Theory of Games and Economic Behavior

Theory of Games and Economic Behavior

John von Neumann
and Oskar Morgenstern

with an Introduction by Harold W. Kuhn
and an Afterword by Ariel Rubinstein

SIXTIETH-ANNIVERSARY EDITION

PRINCETON UNIVERSITY PRESS
Princeton and Oxford

Published by Princeton University Press, 41 William Street, Princeton, New Jersey 08540
In the United Kingdom: Princeton University Press, 3 Market Place,
Woodstock, Oxfordshire OX20 1SY

ISBN: 0-691-11993-7

Printed in the United States of America

Contents

Introduction

HAROLD W. KUHN

Although John von Neumann was without doubt "the father of game theory," the birth took place after a number of miscarriages. From an isolated and amazing minimax solution of a zero-sum two-person game in 1713 [1] to sporadic considerations by E. Zermelo [2], E. Borel [3], and H. Steinhaus [4], nothing matches the path-breaking paper of von Neumann, published in 1928 [5].

This paper, elegant though it is, might have remained a footnote to the history of mathematics were it not for collaboration of von Neumann with Oskar Morgenstern in the early '40s. Their joint efforts led to the publication by the Princeton University Press (with a $4,000 subvention from a source that has been variously identified as being the Carnegie Foundation or the Institute for Advanced Study) of the 616-page *Theory of Games and Economic Behavior* (TGEB).

I will not discuss here the relative contributions of the two authors of this work. Oskar Morgenstern has written his own account [6] of their collaboration, which is reprinted in this volume; I would recommend to the reader the scholarly piece [7] by Robert J. Leonard, who has noted that Morgenstern's "reminiscence sacrifices some of the historical complexity of the run-up to 1944" and has given a superb and historically complete account of the two authors' activities in the relevant period. On balance, I agree with Leonard that "had von Neumann and Morgenstern never met, it seems unlikely that game theory would have been developed." If von Neumann played both father and mother to the theory in an extraordinary act of parthenogenesis, then Morgenstern was the midwife.

In writing this introduction, I have several goals in mind. First, I would like to give the reader a sense of the initial reaction to the publication of this radically new approach to economic theory. Then, we shall survey the subsequent development of the theory of games, attempting to explain the apparent dissonance between the tenor of the book reviews and the response by the communities of economists and mathematicians. As a participant in this response (from the summer of 1948), my account is necessarily colored by subjective and selective recollections; this is a fair warning to the reader.

The book reviews that greeted the publication of TGEB were extraordinary, both in quantity and quality; any author would kill for such reviews. Consider the following partial list of the reviews, paying special attention to the length of these reviews, the quality of the journals, and the prominence of the reviewers:

H. A. Simon, *American Journal of Sociology* (1945) 3 pages*
A. H. Copeland, *Bulletin of the American Mathematical Society* (1945) 7 pages*
L. Hurwicz, *The American Economic Review* (1945) 17 pages*
J. Marschak, *Journal of Political Economy* (1946) 18 pages
T. Barna, *Economica* (1946) 3 pages*
C. Kaysen, *Review of Economic Studies* (1946) 15 pages
D. Hawkins, *Philosophy of Science* (1946) 7 pages
J.R.N. Stone, *Economic Journal* (1948) 16 pages
E. Ruist, *Economisk Tidskrift* (1948) 5 pages
G. Th. Guilbaud, *Economie Appliquée* (1949) 45 pages
E. Justman, *Revue d'Economie Politique* (1949) 18 pages
K. G. Chacko, *Indian Journal of Economics* (1950) 17 pages

The quotes from these reviews are a publisher's dream. Thus:

Simon encouraged "every social scientist who is convinced of the necessity for mathematizing social theory—as well as those unconverted souls who are still open to persuasion on this point—to undertake the task of mastering the Theory of Games."

Copeland asserted: "Posterity may regard this book as one of the major scientific achievements of the first half of the twentieth century."

Hurwicz signaled that "the techniques applied by the authors in tackling economic problems are of sufficient generality to be valid in political science, sociology, or even military strategy" and concluded "the appearance of a book of the caliber of the *Theory of Games* is indeed a rare event."

After praising the "careful and rigorous spirit of the book," Jacob Marschak concludes: "Ten more such books and the progress of economics is assured."

If the quantity of reviews and the quality of the journals in which they were published are impressive, the choice of reviewers and their positions in the social sciences are equally impressive. Two of the reviewers, H. A. Simon and J.R.N. Stone, were awarded Nobel Memorial Prizes in Economics.

The first review to appear was that of Herbert Simon. By his own account [8], he "spent most of [his] 1944 Christmas vacation (days and some nights) reading [the TGEB]." Simon knew of von Neumann's earlier work and was concerned that the TGEB might anticipate results in a book that he was preparing for publication.

*Starred reviews are included in the book.

The first review that was directed at mathematicians was that of A. H. Copeland, a specialist in probability theory and professor at the University of Michigan. Copeland's only significant work in social science is the so-called "Copeland method" for resolving voting problems: simply, it scores 1 for each pairwise win and −1 for each pairwise loss, and declares the alternative with the highest score the winner. His review gave the mathematical community an extremely complete account of the contents of the TGEB. As is typical of almost all of the reviewers, although Copeland pointed to the research challenges opened by the TGEB, he never engaged in research in game theory as such. The only paper in his prolific output that is marginally related to game theory is a joint paper on a one-player game which must be categorized as a game of chance. Copeland's principal contribution to game theory consists in the fact that he was Howard Raiffa's thesis adviser; the book *Games and Decisions,* written by Raiffa with R. Duncan Luce (published by Wiley in 1957 and reprinted by Dover Publications in 1989) was the first nonmathematical exposition that made the theory of games accessible to the broad community of social scientists.

Another reviewer, David Hawkins, is permanently linked to H. A. Simon for their joint discovery of the "Hawkins-Simon conditions," a result that every graduate student in economics must study. Hawkins was a young instructor at the University of California at Berkeley when his friend, J. Robert Oppenheimer, picked him as the "official historian" and "liaison to the military" at Los Alamos, where the first atomic bomb was produced. Hawkins later had a distinguished career at the University of Colorado, where he was chosen in the first class of MacArthur "genius" scholars in 1986. Hawkins did no research in game theory.

The pattern of extravagant praise and no subsequent research is repeated with more significance in the cases of Jacob Marschak and Leonid Hurwicz. Marschak was head of the Cowles Commission at the University of Chicago when he reviewed the TGEB. He had survived a tumultuous early life that took him from Russia, where he was raised, to Berlin, where he trained as an economist, to the United States, where he ran an influential econometric seminar at the New School for Social Research. Leonid Hurwicz preceded Marschak on the staff of the Cowles Commission and continued as a consultant after Tjalling C. Koopmans became director and the commission moved from the University of Chicago to Yale University. Both Marschak and Hurwicz were in a position to influence the research done at the Cowles Commission, but it is an astounding fact that the extensive research output of the commission did not encompass game theory until Martin Shubik joined the Yale faculty in 1963. Eight years after reviewing the TGEB, Hurwicz posed the question: What has happened to the theory of games? His answer [9], published in *The American Economic Review*, contains conclusions that are echoed in this introduction.

Among the reviews and reviewers, the review of G. Th. Guilbaud is surely unique. Occupying 45 pages in the journal, *Economie Appliquée*, it contained not only an account of the main themes of the TGEB, but also went further into consideration of the difficulties that the theory then faced. Guilbaud himself was unique in that he was the only reviewer who has contributed to the theory; his book *Eléments de la Theorie des Jeux* was published by Dunod in Paris in 1968. However, he failed to convince the economic community in France to join him. Guilbaud's seminar in Paris in 1950–51 was attended by such mathematical economists as Allais, Malinvaud, Boiteux, and myself, but none of the French engaged in research in game theory. I am pleased to report that Guilbaud, a very private person, is still with us at 91 years of age, living in St. Germaine-en-Laye. It was he who discovered the minimax solution of 1713 [1], when he purchased the treatise on probability written by Montmort from one of the booksellers whose stalls line the river Seine in Paris.

Given the extravagant praise of these reviewers, one might have expected a flood of research. If nowhere else, surely the Princeton economics department should have been a hotbed of activity. When Martin Shubik arrived in Princeton to do graduate work in economics in the fall of 1949, he expected to find just that. Instead, he found Professor Morgenstern in splendid isolation from the rest of the department, teaching a seminar with four students in attendance [10]. Morgenstern's research project consisted of himself assisted by Maurice Peston, Tom Whitin, and Ed Zabel, who concentrated on areas of operations research such as inventory theory, but did not work on game theory as such. If Shubik had come two years earlier, he would have found the situation in the mathematics department somewhat similar. Samuel Karlin (who received his Ph.D. at Princeton in mathematics in the spring of 1947 then took a faculty position at Cal Tech, and almost immediately started to consult at the RAND Corporation under the tutelage of Frederic Bohnenblust) has written that he never heard game theory mentioned during his graduate studies.

Nevertheless, many observers agree that in the following decade Princeton was one of the two centers in which game theory flourished, the other being the RAND Corporation in Santa Monica. The story of the RAND Corporation and its research sponsored by the Air Force has been told on several occasions (see [11], [12]). We shall concentrate on the activity in the mathematics department at Princeton, a story that illustrates the strong element of chance in human affairs.

The story starts with two visits by George Dantzig to visit John von Neumann in the fall of 1947 and the spring of 1948. In the first visit Dantzig described his new theory of "linear programming" only to be told dismissively by von Neumann that he had encountered similar problems in his study of zero-sum two-person games. In his second visit, Dantzig proposed an academic project to study the relationship between these two fields and asked von Neumann's advice about universities in which such a project might be pursued. Dantzig was driven to the train station for his trip back to Washington by A. W. Tucker (a topologist who was associate chairman of the mathematics department at that time). On the ride, Dantzig gave a quick exposition of his new discoveries, using the Transportation Problem [13] as a lively example. This recalled to Tucker his earlier work on electrical networks and Kirkhoff's Law and planted the idea that the project to study the relationship between linear programming and the theory of games might be established in the mathematics department at Princeton University.

In those halcyon days of no red tape, before a month had elapsed Tucker hired two graduate students, David Gale and myself, and the project was set up through Solomon Lefshetz's project on non-linear differential equations until a formal structure could be established through the Office of Naval Research's Logistics Branch. And so, in the summer of 1948, Gale, Kuhn, and Tucker taught each other the elements of game theory.

How did we do this? We divided up the chapters of the Bible, the TGEB, as handed down by von Neumann and Morgenstern, and lectured to each other in one of the seminar rooms of the old Fine Hall, then the home of the mathematics department at Princeton. By the end of the summer, we had established that, mathematically, linear programming and the theory of zero-sum two-person games are equivalent.

Enthused by the research potential of the subject we had just learned, we wanted to spread the gospel. We initiated a weekly seminar in the department centered on the subjects of game theory and linear programming. To understand the importance of this development, one must contrast the situations today and then. Today, the seminar lists of the university and the Institute of Advanced Studies contain over twenty weekly seminars in subjects such as number theory, topology, analysis, and statistical mechanics. In 1948, there was a weekly colloquium that met alternate weeks at the university and the institute. The topologists and statisticians had weekly seminars and my thesis advisor, Ralph Fox, ran a weekly seminar on knot theory; but that was that. So the addition of a new seminar was an event that raised the visibility of game theory considerably among the graduate students in the department and among the visitors to the institute.

The speakers included von Neumann and Morgenstern, visitors to the institute such as Irving Kaplansky, Ky Fan, and David Bourgin, as well as outside visitors such as Abraham Wald, the Columbia statistician who had made significant connections between game theory and statistical inference. (Wald had done the review of the TGEB for *Mathematical Reviews* and had tutored Morgenstern in mathematics in Vienna.)

More importantly it provided a forum for graduate students in mathematics who were working in this area to present new ideas. As Shubik has reminisced: "The general attitude around Fine Hall was that no one cared who you were or what part of mathematics you worked on as long as you could find some senior member of the faculty and make a case to him that it was interesting and that you did it well. . . . To me the striking thing at that time was not that the mathematics department welcomed game theory with open arms—but that it was open to new ideas and new talent from any source, and that it could convey a sense of challenge and a belief that much new and worthwhile was happening." He did not find that attitude in the economics department.

A crucial fact was that von Neumann's theory was too mathematical for the economists. To illustrate the attitude of a typical economics department of the period and later, more than fifteen years after the publication of TGEB the economists at Princeton voted against instituting a mathematics requirement for undergraduate majors, choosing to run two tracks for students, one which used the calculus and one which avoided it. Richard Lester, who alternated with Lester Chandler as chairman of the department, had carried on a running debate with Fritz Machlup over the validity of marginal product (a calculus notion) as a determinant of wages. Courses that used mathematical terms and which covered mathematical topics such as linear programming were concealed by titles such as "Managerial theory of the firm." Given such prevailing views, there was no incentive or opportunity for graduate students and junior faculty to study the theory of games.

As a consequence, the theory of games was developed almost exclusively by mathematicians in this period. To describe the spirit of the time as seen by another outside observer, we shall paraphrase a section of Robert J. Aumann's magnificent article on game theory from *The New Palgrave Dictionary of Economics* [14].

The period of the late '40s and early '50s was a period of excitement in game theory. The discipline had broken out of its cocoon and was testing its wings. Giants walked the earth. At Princeton, John Nash laid the groundwork for the general non-cooperative theory and for cooperative bargaining theory. Lloyd Shapley defined a value for coalitional games, initiated the theory of stochastic games, coinvented the core with D. B. Gillies, and together with John Milnor developed the first game models with an infinite number of players. Harold Kuhn reformulated the extensive form and introduced the concepts of behavior strategies and perfect recall. A. W. Tucker invented the story of the Prisoner's Dilemma, which has entered popular culture as a crucial example of the interplay between competition and cooperation.

It is important to recognize that the results that Aumann enumerated did not respond to some suggestion of von Neumann; rather they were new ideas that ran counter to von Neumann's preferred version of the theory. In almost every instance, it was a repair of some inadequacy of the theory as presented in the TGEB. Indeed, von Neumann and Morgenstern criticized Nash's non-cooperative theory on a number of occasions. In the case of the extensive form, the book contains the claim that it was impossible to give a useful geometric formulation. Thus, game theory was very much a work in progress, in spite of von Neumann's opinion that the book contained a rather complete theory. Through the efforts at RAND and at Princeton University, many new directions of research had been opened and the way had been paved for the applications to come.

The TGEB was published with unparalleled accolades from the cream of the mathematical economists of the era, then ignored by the economists while mathematicians at the RAND Corporation and at Princeton quietly pushed the boundaries of the subject into new territory. It took nearly a quarter century before reality overcame the stereotypical view that it was merely a theory of zero-sum two-person games and that its usefulness was restricted to military problems. Once these myths were countered, applications came tumbling out and, by the time the Nobel Memorial Prize in Economics was awarded in 1994 to Nash, John Harsanyi, and Reinhard Selten, the theory of games had assumed a central position in academic economic theory. If Oskar Morgenstern had been alive in 1994, he would surely have said, "I told you so!"

In opening this new edition of the TGEB, you are given the opportunity to read for yourselves the revision of the economic theory that it contains and to decide whether it is "one of the major scientific achievements of the twentieth century." Although the subject has enjoyed a spectacular expansion in the sixty years since its publication, everything that followed is based on the foundation laid by von Neumann and Morgenstern in this book.

References

1. Waldegrave, J. (1713) Minimax solution of a 2-person, zero-sum game, reported in a letter from P. de Montmort to N. Bernouilli, transl. and with comments by H. W. Kuhn in W. J. Baumol and S. Goldfeld (eds.), *Precursors of Mathematical Economics* (London: London School of Economics, 1968), 3–9.

2. Zermelo, E. (1913) "Uber eine anwendung der Mengenlehre auf die theorie des Schachspiels," *Proceedings, Fifth International Congress of Mathematicians,* vol. 2, 501–4.

3. Borel, E. (1924) "Sur les jeux ou interviennent l'hasard et l'habilité des joueurs," *Théorie des Probabilités* (Paris: Librarie Scientifique, J Hermann), 204–24.

4. Steinhaus, H. (1925) "Definitions for a theory of games and pursuit" (in Polish), *Mysl Akademika, Lwow* 1, 13–14; E. Rzymovski (trans.) with introduction by H. W. Kuhn, *Naval Research Logistics Quarterly* (1960), 105–8.

5. von Neumann, J. (1928) "Zur theorie der Gesellschaftsspiele," *Math. Ann.* 100, 295–300.

6. Morgenstern, O. (1976) "The collaboration of Oskar Morgerstern and John von Neumann on the theory of games," *Journal of Economic Literature* 14, 805–16.

7. Leonard, R. J. (1995) "From parlor games to social science: von Neumann, Morgenstern, and the creation of game theory 1928–1944," *Journal of Economic Literature* 33, 730–61.

8. Simon, H. A. (1991) *Models of My Life* (New York: Basic Books).

9. Hurwicz, L. (1953) "What has happened to the theory of games?" *American Economic Review* 43, 398–405.

10. Shubik, M. (1992) "Game theory at Princeton, 1949–1955; a personal reminiscence," in E. R. Weintraub (ed.) *Toward a History of Game Theory,* History of Political Economy supplement to vol. 24, (Durham and London: Duke University Press).

11. Poundstone, W. (1992) *Prisoner's Dilemma* (New York: Doubleday).

12. Nasar, S. (1998) *A Beautiful Mind* (New York: Simon & Schuster).

13. Dantzig, G. B. (1963) *Linear Programming and Extensions* (Princeton: Princeton University Press).

14. Aumann, R. J. (1989) "Game theory" in J. Eatwell, M. Milgate, and P. Newman (eds), *The New Palgrave Dictionary of Economics* (New York: W. W. Norton), 1–53.

THEORY OF GAMES AND ECONOMIC BEHAVIOR

By JOHN VON NEUMANN, and

OSKAR MORGENSTERN

PRINCETON

PRINCETON UNIVERSITY PRESS

CONTENTS

CHAPTER III

ZERO-SUM TWO-PERSON GAMES: THEORY

CHAPTER IV

ZERO-SUM TWO-PERSON GAMES: EXAMPLES

CHAPTER V
ZERO-SUM THREE-PERSON GAMES

CHAPTER VI
FORMULATION OF THE GENERAL THEORY: ZERO-SUM *n*-PERSON GAMES

CHAPTER VII

ZERO-SUM FOUR-PERSON GAMES

CHAPTER VIII
SOME REMARKS CONCERNING $n \geqq 5$ PARTICIPANTS

CHAPTER IX
COMPOSITION AND DECOMPOSITION OF GAMES

CHAPTER X
SIMPLE GAMES

CHAPTER XI
GENERAL NON-ZERO-SUM GAMES

CHAPTER XII
EXTENSION OF THE CONCEPTS OF DOMINATION AND SOLUTION

PREFACE TO FIRST EDITION

This book contains an exposition and various applications of a mathematical theory of games. The theory has been developed by one of us since 1928 and is now published for the first time in its entirety. The applications are of two kinds: On the one hand to games in the proper sense, on the other hand to economic and sociological problems which, as we hope to show, are best approached from this direction.

The applications which we shall make to games serve at least as much to corroborate the theory as to investigate these games. The nature of this reciprocal relationship will become clear as the investigation proceeds. Our major interest is, of course, in the economic and sociological direction. Here we can approach only the simplest questions. However, these questions are of a fundamental character. Furthermore, our aim is primarily to show that there is a rigorous approach to these subjects, involving, as they do, questions of parallel or opposite interest, perfect or imperfect information, free rational decision or chance influences.

JOHN VON NEUMANN
OSKAR MORGENSTERN.

PRINCETON, N. J.
January, 1943.

PREFACE TO SECOND EDITION

The second edition differs from the first in some minor respects only. We have carried out as complete an elimination of misprints as possible, and wish to thank several readers who have helped us in that respect. We have added an Appendix containing an axiomatic derivation of numerical utility. This subject was discussed in considerable detail, but in the main qualitatively, in Section 3. A publication of this proof in a periodical was promised in the first edition, but we found it more convenient to add it as an Appendix. Various Appendices on applications to the theory of location of industries and on questions of the four and five person games were also planned, but had to be abandoned because of the pressure of other work.

Since publication of the first edition several papers dealing with the subject matter of this book have appeared.

The attention of the mathematically interested reader may be drawn to the following: *A. Wald* developed a new theory of the foundations of statistical estimation which is closely related to, and draws on, the theory of the zero-sum two-person game ("Statistical Decision Functions Which Minimize the Maximum Risk," Annals of Mathematics, Vol. 46 (1945) pp. 265–280). He also extended the main theorem of the zero-sum two-person games (cf. 17.6.) to certain continuous-infinite-cases, ("Generalization of a Theorem by von Neumann Concerning Zero-Sum Two-Person Games," Annals of Mathematics, Vol. 46 (1945), pp. 281–286.) A new, very simple and elementary proof of this theorem (which covers also the more general theorem referred to in footnote 1 on page 154) was given by *L. H. Loomis*, ("On a Theorem of von Neumann," Proc. Nat. Acad., Vol. 32 (1946) pp. 213–215). Further, interesting results concerning the role of pure and of mixed strategies in the zero-sum two-person game were obtained by *I. Kaplanski*, ("A Contribution to von Neumann's Theory of Games," Annals of Mathematics, Vol. 46 (1945), pp. 474–479). We also intend to come back to various mathematical aspects of this problem. The group theoretical problem mentioned in footnote 1 on page 258 was solved by *C. Chevalley*.

The economically interested reader may find an easier approach to the problems of this book in the expositions of *L. Hurwicz*, ("The Theory of Economic Behavior," American Economic Review, Vol. 35 (1945), pp. 909–925) and of *J. Marschak* ("Neumann's and Morgenstern's New Approach to Static Economics," Journal of Political Economy, Vol. 54, (1946), pp. 97–115).

JOHN VON NEUMANN
OSKAR MORGENSTERN

PRINCETON, N. J.
September, 1946.

PREFACE TO THIRD EDITION

The Third Edition differs from the Second Edition only in the elimination of such further misprints as have come to our attention in the meantime, and we wish to thank several readers who have helped us in that respect.

Since the publication of the Second Edition, the literature on this subject has increased very considerably. A complete bibliography at this writing includes several hundred titles. We are therefore not attempting to give one here. We will only list the following books on this subject:

(1) *H. W. Kuhn* and *A. W. Tucker* (eds.), "Contributions to the Theory of Games, I," Annals of Mathematics Studies, No. 24, Princeton (1950), containing fifteen articles by thirteen authors.

(2) *H. W. Kuhn* and *A. W. Tucker* (eds.), "Contributions to the Theory of Games, II," Annals of Mathematics Studies, No. 28, Princeton (1953), containing twenty-one articles by twenty-two authors.

(3) *J. McDonald*, Strategy in Poker, Business and War, New York (1950).

(4) *J. C. C. McKinsey*, Introduction to the Theory of Games, New York (1952).

(5) *A. Wald*, Statistical Decision Functions, New York (1950).

(6) *J. Williams*, The Compleat Strategyst, Being a Primer on the Theory of Games of Strategy, New York (1953).

Bibliographies on the subject are found in all of the above books except (6). Extensive work in this field has been done during the last years by the staff of the RAND Corporation, Santa Monica, California. A bibliography of this work can be found in the RAND publication RM-950.

In the theory of n-person games, there have been some further developments in the direction of "non-cooperative" games. In this respect, particularly the work of J. F. Nash, "Non-cooperative Games," *Annals of Mathematics*, Vol. 54, (1951), pp. 286–295, must be mentioned. Further references to this work are found in (1), (2), and (4).

Of various developments in economics we mention in particular "linear programming" and the "assignment problem" which also appear to be increasingly connected with the theory of games. The reader will find indications of this again in (1), (2), and (4).

The theory of utility suggested in Section 1.3., and in the Appendix to the Second Edition has undergone considerable development theoretically, as well as experimentally, and in various discussions. In this connection, the reader may consult in particular the following:

M. Friedman and *L. J. Savage*, "The Utility Analysis of Choices Involving Risk," Journal of Political Economy, Vol. 56, (1948), pp. 279–304.

J. Marschak, "Rational Behavior, Uncertain Prospects, and Measurable Utility," Econometrica, Vol. 18, (1950), pp. 111–141.

F. Mosteller and *P. Nogee*, "An Experimental Measurement of Utility," Journal of Political Economy, Vol. 59, (1951), pp. 371–404.

M. Friedman and *L. J. Savage*, "The Expected-Utility Hypothesis and the Measurability of Utility," Journal of Political Economy, Vol. 60, (1952), pp. 463–474.

See also the Symposium on Cardinal Utilities in Econometrica, Vol. 20, (1952):

H. Wold, "Ordinal Preferences or Cardinal Utility?"

A. S. Manne, "The Strong Independence Assumption—Gasoline Blends and Probability Mixtures."

P. A. Samuelson, "Probability, Utility, and the Independence Axiom."

E. Malinvaud, "Note on von Neumann-Morgenstern's Strong Independence Axiom."

In connection with the methodological critique exercised by some of the contributors to the last-mentioned symposium, we would like to mention that we applied the axiomatic method in the customary way with the customary precautions. Thus the strict, axiomatic treatment of the concept of utility (in Section 3.6. and in the Appendix) is complemented by an heuristic preparation (in Sections 3.1.–3.5.). The latter's function is to convey to the reader the viewpoints to evaluate and to circumscribe the validity of the subsequent axiomatic procedure. In particular our discussion and selection of "natural operations" in those sections covers what seems to us the relevant substrate of the Samuelson-Malinvaud "independence axiom."

JOHN VON NEUMANN
OSKAR MORGENSTERN

PRINCETON, N. J.
January, 1953.

TECHNICAL NOTE

The nature of the problems investigated and the techniques employed in this book necessitate a procedure which in many instances is thoroughly mathematical. The mathematical devices used are elementary in the sense that no advanced algebra, or calculus, etc., occurs. (With two, rather unimportant, exceptions: Part of the discussion of an example in 19.7. et sequ. and a remark in A.3.3. make use of some simple integrals.) Concepts originating in set theory, linear geometry and group theory play an important role, but they are invariably taken from the early chapters of those disciplines and are moreover analyzed and explained in special expository sections. Nevertheless the book is not truly elementary because the mathematical deductions are frequently intricate and the logical possibilities are extensively exploited.

Thus no specific knowledge of any particular body of advanced mathematics is required. However, the reader who wants to acquaint himself more thoroughly with the subject expounded here, will have to familiarize himself with the mathematical way of reasoning definitely beyond its routine, primitive phases. The character of the procedures will be mostly that of mathematical logics, set theory and functional analysis.

We have attempted to present the subject in such a form that a reader who is moderately versed in mathematics can acquire the necessary practice in the course of this study. We hope that we have not entirely failed in this endeavour.

In accordance with this, the presentation is not what it would be in a strictly mathematical treatise. All definitions and deductions are considerably broader than they would be there. Besides, purely verbal discussions and analyses take up a considerable amount of space. We have in particular tried to give, whenever possible, a parallel verbal exposition for every major mathematical deduction. It is hoped that this procedure will elucidate in unmathematical language what the mathematical technique signifies—and will also show where it achieves more than can be done without it.

In this, as well as in our methodological stand, we are trying to follow the best examples of theoretical physics.

The reader who is not specifically interested in mathematics should at first omit those sections of the book which in his judgment are too mathematical. We prefer not to give a definite list of them, since this judgment must necessarily be subjective. However, those sections marked with an asterisk in the table of contents are most likely to occur to the average reader in this connection. At any rate he will find that these omissions will little interfere with the comprehension of the early parts, although the logical chain may in the rigorous sense have suffered an interruption. As he proceeds the omissions will gradually assume a more serious character and the lacunae in the deduction will become more and more significant. The reader is then advised to start again from the beginning since the greater familiarity acquired is likely to facilitate a better understanding.

ACKNOWLEDGMENT

The authors wish to express their thanks to Princeton University and to the Institute for Advanced Study for their generous help which rendered this publication possible.

They are also greatly indebted to the Princeton University Press which has made every effort to publish this book in spite of wartime difficulties. The publisher has shown at all times the greatest understanding for the authors' wishes.

CHAPTER I

FORMULATION OF THE ECONOMIC PROBLEM

1. The Mathematical Method in Economics

1.1. Introductory Remarks

1.1.1. The purpose of this book is to present a discussion of some fundamental questions of economic theory which require a treatment different from that which they have found thus far in the literature. The analysis is concerned with some basic problems arising from a study of economic behavior which have been the center of attention of economists for a long time. They have their origin in the attempts to find an exact description of the endeavor of the individual to obtain a maximum of utility, or, in the case of the entrepreneur, a maximum of profit. It is well known what considerable—and in fact unsurmounted—difficulties this task involves given even a limited number of typical situations, as, for example, in the case of the exchange of goods, direct or indirect, between two or more persons, of bilateral monopoly, of duopoly, of oligopoly, and of free competition. It will be made clear that the structure of these problems, familiar to every student of economics, is in many respects quite different from the way in which they are conceived at the present time. It will appear, furthermore, that their exact positing and subsequent solution can only be achieved with the aid of mathematical methods which diverge considerably from the techniques applied by older or by contemporary mathematical economists.

1.1.2. Our considerations will lead to the application of the mathematical theory of "games of strategy" developed by one of us in several successive stages in 1928 and 1940–1941.[1] After the presentation of this theory, its application to economic problems in the sense indicated above will be undertaken. It will appear that it provides a new approach to a number of economic questions as yet unsettled.

We shall first have to find in which way this theory of games can be brought into relationship with economic theory, and what their common elements are. This can be done best by stating briefly the nature of some fundamental economic problems so that the common elements will be seen clearly. It will then become apparent that there is not only nothing artificial in establishing this relationship but that on the contrary this

[1] The first phases of this work were published: *J. von Neumann*, "Zur Theorie der Gesellschaftsspiele," Math. Annalen, vol. 100 (1928), pp. 295–320. The subsequent completion of the theory, as well as the more detailed elaboration of the considerations of loc. cit. above, are published here for the first time.

theory of games of strategy is the proper instrument with which to develop a theory of economic behavior.

One would misunderstand the intent of our discussions by interpreting them as merely pointing out an analogy between these two spheres. We hope to establish satisfactorily, after developing a few plausible schematizations, that the typical problems of economic behavior become strictly identical with the mathematical notions of suitable games of strategy.

1.2. Difficulties of the Application of the Mathematical Method

1.2.1. It may be opportune to begin with some remarks concerning the nature of economic theory and to discuss briefly the question of the role which mathematics may take in its development.

First let us be aware that there exists at present no universal system of economic theory and that, if one should ever be developed, it will very probably not be during our lifetime. The reason for this is simply that economics is far too difficult a science to permit its construction rapidly, especially in view of the very limited knowledge and imperfect description of the facts with which economists are dealing. Only those who fail to appreciate this condition are likely to attempt the construction of universal systems. Even in sciences which are far more advanced than economics, like physics, there is no universal system available at present.

To continue the simile with physics: It happens occasionally that a particular physical theory appears to provide the basis for a universal system, but in all instances up to the present time this appearance has not lasted more than a decade at best. The everyday work of the research physicist is certainly not involved with such high aims, but rather is concerned with special problems which are "mature." There would probably be no progress at all in physics if a serious attempt were made to enforce that super-standard. The physicist works on individual problems, some of great practical significance, others of less. Unifications of fields which were formerly divided and far apart may alternate with this type of work. However, such fortunate occurrences are rare and happen only after each field has been thoroughly explored. Considering the fact that economics is much more difficult, much less understood, and undoubtedly in a much earlier stage of its evolution as a science than physics, one should clearly not expect more than a development of the above type in economics either.

Second we have to notice that the differences in scientific questions make it necessary to employ varying methods which may afterwards have to be discarded if better ones offer themselves. This has a double implication: In some branches of economics the most fruitful work may be that of careful, patient description; indeed this may be by far the largest domain for the present and for some time to come. In others it may be possible to develop already a theory in a strict manner, and for that purpose the use of mathematics may be required.

Mathematics has actually been used in economic theory, perhaps even in an exaggerated manner. In any case its use has not been highly successful. This is contrary to what one observes in other sciences: There mathematics has been applied with great success, and most sciences could hardly get along without it. Yet the explanation for this phenomenon is fairly simple.

1.2.2. It is not that there exists any fundamental reason why mathematics should not be used in economics. The arguments often heard that because of the human element, of the psychological factors etc., or because there is—allegedly—no measurement of important factors, mathematics will find no application, can all be dismissed as utterly mistaken. Almost all these objections have been made, or might have been made, many centuries ago in fields where mathematics is now the chief instrument of analysis. This "might have been" is meant in the following sense: Let us try to imagine ourselves in the period which preceded the mathematical or almost mathematical phase of the development in physics, that is the 16th century, or in chemistry and biology, that is the 18th century. Taking for granted the skeptical attitude of those who object to mathematical economics in principle, the outlook in the physical and biological sciences at these early periods can hardly have been better than that in economics—*mutatis mutandis*—at present.

As to the lack of measurement of the most important factors, the example of the theory of heat is most instructive; before the development of the mathematical theory the possibilities of quantitative measurements were less favorable there than they are now in economics. The precise measurements of the quantity and quality of heat (energy and temperature) were the outcome and not the antecedents of the mathematical theory. This ought to be contrasted with the fact that the quantitative and exact notions of prices, money and the rate of interest were already developed centuries ago.

A further group of objections against quantitative measurements in economics, centers around the lack of indefinite divisibility of economic quantities. This is supposedly incompatible with the use of the infinitesimal calculus and hence (!) of mathematics. It is hard to see how such objections can be maintained in view of the atomic theories in physics and chemistry, the theory of quanta in electrodynamics, etc., and the notorious and continued success of mathematical analysis within these disciplines.

At this point it is appropriate to mention another familiar argument of economic literature which may be revived as an objection against the mathematical procedure.

1.2.3. In order to elucidate the conceptions which we are applying to economics, we have given and may give again some illustrations from physics. There are many social scientists who object to the drawing of such parallels on various grounds, among which is generally found the assertion that economic theory cannot be modeled after physics since it is a science of social, of human phenomena, has to take psychology into account, etc. Such statements are at least premature. It is without doubt reasonable to discover what has led to progress in other sciences, and to investigate whether the application of the same principles may not lead to progress in economics also. Should the need for the application of different principles arise, it could be revealed only in the course of the actual development of economic theory. This would itself constitute a major revolution. But since most assuredly we have not yet reached such a state—and it is by no means certain that there ever will be need for entirely different scientific principles—it would be very unwise to consider anything else than the pursuit of our problems in the manner which has resulted in the establishment of physical science.

1.2.4. The reason why mathematics has not been more successful in economics must, consequently, be found elsewhere. The lack of real success is largely due to a combination of unfavorable circumstances, some of which can be removed gradually. To begin with, the economic problems were not formulated clearly and are often stated in such vague terms as to make mathematical treatment *a priori* appear hopeless because it is quite uncertain what the problems really are. There is no point in using exact methods where there is no clarity in the concepts and issues to which they are to be applied. Consequently the initial task is to clarify the knowledge of the matter by further careful descriptive work. But even in those parts of economics where the descriptive problem has been handled more satisfactorily, mathematical tools have seldom been used appropriately. They were either inadequately handled, as in the attempts to determine a general economic equilibrium by the mere counting of numbers of equations and unknowns, or they led to mere translations from a literary form of expression into symbols, without any subsequent mathematical analysis.

Next, the empirical background of economic science is definitely inadequate. Our knowledge of the relevant facts of economics is incomparably smaller than that commanded in physics at the time when the mathematization of that subject was achieved. Indeed, the decisive break which came in physics in the seventeenth century, specifically in the field of mechanics, was possible only because of previous developments in astronomy. It was backed by several millennia of systematic, scientific, astronomical observation, culminating in an observer of unparalleled caliber, Tycho de Brahe. Nothing of this sort has occurred in economic science. It would have been absurd in physics to expect Kepler and Newton without Tycho,—and there is no reason to hope for an easier development in economics.

These obvious comments should not be construed, of course, as a disparagement of statistical-economic research which holds the real promise of progress in the proper direction.

It is due to the combination of the above mentioned circumstances that mathematical economics has not achieved very much. The underlying vagueness and ignorance has not been dispelled by the inadequate and inappropriate use of a powerful instrument that is very difficult to handle.

In the light of these remarks we may describe our own position as follows: The aim of this book lies not in the direction of empirical research. The advancement of that side of economic science, on anything like the scale which was recognized above as necessary, is clearly a task of vast proportions. It may be hoped that as a result of the improvements of scientific technique and of experience gained in other fields, the development of descriptive economics will not take as much time as the comparison with astronomy would suggest. But in any case the task seems to transcend the limits of any individually planned program.

We shall attempt to utilize only some commonplace experience concerning human behavior which lends itself to mathematical treatment and which is of economic importance.

We believe that the possibility of a mathematical treatment of these phenomena refutes the "fundamental" objections referred to in 1.2.2.

It will be seen, however, that this process of mathematization is not at all obvious. Indeed, the objections mentioned above may have their roots partly in the rather obvious difficulties of any direct mathematical approach. We shall find it necessary to draw upon techniques of mathematics which have not been used heretofore in mathematical economics, and it is quite possible that further study may result in the future in the creation of new mathematical disciplines.

To conclude, we may also observe that part of the feeling of dissatisfaction with the mathematical treatment of economic theory derives largely from the fact that frequently one is offered not proofs but mere assertions which are really no better than the same assertions given in literary form. Very frequently the proofs are lacking because a mathematical treatment has been attempted of fields which are so vast and so complicated that for a long time to come—until much more empirical knowledge is acquired—there is hardly any reason at all to expect progress *more mathematico*. The fact that these fields have been attacked in this way—as for example the theory of economic fluctuations, the time structure of production, etc.—indicates how much the attendant difficulties are being underestimated. They are enormous and we are now in no way equipped for them.

1.2.5. We have referred to the nature and the possibilities of those changes in mathematical technique—in fact, in mathematics itself—which a successful application of mathematics to a new subject may produce. It is important to visualize these in their proper perspective.

It must not be forgotten that these changes may be very considerable. The decisive phase of the application of mathematics to physics—Newton's creation of a rational discipline of mechanics—brought about, and can hardly be separated from, the discovery of the infinitesimal calculus. (There are several other examples, but none stronger than this.)

The importance of the social phenomena, the wealth and multiplicity of their manifestations, and the complexity of their structure, are at least equal to those in physics. It is therefore to be expected—or feared—that mathematical discoveries of a stature comparable to that of calculus will be needed in order to produce decisive success in this field. (Incidentally, it is in this spirit that our present efforts must be discounted.) *A fortiori* it is unlikely that a mere repetition of the tricks which served us so well in physics will do for the social phenomena too. The probability is very slim indeed, since it will be shown that we encounter in our discussions some mathematical problems which are quite different from those which occur in physical science.

These observations should be remembered in connection with the current overemphasis on the use of calculus, differential equations, etc., as the main tools of mathematical economics.

1.3. Necessary Limitations of the Objectives

1.3.1. We have to return, therefore, to the position indicated earlier: It is necessary to begin with those problems which are described clearly, even if they should not be as important from any other point of view. It should be added, moreover, that a treatment of these manageable problems may lead to results which are already fairly well known, but the exact proofs may nevertheless be lacking. Before they have been given the respective theory simply does not exist as a scientific theory. The movements of the planets were known long before their courses had been calculated and explained by Newton's theory, and the same applies in many smaller and less dramatic instances. And similarly in economic theory, certain results—say the indeterminateness of bilateral monopoly—may be known already. Yet it is of interest to derive them again from an exact theory. The same could and should be said concerning practically all established economic theorems.

1.3.2. It might be added finally that we do not propose to raise the question of the practical significance of the problems treated. This falls in line with what was said above about the selection of fields for theory. The situation is not different here from that in other sciences. There too the most important questions from a practical point of view may have been completely out of reach during long and fruitful periods of their development. This is certainly still the case in economics, where it is of utmost importance to know how to stabilize employment, how to increase the national income, or how to distribute it adequately. Nobody can really answer these questions, and we need not concern ourselves with the pretension that there can be scientific answers at present.

The great progress in every science came when, in the study of problems which were modest as compared with ultimate aims, methods were developed which could be extended further and further. The free fall is a very trivial physical phenomenon, but it was the study of this exceedingly simple fact and its comparison with the astronomical material, which brought forth mechanics.

It seems to us that the same standard of modesty should be applied in economics. It is futile to try to explain—and "systematically" at that—everything economic. The sound procedure is to obtain first utmost precision and mastery in a limited field, and then to proceed to another, somewhat wider one, and so on. This would also do away with the unhealthy practice of applying so-called theories to economic or social reform where they are in no way useful.

We believe that it is necessary to know as much as possible about the behavior of the individual and about the simplest forms of exchange. This standpoint was actually adopted with remarkable success by the founders of the marginal utility school, but nevertheless it is not generally accepted. Economists frequently point to much larger, more "burning" questions, and brush everything aside which prevents them from making statements about these. The experience of more advanced sciences, for example physics, indicates that this impatience merely delays progress, including that of the treatment of the "burning" questions. There is no reason to assume the existence of shortcuts.

1.4. Concluding Remarks

1.4. It is essential to realize that economists can expect no easier fate than that which befell scientists in other disciplines. It seems reasonable to expect that they will have to take up first problems contained in the very simplest facts of economic life and try to establish theories which explain them and which really conform to rigorous scientific standards. We can have enough confidence that from then on the science of economics will grow further, gradually comprising matters of more vital importance than those with which one has to begin.[1]

The field covered in this book is very limited, and we approach it in this sense of modesty. We do not worry at all if the results of our study conform with views gained recently or held for a long time, for what is important is the gradual development of a theory, based on a careful analysis of the ordinary everyday interpretation of economic facts. This preliminary stage is necessarily *heuristic,* i.e. the phase of transition from unmathematical plausibility considerations to the formal procedure of mathematics. The theory finally obtained must be mathematically rigorous and conceptually general. Its first applications are necessarily to elementary problems where the result has never been in doubt and no theory is actually required. At this early stage the application serves to corroborate the theory. The next stage develops when the theory is applied

[1] The beginning is actually of a certain significance, because the forms of exchange between a few individuals are the same as those observed on some of the most important markets of modern industry, or in the case of barter exchange between states in international trade.

to somewhat more complicated situations in which it may already lead to a certain extent beyond the obvious and the familiar. Here theory and application corroborate each other mutually. Beyond this lies the field of real success: genuine prediction by theory. It is well known that all mathematized sciences have gone through these successive phases of evolution.

2. Qualitative Discussion of the Problem of Rational Behavior

2.1. The Problem of Rational Behavior

2.1.1. The subject matter of economic theory is the very complicated mechanism of prices and production, and of the gaining and spending of incomes. In the course of the development of economics it has been found, and it is now well-nigh universally agreed, that an approach to this vast problem is gained by the analysis of the behavior of the individuals which constitute the economic community. This analysis has been pushed fairly far in many respects, and while there still exists much disagreement the significance of the approach cannot be doubted, no matter how great its difficulties may be. The obstacles are indeed considerable, even if the investigation should at first be limited to conditions of economics statics, as they well must be. One of the chief difficulties lies in properly describing the assumptions which have to be made about the motives of the individual. This problem has been stated traditionally by assuming that the consumer desires to obtain a maximum of utility or satisfaction and the entrepreneur a maximum of profits.

The conceptual and practical difficulties of the notion of utility, and particularly of the attempts to describe it as a number, are well known and their treatment is not among the primary objectives of this work. We shall nevertheless be forced to discuss them in some instances, in particular in 3.3. and 3.5. Let it be said at once that the standpoint of the present book on this very important and very interesting question will be mainly opportunistic. We wish to concentrate on one problem—which is not that of the measurement of utilities and of preferences—and we shall therefore attempt to simplify all other characteristics as far as reasonably possible. We shall therefore assume that the aim of all participants in the economic system, consumers as well as entrepreneurs, is money, or equivalently a single monetary commodity. This is supposed to be unrestrictedly divisible and substitutable, freely transferable and identical, even in the quantitative sense, with whatever "satisfaction" or "utility" is desired by each participant. (For the quantitative character of utility, cf. 3.3. quoted above.)

It is sometimes claimed in economic literature that discussions of the notions of utility and preference are altogether unnecessary, since these are purely verbal definitions with no empirically observable consequences, i.e., entirely tautological. It does not seem to us that these notions are qualitatively inferior to certain well established and indispensable notions in physics, like force, mass, charge, etc. That is, while they are in their immediate form merely definitions, they become subject to empirical control through the theories which are built upon them—and in no other way. Thus the notion of utility is raised above the status of a tautology by such economic theories as make use of it and the results of which can be compared with experience or at least with common sense.

2.1.2. The individual who attempts to obtain these respective maxima is also said to act "rationally." But it may safely be stated that there exists, at present, no satisfactory treatment of the question of rational behavior. There may, for example, exist several ways by which to reach the optimum position; they may depend upon the knowledge and understanding which the individual has and upon the paths of action open to him. A study of all these questions in qualitative terms will not exhaust them, because they imply, as must be evident, quantitative relationships. It would, therefore, be necessary to formulate them in quantitative terms so that all the elements of the qualitative description are taken into consideration. This is an exceedingly difficult task, and we can safely say that it has not been accomplished in the extensive literature about the topic. The chief reason for this lies, no doubt, in the failure to develop and apply suitable mathematical methods to the problem; this would have revealed that the maximum problem which is supposed to correspond to the notion of rationality is not at all formulated in an unambiguous way. Indeed, a more exhaustive analysis (to be given in 4.3.-4.5.) reveals that the significant relationships are much more complicated than the popular and the "philosophical" use of the word "rational" indicates.

A valuable qualitative preliminary description of the behavior of the individual is offered by the Austrian School, particularly in analyzing the economy of the isolated "Robinson Crusoe." We may have occasion to note also some considerations of Böhm-Bawerk concerning the exchange between two or more persons. The more recent exposition of the theory of the individual's choices in the form of indifference curve analysis builds up on the very same facts or alleged facts but uses a method which is often held to be superior in many ways. Concerning this we refer to the discussions in 2.1.1. and 3.3.

We hope, however, to obtain a real understanding of the problem of exchange by studying it from an altogether different angle; this is, from the perspective of a "game of strategy." Our approach will become clear presently, especially after some ideas which have been advanced, say by Böhm-Bawerk—whose views may be considered only as a prototype of this theory—are given correct quantitative formulation.

2.2. "Robinson Crusoe" Economy and Social Exchange Economy

2.2.1. Let us look more closely at the type of economy which is represented by the "Robinson Crusoe" model, that is an economy of an isolated single person or otherwise organized under a single will. This economy is confronted with certain quantities of commodities and a number of wants which they may satisfy. The problem is to obtain a maximum satisfaction. This is—considering in particular our above assumption of the numerical character of utility—indeed an ordinary maximum problem, its difficulty depending apparently on the number of variables and on the nature of the function to be maximized; but this is more of a practical difficulty than a theoretical one.[1] If one abstracts from continuous production and from the fact that consumption too stretches over time (and often uses durable consumers' goods), one obtains the simplest possible model. It was thought possible to use it as the very basis for economic theory, but this attempt—notably a feature of the Austrian version—was often contested. The chief objection against using this very simplified model of an isolated individual for the theory of a social exchange economy is that it does not represent an individual exposed to the manifold social influences. Hence, it is said to analyze an individual who might behave quite differently if his choices were made in a social world where he would be exposed to factors of imitation, advertising, custom, and so on. These factors certainly make a great difference, but it is to be questioned whether they change the formal properties of the process of maximizing. Indeed the latter has never been implied, and since we are concerned with this problem alone, we can leave the above social considerations out of account.

Some other differences between "Crusoe" and a participant in a social exchange economy will not concern us either. Such is the non-existence of money as a means of exchange in the first case where there is only a standard of calculation, for which purpose any commodity can serve. This difficulty indeed has been ploughed under by our assuming in 2.1.2. a quantitative and even monetary notion of utility. We emphasize again: Our interest lies in the fact that even after all these drastic simplifications Crusoe is confronted with a formal problem quite different from the one a participant in a social economy faces.

2.2.2. Crusoe is given certain physical data (wants and commodities) and his task is to combine and apply them in such a fashion as to obtain a maximum resulting satisfaction. There can be no doubt that he controls exclusively all the variables upon which this result depends—say the allotting of resources, the determination of the uses of the same commodity for different wants, etc.[2]

Thus Crusoe faces an ordinary maximum problem, the difficulties of which are of a purely technical—and not conceptual—nature, as pointed out.

2.2.3. Consider now a participant in a social exchange economy. His problem has, of course, many elements in common with a maximum prob-

[1] It is not important for the following to determine whether its theory is complete in all its aspects.

[2] Sometimes uncontrollable factors also intervene, e.g. the weather in agriculture. These however are purely statistical phenomena. Consequently they can be eliminated by the known procedures of the calculus of probabilities: i.e., by determining the probabilities of the various alternatives and by introduction of the notion of "mathematical expectation." Cf. however the influence on the notion of utility, discussed in 3.3.

lem. But it also contains some, very essential, elements of an entirely different nature. He too tries to obtain an optimum result. But in order to achieve this, he must enter into relations of exchange with others. If two or more persons exchange goods with each other, then the result for each one will depend in general not merely upon his own actions but on those of the others as well. Thus each participant attempts to maximize a function (his above-mentioned "result") of which he does not control all variables. This is certainly no maximum problem, but a peculiar and disconcerting mixture of several conflicting maximum problems. Every participant is guided by another principle and neither determines all variables which affect his interest.

This kind of problem is nowhere dealt with in classical mathematics. We emphasize at the risk of being pedantic that this is no conditional maximum problem, no problem of the calculus of variations, of functional analysis, etc. It arises in full clarity, even in the most "elementary" situations, e.g., when all variables can assume only a finite number of values.

A particularly striking expression of the popular misunderstanding about this pseudo-maximum problem is the famous statement according to which the purpose of social effort is the "greatest possible good for the greatest possible number." A guiding principle cannot be formulated by the requirement of maximizing two (or more) functions at once.

Such a principle, taken literally, is self-contradictory. (In general one function will have no maximum where the other function has one.) It is no better than saying, e.g., that a firm should obtain maximum prices at maximum turnover, or a maximum revenue at minimum outlay. If some order of importance of these principles or some weighted average is meant, this should be stated. However, in the situation of the participants in a social economy nothing of that sort is intended, but all maxima are desired at once—by various participants.

One would be mistaken to believe that it can be obviated, like the difficulty in the Crusoe case mentioned in footnote 2 on p. 10, by a mere recourse to the devices of the theory of probability. Every participant can determine the variables which describe his own actions but not those of the others. Nevertheless those "alien" variables cannot, from his point of view, be described by statistical assumptions. This is because the others are guided, just as he himself, by rational principles—whatever that may mean—and no *modus procedendi* can be correct which does not attempt to understand those principles and the interactions of the conflicting interests of all participants.

Sometimes some of these interests run more or less parallel—then we are nearer to a simple maximum problem. But they can just as well be opposed. The general theory must cover all these possibilities, all intermediary stages, and all their combinations.

2.2.4. The difference between Crusoe's perspective and that of a participant in a social economy can also be illustrated in this way: Apart from those variables which his will controls, Crusoe is given a number of data which are "dead"; they are the unalterable physical background of the situation. (Even when they are apparently variable, cf. footnote 2 on p. 10, they are really governed by fixed statistical laws.) Not a single datum with which he has to deal reflects another person's will or intention of an economic kind—based on motives of the same nature as his own. A participant in a social exchange economy, on the other hand, faces data of this last type as well: they are the product of other participants' actions and volitions (like prices). His actions will be influenced by his expectation of these, and they in turn reflect the other participants' expectation of his actions.

Thus the study of the Crusoe economy and the use of the methods applicable to it, is of much more limited value to economic theory than has been assumed heretofore even by the most radical critics. The grounds for this limitation lie not in the field of those social relationships which we have mentioned before—although we do not question their significance—but rather they arise from the conceptual differences between the original (Crusoe's) maximum problem and the more complex problem sketched above.

We hope that the reader will be convinced by the above that we face here and now a really conceptual—and not merely technical—difficulty. And it is this problem which the theory of "games of strategy" is mainly devised to meet.

2.3. The Number of Variables and the Number of Participants

2.3.1. The formal set-up which we used in the preceding paragraphs to indicate the events in a social exchange economy made use of a number of "variables" which described the actions of the participants in this economy. Thus every participant is allotted a set of variables, "his" variables, which together completely describe his actions, i.e. express precisely the manifestations of his will. We call these sets the partial sets of variables. The partial sets of all participants constitute together the set of all variables, to be called the total set. So the total number of variables is determined first by the number of participants, i.e. of partial sets, and second by the number of variables in every partial set.

From a purely mathematical point of view there would be nothing objectionable in treating all the variables of any one partial set as a single variable, "the" variable of the participant corresponding to this partial set. Indeed, this is a procedure which we are going to use frequently in our mathematical discussions; it makes absolutely no difference conceptually, and it simplifies notations considerably.

For the moment, however, we propose to distinguish from each other the variables within each partial set. The economic models to which one is naturally led suggest that procedure; thus it is desirable to describe for every participant the quantity of every particular good he wishes to acquire by a separate variable, etc.

2.3.2. Now we must emphasize that any increase of the number of variables inside a participant's partial set may complicate our problem technically, but only technically. Thus in a Crusoe economy—where there exists only one participant and only one partial set which then coincides with the total set—this may make the necessary determination of a maximum technically more difficult, but it will not alter the "pure maximum" character of the problem. If, on the other hand, the number of participants—i.e., of the partial sets of variables—is increased, something of a very different nature happens. To use a terminology which will turn out to be significant, that of games, this amounts to an increase in the number of players in the game. However, to take the simplest cases, a three-person game is very fundamentally different from a two-person game, a four-person game from a three-person game, etc. The combinatorial complications of the problem—which is, as we saw, no maximum problem at all—increase tremendously with every increase in the number of players, —as our subsequent discussions will amply show.

We have gone into this matter in such detail particularly because in most models of economics a peculiar mixture of these two phenomena occurs. Whenever the number of players, i.e. of participants in a social economy, increases, the complexity of the economic system usually increases too; e.g. the number of commodities and services exchanged, processes of production used, etc. Thus the number of variables in every participant's partial set is likely to increase. But the number of participants, i.e. of partial sets, has increased too. Thus both of the sources which we discussed contribute *pari passu* to the total increase in the number of variables. It is essential to visualize each source in its proper role.

2.4. The Case of Many Participants: Free Competition

2.4.1. In elaborating the contrast between a Crusoe economy and a social exchange economy in 2.2.2.-2.2.4., we emphasized those features of the latter which become more prominent when the number of participants —while greater than 1—is of moderate size. The fact that every participant is influenced by the anticipated reactions of the others to his own measures, and that this is true for each of the participants, is most strikingly the crux of the matter (as far as the sellers are concerned) in the classical problems of duopoly, oligopoly, etc. When the number of participants becomes really great, some hope emerges that the influence of every particular participant will become negligible, and that the above difficulties may recede and a more conventional theory become possible. These are, of course, the classical conditions of "free competition." Indeed, this was the starting point of much of what is best in economic theory. Compared with this case of great numbers—free competition—the cases of small numbers on the side of the sellers—monopoly, duopoly, oligopoly—were even considered to be exceptions and abnormities. (Even in these cases the number of participants is still very large in view of the competition among the buyers. The cases involving really small numbers are those of bilateral monopoly, of exchange between a monopoly and an oligopoly, or two oligopolies, etc.)

2.4.2. In all fairness to the traditional point of view this much ought to be said: It is a well known phenomenon in many branches of the exact and physical sciences that very great numbers are often easier to handle than those of medium size. An almost exact theory of a gas, containing about 10^{25} freely moving particles, is incomparably easier than that of the solar system, made up of 9 major bodies; and still more than that of a multiple star of three or four objects of about the same size. This is, of course, due to the excellent possibility of applying the laws of statistics and probabilities in the first case.

This analogy, however, is far from perfect for our problem. The theory of mechanics for 2, 3, 4, · · · bodies is well known, and in its general theoretical (as distinguished from its special and computational) form is the foundation of the statistical theory for great numbers. For the social exchange economy—i.e. for the equivalent "games of strategy"—the theory of 2, 3, 4, · · participants was heretofore lacking. It is this need that our previous discussions were designed to establish and that our subsequent investigations will endeavor to satisfy. In other words, only after the theory for moderate numbers of participants has been satisfactorily developed will it be possible to decide whether extremely great numbers of participants simplify the situation. Let us say it again: We share the hope—chiefly because of the above-mentioned analogy in other fields!—that such simplifications will indeed occur. The current assertions concerning free competition appear to be very valuable surmises and inspiring anticipations of results. But they are not results and it is scientifically unsound to treat them as such as long as the conditions which we mentioned above are not satisfied.

There exists in the literature a considerable amount of theoretical discussion purporting to show that the zones of indeterminateness (of rates of exchange)—which undoubtedly exist when the number of participants is small—narrow and disappear as the number increases. This then would provide a continuous transition into the ideal case of free competition—for a very great number of participants—where all solutions would be sharply and uniquely determined. While it is to be hoped that this indeed turns out to be the case in sufficient generality, one cannot concede that anything like this contention has been established conclusively thus far. There is no getting away from it: The problem must be formulated, solved and understood for small numbers of participants before anything can be proved about the changes of its character in any limiting case of large numbers, such as free competition.

2.4.3. A really fundamental reopening of this subject is the more desirable because it is neither certain nor probable that a mere increase in the number of participants will always lead *in fine* to the conditions of free competition. The classical definitions of free competition all involve further postulates besides the greatness of that number. E.g., it is clear that if certain great groups of participants will—for any reason whatsoever—act together, then the great number of participants may not become effective; the decisive exchanges may take place directly between large "coalitions,"[1] few in number, and not between individuals, many in number, acting independently. Our subsequent discussion of "games of strategy" will show that the role and size of "coalitions" is decisive throughout the entire subject. Consequently the above difficulty—though not new—still remains the crucial problem. Any satisfactory theory of the "limiting transition" from small numbers of participants to large numbers will have to explain under what circumstances such big coalitions will or will not be formed—i.e. when the large numbers of participants will become effective and lead to a more or less free competition. Which of these alternatives is likely to arise will depend on the physical data of the situation. Answering this question is, we think, the real challenge to any theory of free competition.

2.5. The "Lausanne" Theory

2.5. This section should not be concluded without a reference to the equilibrium theory of the Lausanne School and also of various other systems which take into consideration "individual planning" and interlocking individual plans. All these systems pay attention to the interdependence of the participants in a social economy. This, however, is invariably done under far-reaching restrictions. Sometimes free competition is assumed, after the introduction of which the participants face fixed conditions and act like a number of Robinson Crusoes—solely bent on maximizing their individual satisfactions, which under these conditions are again independent. In other cases other restricting devices are used, all of which amount to excluding the free play of "coalitions" formed by any or all types of participants. There are frequently definite, but sometimes hidden, assumptions concerning the ways in which their partly parallel and partly opposite interests will influence the participants, and cause them to cooperate or not, as the case may be. We hope we have shown that such a procedure amounts to a *petitio principii*—at least on the plane on which we should like to put the discussion. It avoids the real difficulty and deals with a verbal problem, which is not the empirically given one. Of course we do not wish to question the significance of these investigations—but they do not answer our queries.

3. The Notion of Utility

3.1. Preferences and Utilities

3.1.1. We have stated already in 2.1.1. in what way we wish to describe the fundamental concept of individual preferences by the use of a rather

[1] Such as trade unions, consumers' cooperatives, industrial cartels, and conceivably some organizations more in the political sphere.

far-reaching notion of utility. Many economists will feel that we are assuming far too much (cf. the enumeration of the properties we postulated in 2.1.1.), and that our standpoint is a retrogression from the more cautious modern technique of "indifference curves."

Before attempting any specific discussion let us state as a general excuse that our procedure at worst is only the application of a classical preliminary device of scientific analysis: To divide the difficulties, i.e. to concentrate on one (the subject proper of the investigation in hand), and to reduce all others as far as reasonably possible, by simplifying and schematizing assumptions. We should also add that this high handed treatment of preferences and utilities is employed in the main body of our discussion, but we shall incidentally investigate to a certain extent the changes which an avoidance of the assumptions in question would cause in our theory (cf. 66., 67.).

We feel, however, that one part of our assumptions at least—that of treating utilities as numerically measurable quantities—is not quite as radical as is often assumed in the literature. We shall attempt to prove this particular point in the paragraphs which follow. It is hoped that the reader will forgive us for discussing only incidentally in a condensed form a subject of so great a conceptual importance as that of utility. It seems however that even a few remarks may be helpful, because the question of the measurability of utilities is similar in character to corresponding questions in the physical sciences.

3.1.2. Historically, utility was first conceived as quantitatively measurable, i.e. as a number. Valid objections can be and have been made against this view in its original, naive form. It is clear that every measurement—or rather every claim of measurability—must ultimately be based on some immediate sensation, which possibly cannot and certainly need not be analyzed any further.[1] In the case of utility the immediate sensation of preference—of one object or aggregate of objects as against another—provides this basis. But this permits us only to say when for one person one utility is greater than another. It is not in itself a basis for numerical comparison of utilities for one person nor of any comparison between different persons. Since there is no intuitively significant way to add two utilities for the same person, the assumption that utilities are of non-numerical character even seems plausible. The modern method of indifference curve analysis is a mathematical procedure to describe this situation.

3.2. Principles of Measurement: Preliminaries

3.2.1. All this is strongly reminiscent of the conditions existant at the beginning of the theory of heat: that too was based on the intuitively clear concept of one body feeling warmer than another, yet there was no immediate way to express significantly by how much, or how many times, or in what sense.

[1] Such as the sensations of light, heat, muscular effort, etc., in the corresponding branches of physics.

This comparison with heat also shows how little one can forecast *a priori* what the ultimate shape of such a theory will be. The above crude indications do not disclose at all what, as we now know, subsequently happened. It turned out that heat permits quantitative description not by one number but by two: the quantity of heat and temperature. The former is rather directly numerical because it turned out to be additive and also in an unexpected way connected with mechanical energy which was numerical anyhow. The latter is also numerical, but in a much more subtle way; it is not additive in any immediate sense, but a rigid numerical scale for it emerged from the study of the concordant behavior of ideal gases, and the role of absolute temperature in connection with the entropy theorem.

3.2.2. The historical development of the theory of heat indicates that one must be extremely careful in making negative assertions about any concept with the claim to finality. Even if utilities look very unnumerical today, the history of the experience in the theory of heat may repeat itself, and nobody can foretell with what ramifications and variations.[1] And it should certainly not discourage theoretical explanations of the formal possibilities of a numerical utility.

3.3. Probability and Numerical Utilities

3.3.1. We can go even one step beyond the above double negations—which were only cautions against premature assertions of the impossibility of a numerical utility. It can be shown that under the conditions on which the indifference curve analysis is based very little extra effort is needed to reach a numerical utility.

It has been pointed out repeatedly that a numerical utility is dependent upon the possibility of comparing differences in utilities. This may seem—and indeed is—a more far-reaching assumption than that of a mere ability to state preferences. But it will seem that the alternatives to which economic preferences must be applied are such as to obliterate this distinction.

3.3.2. Let us for the moment accept the picture of an individual whose system of preferences is all-embracing and complete, i.e. who, for any two objects or rather for any two imagined events, possesses a clear intuition of preference.

More precisely we expect him, for any two alternative events which are put before him as possibilities, to be able to tell which of the two he prefers.

It is a very natural extension of this picture to permit such an individual to compare not only events, but even combinations of events with stated probabilities.[2]

By a combination of two events we mean this: Let the two events be denoted by B and C and use, for the sake of simplicity, the probability

[1] A good example of the wide variety of formal possibilities is given by the entirely different development of the theory of light, colors, and wave lengths. All these notions too became numerical, but in an entirely different way.

[2] Indeed this is necessary if he is engaged in economic activities which are explicitly dependent on probability. Cf. the example of agriculture in footnote 2 on p. 10.

50%-50%. Then the "combination" is the prospect of seeing B occur with a probability of 50% and (if B does not occur) C with the (remaining) probability of 50%. We stress that the two alternatives are mutually exclusive, so that no possibility of complementarity and the like exists. Also, that an absolute certainty of the occurrence of either B or C exists.

To restate our position. We expect the individual under consideration to possess a clear intuition whether he prefers the event A to the 50-50 combination of B or C, or conversely. It is clear that if he prefers A to B and also to C, then he will prefer it to the above combination as well; similarly, if he prefers B as well as C to A, then he will prefer the combination too. But if he should prefer A to, say B, but at the same time C to A, then any assertion about his preference of A against the combination contains fundamentally new information. Specifically: If he now prefers A to the 50-50 combination of B and C, this provides a plausible base for the numerical estimate that his preference of A over B is in excess of his preference of C over A.[1,2]

If this standpoint is accepted, then there is a criterion with which to compare the preference of C over A with the preference of A over B. It is well known that thereby utilities—or rather differences of utilities—become numerically measurable.

That the possibility of comparison between A, B, and C only to this extent is already sufficient for a numerical measurement of "distances" was first observed in economics by Pareto. Exactly the same argument has been made, however, by Euclid for the position of points on a line—in fact it is the very basis of his classical derivation of numerical distances.

The introduction of numerical measures can be achieved even more directly if use is made of all possible probabilities. Indeed: Consider three events, C, A, B, for which the order of the individual's preferences is the one stated. Let α be a real number between 0 and 1, such that A is exactly equally desirable with the combined event consisting of a chance of probability $1 - \alpha$ for B and the remaining chance of probability α for C. Then we suggest the use of α as a numerical estimate for the ratio of the preference of A over B to that of C over B.[3] An exact and exhaustive

[1] To give a simple example: Assume that an individual prefers the consumption of a glass of tea to that of a cup of coffee, and the cup of coffee to a glass of milk. If we now want to know whether the last preference—i.e., difference in utilities—exceeds the former, it suffices to place him in a situation where he must decide this: Does he prefer a cup of coffee to a glass the content of which will be determined by a 50 %-50 % chance device as tea or milk.

[2] Observe that we have only postulated an individual intuition which permits decision as to which of two "events" is preferable. But we have not directly postulated any intuitive estimate of the relative sizes of two preferences—i.e. in the subsequent terminology, of two differences of utilities.

This is important, since the former information ought to be obtainable in a reproducible way by mere "questioning."

[3] This offers a good opportunity for another illustrative example. The above technique permits a direct determination of the ratio q of the utility of possessing 1 unit of a certain good to the utility of possessing 2 units of the same good. The individual must

elaboration of these ideas requires the use of the axiomatic method. A simple treatment on this basis is indeed possible. We shall discuss it in 3.5-3.7.

3.3.3. To avoid misunderstandings let us state that the "events" which were used above as the substratum of preferences are conceived as future events so as to make all logically possible alternatives equally admissible. However, it would be an unnecessary complication, as far as our present objectives are concerned, to get entangled with the problems of the preferences between events in different periods of the future.[1] It seems, however, that such difficulties can be obviated by locating all "events" in which we are interested at one and the same, standardized, moment, preferably in the immediate future.

The above considerations are so vitally dependent upon the numerical concept of probability that a few words concerning the latter may be appropriate.

Probability has often been visualized as a subjective concept more or less in the nature of an estimation. Since we propose to use it in constructing an individual, numerical estimation of utility, the above view of probability would not serve our purpose. The simplest procedure is, therefore, to insist upon the alternative, perfectly well founded interpretation of probability as frequency in long runs. This gives directly the necessary numerical foothold.[2]

3.3.4. This procedure for a numerical measurement of the utilities of the individual depends, of course, upon the hypothesis of completeness in the system of individual preferences.[3] It is conceivable—and may even in a way be more realistic—to allow for cases where the individual is neither able to state which of two alternatives he prefers nor that they are equally desirable. In this case the treatment by indifference curves becomes impracticable too.[4]

How real this possibility is, both for individuals and for organizations, seems to be an extremely interesting question, but it is a question of fact. It certainly deserves further study. We shall reconsider it briefly in 3.7.2.

At any rate we hope we have shown that the treatment by indifference curves implies either too much or too little: if the preferences of the indi-

---

be given the choice of obtaining 1 unit with certainty or of playing the chance to get two units with the probability α, or nothing with the probability $1 - \alpha$. If he prefers the former, then $\alpha < q$; if he prefers the latter, then $\alpha > q$; if he cannot state a preference either way, then $\alpha = q$.

[1] It is well known that this presents very interesting, but as yet extremely obscure, connections with the theory of saving and interest, etc.

[2] If one objects to the frequency interpretation of probability then the two concepts (probability and preference) can be axiomatized together. This too leads to a satisfactory numerical concept of utility which will be discussed on another occasion.

[3] We have not obtained any basis for a comparison, quantitatively or qualitatively, of the utilities of different individuals.

[4] These problems belong systematically in the mathematical theory of ordered sets. The above question in particular amounts to asking whether events, with respect to preference, form a completely or a partially ordered set. Cf. 65.3.

vidual are not all comparable, then the indifference curves do not exist.[1] If the individual's preferences are all comparable, then we can even obtain a (uniquely defined) numerical utility which renders the indifference curves superfluous.

All this becomes, of course, pointless for the entrepreneur who can calculate in terms of (monetary) costs and profits.

3.3.5. The objection could be raised that it is not necessary to go into all these intricate details concerning the measurability of utility, since evidently the common individual, whose behavior one wants to describe, does not measure his utilities exactly but rather conducts his economic activities in a sphere of considerable haziness. The same is true, of course, for much of his conduct regarding light, heat, muscular effort, etc. But in order to build a science of physics these phenomena had to be measured. And subsequently the individual has come to use the results of such measurements—directly or indirectly—even in his everyday life. The same may obtain in economics at a future date. Once a fuller understanding of economic behavior has been achieved with the aid of a theory which makes use of this instrument, the life of the individual might be materially affected. It is, therefore, not an unnecessary digression to study these problems.

3.4. Principles of Measurement: Detailed Discussion

3.4.1. The reader may feel, on the basis of the foregoing, that we obtained a numerical scale of utility only by begging the principle, i.e. by really postulating the existence of such a scale. We have argued in 3.3.2. that if an individual prefers A to the 50-50 combination of B and C (while preferring C to A and A to B), this provides a plausible basis for the numerical estimate that this preference of A over B exceeds that of C over A. Are we not postulating here—or taking it for granted—that one preference may exceed another, i.e. that such statements convey a meaning? Such a view would be a complete misunderstanding of our procedure.

3.4.2. We are not postulating—or assuming—anything of the kind. We have assumed only one thing—and for this there is good empirical evidence—namely that imagined events can be combined with probabilities. And therefore the same must be assumed for the utilities attached to them,—whatever they may be. Or to put it in more mathematical language:

There frequently appear in science quantities which are *a priori* not mathematical, but attached to certain aspects of the physical world. Occasionally these quantities can be grouped together in domains within which certain natural, physically defined operations are possible. Thus the physically defined quantity of "mass" permits the operation of addition. The physico-geometrically defined quantity of "distance"[2] permits the same

[1] Points on the same indifference curve must be identified and are therefore no instances of incomparability.

[2] Let us, for the sake of the argument, view geometry as a physical discipline,—a sufficiently tenable viewpoint. By "geometry" we mean—equally for the sake of the argument—Euclidean geometry.

operation. On the other hand, the physico-geometrically defined quantity of "position" does not permit this operation,[1] but it permits the operation of forming the "center of gravity" of two positions.[2] Again other physico-geometrical concepts, usually styled "vectorial"—like velocity and acceleration—permit the operation of "addition."

3.4.3. In all these cases where such a "natural" operation is given a name which is reminiscent of a mathematical operation—like the instances of "addition" above—one must carefully avoid misunderstandings. This nomenclature is not intended as a claim that the two operations with the same name are identical,—this is manifestly not the case; it only expresses the opinion that they possess similar traits, and the hope that some correspondence between them will ultimately be established. This of course—when feasible at all—is done by finding a mathematical model for the physical domain in question, within which those quantities are defined by numbers, so that in the model the mathematical operation describes the synonymous "natural" operation.

To return to our examples: "energy" and "mass" became numbers in the pertinent mathematical models, "natural" addition becoming ordinary addition. "Position" as well as the vectorial quantities became triplets[3] of numbers, called coordinates or components respectively. The "natural" concept of "center of gravity" of two positions $\{x_1, x_2, x_3\}$ and $\{x'_1, x'_2, x'_3\}$,[4] with the "masses" α, $1 - \alpha$ (cf. footnote 2 above), becomes

$$\{\alpha x_1 + (1 - \alpha)x'_1, \alpha x_2 + (1 - \alpha)x'_2, \alpha x_3 + (1 - \alpha)x'_3\}.^5$$

The "natural" operation of "addition" of vectors $\{x_1, x_2, x_3\}$ and $\{x'_1, x'_2, x'_3\}$ becomes $\{x_1 + x'_1, x_2 + x'_2, x_3 + x'_3\}$.[6]

What was said above about "natural" and mathematical operations applies equally to natural and mathematical relations. The various concepts of "greater" which occur in physics—greater energy, force, heat, velocity, etc.—are good examples.

These "natural" relations are the best base upon which to construct mathematical models and to correlate the physical domain with them.[7,8]

[1] We are thinking of a "homogeneous" Euclidean space, in which no origin or frame of reference is preferred above any other.

[2] With respect to two given masses α, β occupying those positions. It may be convenient to normalize so that the total mass is the unit, i.e. $\beta = 1 - \alpha$.

[3] We are thinking of three-dimensional Euclidean space.

[4] We are now describing them by their three numerical coordinates.

[5] This is usually denoted by $\alpha\{x_1, x_2, x_3\} + (1 - \alpha)\{x'_1, x'_2, x'_3\}$. Cf. (16:A:c) in 16.2.1.

[6] This is usually denoted by $\{x_1, x_2, x_3\} + \{x'_1, x'_2, x'_3\}$. Cf. the beginning of 16.2.1.

[7] Not the only one. Temperature is a good counter-example. The "natural" relation of "greater" would not have sufficed to establish the present day mathematical model,—i.e. the absolute temperature scale. The devices actually used were different. Cf. 3.2.1.

[8] We do not want to give the misleading impression of attempting here a complete picture of the formation of mathematical models, i.e. of physical theories. It should be remembered that this is a very varied process with many unexpected phases. An important one is, e.g., the disentanglement of concepts: i.e. splitting up something which at

3.4.4. Here a further remark must be made. Assume that a satisfactory mathematical model for a physical domain in the above sense has been found, and that the physical quantities under consideration have been correlated with numbers. In this case it is not true necessarily that the description (of the mathematical model) provides for a *unique* way of correlating the physical quantities to numbers; i.e., it may specify an entire family of such correlations—the mathematical name is mappings—any one of which can be used for the purposes of the theory. Passage from one of these correlations to another amounts to a *transformation* of the numerical data describing the physical quantities. We then say that in this theory the physical quantities in question are described by numbers *up to* that system of transformations. The mathematical name of such transformation systems is *groups*.[1]

Examples of such situations are numerous. Thus the geometrical concept of distance is a number, up to multiplication by (positive) constant factors.[2] The situation concerning the physical quantity of mass is the same. The physical concept of energy is a number up to any linear transformation,—i.e. addition of any constant and multiplication by any (positive) constant.[3] The concept of position is defined up to an inhomogeneous orthogonal linear transformation.[4,5] The vectorial concepts are defined up to homogeneous transformations of the same kind.[5,6]

3.4.5. It is even conceivable that a physical quantity is a number up to any monotone transformation. This is the case for quantities for which only a "natural" relation "greater" exists—and nothing else. E.g. this was the case for temperature as long as only the concept of "warmer" was known;[7] it applies to the Mohs' scale of hardness of minerals; it applies to

---

superficial inspection seems to be one physical entity into several mathematical notions. Thus the "disentanglement" of force and energy, of quantity of heat and temperature, were decisive in their respective fields.

It is quite unforeseeable how many such differentiations still lie ahead in economic theory.

[1] We shall encounter groups in another context in 28.1.1, where references to the literature are also found.

[2] I.e. there is nothing in Euclidean geometry to fix a unit of distance.

[3] I.e. there is nothing in mechanics to fix a zero or a unit of energy. Cf. with footnote 2 above. Distance has a natural zero,—the distance of any point from itself.

[4] I.e. $\{x_1, x_2, x_3\}$ are to be replaced by $\{x_1{}^*, x_2{}^*, x_3{}^*\}$ where

$$\begin{aligned} x_1{}^* &= a_{11}x_1 + a_{12}x_2 + a_{13}x_3 + b_1, \\ x_2{}^* &= a_{21}x_1 + a_{22}x_2 + a_{23}x_3 + b_2, \\ x_3{}^* &= a_{31}x_1 + a_{32}x_2 + a_{33}x_3 + b_3, \end{aligned}$$

the a_{ij}, b_i being constants, and the matrix (a_{ij}) what is known as orthogonal.

[5] I.e. there is nothing in geometry to fix either origin or the frame of reference when positions are concerned; and nothing to fix the frame of reference when vectors are concerned.

[6] I.e. the $b_i = 0$ in footnote 4 above. Sometimes a wider concept of matrices is permissible,—all those with determinants $\neq 0$. We need not discuss these matters here.

[7] But no quantitatively reproducible method of thermometry.

the notion of utility when this is based on the conventional idea of preference. In these cases one may be tempted to take the view that the quantity in question is not numerical at all, considering how arbitrary the description by numbers is. It seems to be preferable, however, to refrain from such qualitative statements and to state instead objectively up to what system of transformations the numerical description is determined. The case when the system consists of all monotone transformations is, of course, a rather extreme one; various graduations at the other end of the scale are the transformation systems mentioned above: inhomogeneous or homogeneous orthogonal linear transformations in space, linear transformations of one numerical variable, multiplication of that variable by a constant.[1] *In fine*, the case even occurs where the numerical description is absolutely rigorous, i.e. where no transformations at all need be tolerated.[2]

3.4.6. Given a physical quantity, the system of transformations up to which it is described by numbers may vary in time, i.e. with the stage of development of the subject. Thus temperature was originally a number only up to any monotone transformation.[3] With the development of thermometry—particularly of the concordant ideal gas thermometry—the transformations were restricted to the linear ones, i.e. only the absolute zero and the absolute unit were missing. Subsequent developments of thermodynamics even fixed the absolute zero so that the transformation system in thermodynamics consists only of the multiplication by constants. Examples could be multiplied but there seems to be no need to go into this subject further.

For utility the situation seems to be of a similar nature. One may take the attitude that the only "natural" datum in this domain is the relation "greater," i.e. the concept of preference. In this case utilities are numerical up to a monotone transformation. This is, indeed, the generally accepted standpoint in economic literature, best expressed in the technique of indifference curves.

To narrow the system of transformations it would be necessary to discover further "natural" operations or relations in the domain of utility. Thus it was pointed out by Pareto[4] that an equality relation for utility differences would suffice; in our terminology it would reduce the transformation system to the linear transformations.[5] However, since it does not

[1] One could also imagine intermediate cases of greater transformation systems than these but not containing all monotone transformations. Various forms of the theory of relativity give rather technical examples of this.

[2] In the usual language this would hold for physical quantities where an absolute zero as well as an absolute unit can be defined. This is, e.g., the case for the absolute value (not the vector!) of velocity in such physical theories as those in which light velocity plays a normative role: Maxwellian electrodynamics, special relativity.

[3] As long as only the concept of "warmer"—i.e. a "natural" relation "greater"—was known. We discussed this *in extenso* previously.

[4] *V. Pareto*, Manuel d'Economie Politique, Paris, 1907, p. 264.

[5] This is exactly what Euclid did for position on a line. The utility concept of "preference" corresponds to the relation of "lying to the right of" there, and the (desired) relation of the equality of utility differences to the geometrical congruence of intervals.

seem that this relation is really a "natural" one—i.e. one which can be interpreted by reproducible observations—the suggestion does not achieve the purpose.

3.5. Conceptual Structure of the Axiomatic Treatment of Numerical Utilities

3.5.1. The failure of one particular device need not exclude the possibility of achieving the same end by another device. Our contention is that the domain of utility contains a "natural" operation which narrows the system of transformations to precisely the same extent as the other device would have done. This is the combination of two utilities with two given alternative probabilities α, $1 - \alpha$, $(0 < \alpha < 1)$ as described in 3.3.2. The process is so similar to the formation of centers of gravity mentioned in 3.4.3. that it may be advantageous to use the same terminology. Thus we have for utilities u, v the "natural" *relation* $u > v$ (read: u is preferable to v), and the "natural" *operation* $\alpha u + (1 - \alpha)v$, $(0 < \alpha < 1)$, (read: center of gravity of u, v with the respective weights α, $1 - \alpha$; or: combination of u, v with the alternative probabilities α, $1 - \alpha$). If the existence—and reproducible observability —of these concepts is conceded, then our way is clear: We must find a correspondence between utilities and numbers which carries the relation $u > v$ and the operation $\alpha u + (1 - \alpha)v$ for utilities into the synonymous concepts for numbers.

Denote the correspondence by

$$u \to \rho = \mathrm{v}(u),$$

u being the utility and $\mathrm{v}(u)$ the number which the correspondence attaches to it. Our requirements are then:

(3:1:a) $\qquad u > v \quad$ implies $\quad \mathrm{v}(u) > \mathrm{v}(v),$

(3:1:b) $\qquad \mathrm{v}(\alpha u + (1 - \alpha)v) = \alpha\mathrm{v}(u) + (1 - \alpha)\mathrm{v}(v).$[1]

If two such correspondences

(3:2:a) $\qquad u \to \rho = \mathrm{v}(u),$

(3:2:b) $\qquad u \to \rho' = \mathrm{v}'(u),$

should exist, then they set up a correspondence between numbers

(3:3) $\qquad \rho \leftrightarrows \rho',$

for which we may also write

(3:4) $\qquad \rho' = \phi(\rho).$

Since (3:2:a), (3:2:b) fulfill (3:1:a), (3:1:b), the correspondence (3:3), i.e. the function $\phi(\rho)$ in (3:4) must leave the relation $\rho > \sigma$ [2] and the operation

[1] Observe that in in each case the left-hand side has the "natural" concepts for utilities, and the right-hand side the conventional ones for numbers.

[2] Now these are applied to numbers ρ, σ!

$\alpha\rho + (1 - \alpha)\sigma$ unaffected (cf footnote 1 on p. 24). I.e.

(3:5:a) $\rho > \sigma$ implies $\phi(\rho) > \phi(\sigma)$,

(3:5:b) $\phi(\alpha\rho + (1 - \alpha)\sigma) = \alpha\phi(\rho) + (1 - \alpha)\phi(\sigma)$.

Hence $\phi(\rho)$ must be a linear function, i.e.

(3:6) $$\rho' = \phi(\rho) \equiv \omega_0\rho + \omega_1,$$

where ω_0, ω_1 are fixed numbers (constants) with $\omega_0 > 0$.

So we see: If such a numerical valuation of utilities[1] exists at all, then it is determined up to a linear transformation.[2,3] I.e. then utility is a number up to a linear transformation.

In order that a numerical valuation in the above sense should exist it is necessary to postulate certain properties of the relation $u > v$ and the operation $\alpha u + (1 - \alpha)v$ for utilities. The selection of these postulates or axioms and their subsequent analysis leads to problems of a certain mathematical interest. In what follows we give a general outline of the situation for the orientation of the reader; a complete discussion is found in the Appendix.

3.5.2. A choice of axioms is not a purely objective task. It is usually expected to achieve some definite aim—some specific theorem or theorems are to be derivable from the axioms—and to this extent the problem is exact and objective. But beyond this there are always other important desiderata of a less exact nature: The axioms should not be too numerous, their system is to be as simple and transparent as possible, and each axiom should have an immediate intuitive meaning by which its appropriateness may be judged directly.[4] In a situation like ours this last requirement is particularly vital, in spite of its vagueness: we want to make an intuitive concept amenable to mathematical treatment and to see as clearly as possible what hypotheses this requires.

The objective part of our problem is clear: the postulates must imply the existence of a correspondence (3:2:a) with the properties (3:1:a), (3:1:b) as described in 3.5.1. The further heuristic, and even esthetic desiderata, indicated above, do not determine a unique way of finding this axiomatic treatment. In what follows we shall formulate a set of axioms which seems to be essentially satisfactory.

[1] I.e. a correspondence (3:2:a) which fulfills (3:1:a), (3:1:b).

[2] I.e. one of the form (3:6).

[3] Remember the physical examples of the same situation given in 3.4.4. (Our present discussion is somewhat more detailed.) We do not undertake to fix an absolute zero and an absolute unit of utility.

[4] The first and the last principle may represent—at least to a certain extent—opposite influences: If we reduce the number of axioms by merging them as far as technically possible, we may lose the possibility of distinguishing the various intuitive backgrounds. Thus we could have expressed the group (3:B) in 3.6.1. by a smaller number of axioms, but this would have obscured the subsequent analysis of 3.6.2.

To strike a proper balance is a matter of practical—and to some extent even esthetic—judgment.

3.6. The Axioms and Their Interpretation

3.6.1. Our axioms are these:

We consider a system U of entities[1] $u, v, w, \cdots$. In U a *relation* is given, $u > v$, and for any number α, $(0 < \alpha < 1)$, an *operation*

$$\alpha u + (1 - \alpha)v = w.$$

These concepts satisfy the following axioms:

(3:A) $u > v$ *is a complete ordering of* U.[2]

This means: Write $u < v$ when $v > u$. Then:

(3:A:a) For any two u, v one and only one of the three following relations holds:

$$u = v, \qquad u > v, \qquad u < v.$$

(3:A:b) $u > v$, $v > w$ imply $u > w$.[3]

(3:B) *Ordering and combining.*[4]

(3:B:a) $u < v$ implies that $u < \alpha u + (1 - \alpha)v$.

(3:B:b) $u > v$ implies that $u > \alpha u + (1 - \alpha)v$.

(3:B:c) $u < w < v$ implies the existence of an α with

$$\alpha u + (1 - \alpha)v < w.$$

(3:B:d) $u > w > v$ implies the existence of an α with

$$\alpha u + (1 - \alpha)v > w.$$

(3:C) *Algebra of combining.*

(3:C:a) $\alpha u + (1 - \alpha)v = (1 - \alpha)v + \alpha u.$

(3:C:b) $\alpha(\beta u + (1 - \beta)v) + (1 - \alpha)v = \gamma u + (1 - \gamma)v$ where $\gamma = \alpha\beta$.

One can show that these axioms imply the existence of a correspondence (3:2:a) with the properties (3:1:a), (3:1:b) as described in 3.5.1. Hence the conclusions of 3.5.1. hold good: The system U—i.e. in our present interpretation, the system of (abstract) utilities—is one of numbers up to a linear transformation.

The construction of (3:2:a) (with (3:1:a), (3:1:b) by means of the axioms (3:A)-(3:C)) is a purely mathematical task which is somewhat lengthy, although it runs along conventional lines and presents no par-

[1] This is, of course, meant to be the system of (abstract) utilities, to be characterized by our axioms. Concerning the general nature of the axiomatic method, cf. the remarks and references in the last part of 10.1.1.

[2] For a more systematic mathematical discussion of this notion, cf. 65.3.1. The equivalent concept of the completeness of the system of preferences was previously considered at the beginning of 3.3.2. and of 3.4.6.

[3] These conditions (3:A:a), (3:A:b) correspond to (65:A:a), (65:A:b) in 65.3.1.

[4] Remember that the α, β, γ occurring here are always > 0, < 1.

ticular difficulties. (Cf. Appendix.)

It seems equally unnecessary to carry out the usual logistic discussion of these axioms[1] on this occasion.

We shall however say a few more words about the intuitive meaning—i.e. the justification—of each one of our axioms (3:A)-(3:C).

3.6.2. The analysis of our postulates follows:

(3:A:a*) This is the statement of the completeness of the system of individual preferences. It is customary to assume this when discussing utilities or preferences, e.g. in the "indifference curve analysis method." These questions were already considered in 3.3.4. and 3.4.6.

(3:A:b*) This is the "transitivity" of preference, a plausible and generally accepted property.

(3:B:a*) We state here: If v is preferable to u, then even a chance $1 - \alpha$ of v—alternatively to u—is preferable. This is legitimate since any kind of complementarity (or the opposite) has been excluded, cf. the beginning of 3.3.2.

(3:B:b*) This is the dual of (3:B:a*), with "less preferable" in place of "preferable."

(3:B:c*) We state here: If w is preferable to u, and an even more preferable v is also given, then the combination of u with a chance $1 - \alpha$ of v will not affect w's preferability to it if this chance is small enough. I.e.: However desirable v may be in itself, one can make its influence as weak as desired by giving it a sufficiently small chance. This is a plausible "continuity" assumption.

(3:B:d*) This is the dual of (3:B:c*), with "less preferable" in place of "preferable."

(3:C:a*) This is the statement that it is irrelevant in which order the constituents u, v of a combination are named. It is legitimate, particularly since the constituents are alternative events, cf. (3:B:a*) above.

(3:C:b*) This is the statement that it is irrelevant whether a combination of two constituents is obtained in two successive steps,—first the probabilities α, $1 - \alpha$, then the probabilities β, $1 - \beta$; or in one operation,—the probabilities γ, $1 - \gamma$ where $\gamma = \alpha\beta$.[2] The same things can be said for this as for (3:C:a*) above. It may be, however, that this postulate has a deeper significance, to which one allusion is made in 3.7.1. below.

[1] A similar situation is dealt with more exhaustively in 10.; those axioms describe a subject which is more vital for our main objective. The logistic discussion is indicated there in 10.2. Some of the general remarks of 10.3. apply to the present case also.

[2] This is of course the correct arithmetic of accounting for two successive admixtures of v with u.

3.7. General Remarks Concerning the Axioms

3.7.1. At this point it may be well to stop and to reconsider the situation. Have we not shown too much? We can derive from the postulates (3:A)-(3:C) the numerical character of utility in the sense of (3:2:a) and (3:1:a), (3:1:b) in 3.5.1.; and (3:1:b) states that the numerical values of utility combine (with probabilities) like mathematical expectations! And yet the concept of mathematical expectation has been often questioned, and its legitimateness is certainly dependent upon some hypothesis concerning the nature of an "expectation."[1] Have we not then begged the question? Do not our postulates introduce, in some oblique way, the hypotheses which bring in the mathematical expectation?

More specifically: May there not exist in an individual a (positive or negative) utility of the mere act of "taking a chance," of gambling, which the use of the mathematical expectation obliterates?

How did our axioms (3:A)-(3:C) get around this possibility?

As far as we can see, our postulates (3:A)-(3:C) do not attempt to avoid it. Even that one which gets closest to excluding a "utility of gambling" (3:C:b) (cf. its discussion in 3.6.2.), seems to be plausible and legitimate,—unless a much more refined system of psychology is used than the one now available for the purposes of economics. The fact that a numerical utility—with a formula amounting to the use of mathematical expectations—can be built upon (3:A)-(3:C), seems to indicate this: We have practically defined numerical utility as being that thing for which the calculus of mathematical expectations is legitimate.[2] Since (3:A)-(3:C) secure that the necessary construction can be carried out, concepts like a "specific utility of gambling" cannot be formulated free of contradiction on this level.[3]

3.7.2. As we have stated, the last time in 3.6.1., our axioms are based on the relation $u > v$ and on the operation $\alpha u + (1 - \alpha)v$ for utilities. It seems noteworthy that the latter may be regarded as more immediately given than the former: One can hardly doubt that anybody who could imagine two alternative situations with the respective utilities u, v could not also conceive the prospect of having both with the given respective probabilities α, $1 - \alpha$. On the other hand one may question the postulate of axiom (3:A:a) for $u > v$, i.e. the completeness of this ordering.

Let us consider this point for a moment. We have conceded that one may doubt whether a person can always decide which of two alternatives—

[1] Cf. *Karl Menger:* Das Unsicherheitsmoment in der Wertlehre, Zeitschrift für Nationalökonomie, vol. 5, (1934) pp. 459ff. and *Gerhard Tintner:* A contribution to the non-static Theory of Choice, Quarterly Journal of Economics, vol. LVI, (1942) pp. 274ff.

[2] Thus Daniel Bernoulli's well known suggestion to "solve" the "St. Petersburg Paradox" by the use of the so-called "moral expectation" (instead of the mathematical expectation) means defining the utility numerically as the logarithm of one's monetary possessions.

[3] This may seem to be a paradoxical assertion. But anybody who has seriously tried to axiomatize that elusive concept, will probably concur with it.

with the utilities u, v—he prefers.[1] But, whatever the merits of this doubt are, this possibility—i.e. the completeness of the system of (individual) preferences—must be assumed even for the purposes of the "indifference curve method" (cf. our remarks on (3:A:a) in 3.6.2.). But if this property of $u > v$ [2] is assumed, then our use of the much less questionable $\alpha u + (1 - \alpha)v$ [3] yields the numerical utilities too![4]

If the general comparability assumption is not made,[5] a mathematical theory—based on $\alpha u + (1 - \alpha)v$ together with what remains of $u > v$—is still possible.[6] It leads to what may be described as a many-dimensional vector concept of utility. This is a more complicated and less satisfactory set-up, but we do not propose to treat it systematically at this time.

3.7.3. This brief exposition does not claim to exhaust the subject, but we hope to have conveyed the essential points. To avoid misunderstandings, the following further remarks may be useful.

(1) We re-emphasize that we are considering only utilities experienced by one person. These considerations do not imply anything concerning the comparisons of the utilities belonging to different individuals.

(2) It cannot be denied that the analysis of the methods which make use of mathematical expectation (cf. footnote 1 on p. 28 for the literature) is far from concluded at present. Our remarks in 3.7.1. lie in this direction, but much more should be said in this respect. There are many interesting questions involved, which however lie beyond the scope of this work. For our purposes it suffices to observe that the validity of the simple and plausible axioms (3:A)-(3:C) in 3.6.1. for the relation $u > v$ and the operation $\alpha u + (1 - \alpha)v$ makes the utilities numbers up to a linear transformation in the sense discussed in these sections.

3.8. The Role of the Concept of Marginal Utility

3.8.1. The preceding analysis made it clear that we feel free to make use of a numerical conception of utility. On the other hand, subsequent

[1] Or that he can assert that they are precisely equally desirable.

[2] I.e. the completeness postulate (3:A:a).

[3] I.e. the postulates (3:B), (3:C) together with the obvious postulate (3:A:b).

[4] At this point the reader may recall the familiar argument according to which the unnumerical ("indifference curve") treatment of utilities is preferable to any numerical one, because it is simpler and based on fewer hypotheses. This objection might be legitimate if the numerical treatment were based on Pareto's equality relation for utility differences (cf. the end of 3.4.6.). This relation is, indeed, a stronger and more complicated hypothesis, added to the original ones concerning the general comparability of utilities (completeness of preferences).

However, we used the operation $\alpha u + (1 - \alpha)v$ instead, and we hope that the reader will agree with us that it represents an even safer assumption than that of the completeness of preferences.

We think therefore that our procedure, as distinguished from Pareto's, is not open to the objections based on the necessity of artificial assumptions and a loss of simplicity.

[5] This amounts to weakening (3:A:a) to an (3:A:a') by replacing in it "one and only one" by "at most one." The conditions (3:A:a'), (3:A:b) then correspond to (65:B:a), (65:B:b).

[6] In this case some modifications in the groups of postulates (3:B), (3:C) are also necessary.

discussions will show that we cannot avoid the assumption that all subjects of the economy under consideration are completely informed about the physical characteristics of the situation in which they operate and are able to perform all statistical, mathematical, etc., operations which this knowledge makes possible. The nature and importance of this assumption has been given extensive attention in the literature and the subject is probably very far from being exhausted. We propose not to enter upon it. The question is too vast and too difficult and we believe that it is best to "divide difficulties." I.e. we wish to avoid this complication which, while interesting in its own right, should be considered separately from our present problem.

Actually we think that our investigations—although they assume "complete information" without any further discussion—do make a contribution to the study of this subject. It will be seen that many economic and social phenomena which are usually ascribed to the individual's state of "incomplete information" make their appearance in our theory and can be satisfactorily interpreted with its help. Since our theory assumes "complete information," we conclude from this that those phenomena have nothing to do with the individual's "incomplete information." Some particularly striking examples of this will be found in the concepts of "discrimination" in 33.1., of "incomplete exploitation" in 38.3., and of the "transfer" or "tribute" in 46.11., 46.12.

On the basis of the above we would even venture to question the importance usually ascribed to incomplete information in its conventional sense[1] in economic and social theory. It will appear that some phenomena which would *prima facie* have to be attributed to this factor, have nothing to do with it.[2]

3.8.2. Let us now consider an isolated individual with definite physical characteristics and with definite quantities of goods at his disposal. In view of what was said above, he is in a position to determine the maximum utility which can be obtained in this situation. Since the maximum is a well-defined quantity, the same is true for the increase which occurs when a unit of any definite good is added to the stock of all goods in the possession of the individual. This is, of course, the classical notion of the marginal utility of a unit of the commodity in question.[3]

These quantities are clearly of decisive importance in the "Robinson Crusoe" economy. The above marginal utility obviously corresponds to

[1] We shall see that the rules of the games considered may explicitly prescribe that certain participants should not possess certain pieces of information. Cf. 6.3., 6.4. (Games in which this does not happen are referred to in 14.8. and in (15:B) of 15.3.2., and are called games with "perfect information.") We shall recognize and utilize this kind of "incomplete information" (according to the above, rather to be called "imperfect information"). But we reject all other types, vaguely defined by the use of concepts like complication, intelligence, etc.

[2] Our theory attributes these phenomena to the possibility of multiple "stable standards of behavior" cf. 4.6. and the end of 4.7.

[3] More precisely: the so-called "indirectly dependent expected utility."

the maximum effort which he will be willing to make—if he behaves according to the customary criteria of rationality—in order to obtain a further unit of that commodity.

It is not clear at all, however, what significance it has in determining the behavior of a participant in a social exchange economy. We saw that the principles of rational behavior in this case still await formulation, and that they are certainly not expressed by a maximum requirement of the Crusoe type. Thus it must be uncertain whether marginal utility has any meaning at all in this case.[1]

Positive statements on this subject will be possible only after we have succeeded in developing a theory of rational behavior in a social exchange economy,—that is, as was stated before, with the help of the theory of "games of strategy." It will be seen that marginal utility does, indeed, play an important role in this case too, but in a more subtle way than is usually assumed.

4. Structure of the Theory: Solutions and Standards of Behavior

4.1. The Simplest Concept of a Solution for One Participant

4.1.1. We have now reached the point where it becomes possible to give a positive description of our proposed procedure. This means primarily an outline and an account of the main technical concepts and devices.

As we stated before, we wish to find the mathematically complete principles which define "rational behavior" for the participants in a social economy, and to derive from them the general characteristics of that behavior. And while the principles ought to be perfectly general—i.e., valid in all situations—we may be satisfied if we can find solutions, for the moment, only in some characteristic special cases.

First of all we must obtain a clear notion of what can be accepted as a solution of this problem; i.e., what the amount of information is which a solution must convey, and what we should expect regarding its formal structure. A precise analysis becomes possible only after these matters have been clarified.

4.1.2. The immediate concept of a solution is plausibly a set of rules for each participant which tell him how to behave in every situation which may conceivably arise. One may object at this point that this view is unnecessarily inclusive. Since we want to theorize about "rational behavior," there seems to be no need to give the individual advice as to his behavior in situations other than those which arise in a rational community. This would justify assuming rational behavior on the part of the others as well,—in whatever way we are going to characterize that. Such a procedure would probably lead to a unique sequence of situations to which alone our theory need refer.

[1] All this is understood within the domain of our several simplifying assumptions. If they are relaxed, then various further difficulties ensue.

This objection seems to be invalid for two reasons:

First, the "rules of the game,"—i.e. the physical laws which give the factual background of the economic activities under consideration may be explicitly statistical. The actions of the participants of the economy may determine the outcome only in conjunction with events which depend on chance (with known probabilities), cf. footnote 2 on p. 10 and 6.2.1. If this is taken into consideration, then the rules of behavior even in a perfectly rational community must provide for a great variety of situations—some of which will be very far from optimum.[1]

Second, and this is even more fundamental, the rules of rational behavior must provide definitely for the possibility of irrational conduct on the part of others. In other words: Imagine that we have discovered a set of rules for all participants—to be termed as "optimal" or "rational"—each of which is indeed optimal provided that the other participants conform. Then the question remains as to what will happen if some of the participants do not conform. If that should turn out to be advantageous for them—and, quite particularly, disadvantageous to the conformists—then the above "solution" would seem very questionable. We are in no position to give a positive discussion of these things as yet—but we want to make it clear that under such conditions the "solution," or at least its motivation, must be considered as imperfect and incomplete. In whatever way we formulate the guiding principles and the objective justification of "rational behavior," provisos will have to be made for every possible conduct of "the others." Only in this way can a satisfactory and exhaustive theory be developed. But if the superiority of "rational behavior" over any other kind is to be established, then its description must include rules of conduct for all conceivable situations—including those where "the others" behaved irrationally, in the sense of the standards which the theory will set for them.

4.1.3. At this stage the reader will observe a great similarity with the everyday concept of games. We think that this similarity is very essential; indeed, that it is more than that. For economic and social problems the games fulfill—or should fulfill—the same function which various geometrico-mathematical models have successfully performed in the physical sciences. Such models are theoretical constructs with a precise, exhaustive and not too complicated definition; and they must be similar to reality in those respects which are essential in the investigation at hand. To recapitulate in detail: The definition must be precise and exhaustive in order to make a mathematical treatment possible. The construct must not be unduly complicated, so that the mathematical treatment can be brought beyond the mere formalism to the point where it yields complete numerical results. Similarity to reality is needed to make the operation significant. And this similarity must usually be restricted to a few traits

[1] That a unique optimal behavior is at all conceivable in spite of the multiplicity of the possibilities determined by chance, is of course due to the use of the notion of "mathematical expectation." Cf. loc. cit. above.

deemed "essential" *pro tempore*—since otherwise the above requirements would conflict with each other.[1]

It is clear that if a model of economic activities is constructed according to these principles, the description of a game results. This is particularly striking in the formal description of markets which are after all the core of the economic system—but this statement is true in all cases and without qualifications.

4.1.4. We described in 4.1.2. what we expect a solution—i.e. a characterization of "rational behavior"—to consist of. This amounted to a complete set of rules of behavior in all conceivable situations. This holds equivalently for a social economy and for games. The entire result in the above sense is thus a combinatorial enumeration of enormous complexity. But we have accepted a simplified concept of utility according to which all the individual strives for is fully described by one numerical datum (cf. 2.1.1. and 3.3.). Thus the complicated combinatorial catalogue—which we expect from a solution—permits a very brief and significant summarization: the statement of how much[2,3] the participant under consideration can get if he behaves "rationally." This "can get" is, of course, presumed to be a minimum; he may get more if the others make mistakes (behave irrationally).

It ought to be understood that all this discussion is advanced, as it should be, preliminary to the building of a satisfactory theory along the lines indicated. We formulate desiderata which will serve as a gauge of success in our subsequent considerations; but it is in accordance with the usual heuristic procedure to reason about these desiderata—even before we are able to satisfy them. Indeed, this preliminary reasoning is an essential part of the process of finding a satisfactory theory.[4]

4.2. Extension to All Participants

4.2.1. We have considered so far only what the solution ought to be for one participant. Let us now visualize all participants simultaneously. I.e., let us consider a social economy, or equivalently a game of a fixed number of (say n) participants. The complete information which a solution should convey is, as we discussed it, of a combinatorial nature. It was indicated furthermore how a single quantitative statement contains the decisive part of this information, by stating how much each participant

[1] E.g., Newton's description of the solar system by a small number of "masspoints." These points attract each other and move like the stars; this is the similarity in the essentials, while the enormous wealth of the other physical features of the planets has been left out of account.

[2] Utility; for an entrepreneur,—profit; for a player,—gain or loss.

[3] We mean, of course, the "mathematical expectation," if there is an explicit element of chance. Cf. the first remark in 4.1.2. and also the discussion of 3.7.1.

[4] Those who are familiar with the development of physics will know how important such heuristic considerations can be. Neither general relativity nor quantum mechanics could have been found without a "pre-theoretical" discussion of the desiderata concerning the theory-to-be.

obtains by behaving rationally. Consider these amounts which the several participants "obtain." If the solution did nothing more in the quantitative sense than specify these amounts,[1] then it would coincide with the well known concept of imputation: it would just state how the total proceeds are to be distributed among the participants.[2]

We emphasize that the problem of imputation must be solved both when the total proceeds are in fact identically zero and when they are variable. This problem, in its general form, has neither been properly formulated nor solved in economic literature.

4.2.2. We can see no reason why one should not be satisfied with a solution of this nature, providing it can be found: i.e. a single imputation which meets reasonable requirements for optimum (rational) behavior. (Of course we have not yet formulated these requirements. For an exhaustive discussion, cf. loc. cit. below.) The structure of the society under consideration would then be extremely simple: There would exist an absolute state of equilibrium in which the quantitative share of every participant would be precisely determined.

It will be seen however that such a solution, possessing all necessary properties, does not exist in general. The notion of a solution will have to be broadened considerably, and it will be seen that this is closely connected with certain inherent features of social organization that are well known from a "common sense" point of view but thus far have not been viewed in proper perspective. (Cf. 4.6. and 4.8.1.)

4.2.3. Our mathematical analysis of the problem will show that there exists, indeed, a not inconsiderable family of games where a solution can be defined and found in the above sense: i.e. as one single imputation. In such cases every participant obtains at least the amount thus imputed to him by just behaving appropriately, rationally. Indeed, he gets exactly this amount if the other participants too behave rationally; if they do not, he may get even more.

These are the games of two participants where the sum of all payments is zero. While these games are not exactly typical for major economic processes, they contain some universally important traits of all games and the results derived from them are the basis of the general theory of games. We shall discuss them at length in Chapter III.

4.3. The Solution as a Set of Imputations

4.3.1. If either of the two above restrictions is dropped, the situation is altered materially.

[1] And of course, in the combinatorial sense, as outlined above, the procedure how to obtain them.

[2] In games—as usually understood—the total proceeds are always zero; i.e. one participant can gain only what the others lose. Thus there is a pure problem of distribution—i.e. imputation—and absolutely none of increasing the total utility, the "social product." In all economic questions the latter problem arises as well, but the question of imputation remains. Subsequently we shall broaden the concept of a game by dropping the requirement of the total proceeds being zero (cf. Ch. XI).

The simplest game where the second requirement is overstepped is a two-person game where the sum of all payments is variable. This corresponds to a social economy with two participants and allows both for their interdependence and for variability of total utility with their behavior.[1] As a matter of fact this is exactly the case of a bilateral monopoly (cf. 61.2.-61.6.). The well known "zone of uncertainty" which is found in current efforts to solve the problem of imputation indicates that a broader concept of solution must be sought. This case will be discussed loc. cit. above. For the moment we want to use it only as an indicator of the difficulty and pass to the other case which is more suitable as a basis for a first positive step.

4.3.2. The simplest game where the first requirement is disregarded is a three-person game where the sum of all payments is zero. In contrast to the above two-person game, this does not correspond to any fundamental economic problem but it represents nevertheless a basic possibility in human relations. The essential feature is that any two players who combine and cooperate against a third can thereby secure an advantage. The problem is how this advantage should be distributed among the two partners in this combination. Any such scheme of imputation will have to take into account that any two partners can combine; i.e. while any one combination is in the process of formation, each partner must consider the fact that his prospective ally could break away and join the third participant.

Of course the rules of the game will prescribe how the proceeds of a coalition should be divided between the partners. But the detailed discussion to be given in 22.1. shows that this will not be, in general, the final verdict. Imagine a game (of three or more persons) in which two participants can form a very advantageous coalition but where the rules of the game provide that the greatest part of the gain goes to the first participant. Assume furthermore that the second participant of this coalition can also enter a coalition with the third one, which is less effective *in toto* but promises him a greater individual gain than the former. In this situation it is obviously reasonable for the first participant to transfer a part of the gains which he could get from the first coalition to the second participant in order to save this coalition. In other words: One must expect that under certain conditions one participant of a coalition will be willing to pay a compensation to his partner. Thus the apportionment within a coalition depends not only upon the rules of the game but also upon the above principles, under the influence of the alternative coalitions.[2]

Common sense suggests that one cannot expect any theoretical statement as to which alliance will be formed[3] but only information concerning

[1] It will be remembered that we make use of a transferable utility, cf. 2.1.1.

[2] This does not mean that the rules of the game are violated, since such compensatory payments, if made at all, are made freely in pursuance of a rational consideration.

[3] Obviously three combinations of two partners each are possible. In the example to be given in 21., any preference within the solution for a particular alliance will be a

how the partners in a possible combination must divide the spoils in order to avoid the contingency that any one of them deserts to form a combination with the third player. All this will be discussed in detail and quantitatively in Ch. V.

It suffices to state here only the result which the above qualitative considerations make plausible and which will be established more rigorously loc. cit. A reasonable concept of a solution consists in this case of a system of three imputations. These correspond to the above-mentioned three combinations or alliances and express the division of spoils between respective allies.

4.3.3. The last result will turn out to be the prototype of the general situation. We shall see that a consistent theory will result from looking for solutions which are not single imputations, but rather systems of imputations.

It is clear that in the above three-person game no single imputation from the solution is in itself anything like a solution. Any particular alliance describes only one particular consideration which enters the minds of the participants when they plan their behavior. Even if a particular alliance is ultimately formed, the division of the proceeds between the allies will be decisively influenced by the other alliances which each one might alternatively have entered. Thus only the three alliances and their imputations together form a rational whole which determines all of its details and possesses a stability of its own. It is, indeed, this whole which is the really significant entity, more so than its constituent imputations. Even if one of these is actually applied, i.e. if one particular alliance is actually formed, the others are present in a "virtual" existence: Although they have not materialized, they have contributed essentially to shaping and determining the actual reality.

In conceiving of the general problem, a social economy or equivalently a game of n participants, we shall—with an optimism which can be justified only by subsequent success—expect the same thing: A solution should be a system of imputations[1] possessing in its entirety some kind of balance and stability the nature of which we shall try to determine. We emphasize that this stability—whatever it may turn out to be—will be a property of the system as a whole and not of the single imputations of which it is composed. These brief considerations regarding the three-person game have illustrated this point.

4.3.4. The exact criteria which characterize a system of imputations as a solution of our problem are, of course, of a mathematical nature. For a precise and exhaustive discussion we must therefore refer the reader to the subsequent mathematical development of the theory. The exact definition

---

limine excluded by symmetry. I.e. the game will be symmetric with respect to all three participants. Cf. however 33.1.1.

[1] They may again include compensations between partners in a coalition, as described in 4.3.2.

itself is stated in 30.1.1. We shall nevertheless undertake to give a preliminary, qualitative outline. We hope this will contribute to the understanding of the ideas on which the quantitative discussion is based. Besides, the place of our considerations in the general framework of social theory will become clearer.

4.4. The Intransitive Notion of "Superiority" or "Domination"

4.4.1. Let us return to a more primitive concept of the solution which we know already must be abandoned. We mean the idea of a solution as a single imputation. If this sort of solution existed it would have to be an imputation which in some plausible sense was superior to all other imputations. This notion of superiority as between imputations ought to be formulated in a way which takes account of the physical and social structure of the milieu. That is, one should define that an imputation x is superior to an imputation y whenever this happens: Assume that society, i.e. the totality of all participants, has to consider the question whether or not to "accept" a static settlement of all questions of distribution by the imputation y. Assume furthermore that at this moment the alternative settlement by the imputation x is also considered. Then this alternative x will suffice to exclude acceptance of y. By this we mean that a sufficient number of participants prefer in their own interest x to y, and are convinced or can be convinced of the possibility of obtaining the advantages of x. In this comparison of x to y the participants should not be influenced by the consideration of any third alternatives (imputations). I.e. we conceive the relationship of superiority as an elementary one, correlating the two imputations x and y only. The further comparison of three or more—ultimately of all—imputations is the subject of the theory which must now follow, as a superstructure erected upon the elementary concept of superiority.

Whether the possibility of obtaining certain advantages by relinquishing y for x, as discussed in the above definition, can be made convincing to the interested parties will depend upon the physical facts of the situation—in the terminology of games, on the rules of the game.

We prefer to use, instead of "superior" with its manifold associations, a word more in the nature of a *terminus technicus*. When the above described relationship between two imputations x and y exists,[1] then we shall say that x *dominates* y.

If one restates a little more carefully what should be expected from a solution consisting of a single imputation, this formulation obtains: Such an imputation should dominate all others and be dominated by none.

4.4.2. The notion of domination as formulated—or rather indicated—above is clearly in the nature of an ordering, similar to the question of

[1] That is, when it holds in the mathematically precise form, which will be given in 30.1.1.

preference, or of size in any quantitative theory. The notion of a single imputation solution[1] corresponds to that of the first element with respect to that ordering.[2]

The search for such a first element would be a plausible one if the ordering in question, i.e. our notion of domination, possessed the important property of transitivity; that is, if it were true that whenever x dominates y and y dominates z, then also x dominates z. In this case one might proceed as follows: Starting with an arbitrary x, look for a y which dominates x; if such a y exists, choose one and look for a z which dominates y; if such a z exists, choose one and look for a u which dominates z, etc. In most practical problems there is a fair chance that this process either terminates after a finite number of steps with a w which is undominated by anything else, or that the sequence $x, y, z, u, \cdots$, goes on *ad infinitum*, but that these $x, y, z, u, \cdots$ tend to a limiting position w undominated by anything else. And, due to the transitivity referred to above, the final w will in either case dominate all previously obtained $x, y, z, u, \cdots$.

We shall not go into more elaborate details which could and should be given in an exhaustive discussion. It will probably be clear to the reader that the progress through the sequence $x, y, z, u, \cdots$ corresponds to successive "improvements" culminating in the "optimum," i.e. the "first" element w which dominates all others and is not dominated.

All this becomes very different when transitivity does not prevail. In that case any attempt to reach an "optimum" by successive improvements may be futile. It can happen that x is dominated by y, y by z, and z in turn by x.[3]

4.4.3. Now the notion of domination on which we rely is, indeed, not transitive. In our tentative description of this concept we indicated that x dominates y when there exists a group of participants each one of whom prefers his individual situation in x to that in y, and who are convinced that they are able as a group—i.e. as an alliance—to enforce their preferences. We shall discuss these matters in detail in 30.2. This group of participants shall be called the "effective set" for the domination of x over y. Now when x dominates y and y dominates z, the effective sets for these two dominations may be entirely disjunct and therefore no conclusions can be drawn concerning the relationship between z and x. It can even happen that z *dominates* x with the help of a third effective set, possibly disjunct from both previous ones.

[1] We continue to use it as an illustration although we have shown already that it is a forlorn hope. The reason for this is that, by showing what is involved if certain complications did not arise, we can put these complications into better perspective. Our real interest at this stage lies of course in these complications, which are quite fundamental.

[2] The mathematical theory of ordering is very simple and leads probably to a deeper understanding of these conditions than any purely verbal discussion. The necessary mathematical considerations will be found in 65.3.

[3] In the case of transitivity this is impossible because—if a proof be wanted—x never dominates itself. Indeed, if e.g. y dominates x, z dominates y, and x dominates z, then we can infer by transitivity that x dominates x.

This lack of transitivity, especially in the above formalistic presentation, may appear to be an annoying complication and it may even seem desirable to make an effort to rid the theory of it. Yet the reader who takes another look at the last paragraph will notice that it really contains only a circumlocution of a most typical phenomenon in all social organizations. The domination relationships between various imputations $x, y, z, \cdots$—i.e. between various states of society—correspond to the various ways in which these can unstabilize—i.e. upset—each other. That various groups of participants acting as effective sets in various relations of this kind may bring about "cyclical" dominations—e.g., y over x, z over y, and x over z—is indeed one of the most characteristic difficulties which a theory of these phenomena must face.

4.5. The Precise Definition of a Solution

4.5.1. Thus our task is to replace the notion of the optimum—i.e. of the first element—by something which can take over its functions in a static equilibrium. This becomes necessary because the original concept has become untenable. We first observed its breakdown in the specific instance of a certain three-person game in 4.3.2.-4.3.3. But now we have acquired a deeper insight into the cause of its failure: it is the nature of our concept of domination, and specifically its intransitivity.

This type of relationship is not at all peculiar to our problem. Other instances of it are well known in many fields and it is to be regretted that they have never received a generic mathematical treatment. We mean all those concepts which are in the general nature of a comparison of preference or "superiority," or of order, but lack transitivity: e.g., the strength of chess players in a tournament, the "paper form" in sports and races, etc.[1]

4.5.2. The discussion of the three-person game in 4.3.2.-4.3.3. indicated that the solution will be, in general, a set of imputations instead of a single imputation. That is, the concept of the "first element" will have to be replaced by that of a set of elements (imputations) with suitable properties. In the exhaustive discussion of this game in 32. (cf. also the interpretation in 33.1.1. which calls attention to some deviations) the system of three imputations, which was introduced as the solution of the three-person game in 4.3.2.-4.3.3., will be derived in an exact way with the help of the postulates of 30.1.1. These postulates will be very similar to those which characterize a first element. They are, of course, requirements for a set of elements (imputations), but if that set should turn out to consist of a single element only, then our postulates go over into the characterization of the first element (in the total system of all imputations).

We do not give a detailed motivation for those postulates as yet, but we shall formulate them now hoping that the reader will find them to be some-

[1] Some of these problems have been treated mathematically by the introduction of chance and probability. Without denying that this approach has a certain justification, we doubt whether it is conducive to a complete understanding even in those cases. It would be altogether inadequate for our considerations of social organization.

what plausible. Some reasons of a qualitative nature, or rather one possible interpretation, will be given in the paragraphs immediately following.

4.5.3. The postulates are as follows: A set S of elements (imputations) is a solution when it possesses these two properties:

(4:A:a) No y contained in S is dominated by an x contained in S.

(4:A:b) Every y not contained in S is dominated by some x contained in S.

(4:A:a) and (4:A:b) can be stated as a single condition:

(4:A:c) The elements of S are precisely those elements which are undominated by elements of S.[1]

The reader who is interested in this type of exercise may now verify our previous assertion that for a set S which consists of a single element x the above conditions express precisely that x is the first element.

4.5.4. Part of the malaise which the preceding postulates may cause at first sight is probably due to their circular character. This is particularly obvious in the form (4:A:c), where the elements of S are characterized by a relationship which is again dependent upon S. It is important not to misunderstand the meaning of this circumstance.

Since our definitions (4:A:a) and (4:A:b), or (4:A:c), are circular—i.e. implicit—for S, it is not at all clear that there really exists an S which fulfills them, nor whether—if there exists one—the S is unique. Indeed these questions, at this stage still unanswered, are the main subject of the subsequent theory. What is clear, however, is that these definitions tell unambiguously whether any particular S is or is not a solution. If one insists on associating with the concept of a definition the attributes of existence and uniqueness of the object defined, then one must say: We have not given a definition of S, but a definition of a property of S—we have not defined the solution but characterized all possible solutions. Whether the totality of all solutions, thus circumscribed, contains no S, exactly one S, or several S's, is subject for further inquiry.[2]

4.6. Interpretation of Our Definition in Terms of "Standards of Behavior"

4.6.1. The single imputation is an often used and well understood concept of economic theory, while the sets of imputations to which we have been led are rather unfamiliar ones. It is therefore desirable to correlate them with something which has a well established place in our thinking concerning social phenomena.

[1] Thus (4:A:c) is an exact equivalent of (4:A:a) and (4:A:b) together. It may impress the mathematically untrained reader as somewhat involved, although it is really a straightforward expression of rather simple ideas.

[2] It should be unnecessary to say that the circularity, or rather implicitness, of (4:A:a) and (4:A:b), or (4:A:c), does not at all mean that they are tautological. They express, of course, a very serious restriction of S.

Indeed, it appears that the sets of imputations S which we are considering correspond to the "standard of behavior" connected with a social organization. Let us examine this assertion more closely.

Let the physical basis of a social economy be given,—or, to take a broader view of the matter, of a society.[1] According to all tradition and experience human beings have a characteristic way of adjusting themselves to such a background. This consists of not setting up one rigid system of apportionment, i.e. of imputation, but rather a variety of alternatives, which will probably all express some general principles but nevertheless differ among themselves in many particular respects.[2] This system of imputations describes the "established order of society" or "accepted standard of behavior."

Obviously no random grouping of imputations will do as such a "standard of behavior": it will have to satisfy certain conditions which characterize it as a possible order of things. This concept of possibility must clearly provide for conditions of stability. The reader will observe, no doubt, that our procedure in the previous paragraphs is very much in this spirit: The sets S of imputations $x, y, z, \cdots$ correspond to what we now call "standard of behavior," and the conditions (4:A:a) and (4:A:b), or (4:A:c), which characterize the solution S express, indeed, a stability in the above sense.

4.6.2. The disjunction into (4:A:a) and (4:A:b) is particularly appropriate in this instance. Recall that domination of y by x means that the imputation x, if taken into consideration, excludes acceptance of the imputation y (this without forecasting what imputation will ultimately be accepted, cf. 4.4.1. and 4.4.2.). Thus (4:A:a) expresses the fact that the standard of behavior is free from inner contradictions: No imputation y belonging to S—i.e. conforming with the "accepted standard of behavior" —can be upset—i.e. dominated—by another imputation x of the same kind. On the other hand (4:A:b) expresses that the "standard of behavior" can be used to discredit any non-conforming procedure: Every imputation y not belonging to S can be upset—i.e. dominated—by an imputation x belonging to S.

Observe that we have not postulated in 4.5.3. that a y belonging to S should never be dominated by any x.[3] Of course, if this should happen, then x would have to be outside of S, due to (4:A:a). In the terminology of social organizations: An imputation y which conforms with the "accepted

[1] In the case of a game this means simply—as we have mentioned before—that the rules of the game are given. But for the present simile the comparison with a social economy is more useful. We suggest therefore that the reader forget temporarily the analogy with games and think entirely in terms of social organization.

[2] There may be extreme, or to use a mathematical term, "degenerate" special cases where the setup is of such exceptional simplicity that a rigid single apportionment can be put into operation. But it seems legitimate to disregard them as non-typical.

[3] It can be shown, cf. (31:M) in 31.2.3., that such a postulate cannot be fulfilled in general; i.e. that in all really interesting cases it is impossible to find an S which satisfies it together with our other requirements.

standard of behavior" may be upset by another imputation x, but in this case it is certain that x does not conform.[1] It follows from our other requirements that then x is upset in turn by a third imputation z which again conforms. Since y and z both conform, z cannot upset y—a further illustration of the intransitivity of "domination."

Thus our solutions S correspond to such "standards of behavior' as have an inner stability: once they are generally accepted they overrule everything else and no part of them can be overruled within the limits of the accepted standards. This is clearly how things are in actual social organizations, and it emphasizes the perfect appropriateness of the circular character of our conditions in 4.5.3.

4.6.3. We have previously mentioned, but purposely neglected to discuss, an important objection: That neither the existence nor the uniqueness of a solution S in the sense of the conditions (4:A:a) and (4:A:b), or (4:A:c), in 4.5.3. is evident or established.

There can be, of course, no concessions as regards existence. If it should turn out that our requirements concerning a solution S are, in any special case, unfulfillable,—this would certainly necessitate a fundamental change in the theory. Thus a general proof of the existence of solutions S for all particular cases[2] is most desirable. It will appear from our subsequent investigations that this proof has not yet been carried out in full generality but that in all cases considered so far solutions were found.

As regards uniqueness the situation is altogether different. The often mentioned "circular" character of our requirements makes it rather probable that the solutions are not in general unique. Indeed we shall in most cases observe a multiplicity of solutions.[3] Considering what we have said about interpreting solutions as stable "standards of behavior" this has a simple and not unreasonable meaning, namely that given the same physical background different "established orders of society" or "accepted standards of behavior" can be built, all possessing those characteristics of inner stability which we have discussed. Since this concept of stability is admittedly of an "inner" nature—i.e. operative only under the hypothesis of general acceptance of the standard in question—these different standards may perfectly well be in contradiction with each other.

4.6.4. Our approach should be compared with the widely held view that a social theory is possible only on the basis of some preconceived principles of social purpose. These principles would include quantitative statements concerning both the aims to be achieved *in toto* and the apportionments between individuals. Once they are accepted, a simple maximum problem results.

[1] We use the word "conform" (to the "standard of behavior") temporarily as a synonym for *being contained in S*, and the word "upset" as a synonym for *dominate*.

[2] In the terminology of games: for all numbers of participants and for all possible rules of the game.

[3] An interesting exception is 65.8.

Let us note that no such statement of principles is ever satisfactory *per se*, and the arguments adduced in its favor are usually either those of inner stability or of less clearly defined kinds of desirability, mainly concerning distribution.

Little can be said about the latter type of motivation. Our problem is not to determine what ought to happen in pursuance of any set of—necessarily arbitrary—*a priori* principles, but to investigate where the equilibrium of forces lies.

As far as the first motivation is concerned, it has been our aim to give just those arguments precise and satisfactory form, concerning both global aims and individual apportionments. This made it necessary to take up the entire question of inner stability as a problem in its own right. A theory which is consistent at this point cannot fail to give a precise account of the entire interplay of economic interests, influence and power.

4.7. Games and Social Organizations

4.7. It may now be opportune to revive the analogy with games, which we purposely suppressed in the previous paragraphs (cf. footnote 1 on p. 41). The parallelism between the solutions S in the sense of 4.5.3. on one hand and of stable "standards of behavior" on the other can be used for corroboration of assertions concerning these concepts in both directions. At least we hope that this suggestion will have some appeal to the reader. We think that the procedure of the mathematical theory of games of strategy gains definitely in plausibility by the correspondence which exists between its concepts and those of social organizations. On the other hand, almost every statement which we—or for that matter anyone else—ever made concerning social organizations, runs afoul of some existing opinion. And, by the very nature of things, most opinions thus far could hardly have been proved or disproved within the field of social theory. It is therefore a great help that all our assertions can be borne out by specific examples from the theory of games of strategy.

Such is indeed one of the standard techniques of using models in the physical sciences. This two-way procedure brings out a significant function of models, not emphasized in their discussion in 4.1.3.

To give an illustration: The question whether several stable "orders of society" or "standards of behavior" based on the same physical background are possible or not, is highly controversial. There is little hope that it will be settled by the usual methods because of the enormous complexity of this problem among other reasons. But we shall give specific examples of games of three or four persons, where one game possesses several solutions in the sense of 4.5.3. And some of these examples will be seen to be models for certain simple economic problems. (Cf. 62.)

4.8. Concluding Remarks

4.8.1. In conclusion it remains to make a few remarks of a more formal nature.

We begin with this observation: Our considerations started with single imputations—which were originally quantitative extracts from more detailed combinatorial sets of rules. From these we had to proceed to sets S of imputations, which under certain conditions appeared as solutions. Since the solutions do not seem to be necessarily unique, the complete answer to any specific problem consists not in finding a solution, but in determining the set of all solutions. Thus the entity for which we look in any particular problem is really a set of sets of imputations. This may seem to be unnaturally complicated in itself; besides there appears no guarantee that this process will not have to be carried further, conceivably because of later difficulties. Concerning these doubts it suffices to say: First, the mathematical structure of the theory of games of strategy provides a formal justification of our procedure. Second, the previously discussed connections with "standards of behavior" (corresponding to sets of imputations) and of the multiplicity of "standards of behavior" on the same physical background (corresponding to sets of sets of imputations) make just this amount of complicatedness desirable.

One may criticize our interpretation of sets of imputations as "standards of behavior." Previously in 4.1.2. and 4.1.4. we introduced a more elementary concept, which may strike the reader as a direct formulation of a "standard of behavior": this was the preliminary combinatorial concept of a solution as a set of rules for each participant, telling him how to behave in every possible situation of the game. (From these rules the single imputations were then extracted as a quantitative summary, cf. above.) Such a simple view of the "standard of behavior" could be maintained, however, only in games in which coalitions and the compensations between coalition partners (cf. 4.3.2.) play no role, since the above rules do not provide for these possibilities. Games exist in which coalitions and compensations can be disregarded: e.g. the two-person game of zero-sum mentioned in 4.2.3., and more generally the "inessential" games to be discussed in 27.3. and in (31:P) of 31.2.3. But the general, typical game—in particular all significant problems of a social exchange economy—cannot be treated without these devices. Thus the same arguments which forced us to consider sets of imputations instead of single imputations necessitate the abandonment of that narrow concept of "standard of behavior." Actually we shall call these sets of rules the "strategies" of the game.

4.8.2. The next subject to be mentioned concerns the static or dynamic nature of the theory. We repeat most emphatically that our theory is thoroughly static. A dynamic theory would unquestionably be more complete and therefore preferable. But there is ample evidence from other branches of science that it is futile to try to build one as long as the static side is not thoroughly understood. On the other hand, the reader may object to some definitely dynamic arguments which were made in the course of our discussions. This applies particularly to all considerations concerning the interplay of various imputations under the influence of "domination," cf. 4.6.2. We think that this is perfectly legitimate. A static theory deals with equilibria.[1] The essential characteristic of an equilibrium is that it has no tendency to change, i.e. that it is not conducive to dynamic developments. An analysis of this feature is, of course, inconceivable without the use of certain rudimentary dynamic concepts. The important point is that they are rudimentary. In other words: For the real dynamics which investigates the precise motions, usually far away from equilibria, a much deeper knowledge of these dynamic phenomena is required.[2,3]

4.8.3. Finally let us note a point at which the theory of social phenomena will presumably take a very definite turn away from the existing patterns of mathematical physics. This is, of course, only a surmise on a subject where much uncertainty and obscurity prevail.

Our static theory specifies equilibria—i.e. solutions in the sense of 4.5.3. —which are sets of imputations. A dynamic theory—when one is found—will probably describe the changes in terms of simpler concepts: of a single imputation—valid at the moment under consideration—or something similar. This indicates that the formal structure of this part of the theory—the relationship between statics and dynamics—may be generically different from that of the classical physical theories.[4]

All these considerations illustrate once more what a complexity of theoretical forms must be expected in social theory. Our static analysis alone necessitated the creation of a conceptual and formal mechanism which is very different from anything used, for instance, in mathematical physics. Thus the conventional view of a solution as a uniquely defined number or aggregate of numbers was seen to be too narrow for our purposes, in spite of its success in other fields. The emphasis on mathematical methods seems to be shifted more towards combinatorics and set theory—and away from the algorithm of differential equations which dominate mathematical physics.

[1] The dynamic theory deals also with inequilibria—even if they are sometimes called "dynamic equilibria."

[2] The above discussion of statics *versus* dynamics is, of course, not at all a construction *ad hoc*. The reader who is familiar with mechanics for instance will recognize in it a reformulation of well known features of the classical mechanical theory of statics and dynamics. What we do claim at this time is that this is a general characteristic of scientific procedure involving forces and changes in structures.

[3] The dynamic concepts which enter into the discussion of static equilibria are parallel to the "virtual displacements" in classical mechanics. The reader may also remember at this point the remarks about "virtual existence" in 4.3.3.

[4] Particularly from classical mechanics. The analogies of the type used in footnote 2 above, cease at this point.

CHAPTER II

GENERAL FORMAL DESCRIPTION OF GAMES OF STRATEGY

5. Introduction

5.1. Shift of Emphasis from Economics to Games

5.1. It should be clear from the discussions of Chapter I that a theory of rational behavior—i.e. of the foundations of economics and of the main mechanisms of social organization—requires a thorough study of the "games of strategy." Consequently we must now take up the theory of games as an independent subject. In studying it as a problem in its own right, our point of view must of necessity undergo a serious shift. In Chapter I our primary interest lay in economics. It was only after having convinced ourselves of the impossibility of making progress in that field without a previous fundamental understanding of the games that we gradually approached the formulations and the questions which are partial to that subject. But the economic viewpoints remained nevertheless the dominant ones in all of Chapter I. From this Chapter II on, however, we shall have to treat the games as games. Therefore we shall not mind if some points taken up have no economic connections whatever,—it would not be possible to do full justice to the subject otherwise. Of course most of the main concepts are still those familiar from the discussions of economic literature (cf. the next section) but the details will often be altogether alien to it—and details, as usual, may dominate the exposition and overshadow the guiding principles.

5.2. General Principles of Classification and of Procedure

5.2.1. Certain aspects of "games of strategy" which were already prominent in the last sections of Chapter I will not appear in the beginning stages of the discussions which we are now undertaking. Specifically: There will be at first no mention of coalitions between players and the compensations which they pay to each other. (Concerning these, cf. 4.3.2., 4.3.3., in Chapter I.) We give a brief account of the reasons, which will also throw some light on our general disposition of the subject.

An important viewpoint in classifying games is this: Is the sum of all payments received by all players (at the end of the game) always zero; or is this not the case? If it is zero, then one can say that the players pay only to each other, and that no production or destruction of goods is involved. All games which are actually played for entertainment are of this type. But the economically significant schemes are most essentially not such. There the sum of all payments, the total social product, will in general not be zero, and not even constant. I.e., it will depend on the behavior of the players—the participants in the social economy. This distinction was already mentioned in 4.2.1., particularly in footnote 2, p. 34. We shall call games of the first-mentioned type *zero-sum* games, and those of the latter type *non-zero-sum* games.

We shall primarily construct a theory of the zero-sum games, but it will be found possible to dispose, with its help, of all games, without restriction. Precisely: We shall show that the general (hence in particular the variable sum) n-person game can be reduced to a zero-sum $n + 1$-person game. (Cf. 56.2.2.) Now the theory of the zero-sum n-person game will be based on the special case of the zero-sum two-person game. (Cf. 25.2.) Hence our discussions will begin with a theory of these games, which will indeed be carried out in Chapter III.

Now in zero-sum two-person games coalitions and compensations can play no role.[1] The questions which are essential in these games are of a different nature. These are the main problems: How does each player plan his course—i.e. how does one formulate an exact concept of a strategy? What information is available to each player at every stage of the game? What is the role of a player being informed about the other player's strategy? About the entire theory of the game?

5.2.2. All these questions are of course essential in all games, for any number of players, even when coalitions and compensations have come into their own. But for zero-sum two-person games they are the only ones which matter, as our subsequent discussions will show. Again, all these questions have been recognized as important in economics, but we think that in the theory of games they arise in a more elementary—as distinguished from composite—fashion. They can, therefore, be discussed in a precise way and—as we hope to show—be disposed of. But in the process of this analysis it will be technically advantageous to rely on pictures and examples which are rather remote from the field of economics proper, and belong strictly to the field of games of the conventional variety. Thus the discussions which follow will be dominated by illustrations from Chess, "Matching Pennies," Poker, Bridge, etc., and not from the structure of cartels, markets, oligopolies, etc.

At this point it is also opportune to recall that we consider all transactions at the end of a game as purely monetary ones—i.e. that we ascribe to all players an exclusively monetary profit motive. The meaning of this in terms of the utility concept was analyzed in 2.1.1. in Chapter I. For the present—particularly for the "zero-sum two-person games" to be discussed

[1] The only fully satisfactory "proof" of this assertion lies in the construction of a complete theory of all zero-sum two-person games, without use of those devices. This will be done in Chapter III, the decisive result being contained in 17. It ought to be clear by common sense, however, that "understandings" and "coalitions" can have no role here: Any such arrangement must involve at least two players—hence in this case all players—for whom the sum of payments is identically zero. I.e. th re are no opponents left and no possible objectives.

first (cf. the discussion of 5.2.1.)—it is an absolutely necessary simplification. Indeed, we shall maintain it through most of the theory, although variants will be examined later on. (Cf. Chapter XII, in particular 66.)

5.2.3. Our first task is to give an exact definition of what constitutes a game. As long as the concept of a game has not been described with absolute mathematical—combinatorial—precision, we cannot hope to give exact and exhaustive answers to the questions formulated at the end of 5.2.1. Now while our first objective is—as was explained in 5.2.1.—the theory of zero-sum two-person games, it is apparent that the exact description of what constitutes a game need not be restricted to this case. Consequently we can begin with the description of the general n-person game. In giving this description we shall endeavor to do justice to all conceivable nuances and complications which can arise in a game—insofar as they are not of an obviously inessential character. In this way we reach—in several successive steps—a rather complicated but exhaustive and mathematically precise scheme. And then we shall see that it is possible to replace this general scheme by a vastly simpler one, which is nevertheless fully and rigorously equivalent to it. Besides, the mathematical device which permits this simplification is also of an immediate significance for our problem: It is the introduction of the exact concept of a strategy.

It should be understood that the detour—which leads to the ultimate, simple formulation of the problem, over considerably more complicated ones—is not avoidable. It is necessary to show first that all possible complications have been taken into consideration, and that the mathematical device in question does guarantee the equivalence of the involved setup to the simple.

All this can—and must—be done for all games, of any number of players. But after this aim has been achieved in entire generality, the next objective of the theory is—as mentioned above—to find a complete solution for the zero-sum two-person game. Accordingly, this chapter will deal with all games, but the next one with zero-sum two-person games only. After they are disposed of and some important examples have been discussed, we shall begin to re-extend the scope of the investigation—first to zero-sum n-person games, and then to all games.

Coalitions and compensations will only reappear during this latter stage.

6. The Simplified Concept of a Game

6.1. Explanation of the Termini Technici

6.1. Before an exact definition of the combinatorial concept of a game can be given, we must first clarify the use of some termini. There are some notions which are quite fundamental for the discussion of games, but the use of which in everyday language is highly ambiguous. The words which describe them are used sometimes in one sense, sometimes in another, and occasionally—worst of all—as if they were synonyms. We must therefore introduce a definite usage of *termini technici*, and rigidly adhere to it in all that follows.

First, one must distinguish between the abstract concept of a *game*, and the individual *plays* of that game. The *game* is simply the totality of the rules which describe it. Every particular instance at which the game is played—in a particular way—from beginning to end, is a *play*.[1]

Second, the corresponding distinction should be made for the moves, which are the component elements of the game. A move is the occasion of a choice between various alternatives, to be made either by one of the players, or by some device subject to chance, under conditions precisely prescribed by the rules of the game. The *move* is nothing but this abstract "occasion," with the attendant details of description,—i.e. a component of the *game*. The specific alternative chosen in a concrete instance—i.e. in a concrete *play*—is the *choice*. Thus the moves are related to the choices in the same way as the game is to the play. The game consists of a sequence of moves, and the play of a sequence of choices.[2]

Finally, the *rules* of the game should not be confused with the *strategies* of the players. Exact definitions will be given subsequently, but the distinction which we stress must be clear from the start. Each player selects his strategy—i.e. the general principles governing his choices—freely. While any particular strategy may be good or bad—provided that these concepts can be interpreted in an exact sense (cf. 14.5. and 17.8-17.10.)—it is within the player's discretion to use or to reject it. The rules of the game, however, are absolute commands. If they are ever infringed, then the whole transaction by definition ceases to be the game described by those rules. In many cases it is even physically impossible to violate them.[3]

6.2. The Elements of the Game

6.2.1. Let us now consider a game Γ of n players who, for the sake of brevity, will be denoted by $1, \cdots, n$. The conventional picture provides that this game is a sequence of moves, and we assume that both the number and the arrangement of these moves is given *ab initio*. We shall see later that these restrictions are not really significant, and that they can be removed without difficulty. For the present let us denote the (fixed) number of moves in Γ by ν—this is an integer $\nu = 1, 2, \cdots$. The moves themselves we denote by $\mathfrak{M}_1, \cdots, \mathfrak{M}_\nu$, and we assume that this is the chronological order in which they are prescribed to take place.

[1] In most games everyday usage calls a play equally a game; thus in chess, in poker, in many sports, etc. In Bridge a play corresponds to a "rubber," in Tennis to a "set," but unluckily in these games certain components of the play are again called "games." The French terminology is tolerably unambiguous: "game" = "jeu," "play" = "partie."

[2] In this sense we would talk in chess of the first move, and of the choice "E2-E4."

[3] E.g.: In Chess the rules of the game forbid a player to move his king into a position of "check." This is a prohibition in the same absolute sense in which he may not move a pawn sideways. But to move the king into a position where the opponent can "checkmate" him at the next move is merely unwise, but not forbidden.

Every move $\mathfrak{M}_\kappa$, $\kappa = 1, \cdots, \nu$, actually consists of a number of alternatives, among which the choice—which constitutes the move $\mathfrak{M}_\kappa$—takes place. Denote the number of these alternatives by α_κ and the alternatives themselves by $\mathfrak{A}_\kappa(1), \cdots, \mathfrak{A}_\kappa(\alpha_\kappa)$.

The moves are of two kinds. A *move of the first kind*, or a *personal move*, is a choice made by a specific player, i.e. depending on his free decision and nothing else. A *move of the second kind*, or a *chance move*, is a choice depending on some mechanical device, which makes its outcome fortuitous with definite probabilities.[1] Thus for every personal move it must be specified which player's decision determines this move, *whose move* it is. We denote the player in question (i.e. his number) by k_κ. So $k_\kappa = 1, \cdots, n$. For a chance move we put (conventionally) $k_\kappa = 0$. In this case the probabilities of the various alternatives $\mathfrak{A}_\kappa(1), \cdots, \mathfrak{A}_\kappa(\alpha_\kappa)$ must be given. We denote these probabilities by $p_\kappa(1), \cdots, p_\kappa(\alpha_\kappa)$ respectively.[2]

6.2.2. In a move $\mathfrak{M}_\kappa$ the choice consists of selecting an alternative $\mathfrak{A}_\kappa(1), \cdots, \mathfrak{A}_\kappa(\alpha_\kappa)$, i.e. its number $1, \cdots, \alpha_\kappa$. We denote the number so chosen by σ_κ. Thus this choice is characterized by a number $\sigma_\kappa = 1, \cdots, \alpha_\kappa$. And the complete play is described by specifying all choices, corresponding to all moves $\mathfrak{M}_1, \cdots, \mathfrak{M}_\nu$. I.e. it is described by a sequence $\sigma_1, \cdots, \sigma_\nu$.

Now the rule of the game Γ must specify what the outcome of the play is for each player $k = 1, \cdots, n$, if the play is described by a given sequence $\sigma_1, \cdots \sigma_\nu$. I.e. what payments every player receives when the play is completed. Denote the payment to the player k by $\mathfrak{F}_k$ ($\mathfrak{F}_k > 0$ if k receives a payment, $\mathfrak{F}_k < 0$ if he must make one, $\mathfrak{F}_k = 0$ if neither is the case). Thus each $\mathfrak{F}_k$ must be given as a function of the $\sigma_1, \cdots, \sigma_\nu$:

$$\mathfrak{F}_k = \mathfrak{F}_k(\sigma_1, \cdots, \sigma_\nu), \qquad k = 1, \cdots, n.$$

We emphasize again that the rules of the game Γ specify the function $\mathfrak{F}_k(\sigma_1, \cdots, \sigma_\nu)$ only as a function,[3] i.e. the abstract dependence of each $\mathfrak{F}_k$ on the variables $\sigma_1, \cdots, \sigma_\nu$. But all the time each σ_κ is a variable, with the domain of variability $1, \cdots, \alpha_\kappa$. A specification of particular numerical values for the σ_κ, i.e. the selection of a particular sequence $\sigma_1, \cdots, \sigma_\nu$, is no part of the game Γ. It is, as we pointed out above, the definition of a play.

[1] E.g., dealing cards from an appropriately shuffled deck, throwing dice, etc. It is even possible to include certain games of strength and skill, where "strategy" plays a role, e.g. Tennis, Football, etc. In these the actions of the players are up to a certain point personal moves—i.e. dependent upon their free decision—and beyond this point chance moves, the probabilities being characteristics of the player in question.

[2] Since the $p_\kappa(1), \cdots, p_\kappa(\alpha_\kappa)$ are probabilities, they are necessarily numbers $\geqq 0$. Since they belong to disjunct but exhaustive alternatives, their sum (for a fixed κ) must be one. I.e.:

$$p_\kappa(\sigma) \geqq 0, \qquad \sum_{\sigma=1}^{\alpha_\kappa} p_\kappa(\sigma) = 1.$$

[3] For a systematic exposition of the concept of a function cf. 13.1.

6.3. Information and Preliminarity

6.3.1. Our description of the game Γ is not yet complete. We have failed to include specifications about the state of information of every player at each decision which he has to make,—i.e. whenever a personal move turns up which is his move. Therefore we now turn to this aspect of the matter.

This discussion is best conducted by following the moves $\mathfrak{M}_1, \cdots, \mathfrak{M}_\nu$, as the corresponding choices are made.

Let us therefore fix our attention on a particular move $\mathfrak{M}_\kappa$. If this $\mathfrak{M}_\kappa$ is a chance move, then nothing more need be said: the choice is decided by chance; nobody's will and nobody's knowledge of other things can influence it. But if $\mathfrak{M}_\kappa$ is a personal move, belonging to the player k_κ, then it is quite important what k_κ's state of information is when he forms his decision concerning $\mathfrak{M}_\kappa$—i.e. his choice of σ_κ.

The only things he can be informed about are the choices corresponding to the moves preceding $\mathfrak{M}_\kappa$—the moves $\mathfrak{M}_1, \cdots, \mathfrak{M}_{\kappa-1}$. I.e. he may know the values of $\sigma_1, \cdots, \sigma_{\kappa-1}$. But he need not know that much. It is an important peculiarity of Γ, just how much information concerning $\sigma_1, \cdots, \sigma_{\kappa-1}$ the player k_κ is granted, when he is called upon to choose σ_κ. We shall soon show in several examples what the nature of such limitations is.

The simplest type of rule which describes k_κ's state of information at $\mathfrak{M}_\kappa$ is this: a set Λ_κ consisting of some numbers from among $\lambda = 1, \cdots, \kappa - 1$, is given. It is specified that k_κ knows the values of the σ_λ with λ belonging to Λ_κ, and that he is entirely ignorant of the σ_λ with any other λ.

In this case we shall say, when λ belongs to Λ_κ, that λ is *preliminary* to κ. This implies $\lambda = 1, \cdots, \kappa - 1$, i.e. $\lambda < \kappa$, but need not be implied by it. Or, if we consider, instead of λ, κ, the corresponding moves $\mathfrak{M}_\lambda, \mathfrak{M}_\kappa$: Preliminarity implies anteriority,[1] but need not be implied by it.

6.3.2. In spite of its somewhat restrictive character, this concept of preliminarity deserves a closer inspection. In itself, and in its relationship to anteriority (cf. footnote 1 above), it gives occasion to various combinatorial possibilities. These have definite meanings in those games in which they occur, and we shall now illustrate them by some examples of particularly characteristic instances.

6.4. Preliminarity, Transitivity, and Signaling

6.4.1. We begin by observing that there exist games in which preliminarity and anteriority are the same thing. I.e., where the players k_κ who makes the (personal) move $\mathfrak{M}_\kappa$ is informed about the outcome of the choices of all anterior moves $\mathfrak{M}_1, \cdots, \mathfrak{M}_{\kappa-1}$. Chess is a typical representative of this class of games of "perfect" information. They are generally considered to be of a particularly rational character. We shall see in 15., specifically in 15.7., how this can be interpreted in a precise way.

[1] In time, $\lambda < \kappa$ means that $\mathfrak{M}_\lambda$ occurs before $\mathfrak{M}_\kappa$.

Chess has the further feature that all its moves are personal. Now it is possible to conserve the first-mentioned property—the equivalence of preliminarity and anteriority—even in games which contain chance moves. Backgammon is an example of this.[1] Some doubt might be entertained whether the presence of chance moves does not vitiate the "rational character" of the game mentioned in connection with the preceding examples.

We shall see in 15.7.1. that this is not so if a very plausible interpretation of that "rational character" is adhered to. It is not important whether all moves are personal or not; the essential fact is that preliminarity and anteriority coincide.

6.4.2. Let us now consider games where anteriority does not imply preliminarity. I.e., where the player k_κ who makes the (personal) move $\mathfrak{M}_\kappa$ is not informed about everything that happened previously. There is a large family of games in which this occurs. These games usually contain chance moves as well as personal moves. General opinion considers them as being of a mixed character: while their outcome is definitely dependent on chance, they are also strongly influenced by the strategic abilities of the players.

Poker and Bridge are good examples. These two games show, furthermore, what peculiar features the notion of preliminarity can present, once it has been separated from anteriority. This point perhaps deserves a little more detailed consideration.

Anteriority, i.e. the chronological ordering of the moves, possesses the property of transitivity.[2] Now in the present case, preliminarity need not be transitive. Indeed it is neither in Poker nor in Bridge, and the conditions under which this occurs are quite characteristic.

Poker: Let $\mathfrak{M}_\mu$ be the deal of his "hand" to player 1—a chance move; $\mathfrak{M}_\lambda$ the first bid of player 1—a personal move of 1; $\mathfrak{M}_\kappa$ the first (subsequent) bid of player 2—a personal move of 2. Then $\mathfrak{M}_\mu$ is preliminary to $\mathfrak{M}_\lambda$ and $\mathfrak{M}_\lambda$ to $\mathfrak{M}_\kappa$ but $\mathfrak{M}_\mu$ is not preliminary to $\mathfrak{M}_\kappa$.[3] Thus we have intransitivity, but it involves both players. Indeed, it may first seem unlikely that preliminarity could in any game be intransitive among the personal moves of one particular player. It would require that this player should "forget" between the moves $\mathfrak{M}_\lambda$ and $\mathfrak{M}_\kappa$ the outcome of the choice connected with $\mathfrak{M}_\mu$[4]—and it is difficult to see how this "forgetting" could be achieved, and

[1] The chance moves in Backgammon are the dice throws which decide the total number of steps by which each player's men may alternately advance. The personal moves are the decisions by which each player partitions that total number of steps allotted to him among his individual men. Also his decision to double the risk, and his alternative to accept or to give up when the opponent doubles. At every move, however, the outcome of the choices of all anterior moves are visible to all on the board.

[2] I.e.: If $\mathfrak{M}_\mu$ is anterior to $\mathfrak{M}_\lambda$ and $\mathfrak{M}_\lambda$ to $\mathfrak{M}_\kappa$ then $\mathfrak{M}_\mu$ is anterior to $\mathfrak{M}_\kappa$. Special situations where the presence or absence of transitivity was of importance, were analyzed in 4.4.2., 4.6.2. of Chapter I in connection with the relation of domination.

[3] I.e., 1 makes his first bid knowing his own "hand"; 2 makes his first bid knowing 1's (preceding) first bid; but at the same time 2 is ignorant of 1's "hand."

[4] We assume that $\mathfrak{M}_\mu$ is preliminary to $\mathfrak{M}_\lambda$ and $\mathfrak{M}_\lambda$ to $\mathfrak{M}_\kappa$ but $\mathfrak{M}_\mu$ not to $\mathfrak{M}_\kappa$.

even enforced! Nevertheless our next example provides an instance of just this.

Bridge: Although Bridge is played by 4 persons, to be denoted by A,B,C,D, it should be classified as a two-person game. Indeed, A and C form a combination which is more than a voluntary coalition, and so do B and D. For A to cooperate with B (or D) instead of with C would be "cheating," in the same sense in which it would be "cheating" to look into B's cards or failing to follow suit during the play. I.e. it would be a violation of the rules of the game. If three (or more) persons play poker, then it is perfectly permissible for two (or more) of them to cooperate against another player when their interests are parallel—but in Bridge A and C (and similarly B and D) must cooperate, while A and B are forbidden to cooperate. The natural way to describe this consists in declaring that A and C are really one player 1, and that B and D are really one player 2. Or, equivalently: Bridge is a two-person game, but the two players 1 and 2 do not play it themselves. 1 acts through two representatives A and C and 2 through two representatives B and D.

Consider now the representatives of 1, A and C. The rules of the game restrict communication, i.e. the exchange of information, between them. E.g.: let $\mathfrak{M}_\mu$ be the deal of his "hand" to A—a chance move; $\mathfrak{M}_\lambda$ the first card played by A—a personal move of 1; $\mathfrak{M}_\kappa$ the card played into this trick by C—a personal move of 1. Then $\mathfrak{M}_\mu$ is preliminary to $\mathfrak{M}_\lambda$ and $\mathfrak{M}_\lambda$ to $\mathfrak{M}_\kappa$ but $\mathfrak{M}_\mu$ is not preliminary to $\mathfrak{M}_\kappa$.[1] Thus we have again intransitivity, but this time it involves only one player. It is worth noting how the necessary "forgetting" of $\mathfrak{M}_\mu$ between $\mathfrak{M}_\lambda$ and $\mathfrak{M}_\kappa$ was achieved by "splitting the personality" of 1 into A and C.

6.4.3. The above examples show that intransitivity of the relation of preliminarity corresponds to a very well known component of practical strategy: to the possibility of "signaling." If no knowledge of $\mathfrak{M}_\mu$ is available at $\mathfrak{M}_\kappa$, but if it is possible to observe $\mathfrak{M}_\lambda$'s outcome at $\mathfrak{M}_\kappa$ and $\mathfrak{M}_\lambda$ has been influenced by $\mathfrak{M}_\mu$ (by knowledge about $\mathfrak{M}_\mu$'s outcome), then $\mathfrak{M}_\lambda$ is really a signal from $\mathfrak{M}_\mu$ to $\mathfrak{M}_\kappa$—a device which (indirectly) relays information. Now two opposite situations develop, according to whether $\mathfrak{M}_\lambda$ and $\mathfrak{M}_\kappa$ are moves of the same player, or of two different players.

In the first case—which, as we saw, occurs in Bridge—the interest of the player (who is $k_\lambda = k_\kappa$) lies in promoting the "signaling," i.e. the spreading of information "within his own organization." This desire finds its realization in the elaborate system of "conventional signals" in Bridge.[2] These are parts of the strategy, and not of the rules of the game

[1] I.e. A plays his first card knowing his own "hand"; C contributes to this trick knowing the (initiating) card played by A; but at the same time C is ignorant of A's "hand."

[2] Observe that this "signaling" is considered to be perfectly fair in Bridge if it is carried out by actions which are provided for by the rules of the game. E.g. it is correct for A and C (the two components of player 1, cf. 6.4.2.) to agree—before the play begins! —that an "original bid" of two trumps "indicates" a weakness of the other suits. But it is incorrect—i.e. "cheating"—to indicate a weakness by an inflection of the voice at bidding, or by tapping on the table, etc.

(cf. 6.1.), and consequently they may vary,[1] while the game of Bridge remains the same.

In the second case—which, as we saw, occurs in Poker—the interest of the player (we now mean k_λ, observe that here $k_\lambda \neq k_\kappa$) lies in preventing this "signaling," i.e. the spreading of information to the opponent (k_κ). This is usually achieved by irregular and seemingly illogical behavior (when making the choice at $\mathfrak{M}_\lambda$)—this makes it harder for the opponent to draw inferences from the outcome of $\mathfrak{M}_\lambda$ (which he sees) concerning the outcome of $\mathfrak{M}_\mu$ (of which he has no direct news). I.e. this procedure makes the "signal" uncertain and ambiguous. We shall see in 19.2.1. that this is indeed the function of "bluffing" in Poker.[2]

We shall call these two procedures *direct* and *inverted signaling.* It ought to be added that inverted signaling—i.e. misleading the opponent—occurs in almost all games, including Bridge. This is so since it is based on the intransitivity of preliminarity when several players are involved, which is easy to achieve. Direct signaling, on the other hand, is rarer; e.g. Poker contains no vestige of it. Indeed, as we pointed out before, it implies the intransitivity of preliminarity when only one player is involved—i.e. it requires a well-regulated "forgetfulness" of that player, which is obtained in Bridge by the device of "splitting the player up" into two persons.

At any rate Bridge and Poker seem to be reasonably characteristic instances of these two kinds of intransitivity—of direct and of inverted signaling, respectively.

Both kinds of signaling lead to a delicate problem of balancing in actual playing, i.e. in the process of trying to define "good," "rational" playing. Any attempt to signal more or to signal less than "unsophisticated" playing would involve, necessitates deviations from the "unsophisticated" way of playing. And this is usually possible only at a definite cost, i.e. its direct consequences are losses. Thus the problem is to adjust this "extra" signaling so that its advantages—by forwarding or by withholding information—overbalance the losses which it causes directly. One feels that this involves something like the search for an optimum, although it is by no means clearly defined. We shall see how the theory of the two-person game takes care already of this problem, and we shall discuss it exhaustively in one characteristic instance. (This is a simplified form of Poker. Cf. 19.)

Let us observe, finally, that all important examples of intransitive preliminarity are games containing chance moves. This is peculiar, because there is no apparent connection between these two phenomena.[3,4] Our

[1] They may even be different for the two players, i.e. for A and C on one hand and B and D on the other. But "within the organization" of one player, e.g. for A and C, they must agree.

[2] And that "bluffing" is not at all an attempt to secure extra gains—in any direct sense—when holding a weak hand. Cf. loc. cit.

[3] Cf. the corresponding question when preliminarity coincides with anteriority, and thus is transitive, as discussed in 6.4.1. As mentioned there, the presence or absence of chance moves is immaterial in that case.

[4] "Matching pennies" is an example which has a certain importance in this connection. This and other related games will be discussed in 18.

subsequent analysis will indeed show that the presence or absence of chance moves scarcely influences the essential aspects of the strategies in this situation.

7. The Complete Concept of a Game

7.1. Variability of the Characteristics of Each Move

7.1.1. We introduced in 6.2.1. the α_κ alternatives $\mathfrak{A}_\kappa(1), \cdots, \mathfrak{A}_\kappa(\alpha_\kappa)$ of the move $\mathfrak{M}_\kappa$. Also the index k_κ which characterized the move as a personal or chance one, and in the first case the player whose move it is; and in the second case the probabilities $p_\kappa(1), \cdots, p_\kappa(\alpha_\kappa)$ of the above alternatives. We described in 6.3.1. the concept of preliminarity with the help of the sets Λ_κ,—this being the set of all λ (from among the $\lambda = 1, \cdots, \kappa - 1$) which are preliminary to κ. We failed to specify, however, whether all these objects—α_κ, k_κ, Λ_κ and the $\mathfrak{A}_\kappa(\sigma)$, $p_\kappa(\sigma)$ for $\sigma = 1, \cdots, \alpha_\kappa$—depend solely on κ or also on other things. These "other things" can, of course, only be the outcome of the choices corresponding to the moves which are anterior to $\mathfrak{M}_\kappa$. I.e. the numbers $\sigma_1, \cdots, \sigma_{\kappa-1}$. (Cf. 6.2.2.)

This dependence requires a more detailed discussion.

First, the dependence of the alternatives $\mathfrak{A}_\kappa(\sigma)$ themselves (as distinguished from their number α_κ!) on $\sigma_1, \cdots, \sigma_{\kappa-1}$ is immaterial. We may as well assume that the choice corresponding to the move $\mathfrak{M}_\kappa$ is made not between the $\mathfrak{A}_\kappa(\sigma)$ themselves, but between their numbers σ. *In fine*, it is only the σ of $\mathfrak{M}_\kappa$, i.e. σ_κ, which occurs in the expressions describing the outcome of the play,—i.e. in the functions $\mathfrak{F}_k(\sigma_1, \cdots, \sigma_\kappa)$, $k = 1, \cdots, n$.[1] (Cf. 6.2.2.)

Second, all dependences (on $\sigma_1, \cdots, \sigma_{\kappa-1}$) which arise when $\mathfrak{M}_\kappa$ turns out to be a chance move—i.e. when $k_\kappa = 0$ (cf. the end of 6.2.1.)—cause no complications. They do not interfere with our analysis of the behavior of the players. This disposes, in particular, of all probabilities $p_\kappa(\sigma)$, since they occur only in connection with chance moves. (The Λ_κ, on the other hand, never occur in chance moves.)

Third, we must consider the dependences (on $\sigma_1, \cdots, \sigma_{\kappa-1}$) of the α_κ, k_κ, Λ_κ when $\mathfrak{M}_\kappa$ turns out to be a personal move.[2] Now this possibility is indeed a source of complications. And it is a very real possibility.[3] The reason is this.

[1] The form and nature of the alternatives $\mathfrak{A}_\kappa(\sigma)$ offered at $\mathfrak{M}_\kappa$ might, of course, convey to the player k_κ (if $\mathfrak{M}_\kappa$ is a personal move) some information concerning the anterior $\sigma_1, \cdots, \sigma_{\kappa-1}$ values,—if the $\mathfrak{A}_\kappa(\sigma)$ depend on those. But any such information should be specified separately, as information available to k_κ at $\mathfrak{M}_\kappa$. We have discussed the simplest schemes concerning the subject of information in 6.3.1., and shall complete the discussion in 7.1.2. The discussion of α_κ, k_κ, Λ_κ, which follows further below, is characteristic also as far as the role of the $\mathfrak{A}_\kappa(\sigma)$ as possible sources of information is concerned.

[2] Whether this happens for a given κ, will itself depend on k_κ—and hence indirectly on $\sigma_1, \cdots, \sigma_{\kappa-1}$—since it is characterized by $k_\kappa \neq 0$ (cf. the end of 6.2.1.).

[3] E.g.: In Chess the number of possible alternatives α_κ at $\mathfrak{M}_\kappa$ depends on the positions of the men, i.e. the previous course of the play. In Bridge the player who plays the first

7.1.2. The player k_κ must be informed at $\mathfrak{M}_\kappa$ of the values of α_κ, k_κ, Λ_κ—since these are now part of the rules of the game which he must observe. Insofar as they depend upon $\sigma_1, \cdots, \sigma_{\kappa-1}$, he may draw from them certain conclusions concerning the values of $\sigma_1, \cdots, \sigma_{\kappa-1}$. But he is supposed to know absolutely nothing concerning the σ_λ with λ not in Λ_κ! It is hard to see how conflicts can be avoided.

To be precise: There is no conflict in this special case: Let Λ_κ be independent of all $\sigma_1, \cdots, \sigma_{\kappa-1}$, and let α_κ, k_κ depend only on the σ_λ with λ in Λ_κ. Then the player k_κ can certainly not get any information from α_κ, k_κ, Λ_κ beyond what he knows anyhow (i.e. the values of the σ_λ with λ in Λ_κ). If this is the case, we say that we have the *special form of dependence.*

But do we always have the special form of dependence? To take an extreme case: What if Λ_κ is always empty—i.e. k_κ expected to be completely uninformed at $\mathfrak{M}_\kappa$—and yet e.g. α_κ explicitly dependent on some of the $\sigma_1, \cdots, \sigma_{\kappa-1}$!

This is clearly inadmissible. We must demand that all numerical conclusions which can be derived from the knowledge of α_κ, k_κ, Λ_κ, must be explicitly and *ab initio* specified as information available to the player k_κ at $\mathfrak{M}_\kappa$. It would be erroneous, however, to try to achieve this by including in Λ_κ the indices λ of all these σ_λ, on which α_κ, k_κ, Λ_κ explicitly depend. In the first place great care must be exercised in order to avoid circularity in this requirement, as far as Λ_κ is concerned.[1] But even if this difficulty does not arise, because Λ_κ depends only on κ and not on $\sigma_1, \cdots, \sigma_{\kappa-1}$—i.e. if the information available to every player at every moment is independent of the previous course of the play—the above procedure may still be inadmissible. Assume, e.g., that α_κ depends on a certain combination of some σ_λ from among the $\lambda = 1, \cdots, \kappa - 1$, and that the rules of the game do indeed provide that the player k_κ at $\mathfrak{M}_\kappa$ should know the value of this combination, but that it does not allow him to know more (i.e. the values of the individual $\sigma_1, \cdots, \sigma_{\kappa-1}$). E.g.: He may know the value of $\sigma_\mu + \sigma_\lambda$ where μ, λ are both anterior to κ ($\mu, \lambda < \kappa$), but he is not allowed to know the separate values of σ_μ and σ_λ.

One could try various tricks to bring back the above situation to our earlier, simpler, scheme, which describes k_κ's state of information by means of the set Λ_κ.[2] But it becomes completely impossible to disentangle the various components of k_κ's information at $\mathfrak{M}_\kappa$, if they themselves originate from personal moves of different players, or of the same player but in

---

card to the next trick, i.e. k_κ at $\mathfrak{M}_\kappa$, is the one who took the last trick, i.e. again dependent upon the previous course of the play. In some forms of Poker, and some other related games, the amount of information available to a player at a given moment, i.e. Λ_κ at $\mathfrak{M}_\kappa$, depends on what he and the others did previously.

[1] The σ_λ on which, among others, Λ_κ depend are only defined if the totality of all Λ_κ, for all sequences $\sigma_1, \cdots, \sigma_{\kappa-1}$, is considered. Should every Λ_κ contain these λ?

[2] In the above example one might try to replace the move $\mathfrak{M}_\mu$ by a new one in which not σ_μ is chosen, but $\sigma_\mu + \sigma_\lambda$. $\mathfrak{M}_\kappa$ would remain unchanged. Then k_κ at $\mathfrak{M}_\kappa$ would be informed about the outcome of the choice connected with the new $\mathfrak{M}_\mu$ only.

different stages of information. In our above example this happens if $k_\mu \neq k_\lambda$, or if $k_\mu = k_\lambda$ but the state of information of this player is not the same at $\mathfrak{M}_\mu$ and at $\mathfrak{M}_\lambda$.[1]

7.2. The General Description

7.2.1. There are still various, more or less artificial, tricks by which one could try to circumvent these difficulties. But the most natural procedure seems to be to admit them, and to modify our definitions accordingly.

This is done by sacrificing the Λ_κ as a means of describing the state of information. Instead, we describe the state of information of the player k_κ at the time of his personal move $\mathfrak{M}_\kappa$ explicitly: By enumerating those functions of the variable σ_λ anterior to this move—i.e. of the $\sigma_1, \cdots, \sigma_{\kappa-1}$—the numerical values of which he is supposed to know at this moment. This is a system of functions, to be denoted by Φ_κ.

So Φ_κ is a set of functions

$$h(\sigma_1, \cdots, \sigma_{\kappa-1}).$$

Since the elements of Φ_κ describe the dependence on $\sigma_1, \cdots, \sigma_{\kappa-1}$, so Φ_κ itself is fixed, i.e. depending on κ only.[2] α_κ, k_κ may depend on $\sigma_1, \cdots, \sigma_{\kappa-1}$, and since their values are known to k_κ at $\mathfrak{M}_\kappa$, these functions

$$\alpha_\kappa = \alpha_\kappa(\sigma_1, \cdots, \sigma_{\kappa-1}), \qquad k_\kappa = k_\kappa(\sigma_1, \cdots, \sigma_{\kappa-1})$$

must belong to Φ_κ. Of course, whenever it turns out that $k_\kappa = 0$ (for a special set of $\sigma_1, \cdots, \sigma_{\kappa-1}$ values), then the move $\mathfrak{M}_\kappa$ is a chance one (cf. above), and no use will be made of Φ_κ—but this does not matter.

Our previous mode of description, with the Λ_κ, is obviously a special case of the present one, with the Φ_κ.[3]

7.2.2. At this point the reader may feel a certain dissatisfaction about the turn which the discussion has taken. It is true that the discussion was deflected into this direction by complications which arose in actual and typical games (cf. footnote 3 on p. 55). But the necessity of replacing the Λ_κ by the Φ_κ originated in our desire to maintain absolute formal (mathematical) generality. These decisive difficulties, which caused us to take this step (discussed in 7.1.2., particularly as illustrated by the footnotes there) were really extrapolated. I.e. they were not characteristic

[1] In the instance of footnote 2 on p. 56, this means: If $k_\mu \neq k_\lambda$, there is no player to whom the new move $\mathfrak{M}_\mu$ (where $\sigma_\mu + \sigma_\lambda$ is chosen, and which ought to be personal) can be attributed. If $k_\mu = k_\lambda$ but the state of information varies from $\mathfrak{M}_\mu$ to $\mathfrak{M}_\lambda$, then no state of information can be satisfactorily prescribed for the new move $\mathfrak{M}_\mu$.

[2] This arrangement includes nevertheless the possibility that the state of information expressed by Φ_κ depends on $\sigma_1, \cdots, \sigma_{\kappa-1}$. This is the case if, e.g., all functions $h(\sigma_1, \cdots, \sigma_{\kappa-1})$ of Φ_κ show an explicit dependence on σ_μ for one set of values of σ_λ, while being independent of σ_μ for other values of σ_λ. Yet Φ_κ is fixed.

[3] If Φ_κ happens to consist of all functions of certain variables σ_λ—say of those for which λ belongs to a given set M_κ—and of no others, then the Φ_κ description specializes back to the Λ_κ one: Λ_κ being the above set M_κ. But we have seen that we cannot, in general, count upon the existence of such a set.

of the original examples, which are actual games. (E.g. Chess and Bridge can be described with the help of the Λ_κ.)

There exist games which require discussion by means of the Φ_κ. But in most of them one could revert to the Λ_κ by means of various extraneous tricks—and the entire subject requires a rather delicate analysis upon which it does not seem worth while to enter here.[1] There exist unquestionably economic models where the Φ_κ are necessary.[2]

The most important point, however, is this.

In pursuit of the objectives which we have set ourselves we must achieve the certainty of having exhausted all combinatorial possibilities in connection with the entire interplay of the various decisions of the players, their changing states of information, etc. These are problems, which have been dwelt upon extensively in economic literature. We hope to show that they can be disposed of completely. But for this reason we want to be safe against any possible accusation of having overlooked some essential possibility by undue specialization.

Besides, it will be seen that all the formal elements which we are introducing now into the discussion do not complicate it *ultima analysi*. I.e. they complicate only the present, preliminary stage of formal description. The final form of the problem turns out to be unaffected by them. (Cf. 11.2.)

7.2.3. There remains only one more point to discuss: The specializing assumption formulated at the very start of this discussion (at the beginning of 6.2.1.) that both the number and the arrangement of the moves are given (i.e. fixed) *ab initio*. We shall now see that this restriction is not essential.

Consider first the "arrangement" of the moves. The possible variability of the nature of each move—i.e. of its k_κ—has already received full consideration (especially in 7.2.1.). The ordering of the moves $\mathfrak{M}_\kappa$, $k = 1, \cdots, \nu$, was from the start simply the chronological one. Thus there is nothing left to discuss on this score.

Consider next the number of moves ν. This quantity too could be variable, i.e. dependent upon the course of the play.[3] In describing this variability of ν a certain amount of care must be exercised.

[1] We mean card games where players may discard some cards without uncovering them, and are allowed to take up or otherwise use openly a part of their discards later. There exists also a game of double-blind Chess—sometimes called "Kriegsspiel"—which belongs in this class. (For its description cf. 9.2.3. With reference to that description: Each player knows about the "possibility" of the other's anterior choices, without knowing those choices themselves—and this "possibility" is a function of all anterior choices.)

[2] Let a participant be ignorant of the full details of the previous actions of the others, but let him be informed concerning certain statistical resultants of those actions.

[3] It is, too, in most games: Chess, Backgammon, Poker, Bridge. In the case of Bridge this variability is due first to the variable length of the "bidding" phase, and second to the changing number of contracts needed to make a "rubber" (i.e. a play). Examples of games with a fixed ν are harder to find: we shall see that we can make ν fixed in every game by an artifice, but games in which ν is *ab initio* fixed are apt to be monotonous.

The course of the play is characterized by the sequence (of choices) $\sigma_1, \cdots, \sigma_\nu$ (cf. 6.2.2.). Now one cannot state simply that ν may be a function of the variables $\sigma_1, \cdots, \sigma_\nu$, because the full sequence $\sigma_1, \cdots, \sigma_\nu$ cannot be visualized at all, without knowing beforehand what its length ν is going to be.[1] The correct formulation is this: Imagine that the variables $\sigma_1, \sigma_2, \sigma_3, \cdots$ are chosen one after the other.[2] If this succession of choices is carried on indefinitely, then the rules of the game must at some place ν *stop* the procedure. Then ν for which the stop occurs will, of course, depend on all the choices up to that moment. It is the number of moves in that particular play.

Now this *stop rule* must be such as to give a certainty that every conceivable play will be stopped sometime. I.e. it must be impossible to arrange the successive choices of $\sigma_1, \sigma_2, \sigma_3, \cdots$ in such a manner (subject to the restrictions of footnote 2 above) that the stop never comes. The obvious way to guarantee this is to devise a stop rule for which it is certain that the stop will come before a fixed moment, say ν^*. I.e. that while ν may depend on $\sigma_1, \sigma_2, \sigma_3, \cdots$, it is sure to be $\nu \leqq \nu^*$ where ν^* does not depend on $\sigma_1, \sigma_2, \sigma_3, \cdots$. If this is the case we say that the stop rule is *bounded by* ν^*. We shall assume for the games which we consider that they have stop rules bounded by (suitable, but fixed) numbers ν^*.[3,4]

[1] I.e. one cannot say that the length of the game depends on all choices made in connection with all moves, since it will depend on the length of the game whether certain moves will occur at all. The argument is clearly circular.

[2] The domain of variability of σ_1 is $1, \cdots, \alpha_1$. The domain of variability of σ_2 is $1, \cdots, \alpha_2$, and may depend on σ_1: $\alpha_2 = \alpha_2(\sigma_1)$. The domain of variability of σ_3 is $1, \cdots, \alpha_3$, and may depend on σ_1, σ_2: $\alpha_3 = \alpha_3(\sigma_1, \sigma_2)$. Etc., etc.

[3] This stop rule is indeed an essential part of every game. In most games it is easy to find ν's fixed upper bound ν^*. Sometimes, however, the conventional form of the rules of the game does not exclude that the play might—under exceptional conditions—go on *ad infinitum*. In all these cases practical safeguards have been subsequently incorporated into the rules of the game with the purpose of securing the existence of the bound ν^*. It must be said, however, that these safeguards are not always absolutely effective—although the intention is clear in every instance, and even where exceptional infinite plays exist they are of little practical importance. It is nevertheless quite instructive, at least from a mathematical point of view, to discuss a few typical examples.

We give four examples, arranged according to decreasing effectiveness.

Écarté: A play is a "rubber," a "rubber" consists of winning two "games" out of three (cf. footnote 1 on p. 49), a "game" consists of winning five "points," and each "deal" gives one player or the other one or two points. Hence a "rubber" is complete after at most three "games," a "game" after at most nine "deals," and it is easy to verify that a "deal" consists of 13, 14 or 18 moves. Hence $\nu^* = 3 \cdot 9 \cdot 18 = 486$.

Poker: *A priori* two players could keep "overbidding" each other *ad infinitum*. It is therefore customary to add to the rules a proviso limiting the permissible number of "overbids." (The amounts of the bids are also limited, so as to make the number of alternatives α_κ at these personal moves finite.) This of course secures a finite ν^*.

Bridge: The play is a "rubber" and this could go on forever if both sides (players) invariably failed to make their contract. It is not inconceivable that the side which is in danger of losing the "rubber," should in this way permanently prevent a completion of the play by absurdly high bids. This is not done in practice, but there is nothing explicit in the rules of the game to prevent it. In theory, at any rate, some stop rule should be introduced in Bridge.

Chess: It is easy to construct sequences of choices (in the usual terminology:

Now we can make use of this bound ν^* to get entirely rid of the variability of ν.

This is done simply by extending the scheme of the game so that there are always ν^* moves $\mathfrak{M}_1, \cdots, \mathfrak{M}_{\nu^*}$. For every sequence $\sigma_1, \sigma_2, \sigma_3, \cdots$ everything is unchanged up to the move $\mathfrak{M}_\nu$, and all moves beyond $\mathfrak{M}_\nu$ are "dummy moves." I.e. if we consider a move $\mathfrak{M}_\kappa$, $\kappa = 1, \cdots, \nu^*$, for a sequence $\sigma_1, \sigma_2, \sigma_3, \cdots$ for which $\nu < \kappa$, then we make $\mathfrak{M}_\kappa$ a chance move with one alternative only[1]—i.e. one at which nothing happens.

Thus the assumptions made at the beginning of 6.2.1.—particularly that ν is given *ab initio*—are justified *ex post*.

8. Sets and Partitions

8.1. Desirability of a Set-theoretical Description of a Game

8.1. We have obtained a satisfactory and general description of the concept of a game, which could now be restated with axiomatic precision and rigidity to serve as a basis for the subsequent mathematical discussion. It is worth while, however, before doing that, to pass to a different formulation. This formulation is exactly equivalent to the one which we reached in the preceding sections, but it is more unified, simpler when stated in a general form, and it leads to more elegant and transparent notations.

In order to arrive at this formulation we must use the symbolism of the theory of sets—and more particularly of partitions—more extensively than we have done so far. This necessitates a certain amount of explanation and illustration, which we now proceed to give.

"moves")—particularly in the "end game"—which can go on *ad infinitum* without ever ending the play (i.e. producing a "checkmate"). The simplest ones are periodical, i.e. indefinite repetitions of the same cycle of choices, but there exist non-periodical ones as well. All of them offer a very real possibility for the player who is in danger of losing to secure sometimes a "tie." For this reason various "tie rules"—i.e. stop rules—are in use just to prevent that phenomenon.

One well known "tie rule" is this: Any cycle of choices (i.e. "moves"), when three times repeated, terminates the play by a "tie." This rule excludes most but not all infinite sequences, and hence is really not effective.

Another "tie rule" is this: If no pawn has been moved and no officer taken (these are "irreversible" operations, which cannot be undone subsequently) for 40 moves, then the play is terminated by a "tie." It is easy to see that this rule is effective, although the ν^* is enormous.

[4] From a purely mathematical point of view, the following question could be asked: Let the stop rule be effective in this sense only, that it is impossible so to arrange the successive choices $\sigma_1, \sigma_2, \sigma_3, \cdots$ that the stop never comes. I.e. let there always be a finite ν dependent upon $\sigma_1, \sigma_2, \sigma_3, \cdots$. Does this by itself secure the existence of a fixed, finite ν^* bounding the stop rule? I.e. such that all $\nu \leq \nu^*$?

The question is highly academic since all practical game rules aim to establish a ν^* directly. (Cf., however, footnote 3 above.) It is nevertheless quite interesting mathematically.

The answer is "Yes," i.e. ν^* always exists. Cf. e.g. *D. König:* Über eine Schlussweise aus dem Endlichen ins Unendliche, Acta Litt. ac Scient. Univ. Szeged, Sect. Math. Vol. III/II (1927) pp. 121–130; particularly the Appendix, pp. 129–130.

[1] This means, of course, that $\alpha_\kappa = 1$, $k_\kappa = 0$, and $p_\kappa(1) = 1$.

8.2. Sets, Their Properties, and Their Graphical Representation

8.2.1. A *set* is an arbitrary collection of objects, absolutely no restriction being placed on the nature and number of these objects, the *elements* of the set in question. The elements constitute and determine the set as such, without any ordering or relationship of any kind between them. I.e. if two sets A, B are such that every element of A is also one of B and *vice versa*, then they are identical in every respect, $A = B$. The relationship of α being an element of the set A is also expressed by saying that α *belongs* to A.[1]

We shall be interested chiefly, although not always, in *finite sets* only,—i.e. sets consisting of a finite number of elements.

Given any objects $\alpha, \beta, \gamma, \cdots$ we denote the set of which they are the elements by $(\alpha, \beta, \gamma, \cdots)$. It is also convenient to introduce a set which contains no elements at all, the *empty set*.[2] We denote the empty set by $\ominus$. We can, in particular, form sets with precisely one element, *one-element sets*. The one-element set (α), and its unique element α, are not the same thing and should never be confused.[3]

We re-emphasize that any objects can be elements of a set. Of course we shall restrict ourselves to mathematical objects. But the elements can, for instance, perfectly well be sets themselves (cf. footnote 3),—thus leading to sets of sets, etc. These latter are sometimes called by some other —equivalent—name, e.g. systems or aggregates of sets. But this is not necessary.

8.2.2. The main concepts and operations connected with sets are these:

(8:A:a) A is a *subset* of B, or B a *superset* of A, if every element of A is also an element of B. In symbols: $A \subseteq B$ or $B \supseteq A$. A is a *proper subset of* B, or B a *proper superset* of A, if the above is true, but if B contains elements which are not elements of A. In symbols: $A \subset B$ or $B \supset A$. We see: If A is a subset of B and B is a subset of A, then $A = B$. (This is a restatement of the principle formulated at the beginning of 8.2.1.) Also: A is a proper subset of B if and only if A is a subset of B without $A = B$.

[1] The mathematical literature of the theory of sets is very extensive. We make no use of it beyond what will be said in the text. The interested reader will find more information on set theory in the good introduction: *A. Fraenkel:* Einleitung in die Mengenlehre, 3rd Edit. Berlin 1928; concise and technically excellent: *F. Hausdorff:* Mengenlehre, 2nd Edit. Leipzig 1927.

[2] If two sets A, B are both without elements, then we may say that they have the same elements. Hence, by what we said above, $A = B$. I.e. there exists only one empty set.

This reasoning may sound odd, but it is nevertheless faultless.

[3] There are some parts of mathematics where (α) and α can be identified. This is then occasionally done, but it is an unsound practice. It is certainly not feasible in general. E.g., let α be something which is definitely not a one-element set,—i.e. a two-element set (α, β), or the empty set $\ominus$. Then (α) and α must be distinguished, since (α) is a one-element set while α is not.

(8:A:b) The *sum* of two sets A, B is the set of all elements of A together with all elements of B,—to be denoted by $A \cup B$. Similarly the sums of more than two sets are formed.[1]

(8:A:c) The *product*, or *intersection*, of two sets A, B is the set of all common elements of A and of B,—to be denoted by $A \cap B$. Similarly the products of more than two sets are formed.[1]

(8:A:d) The *difference* of two sets A, B (A the *minuend*, B the *subtrahend*) is the set of all those elements of A which do not belong to B,—to be denoted by $A - B$.[1]

(8:A:e) When B is a subset of A, we shall also call $A - B$ the *complement* of B in A. Occasionally it will be so obvious which set A is meant that we shall simply write $-B$ and talk about the *complement* of B without any further specifications.

(8:A:f) Two sets A, B are *disjunct* if they have no elements in common,—i.e. if $A \cap B = \ominus$.

(8:A:g) A system (set) $\mathfrak{A}$ of sets is said to be a system of *pairwise disjunct* sets if all pairs of different elements of $\mathfrak{A}$ are disjunct sets,—i.e. if for A, B belonging to $\mathfrak{A}$, $A \neq B$ implies $A \cap B = \ominus$.

8.2.3. At this point some graphical illustrations may be helpful.

We denote the objects which are elements of sets in these considerations by dots (Figure 1). We denote sets by encircling the dots (elements)

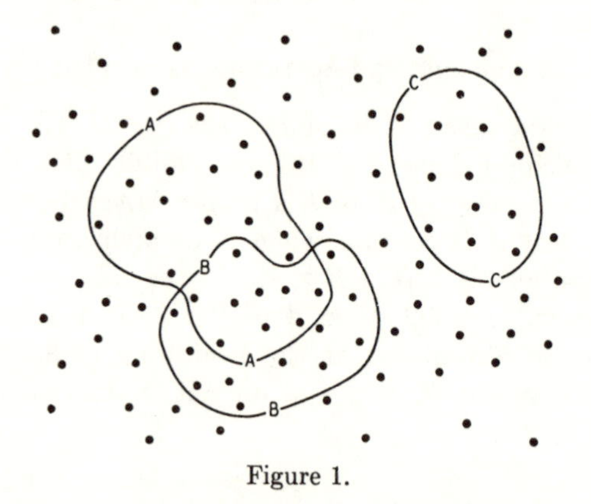

Figure 1.

which belong to them, writing the symbol which denotes the set across the encircling line in one or more places (Figure 1). The sets A, C in this figure are, by the way, disjunct, while A, B are not.

[1] This nomenclature of sums, products, differences, is traditional. It is based on certain algebraic analogies which we shall not use here. In fact, the algebra of these operations $\cup$, $\cap$, also known as *Boolean algebra*, has a considerable interest of its own. Cf. e.g. *A. Tarski:* Introduction to Logic, New York, 1941. Cf. further *Garrett Birkhoff:* Lattice Theory, New York 1940. This book is of wider interest for the understanding of the modern abstract method. Chapt. VI. deals with Boolean Algebras. Further literature is given there.

With this device we can also represent sums, products and differences of sets (Figure 2). In this figure neither A is a subset of B nor B one of A,—hence neither the difference $A - B$ nor the difference $B - A$ is a complement. In the next figure, however, B is a subset of A, and so $A - B$ is the complement of B in A (Figure 3).

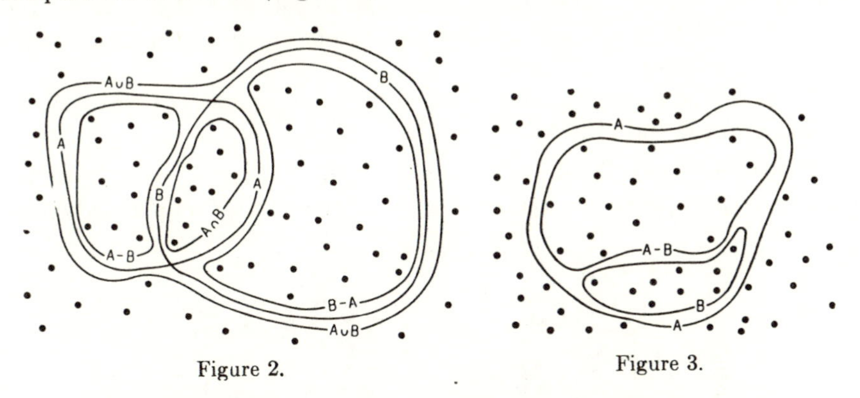

Figure 2. Figure 3.

8.3. Partitions, Their Properties, and Their Graphical Representation

8.3.1. Let a set Ω and a system of sets $\mathcal{A}$ be given. We say that $\mathcal{A}$ is a *partition in* Ω if it fulfills the two following requirements:

(8:B:a) Every element A of $\mathcal{A}$ is a subset of Ω, and not empty.

(8:B:b) $\mathcal{A}$ is a system of pairwise disjunct sets.

This concept too has been the subject of an extensive literature.[1]

We say for two partitions $\mathcal{A}$, $\mathcal{B}$ that $\mathcal{A}$ is a *subpartition* of $\mathcal{B}$, if they fulfill this condition:

(8:B:c) Every element A of $\mathcal{A}$ is a subset of some element B of $\mathcal{B}$.[2]

Observe that if $\mathcal{A}$ is a subpartition of $\mathcal{B}$ and $\mathcal{B}$ a subpartition of $\mathcal{A}$, then $\mathcal{A} = \mathcal{B}$.[3]

Next we define:

(8:B:d) Given two partitions $\mathcal{A}$, $\mathcal{B}$, we form the system of all those intersections $A \cap B$—A running over all elements of $\mathcal{A}$ and B over

[1] Cf. *G. Birkhoff* loc. cit. Our requirements (8:B:a), (8:B:b) are not exactly the customary ones. Precisely:

Ad (8:B:a): It is sometimes not required that the elements A of $\mathcal{A}$ be not empty. Indeed, we shall have to make one exception in 9.1.3. (cf. footnote 4 on p. 69).

Ad (8:B:b): It is customary to require that the sum of all elements of $\mathcal{A}$ be exactly the set Ω. It is more convenient for our purposes to omit this condition.

[2] Since $\mathcal{A}$, $\mathcal{B}$ are also sets, it is appropriate to compare the subset relation (as far as $\mathcal{A}$, $\mathcal{B}$ are concerned) with the subpartition relation. One verifies immediately that if $\mathcal{A}$ is a subset of $\mathcal{B}$ then $\mathcal{A}$ is also a subpartition of $\mathcal{B}$, but that the converse statement is not (generally) true.

[3] *Proof:* Consider an element A of $\mathcal{A}$. It must be subset of an element B of $\mathcal{B}$, and B in turn subset of an element A_1 of $\mathcal{A}$. So A, A_1 have common elements—all those of the not empty set A—i.e. are not disjunct. Since they both belong to the partition $\mathcal{A}$, this necessitates $A = A_1$. So A is a subset of B and B one of A ($= A_1$). Hence $A = B$, and thus A belongs to $\mathcal{B}$.

I.e.: $\mathcal{A}$ is a subset of $\mathcal{B}$. (Cf. footnote 2 above.) Similarly $\mathcal{B}$ is a subset of $\mathcal{A}$. Hence $\mathcal{A} = \mathcal{B}$.

all those of $\mathfrak{B}$—which are not empty. This again is clearly a partition, the *superposition* of $\mathfrak{A}$, $\mathfrak{B}$.[1]

Finally, we also define the above relations for two partitions $\mathfrak{A}$, $\mathfrak{B}$ *within* a given set C.

(8:B:e) $\mathfrak{A}$ is a *subpartition* of $\mathfrak{B}$ *within* C, if every A belonging to $\mathfrak{A}$ which is a subset of C is also subset of some B belonging to $\mathfrak{B}$ which is a subset of C.

(8:B:f) $\mathfrak{A}$ is *equal* to $\mathfrak{B}$ *within* C if the same subsets of C are elements of $\mathfrak{A}$ and of $\mathfrak{B}$.

Clearly footnote 3 on p. 63 applies again, *mutatis mutandis*. Also, the above concepts within Ω are the same as the original unqualified ones.

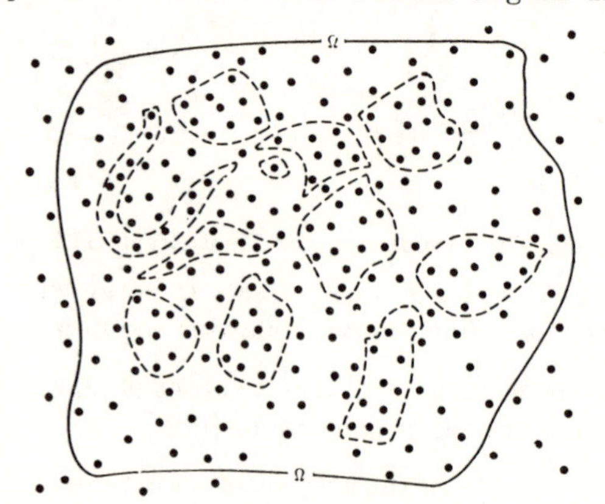

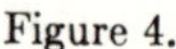
Figure 4.

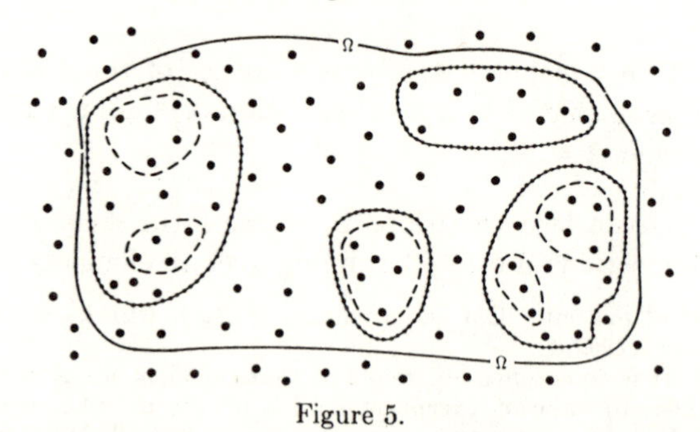

Figure 5.

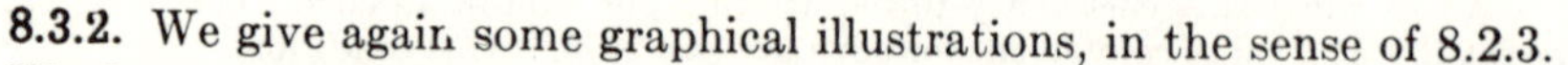

8.3.2. We give again some graphical illustrations, in the sense of 8.2.3.

We begin by picturing a partition. We shall not give the elements of the partition—which are sets—names, but denote each one by an encircling line ——— (Figure 4).

We picture next two partitions $\mathfrak{A}$, $\mathfrak{B}$ distinguishing them by marking the encircling lines of the elements of $\mathfrak{A}$ by – – – and of the elements of $\mathfrak{B}$ by

[1] It is easy to show that the superposition of $\mathfrak{A}$, $\mathfrak{B}$ is a subpartition of both $\mathfrak{A}$ and $\mathfrak{B}$—and that every partition $\mathfrak{C}$ which is a subpartition of both $\mathfrak{A}$ and $\mathfrak{B}$ is also one of their superposition. Hence the name. Cf. *G. Birkhoff*, loc. cit. Chapt. I–II.

-·-·-·- (Figure 5). In this figure $\mathfrak{A}$ is a subpartition of $\mathfrak{B}$. In the following one neither $\mathfrak{A}$ is a subpartition $\mathfrak{B}$ nor is $\mathfrak{B}$ one of $\mathfrak{A}$ (Figure 6). We leave it to the reader to determine the superposition of $\mathfrak{A}$, $\mathfrak{B}$ in this figure.

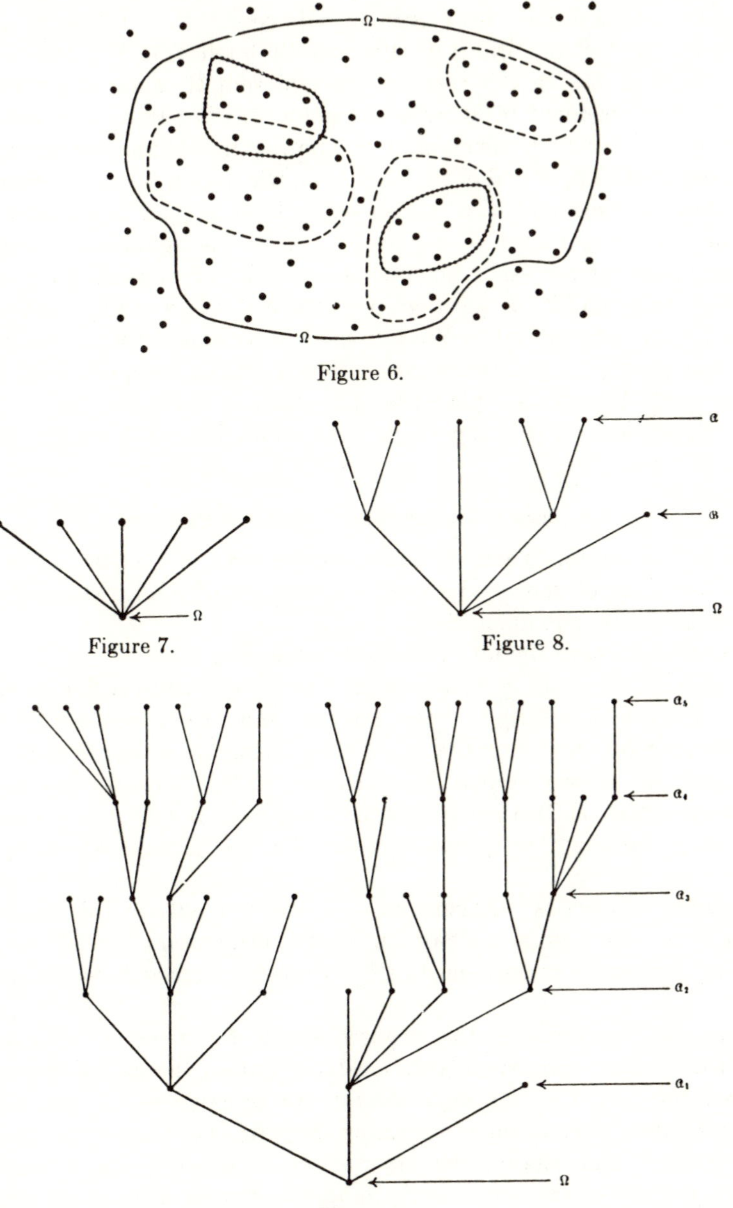

Figure 6.

Figure 7.

Figure 8.

Figure 9.

Another, more schematic, representation of partitions obtains by representing the set Ω by one dot, and every element of the partition—which is a subset of Ω—by a line going upward from this dot. Thus the partition $\mathcal{A}$ of Figure 5 will be represented by a much simpler drawing (Figure 7). This representation does not indicate the elements within the elements of the partition, and it cannot be used to represent several partitions in Ω simultaneously, as was done in Figure 6. However, this deficiency can be removed if the two partitions $\mathcal{A}$, $\mathcal{B}$ in Ω are related as in Figure 5: If $\mathcal{A}$ is a subpartition of $\mathcal{B}$. In this case we can represent Ω again by a dot at the bottom, every element of $\mathcal{B}$ by a line going upward from this dot—as in Figure 7—and every element of $\mathcal{A}$ as another line going further upward, beginning at the upper end of that line of $\mathcal{B}$, which represents the element of $\mathcal{B}$ of which this element of $\mathcal{A}$ is a subset. Thus we can represent the two partitions $\mathcal{A}$, $\mathcal{B}$ of Figure 5 (Figure 8). This representation is again less revealing than the corresponding one of Figure 5. But its simplicity makes it possible to extend it further than pictures in the vein of Figures 4–6 could practically go. Specifically: We can picture by this device a sequence of partitions $\mathcal{A}_1, \cdots, \mathcal{A}_\mu$, where each one is a subpartition of its immediate predecessor. We give a typical example with $\mu = 5$ (Figure 9).

Configurations of this type have been studied in mathematics, and are known as *trees*.

8.4. Logistic Interpretation of Sets and Partitions

8.4.1. The notions which we have described in 8.2.1.-8.3.2. will be useful in the discussion of games which follows, because of the logistic interpretation which can be put upon them.

Let us begin with the interpretation concerning sets.

If Ω is a set of objects of any kind, then every conceivable *property*—which some of these objects may possess, and others not—can be fully characterized by specifying the set of those elements of Ω which have this property. I.e. if two properties correspond in this sense to the same set (the same subset of Ω), then the same elements of Ω will possess these two properties,—i.e. they are *equivalent* within Ω, in the sense in which this term is understood in logic.

Now the properties (of elements of Ω) are not only in this simple correspondence with sets (subsets of Ω), but the elementary logical operations involving properties correspond to the set operations which we discussed in 8.2.2.

Thus the *disjunction* of two properties—i.e. the assertion that at least one of them holds—corresponds obviously to forming the *sum* of their sets,—the operation $A \cup B$. The *conjunction* of two properties —i.e. the assertion that both hold—corresponds to forming the *product* of their sets,—the operation $A \cap B$. And finally, the *negation* of a property—i.e. the assertion of the opposite—corresponds to forming the complement of its set,—the operation $-A$.[1]

[1] Concerning the connection of set theory and of formal logic cf., e.g., *G. Birkhoff*, loc. cit. Chapt. VIII.

Instead of correlating the subsets of Ω to properties in Ω—as done above—we may equally well correlate them with all possible bodies of information concerning an—otherwise undetermined—element of Ω. Indeed, any such information amounts to the assertion that this—unknown—element of Ω possesses a certain—specified—property. It is equivalently represented by the set of all those elements of Ω which possess this property; i.e. to which the given information has narrowed the range of possibilities for the—unknown—element of Ω.

Observe, in particular, that the empty set $\ominus$ corresponds to a property which never occurs, i.e. to an absurd information. And two disjunct sets correspond to two incompatible properties, i.e. to two mutually exclusive bodies of information.

8.4.2. We now turn our attention to partitions.

By reconsidering the definition (8:B:a), (8:B:b) in 8.3.1., and by restating it in our present terminology, we see: A partition is a system of pairwise mutually exclusive bodies of information—concerning an unknown element of Ω—none of which is absurd in itself. In other words: A partition is a preliminary announcement which states how much information will be given later concerning an—otherwise unknown—element of Ω; i.e. to what extent the range of possibilities for this element will be narrowed later. But the actual information is not given by the partition,—that would amount to selecting an element of the partition, since such an element is a subset of Ω, i.e. actual information.

We can therefore say that a partition in Ω is a *pattern of information.* As to the subsets of Ω: we saw in 8.4.1. that they correspond to definite information. In order to avoid confusion with the terminology used for partitions, we shall use in this case—i.e. for a subset of Ω—the words *actual information.*

Consider now the definition (8:B:c) in 8.3.1., and relate it to our present terminology. This expresses for two partitions $\mathfrak{A}$, $\mathfrak{B}$ in Ω the meaning of $\mathfrak{A}$ being a subpartition of $\mathfrak{B}$: it amounts to the assertion that the information announced by $\mathfrak{A}$ includes all the information announced by $\mathfrak{B}$ (and possibly more); i.e. that the pattern of information $\mathfrak{A}$ includes the pattern of information $\mathfrak{B}$.

These remarks put the significance of the Figures 4–9 in 8.3.2. in a new light. It appears, in particular, that the *tree* of Figure 9 pictures a sequence of continually increasing patterns of information.

9. The Set-theoretical Description of a Game

9.1. The Partitions Which Describe a Game

9.1.1. We assume the number of moves—as we now know that we may—to be fixed. Denote this number again by ν, and the moves themselves again by $\mathfrak{M}_1, \cdots, \mathfrak{M}_\nu$.

Consider all possible plays of the game Γ, and form the set Ω of which they are the elements. If we use the description of the preceding sections, then all possible plays are simply all possible sequences $\sigma_1, \cdots, \sigma_\nu$.[1] There exist only a finite number of such sequences,[2] and so Ω is a finite set.

There are, however, also more direct ways to form Ω. We can, e.g., form it by describing each play as the sequence of the $\nu + 1$ consecutive positions[3] which arise during its course. In general, of course, a given position may not be followed by an arbitrary position, but the positions which are possible at a given moment are restricted by the previous positions, in a way which must be precisely described by the rules of the game.[4] Since our description of the rules of the game begins by forming Ω, it may be undesirable to let Ω itself depend so heavily on all the details of those rules. We observe, therefore, that there is no objection to including in Ω absurd sequences of positions as well.[5] Thus it would be perfectly acceptable even to let Ω consist of all sequences of $\nu + 1$ successive positions, without any restrictions whatsoever.

Our subsequent descriptions will show how the really possible plays are to be selected from this, possibly redundant, set Ω.

9.1.2. ν and Ω being given, we enter upon the more elaborate details of the course of a play.

Consider a definite moment during this course, say that one which immediately precedes a given move $\mathfrak{M}_\kappa$. At this moment the following general specifications must be furnished by the rules of the game.

First it is necessary to describe to what extent the events which have led up to the move $\mathfrak{M}_\kappa$[6] have determined the course of the play. Every particular sequence of these events narrows the set Ω down to a subset A_κ: this being the set of all those plays from Ω, the course of which is, up to $\mathfrak{M}_\kappa$, the particular sequence of events referred to. In the terminology of the earlier sections, Ω is—as pointed out in 9.1.1.—the set of all sequences $\sigma_1, \cdots, \sigma_\nu$; then A_κ would be the set of those sequences $\sigma_1, \cdots, \sigma_\nu$ for which the $\sigma_1, \cdots, \sigma_{\kappa-1}$ have given numerical values (cf. footnote 6 above). But from our present broader point of view we need only say that A_κ must be a subset of Ω.

Now the various possible courses the game may have taken up to $\mathfrak{M}_\kappa$ must be represented by different sets A_κ. Any two such courses, if they are different from each other, initiate two entirely disjunct sets of plays; i.e. no play can have begun (i.e. run up to $\mathfrak{M}_\kappa$) both ways at once. This means that any two different sets A_κ must be disjunct.

[1] Cf. in particular 6.2.2. The range of the $\sigma_1, \cdots, \sigma_\nu$ is described in footnote 2 on p. 59.

[2] Verification by means of the footnote referred to above is immediate.

[3] Before $\mathfrak{M}_1$, between $\mathfrak{M}_1$ and $\mathfrak{M}_2$, between $\mathfrak{M}_2$ and $\mathfrak{M}_3$, etc., etc., between $\mathfrak{M}_{\nu-1}$ and $\mathfrak{M}_\nu$, after $\mathfrak{M}_\nu$.

[4] This is similar to the development of the sequence $\sigma_1, \cdots, \sigma_\nu$, as described in footnote 2 on p. 59.

[5] I.e. ones which will ultimately be found to be disallowed by the fully formulated rules of the game.

[6] I.e. the choices connected with the anterior moves $\mathfrak{M}_1, \cdots, \mathfrak{M}_{\kappa-1}$—i.e. the numerical values $\sigma_1, \cdots, \sigma_{\kappa-1}$.

Thus the complete formal possibilities of the course of all conceivable plays of our game up to $\mathfrak{M}_\kappa$ are described by a family of pairwise disjunct subsets of Ω. This is the family of all the sets A_κ mentioned above. We denote this family by $\mathfrak{A}_\kappa$.

The sum of all sets A_κ contained in $\mathfrak{A}_\kappa$ must contain all possible plays. But since we explicitly permitted a redundancy of Ω (cf. the end of 9.1.1.), this sum need nevertheless not be equal to Ω. Summing up:

(9:A) $\mathfrak{A}_\kappa$ is a *partition in* Ω.

We could also say that the partition $\mathfrak{A}_\kappa$ describes the pattern of information of a person who knows everything that happened up to $\mathfrak{M}_\kappa$;[1] e.g. of an umpire who supervises the course of the play.[2]

9.1.3. Second, it must be known what the nature of the move $\mathfrak{M}_\kappa$ is going to be. This is expressed by the k_κ of 6.2.1.: $k_\kappa = 1, \cdots, n$ if the move is personal and belongs to the player k_κ; $k_\kappa = 0$ if the move is chance. k_κ may depend upon the course of the play up to $\mathfrak{M}_\kappa$, i.e. upon the information embodied in $\mathfrak{A}_\kappa$.[3] This means that k_κ must be a constant within each set A_κ of $\mathfrak{A}_\kappa$, but that it may vary from one A_κ to another.

Accordingly we may form for every $k = 0, 1, \cdots, n$ a set $B_\kappa(k)$, which contains all sets A_κ with $k_\kappa = k$, the various $B_\kappa(k)$ being disjunct. Thus the $B_\kappa(k)$, $k = 0, 1, \cdots, n$, form a family of disjunct subsets of Ω. We denote this family by $\mathfrak{B}_\kappa$.

(9:B) $\mathfrak{B}_\kappa$ is again a *partition in* Ω. Since every A_κ of $\mathfrak{A}_\kappa$ is a subset of some $B_\kappa(k)$ of $\mathfrak{B}_\kappa$, therefore $\mathfrak{A}_\kappa$ is a *subpartition* of $\mathfrak{B}_\kappa$.

But while there was no occasion to specify any particular enumeration of the sets A_κ of $\mathfrak{A}_\kappa$, it is not so with $\mathfrak{B}_\kappa$. $\mathfrak{B}_\kappa$ consists of exactly $n + 1$ sets $B_\kappa(k)$, $k = 0, 1, \cdots, n$, which in this way appear in a fixed enumeration by means of the $k = 0, 1, \cdots, n$.[4] And this enumeration is essential since it replaces the function k_κ (cf. footnote 3 above).

9.1.4. Third, the conditions under which the choice connected with the move $\mathfrak{M}_\kappa$ is to take place must be described in detail.

Assume first that $\mathfrak{M}_\kappa$ is a chance move, i.e. that we are within the set $B_\kappa(0)$. Then the significant quantities are: the number of alternatives α_κ and the probabilities $p_\kappa(1), \cdots, p_\kappa(\alpha_\kappa)$ of these various alternatives (cf. the end of 6.2.1.). As was pointed out in 7.1.1. (this was the second item

[1] I.e. the outcome of all choices connected with the moves $\mathfrak{M}_1, \cdots, \mathfrak{M}_{\kappa-1}$. In our earlier terminology: the values of $\sigma_1, \cdots, \sigma_{\kappa-1}$.

[2] It is necessary to introduce such a person since, in general, no player will be in possession of the full information embodied in $\mathfrak{A}_\kappa$.

[3] In the notations of 7.2.1., and in the sense of the preceding footnotes: $k_\kappa = k_\kappa(\sigma_1, \cdots, \sigma_{\kappa-1})$.

[4] Thus $\mathfrak{B}_\kappa$ is really not a set and not a partition, but a more elaborate concept: it consists of the sets $\mathfrak{B}_\kappa(k)$, $k = 0, 1, \cdots, n$, in this enumeration.

It possesses, however, the properties (8:B:a), (8:B:b) of 8.3.1., which characterize a partition. Yet even there an exception must be made: among the sets $\mathfrak{B}_\kappa(k)$ there can be empty ones.

of the discussion there), all these quantities may depend upon the entire information embodied in $\mathcal{A}_\kappa$ (cf. footnote 3 on p. 69), since $\mathfrak{M}_\kappa$ is now a chance move. I.e. α_κ and the $p_\kappa(1), \cdots, p_\kappa(\alpha_\kappa)$ must be constant within each set A_κ of $\mathcal{A}_\kappa$[1] but they may vary from one A_κ to another.

Within each one of these A_κ the choice among the alternatives $\mathcal{A}_\kappa(1), \cdots, \mathcal{A}_\kappa(\alpha_\kappa)$ takes place, i.e. the choice of a $\sigma_\kappa = 1, \cdots, \alpha_\kappa$ (cf. 6.2.2.). This can be described by specifying α_κ disjunct subsets of A_κ which correspond to the restriction expressed by A_κ, plus the choice of σ_κ which has taken place. We call these sets C_κ, and their system—consisting of all C_κ in all the A_κ which are subsets of $B_\kappa(0)$— $\mathcal{C}_\kappa(0)$. Thus $\mathcal{C}_\kappa(0)$ is a *partition in* $B_\kappa(0)$. And since every C_κ of $\mathcal{C}_\kappa(0)$ is a subset of some A_κ of $\mathcal{A}_\kappa$, therefore $\mathcal{C}_\kappa(0)$ is a *subpartition* of $\mathcal{A}_\kappa$.

The α_κ are determined by $\mathcal{C}_\kappa(0)$;[2] hence we need not mention them any more. For the $p_\kappa(1), \cdots, p_\kappa(\alpha_\kappa)$ this description suggests itself: with every C_κ of $\mathcal{C}_\kappa(0)$ a number $p_\kappa(C_\kappa)$ (its probability) must be associated, subject to the equivalents of footnote 2 on p. 50.[3]

9.1.5. Assume, secondly, that $\mathfrak{M}_\kappa$ is a personal move, say of the player $k = 1, \cdots, n$, i.e. that we are within the set $B_\kappa(k)$. In this case we must specify the state of information of the player k at $\mathfrak{M}_\kappa$. In 6.3.1. this was described by means of the set Λ_κ, in 7.2.1. by means of the family of functions Φ_κ, the latter description being the more general and the final one. According to this description k knows at $\mathfrak{M}_\kappa$ the values of all functions $h(\sigma_1, \cdots, \sigma_{\kappa-1})$ of Φ_κ and no more. This amount of information operates a subdivision of $B_\kappa(k)$ into several disjunct subsets, corresponding to the various possible contents of k's information at $\mathfrak{M}_\kappa$. We call these sets D_κ, and their system $\mathfrak{D}_\kappa(k)$. Thus $\mathfrak{D}_\kappa(k)$ is a *partition in* $B_\kappa(k)$.

Of course k's information at $\mathfrak{M}_\kappa$ is part of the total information existing at that moment—in the sense of 9.1.2.—which is embodied in $\mathcal{A}_\kappa$—Hence in an A_κ of $\mathcal{A}_\kappa$, which is a subset of $B_\kappa(k)$, no ambiguity can exist, i.e. this A_κ cannot possess common elements with more than one D_κ of $\mathfrak{D}_\kappa(k)$. This means that the A_κ in question must be a subset of a D_κ of $\mathfrak{D}_\kappa(k)$. In other words: within $B_\kappa(k)$ $\mathcal{A}_\kappa$ is a subpartition of $\mathfrak{D}_\kappa(k)$.

In reality the course of the play is narrowed down at $\mathfrak{M}_\kappa$ within a set A_κ of $\mathcal{A}_\kappa$. But the player k whose move $\mathfrak{M}_\kappa$ is, does not know as much: as far as he is concerned, the play is merely within a set D_κ of $\mathfrak{D}_\kappa(k)$. He must now make the choice among the alternatives $\mathcal{A}_\kappa(1), \cdots, \mathcal{A}_\kappa(\alpha_\kappa)$, i.e. the choice of a $\sigma_\kappa = 1, \cdots, \alpha_\kappa$. As was pointed out in 7.1.2. and 7.2.1. (particularly at the end of 7.2.1.), α_κ may well be variable, but it can only depend upon the information embodied in $\mathfrak{D}_\kappa(k)$. I.e. it must be a constant within the set D_κ of $\mathfrak{D}_\kappa(k)$ to which we have restricted ourselves. Thus the choice of a $\sigma_\kappa = 1, \cdots, \alpha_\kappa$ can be described by specifying α_κ disjunct subsets of D_κ, which correspond to the restriction expressed by D_κ, plus the

[1] We are within $B_\kappa(0)$, hence all this refers only to A_κ's which are subsets of $B_\kappa(0)$.

[2] α_κ is the number of those C_κ of $\mathcal{C}_\kappa(0)$ which are subsets of the given A_κ.

[3] I.e. every $p_\kappa(C_\kappa) \geq 0$, and for each A_κ, and the sum extended over all C_κ of $\mathcal{C}_\kappa(0)$ which are subsets of A_κ, we have $\Sigma p_\kappa(C_\kappa) = 1$.

choice of σ_κ which has taken place. We call these sets C_κ, and their system—consisting of all C_κ in all the D_κ of $\mathfrak{D}_\kappa(k)$ — $\mathfrak{C}_\kappa(k)$. Thus $\mathfrak{C}_\kappa(k)$ is a partition in $B_\kappa(k)$. And since every C_κ of $\mathfrak{C}_\kappa(k)$ is a subset of some D_κ of $\mathfrak{D}_\kappa(k)$, therefore $\mathfrak{C}_\kappa(k)$ is a *subpartition* of $\mathfrak{D}_\kappa(k)$.

The α_κ are determined by $\mathfrak{C}_\kappa(k)$;[1] hence we need not mention them any more. α_κ must not be zero,—i.e., given a D_κ of $\mathfrak{D}_\kappa(k)$, some C_κ of $\mathfrak{C}_\kappa(k)$, which is a subset of D_κ, must exist.[2]

9.2. Discussion of These Partitions and Their Properties

9.2.1. We have completely described in the preceding sections the situation at the moment which precedes the move $\mathfrak{M}_\kappa$. We proceed now to discuss what happens as we go along these moves $\kappa = 1, \cdots, \nu$. It is convenient to add to these a $\kappa = \nu + 1$, too, which corresponds to the conclusion of the play, i.e. follows after the last move $\mathfrak{M}_\nu$.

For $\kappa = 1, \cdots, \nu$ we have, as we discussed in the preceding sections, the partitions

$$\mathfrak{A}_\kappa,\ \mathfrak{B}_\kappa = (B_\kappa(0), B_\kappa(1), \cdots, B_\kappa(n)),\ \mathfrak{C}_\kappa(0), \mathfrak{C}_\kappa(1), \cdots, \mathfrak{C}_\kappa(n),$$
$$\mathfrak{D}_\kappa(1), \cdots, \mathfrak{D}_\kappa(n).$$

All of these, with the sole exception of $\mathfrak{A}_\kappa$, refer to the move $\mathfrak{M}_\kappa$,—hence they need not and cannot be defined for $\kappa = \nu + 1$. But $\mathfrak{A}_{\nu+1}$ has a perfectly good meaning, as its discussion in 9.1.2. shows: It represents the full information which can conceivably exist concerning a play,—i.e. the individual identity of the play.[3]

At this point two remarks suggest themselves: In the sense of the above observations $\mathfrak{A}_1$ corresponds to a moment at which no information is available at all. Hence $\mathfrak{A}_1$ should consist of the one set Ω. On the other hand, $\mathfrak{A}_{\nu+1}$ corresponds to the possibility of actually identifying the play which has taken place. Hence $\mathfrak{A}_{\nu+1}$ is a system of one-element sets.

We now proceed to describe the transition from κ to $\kappa + 1$, when $\kappa = 1, \cdots, \nu$.

9.2.2. Nothing can be said about the change in the $\mathfrak{B}_\kappa$, $\mathfrak{C}_\kappa(k)$, $\mathfrak{D}_\kappa(k)$ when κ is replaced by $\kappa + 1$,—our previous discussions have shown that when this replacement is made anything may happen to those objects, i.e. to what they represent.

It is possible, however, to tell how $\mathfrak{A}_{\kappa+1}$ obtains from $\mathfrak{A}_\kappa$.

The information embodied in $\mathfrak{A}_{\kappa+1}$ obtains from that one embodied in $\mathfrak{A}_\kappa$ by adding to it the outcome of the choice connected with the move $\mathfrak{M}_\kappa$.[4] This ought to be clear from the discussions of 9.1.2. Thus the

[1] α_κ is the number of those C_κ of $\mathfrak{C}_\kappa(k)$ which are subsets of the given A_κ.

[2] We required this for $k = 1, \cdots, n$ only, although it must be equally true for $k = 0$—with an A_κ, subset of $B_\kappa(0)$, in place of our D_κ of $\mathfrak{D}_\kappa(k)$. But it is unnecessary to state it for that case, because it is a consequence of footnote 3 on p. 70; indeed, if no C_κ of the desired kind existed, the $\Sigma p_\kappa(C_\kappa)$ of loc. cit. would be 0 and not 1.

[3] In the sense of footnote 1 on p. 69, the values of all $\sigma_1, \cdots, \sigma_\nu$. And the sequence $\sigma_1, \cdots, \sigma_\nu$ characterizes, as stated in 6.2.2., the play itself.

[4] In our earlier terminology: the value of σ_κ.

information in $\mathcal{A}_{\kappa+1}$ which goes beyond that in $\mathcal{A}_\kappa$ is precisely the information embodied in the $\mathcal{C}_\kappa(0), \mathcal{C}_\kappa(1), \cdots, \mathcal{C}_\kappa(n)$.

This means that the partitions $\mathcal{A}_{\kappa+1}$ obtains by *superposing* the partition $\mathcal{A}_\kappa$ with all partitions $\mathcal{C}_\kappa(0), \mathcal{C}_\kappa(1), \cdots, \mathcal{C}_\kappa(k)$. I.e. by forming the intersection of every A_κ in $\mathcal{A}_\kappa$ with every C_κ in any $\mathcal{C}_\kappa(0), \mathcal{C}_\kappa(1), \cdots, \mathcal{C}_\kappa(n)$, and then throwing away the empty sets.

Owing to the relationship of $\mathcal{A}_\kappa$ and of the $\mathcal{C}_\kappa(k)$ to the sets $B_\kappa(k)$—as discussed in the preceding sections—we can say a little more about this process of superposition.

In $B_\kappa(0)$, $\mathcal{C}_\kappa(0)$ is a subpartition of $\mathcal{A}_\kappa$ (cf. the discussion in 9.1.4.). Hence there $\mathcal{A}_{\kappa+1}$ simply coincides with $\mathcal{C}_\kappa(0)$. In $B_\kappa(k)$, $k = 1, \cdots, n$, $\mathcal{C}_\kappa(k)$ and $\mathcal{A}_\kappa$ are both subpartitions of $\mathcal{D}_\kappa(k)$ (cf. the discussion in 9.1.5.). Hence there $\mathcal{A}_{\kappa+1}$ obtains by first taking every D_κ of $\mathcal{D}_\kappa(k)$, then for every such D_κ all A_κ of $\mathcal{A}_\kappa$ and all C_κ of $\mathcal{C}_\kappa(k)$ which are subsets of this D_κ, and forming all intersections $A_\kappa \cap C_\kappa$.

Every such set $A_\kappa \cap C_\kappa$ represents those plays which arise when the player k, with the information of D_κ before him, but in a situation which is really in A_κ (a subset of D_κ), makes the choice C_κ at the move $\mathcal{M}_\kappa$ so as to restrict things to C_κ.

Since this choice, according to what was said before, is a possible one, there exist such plays. I.e. the set $A_\kappa \cap C_\kappa$ must not be empty. We restate this:

(9:C) If A_κ of $\mathcal{A}_\kappa$ and C_κ of $\mathcal{C}_\kappa(k)$ are subsets of the same D_κ of $\mathcal{D}_\kappa(k)$, then the intersection $A_\kappa \cap C_\kappa$ must not be empty.

9.2.3. There are games in which one might be tempted to set this requirement aside. These are games in which a player may make a legitimate choice which turns out subsequently to be a forbidden one; e.g. the double-blind Chess referred to in footnote 1 on p. 58: here a player can make an apparently possible choice ("move") on his own board, and will (possibly) be told only afterwards by the "umpire" that it is an "impossible" one.

This example is, however, spurious. The move in question is best resolved into a sequence of several alternative ones. It seems best to give the contemplated rules of double-blind Chess in full.

The game consists of a sequence of moves. At each move the "umpire" announces to both players whether the preceding move was a "possible" one. If it was not, the next move is a personal move of the same player as the preceding one; if it was, then the next move is the other player's personal move. At each move the player is informed about all of his own anterior choices, about the entire sequence of "possibility" or "impossibility" of all anterior choices of both players, and about all anterior instances where either player threatened check or took anything. But he knows the identity of his own losses only. In determining the course of the game, the "umpire" disregards the "impossible" moves. Otherwise the game is played like Chess, with a stop rule in the sense of footnote 3 on p. 59, amplified by the further requirement that no player may make ("try") the same choice twice in any one uninterrupted sequence of his own personal moves. (In practice, of course, the players need two chessboards—out of each other's view but both in the "umpire's" view—to obtain these conditions of information.)

At any rate we shall adhere to the requirement stated above. It will be seen that it is very convenient for our subsequent discussion (cf. 11.2.1.).

9.2.4. Only one thing remains: to reintroduce in our new terminology, the quantities $\mathfrak{F}_k$, $k = 1, \cdots, n$, of 6.2.2. $\mathfrak{F}_k$ is the outcome of the play for the player k. $\mathfrak{F}_k$ must be a function of the actual play which has taken place.[1] If we use the symbol π to indicate that play, then we may say: $\mathfrak{F}_k$ is a function of a variable π with the domain of variability Ω. I.e.:

$$\mathfrak{F}_k = \mathfrak{F}_k(\pi), \qquad \pi \text{ in } \Omega, \qquad k = 1, \cdots, n.$$

10. Axiomatic Formulation

10.1. The Axioms and Their Interpretations

10.1.1. Our description of the general concept of a game, with the new technique involving the use of sets and of partitions, is now complete. All constructions and definitions have been sufficiently explained in the past sections, and we can therefore proceed to a rigorous axiomatic definition of a game. This is, of course, only a concise restatement of the things which we discussed more broadly in the preceding sections.

We give first the precise definition, without any commentary:[2]

An n-person game Γ, i.e. the complete system of its rules, is determined by the specification of the following data:

(10:A:a) A number ν.

(10:A:b) A finite set Ω.

(10:A:c) For every $k = 1, \cdots, n$: A function

$$\mathfrak{F}_k = \mathfrak{F}_k(\pi), \qquad \pi \text{ in } \Omega.$$

(10:A:d) For every $\kappa = 1, \cdots, \nu, \nu + 1$: A partition $\mathfrak{A}_\kappa$ in Ω.

(10:A:e) For every $\kappa = 1, \cdots, \nu$: A partition $\mathfrak{B}_\kappa$ in Ω. $\mathfrak{B}_\kappa$ consists of $n + 1$ sets $B_\kappa(k)$, $k = 0, 1, \cdots, n$, enumerated in this way.

(10:A:f) For every $\kappa = 1, \cdots, \nu$ and every $k = 0, 1, \cdots, n$: A partition $\mathfrak{C}_\kappa(k)$ in $B_\kappa(k)$.

(10:A:g) For every $\kappa = 1, \cdots, \nu$ and every $k = 1, \cdots, n$: A partition $\mathfrak{D}_\kappa(k)$ in $B_\kappa(k)$.

(10:A:h) For every $\kappa = 1, \cdots, \nu$ and every C_κ of $\mathfrak{C}_\kappa(0)$: A number $p_\kappa(C_\kappa)$.

These entities must satisfy the following requirements:

(10:1:a) $\mathfrak{A}_\kappa$ is a subpartition of $\mathfrak{B}_\kappa$.

(10:1:b) $\mathfrak{C}_\kappa(0)$ is a subpartition of $\mathfrak{A}_\kappa$.

[1] In the old terminology, accordingly, we had $\mathfrak{F}_k = \mathfrak{F}_k(\sigma_1, \cdots, \sigma_\nu)$. Cf. 6.2.2.

[2] For "explanations" cf. the end of 10.1.1. and the discussion of 10.1.2.

(10:1:c) For $k = 1, \cdots, n$: $\mathfrak{C}_\kappa(k)$ is a subpartition of $\mathfrak{D}_\kappa(k)$.

(10:1:d) For $k = 1, \cdots, n$: Within $B_\kappa(k)$, $\mathfrak{A}_\kappa$ is a subpartition of $\mathfrak{D}_\kappa(k)$.

(10:1:e) For every $\kappa = 1, \cdots, \nu$ and every A_κ of $\mathfrak{A}_\kappa$ which is a subset of $B_\kappa(0)$: For all C_κ of $\mathfrak{C}_\kappa(0)$ which are subsets of this A_κ, $p_\kappa(C_\kappa) \geqq 0$, and for the sum extended over them $\Sigma p_\kappa(C_\kappa) = 1$.

(10:1:f) $\mathfrak{A}_1$ consists of the one set Ω.

(10:1:g) $\mathfrak{A}_{\nu+1}$ consists of one-element sets.

(10:1:h) For $\kappa = 1, \cdots, \nu$: $\mathfrak{A}_{\kappa+1}$ obtains from $\mathfrak{A}_\kappa$ by superposing it with all $\mathfrak{C}_\kappa(k)$, $k = 0, 1, \cdots, n$. (For details, cf. 9.2.2.)

(10:1:i) For $\kappa = 1, \cdots, \nu$: If A_κ of $\mathfrak{A}_\kappa$ and C_κ of $\mathfrak{C}_\kappa(k)$, $k = 1, \cdots, n$ are subsets of the same D_κ of $\mathfrak{D}_\kappa(k)$, then the intersection $A_\kappa \cap C_\kappa$ must not be empty.

(10:1:j) For $\kappa = 1, \cdots, \nu$ and $k = 1, \cdots, n$ and every D_κ of $\mathfrak{D}_\kappa(k)$: Some $C_\kappa(k)$ of $\mathfrak{C}_\kappa$, which is a subset of D_κ, must exist.

This definition should be viewed primarily in the spirit of the modern axiomatic method. We have even avoided giving names to the mathematical concepts introduced in (10:A:a)-(10:A:h) above, in order to establish no correlation with any meaning which the verbal associations of names may suggest. In this absolute "purity" these concepts can then be the objects of an exact mathematical investigation.[1]

This procedure is best suited to develop sharply defined concepts. The application to intuitively given subjects follows afterwards, when the exact analysis has been completed. Cf. also what was said in 4.1.3. in Chapter I about the role of models in physics: The axiomatic models for intuitive systems are analogous to the mathematical models for (equally intuitive) physical systems.

Once this is understood, however, there can be no harm in recalling that this axiomatic definition was distilled out of the detailed empirical discussions of the sections which precede it. And it will facilitate its use, and make its structure more easily understood, if we give the intervening concepts appropriate names,—which indicate, as much as possible, the intuitive background. And it is further useful to express, in the same spirit, the "meaning" of our postulates (10:1:a)-(10:1:j)—i.e. the intuitive considerations from which they sprang.

All this will be, of course, merely a concise summary of the intuitive considerations of the preceding sections, which lead up to this axiomatization.

10.1.2. We state first the technical names for the concepts of (10:A:a)-(10:A:h) in 10.1.1.

[1] This is analogous to the present attitude in axiomatizing such subjects as logic, geometry, etc. Thus, when axiomatizing geometry, it is customary to state that the notions of points, lines, and planes are not to be a *priori* identified with anything intuitive,—they are only notations for things about which only the properties expressed in the axioms are assumed. Cf., e.g., *D. Hilbert:* Die Grundlagen der Geometrie, Leipzig 1899, 2nd Engl. Edition Chicago 1910.

(10:A:a*) ν is the *length* of the game Γ.

(10:A:b*) Ω is the *set of all plays* of Γ.

(10:A:c*) $\mathfrak{F}_k(\pi)$ is the *outcome* of the play π for the player k.

(10:A:d*) $\mathfrak{A}_\kappa$ is the *umpire's pattern of information*, an A_κ of $\mathfrak{A}_\kappa$ is the *umpire's actual information* at (i.e. immediately preceding) the move $\mathfrak{M}_\kappa$. (For $\kappa = \nu + 1$: At the end of the game.)

(10:A:e*) $\mathfrak{B}_\kappa$ is the *pattern of assignment*, a $B_\kappa(k)$ of $\mathfrak{B}_\kappa$ is the *actual assignment*, of the move $\mathfrak{M}_\kappa$.

(10:A:f*) $\mathfrak{C}_\kappa(k)$ is the *pattern of choice*, a C_κ of $\mathfrak{C}_\kappa(k)$ is the *actual choice*, of the player k at the move $\mathfrak{M}_\kappa$. (For $k = 0$: Of chance.)

(10:A:g*) $\mathfrak{D}_\kappa(k)$ is the *player k's pattern of information*, a D_κ of $\mathfrak{D}_\kappa(k)$ the *player k's actual information*, at the move $\mathfrak{M}_\kappa$.

(10:A:h*) $p_\kappa(C_\kappa)$ is the probability of the actual choice C_κ at the (chance) move $\mathfrak{M}_\kappa$.

We now formulate the "meaning" of the requirements (10:1:a)-(10:1:j)—in the sense of the concluding discussion of 10.1.1—with the use of the above nomenclature.

(10:1:a*) The umpire's pattern of information at the move $\mathfrak{M}_\kappa$ includes the assignment of that move.

(10:1:b*) The pattern of choice at a chance move $\mathfrak{M}_\kappa$ includes the umpire's pattern of information at that move.

(10:1:c*) The pattern of choice at a personal move $\mathfrak{M}_\kappa$ of the player k includes the player k's pattern of information at that move.

(10:1:d*) The umpire's pattern of information at the move $\mathfrak{M}_\kappa$ includes—to the extent to which this is a personal move of the player k—the player k's pattern of information at that move.

(10:1:e*) The probabilities of the various alternative choices at a chance move $\mathfrak{M}_\kappa$ behave like probabilities belonging to disjunct but exhaustive alternatives.

(10:1:f*) The umpire's pattern of information at the first move is void.

(10:1:g*) The umpire's pattern of information at the end of the game determines the play fully.

(10:1:h*) The umpire's pattern of information at the move $\mathfrak{M}_{\kappa+1}$ (for $\kappa = \nu$: at the end of the game) obtains from that one at the move $\mathfrak{M}_\kappa$ by superposing it with the pattern of choice at the move $\mathfrak{M}_\kappa$.

(10:1:i*) Let a move $\mathfrak{M}_\kappa$ be given, which is a personal move of the player k, and any actual information of the player k at that move also be given. Then any actual information of the umpire at that move and any actual choice of the player k at that move, which are both within (i.e. refinements of) this actual (player's) information, are also compatible with each other. I.e. they occur in actual plays.

(10:1:j*) Let a move $\mathfrak{M}_\kappa$ be given, which is a personal move of the player k, and any actual information of the player k at that move also be given. Then the number of alternative actual choices, available to the player k, is not zero.

This concludes our formalization of the general scheme of a game.

10.2. Logistic Discussion of the Axioms

10.2. We have not yet discussed those questions which are conventionally associated in formal logics with every axiomatization: freedom from contradiction, categoricity (completeness), and independence of the axioms.[1] Our system possesses the first and the last-mentioned properties, but not the second one. These facts are easy to verify, and it is not difficult to see that the situation is exactly what it should be. *In summa:*

Freedom from contradiction: There can be no doubt as to the existence of games, and we did nothing but give an exact formalism for them. We shall discuss the formalization of several games later in detail, cf. e.g. the examples of 18., 19. From the strictly mathematical—logistic—point of view, even the simplest game can be used to establish the fact of freedom from contradiction. But our real interest lies, of course, with the more involved games, which are the really interesting ones.[2]

Categoricity (completeness): This is not the case, since there exist many different games which fulfill these axioms. Concerning effective examples, cf. the preceding reference.

The reader will observe that categoricity is not intended in this case, since our axioms have to define a class of entities (games) and not a unique entity.[3]

Independence: This is easy to establish, but we do not enter upon it.

10.3. General Remarks Concerning the Axioms

10.3. There are two more remarks which ought to be made in connection with this axiomatization.

First, our procedure follows the classical lines of obtaining an exact formulation for intuitively—empirically—given ideas. The notion of a game exists in general experience in a practically satisfactory form, which is nevertheless too loose to be fit for exact treatment. The reader who has followed our analysis will have observed how this imprecision was gradually

[1] Cf. *D. Hilbert*, loc. cit.; *O. Veblen & J. W. Young:* Projective Geometry, New York 1910; *H. Weyl:* Philosophie der Mathematik und Naturwissenschaften, in Handbuch der Philosophie, Munich, 1927.

[2] This is the simplest game: $\nu = 0$, Ω has only one element, say π_0. Consequently no $\mathfrak{B}_\kappa$, $\mathfrak{C}_\kappa(k)$, $\mathfrak{D}_\kappa(k)$, exist, while the only $\mathfrak{A}_\kappa$ is $\mathfrak{A}_1$, consisting of Ω alone. Define $\mathfrak{F}(\pi_0) = 0$ for $k = 1, \cdots, n$. An obvious description of this game consists in the statement that nobody does anything and that nothing happens. This also indicates that the freedom from contradiction is not in this case an interesting question.

[3] This is an important distinction in the general logistic approach to axiomatization. Thus the axioms of Euclidean geometry describe a unique object—while those of group theory (in mathematics) or of rational mechanics (in physics) do not, since there exist many different groups and many different mechanical systems.

removed, the "zone of twilight" successively reduced, and a precise formulation obtained eventually.

Second, it is hoped that this may serve as an example of the truth of a much disputed proposition: That it is possible to describe and discuss mathematically human actions in which the main emphasis lies on the psychological side. In the present case the psychological element was brought in by the necessity of analyzing decisions, the information on the basis of which they are taken, and the interrelatedness of such sets of information (at the various moves) with each other. This interrelatedness originates in the connection of the various sets of information in time, causation, and by the speculative hypotheses of the players concerning each other.

There are of course many—and most important—aspects of psychology which we have never touched upon, but the fact remains that a primarily psychological group of phenomena has been axiomatized.

10.4. Graphical Representation

10.4.1. The graphical representation of the numerous partitions which we had to use to represent a game is not easy. We shall not attempt to treat this matter systematically: even relatively simple games seem to lead to complicated and confusing diagrams, and so the usual advantages of graphical representation do not obtain.

There are, however, some restricted possibilities of graphical representation, and we shall say a few words about these.

In the first place it is clear from (10:1:h) in 10.1.1., (or equally by (10:1:h*) in 10.1.2., i.e. by remembering the "meaning,") that $\mathcal{A}_{\kappa+1}$ is a subpartition of $\mathcal{A}_\kappa$. I.e. in the sequence of partitions $\mathcal{A}_1, \cdots, \mathcal{A}_\nu, \mathcal{A}_{\nu+1}$ each one is a subpartition of its immediate predecessor. Consequently this much can be pictured with the devices of Figure 9 in 8.3.2., i.e. by a tree. (Figure 9 is not characteristic in one way: since the length of the game Γ is assumed to be fixed, all branches of the tree must continue to its full height. Cf. Figure 10 in 10.4.2. below.) We shall not attempt to add the $B_\kappa(k)$, $\mathcal{C}_\kappa(k)$, $\mathcal{D}_\kappa(k)$ to this picture.

There is, however, a class of games where the sequence $\mathcal{A}_1, \cdots, \mathcal{A}_\nu, \mathcal{A}_{\nu+1}$ tells practically the entire story. This is the important class—already discussed in 6.4.1., and about which more will be said in 15.—where preliminarity and anteriority are equivalent. Its characteristics find a simple expression in our present formalism.

10.4.2. Preliminarity and anteriority are equivalent—as the discussions of 6.4.1., 6.4.2. and the interpretation of 6.4.3. show—if and only if every player who makes a personal move knows at that moment the entire anterior history of the play. Let the player be k, the move $\mathfrak{M}_\kappa$. The assertion that $\mathfrak{M}_\kappa$ is k's personal move means, then, that we are within $B_\kappa(k)$. Hence the assertion is that within $B_\kappa(k)$ the player k's pattern of information coincides with the umpire's pattern of information; i.e. that $\mathcal{D}_\kappa(k)$ is equal to

$\mathcal{A}_\kappa$ within $B_\kappa(k)$. But $\mathcal{D}_\kappa(k)$ is a partition in $B_\kappa(k)$; hence the above statement means that $\mathcal{D}_\kappa(k)$ simply is that part of $\mathcal{A}_\kappa$ which lies in $B_\kappa(k)$.

We restate this:

(10:B) Preliminarity and anteriority coincide—i.e. every player who makes a personal move is at that moment fully informed about the entire anterior history of the play—if and only if $\mathcal{D}_\kappa(k)$ is that part of $\mathcal{A}_\kappa$ which lies in $B_\kappa(k)$.

If this is the case, then we can argue on as follows: By (10:1:c) in 10.1.1. and the above, $\mathcal{C}_\kappa(k)$ must now be a subpartition of $\mathcal{A}_\kappa$. This holds for personal moves, i.e. for $k = 1, \cdots, n$, but for $k = 0$ it follows immedi-

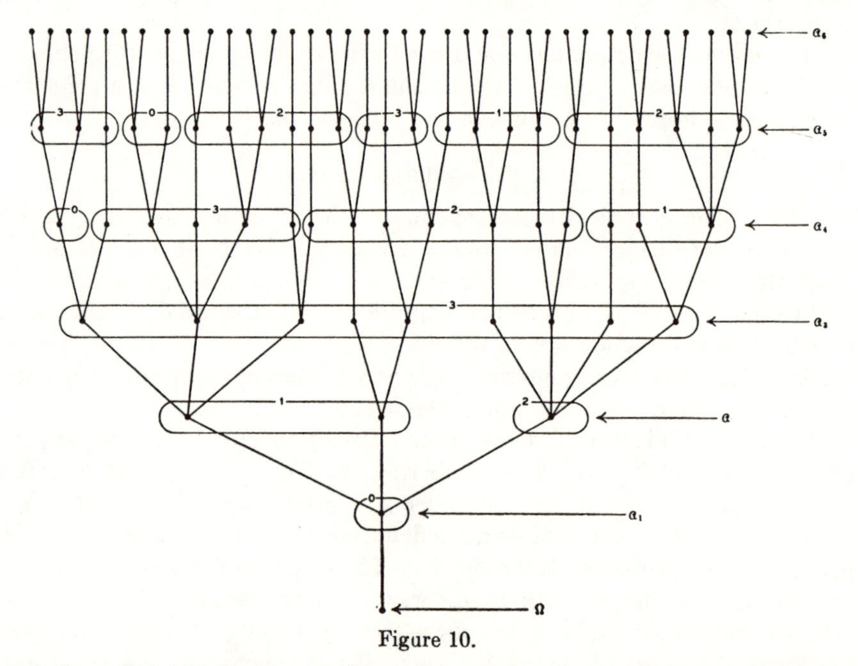

Figure 10.

ately from (10:1:b) in 10.1.1. Now (10:1:h) in 10.1.1. permits the inference from this (for details cf. 9.2.2.) that $\mathcal{A}_{\kappa+1}$ coincides with $\mathcal{C}_\kappa(k)$ in $B_\kappa(k)$—for all $k = 0, 1, \cdots, n$. (We could equally have used the corresponding points in 10.1.2., i.e. the "meaning" of these concepts. We leave the verbal expression of the argument to the reader.) But $\mathcal{C}_\kappa(k)$ is a partition in $B_\kappa(k)$; hence the above statement means that $\mathcal{C}_\kappa(k)$ simply is that part of $\mathcal{A}_{\kappa+1}$ which lies in $B_\kappa(k)$.

We restate this:

(10:C) If the condition of (10:B) is fulfilled, then $\mathcal{C}_\kappa(k)$ is that part of $\mathcal{A}_{\kappa+1}$ which lies in $B_\kappa(k)$.

Thus when preliminarity and anteriority coincide, then in our present formalism the sequence $\mathcal{A}_1, \cdots, \mathcal{A}_\nu, \mathcal{A}_{\nu+1}$ and the sets $B_\kappa(k)$, $k = 0, 1, \cdots, n$, for each $\kappa = 1, \cdots, \nu$, describe the game fully. I.e. the picture of Figure 9 in 8.3.2. must be amplified only by bracketing together those elements of each $\mathfrak{A}_\kappa$, which belong to the same set $\mathfrak{B}_\kappa(k)$. (Cf. however, the remark made in 10.4.1.) We can do this by encircling them with a line, across which the number k of $B_\kappa(k)$ is written. Such $B_\kappa(k)$ as are empty can be omitted. We give an example of this for $\nu = 5$ and $n = 3$ (Figure 10).

In many games of this class even this extra device is not necessary, because for every κ only one $B_\kappa(k)$ is not empty. I.e. the character of each move $\mathfrak{M}_\kappa$ is independent of the previous course of the play.[1] Then it suffices to indicate at each $\mathfrak{A}_\kappa$ the character of the move $\mathfrak{M}_\kappa$—i.e. the unique $k = 0, 1, \cdots, n$ for which $B_\kappa(k) \neq \ominus$.

11. Strategies and the Final Simplification of the Description of a Game

11.1. The Concept of a Strategy and Its Formalization

11.1.1. Let us return to the course of an actual play π of the game Γ.

The moves $\mathfrak{M}_\kappa$ follow each other in the order $\kappa = 1, \cdots, \nu$. At each move $\mathfrak{M}_\kappa$ a choice is made, either by chance—if the play is in $B_\kappa(0)$—or by a player $k = 1, \cdots, n$—if the play is in $B_\kappa(k)$. The choice consists in the selection of a C_κ from $\mathfrak{C}_\kappa(k)$ ($k = 0$ or $k = 1, \cdots, n$, cf. above), to which the play is then restricted. If the choice is made by a player k, then precautions must be taken that this player's pattern of information should be at this moment $\mathfrak{D}_\kappa(k)$, as required. (That this can be a matter of some practical difficulty is shown by such examples as Bridge [cf. the end of 6.4.2.] and double-blind Chess [cf. 9.2.3.].)

Imagine now that each player $k = 1, \cdots, n$, instead of making each decision as the necessity for it arises, makes up his mind in advance for all possible contingencies; i.e. that the player k begins to play with a complete plan: a plan which specifies what choices he will make in every possible situation, for every possible actual information which he may possess at that moment in conformity with the pattern of information which the rules of the game provide for him for that case. We call such a plan a *strategy*.

Observe that if we require each player to start the game with a complete plan of this kind, i.e. with a strategy, we by no means restrict his freedom of action. In particular, we do not thereby force him to make decisions on the basis of less information than there would be available for him in each practical instance in an actual play. This is because the strategy is supposed to specify every particular decision only as a function of just that amount of actual information which would be available for this purpose in an actual play. The only extra burden our assumption puts on the player is the intellectual one to be prepared with a rule of behavior for all eventualities,—although he is to go through one play only. But this is an innocuous assumption within the confines of a mathematical analysis. (Cf. also 4.1.2.)

[1] This is true for Chess. The rules of Backgammon permit interpretations both ways.

11.1.2. The chance component of the game can be treated in the same way.

It is indeed obvious that it is not necessary to make the choices which are left to chance, i.e. those of the chance moves, only when those moves come along. An umpire could make them all in advance, and disclose their outcome to the players at the various moments and to the varying extent, as the rules of the game provide about their information.

It is true that the umpire cannot know in advance which moves will be chance ones, and with what probabilities; this will in general depend upon the actual course of the play. But—as in the strategies which we considered above—he could provide for all contingencies: He could decide in advance what the outcome of the choice in every possible chance move should be, for every possible anterior course of the play,—i.e. for every possible actual umpire's information at the move in question. Under these conditions the probabilities prescribed by the rules of the game for each one of the above instances would be fully determined—and so the umpire could arrange for each one of the necessary choices to be effected by chance, with the appropriate probabilities.

The outcomes could then be disclosed by the umpire to the players—at the proper moments and to the proper extent—as described above.

We call such a preliminary decision of the choices of all conceivable chance moves an *umpire's choice*.

We saw in the last section that the replacement of the choices of all personal moves of the player k by the strategy of the player k is legitimate; i.e. that it does not modify the fundamental character of the game Γ. Clearly our present replacement of the choices of all chance moves by the umpire's choice is legitimate in the same sense.

11.1.3. It remains for us to formalize the concepts of a strategy and of the umpire's choice. The qualitative discussion of the two last sections makes this an unambiguous task.

A strategy of the player k does this: Consider a move $\mathfrak{M}_\kappa$. Assume that it has turned out to be a personal move of the player k,—i.e. assume that the play is within $B_\kappa(k)$. Consider a possible actual information of the player k at that moment,—i.e. consider a D_κ of $\mathfrak{D}_\kappa(k)$. Then the strategy in question must determine his choice at this juncture,—i.e. a C_κ of $\mathfrak{C}_\kappa(k)$ which is a subset of the above D_κ.

Formalized:

(11:A) A *strategy* of the player k is a function $\Sigma_k(\kappa; D_\kappa)$ which is defined for every $\kappa = 1, \cdots, \nu$ and every D_κ of $\mathfrak{D}_\kappa(k)$, and whose value

$$\Sigma_k(\kappa; D_\kappa) = C_\kappa$$

has always these properties: C_κ belongs to $\mathfrak{C}_\kappa(k)$ and is a subset of D_κ.

That strategies—i.e. functions $\Sigma_k(\kappa; D_\kappa)$ fulfilling the above requirement—exist at all, coincides precisely with our postulate (10:1:j) in 10.1.1.

An umpire's choice does this:

Consider a move $\mathfrak{M}_\kappa$. Assume that it has turned out to be a chance move,—i.e. assume that the play is within $B_\kappa(0)$. Consider a possible actual information of the umpire at this moment; i.e. consider an A_κ of $\mathfrak{A}_\kappa$ which is a subset of $B_\kappa(0)$. Then the umpire's choice in question must determine the chance choice at this juncture,—i.e. a C_κ of $\mathfrak{C}_\kappa(0)$ which is a subset of the above A_κ.

Formalized:

(11:B) An *umpire's choice* is a function $\Sigma_0(\kappa; A_\kappa)$ which is defined for every $\kappa = 1, \cdots, \nu$ and every A_κ of $\mathfrak{A}_\kappa$ which is a subset of $B_\kappa(0)$ and whose value

$$\Sigma_0(\kappa; A_\kappa) = C_\kappa$$

has always these properties: C_κ belongs to $\mathfrak{C}_\kappa(0)$ and is a subset of A_κ.

Concerning the existence of umpire's choices—i.e. of functions $\Sigma_0(\kappa; A_\kappa)$ fulfilling the above requirement—cf. the remark after (11:A) above, and footnote 2 on p. 71.

Since the outcome of the umpire's choice depends on chance, the corresponding probabilities must be specified. Now the umpire's choice is an aggregate of independent chance events. There is such an event, as described in 11.1.2., for every $\kappa = 1, \cdots, \nu$ and every A_κ of $\mathfrak{A}_\kappa$ which is a subset of $B_\kappa(0)$. I.e. for every pair κ, A_κ in the domain of definition of $\Sigma_0(\kappa; A_\kappa)$. As far as this event is concerned the probability of the particular outcome $\Sigma_0(\kappa; A_\kappa) = C_\kappa$ is $p_\kappa(C_\kappa)$. Hence the probability of the entire umpire's choice, represented by the function $\Sigma_0(\kappa; A_\kappa)$ is the product of the individual probabilities $p_\kappa(C_\kappa)$.[1]

Formalized:

(11:C) The probability of the *umpire's choice*, represented by the function $\Sigma_0(\kappa; A_\kappa)$ is the product of the probabilities $p_\kappa(C_\kappa)$, where $\Sigma_0(\kappa; A_\kappa) = C_\kappa$, and κ, A_κ run over the entire domain of definition of $\Sigma_0(\kappa; A_\kappa)$ (cf. (11:B) above).

If we consider the conditions of (10:1:e) in 10.1.1. for all these pairs κ, A_κ, and multiply them all with each other, then these facts result: The probabilities of (11:C) above are all $\geqq 0$, and their sum (extended over all umpire's choices) is one. This is as it should be, since the totality of all umpire's choices is a system of disjunct but exhaustive alternatives.

11.2. The Final Simplification of the Description of a Game

11.2.1. If a definite strategy has been adopted by each player $k = 1, \cdots, n$, and if a definite umpire's choice has been selected, then these determine the entire course of the play uniquely,—and accordingly its

[1] The chance events in question must be treated as independent.

outcome too, for each player $k = 1, \cdots, n$. This should be clear from the verbal description of all these concepts, but an equally simple formal proof can be given.

Denote the strategies in question by $\Sigma_k(\kappa; D_\kappa)$, $k = 1, \cdots, n$, and the umpire's choice by $\Sigma_0(\kappa; A_\kappa)$. We shall determine the umpire's actual information at all moments $\kappa = 1, \cdots, \nu, \nu + 1$. In order to avoid confusing it with the above variable A_κ, we denote it by $\bar{A}_\kappa$.

$\bar{A}_1$ is, of course, equal to Ω itself. (Cf. (10:1:f) in 10.1.1.)

Consider now a $\kappa = 1, \cdots, \nu$, and assume that the corresponding $\bar{A}_\kappa$ is already known. Then $\bar{A}_\kappa$ is a subset of precisely one $B_\kappa(k)$, $k = 0, 1, \cdots, n$. (Cf. (10:1:a) in 10.1.1.) If $k = 0$, then $\mathfrak{M}_\kappa$ is a chance move, and so the outcome of the choice is $\Sigma_0(\kappa; \bar{A}_\kappa)$. Accordingly $\bar{A}_{\kappa+1} = \Sigma_0(\kappa; \bar{A}_\kappa)$. (Cf. (10:1:h) in 10.1.1. and the details in 9.2.2.) If $k = 1, \cdots, n$, then $\mathfrak{M}_\kappa$ is a personal move of the player k. $\bar{A}_\kappa$ is a subset of precisely one $\bar{D}_\kappa$ of $\mathfrak{D}_\kappa(k)$. (Cf. (10:1:d) in 10.1.1.) So the outcome of the choice is $\Sigma_k(\kappa; \bar{D}_\kappa)$. Accordingly $\bar{A}_{\kappa+1} = \bar{A}_\kappa \cap \Sigma_k(\kappa; \bar{D}_\kappa)$. (Cf. (10:1:h) in 10.1.1. and the details in 9.2.2.)

Thus we determine inductively $\bar{A}_1, \bar{A}_2, \bar{A}_3, \cdots, \bar{A}_\nu, \bar{A}_{\nu+1}$ in succession. But $\bar{A}_{\nu+1}$ is a one-element set (cf. (10:1:g) in 10.1.1.); denote its unique element by $\bar{\pi}$.

This $\bar{\pi}$ is the actual play which took place.[1] Consequently the outcome of the play is $\mathfrak{F}_k(\bar{\pi})$ for the player $k = 1, \cdots, n$.

11.2.2. The fact that the strategies of all players and the umpire's choice determine together the actual play—and so its outcome for each player—opens up the possibility of a new and much simpler description of the game Γ.

Consider a given player $k = 1, \cdots, n$. Form all possible strategies of his, $\Sigma_k(\kappa; D_\kappa)$, or for short Σ_k. While their number is enormous, it is obviously finite. Denote it by β_k, and the strategies themselves by $\Sigma_k^1, \cdots, \Sigma_k^{\beta_k}$.

Form similarly all possible umpire's choices, $\Sigma_0(\kappa; A_\kappa)$, or for short Σ_0. Again their number is finite. Denote it by β_0, and the umpire's choices by $\Sigma_0^1, \cdots, \Sigma_0^{\beta_0}$. Denote their probabilities by $p^1, \cdots, p^{\beta_0}$ respectively. (Cf. (11:C) in 11.1.3.) All these probabilities are $\geqq 0$ and their sum is one. (Cf. the end of 11.1.3.)

A definite choice of all strategies and of the umpire's choices, say $\Sigma_k^{\tau_k}$ for $k = 1, \cdots, n$ and for $k = 0$ respectively, where

$$\tau_k = 1, \cdots, \beta_k \qquad \text{for} \qquad k = 0, 1, \cdots, n,$$

determines the play $\bar{\pi}$ (cf. the end of 11.2.1.), and its outcome $\mathfrak{F}_k(\bar{\pi})$ for each player $k = 1, \cdots, n$. Write accordingly

$$(11:1) \qquad \mathfrak{F}_k(\bar{\pi}) = \mathfrak{G}_k(\tau_0, \tau_1, \cdots, \tau_n) \qquad \text{for} \qquad k = 1, \cdots, n.$$

[1] The above inductive derivation of the $\bar{A}_1, \bar{A}_2, \bar{A}_3, \cdots, \bar{A}_\nu, \bar{A}_{\nu+1}$ is just a mathematical reproduction of the actual course of the play. The reader should verify the parallelism of the steps involved.

The entire play now consists of each player k choosing a strategy $\Sigma_k^{\tau_k}$, i.e. a number $\tau_k = 1, \cdots, \beta_k$; and of the chance umpire's choice of $\tau_0 = 1, \cdots, \beta_0$, with the probabilities $p^1, \cdots, p^{\beta_0}$ respectively.

The player k must choose his strategy, i.e. his τ_k, without any information concerning the choices of the other players, or of the chance events (the umpire's choice). This must be so since all the information he can at any time possess is already embodied in his strategy $\Sigma_k = \Sigma_k^{\tau_k}$ i.e. in the function $\Sigma_k = \Sigma_k(\kappa; D_\kappa)$. (Cf. the discussion of 11.1.1.) Even if he holds definite views as to what the strategies of the other players are likely to be, they must be already contained in the function $\Sigma_k\,(\kappa; D_\kappa)$.

11.2.3. All this means, however, that Γ has been brought back to the very simplest description, within the least complicated original framework of the sections 6.2.1.-6.3.1. We have $n + 1$ moves, one chance and one personal for each player $k = 1, \cdots, n$—each move has a fixed number of alternatives, β_0 for the chance move and $\beta_1, \cdots, \beta_n$ for the personal ones—and every player has to make this choice with absolutely no information concerning the outcome of all other choices.[1]

Now we can get rid even of the chance move. If the choices of the players have taken place, the player k having chosen τ_k, then the total influence of the chance move is this: The outcome of the play for the player k may be any one of the numbers

$$\mathcal{G}_k(\tau_0, \tau_1, \cdots, \tau_n), \qquad \tau_0 = 1, \cdots, \beta_0,$$

with the probabilities $p^1, \cdots, p^{\beta_0}$ respectively. Consequently his "mathematical expectation" of the outcome is

$$(11{:}2) \qquad \mathcal{H}_k(\tau_1, \cdots, \tau_n) = \sum_{\tau_0=1}^{\beta_0} p^{\tau_0}\mathcal{G}_k(\tau_0, \tau_1, \cdots, \tau_n).$$

The player's judgment must be directed solely by this "mathematical expectation,"—because the various moves, and in particular the chance move, are completely isolated from each other.[2] Thus the only moves which matter are the n personal moves of the players $k = 1, \cdots n$.

The final formulation is therefore this:

(11:D) The n person game Γ, i.e. the complete system of its rules, is determined by the specification of the following data:

(11:D:a) For every $k = 1, \cdots, n$: A number β_k.

(11:D:b) For every $k = 1, \cdots, n$: A function

$$\mathcal{H}_k = \mathcal{H}_k(\tau_1, \cdots, \tau_n),$$
$$\tau_j = 1, \cdots, \beta_j \qquad \text{for} \qquad j = 1, \cdots, n.$$

[1] Owing to this complete disconnectedness of the $n + 1$ moves, it does not matter in what chronological order they are placed.

[2] We are entitled to use the unmodified "mathematical expectation" since we are satisfied with the simplified concept of utility, as stressed at the end of 5.2.2. This excludes in particular all those more elaborate concepts of "expectation," which are really attempts at improving that naive concept of utility. (E.g. D. Bernoulli's "moral expectation" in the "St. Petersburg Paradox.")

The course of a play of Γ is this:

Each player k chooses a number $\tau_k = 1, \cdots, \beta_k$. Each player must make his choice in absolute ignorance of the choices of the others. After all choices have been made, they are submitted to an umpire who determines that the outcome of the play for the player k is $\mathcal{H}_k(\tau_1, \cdots, \tau_n)$.

11.3. The Role of Strategies in the Simplified Form of a Game

11.3. Observe that in this scheme no space is left for any kind of further "strategy." Each player has one move, and one move only; and he must make it in absolute ignorance of everything else.[1] This complete crystallization of the problem in this rigid and final form was achieved by our manipulations of the sections from 11.1.1. on, in which the transition from the original moves to strategies was effected. Since we now treat these strategies themselves as moves, there is no need for strategies of a higher order.

11.4. The Meaning of the Zero-sum Restriction

11.4. We conclude these considerations by determining the place of the zero-sum games (cf. 5.2.1.) within our final scheme.

That Γ is a zero-sum game means, in the notation of 10.1.1., this:

$$\text{(11:3)} \qquad \sum_{k=1}^{n} \mathcal{F}_k(\pi) = 0 \qquad \text{for all } \pi \text{ of } \Omega.$$

If we pass from $\mathcal{F}_k(\pi)$ to $\mathcal{G}_k(\tau_0, \tau_1, \cdots, \tau_n)$, in the sense of 11.2.2., then this becomes

$$\text{(11:4)} \qquad \sum_{k=1}^{n} \mathcal{G}_k(\tau_0, \tau_1, \cdots, \tau_n) = 0 \qquad \text{for all } \tau_0, \tau_1, \cdots, \tau_n.$$

And if we finally introduce $\mathcal{H}_k(\tau_1, \cdots, \tau_n)$, in the sense of 11.2.3., we obtain

$$\text{(11:5)} \qquad \sum_{k=1}^{n} \mathcal{H}_k(\tau_1, \cdots, \tau_n) = 0 \qquad \text{for all } \tau_1, \cdots, \tau_n.$$

Conversely, it is clear that the condition (11:5) makes the game Γ, which we defined in 11.2.3., one of zero sum.

[1] Reverting to the definition of a strategy as given in 11.1.1.: In this game a player k has one and only one personal move, and this independently of the course of the play, —the move $\mathfrak{M}_k$. And he must make his choice at $\mathfrak{M}_k$ with nil information. So his strategy is simply a definite choice for the move $\mathfrak{M}_k$,—no more and no less; i.e. precisely $\tau_k = 1, \cdots, \beta_k$.

We leave it to the reader to describe this game in terms of partitions, and to compare the above with the formalistic definition of a strategy in (11:A) in 11.1.3.

CHAPTER III

ZERO-SUM TWO-PERSON GAMES: THEORY

12. Preliminary Survey

12.1. General Viewpoints

12.1.1. In the preceding chapter we obtained an all-inclusive formal characterization of the general game of n persons (cf. 10.1.). We followed up by developing an exact concept of strategy which permitted us to replace the rather complicated general scheme of a game by a much more simple special one, which was nevertheless shown to be fully equivalent to the former (cf. 11.2.). In the discussion which follows it will sometimes be more convenient to use one form, sometimes the other. It is therefore desirable to give them specific technical names. We will accordingly call them the *extensive* and the *normalized* form of the game, respectively.

Since these two forms are strictly equivalent, it is entirely within our province to use in each particular case whichever is technically more convenient at that moment. We propose, indeed, to make full use of this possibility, and must therefore re-emphasize that this does not in the least affect the absolutely general validity of all our considerations.

Actually the normalized form is better suited for the derivation of general theorems, while the extensive form is preferable for the analysis of special cases; i.e., the former can be used advantageously to establish properties which are common to all games, while the latter brings out characteristic differences of games and the decisive structural features which determine these differences. (Cf. for the former 14., 17., and for the latter e.g. 15.)

12.1.2. Since the formal description of all games has been completed, we must now turn to build up a positive theory. It is to be expected that a systematic procedure to this end will have to proceed from simpler games to more complicated games. It is therefore desirable to establish an ordering for all games according to their increasing degree of complication.

We have already classified games according to the number of participants—a game with n participants being called an n-person game—and also according to whether they are or are not of zero-sum. Thus we must distinguish *zero-sum n-person games* and *general n-person games.* It will be seen later that the general n-person game is very closely related to the zero-sum $(n + 1)$-person game,—in fact the theory of the former will obtain as a special case of the theory of the latter. (Cf. 56.2.2.)

12.2. The One-person Game

12.2.1. We begin with some remarks concerning the one-person game. In the normalized form this game consists of the choice of a number

$\tau = 1, \cdots, \beta$, after which the (only) player 1 gets the amount $\mathcal{H}(\tau)$.[1] The zero-sum case is obviously void[2] and there is nothing to say concerning it. The general case corresponds to a general function $\mathcal{H}(\tau)$ and the "best" or "rational" way of acting—i.e. of playing—consists obviously of this: The player 1 will choose $\tau = 1, \cdots, \beta$ so as to make $\mathcal{H}(\tau)$ a maximum.

This extreme simplification of the one-person game is, of course, due to the fact that our variable τ represents not a choice (in a move) but the player's strategy; i.e., it expresses his entire "theory" concerning the handling of all conceivable situations which may occur in the course of the play. It should be remembered that even a one-person game can be of a very complicated pattern: It may contain chance moves as well as personal moves (of the only player), each one possibly with numerous alternatives, and the amount of information available to the player at any particular personal move may vary in any prescribed way.

12.2.2. Numerous good examples of many complications and subtleties which may arise in this way are given by the various games of "Patience" or "Solitaire." There is, however, an important possibility for which, to the best of our knowledge, examples are lacking among the customary one-person games. This is the case of incomplete information, i.e. of non-equivalence of anteriority and preliminarity of personal moves of the unique player (cf. 6.4.). For such an absence of equivalence it would be necessary that the player have two personal moves $\mathfrak{M}_\kappa$ and $\mathfrak{M}_\lambda$ at neither of which he is informed about the outcome of the choice of the other. Such a state of lack of information is not easy to achieve, but we discussed in 6.4.2. how it can be brought about by "splitting" the player into two or more persons of identical interest and with imperfect communications. We saw loc. cit. that Bridge is an example of this in a two-person game; it would be easy to construct an analogous one-person game—but unluckily the known forms of "solitaire" are not such.[3]

This possibility is nevertheless a practical one for certain economic setups: A rigidly established communistic society, in which the structure of the distribution scheme is beyond dispute (i.e. where there is no exchange, but only one unalterable imputation) would be such—since the interests of all the members of such a society are strictly identical[4] this setup must be treated as a one-person game. But owing to the conceivable imperfections of communications among the members, all sorts of incomplete information can occur.

This is then the case which, by a consistent use of the concept of strategy (i.e. of planning), is naturally reduced to a simple maximum problem. On the basis of our previous discussions it will therefore be apparent now

[1] Cf. (11:D:a), (11:D:b) at the end of 11.2.3. We suppress the index 1.

[2] Then $\mathcal{H}(\tau) = 0$, cf. 11.4.

[3] The existing "double solitaires" are competitive games between the two participants, i.e. two-person games.

[4] The individual members themselves cannot be considered as players, since all possibilities of conflict among them, as well as coalitions of some of them against the others, are excluded.

that this—and this only—is the case in which the simple maximum formulation—i.e. the "Robinson Crusoe" form—of economics is appropriate.

12.2.3. These considerations show also the limitations of the pure maximum—i.e. the "Robinson Crusoe"—approach. The above example of a society of a rigidly established and unquestioned distribution scheme shows that on this plane a rational and critical appraisal of the distribution scheme itself is impossible. In order to get a maximum problem it was necessary to place the entire scheme of distribution among the rules of the game, which are absolute, inviolable and above criticism. In order to bring them into the sphere of combat and competition—i.e. the strategy of the game—it is necessary to consider n-person games with $n \geqq 2$ and thereby to sacrifice the simple maximum aspect of the problem.

12.3. Chance and Probability

12.3. Before going further, we wish to mention that the extensive literature of "mathematical games"—which was developed mainly in the 18th and 19th centuries—deals essentially only with an aspect of the matter which we have already left behind. This is the appraisal of the influence of chance. This was, of course, effected by the discovery and appropriate application of the calculus of probability and particularly of the concept of mathematical expectations. In our discussions, the operations necessary for this purpose were performed in 11.2.3.[1,2]

Consequently we are no longer interested in these games, where the mathematical problem consists only in evaluating the role of chance—i.e. in computing probabilities and mathematical expectations. Such games lead occasionally to interesting exercises in probability theory;[3] but we hope that the reader will agree with us that they do not belong in the theory of games proper.

12.4. The Next Objective

12.4. We now proceed to the analysis of more complicated games. The general one-person game having been disposed of, the simplest one of the remaining games is the zero-sum two-person game. Accordingly we are going to discuss it first.

Afterwards there is a choice of dealing either with the general two-person game or with the zero sum three-person game. It will be seen that our technique of discussion necessitates taking up the zero-sum three-person

[1] We do not in the least intend, of course, to detract from the enormous importance of those discoveries. It is just because of their great power that we are now in a position to treat this side of the matter as briefly as we do. We are interested in those aspects of the problem which are not settled by the concept of probability alone; consequently these and not the satisfactorily settled ones must occupy our attention.

[2] Concerning the important connection between the use of mathematical expectation and the concept of numerical utility, cf. 3.7. and the considerations which precede it.

[3] Some games like Roulette are of an even more peculiar character. In Roulette the mathematical expectation of the players is clearly negative. Thus the motives for participating in that game cannot be understood if one identifies the monetary return with utility.

game first. After that we shall extend the theory to the zero-sum n-person game (for all $n = 1, 2, 3, \cdots$) and only subsequently to this will it be found convenient to investigate the general n-person game.

13. Functional Calculus

13.1. Basic Definitions

13.1.1. Our next objective is—as stated in 12.4.—the exhaustive discussion of the zero-sum two-person games. In order to do this adequately, it will be necessary to use the symbolism of the functional calculus—or at least of certain parts of it—more extensively than we have done thus far. The concepts which we need are those of *functions*, of *variables*, of *maxima* and *minima*, and of the use of the two latter as functional operations. All this necessitates a certain amount of explanation and illustration, which will be given here.

After that is done, we will prove some theorems concerning maxima, minima, and a certain combination of these two, the *saddle value*. These theorems will play an important part in the theory of the zero-sum two-person games.

13.1.2. A *function* ϕ is a dependence which states how certain entities $x, y, \cdots$—called the *variables of* ϕ—determine an entity u—called the *value of* ϕ. Thus u is determined by ϕ and by the $x, y, \cdots$, and this determination—i.e. dependence—will be indicated by the symbolic equation

$$u = \phi(x, y, \cdots).$$

In principle it is necessary to distinguish between the function ϕ itself—which is an abstract entity, embodying only the general dependence of $u = \phi(x, y, \cdots)$ on the $x, y, \cdots$—and its value $\phi(x, y, \cdots)$ for any specific $x, y, \cdots$. In practical mathematical usage, however, it is often convenient to write $\phi(x, y, \cdots)$—but with $x, y, \cdots$ indeterminate—instead of ϕ (cf. the examples (c)–(e) below; (a), (b) are even worse, cf. footnote 1 below).

In order to describe the function ϕ it is of course necessary—among other things—to specify the number of its variables $x, y, \cdots$. Thus there exist *one-variable functions* $\phi(x)$, *two-variable functions* $\phi(x, y)$, etc.

Some examples:

(a) The arithmetical operations $x + 1$ and x^2 are one-variable functions.[1]

(b) The arithmetical operations of addition and of multiplication $x + y$ and xy, are two-variable functions.[1]

(c) For any fixed k the $\mathfrak{F}_k(\pi)$ of 9.2.4. is a one-variable function (of π). But it can also be viewed as a two-variable function (of k, π).

(d) For any fixed k the $\Sigma_k(\kappa, D_\kappa)$ of (11:A) in 11.1.3. is a two-variable function (of κ, D_κ).[2]

(e) For any fixed k the $\mathcal{H}_k(\tau_1, \cdots, \tau_n)$ of 11.2.3. is a n-variable function (of $\tau_1, \cdots, \tau_n$).[1]

[1] Although they do not appear in the above canonical forms $\phi(x)$, $\phi(x, y)$.

[2] We could also treat k in (d) and k in (e) like k in (c), i.e. as a variable.

13.1.3. It is equally necessary, in order to describe a function ϕ to specify for which specific choices of its variables $x, y, \cdots$ the value $\phi(x, y, \cdots)$ is defined at all. These choices—i.e. these combinations—of $x, y, \cdots$ form the *domain of* ϕ.

The examples (a)–(e) show some of the many possibilities for the domains of functions: They may consist of arithmetical or of analytical entities, as well as of others. Indeed:

(a) We may consider the domain to consist of all integer numbers,—or equally well of all real numbers.

(b) All pairs of either category of numbers used in (a), form the domain.

(c) The domain is the set Ω of all objects π which represent the plays of the game Γ (cf. 9.1.1. and 9.2.4.).

(d) The domain consists of pairs of a positive integer κ and a set D_κ.

(e) The domain consists of certain systems of positive integers.

A function ϕ is an *arithmetical function* if its variables are positive integers; it is a *numerical function* if its variables are real numbers; it is a *set-function* if its variables are sets (as, e.g., D_κ in (d)).

For the moment we are mainly interested in arithmetical and numerical functions.

We conclude this section by an observation which is a natural consequence of our view of the concept of a function. This is, that the number of variables, the domain, and the dependence of the value on the variables, constitute the function as such: i.e., if two functions ϕ, ψ have the same variables $x, y, \cdots$ and the same domain, and if $\phi(x, y, \cdots) = \psi(x, y, \cdots)$ throughout this domain, then ϕ, ψ are identical in all respects.[1]

13.2. The Operations Max and Min

13.2.1. Consider a function ϕ which has real numbers for values

$$\phi(x, y, \cdots).$$

Assume first that ϕ is a one-variable function. If its variable can be chosen, say as $x = x_0$ so that $\phi(x_0) \geqq \phi(x')$ for all other choices x', then we say that ϕ has the *maximum* $\phi(x_0)$ and *assumes* it at $x = x_0$.

Observe that this maximum $\phi(x_0)$ is uniquely determined; i.e., the maximum may be assumed at $x = x_0$ for several x_0, but they must all furnish the same value $\phi(x_0)$.[2] We denote this value by Max $\phi(x)$, the *maximum value* of $\phi(x)$.

If we replace $\geqq$ by $\leqq$, then the concept of ϕ's *minimum*, $\phi(x_0)$, obtains, and of x_0 where ϕ *assumes* it. Again there may be several such x_0, but they must all furnish the same value $\phi(x_0)$. We denote this value by Min $\phi(x)$, the *minimum value* of ϕ.

[1] The concept of a function is closely allied to that of a set, and the above should be viewed in parallel with the exposition of 8.2.

[2] Proof: Consider two such x_0, say x_0' and x_0''. Then $\phi(x_0') \geqq \phi(x_0'')$ and $\phi(x_0'') \geqq \phi(x_0')$. Hence $\phi(x_0') = \phi(x_0'')$.

Observe that there is no *a priori* guarantee that either Max $\phi(x)$ or Min $\phi(x)$ exist.[1]

If, however, the domain of ϕ—over which the variable x may run—consists only of a finite number of elements, then the existence of both Max $\phi(x)$ and Min $\phi(x)$ is obvious. This will actually be the case for most functions which we shall discuss.[2] For the remaining ones it will be a consequence of their continuity together with the geometrical limitations of their domains.[3] At any rate we are restricting our considerations to such functions, for which Max and Min exist.

13.2.2. Let now ϕ have any number of variables $x, y, z, \cdots$. By singling out one of the variables, say x, and treating the others, $y, z, \cdots$, as constants, we can view $\phi(x, y, z, \cdots)$ as a one-variable function, of the variable x. Hence we may form Max $\phi(x, y, z, \cdots)$, Min $\phi(x, y, z, \cdots)$ as in 13.2.1., of course with respect to this x.

But since we could have done this equally well for any one of the other variables $y, z, \cdots$ it becomes necessary to indicate that the operations Max, Min were performed with respect to the variable x. We do this by writing $\text{Max}_x\ \phi(x, y, z, \cdots)$, $\text{Min}_x\ \phi(x, y, z, \cdots)$ instead of the incomplete expressions Max ϕ, Min ϕ. Thus we can now apply to the function $\phi(x, y, z, \cdots)$ any one of the operators Max_x, Min_x, Max_y, Min_y, Max_z, Min_z, $\cdots$. They are all distinct and our notation is unambiguous.

This notation is even advantageous for one variable functions, and we will use it accordingly; i.e. we write $\text{Max}_x\ \phi(x)$, $\text{Min}_x\ \phi(x)$ instead of the Max $\phi(x)$, Min $\phi(x)$ of 13.2.1.

Sometimes it will be convenient—or even necessary—to indicate the domain S for a maximum or a minimum explicitly. E.g. when the function $\phi(x)$ is defined also for (some) x outside of S, but it is desired to form the maximum or minimum within S only. In such a case we write

$$\text{Max}_{x \text{ in } S}\ \phi(x), \qquad \text{Min}_{x \text{ in } S}\ \phi(x)$$

instead of $\text{Max}_x\ \phi(x)$, $\text{Min}_x\ \phi(x)$.

In certain other cases it may be simpler to enumerate the values of $\phi(x)$—say $a, b, \cdots$—than to express $\phi(x)$ as a function. We may then write

[1] E.g. if $\phi(x) \equiv x$ with all real numbers as domain, then neither Max $\phi(x)$ nor Min $\phi(x)$ exist.

[2] Typical examples: The functions $\mathcal{H}_k(\tau_1, \cdots, \tau_n)$ of 11.2.3. (or of (e) in 13.1.2.), the function $\mathcal{H}(\tau_1, \tau_2)$ of 14.1.1.

[3] Typical examples: The functions $K(\vec{\xi}, \vec{\eta})$, $\text{Max}_{\vec{\xi}}\ K(\vec{\xi}, \vec{\eta})$, $\text{Min}_{\vec{\eta}}\ K(\vec{\xi}, \vec{\eta})$ in 17.4., the functions $\text{Min}_{\tau_2} \sum_{\tau_1=1}^{\beta_1} \mathcal{H}(\tau_1, \tau_2)\xi_{\tau_1}$, $\text{Max}_{\tau_1} \sum_{\tau_2=1}^{\beta_2} \mathcal{H}(\tau_1, \tau_2)\eta_{\tau_2}$ in 17.5.2. The variables of all these functions are $\vec{\xi}$ or $\vec{\eta}$ or both, with respect to which subsequent maxima and minima are formed.

Another instance is discussed in 46.2.1. espec. footnote 1 on p. 384, where the mathematical background of this subject and its literature are considered. It seems unnecessary to enter upon these here, since the above examples are entirely elementary.

Max $(a, b, \cdots)$, [Min $(a, b, \cdots)$] instead of $\text{Max}_x\ \phi(x)$, [$\text{Min}_x\ \phi(x)$].[1]

13.2.3. Observe that while $\phi(x, y, z, \cdots)$ is a function of the variables $x, y, z, \cdots$, $\text{Max}_x\ \phi(x, y, z, \cdots)$, $\text{Min}_x\ \phi(x, y, z, \cdots)$ are still functions, but of the variables $y, z, \cdots$ only. Purely typographically, x is still present in $\text{Max}_x\ \phi(x, y, z, \cdots)$, $\text{Min}_x\ \phi(x, y, z, \cdots)$, but it is no longer a variable of these functions. We say that the operations Max_x, Min_x *kill* the variable x which appears as their index.[2]

Since $\text{Max}_x\ \phi(x, y, z, \cdots)$, $\text{Min}_x\ \phi(x, y, z, \cdots)$ are still functions of the variables $y, z, \cdots$,[3] we can go on and form the expressions

$$\text{Max}_y\ \text{Max}_x\ \phi(x, y, z, \cdots), \qquad \text{Max}_y\ \text{Min}_x\ \phi(x, y, z, \cdots),$$
$$\text{Min}_y\ \text{Max}_x\ \phi(x, y, z, \cdots), \qquad \text{Min}_y\ \text{Min}_x\ \phi(x, y, z, \cdots),$$

We could equally form

$$\text{Max}_x\ \text{Max}_y\ \phi(x, y, z, \cdots), \qquad \text{Max}_x\ \text{Min}_y\ \phi(x, y, z, \cdots)$$

etc.;[4] or use two other variables than x, y (if there are any); or use more variables than two (if there are any).

In fine, after having applied as many operations Max or Min as there are variables of $\phi(x, y, z, \cdots)$—in any order and combination, but precisely one for each variable $x, y, z, \cdots$—we obtain a function of no variables at all, i.e. a *constant*.

13.3. Commutativity Questions

13.3.1. The discussions of 13.2.3. provide the basis for viewing the Max_x, Min_x, Max_y, Min_y, Max_z, Min_z, $\cdots$ entirely as *functional operations*, each one of which carries a function into another function.[5] We have seen that we can apply several of them successively. In this latter case it is *prima facie* relevant, in which order the successive operations are applied.

But is it really relevant? Precisely: Two operations are said to *commute* if, in case of their successive application (to the same object), the order in which this is done does not matter. Now we ask: Do the operations Max_x, Min_x, Max_y, Min_y, Max_z, Min_z, $\cdots$ all commute with each other or not?

We shall answer this question. For this purpose we need use only two variables, say x, y and then it is not necessary that ϕ be a function of further variables besides x, y.[6]

[1] Of course Max $(a, b, \cdots)$ [Min $(a, b, \cdots)$] is simply the greatest [smallest] one among the numbers $a, b, \cdots$.

[2] A well known operation in analysis which kills a variable x is the definite integral: $\phi(x)$ is a function of x, but $\int_0^1 \phi(x)dx$ is a constant.

[3] We treated $y, z, \cdots$ as constant parameters in 13.2.2. But now that x has been killed we release the variables $y, z, \cdots$.

[4] Observe that if two or more operations are applied, the innermost applies first and kills its variable; then the next one follows suit, etc

[5] With one variable less, since these operations kill one variable each.

[6] Further variables of ϕ, if any, may be treated as constants for the purpose of this analysis.

So we consider a two-variable function $\phi(x, y)$. The significant questions of commutativity are then clearly these:

Which of the three equations which follow are generally true:

$$\text{(13:1)} \qquad \text{Max}_x \text{Max}_y \phi(x, y) = \text{Max}_y \text{Max}_x \phi(x, y),$$
$$\text{(13:2)} \qquad \text{Min}_x \text{Min}_y \phi(x, y) = \text{Min}_y \text{Min}_x \phi(x, y),$$
$$\text{(13:3)} \qquad \text{Max}_x \text{Min}_y \phi(x, y) = \text{Min}_y \text{Max}_x \phi(x, y).$$ [1]

We shall see that (13:1), (13:2) are true, while (13:3) is not; i.e., any two Max or any two Min commute, while a Max and a Min do not commute in general. We shall also obtain a criterion which determines in which special cases Max and Min commute.

This question of commutativity of Max and Min will turn out to be decisive for the zero-sum two-person game (cf. 14.4.2. and 17.6.).

13.3.2. Let us first consider (13:1). It ought to be intuitively clear that $\text{Max}_x \text{Max}_y \phi(x, y)$ is the maximum of $\phi(x, y)$ if we treat x, y together as one variable; i.e. that for some suitable x_0, y_0, $\phi(x_0, y_0) = \text{Max}_x \text{Max}_y \phi(x, y)$ and that for all x', y', $\phi(x_0, y_0) \geqq \phi(x', y')$.

If a mathematical proof is nevertheless wanted, we give it here: Choose x_0 so that $\text{Max}_y \phi(x, y)$ assumes its x-maximum at $x = x_0$, and then choose y_0 so that $\phi(x_0, y)$ assumes its y-maximum at $y = y_0$. Then

$$\phi(x_0, y_0) = \text{Max}_y \phi(x_0, y) = \text{Max}_x \text{Max}_y \phi(x, y),$$

and for all x', y'

$$\phi(x_0, y_0) = \text{Max}_y \phi(x_0, y) \geqq \text{Max}_y \phi(x', y) \geqq \phi(x', y').$$

This completes the proof.

Now by interchanging x, y we see that $\text{Max}_y \text{Max}_x \phi(x, y)$ is equally the maximum of $\phi(x, y)$ if we treat x, y as one variable.

Thus both sides of (13:1) have the same characteristic property, and therefore they are equal to each other. This proves (13:1).

Literally the same arguments apply to Min in place of Max: we need only use $\leqq$ consistently in place of $\geqq$. This proves (13:2).

This device of treating two variables x, y as one, is occasionally quite convenient in itself. When we use it (as, e.g., in 18.2.1., with $\tau_1, \tau_2, \mathcal{H}(\tau_1, \tau_2)$ in place of our present x, y, $\phi(x, y)$), we shall write $\text{Max}_{x,y} \phi(x, y)$ and $\text{Min}_{x,y} \phi(x, y)$.

13.3.3. At this point a graphical illustration may be useful. Assume that the domain of ϕ for x, y is a finite set. Denote, for the sake of simplicity, the possible values of x (in this domain) by $1, \cdots, t$ and those of y by $1, \cdots, s$. Then the values of $\phi(x, y)$ corresponding to all x, y in this domain—i.e. to all combinations of $x = 1, \cdots, t$, $y = 1, \cdots, s$—can be arranged in a rectangular scheme: We use a rectangle of t rows and s

[1] The combination $\text{Min}_x \text{Max}_y$ requires no treatment of its own, since it obtains from the above—$\text{Max}_x \text{Min}_y$—by interchanging x, y.

columns, using the number $x = 1, \cdots, t$ to enumerate the former and the number $y = 1, \cdots, s$ to enumerate the latter. Into the field of intersection of row x and column y—to be known briefly as the field x, y—we write the value $\phi(x, y)$ (Fig. 11). This arrangement, known in mathematics as a *rectangular matrix*, amounts to a complete characterization of the function $\phi(x, y)$. The specific values $\phi(x, y)$ are the *matrix elements*.

	1	2	....	y	....	s
1	$\phi(1, 1)$	$\phi(1, 2)$	....	$\phi(1, y)$	....	$\phi(1, s)$
2	$\phi(2, 1)$	$\phi(2, 2)$	....	$\phi(2, y)$	....	$\phi(2, s)$
⋮	⋮	⋮		⋮		⋮
x	$\phi(x, 1)$	$\phi(x, 2)$	....	$\phi(x, y)$	....	$\phi(x, s)$
⋮	⋮	⋮		⋮		⋮
t	$\phi(t, 1)$	$\phi(t, 2)$	....	$\phi(t, y)$	....	$\phi(t, s)$

Figure 11.

Now $\text{Max}_y \, \phi(x, y)$ is the maximum of $\phi(x, y)$ in the row x.

$$\text{Max}_x \, \text{Max}_y \, \phi(x, y)$$

is therefore the maximum of the row maxima. On the other hand,

$$\text{Max}_x \, \phi(x, y)$$

is the maximum of $\phi(x, y)$ in the column y. $\text{Max}_y \, \text{Max}_x \, \phi(x, y)$ is therefore the maximum of the column maxima. Our assertions in 13.3.2. concerning (13:1) can now be stated thus: The maximum of the row maxima is the same as the maximum of the column maxima; both are the absolute maximum of $\phi(x, y)$ in the matrix. In this form, at least, these assertions should be intuitively obvious. The assertions concerning (13:2) obtain in the same way if Min is put in place of Max.

13.4. The Mixed Case. Saddle Points

13.4.1. Let us now consider (13:3). Using the terminology of 13.3.3. the left-hand side of (13:3) is the maximum of the row minima and the right-hand side is the minimum of the column maxima. These two numbers are neither absolute maxima, nor absolute minima, and there is no *prima facie* evidence why they should be generally equal. Indeed, they are not. Two functions for which they are different are given in Figs. 12, 13. A function for which they are equal is given in Fig. 14. (All these figures should be read in the sense of the explanations of 13.3.3. and of Fig. 11.)

These figures—as well as the entire question of commutativity of Max and Min—will play an essential role in the theory of zero-sum two-person games. Indeed, it will be seen that they represent certain games which are typical for some important possibilities in that theory (cf. 18.1.2.). But for the moment we want to discuss them on their own account, without any reference to those applications.

$t = s = 2$

	1	2	row minima
1	1	−1	−1
2	−1	1	−1
column maxima	1	1	

Maximum of row minima = −1
Minimum of column maxima = 1
Figure 12.

$t = s = 3$

	1	2	3	row minima
1	0	−1	1	−1
2	1	0	−1	−1
3	−1	1	0	−1
column maxima	1	1	1	

Maximum of row minima = −1
Minimum of column maxima = 1
Figure 13.

$t = s = 2$

	1	2	row minima
1	−2	1	−2
2	−1	2	−1
column maxima	−1	2	

Maximum of row minima = −1
Minimum of column maxima = −1
Figure 14.

13.4.2. Since (13:3) is neither generally true, nor generally false, it is desirable to discuss the relationship of its two sides

(13:4) $\quad \text{Max}_x \text{Min}_y \phi(x, y), \qquad \text{Min}_y \text{Max}_x \phi(x, y),$

more fully. Figs. 12–14, which illustrated the behavior of (13:3) to a certain degree, give some clues as to what this behavior is likely to be. Specifically:

(13:A) In all three figures the left-hand side of (13:3) (i.e. the first expression of (13:4)) is $\leqq$ the right-hand side of (13:3) (i.e. the second expression in (13:4)).

(13:B) In Fig. 14—where (13:3) is true—there exists a field in the matrix which contains simultaneously the minimum of its row and the maximum of its column. (This happens to be the element -1—the left lower corner field of the matrix.) In the other figures 12, 13, where (13:3) is not true, there exists no such field.

It is appropriate to introduce a general concept which describes the behavior of the field mentioned in (13:B). We define accordingly:

Let $\phi(x, y)$ be any two-variable function. Then x_0, y_0 is a *saddle point* of ϕ if at the same time $\phi(x, y_0)$ assumes its maximum at $x = x_0$ and $\phi(x_0, y)$ assumes its minimum at $y = y_0$.

The reason for the use of the name *saddle point* is this: Imagine the matrix of all x, y elements ($x = 1, \cdots, t$, $y = 1, \cdots s$; cf. Fig. 11) as an oreographical map, the height of the mountain over the field x, y being the value of $\phi(x, y)$ there. Then the definition of a saddle point x_0, y_0 describes indeed essentially a *saddle* or *pass* at that point (i.e. over the field x_0, y_0); the row x_0 is the ridge of the mountain, and the column y_0 is the road (from valley to valley) which crosses this ridge.

The formula (13:C*) in 13.5.2. also falls in with this interpretation.[1]

13.4.3. Figs. 12, 13 show that a ϕ may have no saddle points at all. On the other hand it is conceivable that ϕ possesses several saddle points. But all saddle points x_0, y_0—if they exist at all—must furnish the same value $\phi(x_0, y_0)$.[2] We denote this value—if it exists at all—by $\mathrm{Sa}_{x/y}\phi(x, y)$, the *saddle value* of $\phi(x, y)$.[3]

We now formulate the theorems which generalize the indications of (13:A), (13:B). We denote them by (13:A*), (13:B*), and emphasize that they are valid for all functions $\phi(x, y)$.

(13:A*) Always

$$\mathrm{Max}_x \mathrm{Min}_y \phi(x, y) \leqq \mathrm{Min}_y \mathrm{Max}_x \phi(x, y).$$

(13:B*) We have

$$\mathrm{Max}_x \mathrm{Min}_y \phi(x, y) = \mathrm{Min}_y \mathrm{Max}_x \phi(x, y)$$

if and only if a saddle point x_0, y_0 of ϕ exists.

13.5. Proofs of the Main Facts

13.5.1. We define first two sets A^ϕ, B^ϕ for every function $\phi(x, y)$. $\mathrm{Min}_y \phi(x, y)$ is a function of x; let A^ϕ be the set of all those x_0 for which

[1] All this is closely connected with—although not precisely a special case of—certain more general mathematical theories involving extremal problems, calculus of variations, etc. Cf. *M. Morse:* The Critical Points of Functions and the Calculus of Variations in the Large, Bull. Am. Math. Society, Jan.–Feb. 1929, pp. 38 cont., and What is Analysis in the Large?, Am. Math. Monthly, Vol. XLIX, 1942, pp. 358 cont.

[2] This follows from (13:C*) in 13.5.2. There exists an equally simple direct proof: Consider two saddle points x_0, y_0, say x_0', y_0' and x_0'', y_0''. Then:

$$\phi(x_0', y_0') = \mathrm{Max}_x \phi(x, y_0') \geqq \phi(x_0'', y_0') \geqq \mathrm{Min}_y \phi(x_0'', y) = \phi(x_0'', y_0''),$$

i.e.: $\phi(x_0', y_0') \geqq \phi(x_0'', y_0'')$. Similarly $\phi(x_0'', y_0'') \geqq \phi(x_0', y_0')$.
Hence $\phi(x_0', y_0') = \phi(x_0'', y_0'')$.

[3] Clearly the operation $\mathrm{Sa}_{x/y}\phi(x, y)$ kills both variables x, y. Cf. 13.2.3.

this function assumes its maximum at $x = x_0$. $\mathrm{Max}_x \phi(x, y)$ is a function of y; let B^ϕ be the set of all those y_0 for which this function assumes its minimum at $y = y_0$.

We now prove (13:A*), (13:B*).

Proof of (13:A*): Choose x_0 in A^ϕ and y_0 in B^ϕ. Then

$$\begin{aligned}\mathrm{Max}_x \mathrm{Min}_y \phi(x, y) &= \mathrm{Min}_y \phi(x_0, y) \leqq \phi(x_0, y_0) \\ &\leqq \mathrm{Max}_x \phi(x, y_0) = \mathrm{Min}_y \mathrm{Max}_x \phi(x, y),\end{aligned}$$

i.e.: $\mathrm{Max}_x \mathrm{Min}_y \phi(x, y) \leqq \mathrm{Min}_y \mathrm{Max}_x \phi(x, y)$ as desired.

Proof of the necessity of the existence of a saddle point in (13:B*): Assume that

$$\mathrm{Max}_x \mathrm{Min}_y \phi(x, y) = \mathrm{Min}_y \mathrm{Max}_x \phi(x, y).$$

Choose x_0 in A^ϕ and y_0 in B^ϕ; then we have

$$\begin{aligned}\mathrm{Max}_x \phi(x, y_0) &= \mathrm{Min}_y \mathrm{Max}_x \phi(x, y) \\ &= \mathrm{Max}_x \mathrm{Min}_y \phi(x, y) = \mathrm{Min}_y \phi(x_0, y).\end{aligned}$$

Hence for every x'

$$\phi(x', y_0) \leqq \mathrm{Max}_x \phi(x, y_0) = \mathrm{Min}_y \phi(x_0, y) \leqq \phi(x_0, y_0),$$

i.e. $\phi(x_0, y_0) \geqq \phi(x', y_0)$—so $\phi(x, y_0)$ assumes its maximum at $x = x_0$. And for every y'

$$\phi(x_0, y') \geqq \mathrm{Min}_y \phi(x_0, y) = \mathrm{Max}_x \phi(x, y_0) \geqq \phi(x_0, y_0),$$

i.e. $\phi(x_0, y_0) \leqq \phi(x_0, y')$—so $\phi(x_0, y)$ assumes its minimum at $y = y_0$.

Consequently these x_0, y_0 form a saddle point.

Proof of the sufficiency of the existence of a saddle point in (13:B*): Let x_0, y_0 be a saddle point. Then

$$\begin{aligned}\mathrm{Max}_x \mathrm{Min}_y \phi(x, y) &\geqq \mathrm{Min}_y \phi(x_0, y) = \phi(x_0, y_0), \\ \mathrm{Min}_y \mathrm{Max}_x \phi(x, y) &\leqq \mathrm{Max}_x \phi(x, y_0) = \phi(x_0, y_0),\end{aligned}$$

hence

$$\mathrm{Max}_x \mathrm{Min}_y \phi(x, y) \geqq \phi(x_0, y_0) \geqq \mathrm{Min}_y \mathrm{Max}_x \phi(x, y).$$

Combining this with (13:A*) gives

$$\mathrm{Max}_x \mathrm{Min}_y \phi(x, y) = \phi(x_0, y_0) = \mathrm{Min}_y \mathrm{Max}_x \phi(x, y),$$

and hence the desired equation.

13.5.2. The considerations of 13.5.1. yield some further results which are worth noting. We assume now the existence of saddle points,—i.e. the validity of the equation of (13:B*).

For every saddle point x_0, y_0

(13:C*) $\quad \phi(x_0, y_0) = \mathrm{Max}_x \mathrm{Min}_y \phi(x, y) = \mathrm{Min}_y \mathrm{Max}_x \phi(x, y).$

Proof: This coincides with the last equation of the sufficiency proof of (13:B*) in 13.5.1.

(13:D*) x_0, y_0 is a saddle point if and only if x_0 belongs to A^ϕ and y_0 belongs to B^ϕ.[1]

Proof of sufficiency: Let x_0 belong to A^ϕ and y_0 belong to B^ϕ. Then the necessity proof of (13:B*) in 13.5.1. shows exactly that this x_0, y_0 is a saddle point.

Proof of necessity: Let x_0, y_0 be a saddle point. We use (13:C*). For every x'

$$\mathrm{Min}_y\, \phi(x', y) \leqq \mathrm{Max}_x\, \mathrm{Min}_y\, \phi(x, y) = \phi(x_0, y_0) = \mathrm{Min}_y\, \phi(x_0, y),$$

i.e. $\mathrm{Min}_y\, \phi(x_0, y) \geqq \mathrm{Min}_y\, \phi(x', y)$—so $\mathrm{Min}_y\, \phi(x, y)$ assumes its maximum at $x = x_0$. Hence x_0 belongs to A^ϕ. Similarly for every y'

$$\mathrm{Max}_x\, \phi(x, y') \geqq \mathrm{Min}_y\, \mathrm{Max}_x\, \phi(x, y) = \phi(x_0, y_0) = \mathrm{Max}_x\, \phi(x, y_0),$$

i.e. $\mathrm{Max}_x\, \phi(x, y_0) \leqq \mathrm{Max}_x\, \phi(x, y')$—so $\mathrm{Max}_x\, \phi(x, y)$ assumes its minimum at $y = y_0$. Hence y belongs to B^ϕ. This completes the proof.

The theorems (13:C*), (13:D*) indicate, by the way, the limitations of the analogy described at the end of 13.4.2.; i.e. they show that our concept of a saddle point is narrower than the everyday (oreographical) idea of a saddle or a pass. Indeed, (13:C*) states that all saddles—provided that they exist—are at the same altitude. And (13:D*) states—if we depict the sets A^ϕ, B^ϕ as two intervals of numbers[2]—that all saddles together are an area which has the form of a rectangular plateau.[3]

13.5.3. We conclude this section by proving the existence of a saddle point for a special kind of x, y and $\phi(x, y)$. This special case will be seen to be of a not inconsiderable generality. Let a function $\psi(x, u)$ of two variables x, u be given. We consider all functions $f(x)$ of the variable which have values in the domain of u. Now we keep the variable x but in place of the variable u we use the function f itself.[4] The expression $\psi(x, f(x))$ is determined by x, f; hence we may treat $\psi(x, f(x))$ as a function of the variables x, f and let it take the place of $\phi(x, y)$.

We wish to prove that for these x, f and $\psi(x, f(x))$—in place of x, y and $\phi(x, y)$—a saddle point exists; i.e. that

(13:E) $$\mathrm{Max}_x\, \mathrm{Min}_f\, \psi(x, f(x)) = \mathrm{Min}_f\, \mathrm{Max}_x\, \psi(x, f(x)).$$

Proof: For every x choose a u_0 with $\psi(x, u_0) = \mathrm{Min}_u\, \psi(x, u)$. This u_0 depends on x, hence we can define a function f_0 by $u_0 = f_0(x)$. Thus $\psi(x, f_0(x)) = \mathrm{Min}_u\, \psi(x, u)$. Consequently

$$\mathrm{Max}_x\, \psi(x, f_0(x)) = \mathrm{Max}_x\, \mathrm{Min}_u\, \psi(x, u).$$

[1] Only under our hypothesis at the beginning of this section! Otherwise there exist no saddle points at all.

[2] If the x, y are positive integers, then this can certainly be brought about by two appropriate permutations of their domain.

[3] The general mathematical concepts alluded to in footnote 1 on p. 95 are free from these limitations. They correspond precisely to the everyday idea of a pass.

[4] The reader is asked to visualize this: Although itself a function, f may perfectly well be the variable of another function.

A fortiori,

(13:F) $$\mathrm{Min}_f \, \mathrm{Max}_x \, \psi(x, f(x)) \leqq \mathrm{Max}_x \, \mathrm{Min}_u \, \psi(x, u).$$

Now $\mathrm{Min}_f \, \psi(x, f(x))$ is the same thing as $\mathrm{Min}_u \, \psi(x, u)$ since f enters into this expression only via its value at the one place x, i.e. $f(x)$, for which we may write u. So $\mathrm{Min}_f \, \psi(x, f(x)) = \mathrm{Min}_u \, \psi(x, u)$ and consequently,

(13:G) $$\mathrm{Max}_x \, \mathrm{Min}_f \, \psi(x, f(x)) = \mathrm{Max}_x \, \mathrm{Min}_u \, \psi(x, u).$$

(13:F), (13:G) together establish the validity of a $\geqq$ in (13:E). The $\leqq$ in (13:E) holds owing to (13:A*). Hence we have $=$ in (13:E), i.e. the proof is completed.

14. Strictly Determined Games

14.1. Formulation of the Problem

14.1.1. We now proceed to the consideration of the zero-sum two-person game. Again we begin by using the normalized form.

According to this the game consists of two moves: Player 1 chooses a number $\tau_1 = 1, \cdots, \beta_1$, player 2 chooses a number $\tau_2 = 1, \cdots, \beta_2$, each choice being made in complete ignorance of the other, and then players 1 and 2 get the amounts $\mathcal{H}_1(\tau_1, \tau_2)$ and $\mathcal{H}_2(\tau_1, \tau_2)$, respectively.[1]

Since the game is zero-sum, we have, by 11.4.

$$\mathcal{H}_1(\tau_1, \tau_2) + \mathcal{H}_2(\tau_1, \tau_2) \equiv 0.$$

We prefer to express this by writing

$$\mathcal{H}_1(\tau_1, \tau_2) \equiv \mathcal{H}(\tau_1, \tau_2), \qquad \mathcal{H}_2(\tau_1, \tau_2) \equiv -\mathcal{H}(\tau_1, \tau_2).$$

We shall now attempt to understand how the obvious desires of the players 1, 2 will determine the events, i.e. the choices τ_1, τ_2.

It must again be remembered, of course, that τ_1, τ_2 stand *ultima analysi* not for a choice (in a move) but for the players' strategies; i.e. their entire "theory" or "plan" concerning the game.

For the moment we leave it at that. Subsequently we shall also go "behind" the τ_1, τ_2 and analyze the course of a play.

14.1.2. The desires of the players 1, 2, are simple enough. 1 wishes to make $\mathcal{H}_1(\tau_1, \tau_2) \equiv \mathcal{H}(\tau_1, \tau_2)$ a maximum, 2 wishes to make $\mathcal{H}_2(\tau_1, \tau_2) \equiv -\mathcal{H}(\tau_1, \tau_2)$ a maximum; i.e. 1 wants to maximize and 2 wants to minimize $\mathcal{H}(\tau_1, \tau_2)$.

So the interests of the two players concentrate on the same object: the one function $\mathcal{H}(\tau_1, \tau_2)$. But their intentions are—as is to be expected in a zero-sum two-person game—exactly opposite: 1 wants to maximize, 2 wants to minimize. Now the peculiar difficulty in all this is that neither player has full control of the object of his endeavor—of $\mathcal{H}(\tau_1, \tau_2)$—i.e. of both its variables τ_1, τ_2. 1 wants to maximize, but he controls only τ_1; 2 wants to minimize, but he controls only τ_2: What is going to happen?

[1] Cf. (11:D) in 11.2.3.

The difficulty is that no particular choice of, say τ_1, need in itself make $\mathcal{H}(\tau_1, \tau_2)$ either great or small. The influence of τ_1 on $\mathcal{H}(\tau_1, \tau_2)$ is, in general, no definite thing; it becomes that only in conjunction with the choice of the other variable, in this case τ_2. (Cf. the corresponding difficulty in economics as discussed in 2.2.3.)

Observe that from the point of view of the player 1 who chooses a variable, say τ_1, the other variable can certainly not be considered as a chance event. The other variable, in this case τ_2, is dependent upon the will of the other player, which must be regarded in the same light of "rationality" as one's own. (Cf. also the end of 2.2.3. and 2.2.4.)

14.1.3. At this point it is convenient to make use of the graphical representation developed in 13.3.3. We represent $\mathcal{H}(\tau_1, \tau_2)$ by a rectangular matrix: We form a rectangle of β_1 rows and β_2 columns, using the number $\tau_1 = 1, \cdots, \beta_1$ to enumerate the former, and the number $\tau_2 = 1, \cdots, \beta_2$ to enumerate the latter; and into the field τ_1, τ_2 we write the matrix element $\mathcal{H}(\tau_1, \tau_2)$. (Cf. with Figure 11 in 13.3.3. The ϕ, x, y, t, s there correspond to our $\mathcal{H}, \tau_1, \tau_2, \beta_1, \beta_2$ (Figure 15).)

	1	2	....	τ_2	....	β_2
1	$\mathcal{H}(1, 1)$	$\mathcal{H}(1, 2)$	....	$\mathcal{H}(1, \tau_2)$	....	$\mathcal{H}(1, \beta_2)$
2	$\mathcal{H}(2, 1)$	$\mathcal{H}(2, 2)$	....	$\mathcal{H}(2, \tau_2)$	....	$\mathcal{H}(2, \beta_2)$
⋮	⋮	⋮		⋮		⋮
τ_1	$\mathcal{H}(\tau_1, 1)$	$\mathcal{H}(\tau_1, 2)$	....	$\mathcal{H}(\tau_1, \tau_2)$	....	$\mathcal{H}(\tau_1, \beta_2)$
⋮	⋮	⋮		⋮		⋮
β_1	$\mathcal{H}(\beta_1, 1)$	$\mathcal{H}(\beta_1, 2)$	....	$\mathcal{H}(\beta_1, \tau_2)$	....	$\mathcal{H}(\beta_1, \beta_2)$

Figure 15.

It ought to be understood that the function $\mathcal{H}(\tau_1, \tau_2)$ is subject to no restrictions whatsoever; i.e., we are free to choose it absolutely at will.[1] Indeed, any given function $\mathcal{H}(\tau_1, \tau_2)$ defines a zero-sum two-person game in the sense of (11:D) of 11.2.3. by simply defining

$$\mathcal{H}_1(\tau_1, \tau_2) \equiv \mathcal{H}(\tau_1, \tau_2), \qquad \mathcal{H}_2(\tau_1, \tau_2) \equiv -\mathcal{H}(\tau_1, \tau_2)$$

(cf. 14.1.1.). The desires of the players 1, 2, as described above in the last section, can now be visualized as follows: Both players are solely

[1] The domain, of course, is prescribed: It consists of all pairs τ_1, τ_2 with $\tau_1 = 1, \cdots, \beta_1$; $\tau_2 = 1, \cdots, \beta_2$. This is a finite set, so all Max and Min exist, cf. the end of 13.2.1.

interested in the value of the matrix element $\mathcal{H}(\tau_1, \tau_2)$. Player 1 tries to maximize it, but he controls only the row,—i.e. the number τ_1. Player 2 tries to minimize it, but he controls only the column,—i.e. the number τ_2.

We must now attempt to find a satisfactory interpretation for the outcome of this peculiar tug-of-war.[1]

14.2. The Minorant and the Majorant Games

14.2. Instead of attempting a direct attack on the game Γ itself—for which we are not yet prepared—let us consider two other games, which are closely connected with Γ and the discussion of which is immediately feasible.

The difficulty in analyzing Γ is clearly that the player 1, in choosing τ_1 does not know what choice τ_2 of the player 2 he is going to face and *vice versa*. Let us therefore compare Γ with other games where this difficulty does not arise.

We define first a game Γ_1, which agrees with Γ in every detail except that player 1 has to make his choice of τ_1 before player 2 makes his choice of τ_2, and that player 2 makes his choice in full knowledge of the value given by player 1 to τ_1 (i.e. 1's move is preliminary to 2's move).[2] In this game Γ_1 player 1 is obviously at a disadvantage as compared to his position in the original game Γ. We shall therefore call Γ_1 the *minorant* game of Γ.

We define similarly a second game Γ_2 which again agrees with Γ in every detail except that now player 2 has to make his choice of τ_2 before player 1 makes his choice of τ_1 and that 1 makes his choice in full knowledge of the value given by 2 to τ_2 (i.e. 2's move is preliminary to 1's move).[3] In this game Γ_2 the player 1 is obviously at an advantage as compared to his position in the game Γ. We shall therefore call Γ_2 the *majorant* game of Γ.

The introduction of these two games Γ_1, Γ_2 achieves this: It ought to be evident by common sense—and we shall also establish it by an exact discussion—that for Γ_1, Γ_2 the "best way of playing"—i.e. the concept of rational behavior—has a clear meaning. On the other hand, the game Γ lies clearly "between" the two games Γ_1, Γ_2; e.g. from 1's point of view Γ_1 is always less and Γ_2 is always more advantageous than Γ.[4] Thus Γ_1, Γ_2 may be expected to provide lower and upper bounds for the significant quantities concerning Γ. We shall, of course, discuss all this in an entirely precise form. *A priori*, these "bounds" could differ widely and leave a considerable uncertainty as to the understanding of Γ. Indeed, *prima facie* this will seem to be the case for many games. But we shall succeed in manipulating this technique in such a way—by the introduction of certain

[1] The point is, of course, that this is not a tug-of-war. The two players have opposite interests, but the means by which they have to promote them are not in opposition to each other. On the contrary, these "means"—i.e. the choices of τ_1, τ_2—are apparently independent. This discrepancy characterizes the entire problem.

[2] Thus Γ_1—while extremely simple—is no longer in the normalized form.

[3] Thus Γ_2—while extremely simple—is no longer in the normalized form.

[4] Of course, to be precise we should say "less than or equal to" instead of "less," and "more than or equal to" instead of "more."

further devices—as to obtain in the end a precise theory of Γ, which gives complete answers to all questions.

14.3. Discussion of the Auxiliary Games

14.3.1. Let us first consider the minorant game Γ_1. After player 1 has made his choice τ_1 the player 2 makes his choice τ_2 in full knowledge of the value of τ_1. Since 2's desire is to minimize $\mathcal{H}(\tau_1, \tau_2)$, it is certain that he will choose τ_2 so as to make the value of $\mathcal{H}(\tau_1, \tau_2)$ a minimum for this τ_1. In other words: When 1 chooses a particular value of τ_1 he can already foresee with certainty what the value of $\mathcal{H}(\tau_1, \tau_2)$ will be. This will be $\mathrm{Min}_{\tau_2} \mathcal{H}(\tau_1, \tau_2)$.[1] This is a function of τ_1 alone. Now 1 wishes to maximize $\mathcal{H}(\tau_1, \tau_2)$ and since his choice of τ_1 is conducive to the value $\mathrm{Min}_{\tau_2} \mathcal{H}(\tau_1, \tau_2)$—which depends on τ_1 only, and not at all on τ_2—so he will choose τ_1 so as to maximize $\mathrm{Min}_{\tau_2} \mathcal{H}(\tau_1, \tau_2)$. Thus the value of this quantity will finally be

$$\mathrm{Max}_{\tau_1} \mathrm{Min}_{\tau_2} \mathcal{H}(\tau_1, \tau_2).^{2}$$

Summing up:

(14:A:a) The good way (strategy) for 1 to play the minorant game Γ_1 is to choose τ_1, belonging to the set A,— A being the set of those τ_1 for which $\mathrm{Min}_{\tau_2} \mathcal{H}(\tau_1, \tau_2)$ assumes its maximum value $\mathrm{Max}_{\tau_1} \mathrm{Min}_{\tau_2} \mathcal{H}(\tau_1, \tau_2)$.

(14:A:b) The good way (strategy) for 2 to play is this: If 1 has chosen a definite value of τ_1,[3] then τ_2 should be chosen belonging to the set B_{τ_1},— B_{τ_1} being the set of those τ_2 for which $\mathcal{H}(\tau_1, \tau_2)$ assumes its minimum value $\mathrm{Min}_{\tau_2} \mathcal{H}(\tau_1, \tau_2)$.[4]

On the basis of this we can state further:

(14:A:c) If both players 1 and 2 play the minorant game Γ_1 well, i.e. if τ_1 belongs to A and τ_2 belongs to B_{τ_1} then the value of $\mathcal{H}(\tau_1, \tau_2)$ will be equal to

$$v_1 = \mathrm{Max}_{\tau_1} \mathrm{Min}_{\tau_2} \mathcal{H}(\tau_1, \tau_2).$$

[1] Observe that τ_2 may not be uniquely determined: For a given τ_1 the τ_2-function $\mathcal{H}(\tau_1, \tau_2)$ may assume its τ_2-minimum for several values of τ_2. The value of $\mathcal{H}(\tau_1, \tau_2)$ will, however, be the same for all these τ_2, namely the uniquely defined minimum value $\mathrm{Min}_{\tau_2} \mathcal{H}(\tau_1, \tau_2)$. (Cf. 13.2.1.)

[2] For the same reason as in footnote 1 above, the value of τ_1 may not be unique, but the value of $\mathrm{Min}_{\tau_2} \mathcal{H}(\tau_1, \tau_2)$ is the same for all τ_1 in question, namely the uniquely-defined maximum value

$$\mathrm{Max}_{\tau_1} \mathrm{Min}_{\tau_2} \mathcal{H}(\tau_1 \tau_2).$$

[3] 2 is informed of the value of τ_1 when called upon to make his choice of τ_2,—this is the rule of Γ_1. It follows from our concept of a strategy (cf. 4.1.2. and end of 11.1.1.) that at this point a rule must be provided for 2's choice of τ_2 for every value of τ_1,—irrespective of whether 1 has played well or not, i.e. whether or not the value chosen belongs to A.

[4] In all, this τ_1 is treated as a known parameter on which everything depends,—including the set B_{τ_1} from which τ_2 ought to be chosen.

The truth of the above assertion is immediately established in the mathematical sense by remembering the definitions of the sets A and B_{τ_1}, and by substituting accordingly in the assertion. We leave this exercise—which is nothing but the classical operation of "substituting the defining for the defined"—to the reader. Moreover, the statement ought to be clear by common sense.

The entire discussion should make it clear that every play of the game Γ_1 has a definite *value* for each player. This value is the above v_1 for the player 1 and therefore $-v_1$ for the player 2.

An even more detailed idea of the significance of v_1 is obtained in this way:

(14:A:d) Player 1 can, by playing appropriately, secure for himself a gain $\geqq v_1$ irrespective of what player 2 does. Player 2 can, by playing appropriately, secure for himself a gain $\geqq -v_1$, irrespective of what player 1 does.

(*Proof:* The former obtains by any choice of τ_1 in A. The latter obtains by any choice of τ_2 in B_{τ_1}.[1] Again we leave the details to the reader; they are altogether trivial.)

The above can be stated equivalently thus:

(14:A:e) Player 2 can, by playing appropriately, make it sure that the gain of player 1 is $\leqq v_1$, i.e. prevent him from gaining $> v_1$ irrespective of what player 1 does. Player 1 can, by playing appropriately, make it sure that the gain of player 2 is $\leqq -v_1$, i.e. prevent him from gaining $> -v_1$ irrespective of what player 2 does.

14.3.2. We have carried out the discussion of Γ_1 in rather profuse detail although the "solution" is a rather obvious one. That is, it is very likely that anybody with a clear vision of the situation will easily reach the same conclusions "unmathematically," just by exercise of common sense. Nevertheless we felt it necessary to discuss this case so minutely because it is a prototype of several others to follow where the situation will be much less open to "unmathematical" vision. Also, all essential elements of complication as well as the bases for overcoming them are really present in this very simplest case. By seeing their respective positions clearly in this case, it will be possible to visualize them in the subsequent, more complicated, ones. And it will be possible, in this way only, to judge precisely how much can be achieved by every particular measure.

14.3.3. Let us now consider the majorant game Γ_2.

Γ_2 differs from Γ_1 only in that the roles of players 1 and 2 are interchanged: Now player 2 must make his choice τ_2 first and then the player 1 makes his choice of τ_1 in full knowledge of the value of τ_2.

[1] Recall that τ_1 must be chosen without knowledge of τ_2, while τ_2 is chosen in full knowledge of τ_1.

But in saying that Γ_2 arises from Γ_1 by interchanging the players 1 and 2, it must be remembered that these players conserve in the process their respective functions $\mathcal{H}_1(\tau_1, \tau_2)$, $\mathcal{H}_2(\tau_1, \tau_2)$, i.e. $\mathcal{H}(\tau_1, \tau_2)$, $-\mathcal{H}(\tau_1, \tau_2)$. That is, 1 still desires to maximize and 2 still desires to minimize $\mathcal{H}(\tau_1, \tau_2)$.

These being understood, we can leave the practically literal repetition of the considerations of 14.3.1. to the reader. We confine ourselves to restating the significant definitions, in the form in which they apply to Γ_2.

(14:B:a) The good way (strategy) for 2 to play the majorant game Γ_2 is to choose τ_2 belonging to the set B,—B being the set of those τ_2 for which $\text{Max}_{\tau_1} \mathcal{H}(\tau_1, \tau_2)$ assumes its minimum value $\text{Min}_{\tau_2} \text{Max}_{\tau_1} \mathcal{H}(\tau_1, \tau_2)$.

(14:B:b) The good way (strategy) for 1 to play is this: If 2 has chosen a definite value of τ_2,[1] then τ_1 should be chosen belonging to the set A_{τ_2},—A_{τ_2} being the set of those τ_1 for which $\mathcal{H}(\tau_1, \tau_2)$ assumes its maximum value $\text{Max}_{\tau_1} \mathcal{H}(\tau_1, \tau_2)$.[2]

On the basis of this we can state further:

(14:B:c) If both players 1 and 2 play the majorant game Γ_2 well, i.e. if τ_2 belongs to B and τ_1 belongs to A_{τ_2} then the value of $\mathcal{H}(\tau_1, \tau_2)$ will be equal to

$$v_2 = \text{Min}_{\tau_2} \text{Max}_{\tau_1} \mathcal{H}(\tau_1, \tau_2).$$

The entire discussion should make it clear that every play of the game Γ_2 has a definite *value* for each player. This value is the above v_2 for the player 1 and therefore $-v_2$ for the player 2.

In order to stress the symmetry of the entire arrangement, we repeat, *mutatis mutandis*, the considerations which concluded 14.3.1. They now serve to give a more detailed idea of the significance of v_2.

(14:B:d) Player 1 can, by playing appropriately, secure for himself a gain $\geqq v_2$, irrespective of what player 2 does. Player 2 can, by playing appropriately, secure for himself a gain $\geqq -v_2$, irrespective of what player 1 does.

(*Proof:* The latter obtains by any choice of τ_2 in B. The former obtains by any choice of τ_1 in A_{τ_2}.[3] Cf. with the proof, loc. cit.)

The above can again be stated equivalently thus:

(14:B:e) Player 2 can, by playing appropriately, make it sure that the gain of player 1 is $\leqq v_2$, i.e. prevent him from gaining

[1] 1 is informed of the value of τ_2 when called upon to make his choice of τ_1—this is the rule of Γ_2 (Cf. footnote 3 on p. 101).

[2] In all this τ_2 is treated as a known parameter on which everything depends, including the set A_{τ_2} from which τ_1 ought to be chosen.

[3] Remember that τ_2 must be chosen without any knowledge of τ_1, while τ_1 is chosen with full knowledge of τ_2.

> $> v_2$, irrespective of what player 1 does. Player 1 can, by playing appropriately, make it sure that the gain of player 2 is $\leqq -v_2$, i.e. prevent him from gaining $> -v_2$, irrespective of what player 2 does.

14.3.4. The discussions of Γ_1 and Γ_2, as given in 14.3.1. and 14.3.3., respectively, are in a relationship of *symmetry* or duality to each other; they obtain from each other, as was pointed out previously (at the beginning of 14.3.3.) by interchanging the roles of the players 1 and 2. In itself neither game Γ_1 nor Γ_2 is symmetric with respect to this interchange; indeed, this is nothing but a restatement of the fact that the interchange of the players 1 and 2 also interchanges the two games Γ_1 and Γ_2, and so modifies both. It is in harmony with this that the various statements which we made in 14.3.1. and 14.3.3. concerning the good strategies of Γ_1 and Γ_2, respectively—i.e. (14:A:a), (14:A:b), (14:B:a), (14:B:b), loc. cit.—were not symmetric with respect to the players 1 and 2 either. Again we see: An interchange of the players 1 and 2 interchanges the pertinent definitions for Γ_1 and Γ_2, and so modifies both.[1]

It is therefore very remarkable that the characterization of the *value* of a play (v_1 for Γ_1, v_2 for Γ_2), as given at the end of 14.3.1. and 14.3.3.—i.e. (14:A:c), (14:A:d), (14:A:e), (14:B:c), (14:B:d), (14:B:e), loc. cit. (except for the formulae at the end of (14:A:c) and of (14:B:c))—are fully symmetric with respect to the players 1 and 2. According to what was said above, this is the same thing as asserting that these characterizations are stated exactly the same way for Γ_1 and Γ_2.[2] All this is, of course, equally clear by immediate inspection of the relevant passages.

Thus we have succeeded in defining the value of a play in the same way for the games Γ_1 and Γ_2, and symmetrically for the players 1 and 2: in (14:A:c), (14:A:d), (14:A:e), (14:B:c), (14:B:d) and (14:B:e) in 14.3.1. and in 14.3.3.,—this in spite of the fundamental difference of the individual role of each player in these two games. From this we derive the hope that the definition of the value of a play may be used in the same form for other games as well—in particular for the game Γ—which, as we know, occupies a middle position between Γ_1 and Γ_2. This hope applies, of course, only to the concept of value itself, but not to the reasonings which lead to it; those were specific to Γ_1 and Γ_2, indeed different for Γ_1 and for Γ_2, and altogether impracticable for Γ itself; i.e., we expect for the future more from (14:A:d), (14:A:e), (14:B:d), (14:B:e) than from (14:A:a), (14:A:b), (14:B:a), (14:B:b).

[1] Observe that the original game Γ was symmetric with respect to the two players 1 and 2, if we let each player take his function $\mathcal{H}_1(\tau_1, \tau_2)$, $\mathcal{H}_2(\tau_1, \tau_2)$ with him in an interchange; i.e. the personal moves of 1 and 2 had both the same character in Γ.

For a narrower *concept* of symmetry, where the functions $\mathcal{H}_1(\tau_1, \tau_2)$, $\mathcal{H}_2(\tau_1, \tau_2)$ are held fixed, cf. 14.6.

[2] This point deserves careful consideration: Naturally these two characterizations must obtain from each other by interchanging the roles of the players 1 and 2. But in this case the statements coincide also directly when no interchange of the players is made at all. This is due to their individual symmetry.

These are clearly only heuristic indications. Thus far we have not even attempted the proof that a numerical *value* of a play can be defined in this manner for Γ. We shall now begin the detailed discussion by which this gap will be filled. It will be seen that at first definite and serious difficulties seem to limit the applicability of this procedure, but that it will be possible to remove them by the introduction of a new device (Cf. 14.7.1. and 17.1.-17.3., respectively).

14.4. Conclusions

14.4.1. We have seen that a perfectly plausible interpretation of the *value* of a play determines this quantity as

$$v_1 = \text{Max}_{\tau_1} \text{Min}_{\tau_2} \mathcal{H}(\tau_1, \tau_2),$$
$$v_2 = \text{Min}_{\tau_2} \text{Max}_{\tau_1} \mathcal{H}(\tau_1, \tau_2),$$

for the games Γ_1, Γ_2, respectively, as far as the player 1 is concerned.[1]

Since the game Γ_1 is less advantageous for 1 than the game Γ_2—in Γ_1 he must make his move prior to, and in full view of, his adversary, while in Γ_2 the situation is reversed—it is a reasonable conclusion that the value of Γ_1 is less than, or equal to (i.e. certainly not greater than) the value of Γ_2. One may argue whether this is a rigorous "proof." The question whether it is, is hard to decide, but at any rate a close analysis of the verbal argument involved shows that it is entirely parallel to the mathematical proof of the same proposition which we already possess. Indeed, the proposition in question,

$$v_1 \leqq v_2$$

coincides with (13:A*) in 13.4.3. (The ϕ, x, y there correspond to our $\mathcal{H}$, τ_1, τ_2.)

Instead of ascribing v_1, v_2 as values to two games Γ_1 and Γ_2 different from Γ we may alternatively correlate them with Γ itself, under suitable assumptions concerning the "intellect" of the players 1 and 2.

Indeed, the rules of the game Γ prescribe that each player must make his choice (his personal move) in ignorance of the outcome of the choice of his adversary. It is nevertheless conceivable that one of the players, say 2, "finds out" his adversary; i.e., that he has somehow acquired the knowledge as to what his adversary's strategy is.[2] The basis for this knowledge does not concern us; it may (but need not) be experience from previous plays. At any rate we assume that the player 2 possesses this knowledge. It is possible, of course, that in this situation 1 will change his strategy; but again let us assume that, for any reason whatever, he does not do it.[3] Under these assumptions we may then say that player 2 has "found out" his adversary.

[1] For player 2 the values are consequently $-v_1$, $-v_2$.

[2] In the game Γ—which is in the normalized form—the strategy is just the actual choice at the unique personal move of the player. Remember how this normalized form was derived from the original extensive form of the game; consequently it appears that this choice corresponds equally to the strategy in the original game.

[3] For an interpretation of all these assumptions, cf. 17.3.1.

In this case, conditions in Γ become exactly the same as if the game were Γ_1, and hence all discussions of 14.3.1. apply literally.

Similarly, we may visualize the opposite possibility, that player 1 has "found out" his adversary. Then conditions in Γ become exactly the same as if the game were Γ_2; and hence all discussions of 14.3.3. apply literally.

In the light of the above we can say:

The value of a play of the game Γ is a well-defined quantity if one of the following two extreme assumptions is made: Either that player 2 "finds out" his adversary, or that player 1 "finds out" his adversary. In the first case the value of a play is v_1 for 1, and $-v_1$ for 2; in the second case the value of a play is v_2 for 1 and $-v_2$ for 2

14.4.2. This discussion shows that if the value of a play of Γ itself—without any further qualifications or modifications—can be defined at all, then it must lie between the values of v_1 and v_2. (We mean the values for the player 1.) I.e. if we write v for the hoped-for value of a play of Γ itself (for player 1), then there must be

$$v_1 \leqq v \leqq v_2.$$

The length of this interval, which is still available for v, is

$$\Delta = v_2 - v_1 \geqq 0.$$

At the same time Δ expresses the advantage which is gained (in the game Γ) by "finding out" one's adversary instead of being "found out" by him.[1]

Now the game may be such that it does not matter which player "finds out" his opponent; i.e., that the advantage involved is zero. According to the above, this is the case if and only if

$$\Delta = 0$$

or equivalently

$$v_1 = v_2$$

Or, if we replace v_1, v_2 by their definitions:

$$\text{Max}_{\tau_1} \text{Min}_{\tau_2} \mathcal{H}(\tau_1, \tau_2) = \text{Min}_{\tau_2} \text{Max}_{\tau_1} \mathcal{H}(\tau_1, \tau_2).$$

If the game Γ possesses these properties, then we call it *strictly determined.*

The last form of this condition calls for comparison with (13:3) in 13.3.1. and with the discussions of 13.4.1.-13.5.2. (The ϕ, x, y there again correspond to our $\mathcal{H}$, τ_1, τ_2). Indeed, the statement of (13:B*) in 13.4.3. says that the game Γ is strictly determined if and only if a saddle point of $\mathcal{H}(\tau_1, \tau_2)$ exists.

14.5. Analysis of Strict Determinateness

14.5.1. Let us assume the game Γ to be strictly determined; i.e. that a saddle point of $\mathcal{H}(\tau_1, \tau_2)$ exists.

[1] Observe that this expression for the advantage in question applies for both players: The advantage for the player 1 is $v_2 - v_1$; for the player 2 it is $(-v_1) - (-v_2)$ and these two expressions are equal to each other, i.e. to Δ.

In this case it is to be hoped—considering the analysis of 14.4.2.—that it will be possible to interpret the quantity

$$v = v_1 = v_2$$

as the value of a play of Γ (for the player 1). Recalling the definitions of v_1, v_2 and the definition of the saddle value in 13.4.3. and using (13:C*) in 13.5.2., we see that the above equation may also be written as

$$v = \mathrm{Max}_{\tau_1} \mathrm{Min}_{\tau_2} \mathcal{H}(\tau_1, \tau_2) = \mathrm{Min}_{\tau_2} \mathrm{Max}_{\tau_1} \mathcal{H}(\tau_1, \tau_2)$$
$$= \mathrm{Sa}_{\tau_1/\tau_2} \mathcal{H}(\tau_1, \tau_2).$$

By retracing the steps made at the end of 14.3.1. and at the end of 14.3.3. it is indeed not difficult to establish that the above can be interpreted as the value of a play of Γ (for player 1).

Specifically: (14:A:c), (14:A:d), (14:A:e), (14:B:c), (14:B:d), (14:B:e) of 14.3.1. and 14.3.3. where they apply to Γ_1 and Γ_2 respectively, can now be obtained for Γ itself. We restate first the equivalent of (14:A:d) and (14:B:d):

(14:C:d) Player 1 can, by playing appropriately, secure for himself a gain $\geqq$ v, irrespective of what player 2 does.

Player 2 can, by playing appropriately, secure for himself a gain $\geqq -v$ irrespective of what player 1 does.

In order to prove this, we form again the set A of (14:A:a) in 14.3.1. and the set B of (14:B:a) in 14.3.3. These are actually the sets A^ϕ, B^ϕ of 13.5.1. (the ϕ corresponds to our $\mathcal{H}$). We repeat:

(14:D:a) A is the set of those τ_1 for which $\mathrm{Min}_{\tau_2} \mathcal{H}(\tau_1, \tau_2)$ assumes its maximum value; i.e. for which

$$\mathrm{Min}_{\tau_2} \mathcal{H}(\tau_1, \tau_2) = \mathrm{Max}_{\tau_1} \mathrm{Min}_{\tau_2} \mathcal{H}(\tau_1, \tau_2) = v.$$

(14:D:b) B is the set of those τ_2 for which $\mathrm{Max}_{\tau_1} \mathcal{H}(\tau_1, \tau_2)$ assumes its minimum value; i.e. for which

$$\mathrm{Max}_{\tau_1} \mathcal{H}(\tau_1, \tau_2) = \mathrm{Min}_{\tau_2} \mathrm{Max}_{\tau_1} \mathcal{H}(\tau_1, \tau_2) = v.$$

Now the demonstration of (14:C:d) is easy:

Let player 1 choose τ_1 from A. Then irrespective of what player 2 does, i.e. for every τ_2, we have $\mathcal{H}(\tau_1, \tau_2) \geqq \mathrm{Min}_{\tau_2} \mathcal{H}(\tau_1, \tau_2) = v$, i.e., 1's gain is $\geqq$ v.

Let player 2 choose τ_2 from B. Then, irrespective of what player 1 does, i.e. for every τ_1, we have $\mathcal{H}(\tau_1, \tau_2) \leqq \mathrm{Max}_{\tau_1} \mathcal{H}(\tau_1, \tau_2) = v$, i.e. 1's gain is $\leqq$ v and so 2's gain is $\geqq -v$.

This completes the proof.

We pass now to the equivalent of (14:A:e) and (14:B:e). Indeed, (14:C:d) as formulated above can be equivalently formulated thus:

(14:C:e) Player 2 can, by playing appropriately, make it sure that the gain of player 1 is $\leqq$ v, i.e. prevent him from gaining $>$ v irrespective of what player 1 does.

Player 1 can, by playing appropriately, make it sure that the gain of player 2 is $\leqq -v$ i.e. present him from gaining $> -v$ irrespective of what player 2 does.

(14:C:d) and (14:C:e) establish satisfactorily our interpretation of v as the value of a play of Γ for the player 1, and of $-v$ for the player 2.

14.5.2. We consider now the equivalents of (14:A:a), (14:A:b), (14:B:a), (14:B:b).

Owing to (14:C:d) in 14.5.1. it is reasonable to define a good way for 1 to play the game Γ as one which guarantees him a gain which is greater than or equal to the value of a play for 1, irrespective of what 2 does; i.e. a choice of τ_1 for which $\mathcal{H}(\tau_1, \tau_2) \geqq v$ for all τ_2. This may be equivalently stated as $\text{Min}_{\tau_2} \mathcal{H}(\tau_1, \tau_2) \geqq v$.

Now we have always $\text{Min}_{\tau_2} \mathcal{H}(\tau_1, \tau_2) \leqq \text{Max}_{\tau_1} \text{Min}_{\tau_2} \mathcal{H}(\tau_1, \tau_2) = v$.

Hence the above condition for τ_1 amounts to $\text{Min}_{\tau_2} \mathcal{H}(\tau_1, \tau_2) = v$, i.e. (by (14:D:a) in 14.5.1.) to τ_1 being in A.

Again, by (14:C:d) in 14.5.1. it is reasonable to define the good way for 2 to play the game Γ as one which guarantees him a gain which is greater than or equal to the value of a play for 2, irrespective of what 1 does; i.e. a choice of τ_2 for which $-\mathcal{H}(\tau_1, \tau_2) \geqq -v$ for all τ_1. That is, $\mathcal{H}(\tau_1, \tau_2) \leqq v$ for all τ_1. This may be equivalently stated as $\text{Max}_{\tau_1} \mathcal{H}(\tau_1, \tau_2) \leqq v$.

Now we have always $\text{Max}_{\tau_1} \mathcal{H}(\tau_1, \tau_2) \geqq \text{Min}_{\tau_2} \text{Max}_{\tau_1} \mathcal{H}(\tau_1, \tau_2) = v$. Hence the above conditions for τ_2 amounts to $\text{Max}_{\tau_1} \mathcal{H}(\tau_1, \tau_2) = v$, i.e. (by (14:D:b) in 14.5.1.) to τ_2 being in B.

So we have:

(14:C:a) The good way (strategy) for 1 to play the game Γ is to choose any τ_1 belonging to A,—A being the set of (14:D:a) in 14.5.1.

(14:C:b) The good way (strategy) for 2 to play the game Γ is to choose any τ_2 belonging to B,—B being the set of (14:D:b) in 14.5.1.[1]

Finally our definition of the good way of playing, as stated at the beginning of this section, yields immediately the equivalent of (14:A:c) or (14:B:c):

(14:C:c) If both players 1 and 2 play the game Γ well—i.e. if τ_1 belongs to A and τ_2 belongs to B—then the value of $\mathcal{H}(\tau_1, \tau_2)$ will be equal to the value of a play (for 1), i.e. to v.

We add the observation that (13:D*) in 13.5.2. and the remark concerning the sets A, B before (14:D:a), (14:D:b) in 14.5.1. together give this:

(14:C:f) Both players 1 and 2 play the game Γ well—i.e. τ_1 belongs to A and τ_2 belongs to B—if and only if τ_1, τ_2 is a saddle point of $\mathcal{H}(\tau_1, \tau_2)$.

[1] Since this is the game Γ each player must make his choice (of τ_1 or τ_2) without knowledge of the other player's choice (of τ_2 or τ_1). Contrast this with (14:A:b) in 14.3.1. for Γ_1 and with (14:B:b) in 14.3.3. for Γ_2.

14.6. The Interchange of Players. Symmetry

14.6. (14:C:a)-(14:C:f) in 14.5.1. and 14.5.2. settle everything as far as the strictly determined two-person games are concerned. In this connection let us remark that in 14.3.1., 14.3.3.—for Γ_1, Γ_2—we derived (14:A:d), (14:A:e), (14:B:d), (14:B:e) from (14:A:a), (14:A:b), (14:B:a), (14:B:b) while in 14.5.1., 14.5.2.—for Γ itself—we obtained (14:C:a), (14:C:b) from (14:C:d), (14:C:e). This is an advantage since the arguments of 14.3.1., 14.3.3. in favor of (14:A:a), (14:A:b), (14:B:a), (14:B:b) were of a much more heuristic character than those of 14.5.1., 14.5.2. in favor of (14:C:d), (14:C:e).

Our use of the function $\mathcal{H}(\tau_1, \tau_2) \equiv \mathcal{H}_1(\tau_1, \tau_2)$ implies a certain asymmetry of the arrangement; the player 1 is thereby given a special role. It ought to be intuitively clear, however, that equivalent results would be obtained if we gave this special role to the player 2 instead. Since interchanging the players 1 and 2 will play a certain role later, we shall nevertheless give a brief mathematical discussion of this point also.

Interchanging the players 1 and 2 in the game Γ—of which we need not assume now that it is strictly determined—amounts to replacing the functions $\mathcal{H}_1(\tau_1, \tau_2)$, $\mathcal{H}_2(\tau_1, \tau_2)$ by $\mathcal{H}_2(\tau_2, \tau_1)$, $\mathcal{H}_1(\tau_2, \tau_1)$.[1,2] It follows, therefore, that this interchange means replacing the function $\mathcal{H}(\tau_1, \tau_2)$ by $-\mathcal{H}(\tau_2, \tau_1)$.

Now the change of sign has the effect of interchanging the operations Max and Min. Consequently the quantities

$$\begin{aligned} \mathrm{Max}_{\tau_1} \mathrm{Min}_{\tau_2} \mathcal{H}(\tau_1, \tau_2) &= v_1, \\ \mathrm{Min}_{\tau_2} \mathrm{Max}_{\tau_1} \mathcal{H}(\tau_1, \tau_2) &= v_2, \end{aligned}$$

as defined in 14.4.1. become now

$$\begin{aligned} \mathrm{Max}_{\tau_1} \mathrm{Min}_{\tau_2} [-\mathcal{H}(\tau_2, \tau_1)] &= -\mathrm{Min}_{\tau_1} \mathrm{Max}_{\tau_2} \mathcal{H}(\tau_2, \tau_1) \\ &= -\mathrm{Min}_{\tau_2} \mathrm{Max}_{\tau_1} \mathcal{H}(\tau_1, \tau_2)\,^{3} = -v_2. \\ \mathrm{Min}_{\tau_2} \mathrm{Max}_{\tau_1} [-\mathcal{H}(\tau_2, \tau_1)] &= -\mathrm{Max}_{\tau_2} \mathrm{Min}_{\tau_1} \mathcal{H}(\tau_2, \tau_1) \\ &= -\mathrm{Max}_{\tau_1} \mathrm{Min}_{\tau_2} \mathcal{H}(\tau_1, \tau_2)\,^{3} = -v_1. \end{aligned}$$

So v_1, v_2 become $-v_2$, $-v_1$.[4] Hence the value of

$$\Delta = v_2 - v_1 = (-v_1) - (-v_2)$$

[1] This is no longer the operation of interchanging the players used in 14.3.4. There we were only interested in the arrangement and the state of information at each move, and the players 1 and 2 were considered as taking their functions $\mathcal{H}_1(\tau_1, \tau_2)$ and $\mathcal{H}_2(\tau_1, \tau_2)$ with them (cf. footnote 1 on p. 104). In this sense Γ was symmetric, i.e. unaffected by that interchange (id.).

At present we interchange the roles of the players 1 and 2 completely, even in their functions $\mathcal{H}_1(\tau_1, \tau_2)$ and $\mathcal{H}_2(\tau_1, \tau_2)$.

[2] We had to interchange the variables τ_1, τ_2 since τ_1 represents the choice of player 1 and τ_2 the choice of player 2. Consequently it is now τ_2 which has the domain $1, \cdots, \beta_1$. Thus it is again true for $\mathcal{H}_k(\tau_2, \tau_1)$—as it was before for $\mathcal{H}_k(\tau_1, \tau_2)$—that the variable before the comma has the domain $1, \cdots, \beta_1$ and the variable after the comma, the domain $1, \cdots, \beta_2$.

[3] This is a mere change of notations: The variables τ_1, τ_2 are changed around to τ_2, τ_1.

[4] This is in harmony with footnote 1 on p. 105, as it should be.

is unaffected,[1] and if Γ is strictly determined, it remains so, since this property is equivalent to $\Delta = 0$. In this case $v = v_1 = v_2$ becomes

$$-v = -v_1 = -v_2.$$

It is now easy to verify that all statements (14:C:a)-(14:C:f) in 14.5.1., 14.5.2. remain the same when the players 1 and 2 are interchanged.

14.7. Non-strictly Determined Games

14.7.1. All this disposes completely of the strictly determined games, but of no others. For a game I which is not strictly determined we have $\Delta > 0$ i.e. in such a game it involves a positive advantage to "find out" one's adversary. Hence there is an essential difference between the results, i.e. the values in Γ_1 and in Γ_2, and therefore also between the good ways of playing these games. The considerations of 14.3.1., 14.3.3. provide therefore no guidance for the treatment of Γ. Those of 14.5.1., 14.5.2. do not apply either, since they make use of the existence of saddle points of $\mathcal{K}(\tau_1, \tau_2)$ and of the validity of

$$\text{Max}_{\tau_1} \text{Min}_{\tau_2} \mathcal{K}(\tau_1, \tau_2) = \text{Min}_{\tau_2} \text{Max}_{\tau_1} \mathcal{K}(\tau_1, \tau_2),$$

i.e. of Γ being strictly determined. There is, of course, some plausibility in the inequality at the beginning of 14.4.2. According to this, the value v of a play of Γ (for the player 1)—if such a concept can be formed at all in this generality, for which we have no evidence as yet[2]—is restricted by

$$v_1 \leqq v \leqq v_2.$$

But this still leaves an interval of length $\Delta = v_2 - v_1 > 0$ open to v; and, besides, the entire situation is conceptually most unsatisfactory.

One might be inclined to give up altogether: Since there is a positive advantage in "finding out" one's opponent in such a game Γ, it seems plausible to say that there is no chance to find a solution unless one makes some definite assumption as to "who finds out whom," and to what extent.[3]

We shall see in 17. that this is not so, and that in spite of $\Delta > 0$ a solution can be found along the same lines as before. But we propose first, without attacking that difficulty, to enumerate certain games Γ with $\Delta > 0$, and others with $\Delta = 0$. The first—which are not strictly determined—will be dealt with briefly now; their detailed investigation will be undertaken in 17.1. The second—which are strictly determined—will be analyzed in considerable detail.

14.7.2. Since there exist functions $\mathcal{K}(\tau_1, \tau_2)$ without saddle points (cf. 13.4.1., 13.4.2.; the $\phi(x, y)$ there, is our $\mathcal{K}(\tau_1, \tau_2)$) there exist not strictly determined games Γ. It is worth while to re-examine those examples—i.e.

[1] This is in harmony with footnote 1 on p. 106, as it should be.

[2] Cf. however, 17.8.1.

[3] In plainer language: $\Delta > 0$ means that it is not possible in this game for each player simultaneously to be cleverer than his opponent. Consequently it seems desirable to know just how clever each particular player is.

the functions described by the matrices of Figs. 12, 13 on p. 94—in the light of our present application. That is, to describe explicitly the games to which they belong. (In each case, replace $\phi(x, y)$ by our $\mathcal{H}(\tau_1, \tau_2)$, τ_2 being the column number and τ_1 the row number in every matrix. Cf. also Fig. 15 on p. 99).

Fig. 12: This is the game of "Matching Pennies." Let—for τ_1 and for τ_2—1 be "heads" and 2 be "tails," then the matrix element has the value 1 if τ_1, τ_2 "match"—i.e. are equal to each other—and -1, if they do not. So player 1 "matches" player 2: He wins (one unit) if they "match" and he loses (one unit), if they do not.

Fig. 13: This is the game of "Stone, Paper, Scissors." Let—for τ_1 and for τ_2—1 be "stone," 2 be "paper," and 3 be "scissors." The distribution of elements 1 and -1 over the matrix expresses that "paper" defeats "stone," "scissors" defeat "paper," "stone" defeats "scissors."[1] Thus player 1 wins (one unit) if he defeats player 2, and he loses (one unit) if he is defeated. Otherwise (if both players make the same choice) the game is tied.

14.7.3. These two examples show the difficulties which we encounter in a not strictly determined game, in a particularly clear form; just because of their extreme simplicity the difficulty is perfectly isolated here, *in vitro*. The point is that in "Matching Pennies" and in "Stone, Paper, Scissors," any way of playing—i.e. any τ_1 or any τ_2 —is just as good as any other: There is no intrinsic advantage or disadvantage in "heads" or in "tails" *per se*, nor in "stone," "paper" or "scissors" *per se*. The only thing which matters is to guess correctly what the adversary is going to do; but how are we going to describe that without further assumptions about the players' "intellects"?[2]

There are, of course, more complicated games which are not strictly determined and which are important from various more subtle, technical viewpoints (cf. 18., 19.). But as far as the main difficulty is concerned, the simple games of "Matching Pennies" and of "Stone, Paper, Scissors" are perfectly characteristic.

14.8. Program of a Detailed Analysis of Strict Determinateness

14.8. While the strictly determined games Γ—for which our solution is valid—are thus a special case only, one should not underestimate the size of the territory they cover. The fact that we are using the normalized form for the game Γ may tempt to such an underestimation: It makes things look more elementary than they really are. One must remember that the τ_1, τ_2 represent strategies in the extensive form of the game, which may be of a very complicated structure, as mentioned in 14.1.1.

In order to understand the meaning of strict determinateness, it is therefore necessary to investigate it in relation to the extensive form of the game. This brings up questions concerning the detailed nature of the moves,

[1] "Paper covers the stone, scissors cut the paper, stone grinds the scissors."
[2] As mentioned before, we shall show in 17.1. that it can be done.

—chance or personal—the state of information of the players, etc.; i.e. we come to the structural analysis based on the extensive form, as mentioned in 12.1.1.

We are particularly interested in those games in which each player who makes a personal move is perfectly informed about the outcome of the choices of all anterior moves. These games were already mentioned in 6.4.1. and it was stated there that they are generally considered to be of a particular rational character. We shall now establish this in a precise sense, by proving that all such games are strictly determined. And this will be true not only when all moves are personal, but also when chance moves too are present.

15. Games with Perfect Information

15.1. Statement of Purpose. Induction

15.1.1. We wish to investigate the zero-sum two-person games somewhat further, with the purpose of finding as wide a subclass among them as possible in which only strictly determined games occur; i.e. where the quantities

$$v_1 = \mathrm{Max}_{\tau_1} \mathrm{Min}_{\tau_2} \mathcal{K}(\tau_1, \tau_2),$$
$$v_2 = \mathrm{Min}_{\tau_2} \mathrm{Max}_{\tau_1} \mathcal{K}(\tau_1, \tau_2)$$

of 14.4.1.—which turned out to be so important for the appraisal of the game—fulfill

$$v_1 = v_2 = v.$$

We shall show that when perfect information prevails in Γ—i.e. when preliminarity is equivalent to anteriority (cf. 6.4.1. and the end of 14.8.)—then Γ is strictly determined. We shall also discuss the conceptual significance of this result (cf. 15.8.). Indeed, we shall obtain this as a special case of a more general rule concerning v_1, v_2, (cf. 15.5.3.).

We begin our discussions in even greater generality, by considering a perfectly unrestricted general n-person game Γ. The greater generality will be useful in a subsequent instance.

15.1.2. Let Γ be a general n-person game, given in its extensive form. We shall consider certain aspects of Γ, first in our original pre-set-theoretical terminology of 6., 7., (cf. 15.1.), and then translate everything into the partition and set terminology of 9., 10. (cf. 15.2., et sequ.). The reader will probably obtain a full understanding with the help of the first discussion alone; and the second, with its rather formalistic machinery, is only undertaken for the sake of absolute rigor, in order to show that we are really proceeding strictly on the basis of our axioms of 10.1.1.

We consider the sequence of all moves in Γ: $\mathfrak{M}_1, \mathfrak{M}_2, \cdots, \mathfrak{M}_\nu$. Let us fix our attention on the first move, $\mathfrak{M}_1$, and the situation which exists at the moment of this move.

Since nothing is anterior to this move, nothing is preliminary to it either; i.e. the characteristics of this move depend on nothing,—they are constants. This applies in the first place to the fact, whether $\mathfrak{M}_1$ is a chance move or a personal move; and in the latter case, to which player $\mathfrak{M}_1$ belongs,—i.e. to the value of $k_1 = 0, 1, \cdots, n$ respectively, in the sense of 6.2.1. And it applies also to the number of alternatives α_1 at $\mathfrak{M}_1$ and for a chance move (i.e. when $k_1 = 0$) to the values of the probabilities $p_1(1), \cdots, p_1(\alpha_1)$. The result of the choice at $\mathfrak{M}_1$—chance or personal—is a $\sigma_1 = 1, \cdots, \alpha_1$.

Now a plausible step suggests itself for the mathematical analysis of the game Γ, which is entirely in the spirit of the method of "complete induction" widely used in all branches of mathematics. It replaces, if successful, the analysis of Γ by the analysis of other games which contain one move less than Γ.[1] This step consists in choosing a $\bar{\sigma}_1 = 1, \cdots, \alpha_1$ and denoting by $\Gamma_{\bar{\sigma}_1}$ a game which agrees with Γ in every detail except that the move $\mathfrak{M}_1$ is omitted, and instead the choice σ_1 is dictated (by the rules of the new game) the value $\sigma_1 = \bar{\sigma}_1$.[2] $\Gamma_{\bar{\sigma}_1}$ has, indeed, one move less than Γ: Its moves are $\mathfrak{M}_2, \cdots, \mathfrak{M}_\nu$.[3] And our "inductive" step will have been successful if we can derive the essential characteristics of Γ from those of all $\Gamma_{\bar{\sigma}_1}$, $\bar{\sigma}_1 = 1, \cdots, \alpha_1$.

15.1.3. It must be noted, however, that the possibilities of forming $\Gamma_{\bar{\sigma}_1}$ are dependent upon a certain restriction on Γ. Indeed, every player who makes a certain personal move in the game $\Gamma_{\bar{\sigma}_1}$ must be fully informed about the rules of this game. Now this knowledge consists of the knowledge of the rules of the original game Γ plus the value of the dictated choice at $\mathfrak{M}_1$, i.e. $\bar{\sigma}_1$. Hence $\Gamma_{\bar{\sigma}_1}$ can be formed out of Γ—without modifying the rules which govern the player's state of information in Γ—only if the outcome of the choice at $\mathfrak{M}_1$, by virtue of the original rules Γ, is known to every player at any personal move of his $\mathfrak{M}_2, \cdots, \mathfrak{M}_\nu$; i.e. $\mathfrak{M}_1$ must be preliminary to all personal moves $\mathfrak{M}_2, \cdots, \mathfrak{M}_\nu$. We restate this:

(15:A) $\Gamma_{\bar{\sigma}_1}$ can be formed—without essentially modifying the structure of Γ for that purpose—only if Γ possess the following property:

(15:A:a) $\mathfrak{M}_1$ is preliminary to all personal moves $\mathfrak{M}_2, \cdots, \mathfrak{M}_\nu$.[4]

[1] I.e. have $\nu - 1$ instead of ν. Repeated application of this "inductive" step—if feasible at all—will reduce the game Γ to one with 0 steps; i.e. to one of fixed, unalterable outcome. And this means, of course, a complete solution for Γ. (Cf. (15:C:a) in 15.6.1.)

[2] E.g. Γ is the game of Chess, and $\bar{\sigma}_1$ a particular opening move—i.e. choice at $\mathfrak{M}_1$—of "white," i.e. player 1. Then $\Gamma_{\bar{\sigma}_1}$ is again Chess, but beginning with a move of the character of the second move in ordinary Chess—a "black," player 2— and in the position created by the "opening move" $\bar{\sigma}_1$. This dictated "opening move" may, but need not, be a conventional one (like E2-E4).

The same operation is exemplified by forms of Tournament Bridge where the "umpires" assign the players definite—known and previously selected—"hands." (This is done, e.g., in Duplicate Bridge.)

In the first example, the dictated move $\mathfrak{M}_1$ was originally personal (of "white," player 1); in the second example it was originally chance (the "deal").

In some games occasionally "handicaps" are used which amount to one or more such operations.

[3] We should really use the indices $1, \cdots, \nu - 1$ and indicate the dependence on $\bar{\sigma}_1$; e.g. by writing $\mathfrak{M}_1^{\bar{\sigma}_1}, \cdots, \mathfrak{M}_{\nu-1}^{\bar{\sigma}_1}$. But we prefer the simpler notation $\mathfrak{M}_2, \cdots, \mathfrak{M}_\nu$.

[4] This is the terminology of 6.3.; i.e. we use the special form of dependence in the sense of 7.2.1. Using the general description of 7.2.1. we must state (15:A:a) like this: For every personal move $\mathfrak{M}_\kappa$, $\kappa = 2, \cdots, \nu$, the set Φ_κ contains the function σ_1.

15.2. The Exact Condition (First Step)

15.2.1. We now translate 15.1.2., 15.1.3. into the partition and set terminology of 9., 10., (cf. also the beginning of 15.1.2.). The notations of 10.1. will therefore be used.

$\mathfrak{A}_1$ consists of the one set Ω ((10:1:f) in 10.1.1.), and it is a subpartition of $\mathfrak{B}_1$ ((10:1:a) in 10.1.1.); hence $\mathfrak{B}_1$ too consists of the one set Ω (the others being empty).[1,2] That is:

$$B_1(k) = \begin{cases} \Omega & \text{for precisely one } k, \text{ say } k = k_1, \\ \ominus & \text{for all } k \neq k_1. \end{cases}$$

This $k_1 = 0, 1, \cdots, n$ determines the character of $\mathfrak{M}_1$; it is the k_1 of 6.2.1. If $k_1 = 1, \cdots, n$—i.e. if the move is personal—then $\mathfrak{A}_1$ is also a subpartition of $\mathfrak{D}_1(k_1)$, ((10:1:d) in 10.1.1. This was only postulated within $B_1(k_1)$, but $B_1(k_1) = \Omega$). Hence $\mathfrak{D}_1(k_1)$ too consists of the one set Ω.[3] And for $k \neq k_1$, the $\mathfrak{D}_1(k)$ which is a partition in $B_1(k) = \ominus$ ((10:A:g) in 10.1.1.) must be empty.

So we have precisely one A_1 of $\mathfrak{A}_1$, which is Ω, and for $k_1 = 1, \cdots, n$ precisely one D_1 in all $\mathfrak{D}_1(k)$, which is also Ω; while for $k_1 = 0$ there are no D_1 in all $\mathfrak{D}_1(k)$.

The move $\mathfrak{M}_1$ consists of the choice of a C_1 from $\mathfrak{C}_1(k_1)$; by chance if $k_1 = 0$; by the player k_1 if $k_1 = 1, \cdots, n$. C_1 is automatically a subset of the unique $A_1(= \Omega)$ in the former case, and of the unique $D_1(= \Omega)$ in the latter. The number of these C_1 is α_1 (cf. 9.1.5., particularly footnote 2 on p. 70); and since the A_1 or D_1 in question is fixed, this α_1 is a definite constant. α_1 is the number of alternatives at $\mathfrak{M}_1$, the α_1 of 6.2.1. and 15.1.2.

These C_1 correspond to the $\sigma_1 = 1, \cdots, \alpha_1$ of 15.1.2., and we denote them accordingly by $C_1(1), \cdots, C_1(\alpha_1)$.[4] Now (10:1:h) in 10.1.1. shows—as is readily verified—that $\mathfrak{A}_2$ is also the set of the $C_1(1), \cdots, C_1(\alpha_1)$, i.e. equal to $\mathfrak{C}_1$.

So far our analysis has been perfectly general,—valid for $\mathfrak{M}_1$ (and to a certain extent for $\mathfrak{M}_2$) of any game Γ. The reader should translate these properties into everyday terminology in the sense of 8.4.2. and 10.4.2.

We pass now to $\Gamma_{\bar{\sigma}_1}$. This should obtain from Γ by dictating the move $\mathfrak{M}_1$—as described in 15.1.2.—by putting $\sigma_1 = \bar{\sigma}_1$. At the same time the moves of the game are restricted to $\mathfrak{M}_2, \cdots, \mathfrak{M}_\nu$. This means that the

[1] This $\mathfrak{B}_1$ is an exception from (8:B:a) in 8.3.1.; cf. the remark concerning this (8:B:a) in footnote 1 on p. 63, and also footnote 4 on p. 69.

[2] *Proof:* Ω belongs to $\mathfrak{A}_1$, which is a subpartition of $\mathfrak{B}_1$; hence Ω is a subset of an element of $\mathfrak{B}_1$. This element is necessarily equal to Ω. All other elements of $\mathfrak{B}_1$ are therefore disjunct from Ω (cf. 8.3.1.), i.e. empty.

[3] $\mathfrak{A}_1$, $\mathfrak{D}_1(k_1)$ unlike $\mathfrak{B}_1$ (cf. above) must fulfill both (8:B:a), (8:B:b) in 8.3.1.; hence both have no further elements besides Ω.

[4] They represent the alternatives $\mathfrak{A}_1(1), \cdots, \mathfrak{A}_1(\alpha_1)$ of 6.2. and 9.1.4., 9.1.5.

element π—which represents the actual play—can no longer vary over all Ω, but is restricted to $C_1(\bar{\sigma}_1)$. And the partitions enumerated in 9.2.1. are restricted to those with $\kappa = 2, \cdots, \nu$,[1] (and $\kappa = \nu + 1$ for $\mathcal{A}_\kappa$).

15.2.2. We now come to the equivalent of the restriction of 15.1.3.

The possibility of carrying out the changes formulated at the end of 15.2.1. is dependent upon a certain restriction on Γ.

As indicated, we wish to restrict the play—i.e. π—within $C_1(\bar{\sigma}_1)$. Therefore all those sets which figured in the description of Γ and which were subsets of Ω, must be made over into subsets of $C_1(\bar{\sigma}_1)$—and the partitions into partitions within $C_1(\bar{\sigma}_1)$ (or within subsets of $C_1(\bar{\sigma}_1)$). How is this to be done?

The partitions which make up the descriptions of Γ (cf. 9.2.1.) fall into two classes: those which represent objective facts—the $\mathcal{A}_\kappa$, the $\mathcal{B}_\kappa = (B_\kappa(0), B_\kappa(1), \cdots, B_\kappa(n))$ and the $\mathcal{C}_\kappa(k)$, $k = 0, 1, \cdots, n$—and those which represent only the player's state of information,[2] the $\mathcal{D}_\kappa(k)$, $k = 1, \cdots, n$. We assume, of course $\kappa \geqq 2$ (cf. the end of 15.2.1.).

In the first class of partitions we need only replace each element by its intersection with $C_1(\bar{\sigma}_1)$. Thus $\mathcal{B}_\kappa$ is modified by replacing its elements $B_\kappa(0), B_\kappa(1), \cdots, B_\kappa(n)$ by $C_1(\bar{\sigma}_1) \cap B_\kappa(0), C_1(\bar{\sigma}_1) \cap B_\kappa(1), \cdots, C_1(\bar{\sigma}_1) \cap B_\kappa(n)$. In $\mathcal{A}_\kappa$ even this is not necessary: It is a subpartition of $\mathcal{A}_2$ (since $\kappa \geqq 2$, cf. 10.4.1.), i.e. of the system of pairwise disjunct sets $(C_1(1), \cdots, C_1(\alpha_1))$ (cf. 15.2.1.); hence we keep only those elements of $\mathcal{A}_\kappa$ which are subsets of $C_1(\bar{\sigma}_1)$, i.e. that part of $\mathcal{A}_\kappa$ which lies in $C_1(\bar{\sigma}_1)$. The $\mathcal{C}_\kappa(k)$ should be treated like $\mathcal{B}_\kappa$ but we prefer to postpone this discussion.

In the second class of partitions—i.e. for the $\mathcal{D}_\kappa(k)$—we cannot do anything like it. Replacing the elements of $\mathcal{D}_\kappa(k)$ by their intersections with $C_1(\bar{\sigma}_1)$ would involve a modification of a player's state of information[3] and should therefore be avoided. The only permissible procedure would be that which was feasible in the case of $\mathcal{A}_\kappa$: replacement—of $\mathcal{D}_\kappa(k)$—by that part of itself which lies in $C_1(\bar{\sigma}_1)$. But this is applicable only if $\mathcal{D}_\kappa(k)$—like $\mathcal{A}_\kappa$ before—is a subpartition of $\mathcal{A}_2$ (for $\kappa \geqq 2$). So we must postulate this.

Now $\mathcal{C}_\kappa(k)$ takes care of itself: It is a subpartition of $\mathcal{D}_\kappa(k)$ ((10:1:c) in 10.1.1.), hence of $\mathcal{A}_2$ (by the above assumption); and so we can replace it by that part of itself which lies in $C_1(\bar{\sigma}_1)$.

So we see: The necessary restriction of Γ is that every $\mathcal{D}_\kappa(k)$ (with $\kappa \geqq 2$) must be a subpartition of $\mathcal{A}_2$. Recall now the interpretation of 8.4.2. and of (10:A:d*), (10:A:g*) in 10.1.2. They give to this restriction the meaning that every player at a personal move $\mathfrak{M}_2, \cdots, \mathfrak{M}_\nu$ is fully

[1] We do not wish to change the enumeration to $\kappa = 1, \cdots, \nu - 1$, cf. footnote 3 on p. 113.

[2] $\mathcal{A}_\kappa$ represents the umpire's state of information, but this is an objective fact: the events up to that moment have determined the course of the play precisely to that extent (cf. 9.1.2.).

[3] Namely, giving him additional information.

informed about the state of things after the move $\mathfrak{M}_1$ (i.e. before the move $\mathfrak{M}_2$) expressed by $\mathfrak{A}_2$. (Cf. also the discussion before (10:B) in 10.4.2.) That is, $\mathfrak{M}_1$ must be preliminary to all moves $\mathfrak{M}_2, \cdots, \mathfrak{M}_\nu$.

Thus we have again obtained the condition (15:A:a) of 15.1.3. We leave to the reader the simple verification that the game $\Gamma_{\bar{\sigma}_1}$ fulfills the requirements of 10.1.1.

15.3. The Exact Condition (Entire Induction)

15.3.1. As indicated at the end of 15.1.2., we wish to obtain the characteristics of Γ from those of all $\Gamma_{\bar{\sigma}_1}$, $\bar{\sigma}_1 = 1, \cdots, \alpha_1$, since this—if successful—would be a typical step of a "complete induction."

For the moment, however, the only class of games for which we possess any kind of (mathematical) characteristics consists of the zero-sum two-person games: for these we have the quantities v_1, v_2 (cf. 15.1.1.). Let us therefore assume that Γ is a zero-sum two-person game.

Now we shall see that the v_1, v_2 of Γ can indeed be expressed with the help of those of the $\Gamma_{\bar{\sigma}_1}$, $\bar{\sigma}_1 = 1, \cdots, \alpha_1$ (cf. 15.1.2.). This circumstance makes it desirable to push the "induction" further, to its conclusion: i.e., to form in the same way $\Gamma_{\bar{\sigma}_1,\bar{\sigma}_2}$, $\Gamma_{\bar{\sigma}_1,\bar{\sigma}_2,\bar{\sigma}_3}$, $\cdots$, $\Gamma_{\bar{\sigma}_1,\bar{\sigma}_2,\ldots,\bar{\sigma}_\nu}$.[1] The point is that the number of steps in these games decreases successively from ν (for Γ), $\nu - 1$ (for $\Gamma_{\bar{\sigma}_1}$), over $\nu - 2$, $\nu - 3$, $\cdots$ to 0 (for $\Gamma_{\bar{\sigma}_1,\bar{\sigma}_2,\ldots,\bar{\sigma}_\nu}$); i.e. $\Gamma_{\bar{\sigma}_1,\bar{\sigma}_2,\ldots,\bar{\sigma}_\nu}$ is a "vacuous" game (like the one mentioned in the footnote 2 on p. 76). There are no moves; the player k gets the fixed amount $\mathfrak{F}_k(\bar{\sigma}_1, \cdots, \bar{\sigma}_\nu)$.

This is the terminology of 15.1.2., 15.1.3.,—i.e. of 6., 7. In that of 15.2.1., 15.2.2.—i.e. of 9., 10.—we would say that Ω (for Γ) is gradually restricted to a $C_1(\bar{\sigma}_1)$ of $\mathfrak{A}_2$ (for $\Gamma_{\bar{\sigma}_1}$), a $C_2(\bar{\sigma}_1, \bar{\sigma}_2)$ of $\mathfrak{A}_3$ (for $\Gamma_{\bar{\sigma}_1,\bar{\sigma}_2}$), a $C_3(\bar{\sigma}_1, \bar{\sigma}_2, \bar{\sigma}_3)$ of $\mathfrak{A}_4$ (for $\Gamma_{\bar{\sigma}_1,\bar{\sigma}_2,\bar{\sigma}_3}$), etc., etc., and finally to a $C_\nu(\bar{\sigma}_1, \bar{\sigma}_2, \cdots, \bar{\sigma}_\nu)$ of $\mathfrak{A}_{\nu+1}$ (for $\Gamma_{\bar{\sigma}_1,\bar{\sigma}_2,\ldots,\bar{\sigma}_\nu}$). And this last set has a unique element ((10:1:g) in 10.1.1.), say $\bar{\pi}$. Hence the outcome of the game $\Gamma_{\bar{\sigma}_1,\bar{\sigma}_2,\ldots,\bar{\sigma}_\nu}$ is fixed: The player k gets the fixed amount $\mathfrak{F}_k(\bar{\pi})$.

Consequently, the nature of the game $\Gamma_{\bar{\sigma}_1,\bar{\sigma}_2,\ldots,\bar{\sigma}_\nu}$ is—trivially—clear; it is clear what this game's value is for every player. Therefore the process which leads from the $\Gamma_{\bar{\sigma}_1}$ to Γ—if established—can be used to work backwards from $\Gamma_{\bar{\sigma}_1,\bar{\sigma}_2,\ldots,\bar{\sigma}_\nu}$ to $\Gamma_{\bar{\sigma}_1,\bar{\sigma}_2,\ldots,\bar{\sigma}_{\nu-1}}$ to $\Gamma_{\bar{\sigma}_1,\bar{\sigma}_2,\ldots,\bar{\sigma}_{\nu-2}}$ etc., etc., to $\Gamma_{\bar{\sigma}_1,\bar{\sigma}_2}$ to $\Gamma_{\bar{\sigma}_1}$ and finally to Γ.

But this is feasible only if we are able to form all games of the sequence $\Gamma_{\bar{\sigma}_1}$, $\Gamma_{\bar{\sigma}_1,\bar{\sigma}_2}$, $\Gamma_{\bar{\sigma}_1,\bar{\sigma}_2,\bar{\sigma}_3}$, $\cdots$, $\Gamma_{\bar{\sigma}_1,\bar{\sigma}_2,\ldots,\bar{\sigma}_\nu}$, i.e. if the final condition of 15.1.3. or 15.2.2. is fulfilled for all these games. This requirement may again be formulated for any general n-person game Γ; so we return now to those Γ.

15.3.2. The requirement then is, in the terminology of 15.1.2., 15.1.3. (i.e. of 6., 7.) that $\mathfrak{M}_1$ must be preliminary to all $\mathfrak{M}_2, \mathfrak{M}_3, \cdots, \mathfrak{M}_\nu$; that

[1] $\bar{\sigma}_1 = 1, \cdots, \alpha_1$; $\bar{\sigma}_2 = 1, \cdots, \alpha_2$ where $\alpha_2 = \alpha_2(\bar{\sigma}_1)$; $\bar{\sigma}_3 = 1, \cdots, \alpha_3$ where $\alpha_3 = \alpha_3(\bar{\sigma}_1, \bar{\sigma}_2)$; etc., etc.

$\mathfrak{M}_2$ must be preliminary to all $\mathfrak{M}_3, \mathfrak{M}_4, \cdots, \mathfrak{M}_\nu$; etc., etc.; i.e. that preliminarity must coincide with anteriority.

In the terminology of 15.2.1., 15.2.2.—i.e. of 9., 10.—of course the same is obtained: All $\mathfrak{D}_\kappa(k)$, $\kappa \geqq 2$ must be subpartitions of $\mathfrak{A}_2$; all $\mathfrak{D}_\kappa(k)$, $\kappa \geqq 3$ must be subpartitions of $\mathfrak{A}_3$, etc., etc.; i.e. all $\mathfrak{D}_\kappa(k)$ must be subpartitions of $\mathfrak{A}_\lambda$ if $\kappa \geqq \lambda$.[1] Since $\mathfrak{A}_\kappa$ is a subpartition of $\mathfrak{A}_\lambda$ in any case (cf. 10.4.1.), it suffices to require that all $\mathfrak{D}_\kappa(k)$ be subpartitions of $\mathfrak{A}_\kappa$. However $\mathfrak{A}_\kappa$ is a subpartition of $\mathfrak{D}_\kappa(k)$ within $\mathfrak{B}_\kappa(k)$ ((10:1:d) in 10.1.1.); consequently our requirement is equivalent to saying that $\mathfrak{D}_\kappa(k)$ is that part of $\mathfrak{A}_\kappa$ which lies in $B_\kappa(k)$.[2] By (10:B) in 10.4.2. this means precisely that preliminarity and anteriority coincide in Γ.

By all these considerations we have established this:

(15:B) In order to be able to form the entire sequence of games

(15:1) $$\Gamma, \Gamma_{\bar{\sigma}_1}, \Gamma_{\bar{\sigma}_1,\bar{\sigma}_2}, \Gamma_{\bar{\sigma}_1,\bar{\sigma}_2,\bar{\sigma}_3}, \cdots, \Gamma_{\bar{\sigma}_1,\bar{\sigma}_2,\ldots,\bar{\sigma}_\nu}$$

of

$$\nu, \nu - 1, \nu - 2, \cdots, 0$$

moves respectively, it is necessary and sufficient that in the game Γ preliminarity and anteriority should coincide,—i.e. that perfect information should prevail. (Cf. 6.4.1. and the end of 14.8.)

If Γ is a zero-sum two-person game, then this permits the elucidation of Γ by going through the sequence (15:1) backwards—from the trivial game $\Gamma_{\bar{\sigma}_1,\bar{\sigma}_2,\ldots,\bar{\sigma}_\nu}$ to the significant game Γ—performing each step with the help of the device which leads from the $\Gamma_{\bar{\sigma}_1}$ to Γ as will be shown in 15.6.2.

15.4. Exact Discussion of the Inductive Step

15.4.1. We now proceed to carry out the announced step from the Γ_{σ_1}[3] to Γ, the "inductive step." Γ need therefore fullfill only the final condition of 15.1.3. or 15.2.2., but it must be a zero-sum two-person game.

Hence we can form all $\Gamma_{\sigma_1}, \sigma_1 = 1, \cdots, \alpha_1$, and they also are zero-sum two-person games. We denote the two players' strategies in Γ by $\Sigma'_1, \cdots, \Sigma_1^{\beta_1}$ and $\Sigma'_2, \cdots, \Sigma_2^{\beta_2}$; and the "mathematical expectation" of the outcome of the play for the two players, if the strategies $\Sigma_1^{\tau_1}, \Sigma_2^{\tau_2}$ are used, by

$$\mathcal{K}_1(\tau_1, \tau_2) \equiv \mathcal{K}(\tau_1, \tau_2), \qquad \mathcal{K}_2(\tau_1, \tau_2) \equiv -\mathcal{K}(\tau_1, \tau_2)$$

(cf. 11.2.3. and 14.1.1.). We denote the corresponding quantities in Γ_{σ_1}, by $\Sigma'_{\sigma_1/1}, \cdots, \Sigma_{\sigma_1/1}^{\beta_{\sigma_1/1}}$ and $\Sigma'_{\sigma_1/2}, \cdots, \Sigma_{\sigma_1/2}^{\beta_{\sigma_1/2}}$, and if the strategies $\Sigma_{\sigma_1/1}^{\tau_{\sigma_1/1}}$, $\Sigma_{\sigma_1/2}^{\tau_{\sigma_1/2}}$ are used, by

$$\mathcal{K}_{\sigma_1/1}(\tau_{\sigma_1/1}, \tau_{\sigma_1/2}) \equiv \mathcal{K}_{\sigma_1}(\tau_{\sigma_1/1}, \tau_{\sigma_1/2}), \qquad \mathcal{K}_{\sigma_1/2}(\tau_{\sigma_1/1}, \tau_{\sigma_1/2}) \equiv -\mathcal{K}_{\sigma_1}(\tau_{\sigma_1/1}, \tau_{\sigma_1/2}).$$

[1] We stated this above for $\lambda = 2, 3, \cdots$; for $\lambda = 1$ it is automatically true: every partition is a subpartition of $\mathfrak{A}_1$ since $\mathfrak{A}_1$ consists of the one set Ω ((10:1:f) in 10.1.1.).

[2] For the motivation—if one is wanted—cf. the argument of footnote 3 on p. 63.

[3] From now on we write $\sigma_1, \sigma_2, \cdots, \sigma_\nu$ instead of $\bar{\sigma}_1, \bar{\sigma}_2, \cdots, \bar{\sigma}_\nu$ because no misunderstandings will be possible.

We form the v_1, v_2 of 14.4.1. for Γ and for Γ_{σ_1} denoting them in the latter case by $v_{\sigma_1/1}$, $v_{\sigma_1/2}$. So

$$v_1 = \mathrm{Max}_{\tau_1} \mathrm{Min}_{\tau_2} \mathcal{K}(\tau_1, \tau_2),$$
$$v_2 = \mathrm{Min}_{\tau_2} \mathrm{Max}_{\tau_1} \mathcal{K}(\tau_1, \tau_2),$$

and

$$v_{\sigma_1/1} = \mathrm{Max}_{\tau_{\sigma_1/1}} \mathrm{Min}_{\tau_{\sigma_1/2}} \mathcal{K}_{\sigma_1}(\tau_{\sigma_1/1}, \tau_{\sigma_1/2}),$$
$$v_{\sigma_1/2} = \mathrm{Min}_{\tau_{\sigma_1/2}} \mathrm{Max}_{\tau_{\sigma_1/1}} \mathcal{K}_{\sigma_1}(\tau_{\sigma_1/1}, \tau_{\sigma_1/2}).$$

Our aim is to express the v_1, v_2 in terms of the $v_{\sigma_1/1}$, $v_{\sigma_1/2}$.

The k_1 of 15.1.2., 15.2.1. which determine the character of the move $\mathfrak{M}_1$ will play an essential role. Since $n = 2$, its possible values are $k_1 = 0, 1, 2$. We must consider these three alternatives separately.

15.4.2. Consider first the case $k_1 = 0$; i.e. let $\mathfrak{M}_1$ be a chance move. The probabilities of its alternatives $\sigma_1 = 1, \cdots, \alpha_1$ are the $p_1(1), \cdots, p_1(\alpha_1)$ mentioned in 15.1.2. ($p_1(\sigma_1)$ is the $p_1(C_1)$ of (10:A:h) in 10.1.1. with $C_1 = C_1(\sigma_1)$ in 15.2.1.).

Now a strategy of player 1 in Γ, $\Sigma_1^{\tau_1}$, consists obviously in specifying a strategy of player 1 in Γ_{σ_1}, $\Sigma_{\sigma_1/1}^{\tau_{\sigma_1/1}}$ for every value of the chance variable $\sigma_1 = 1, \cdots, \alpha_1$,[1] i.e., the $\Sigma_1^{\tau_1}$ correspond to the aggregates $\Sigma_{1/1}^{\tau_{1/1}}, \cdots, \Sigma_{\alpha_1/1}^{\tau_{\alpha_1/1}}$ for all possible combinations $\tau_{1/1}, \cdots, \tau_{\alpha_1/1}$.

Similarly a strategy of player 2 in Γ, $\Sigma_2^{\tau_2}$ consists in specifying a strategy of player 2 in Γ_{σ_1}, $\Sigma_{\sigma_1/2}^{\tau_{\sigma_1/2}}$ for every value of the chance variable $\sigma_1 = 1, \cdots, \alpha_1$; i.e. the $\Sigma_2^{\tau_2}$ correspond to the aggregates $\Sigma_{1/2}^{\tau_{1/2}}, \cdots, \Sigma_{\alpha_1/2}^{\tau_{\alpha_1/2}}$ for all possible combinations $\tau_{1/2}, \cdots, \tau_{\alpha_1/2}$.

Now the "mathematical expectations" of the outcomes in Γ and in Γ_{σ_1} are connected by the obvious formula

$$\mathcal{K}(\tau_1, \tau_2) = \sum_{\sigma_1=1}^{\alpha_1} p_1(\sigma_1)\mathcal{K}_{\sigma_1}(\tau_{\sigma_1/1}, \tau_{\sigma_1/2}).$$

Therefore our formula for v_1 gives

$$v_1 = \mathrm{Max}_{\tau_1} \mathrm{Min}_{\tau_2} \mathcal{K}(\tau_1, \tau_2)$$
$$= \mathrm{Max}_{\tau_{1/1}, \cdots, \tau_{\alpha_1/1}} \mathrm{Min}_{\tau_{1/2}, \cdots, \tau_{\alpha_1/2}} \sum_{\sigma_1=1}^{\alpha_1} p_1(\sigma_1)\mathcal{K}_{\sigma_1}(\tau_{\sigma_1/1}, \tau_{\sigma_1/2}).$$

The σ_1-term of the sum $\sum_{\sigma_1=1}^{\alpha_1}$ on the extreme right-hand side

$$p_1(\sigma_1)\mathcal{K}_{\sigma_1}(\tau_{\sigma_1/1}, \tau_{\sigma_1/2})$$

[1] This is clear intuitively. The reader may verify it from the formalistic point of view by applying the definitions of 11.1.1. and (11:A) in 11.1.3. to the situation described in 15.2.1.

contains only the two variables $\tau_{\sigma_1/1}, \tau_{\sigma_1/2}$. Thus the variable pairs

$$\tau_{1/1}, \tau_{1/2}; \cdots ; \tau_{\alpha_1/1}, \tau_{\alpha_1/2}$$

occur separately, in the separate σ_1-terms

$$\sigma_1 = 1, \qquad \cdots ; \qquad \sigma_1 = \alpha_1.$$

Hence in forming the $\mathrm{Min}_{\tau_{1/2}, \ldots, \tau_{\alpha_1/2}}$ we can minimize each σ_1-term separately, and in forming the $\mathrm{Max}_{\tau_{1/1}, \ldots, \tau_{\alpha_1/1}}$ we can again maximize each σ_1-term separately. Accordingly, our expression becomes

$$\sum_{\sigma_1=1}^{\alpha_1} p_1(\sigma_1) \, \mathrm{Max}_{\tau_{\sigma_1/1}} \, \mathrm{Min}_{\tau_{\sigma_1/2}} \mathcal{K}_{\sigma_1}(\tau_{\sigma_1/1}, \tau_{\sigma_1/2}) = \sum_{\sigma_1=1}^{\alpha_1} p_1(\sigma_1) \mathrm{v}_{\sigma_1/1}.$$

Thus we have shown

$$\text{(15:2)} \qquad \mathrm{v}_1 = \sum_{\sigma_1=1}^{\alpha_1} p_1(\sigma_1) \mathrm{v}_{\sigma_1/1}.$$

If the positions of Max and Min are interchanged, then literally the same argument yields

$$\text{(15:3)} \qquad \mathrm{v}_2 = \sum_{\sigma_1=1}^{\alpha_1} p_1(\sigma_1) \mathrm{v}_{\sigma_1/2}.$$

15.4.3. The case of $k_1 = 1$ comes next, and in this case we shall have to make use of the result of 13.5.3. Considering the highly formal character of this result, it seems desirable to bring it somewhat nearer to the reader's imagination by showing that it is the formal statement of an intuitively plausible fact concerning games. This will also make it clearer why this result must play a role at this particular juncture.

The interpretation which we are now going to give to the result of 13.5.3. is based on our considerations of 14.2.-14.5.—particularly those of 14.5.1., 14.5.2.—and for this reason we could not propose it in 13.5.3.

For this purpose we shall consider a zero-sum two-person game Γ in its normalized form (cf. 14.1.1.) and also its minorant and majorant games Γ_1, Γ_2 (cf. 14.2.).

If we decided to treat the normalized form of Γ as if it were an extensive form, and introduced strategies etc. with the aim of reaching a (new) normalized form by the procedure of 11.2.2., 11.2.3. then nothing would happen, as described in 11.3. and particularly in footnote 1 on p. 84. The situation is different, however, for the majorant and minorant games Γ_1, Γ_2; these are not given in the normalized form, as mentioned in footnotes 2 and 3 on p. 100. Consequently it is appropriate and necessary to bring them into their normalized forms—which we are yet to find—by the procedure of 11.2.2., 11.2.3.

Since complete solutions of Γ_1, Γ_2 were found in 14.3.1., 14.3.3., it is to be expected that they will turn out to be strictly determined.[1]

It suffices to consider Γ_1 (cf. the beginning of 14.3.4.), and this we now proceed to do.

We use the notations $\tau_1, \tau_2, \mathcal{H}(\tau_1, \tau_2)$ and v_1, v_2 for Γ and we denote the corresponding concepts for Γ_1 by $\tau_1', \tau_2', \mathcal{H}'(\tau_1', \tau_2')$ and v_1', v_2'.

A strategy of player 1 in Γ_1 consists in specifying a (fixed) value $\tau_1(= 1, \cdots, \beta_1)$ while a strategy of player 2 in Γ_1 consists in specifying a value of $\tau_2(= 1, \cdots, \beta_2)$ depending on τ_1 for every value of $\tau_1(= 1, \cdots, \beta_1)$.[2] So it is a function of τ_1: $\tau_2 = \mathfrak{I}_2(\tau_1)$.

Thus τ_1' is τ_1, while τ_2' corresponds to the functions $\mathfrak{I}_2$, and $\mathcal{H}'(\tau_1', \tau_2')$ to $\mathcal{H}(\tau_1, \mathfrak{I}_2(\tau_1))$. Accordingly

$$v_1' = \mathrm{Max}_{\tau_1} \mathrm{Min}_{\mathfrak{I}_2} \mathcal{H}(\tau_1, \mathfrak{I}_2(\tau_1)),$$
$$v_2' = \mathrm{Min}_{\mathfrak{I}_2} \mathrm{Max}_{\tau_1} \mathcal{H}(\tau_1, \mathfrak{I}_2(\tau_1)).$$

Hence the assertion that Γ_1 is strictly determined, i.e. the validity of $v_1' = v_2'$ coincides precisely with (13:E) in 13.5.3.; there we need only replace the $x, u, f(x), \psi(x, f(x))$ by our $\tau_1, \tau_2, \mathfrak{I}_2(\tau_1), \mathcal{H}(\tau_1, \mathfrak{I}_2(\tau_1))$.

This equivalence of the result of 13.5.3. to the strictly determined character of Γ_1 makes intelligible why 13.5.3. will play an essential role in the discussion which follows. Γ_1 is a very simple example of a game in which perfect information prevails,—and these are the games which are the ultimate goal of our present discussions (cf. the end of 15.3.2.). And the first move in Γ_1 is precisely of the kind which is coming up for discussion now: It is a personal move of player 1,—i.e. $k_1 = 1$.

15.5. Exact Discussion of the Inductive Step (Continuation)

15.5.1. Consider now the case $k_1 = 1$; i.e. let $\mathfrak{M}_1$ be a personal move of the player 1.

A strategy of player 1 in Γ, $\Sigma_1^{\tau_1}$ consists obviously in specifying a (fixed) value $\sigma_1^0(= 1, \cdots, \alpha_1)$ and a (fixed) strategy of player 1 in $\Gamma_{\sigma_1^0}$, $\Sigma_{\sigma_1^0/1}^{\tau_{\sigma_1^0}/1}$ [3]; i.e. the $\Sigma_1^{\tau_1}$ correspond to the pairs $\sigma_1^0, \tau_{\sigma_1^0/1}$.

[1] This is merely a heuristic argument, since the principles on which the "solutions" of 14.3.1., 14.3.3. are based are not entirely the same as those by which we disposed of the strictly determined case in 14.5.1., 14.5.2., although the former principles were a stepping stone to the latter. It is true that the argument could be made pretty convincing by an "unmathematical," purely verbal, amplification. We prefer to settle the matter mathematically, the reasons being the same as given in a similar situation in 14.3.2.

[2] This is clear intuitively. The reader may verify it from the formalistic point of view, by reformulating the definition of Γ_1 as given in 14.2. in the partition and set terminology, and then applying the definitions of 11.1.1. and (11:A) in 11.1.3.

The essential fact is, at any rate, that in Γ_1 the personal move of player 1 is preliminary to that of player 2.

[3] Cf. footnote 1 on p. 118 or footnote 2 above.

A strategy of player 2 in Γ, $\Sigma_2^{\tau_2}$, on the other hand, consists in specifying a strategy of player 2 in $\Gamma_{\sigma_1^0}$, $\Sigma_{\sigma_1^0/2}^{\tau_{\sigma_1^0}/2}$, for every value of the variable $\sigma_1^0 = 1, \cdots, \alpha_1$.[1] So $\tau_{\sigma_1^0}/2$ is a function of $\sigma_1^0 : \tau_{\sigma_1^0}/2 = \mathfrak{J}_2(\sigma_1^0)$; i.e. the $\Sigma_2^{\tau_2}$ correspond to the functions $\mathfrak{J}_2$ and clearly

$$\mathcal{K}(\tau_1, \tau_2) = \mathcal{K}_{\sigma_1^0}(\tau_{\sigma_1^0}/1, \mathfrak{J}_2(\sigma_1^0)).$$

Therefore our formula for v_1 gives:

$$\begin{aligned} v_1 &= \mathrm{Max}_{\sigma_1^0, \tau_{\sigma_1^0}/1} \mathrm{Min}_{\mathfrak{J}_2} \mathcal{K}_{\sigma_1^0}(\tau_{\sigma_1^0}/1, \mathfrak{J}_2(\sigma_1^0)) \\ &= \mathrm{Max}_{\tau_{\sigma_1^0}/1} \mathrm{Max}_{\sigma_1^0} \mathrm{Min}_{\mathfrak{J}_2} \mathcal{K}_{\sigma_1^0}(\tau_{\sigma_1^0}/1, \mathfrak{J}_2(\sigma_1^0)). \end{aligned}$$

Now

$$\mathrm{Max}_{\sigma_1^0} \mathrm{Min}_{\mathfrak{J}_2} \mathcal{K}_{\sigma_1^0}(\tau_{\sigma_1^0}/1, \mathfrak{J}_2(\sigma_1^0)) = \mathrm{Max}_{\sigma_1^0} \mathrm{Min}_{\tau_{\sigma_1^0}/2} \mathcal{K}_{\sigma_1^0}(\tau_{\sigma_1^0}/1, \tau_{\sigma_1^0}/2)$$

owing to (13:G) in 13.5.3.; there we need only replace the x, u, $f(x)$, $\psi(x, u)$ by our σ_1^0, $\tau_{\sigma_1^0}/2$, $\mathfrak{J}_2(\sigma_1^0)$, $\mathcal{K}_{\sigma_1^0}(\tau_{\sigma_1^0}/1, \tau_{\sigma_1^0}/2)$.[2] Consequently

$$\begin{aligned} v_1 &= \mathrm{Max}_{\tau_{\sigma_1^0}/1} \mathrm{Max}_{\sigma_1^0} \mathrm{Min}_{\tau_{\sigma_1^0}/2} \mathcal{K}_{\sigma_1^0}(\tau_{\sigma_1^0}/1, \tau_{\sigma_1^0}/2) \\ &= \mathrm{Max}_{\sigma_1^0} \mathrm{Max}_{\tau_{\sigma_1^0}/1} \mathrm{Min}_{\tau_{\sigma_1^0}/2} \mathcal{K}_{\sigma_1^0}(\tau_{\sigma_1^0}/1, \tau_{\sigma_1^0}/2) \\ &= \mathrm{Max}_{\sigma_1^0} v_{\sigma_1^0}/1 \end{aligned}$$

And our formula for v_2 gives:[3]

$$\begin{aligned} v_2 &= \mathrm{Min}_{\mathfrak{J}_2} \mathrm{Max}_{\sigma_1^0, \tau_{\sigma_1^0}/1} \mathcal{K}_{\sigma_1^0}(\tau_{\sigma_1^0}/1, \mathfrak{J}_2(\sigma_1^0)) \\ &= \mathrm{Min}_{\mathfrak{J}_2} \mathrm{Max}_{\sigma_1^0} \mathrm{Max}_{\tau_{\sigma_1^0}/1} \mathcal{K}_{\sigma_1^0}(\tau_{\sigma_1^0}/1, \mathfrak{J}_2(\sigma_1^0)). \end{aligned}$$

Now

$$\begin{aligned} &\mathrm{Min}_{\mathfrak{J}_2} \mathrm{Max}_{\sigma_1^0} \mathrm{Max}_{\tau_{\sigma_1^0}/1} \mathcal{K}_{\sigma_1^0}(\tau_{\sigma_1^0}/1, \mathfrak{J}_2(\sigma_1^0)) \\ &\qquad = \mathrm{Max}_{\sigma_1^0} \mathrm{Min}_{\mathfrak{J}_2} \mathrm{Max}_{\tau_{\sigma_1^0}/1} \mathcal{K}_{\sigma_1^0}(\tau_{\sigma_1^0}/1, \mathfrak{J}_2(\sigma_1^0)) \\ &\qquad\qquad = \mathrm{Max}_{\sigma_1^0} \mathrm{Min}_{\tau_{\sigma_1^0}/2} \mathrm{Max}_{\tau_{\sigma_1^0}/1} \mathcal{K}_{\sigma_1^0}(\tau_{\sigma_1^0}/1, \tau_{\sigma_1^0}/2) \end{aligned}$$

owing to (13:E) and (13:G) in 13.5.3.; there we need only replace the x, u, $f(x)$, $\psi(x, u)$ by our σ_1^0, $\tau_{\sigma_1^0}/2$, $\mathfrak{J}_2(\sigma_1^0)$, $\mathrm{Max}_{\tau_{\sigma_1^0}/1} \mathcal{K}_{\sigma_1^0}(\tau_{\sigma_1^0}/1, \tau_{\sigma_1^0}/2)$.[4] Consequently

$$\begin{aligned} v_2 &= \mathrm{Max}_{\sigma_1^0} \mathrm{Min}_{\tau_{\sigma_1^0}/2} \mathrm{Max}_{\tau_{\sigma_1^0}/1} \mathcal{K}_{\sigma_1^0}(\tau_{\sigma_1^0}/1, \tau_{\sigma_1^0}/2) \\ &= \mathrm{Max}_{\sigma_1^0} v_{\sigma_1^0}/2. \end{aligned}$$

Summing up (and writing σ_1 instead of σ_1^0):

[1] Cf. footnote 1 on p. 118 or footnote 2 on p. 120.

[2] $\tau_{\sigma_1^0}/1$ must be treated in this case as a constant.

This step is of course a rather trivial one,—cf. the argument loc. cit.

[3] In contrast to 15.4.2., there is now an essential difference between the treatments of v_1 and v_2.

[4] $\tau_{\sigma_1^0}/1$ is killed in this case by the operation $\mathrm{Max}_{\tau_{\sigma_1^0}/1}$.

This step is not trivial. It makes use of (13:E) in 13.5.3., i.e. of the essential result of that paragraph, as stated in 15.4.3.

(15:4) $$v_1 = \mathrm{Max}_{\sigma_1} v_{\sigma_1/1},$$

(15:5) $$v_2 = \mathrm{Max}_{\sigma_1} v_{\sigma_1/2}.$$

15.5.2. Consider finally the case $k_1 = 2$; i.e. let $\mathfrak{M}_1$ be a personal move of player 2.

Interchanging players 1 and 2 carries this into the preceding case ($k_1 = 1$). As discussed in 14.6., this interchange replaces v_1, v_2 by $-v_2$, $-v_1$ and hence equally $v_{\sigma_1/1}$, $v_{\sigma_1/2}$ by $-v_{\sigma_1/2}$, $-v_{\sigma_1/1}$. Substituting these changes into the above formulae (15:4), (15:5), it becomes clear that these formulae must be modified only by replacing Max in them by Min. So we have:

(15:6) $$v_1 = \mathrm{Min}_{\sigma_1} v_{\sigma_1/1},$$

(15:7) $$v_2 = \mathrm{Min}_{\sigma_1} v_{\sigma_1/2}.$$

15.5.3. We may sum up the formulae (15:2)-(15:7) of 15.4.2., 15.5.1., 15.5.2., as follows:

For all functions $f(\sigma_1)$ of the variable $\sigma_1(= 1, \cdots, \alpha_1)$ define three operations $M^{k_1}_{\sigma_1}$, $k_1 = 0, 1, 2$ as follows:

(15:8) $$M^{k_1}_{\sigma_1} f(\sigma_1) = \begin{cases} \sum_{\sigma_1=1}^{\alpha_1} p_1(\sigma_1) f(\sigma_1) & \text{for} \quad k_1 = 0, \\ \mathrm{Max}_{\sigma_1} f(\sigma_1) & \text{for} \quad k_1 = 1, \\ \mathrm{Min}_{\sigma_1} f(\sigma_1) & \text{for} \quad k_1 = 2. \end{cases}$$

Then

$$v_k = M^{k_1}_{\sigma_1} v_{\sigma_1/k} \quad \text{for} \quad k = 1, 2.$$

We wish to emphasize some simple facts concerning these operations $M^{k_1}_{\sigma_1}$.

First, $M^{k_1}_{\sigma_1}$ kills the variable σ_1; i.e. $M^{k_1}_{\sigma_1} f(\sigma_1)$ no longer depends on σ_1. For $k_1 = 1, 2$—i.e. for Max_{σ_1}, Min_{σ_1}—this was pointed out in 13.2.3. For $k_1 = 0$ it is obvious; and this operation is, by the way, analogous to the integral used as an illustration in footnote 2 on p. 91.

Second, $M^{k_1}_{\sigma_1}$ depends explicitly on the game Γ. This is evident since k_1 occurs in it and σ_1 has the range $1, \cdots, \alpha_1$. But a further dependence is due to the use of the $p_1(1), \cdots, p_1(\alpha_1)$, in the case of $k_1 = 0$.

Third, the dependence of v_k on $v_{\sigma_1/k}$ is the same for $k = 1, 2$ for each value of k_1.

We conclude by observing that it would have been easy to make these formulae—involving the average $\sum_{\sigma_1=1}^{\alpha_1} p_1(\sigma_1) f(\sigma_1)$ for a chance move, the maximum for a personal move of the first player, and the minimum for one of his opponent—plausible by a purely verbal (unmathematical) argument. It seemed nevertheless necessary to give an exact mathematical treatment in order to do full justice to the precise position of v_1 and of v_2. A purely verbal argument attempting this would unavoidably become so involved—if not obscure—as to be of little value.

15.6. The Result in the Case of Perfect Information

15.6.1. We return now to the situation described at the end of 15.3.2. and make all the hypotheses mentioned there; i.e. we assume that perfect information prevails in the game Γ and also that it is a zero-sum two-person game. The scheme indicated loc. cit., together with the formula (15:8) of 15.5.3. which takes care of the "inductive" step, enable us to determine the essential properties of Γ.

We prove first—without going any further into details—that such a Γ is always strictly determined. We do this by "complete induction" with respect to the length ν of the game (cf. 15.1.2.). This consists of proving two things:

(15:C:a) That this is true for all games of minimum length; i.e. for $\nu = 0$.

(15:C:b) That if it is true for all games of length $\nu - 1$, for a given $\nu = 1, 2, \cdots$, then it is also true for all games of length ν.

Proof of (15:C:a): If the length ν is zero, then the game has no moves at all; it consists of paying fixed amounts to the players 1, 2,—say the amounts w, $-$w.[1] Hence $\beta_1 = \beta_2 = 1$ so $\tau_1 = \tau_2 = 1$, $\mathcal{H}(\tau_1, \tau_2) = \mathrm{w}$,[2] and so

$$\mathrm{v}_1 = \mathrm{v}_2 = \mathrm{w};$$

i.e. Γ is strictly determined, and its $\mathrm{v} = \mathrm{w}$.[3]

Proof of (15:C:b): Let Γ be of length ν. Then every Γ_{σ_1} is of length $\nu - 1$; hence by assumption every Γ_{σ_1} is strictly determined. Therefore $\mathrm{v}_{\sigma_1/1} \equiv \mathrm{v}_{\sigma_1/2}$. Now the formula (15:8) of 15.5.3. shows[4] that $\mathrm{v}_1 = \mathrm{v}_2$. Hence Γ is also strictly determined and the proof is completed.

15.6.2. We shall now go more into detail and determine the $\mathrm{v}_1 = \mathrm{v}_2 = \mathrm{v}$ of Γ explicitly. For this we do not even need the above result of 15.6.1.

We form, as at the end of 15.3.2., the sequence of games

$$(15:9) \qquad \Gamma, \Gamma_{\sigma_1}, \Gamma_{\sigma_1, \sigma_2}, \cdots, \Gamma_{\sigma_1, \sigma_2, \ldots, \sigma_\nu}{}^{5}$$

of the respective lengths

$$\nu, \nu - 1, \nu - 2, \cdots, 0.$$

Denote the v_1, v_2 of these games by

$$\mathrm{v}_k, \mathrm{v}_{\sigma_1/k}, \mathrm{v}_{\sigma_1, \sigma_2/k}, \cdots, \mathrm{v}_{\sigma_1, \sigma_2, \ldots, \sigma_\nu/k}.$$

[1] Cf. the game in footnote 2 on p. 76 or $\Gamma_{\bar{\sigma}_1, \bar{\sigma}_2, \ldots, \bar{\sigma}_\nu}$ in 15.3.1. In the partition and set terminology: For $\nu = 0$ (10:1:f), (10:1:g) in 10.1.1. show that Ω has only one element, say $\bar{\pi}: \Omega = (\bar{\pi})$. So $\mathrm{w} = \mathfrak{F}_1(\bar{\pi})$, $-\mathrm{w} = \mathfrak{F}_2(\bar{\pi})$ play the role indicated above.

[2] I.e. each player has only one strategy, which consists of doing nothing.

[3] This is, of course, rather obvious. The essential step is (15:C:b).

[4] I.e. the fact mentioned at the end of 15.5.3., that the formula is the same for $k = 1, 2$ for each value of k_1.

[5] Cf. footnote 3 on p. 117.

Let us apply (15:8) of 15.5.3. for the "inductive" step described at the end of 15.3.2.; i.e. let us replace the $\sigma_1, \Gamma, \Gamma_{\sigma_1}$ of 15.5.3. by $\sigma_\kappa, \Gamma_{\sigma_1, \ldots, \sigma_{\kappa-1}}, \Gamma_{\sigma_1, \ldots, \sigma_{\kappa-1}, \sigma_\kappa}$ for each $\kappa = 1, \cdots, \nu$. The k_1 of 15.5.3. then refers to the first move of $\Gamma_{\sigma_1, \ldots, \sigma_{\kappa-1}}$; i.e. to the move $\mathfrak{M}_\kappa$ in Γ. It is therefore convenient to denote it by $k_\kappa(\sigma_1, \cdots, \sigma_{\kappa-1})$. (Cf. 7.2.1.) Accordingly we form the operation $M_{\sigma_\kappa}^{k_\kappa(\sigma_1, \ldots, \sigma_{\kappa-1})}$, replacing the $M_{\sigma_1}^{k_1}$ of 15.5.3. In this way we obtain

$$(15:10) \qquad v_{\sigma_1, \ldots, \sigma_{\kappa-1}/k} = M_{\sigma_\kappa}^{k_\kappa(\sigma_1, \ldots, \sigma_{\kappa-1})} v_{\sigma_1, \ldots, \sigma_\kappa/k} \qquad \text{for} \qquad k = 1, 2.$$

Consider now the last element of the sequence (15:9), the game $\Gamma_{\sigma_1, \ldots, \sigma_\nu}$. This falls under the discussion of (15:C:a) in 15.6.1.; it has no moves at all. Denote its unique play[1] by $\bar{\pi} = \bar{\pi}(\sigma_1, \cdots, \sigma_\nu)$. Hence its fixed w [2] is equal to $\mathfrak{F}_1(\bar{\pi}(\sigma_1, \cdots, \sigma_\nu))$. So we have:

$$(15:11) \qquad v_{\sigma_1, \ldots, \sigma_\nu/1} = v_{\sigma_1, \ldots, \sigma_\nu/2} = \mathfrak{F}_1(\bar{\pi}(\sigma_1, \cdots, \sigma_\nu)).$$

Now apply (15:10) with $\kappa = \nu$ to (15:11) and then to the result, with $\kappa = \nu - 1, \cdots, 2, 1$ successively. In this manner

$$(15:12) \quad v_1 = v_2 = v = M_{\sigma_1}^{k_1} M_{\sigma_2}^{k_2(\sigma_1)} \cdots M_{\sigma_\nu}^{k_\nu(\sigma_1, \ldots, \sigma_{\nu-1})} \mathfrak{F}_1(\bar{\pi}(\sigma_1, \cdots, \sigma_\nu)).$$

obtains.

This proves once more that Γ is strictly determined, and also gives an explicit formula for its value.

15.7. Application to Chess

15.7.1. The allusions of 6.4.1. and the assertions of 14.8. concerning those zero-sum two-person games in which preliminarity and anteriority coincide—i.e. where perfect information prevails—are now established. We referred there to the general opinion that these games are of a particularly rational character, and we have now given this vague view a precise meaning by showing that the games in question are strictly determined. And we have also shown—a fact much less founded on any "general opinion"—that this is also true when the game contains chance moves.

Examples of games with perfect information were already given in 6.4.1.: Chess (without chance moves) and Backgammon (with chance moves). Thus we have established for all these games the existence of a definite value (of a play) and of definite best strategies. But we have established their existence only in the abstract, while our method for their construction is in most cases too lengthy for effective use.[3]

In this connection it is worth while to consider Chess in a little more detail.

[1] Cf. the remarks concerning $\Gamma_{\bar{\sigma}_1, \ldots, \bar{\sigma}_\nu}$ in 15.3.1.

[2] Cf. (15:C:a) in 15.6.1., particularly footnote 1 on p. 123.

[3] This is due mainly to the enormous value of ν. For Chess, cf. the pertinent part of footnote 3 on p. 59. (The ν^* there is our ν, cf. the end of 7.2.3.)

The outcome of a play in Chess—i.e. every value of the functions $\mathfrak{F}_k$ of 6.2.2. or 9.2.4.—is restricted to the numbers 1, 0, -1.[1] Thus the functions $\mathfrak{G}_k$ of 11.2.2. have the same values, and since there are no chance moves in Chess, the same is true for the function $\mathcal{H}_k$ of 11.2.3.[2] In what follows we shall use the function $\mathcal{H} = \mathcal{H}_1$ of 14.1.1.

Since $\mathcal{H}$ has only the values, 1, 0, -1, the number

$$\text{(15:13)} \qquad v = \text{Max}_{\tau_1} \text{Min}_{\tau_2} \mathcal{H}(\tau_1, \tau_2) = \text{Min}_{\tau_2} \text{Max}_{\tau_1} \mathcal{H}(\tau_1, \tau_2)$$

has necessarily one of these values

$$v = 1, 0, -1.$$

We leave to the reader the simple discussion that (15:13) means this:

(15:D:a) If $v = 1$ then player 1 ("white") possesses a strategy with which he "wins," irrespective of what player 2 ("black") does.

(15:D:b) If $v = 0$ then both players possess a strategy with which each one can "tie" (and possibly "win"), irrespective of what the other player does.

(15:D:c) If $v = -1$ then player 2 ("black") possesses a strategy with which he "wins," irrespective of what player 1 ("white") does.[3]

15.7.2. This shows that if the theory of Chess were really fully known there would be nothing left to play. The theory would show which of the three possibilities (15:D:a), (15:D:b), (15:D:c) actually holds, and accordingly the play would be decided before it starts: The decision would be in case (15:D:a) for "white," in case (15:D:b) for a "tie," in case (15:D:c) for "black."

But our proof, which guarantees the validity of one (and only one) of these three alternatives, gives no practically usable method to determine the true one. This relative, human difficulty necessitates the use of those incomplete, heuristic methods of playing, which constitute "good" Chess; and without it there would be no element of "struggle" and "surprise" in that game.

[1] This is the simplest way to interpret a "win," "tie," or "loss" of a play by the player k.

[2] Every value of $\mathfrak{G}_k$ is one of $\mathfrak{F}_k$; every value of $\mathcal{H}_k$—in the absence of chance moves—is one of $\mathfrak{G}_k$, cf. loc. cit. If there were chance moves, then the value of $\mathcal{H}_k$ would be the probability of a "win" minus that of a "loss," i.e. a number which may lie anywhere between 1 and -1.

[3] When there are chance moves, then $\mathcal{H}(\tau_1, \tau_2)$ is the excess probability of a "win" over a "loss," cf. footnote 2 above. The players try to maximize or to minimize this number, and the sharp trichotomy of (15:D:a)—(15:D:c) above does not, in general, obtain.

Although Backgammon is a game in which complete information prevails, and which contains chance moves, it is not a good example for the above possibility; Backgammon is played for varying payments, and not for simple "win," "tie" or "loss,"—i.e. the values of the $\mathfrak{F}_k$ are not restricted to the numbers 1, 0, -1.

15.8. The Alternative, Verbal Discussion

15.8.1. We conclude this chapter by an alternative, simpler, less formalistic approach to our main result,—that all zero-sum two-person games, in which perfect information prevails, are strictly determined.

It can be questioned whether the argumentation which follows is really a proof; i.e., we prefer to formulate it as a plausibility argument by which a value can be ascribed to each play of any game Γ of the above type, but this is still open to criticism. It is not necessary to show in detail how those criticisms can be invalidated, since we obtain the same value v of a play of Γ as in 15.4.-15.6., and there we gave an absolutely rigorous proof using precisely defined concepts. The value of the present plausibility argument is that it is easier to grasp and that it may be repeated for other games, in which perfect information prevails, which are not subject to the zero-sum two-person restriction. The point which we wish to bring out is that the same criticisms apply in the general case too, and that they can no longer be invalidated there. Indeed, the solution there will be found (even in games where perfect information prevails) along entirely different lines. This will make clearer the nature of the difference between the zero-sum two-person case and the general case. That will be rather important for the justification of the fundamentally different methods which will have to be used for the treatment of the general case (cf. 24.).

15.8.2. Consider a zero-sum two-person game Γ in which perfect information prevails. We use the notations of 15.6.2. in all respects: For the $\mathfrak{M}_1, \mathfrak{M}_2, \cdots, \mathfrak{M}_\nu$; the $\sigma_1, \sigma_2, \cdots, \sigma_\nu$; the $k_1, k_2(\sigma_1), \cdots, k_\nu(\sigma_1, \sigma_2, \cdots, \sigma_{\nu-1})$; the probabilities; the operators $M^{k_1}_{\sigma_1}, M^{k_2(\sigma_1)}_{\sigma_2}, \cdots, M^{k_\nu(\sigma_1, \sigma_2, \cdots, \sigma_{\nu-1})}_{\sigma_\nu}$; the sequence (15:9) of games derived from Γ; and the function $\mathfrak{F}_1(\bar{\pi}(\sigma_1, \cdots, \sigma_\nu))$.

We proceed to discuss the game Γ by starting with the last move $\mathfrak{M}_\nu$ and then going backward from there through the moves $\mathfrak{M}_{\nu-1}, \mathfrak{M}_{\nu-2}, \cdots$. Assume first that the choices $\sigma_1, \sigma_2, \cdots, \sigma_{\nu-1}$ (of the moves $\mathfrak{M}_1, \mathfrak{M}_2, \cdots, \mathfrak{M}_{\nu-1}$) have already been made, and that the choice σ_ν (of the move $\mathfrak{M}_\nu$) is now to be made.

If $\mathfrak{M}_\nu$ is a chance move, i.e. if $k_\nu(\sigma_1, \sigma_2, \cdots, \sigma_{\nu-1}) = 0$, then σ_ν will have the values $1, 2, \cdots, \alpha_\nu(\sigma_1, \cdots, \sigma_{\nu-1})$ with the respective probabilities $p_\nu(1), p_\nu(2), \cdots, p_\nu(\alpha_\nu(\sigma_1, \cdots, \sigma_{\nu-1}))$. So the mathematical expectation of the final payment (for the player 1) $\mathfrak{F}_1(\bar{\pi}(\sigma_1, \cdots, \sigma_{\nu-1}, \sigma_\nu))$ is

$$\sum_{\sigma_\nu = 1}^{\alpha_\nu(\sigma_1, \cdots, \sigma_{\nu-1})} p_\nu(\sigma_\nu)\mathfrak{F}_1(\bar{\pi}(\sigma_1, \cdots, \sigma_{\nu-1}, \sigma_\nu)).$$

If $\mathfrak{M}_\nu$ is a personal move of players 1 or 2, i.e. if $k_\nu(\sigma_1, \cdots, \sigma_{\nu-1}) = 1$ or 2, then that player can be expected to maximize or to minimize $\mathfrak{F}_1(\bar{\pi}(\sigma_1, \cdots, \sigma_{\nu-1}, \sigma_\nu))$ by his choice of σ_ν; i.e. the outcome $\text{Max}_{\sigma_\nu} \mathfrak{F}_1(\bar{\pi}(\sigma_1, \cdots, \sigma_{\nu-1}, \sigma_\nu))$ or $\text{Min}_{\sigma_\nu} \mathfrak{F}_1(\bar{\pi}(\sigma_1, \cdots, \sigma_{\nu-1}, \sigma_\nu))$, respectively is to be expected.

I.e., the outcome to be expected for the play—after the choices $\sigma_1, \cdots, \sigma_{\nu-1}$ have been made—is at any rate

$$M_{\sigma_\nu}^{k_\nu(\sigma_1, \cdots, \sigma_{\nu-1})} \mathfrak{F}_1(\pi(\sigma_1, \cdots, \sigma_\nu)).$$

Assume next that only the choices $\sigma_1, \cdots, \sigma_{\nu-2}$ (of the moves $\mathfrak{M}_1, \cdots, \mathfrak{M}_{\nu-2}$) have been made and that the choice $\sigma_{\nu-1}$ (of the move $\mathfrak{M}_{\nu-1}$) is now to be made.

Since a definite choice of $\sigma_{\nu-1}$ entails, as we have seen, the outcome $M_{\sigma_\nu}^{k_\nu(\sigma_1, \cdots, \sigma_{\nu-1})} \mathfrak{F}_1(\pi(\sigma_1, \cdots, \sigma_\nu))$—which is a function of $\sigma_1, \cdots, \sigma_{\nu-1}$ only, since the operation $M_{\sigma_\nu}^{k_\nu(\sigma_1, \cdots, \sigma_{\nu-1})}$ kills σ_ν—we can proceed as above. We need only replace $\nu; \sigma_1, \cdots, \sigma_\nu; M_{\sigma_\nu}^{k_\nu(\sigma_1, \cdots, \sigma_{\nu-1})} \mathfrak{F}_1(\pi(\sigma_1, \cdots, \sigma_\nu))$ by $\nu - 1; \sigma_1, \cdots, \sigma_{\nu-1}; M_{\sigma_{\nu-1}}^{k_{\nu-1}(\sigma_1, \cdots, \sigma_{\nu-2})} M_{\sigma_\nu}^{k_\nu(\sigma_1, \cdots, \sigma_{\nu-1})} \mathfrak{F}_1(\pi(\sigma_1, \cdots, \sigma_\nu))$. Consequently the outcome to be expected for the play—after the choices $\sigma_1, \cdots, \sigma_{\nu-2}$ have been made—is

$$M_{\sigma_{\nu-1}}^{k_{\nu-1}(\sigma_1, \cdots, \sigma_{\nu-2})} M_{\sigma_\nu}^{k_\nu(\sigma_1, \cdots, \sigma_{\nu-1})} \mathfrak{F}_1(\pi(\sigma_1, \cdots, \sigma_\nu)).$$

Similarly the outcome to be expected for the play—after the choices $\sigma_1, \cdots, \sigma_{\nu-3}$ have been made—is

$$M_{\sigma_{\nu-2}}^{k_{\nu-2}(\sigma_1, \cdots, \sigma_{\nu-3})} M_{\sigma_{\nu-1}}^{k_{\nu-1}(\sigma_1, \cdots, \sigma_{\nu-2})} M_{\sigma_\nu}^{k_\nu(\sigma_1, \cdots, \sigma_{\nu-1})} \mathfrak{F}_1(\pi(\sigma_1, \cdots, \sigma_\nu)).$$

Finally, the outcome to be expected for the play outright—before it has begun—is

$$M_{\sigma_1}^{k_1} M_{\sigma_2}^{k_2(\sigma_1)} \cdots M_{\sigma_{\nu-1}}^{k_{\nu-1}(\sigma_1, \cdots, \sigma_{\nu-2})} M_{\sigma_\nu}^{k_\nu(\sigma_1, \cdots, \sigma_{\nu-1})} \mathfrak{F}_1(\pi(\sigma_1, \cdots, \sigma_\nu)).$$

And this is precisely the v of (15:12) in 15.6.2.[1]

15.8.3. The objection against the procedure of 15.8.2. is that this approach to the "value" of a play of Γ presupposes "rational" behavior of all players; i.e. player 1's strategy is based upon the assumption that player 2's strategy is optimal and *vice-versa.*

Specifically: Assume $k_{\nu-1}(\sigma_1, \cdots, \sigma_{\nu-2}) = 1$, $k_\nu(\sigma_1, \cdots, \sigma_{\nu-1}) = 2$. Then player 1, whose personal move is $\mathfrak{M}_{\nu-1}$ chooses his $\sigma_{\nu-1}$ in the conviction that player 2, whose personal move is $\mathfrak{M}_\nu$ chooses his σ_ν "rationally." Indeed, this is his sole excuse for assuming that his choice of $\sigma_{\nu-1}$ entails the outcome $\text{Min}_{\sigma_\nu} \mathfrak{F}_1(\pi(\sigma_1, \cdots, \sigma_\nu))$, i.e. $M_{\sigma_\nu}^{k_\nu(\sigma_1, \cdots, \sigma_{\nu-1})} \mathfrak{F}_1(\pi(\sigma_1, \cdots, \sigma_\nu))$, of the play. (Cf. the discussion of $\mathfrak{M}_{\nu-1}$ in 15.8.2.)

[1] In imagining the application of this procedure to any specific game it must be remembered that we assume the length ν of Γ to be fixed. If ν is actually variable—and it is so in most games (cf. footnote 3 on p. 58)—then we must first make it constant, by the device of adding "dummy moves" to Γ as described at the end of 7.2.3. It is only after this has been done that the above regression through $\mathfrak{M}_\nu, \mathfrak{M}_{\nu-1}, \cdots, \mathfrak{M}_1$ becomes feasible.

For practical construction this procedure is of course no better than that of 15.4.-15.6.

Possibly some very simple games, like Tit-tat-toe, could be effectively treated in either manner.

Now in the second part of 4.1.2. we came to the conclusion that the hypothesis of "rationality" in others must be avoided. The argumentation of 15.8.2. did not meet this requirement.

It is possible to argue that in a zero-sum two-person game the rationality of the opponent can be assumed, because the irrationality of his opponent can never harm a player. Indeed, since there are only two players and since the sum is zero, every loss which the opponent—irrationally—inflicts upon himself, necessarily causes an equal gain to the other player.[1] As it stands, this argument is far from complete, but it could be elaborated considerably. However, we do not need to be concerned with its stringency: We have the proof of 15.4.-15.6. which is not open to these criticisms.[2]

But the above discussion is probably nevertheless significant for an essential aspect of this matter. We shall see how it affects the modified conditions in the more general case—not subject to the zero-sum two-person restriction—referred to at the end of 15.8.1.

16. Linearity and Convexity

16.1. Geometrical Background

16.1.1. The task which confronts us next is that of finding a solution which comprises all zero-sum two-person games,—i.e. which meets the difficulties of the non-strictly determined case. We shall succeed in doing this with the help of the same ideas with which we mastered the strictly determined case: It will appear that they can be extended so as to cover all zero-sum two-person games. In order to do this we shall have to make use of certain possibilities of probability theory (cf. 17.1., 17.2.). And it will be necessary to use some mathematical devices which are not quite the usual ones. Our analysis of 13. provides one part of the tools; for the remainder it will be most convenient to fall back on the mathematico-geometrical theory of *linearity* and *convexity*. Two theorems on convex bodies[3] will be particularly significant.

For these reasons we are now going to discuss—to the extent to which they are needed—the concepts of linearity and convexity.

16.1.2. It is not necessary for us to analyze in a fundamental way the notion of n-dimensional linear (Euclidean) space. All we need to say is that this space is described by n numerical coordinates. Accordingly we define for each $n = 1, 2, \cdots$, the *n-dimensional linear space* L_n as the set of all n-uplets of real numbers $\{x_1, \cdots, x_n\}$. These n-uplets can also be looked upon as functions x_i of the variable i, with the domain $(1, \cdots, n)$

[1] This is not necessarily true if the sum is not constantly zero, or if there are more than two players. For details cf. 20.1., 24.2.2., 58.3.

[2] Cf. in this respect particularly (14:D:a), (14:D:b), (14:C:d), (14:C:e) in 14.5.1. and (14:C:a), (14:C:b) in 14.5.2.

[3] Cf. *T. Bonessen* and *W. Fenchel:* Theorie der konvexen Körper, in Ergebnisse der Mathematik und ihrer Grenzgebiete, Vol. III/1, Berlin 1934. Further investigations in *H. Weyl:* Elementare Theorie der konvexen Polyeder. Commentarii Mathematici Helvetici, Vol. VII, 1935, pp. 290–306.

in the sense of 13.1.2., 13.1.3.[1] We shall—in conformity with general usage—call i an index and not a variable; but this does not alter the nature of the case. In particular we have

$$\{x_1, \cdots, x_n\} = \{y_1, \cdots, y_n\}$$

if and only if $x_i = y_i$ for all $i = 1, \cdots, n$ (cf. the end of 13.1.3.). One could even take the view that L_n is the simplest possible space of (numerical) functions, where the domain is a fixed finite set—the set $(1, \cdots, n)$.[2]

We shall also call these n-uplets—or functions—of L_n *points* or *vectors* of L_n and write

$$\overrightarrow{x} = \{x_1, \cdots, x_n\}. \tag{16:1}$$

The x_i for the specific $i = 1, \cdots, n$—the values of the function x_i—are the *components* of the vector $\overrightarrow{x}$.

16.1.3. We mention—although this is not essential for our further work—that L_n is not an *abstract Euclidean space*, but one in which a *frame of reference* (system of coordinates) has already been chosen.[3] This is due to the possibility of specifying the origin and the coordinate vectors of L_n numerically (cf. below)—but we do not propose to dwell upon this aspect of the matter.

The *zero vector* or *origin* of L_n is

$$\overrightarrow{0} = \{0, \cdots, 0\}.$$

The n *coordinate vectors* of L_n are the

$$\overrightarrow{\delta^j} = \{0, \cdots, 1, \cdots, 0\} = \{\delta_{1j}, \cdots, \delta_{nj}\} \qquad j = 1, \cdots, n,$$

where

$$\delta_{ij} = \begin{cases} 1 & \text{for} \quad i = j, \\ 0 & \text{for} \quad i \neq j. \end{cases}$$ [4,5]

After these preliminaries we can now describe the fundamental operations and properties of vectors in L_n.

16.2. Vector Operations

16.2.1. The main operations involving vectors are those of *scalar multiplication*, i.e. the multiplication of a vector $\overrightarrow{x}$ by a number t, and of

[1] I.e. the n-uplets $\{x_1, \cdots, x_n\}$ are not merely sets in the sense of 8.2.1. The effective enumeration of the x_i by means of the index $i = 1, \cdots, n$ is just as essential as the aggregate of their values. Cf. the similar situation in footnote 4 on p. 69.

[2] Much in modern analysis tends to corroborate this attitude.

[3] This at least is the orthodox geometrical standpoint.

[4] Thus the zero vector has all components 0, while the coordinate vectors have all components but one 0—that one component being 1, and its index j for the j-th coordinate vector.

[5] δ_{ij} is the "symbol of Kronecker and Weierstrass," which is quite useful in many respects.

vector addition, i.e. addition of two vectors. The two operations are defined by the corresponding operations, i.e. multiplication and addition, on the components of the vector in question. More precisely:

Scalar multiplication: $t\{x_1, \cdots, x_n\} = \{tx_1, \cdots, tx_n\}$.
Vector addition:

$$\{x_1, \cdots, x_n\} + \{y_1, \cdots, y_n\} = \{x_1 + y_1, \cdots, x_n + y_n\}.$$

The algebra of these operations is so simple and obvious that we forego its discussion. We note, however, that they permit the expression of any vector $\vec{x} = \{x_1, \cdots, x_n\}$ with the help of its components and the coordinate vectors of L_n

$$\vec{x} = \sum_{j=1}^{n} x_j \vec{\delta}^{\,j}.$$[1]

Some important subsets of L_n:

(16:A:a) Consider a (linear, inhomogeneous) equation

$$\sum_{i=1}^{n} a_i x_i = b \tag{16:2:a}$$

($a_1, \cdots, a_n, b$ are constants). We exclude

$$a_1 = \cdots = a_n = 0$$

since in that case there would be no equation at all. All points (vectors) $\vec{x} = \{x_1, \cdots, x_n\}$ which fulfill this equation, form a *hyperplane*.[2]

(16:A:b) Given a hyperplane

$$\sum_{i=1}^{n} a_i x_i = b, \tag{16:2:a}$$

it defines two parts of L_n. It cuts L_n into these two parts:

$$\sum_{i=1}^{n} a_i x_i > b, \tag{16:2:b}$$

and

$$\sum_{i=1}^{n} a_i x_i < b. \tag{16:2:c}$$

These are the two *half-spaces* produced by the hyperplane.

[1] The x_j are numbers, and hence they act in $x_j \vec{\delta}^{\,j}$ as scalar multipliers. $\sum_{j=1}^{n}$ is a vector summation.

[2] For $n = 3$, i.e. in ordinary (3-dimensional Euclidean) space, these are just the ordinary (2-dimensional) planes. In our general case they are the (($n - 1$)-dimensional) analogues; hence the name.

Observe that if we replace $a_1, \cdots, a_n, b$ by $-a_1, \cdots, -a_n, -b$, then the hyperplane (16:2:a) remains unaffected, but the two half-spaces (16:2:b), (16:2:c) are interchanged. Hence we may always assume a half space to be given in the form (16:2:b).

(16:A:c) Given two points (vectors) $\vec{x}, \vec{y}$ and a $t \geqq 0$ with $1 - t \geqq 0$; then the *center of gravity* of $\vec{x}, \vec{y}$ with the respective weights $t, 1 - t$—in the sense of mechanics— is $t\vec{x} + (1 - t)\vec{y}$.

The equations

$$\vec{x} = \{x_1, \cdots, x_n\}, \qquad \vec{y} = \{y_1, \cdots, y_n\},$$
$$t\vec{x} + (1 - t)\vec{y} = \{tx_1 + (1 - t)y_1, \cdots, tx_n + (1 - t)y_n\}$$

should make this amply clear.

A subset, C, of L_n which contains all centers of gravity of all its points—i.e. which contains with $\vec{x}, \vec{y}$ all $t\vec{x} + (1 - t)\vec{y}$, $0 \leqq t \leqq 1$—is *convex*.

The reader will note that for $n = 2$, 3—i.e. in the ordinary plane or space—this is the customary concept of convexity. Indeed, the set of all points $t\vec{x} + (1 - t)\vec{y}$, $0 \leqq t \leqq 1$ is precisely the linear (straight) interval connecting the points $\vec{x}$ and $\vec{y}$, the *interval* $[\vec{x}, \vec{y}]$. And so a convex set is one which, with any two of its points $\vec{x}$, $\vec{y}$, also contains their interval $[\vec{x}, \vec{y}]$. Figure 16 shows the conditions for $n = 2$, i.e. in the plane.

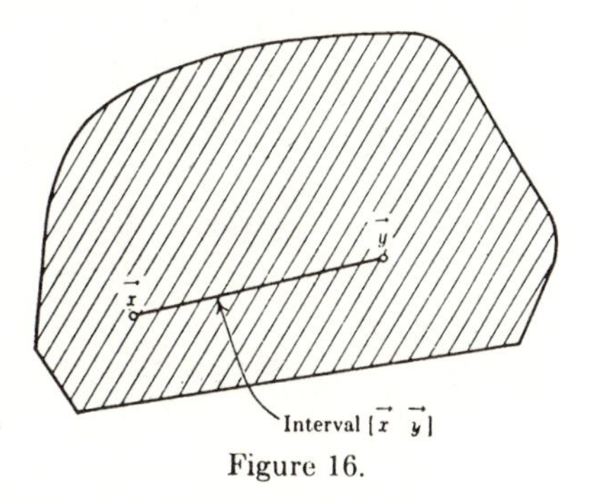

Figure 16.

16.2.2. Clearly the intersection of any number of convex sets is again convex. Hence if any number of points (vectors) $\vec{x}^{\,\prime}, \cdots, \vec{x}^{\,p}$ is given, there exists a smallest convex set containing them all: the intersection of all convex sets which contain $\vec{x}^{\,\prime}, \cdots, \vec{x}^{\,p}$. This is the convex set *spanned* by $\vec{x}^{\,\prime}, \cdots, \vec{x}^{\,p}$. It is again useful to visualize the case $n = 2$ (plane). Cf. Fig. 17, where $p = 6$. It is easy to verify that this set consists of all points (vectors)

(16:2:d) $$\sum_{j=1}^{p} t_j \vec{x}^{\,j} \text{ for all } t_1 \geqq 0, \cdots, t_p \geqq 0 \quad \text{with} \quad \sum_{j=1}^{p} t_j = 1.$$

Proof: The points (16:2:d) form a set containing all $\vec{x}^{\,\prime}, \cdots, \vec{x}^{\,p}$. $\vec{x}^{\,i}$ is such a point: put $t_j = 1$ and all other $t_i = 0$.

The points (16:2:d) form a convex set: If $\vec{x} = \sum_{j=1}^{p} t_j \vec{x}^{\,j}$ and $\vec{y} = \sum_{j=1}^{p} s_j \vec{x}^{\,j}$, then $t\vec{x} + (1-t)\vec{y} = \sum_{j=1}^{p} u_j \vec{x}^{\,j}$ with $u_j = t t_j + (1-t)s_j$.

Any convex set, D, containing $\vec{x}^{\,\prime}, \cdots, \vec{x}^{\,p}$ contains also all points of (16:2:d): We prove this by induction for all $p = 1, 2, \cdots$.

Proof: For $p = 1$ it is obvious; since then $t_1 = 1$ and so $\vec{x}^{\,\prime}$ is the only point of (16:2:d).

Assume that it is true for $p - 1$. Consider p itself. If $\sum_{j=1}^{p-1} t_j = 0$ then $t_1 = \cdots = t_{p-1} = 0$, the point of (16:2:d) is $\vec{x}^{\,p}$ and thus belongs to D. If

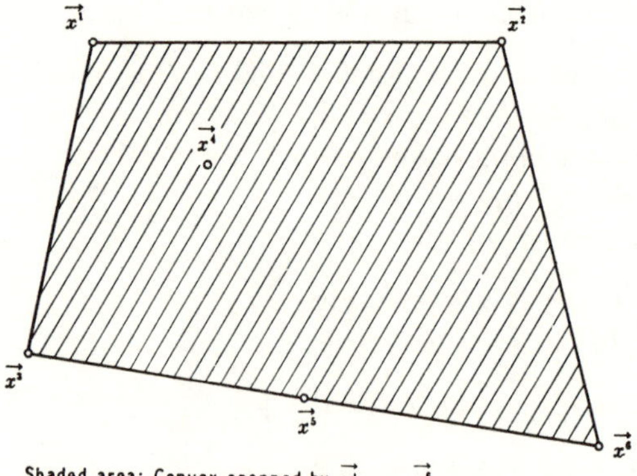

Shaded area: Convex spanned by $\vec{x}^{\,1}, \ldots, \vec{x}^{\,6}$

Figure 17.

$\sum_{j=1}^{p-1} t_j > 0$, then put $t = \sum_{j=1}^{p-1} t_j$, so $1 - t = \sum_{j=1}^{p} t_j - \sum_{j=1}^{p-1} t_j = t_p$. Hence $0 < t \leqq 1$.

Put $s_j = t_j/t$ for $j = 1, \cdots, p-1$. So $\sum_{j=1}^{p-1} s_j = 1$. Hence, by our assumption for $p - 1$, $\sum_{j=1}^{p-1} s_j \vec{x}^{\,j}$ is in D. D is convex, hence

$$t \sum_{j=1}^{p-1} s_j \vec{x}^{\,j} + (1-t)\vec{x}^{\,p}$$

is also in D; but this vector is equal to

$$\sum_{j=1}^{p-1} t_j \vec{x}^{\,j} + t_p \vec{x}^{\,p} = \sum_{j=1}^{p} t_j \vec{x}^{\,j}$$

which thus belongs to D.

The proof is therefore completed.

The $t_1, \cdots, t_p$ of (16:2:d) may themselves be viewed as the components of a vector $\overrightarrow{t} = \{t_1, \cdots, t_p\}$ in L_p. It is therefore appropriate to give a name to the set to which they are restricted, defined by

$$t_1 \geqq 0, \qquad \cdots, \qquad t_p \geqq 0,$$

and

$$\sum_{j=1}^{p} t_j = 1.$$

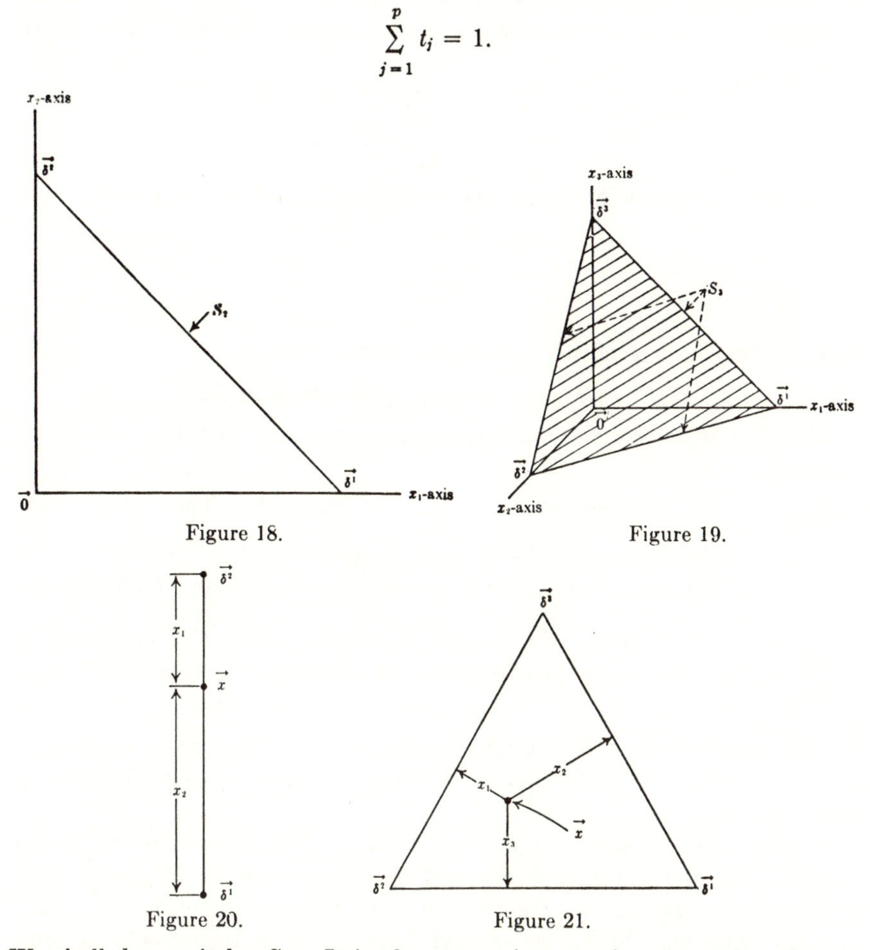

Figure 18.

Figure 19.

Figure 20.

Figure 21.

We shall denote it by S_p. It is also convenient to give a name to the set which is described by the first line of conditions above alone, i.e. by $t_1 \geqq 0, \cdots, t_p \geqq 0$. We shall denote it by P_p. Both sets S_p, P_p are convex.

Let us picture the cases $p = 2$ (plane) and $p = 3$ (space). P_2 is the *positive quadrant*, the area between the positive x_1 and x_2 axes (Figure 18). P_3 is the *positive octant*, the space between the positive x_1, x_2 and x_3 axes,—i.e. between the plane quadrants limited by the pairs x_1, x_2; x_1, x_3; x_2, x_3 of these (Fig. 19). S_2 is a linear interval crossing P_2 (Figure 18). S_3 is a plane triangle, likewise crossing P_3 (Figure 19). It is useful to draw S_2, S_3 separately, without the P_2, P_3 (or even the L_2, L_3) into which they are naturally immersed (Figures 20, 21). We have indicated on these figures those distances which are proportional to x_1, x_2 or x_1, x_2, x_3, respectively.

(We re-emphasize: The distances marked x_1, x_2, x_3 in Figures 20, 21 are not the coordinates x_1, x_2, x_3 themselves. These lie in L_2 or L_3 outside of S_2 or S_3, and therefore cannot be pictured in S_2 or S_3; but they are easily seen to be proportional to those coordinates.)

16.2.3. Another important notion is the *length* of a vector. The length of $\overrightarrow{x} = \{x_1, \cdots, x_n\}$ is

$$|\overrightarrow{x}| = \sqrt{\sum_{i=1}^{n} x_i^2}.$$

The *distance* of two points (vectors) is the length of their difference:

$$|\overrightarrow{x} - \overrightarrow{y}| = \sqrt{\sum_{i=1}^{n} (x_i - y_i)^2}.$$

Thus the length of $\overrightarrow{x}$ is the distance from the origin $\overrightarrow{0}$.[1]

16.3. The Theorem of the Supporting Hyperplanes

16.3. We shall now establish an important general property of convex sets:

(16:B) Let p vectors $\overrightarrow{x}^1, \cdots, \overrightarrow{x}^p$ be given. Then a vector $\overrightarrow{y}$ either belongs to the convex C spanned by $\overrightarrow{x}', \cdots, \overrightarrow{x}^p$ (cf. (16:A:c) in 16.2.1.), or there exists a hyperplane which contains $\overrightarrow{y}$ (cf. (16:2:a) in 16.2.1.) such that all of C is contained in one half-space produced by that hyperplane (say (16:2:b) in 16.2.1.; cf. (16:A:b) id.).

This is true even if the convex spanned by $\overrightarrow{x}', \cdots, \overrightarrow{x}^p$ is replaced by any convex set. In this form it is a fundamental tool in the modern theory of convex sets.

A picture in the case $n = 2$ (plane) follows: Figure 22 uses the convex set C of Figure 17 (which is spanned by a finite number of points, as in the assertion above), while Figure 23 shows a general convex set C.[2]

Before proving (16:B), we observe that the second alternative clearly excludes the first, since $\overrightarrow{y}$ belongs to the hyperplane, hence not to the half space. (I.e. it fulfills (16:2:a) and not (16:2:b) in (16:A:b) above.)

We now give the proof:

Proof: Assume that $\overrightarrow{y}$ does not belong to C. Then consider a point of C which lies as near to $\overrightarrow{y}$ as possible,—i.e. for which

[1] The Euclidean—Pythagorean—meaning of these notions is immediate.

[2] For the reader who is familiar with topology, we add: To be exact, this sentence should be qualified—the statement is meant for closed convex sets. This guarantees the existence of the minimum that we use in the proof that follows. Regarding these concepts, cf. footnote[1] on page 384.

$$|\vec{z} - \vec{y}|^2 = \sum_{i=1}^{n} (z_i - y_i)^2$$

assumes its minimum value.

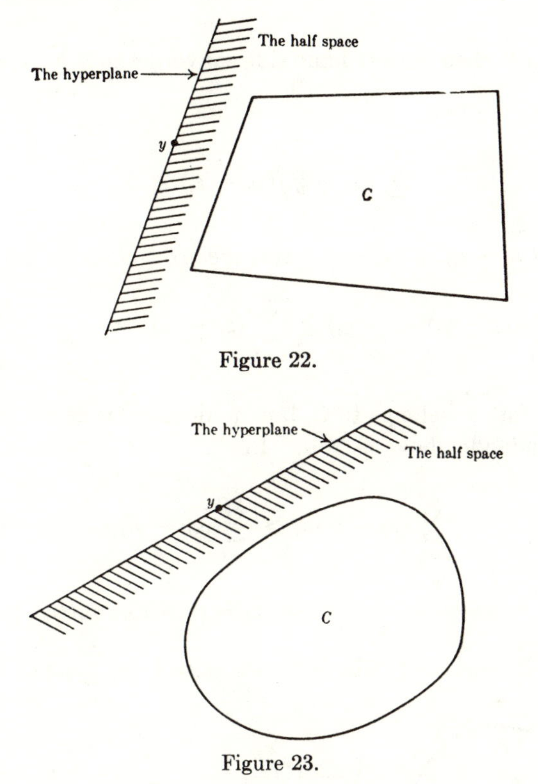

Figure 22.

Figure 23.

Consider any other points $\vec{u}$ of C. Then for every t with $0 \leqq t \leqq 1$, $t\vec{u} + (1-t)\vec{z}$ also belongs to the convex C. By virtue of the minimum property of $\vec{z}$ (cf. above) this necessitates

$$|t\vec{u} + (1-t)\vec{z} - \vec{y}|^2 \geqq |\vec{z} - \vec{y}|^2,$$

i.e.

$$|(\vec{z} - \vec{y}) + t(\vec{u} - \vec{z})|^2 \geqq |\vec{z} - \vec{y}|^2,$$

i.e.

$$\sum_{i=1}^{n} \{(z_i - y_i) + t(u_i - z_i)\}^2 \geqq \sum_{i=1}^{n} (z_i - y_i)^2.$$

By elementary algebra this means

$$2 \sum_{i=1}^{n} (z_i - y_i)(u_i - z_i)t + \sum_{i=1}^{n} (u_i - z_i)^2 t^2 \geqq 0.$$

So for $t > 0$ (but of course $t \leqq 1$) even

$$2 \sum_{i=1}^{n} (z_i - y_i)(u_i - z_i) + \sum_{i=1}^{n} (u_i - z_i)^2 t \geqq 0.$$

If t converges to 0, then the left-hand side converges to $2 \sum_{i=1}^{n} (z_i - y_i)(u_i - z_i)$. Hence

$$(16{:}3) \qquad \sum_{i=1}^{n} (z_i - y_i)(u_i - z_i) \geqq 0.$$

As $u_i - y_i = (u_i - z_i) + (z_i - y_i)$, this means

$$\sum_{i=1}^{n} (z_i - y_i)(u_i - y_i) \geqq \sum_{i=1}^{n} (z_i - y_i)^2 = |\overrightarrow{z} - \overrightarrow{y}|^2.$$

Now $\overrightarrow{z} \neq \overrightarrow{y}$ (as $\overrightarrow{z}$ belongs to C, but $\overrightarrow{y}$ does not); hence $|\overrightarrow{z} - \overrightarrow{y}|^2 > 0$. So the left-hand side above is > 0. I.e.

$$(16{:}4) \qquad \sum_{i=1}^{n} (z_i - y_i)u_i > \sum_{i=1}^{n} (z_i - y_i)y_i.$$

Put $a_i = z_i - y_i$ then $a_1 = \cdots = a_n = 0$ is excluded by $\overrightarrow{z} \neq \overrightarrow{y}$ (cf. above). Put also $b = \sum_{i=1}^{n} a_i y_i$. Thus

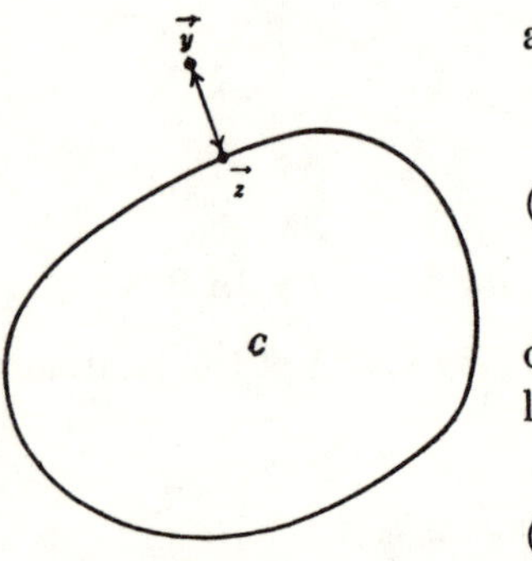

Figure 24.

$$(16{:}2{:}a^*) \qquad \sum_{i=1}^{n} a_i x_i = b$$

defines a hyperplane, to which $\overrightarrow{y}$ clearly belongs. Next

$$(16{:}2{:}b^*) \qquad \sum_{i=1}^{n} a_i x_i > b$$

is a half space produced by this hyperplane, and (16:4) states precisely that $\overrightarrow{u}$ belongs to this half space.

Since $\overrightarrow{u}$ was an arbitrary element of C this completes the proof.

This algebraic proof can also be stated in the geometrical language. Let us do this for the case $n = 2$ (plane) first. The situation is pictured in Figure 24: $\overrightarrow{z}$ is a point of C which is as near to the given point $\overrightarrow{y}$ as possible; i.e. for which the distance of $\overrightarrow{y}$ and $\overrightarrow{z}$, $|\overrightarrow{z} - \overrightarrow{y}|$ assumes its minimum value.

Since $\vec{y}$, $\vec{z}$ are fixed, and $\vec{u}$ is a variable point (of C), therefore (16:3) defined a hyperplane and one of the half spaces produced by it. And it is easy to verify that $\vec{z}$ belongs to this hyperplane, and that it consists of those points $\vec{u}$ for which the angle formed by the three points is a right-angle (i.e. for which the vectors $\vec{z} - \vec{y}$ and $\vec{u} - \vec{z}$ are orthogonal). This means,

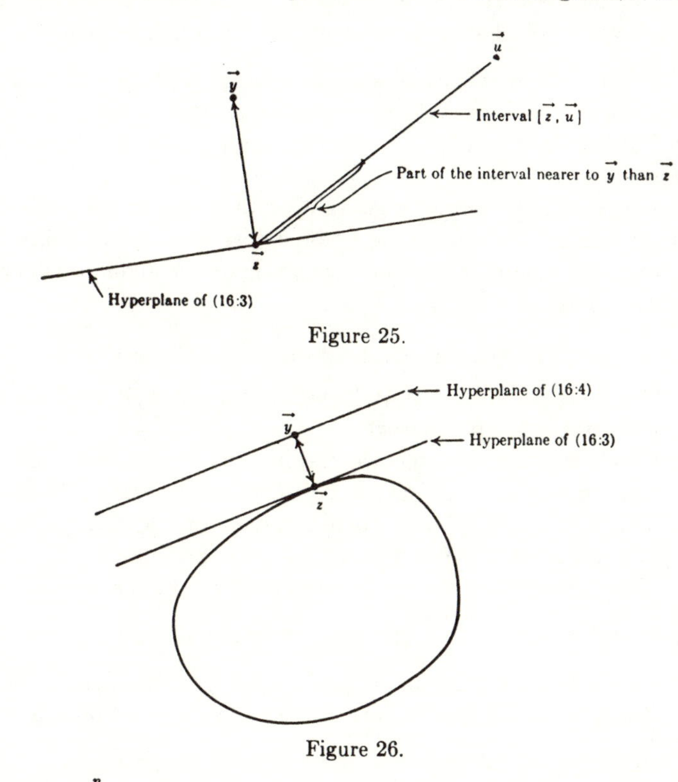

Figure 25.

Figure 26.

indeed, that $\sum_{i=1}^{n} (z_i - y_i)(u_i - z_i) = 0$. Clearly all of C must lie on this hyperplane, or on that side of it which is away from $\vec{y}$. If any point $\vec{u}$ of C did lie on the $\vec{y}$ side, then some points of the interval $[\vec{z}, \vec{u}]$ would be nearer to $\vec{y}$ than $\vec{z}$ is. (Cf. Figure 25. The computation on pp. 135–136—properly interpreted—shows precisely this.) Since C contains $\vec{z}$ and $\vec{u}$, and so all of $[\vec{z}, \vec{u}]$, this would contradict the statement that $\vec{z}$ is as near to $\vec{y}$ as possible in C.

Now our passage from (16:3) to (16:4) amounts to a parallel shift of this hyperplane from $\vec{z}$ to $\vec{y}$ (parallel, because the coefficients $a_i = z_i - y_i$

of u_i, $i = 1, \cdots, n$ are unaltered). Now $\overrightarrow{y}$ lies on the hyperplane, and all of C in one half-space produced by it (Figure 26).

The case $n = 3$ (space) could be visualized in a similar way.

It is even possible to account for a general n in this geometrical manner. If the reader can persuade himself that he possesses n-dimensional "geometrical intuition" he may accept the above as a proof which is equally valid in n dimensions. It is even possible to avoid this by arguing as follows: Whatever n, the entire proof deals with only three points at once, namely $\overrightarrow{y}$, $\overrightarrow{z}$, $\overrightarrow{u}$. Now it is always possible to lay a (2-dimensional) plane through three given points. If we consider only the situation in this plane, then Figures 24–26 and the associated argument can be used without any re-interpretation.

Be this as it may, the purely algebraic proof given above is absolutely rigorous at any rate. We gave the geometrical analogies mainly in the hope that they may facilitate the understanding of the algebraic operations performed in that proof.

16.4. The Theorem of the Alternative for Matrices

16.4.1. The theorem (16:B) of 16.3. permits an inference which will be fundamental for our subsequent work.

We start by considering a rectangular matrix in the sense of 13.3.3. with n rows and m columns, and the matrix element $a(i, j)$. (Cf. Figure 11 in 13.3.3. The ϕ, x, y, t, s there correspond to our a, i, j, n, m.) I.e. $a(i, j)$ is a perfectly arbitrary function of the two variables $i = 1, \cdots, n$; $j = 1, \cdots, m$. Next we form certain vectors in L_n: For each $j = 1, \cdots, m$ the vector $\overrightarrow{x}^j = \{x_1^j, \cdots, x_n^j\}$ with $x_i^j = a(i, j)$ and for each $l = 1, \cdots, n$ the coordinate vector $\overrightarrow{\delta}^l = \{\delta_{il}\}$. (Cf. for the latter the end of 16.1.3.; we have replaced the j there by our l.) Let us now apply the theorem (16:B) of 16.3. for $p = n + m$ to these $n + m$ vectors $\overrightarrow{x}^1, \cdots, \overrightarrow{x}^m, \overrightarrow{\delta}^1, \cdots, \overrightarrow{\delta}^n$. (They replace the $\overrightarrow{x}^1, \cdots, \overrightarrow{x}^p$ loc. cit.) We put $\overrightarrow{y} = \overrightarrow{0}$.

The convex C spanned by $\overrightarrow{x}^1, \cdots, \overrightarrow{x}^m, \overrightarrow{\delta}^1, \cdots, \overrightarrow{\delta}^n$ may contain $\overrightarrow{0}$. If this is the case, then we can conclude from (16:2:d) in 16.2.2. that

$$\sum_{j=1}^{m} t_j \overrightarrow{x}^j + \sum_{l=1}^{n} s_l \overrightarrow{\delta}^l = \overrightarrow{0},$$

with

(16:5) $$t_1 \geqq 0, \cdots, t_m \geqq 0, \ s_1 \geqq 0, \cdots, s_n \geqq 0.$$

(16:6) $$\sum_{j=1}^{m} t_j + \sum_{l=1}^{n} s_l = 1.$$

$t_1, \cdots, t_m, s_1, \cdots, s_n$ replace the $t_1, \cdots, t_p$ (loc. cit.). In terms of the components this means

$$\sum_{j=1}^{m} t_j a(i, j) + \sum_{l=1}^{n} s_l \delta_{il} = 0.$$

The second term on the left-hand side is equal to s_i, so we write

$$(16{:}7) \qquad \sum_{j=1}^{m} a(i, j) t_j = -s_i.$$

If we had $\sum_{j=1}^{m} t_j = 0$, then $t_1 = \cdots = t_m = 0$, hence by (16:7) $s_1 = \cdots = s_n = 0$, thus contradicting (16:6). Hence $\sum_{j=1}^{m} t_j > 0$. We replace (16:7) by its corollary

$$(16{:}8) \qquad \sum_{j=1}^{m} a(i, j) t_j \leqq 0.$$

Now put $x_j = t_j \Big/ \sum_{j=1}^{m} t_j$ for $j = 1, \cdots, m$. Then we have $\sum_{j=1}^{m} x_j = 1$ and (16:5) gives $x_1 \geqq 0, \cdots, x_m \geqq 0$. Hence

$$(16{:}9) \qquad \vec{x} = \{x_1, \cdots, x_m\} \quad \text{belongs to } S_m$$

and (16:8) gives

$$(16{:}10) \qquad \sum_{j=1}^{m} a(i, j) x_j \leqq 0 \quad \text{for} \quad i = 1, \cdots, n.$$

Consider, on the other hand, the possibility that C does not contain $\vec{0}$. Then the theorem (16:B) of 16.3. permits us to infer the existence of a hyperplane which contains $\vec{y}$ (cf. (16:2:a) in 16.2.1.), such that all of C is contained in one half-space produced by that hyperplane (cf. (16:2:b) in 16.2.1.). Denote this hyperplane by

$$\sum_{i=1}^{n} a_i x_i = b.$$

Since $\vec{0}$ belongs to it, therefore $b = 0$. So the half space in question is

$$(16{:}11) \qquad \sum_{i=1}^{n} a_i x_i > 0.$$

$\overrightarrow{x}'$, $\cdots$, $\overrightarrow{x}^m$, $\overrightarrow{\delta}'$, $\cdots$, $\overrightarrow{\delta}^n$ belong to this half space. Stating this for $\overrightarrow{\delta}^l$, (16:11) becomes $\sum_{i=1}^{n} a_i\delta_{il} > 0$, i.e. $a_l > 0$. So we have

(16:12) $$a_1 > 0, \cdots, a_n > 0.$$

Stating it for $\overrightarrow{x}^i$, (16:11) becomes

(16:13) $$\sum_{i=1}^{n} a(i, j)a_i > 0.$$

Now put $w_i = a_i \Big/ \sum_{i=1}^{n} a_i$ for $i = 1, \cdots, n$. Then we have $\sum_{i=1}^{n} w_i = 1$ and (16:12) gives $w_1 > 0, \cdots, w_n > 0$. Hence

(16:14) $$\overrightarrow{w} = \{w_1, \cdots, w_n\} \qquad \text{belongs to } S_n.$$

And (16:13) gives

(16:15) $$\sum_{i=1}^{m} a(i, j)w_i > 0 \qquad \text{for} \qquad j = 1, \cdots, m.$$

Summing up (16:9), (16:10), (16:14), (16:15), we may state:

(16:C) Let a rectangular matrix with n rows and m columns be given. Denote its matrix element by $a(i, j)$, $i = 1, \cdots, n$; $j = 1, \cdots, m$. Then there exists either a vector $\overrightarrow{x} = \{x_1, \cdots, x_m\}$ in S_m with

(16:16:a) $$\sum_{j=1}^{m} a(i, j)x_j \leqq 0 \qquad \text{for} \qquad i = 1, \cdots, n,$$

or a vector $\overrightarrow{w} = \{w_1, \cdots, w_n\}$ in S_n with

(16:16:b) $$\sum_{i=1}^{n} a(i, j)w_i > 0 \qquad \text{for} \qquad j = 1, \cdots, m.$$

We observe further:

The two alternatives (16:16:a), (16:16:b) exclude each other.

Proof: Assume both (16:16:a) and (16:16:b). Multiply each (16:16:a) by w_i and sum over $i = 1, \cdots, n$; this gives $\sum_{i=1}^{n}\sum_{j=1}^{m} a(i, j)w_ix_j \leqq 0$. Multiply each (16:16:b) by x_j and sum over $j = 1, \cdots, m$; this gives

$$\sum_{i=1}^{n} \sum_{j=1}^{m} a(i, j) w_i x_j > 0.$$ [1]

Thus we have a contradiction.

16.4.2. We replace the matrix $a(i, j)$ by its *negative transposed matrix;* i.e. we denote the columns (and not, as before, the rows) by $i = 1, \cdots, n$ and the rows (and not, as before, the columns) by $j = 1, \cdots, m$. And we let the matrix element be $-a(i, j)$ (and not, as before $a(i, j)$). (Thus n, m too are interchanged.)

We restate now the final results of 16.4.1. as applied to this new matrix. But in formulating them, we let $\overrightarrow{x}' = \{x_1', \cdots, x_m'\}$ play the role which $\overrightarrow{w} = \{w_1, \cdots, w_n\}$ had, and $\overrightarrow{w}' = \{w_1', \cdots, w_n'\}$ the role which $\overrightarrow{x} = \{x_1, \cdots, x_m\}$ had. And we announce the result in terms of the original matrix.

Then we have:

(16:D) Let a rectangular matrix with n rows and m columns be given. Denote its matrix element by $a(i, j)$, $i = 1, \cdots, n$; $j = 1, \cdots, m$. Then there exists either a vector $\overrightarrow{x}' = \{x_1', \cdots, x_m'\}$ in S_m with

(16:17:a) $$\sum_{j=1}^{m} a(i, j) x_j' < 0 \quad \text{for} \quad i = 1, \cdots, n,$$

or a vector $\overrightarrow{w}' = \{w_1', \cdots, w_n'\}$ in S_n with

(16:17:b) $$\sum_{i=1}^{n} a(i, j) w_i' \geqq 0 \quad \text{for} \quad j = 1, \cdots, m.$$

And the two alternatives exclude each other.

16.4.3. We now combine the results of 16.4.1. and 16.4.2. They imply that we must have (16:17:a), or (16:16:b), or (16:16:a) and (16:17:b) simultaneously; and also that these three possibilities exclude each other.

Using the same matrix $a(i, j)$ but writing $\overrightarrow{x}, \overrightarrow{w}, \overrightarrow{x}', \overrightarrow{w}'$ for the vectors $\overrightarrow{x}', \overrightarrow{w}, \overrightarrow{x}, \overrightarrow{w}'$ in 16.4.1., 16.4.2. we obtain this:

[1] > 0 and not only $\geqq 0$. Indeed, $= 0$ would necessitate $x_1 = \cdots = x_m = 0$ which is impossible since $\sum_{j=1}^{m} x_j = 1$.

(16:E) There exists either a vector $\vec{x} = \{x_1, \cdots, x_m\}$ in S_m with

(16:18:a) $$\sum_{j=1}^{m} a(i, j)x_j < 0 \quad \text{for} \quad i = 1, \cdots, n,$$

or a vector $\vec{w} = \{w_1, \cdots, w_n\}$ in S_n with

(16:18:b) $$\sum_{i=1}^{n} a(i, j)w_i > 0 \quad \text{for} \quad j = 1, \cdots, m,$$

or two vectors $\vec{x}' = \{x'_1, \cdots, x'_m\}$ in S_m and $\vec{w}' = \{w'_1, \cdots, w'_n\}$ in S_n with

(16:18:c) $$\sum_{j=1}^{m} a(i, j)x'_j \leqq 0 \quad \text{for} \quad i = 1, \cdots, n,$$
$$\sum_{i=1}^{n} a(i, j)w'_i \geqq 0 \quad \text{for} \quad j = 1, \cdots, m.$$

The three alternatives (16:18:a), (16:8:b), (16:8:c) exclude each other.

By combining (16:18:a) and (16:18:c) on one hand and (16:18:b) and (16:18:c) on the other, we get this simpler but weaker statement.[1,2]

(16:F) There exists either a vector $\vec{x} = \{x_1, \cdots, x_m\}$ in S_m with

(16:19:a) $$\sum_{j=1}^{m} a(i, j)x_j \leqq 0 \quad \text{for} \quad i = 1, \cdots, n,$$

or a vector $\vec{w} = \{w_1, \cdots, w_n\}$ in S_n with

(16:19:b) $$\sum_{i=1}^{n} a(i, j)w_i \geqq 0 \quad \text{for} \quad j = 1, \cdots, m.$$

16.4.4. Consider now a *skew symmetric matrix* $a(i, j)$, i.e. one which coincides with its negative transposed in the sense of 16.4.2.; i.e. $n = m$ and

$$a(i, j) = -a(j, i) \quad \text{for} \quad i, j = 1, \cdots, n.$$

[1] The two alternatives (16:19:a), (16:19:b) do not exclude each other: Their conjunction is precisely (16:18:c).

[2] This result could also have been obtained directly from the final result of 16.4.1.: (16:19:a) is precisely (16:16:a) there, and (16:19:b) is a weakened form of (16:16:b) there. We gave the above more detailed discussion because it gives a better insight into the entire situation.

Then the conditions (16:19:a) and (16:19:b) in 16.4.3. express the same thing: Indeed, (16:19:b) is

$$\sum_{i=1}^{n} a(i, j)w_i \geqq 0;$$

this may be written

$$-\sum_{i=1}^{n} a(j, i)w_i \geqq 0, \qquad \text{or} \qquad \sum_{i=1}^{n} a(j, i)w_i \leqq 0.$$

We need only write j, i for i, j [1] so that this becomes $\sum_{j=1}^{n} a(i, j)w_j \leqq 0$, and then $\vec{x}$ for $\vec{w}$,[1] so that we have $\sum_{j=1}^{n} a(i, j)x_j \leqq 0$. And this is precisely (16:19:a).

Therefore we can replace the disjunction of (16:19:a) and (16:19:b) by either one of them,—say by (16:19:b). So we obtain:

(16:G) If the matrix $a(i, j)$ is skew-symmetric (and therefore $n = m$ cf. above), then there exists a vector $\vec{w} = \{w_1, \cdots, w_n\}$ in S_n with

$$\sum_{i=1}^{n} a(i, j)w_i \geqq 0 \qquad \text{for} \qquad j = 1, \cdots, n.$$

17. Mixed Strategies. The Solution for All Games

17.1. Discussion of Two Elementary Examples

17.1.1. In order to overcome the difficulties in the non-strictly determined case—which we observed particularly in 14.7.—it is best to reconsider the simplest examples of this phenomenon. These are the games of Matching Pennies and of Stone, Paper, Scissors (cf. 14.7.2., 14.7.3.). Since an empirical, common-sense attitude with respect to the "problems" of these games exists, we may hope to get a clue for the solution of non-strictly determined (zero-sum two-person) games by observing and analyzing these attitudes.

It was pointed out that, e.g. in Matching Pennies, no particular way of playing—i.e. neither playing "heads" nor playing "tails"—is any better than the other, and all that matters is to find out the opponent's intentions. This seems to block the way to a solution, since the rules of the game in question explicitly bar each player from the knowledge about the opponent's actions, at the moment when he has to make his choice. But

[1] Observe that now, with $n = m$ this is only a change in notation!

the above observation does not correspond fully to the realities of the case: In playing Matching Pennies against an at least moderately intelligent opponent, the player will not attempt to find out the opponent's intentions but will concentrate on avoiding having his own intentions found out, by playing irregularly "heads" and "tails" in successive games. Since we wish to describe the strategy in one play—indeed we must discuss the course in one play and not that of a sequence of successive plays—it is preferable to express this as follows: The player's strategy consists neither of playing "tails" nor of playing "heads," but of playing "tails" with the probability of $\frac{1}{2}$ and "heads" with the probability of $\frac{1}{2}$.

17.1.2. One might imagine that in order to play Matching Pennies in a rational way the player will—before his choice in each play—decide by some 50:50 chance device whether to play "heads" or "tails."[1] The point is that this procedure protects him from loss. Indeed, whatever strategy the opponent follows, the player's expectation for the outcome of the play will be zero.[2] This is true in particular if with certainty the opponent plays "tails," and also if with certainty he plays "heads"; and also, finally, if he—like the player himself—may play both "heads" and "tails," with certain probabilities.[3]

Thus, if we permit a player in Matching Pennies to use a "statistical" strategy, i.e. to "mix" the possible ways of playing with certain probabilities (chosen by him), then he can protect himself against loss. Indeed, we specified above such a statistical strategy with which he cannot lose, irrespective of what his opponent does. The same is true for the opponent, i.e. the opponent can use a statistical strategy which prevents the player from winning, irrespective of what the player does.[4]

The reader will observe the great similarity of this with the discussions of 14.5.[5] In the spirit of those discussions it seems legitimate to consider zero as the value of a play of Matching Pennies and the 50:50 statistical mixture of "heads" and "tails" as a good strategy.

The situation in Paper, Stone, Scissors is entirely similar. Common sense will tell that the good way of playing is to play all three alternatives with the probabilities of $\frac{1}{3}$ each.[6] The value of a play as well as the inter-

[1] E.g. he could throw a die—of course without letting the opponent see the result—and then play "tails" if the number of spots showing is even, and "heads" if that number is odd.

[2] I.e. his probability of winning equals his probability of losing, because under these conditions the probability of matching as well as that of not matching will be $\frac{1}{2}$, whatever the opponent's conduct.

[3] Say p, $1 - p$. For the player himself we used the probabilities $\frac{1}{2}$, $\frac{1}{2}$.

[4] All this, of course, in the statistical sense: that the player cannot lose, means that his probability of losing is $\leqq$ his probability of winning. That he cannot win, means that the former is $\geqq$ to the latter. Actually each play will be won or lost, since Matching Pennies knows no ties.

[5] We mean specifically (14:C:d), (14:C:e) in 14.5.1.

[6] A chance device could be introduced as before. The die mentioned in footnote 1, above, would be a possible one. E.g. the player could decide "stone" if 1 or 2 spots show, "paper" is 3 or 4 spots show, "scissors" if 5 or 6 show.

pretation of the above strategy as a good one can be motivated as before, again in the sense of the quotation there.[1]

17.2. Generalization of This Viewpoint

17.2.1. It is plausible to try to extend the results found for Matching Pennies and Stone, Paper, Scissors to all zero-sum two-person games.

We use the normalized form, the possible choices of the two players being $\tau_1 = 1, \cdots, \beta_1$ and $\tau_2 = 1, \cdots, \beta_2$, and the outcome for player 1 $\mathcal{H}(\tau_1, \tau_2)$, as formerly. We make no assumption of strict determinateness.

Let us now try to repeat the procedure which was successful in 17.1.; i.e. let us again visualize players whose "theory" of the game consists not in the choice of definite strategies but rather in the choice of several strategies with definite probabilities.[2] Thus player 1 will not choose a number $\tau_1 = 1, \cdots, \beta_1$—i.e. the corresponding strategy $\Sigma_1^{\tau_1}$—but β_1 numbers $\xi_1, \cdots, \xi_{\beta_1}$—the probabilities of these strategies $\Sigma_1', \cdots, \Sigma_1^{\beta_1}$, respectively. Equally player 2 will not choose a number $\tau_2 = 1, \cdots, \beta_2$—i.e. the corresponding strategy $\Sigma_2^{\tau_2}$—but β_2 numbers $\eta_1, \cdots, \eta_{\beta_2}$—the probabilities of these strategies $\Sigma_2', \cdots, \Sigma_2^{\beta_2}$, respectively. Since these probabilities belong to disjoint but exhaustive alternatives, the numbers ξ_{τ_1}, η_{τ_2} are subject to the conditions

$$(17:1:a) \qquad \text{all } \xi_{\tau_1} \geqq 0, \qquad \sum_{\tau_1=1}^{\beta_1} \xi_{\tau_1} = 1;$$

$$(17:1:b) \qquad \text{all } \eta_{\tau_2} \geqq 0, \qquad \sum_{\tau_2=1}^{\beta_2} \eta_{\tau_2} = 1.$$

and to no others.

We form the vectors $\vec{\xi} = \{\xi_1, \cdots, \xi_{\beta_1}\}$ and $\vec{\eta} = \{\eta_1, \cdots, \eta_{\beta_2}\}$. Then the above conditions state that $\vec{\xi}$ must belong to S_{β_1} and $\vec{\eta}$ to S_{β_2} in the sense of 16.2.2.

In this setup a player does not, as previously, choose his strategy, but he plays all possible strategies and chooses only the probabilities with which he is going to play them respectively. This generalization meets the major difficulty of the not strictly determined case to a certain point: We have seen that the characteristic of that case was that it constituted a definite disadvantage[3] for each player to have his intentions found out by his

[1] In Stone, Paper, Scissors there exists a tie, but no loss still means that the probability of losing is $\leqq$ the probability of winning, and no gain means the reverse. Cf. footnote 4 on p. 144.

[2] That these probabilities were the same for all strategies ($\frac{1}{2}$, $\frac{1}{2}$ or $\frac{1}{3}$, $\frac{1}{3}$, $\frac{1}{3}$ in the examples of the last paragraph) was, of course accidental. It is to be expected that this equality was due to the symmetric way in which the various alternatives appeared in those games. We proceed now on the assumption that the appearance of probabilities in formulating a strategy was the essential thing, while the particular values were accidental.

[3] The $\Delta > 0$ of 14.7.1.

opponent. Thus one important consideration[1] for a player in such a game is to protect himself against having his intentions found out by his opponent. Playing several different strategies at random, so that only their probabilities are determined, is a very effective way to achieve a degree of such protection: By this device the opponent cannot possibly find out what the player's strategy is going to be, since the player does not know it himself.[2] Ignorance is obviously a very good safeguard against disclosing information directly or indirectly.

17.2.2. It may now seem that we have incidentally restricted the player's freedom of action. It may happen, after all, that he wishes to play one definite strategy to the exclusion of all others; or that, while desiring to use certain strategies with certain probabilities, he wants to exclude absolutely the remaining ones.[3] We emphasize that these possibilities are perfectly within the scope of our scheme. A player who does not wish to play certain strategies at all will simply choose for them the probabilities zero. A player who wishes to play one strategy to the exclusion of all others will choose for this strategy the probability 1 and for all other strategies the probability zero.

Thus if player 1 wishes to play the strategy $\Sigma_1^{\tau_1}$ only, he will choose for $\overrightarrow{\xi}$ the coordinate vector $\overrightarrow{\delta}^{\tau_1}$ (cf. 16.1.3.). Similarly for player 2, the strategy $\Sigma_2^{\tau_2}$ and the vectors $\overrightarrow{\eta}$ and $\overrightarrow{\delta}^{\tau_2}$.

In view of all these considerations we call a vector $\overrightarrow{\xi}$ of S_{β_1} or a vector $\overrightarrow{\eta}$ of S_{β_2} a *statistical* or *mixed strategy* of player 1 or 2, respectively. The coordinate vectors $\overrightarrow{\delta}^{\tau_1}$ or $\overrightarrow{\delta}^{\tau_2}$ correspond, as we saw, to the original strategies τ_1 or τ_2—i.e. $\Sigma_1^{\tau_1}$ or $\Sigma_2^{\tau_2}$—of player 1 or 2, respectively. We call them *strict* or *pure strategies*.

17.3. Justification of the Procedure As Applied to an Individual Play

17.3.1. At this stage the reader may have become uneasy and perceive a contradiction between two viewpoints which we have stressed as equally vital throughout our discussions. On the one hand we have always insisted that our theory is a static one (cf. 4.8.2.), and that we analyze the course

[1] But not necessarily the only one.

[2] If the opponent has enough statistical experience about the player's "style," or if he is very shrewd in rationalizing his expected behavior, he may discover the probabilities —frequencies—of the various strategies. (We need not discuss whether and how this may happen. Cf. the argument of 17.3.1.) But by the very concept of probability and randomness nobody under any conditions can foresee what will actually happen in any particular case. (Exception must be made for such probabilities as may vanish; cf. below.)

[3] In this case he clearly increases the danger of having his strategy found out by the opponent. But it may be that the strategy or strategies in question have such intrinsic advantages over the others as to make this worth while. This happens,—e.g. in an extreme form for the "good" strategies of the strictly determined case (cf. 14.5., particularly (14:C:a), (14:C:b) in 14.5.2.).

of one play and not that of a sequence of successive plays (cf. 17.1.). But on the other hand we have placed considerations concerning the danger of one's strategy being found out by the opponent into an absolutely central position (cf. 14.4., 14.7.1. and again the last part of 17.2.). How can the strategy of a player—particularly one who plays a random mixture of several different strategies—be found out if not by continued observation! We have ruled out that this observation should extend over many plays. Thus it would seem necessary to carry it out in a single play. Now even if the rules of the game should be such as to make this possible—i.e. if they lead to long and repetitious plays—the observation would be effected only gradually and successively in the course of the play. It would not be available at the beginning. And the whole thing would be tied up with various dynamical considerations,—while we insisted on a static theory! Besides, the rules of the game may not even give such opportunities for observation;[1] they certainly do not in our original examples of Matching Pennies, and Stone, Paper, Scissors. These conflicts and contradictions occur both in the discussions of 14.—where we used no probabilities in connection with the choice of a strategy—and in our present discussions of 17. where probabilities will be used.

How are they to be solved?

17.3.2. Our answer is this:

To begin with, the ultimate proof of the results obtained in 14. and 17.—i.e. the discussions of 14.5. and of 17.8.—do not contain any of these conflicting elements. So we could answer that our final proofs are correct even though the heuristic procedures which lead to them are questionable.

But even these procedures can be justified. We make no concessions: Our viewpoint is static and we are analyzing only a single play. We are trying to find a satisfactory theory,—at this stage for the zero-sum two-person game. Consequently we are not arguing deductively from the firm basis of an existing theory—which has already stood all reasonable tests—but we are searching for such a theory.[2] Now in doing this, it is perfectly legitimate for us to use the conventional tools of logics, and in particular that of the *indirect proof*. This consists in imagining that we have a satisfactory theory of a certain desired type,[3] trying to picture the consequences of this imaginary intellectual situation, and then in drawing conclusions from this as to what the hypothetical theory must be like in detail. If this process is applied successfully, it may narrow the possibilities for the hypothetical theory of the type in question to such an extent that only one

[1] I.e. "gradual," "successive" observations of the behavior of the opponent within one play.

[2] Our method is, of course, the empirical one: We are trying to understand, formalize and generalize those features of the simplest games which impress us as typical. This is, after all, the standard method of all sciences with an empirical basis.

[3] This is full cognizance of the fact that we do not (yet) possess one, and that we cannot imagine (yet) what it would be like, if we had one.

All this is—in its own domain—no worse than any other indirect proof in any part of science (e.g. the *per absurdum* proofs in mathematics and in physics).

possibility is left,—i.e. that the theory is determined, discovered by this device.[1] Of course, it can happen that the application is even more "successful," and that it narrows the possibilities down to nothing—i.e. that it demonstrates that a consistent theory of the desired kind is inconceivable.[2]

17.3.3. Let us now imagine that there exists a complete theory of the zero-sum two-person game which tells a player what to do, and which is absolutely convincing. If the players knew such a theory then each player would have to assume that his strategy has been "found out" by his opponent. The opponent knows the theory, and he knows that a player would be unwise not to follow it.[3] Thus the hypothesis of the existence of a satisfactory theory legitimatizes our investigation of the situation when a player's strategy is "found out" by his opponent. And a satisfactory theory[4] can exist only if we are able to harmonize the two extremes Γ_1 and Γ_2,—strategies of player 1 "found out" or of player 2 "found out."

For the original treatment—free from probability (i.e. with pure strategies)—the extent to which this can be done was determined in 14.5.

We saw that the strictly determined case is the one where there exists a theory satisfactory on that basis. We are now trying to push further, by using probabilities (i.e. with mixed strategies). The same device which we used in 14.5. when there were no probabilities will do again,—the analysis of "finding out" the strategy of the other player.

It will turn out that this time the hypothetical theory can be determined completely and in all cases (not merely for the strictly determined case—cf. 17.5.1., 17.6.).

After the theory is found we must justify it independently by a *direct argument*.[5] This was done for the strictly determined case in 14.5., and we shall do it for the present complete theory in 17.8.

[1] There are several important examples of this performance in physics. The successive approaches to Special and to General Relativity or to Wave Mechanics may be viewed as such. Cf. *A. D'Abro:* The Decline of Mechanism in Modern Physics, New York 1939.

[2] This too occurs in physics. The N. Bohr-Heisenberg analysis of "quantities which are not simultaneously observable" in Quantum Mechanics permits this interpretation. Cf. *N. Bohr:* Atomic Theory and the Description of Nature, Cambridge 1934 and *P. A. M. Dirac:* The Principles of Quantum Mechanics, London 1931, Chap. I.

[3] Why it would be unwise not to follow it is none of our concern at present; we have assumed that the theory is absolutely convincing.

That this is not impossible will appear from our final result. We shall find a theory which is satisfactory; nevertheless it implies that the player's strategy is found out by his opponent. But the theory gives him the directions which permit him to adjust himself so that this causes no loss. (Cf. the theorem of 17.6. and the discussion of our complete solution in 17.8.)

[4] I.e. a theory using our present devices only. Of course we do not pretend to be able to make "absolute" statements. If our present requirements should turn out to be unfulfillable we should have to look for another basis for a theory. We have actually done this once by passing from 14. (with pure strategies) to 17. (with mixed strategies).

[5] The indirect argument, as outlined above, gives only necessary conditions. Hence it may establish absurdity (*per absurdum* proof), or narrow down the possibilities to one; but in the latter case it is still necessary to show that the one remaining possibility is satisfactory.

17.4. The Minorant and the Majorant Games (For Mixed Strategies)

17.4.1. Our present picture is then that player 1 chooses an arbitrary element $\vec{\xi}$ from S_{β_1} and that player 2 chooses an arbitrary element $\vec{\eta}$ from S_{β_2}.

Thus if player 1 wishes to play the strategy $\Sigma_1^{\tau_1}$ only, he will choose for $\vec{\xi}$ the coordinate vector $\vec{\delta}^{\tau_1}$ (cf. 16.1.3.); similarly for player 2, the strategy $\Sigma_2^{\tau_2}$ and the vectors $\vec{\eta}$ and $\vec{\delta}^{\tau_2}$.

We imagine again that player 1 makes his choice of $\vec{\xi}$ in ignorance of player 2's choice of $\vec{\eta}$ and *vice versa*.

The meaning is, of course, that when these choices have been made player 1 will actually use (every) $\tau_1 = 1, \cdots, \beta_1$ with the probabilities ξ_{τ_1} and the player 2 will use (every) $\tau_2 = 1, \cdots, \beta_2$ with the probabilities η_{τ_2}. Since their choices are independent, the mathematical expectation of the outcome is

$$\text{(17:2)} \qquad K(\vec{\xi}, \vec{\eta}) = \sum_{\tau_1=1}^{\beta_1} \sum_{\tau_2=1}^{\beta_2} \mathcal{H}(\tau_1, \tau_2)\xi_{\tau_1}\eta_{\tau_2}.$$

In other words, we have replaced the original game Γ by a new one of essentially the same structure, but with the following formal differences: The numbers τ_1, τ_2—the choices of the players—are replaced by the vectors $\vec{\xi}, \vec{\eta}$. The function $\mathcal{H}(\tau_1, \tau_2)$—the outcome, or rather the "mathematical expectation" of the outcome of a play—is replaced by $K(\vec{\xi}, \vec{\eta})$. All these considerations demonstrate the identity of structure of our present view of Γ with that of 14.1.2.,—the sole difference being the replacement of $\tau_1, \tau_2, \mathcal{H}(\tau_1, \tau_2)$ by $\vec{\xi}, \vec{\eta}, K(\vec{\xi}, \vec{\eta})$, described above. This isomorphism suggests the application of the same devices which we used on the original Γ, the comparison with the majorant and minorant games Γ_1 and Γ_2 as described in 14.2., 14.3.1., 14.3.3.

17.4.2. Thus in Γ_1 player 1 chooses his $\vec{\xi}$ first and player 2 chooses his $\vec{\eta}$ afterwards in full knowledge of the $\vec{\xi}$ chosen by his opponent. In Γ_2 the order of their choices is reversed. So the discussion of 14.3.1. applies literally. Player 1, choosing a certain $\vec{\xi}$, may expect that player 2 will choose his $\vec{\eta}$, so as to minimize $K(\vec{\xi}, \vec{\eta})$; i.e. player 1's choice of $\vec{\xi}$ leads to the value $\text{Min}_{\vec{\eta}} K(\vec{\xi}, \vec{\eta})$. This is a function of $\vec{\xi}$ alone; hence player 1 should choose his $\vec{\xi}$ so as to maximize $\text{Min}_{\vec{\eta}} K(\vec{\xi}, \vec{\eta})$. Thus the value of a play of Γ_1 is (for player 1)

$$v_1' = \mathrm{Max}_{\vec{\xi}}\ \mathrm{Min}_{\vec{\eta}}\ K(\vec{\xi}, \vec{\eta}).$$

Similarly the value of a play of Γ_2 (for player 1) turns out to be

$$v_2' = \mathrm{Min}_{\vec{\eta}}\ \mathrm{Max}_{\vec{\xi}}\ K(\vec{\xi}, \vec{\eta}).$$

(The apparent assumption of rational behavior of the opponent does not really matter, since the justifications (14:A:a)-(14:A:e), (14:B:a)-(14:B:e) of 14.3.1. and 14.3.3. again apply literally.)

As in 14.4.1. we can argue that the obvious fact that Γ_1 is less favorable for player 1 than Γ_2 constitutes a proof of

$$v_1' \leqq v_2',$$

and that if this is questioned, a rigorous proof is contained in (13:A*) in 13.4.3. The x, y, ϕ there correspond to our $\vec{\xi}$, $\vec{\eta}$, K.[1] If it should happen that

$$v_1' = v_2',$$

then the considerations of 14.5. apply literally. The arguments (14:C:a)-(14:C:f), (14:D:a), (14:D:b) loc. cit., determine the concept of a "good" $\vec{\xi}$ and $\vec{\eta}$ and fix the "value" of a play of (for the player 1) at

$$v' = v_1' = v_2'.$$[2]

All this happens by (13:B*) in 13.4.3. if and only if a saddle point of K exists. (The x, y, ϕ there correspond to our $\vec{\xi}$, $\vec{\eta}$, K.)

17.5. General Strict Determinateness

17.5.1. We have replaced the v_1, v_2 of (14:A:c) and (14:B:c) by our present v_1', v_2', and the above discussion shows that the latter can perform the functions of the former. But we are just as much dependent upon $v_1' = v_2'$ as we were then upon $v_1 = v_2$. It is natural to ask, therefore, whether there is any gain in this substitution.

Evidently this is the case if, as and when there is a better prospect of having $v_1' = v_2'$ (for any given Γ) than of having $v_1 = v_2$. We called Γ *strictly determined* when $v_1 = v_2$; it now seems preferable to make a distinction and to designate Γ for $v_1 = v_2$ as *specially strictly determined*, and for $v_1' = v_2'$ as *generally strictly determined*. This nomenclature is justified only provided we can show that the former implies the latter.

[1] Although $\vec{\xi}$, $\vec{\eta}$ are vectors, i.e. sequences of real numbers ($\xi_1, \cdots, \xi_{\beta_1}$ and $\eta_1, \cdots, \eta_{\beta_2}$) it is perfectly admissible to view each as a single variable in the maxima and minima which we are now forming. Their domains are, of course, the sets S_{β_1}, S_{β_2} which we introduced in 17.2.

[2] For an exhaustive repetition of the arguments in question cf. 17.8.

This implication is plausible by common sense: Our introduction of mixed strategies has increased the player's ability to defend himself against having his strategy found out; so it may be expected that v'_1, v'_2 actually lie between v_1, v_2. For this reason one may even assert that

(17:3) $$v_1 \leqq v'_1 \leqq v'_2 \leqq v_2.$$

(This inequality secures, of course, the implication just mentioned.)

To exclude all possibility of doubt we shall give a rigorous proof of (17:3). It is convenient to prove this as a corollary of another lemma.

17.5.2. First we prove this lemma:

(17:A) For every $\vec{\xi}$ in S_{β_1}

$$\text{Min}_{\vec{\eta}} \text{K}(\vec{\xi}, \vec{\eta}) = \text{Min}_{\vec{\eta}} \sum_{\tau_1=1}^{\beta_1} \sum_{\tau_2=1}^{\beta_2} \mathcal{H}(\tau_1, \tau_2)\xi_{\tau_1}\eta_{\tau_2}$$
$$= \text{Min}_{\tau_2} \sum_{\tau_1=1}^{\beta_1} \mathcal{H}(\tau_1, \tau_2)\xi_{\tau_1}.$$

For every $\vec{\eta}$ in S_{β_2}

$$\text{Max}_{\vec{\xi}} \text{K}(\vec{\xi}, \vec{\eta}) = \text{Max}_{\vec{\xi}} \sum_{\tau_1=1}^{\beta_1} \sum_{\tau_2=1}^{\beta_2} \mathcal{H}(\tau_1, \tau_2)\xi_{\tau_1}\eta_{\tau_2}$$
$$= \text{Max}_{\tau_1} \sum_{\tau_2=1}^{\beta_2} \mathcal{H}(\tau_1, \tau_2)\eta_{\tau_2}.$$

Proof: We prove the first formula only; the proof of the second is exactly the same, only interchanging Max and Min as well as $\leqq$ and $\geqq$.

Consideration of the special vector $\vec{\eta} = \vec{\delta}^{\tau'_2}$ (cf. 16.1.3. and the end of 17.2.) gives

$$\text{Min}_{\vec{\eta}} \sum_{\tau_1=1}^{\beta_1} \sum_{\tau_2=1}^{\beta_2} \mathcal{H}(\tau_1, \tau_2)\xi_{\tau_1}\eta_{\tau_2} \leqq \sum_{\tau_1=1}^{\beta_1} \sum_{\tau_2=1}^{\beta_2} \mathcal{H}(\tau_1, \tau_2)\xi_{\tau_1}\delta_{\tau_2\tau'_2} = \sum_{\tau_1=1}^{\beta_1} \mathcal{H}(\tau_1, \tau'_2)\xi_{\tau_1}.$$

Since this is true for all τ'_2, so

(17:4:a) $$\text{Min}_{\vec{\eta}} \sum_{\tau_1=1}^{\beta_1} \sum_{\tau_2=1}^{\beta_2} \mathcal{H}(\tau_1, \tau_2)\xi_{\tau_1}\eta_{\tau_2} \leqq \text{Min}_{\tau'_2} \sum_{\tau_1=1}^{\beta_1} \mathcal{H}(\tau_1, \tau'_2)\xi_{\tau_1}.$$

On the other hand, for all τ_2

$$\sum_{\tau_1=1}^{\beta_1} \mathcal{H}(\tau_1, \tau_2)\xi_{\tau_1} \geqq \text{Min}_{\tau_2} \sum_{\tau_1=1}^{\beta_1} \mathcal{H}(\tau_1, \tau_2)\xi_{\tau_1}.$$

Given any $\vec{\eta}$ in S_{β_2}, multiply this by η_{τ_2} and sum over $\tau_2 = 1, \cdots, \beta_2$. Since $\sum_{\tau_2=1}^{\beta} \eta_{\tau_2} = 1$, therefore

$$\sum_{\tau_1=1}^{\beta_1}\sum_{\tau_2=1}^{\beta_2}\mathcal{H}(\tau_1,\tau_2)\xi_{\tau_1}\eta_{\tau_2} \geqq \text{Min}_{\tau_2}\sum_{\tau_1=1}^{\beta_1}\mathcal{H}(\tau_1,\tau_2)\xi_{\tau_1}$$

results. Since this is true for all $\overrightarrow{\eta}$, so

$$\text{(17:4:b)}\qquad \text{Min}_{\overrightarrow{\eta}}\sum_{\tau_1=1}^{\beta_1}\sum_{\tau_2=1}^{\beta_2}\mathcal{H}(\tau_1,\tau_2)\xi_{\tau_1}\eta_{\tau_2} \geqq \text{Min}_{\tau_2}\sum_{\tau_1=1}^{\beta_1}\mathcal{H}(\tau_1,\tau_2)\xi_{\tau_1}.$$

(17:4:a), (17:4:b) yield together the desired relation.

If we combine the above formulae with the definition of v_1', v_2' in 17.4., then we obtain

$$\text{(17:5:a)}\qquad v_1' = \text{Max}_{\overrightarrow{\xi}}\ \text{Min}_{\tau_2}\sum_{\tau_1=1}^{\beta_1}\mathcal{H}(\tau_1,\tau_2)\xi_{\tau_1},$$

$$\text{(17:5:b)}\qquad v_2' = \text{Min}_{\overrightarrow{\eta}}\ \text{Max}_{\tau_1}\sum_{\tau_2=1}^{\beta_2}\mathcal{H}(\tau_1,\tau_2)\eta_{\tau_2}.$$

These formulae have a simple verbal interpretation: In computing v_1' we need only to give player 1 the protection against having his strategy found out which lies in the use of $\overrightarrow{\xi}$ (instead of τ_1); player 2 might as well proceed in the old way and use τ_2 (and not $\overrightarrow{\eta}$). In computing v_2' the roles are interchanged. This is plausible by common-sense: v_1' belongs to the game Γ_1 (cf. 17.4. and 14.2.); there player 2 chooses after player 1 and is fully informed about the choice of player 1,—hence he needs no protection against having his strategy found out by player 1. For v_2' which belongs to the game Γ_2 (cf. id.) the roles are interchanged.

Now the value of v_1' becomes $\leqq$ if we restrict the variability of $\overrightarrow{\xi}$ in the $\text{Max}_{\overrightarrow{\xi}}$ of the above formula. Let us restrict it to the vectors $\overrightarrow{\xi} = \overrightarrow{\delta}^{\tau_1'}$ ($\tau_1' = 1, \cdots, \beta_1$, cf. 16.1.3. and the end of 17.2.). Since

$$\sum_{\tau_1=1}^{\beta_1}\mathcal{H}(\tau_1,\tau_2)\delta_{\tau_1\tau_1'} = \mathcal{H}(\tau_1',\tau_2),$$

this replaces our expression by

$$\text{Max}_{\tau_1'}\ \text{Min}_{\tau_2}\mathcal{H}(\tau_1',\tau_2) = v_1.$$

So we have shown that

$$v_1 \leqq v_1'.$$

Similarly (cf. the remark at the beginning of the proof of our lemma above) restriction of $\overrightarrow{\eta}$ to the $\overrightarrow{\eta} = \overrightarrow{\delta}^{\tau_2'}$ establishes

$$v_2 \geqq v_2'.$$

Together with $v'_1 \leqq v'_2$ (cf. 17.4.), these inequalities prove

$$(17{:}3) \qquad v_1 \leqq v'_1 \leqq v'_2 \leqq v_2,$$

as desired.

17.6. Proof of the Main Theorem

17.6. We have established that general strict determinateness ($v'_1 = v'_2$) holds in all cases of special strict determinateness ($v_1 = v_2$) as is to be expected. That it holds in some further cases as well—i.e. that we can have $v'_1 = v'_2$ but not $v_1 = v_2$ —is clear from our discussions of Matching Pennies and Stone, Paper, Scissors.[1] Thus we may say, in the sense of 17.5.1. that the passage from special to general strict determinateness does constitute an advance. But for all we know at this moment this advance may not cover the entire ground which should be controlled; it could happen that certain games Γ are not even generally strictly determined,—i.e. we have not yet excluded the possibility

$$v'_1 < v'_2.$$

If this possibility should occur, then all that was said in 14.7.1. would apply again and to an increased extent: finding out one's opponent's strategy would constitute a definite advantage

$$\Delta' = v'_2 - v'_1 > 0,$$

and it would be difficult to see how a theory of the game should be constructed without some additional hypotheses as to "who finds out whose strategy."

The decisive fact is, therefore, that it can be shown that this never happens. For all games Γ

$$v'_1 = v'_2$$

i.e.

$$(17{:}6) \qquad \operatorname{Max}_{\vec{\xi}} \operatorname{Min}_{\vec{\eta}} K(\vec{\xi}, \vec{\eta}) = \operatorname{Min}_{\vec{\eta}} \operatorname{Max}_{\vec{\xi}} K(\vec{\xi}, \vec{\eta}),$$

or equivalently (again use (13:B*) in 13.4.3. the x, y, ϕ there corresponding to our $\vec{\xi}$, $\vec{\eta}$, K): A saddle point of $K(\vec{\xi}, \vec{\eta})$ exists.

This is a general theorem valid for all functions $K(\vec{\xi}, \vec{\eta})$ of the form

$$(17{:}2) \qquad K(\vec{\xi}, \vec{\eta}) = \sum_{\tau_1=1}^{\beta_1} \sum_{\tau_2=1}^{\beta_2} \mathcal{H}(\tau_1, \tau_2)\xi_{\tau_1}\eta_{\tau_2}.$$

The coefficients $\mathcal{H}(\tau_1, \tau_2)$ are absolutely unrestricted; they form, as described in 14.1.3. a perfectly arbitrary matrix. The variables $\vec{\xi}$, $\vec{\eta}$ are really

[1] In both games $v_1 = -1$, $v_2 = 1$ (cf. 14.7.2., 14.7.3.), while the discussion of 17.1. can be interpreted as establishing $v'_1 = v'_2 = 0$.

sequences of real numbers $\xi_1, \cdots, \xi_{\beta_1}$ and $\eta_1, \cdots, \eta_{\beta_2}$; their domains being the sets S_{β_1}, S_{β_2} (cf. footnote 1 on p. 150). The functions $K(\vec{\xi}, \vec{\eta})$ of the form (17:2) are called *bilinear forms*.

With the help of the results of 16.4.3. the proof is easy.[1] This is it:

We apply (16:19:a), (16:19:b) in 16.4.3. replacing the $i, j, n, m, a(i, j)$ there by our $\tau_1, \tau_2, \beta_1, \beta_2, \mathcal{H}(\tau_1, \tau_2)$ and the vectors $\vec{w}, \vec{x}$ there by our $\vec{\xi}, \vec{\eta}$.

If (16:19:b) holds, then we have a $\vec{\xi}$ in S_{β_1} with

$$\sum_{\tau_1=1}^{\beta_1} \mathcal{H}(\tau_1, \tau_2)\xi_{\tau_1} \geqq 0 \qquad \text{for} \qquad \tau_2 = 1, \cdots, \beta_2,$$

i.e. with

$$\text{Min}_{\tau_2} \sum_{\tau_1=1}^{\beta_1} \mathcal{H}(\tau_1, \tau_2)\xi_{\tau_1} \geqq 0.$$

Therefore the formula (17:5:a) of 17.5.2. gives

$$v_1' \geqq 0.$$

If (16:19:a) holds, then we have an $\vec{\eta}$ in S_{β_2} with

$$\sum_{\tau_2=1}^{\beta_2} \mathcal{H}(\tau_1, \tau_2)\eta_{\tau_2} \leqq 0 \qquad \text{for} \qquad \tau_1 = 1, \cdots, \beta_1,$$

[1] This theorem occurred and was proved first in the original publication of one of the authors on the theory of games: *J. von Neumann:* "Zur Theorie der Gesellschaftsspiele," Math. Annalen, Vol. 100 (1928), pp. 295–320.

A slightly more general form of this Min-Max problem arises in another question of mathematical economics in connection with the equations of production:

J. von Neumann: "Über ein ökonomisches Gleichungssystem und eine Verallgemeinerung des Brouwer'schen Fixpunktsatzes," Ergebnisse eines Math. Kolloquiums, Vol. 8 (1937), pp. 73–83.

It seems worth remarking that two widely different problems related to mathematical economics—although discussed by entirely different methods—lead to the same mathematical problem,—and at that to one of a rather uncommon type: The "Min-Max type." There may be some deeper formal connections here, as well as in some other directions, mentioned in the second paper. The subject should be clarified further.

The proof of our theorem, given in the first paper, made a rather involved use of some topology and of functional calculus. The second paper contained a different proof, which was fully topological and connected the theorem with an important device of that discipline: the so-called "Fixed Point Theorem" of *L. E. J. Brouwer*. This aspect was further clarified and the proof simplified by *S. Kakutani:* "A Generalization of Brouwer's Fixed Point Theorem," Duke Math. Journal, Vol. 8 (1941), pp. 457–459.

All these proofs are definitely non-elementary. The first elementary one was given by *J. Ville* in the collection by *E. Borel* and collaborators, "Traité du Calcul des Probabilités et de ses Applications," Vol. IV, 2: "Applications aux Jeux de Hasard," Paris (1938), Note by *J. Ville:* "Sur la Théorie Générale des Jeux où intervient l'Habileté des Joueurs," pp. 105–113.

The proof which we are going to give carries the elementarization initiated by *J. Ville* further, and seems to be particularly simple. The key to the procedure is, of course, the connection with the theory of convexity in 16. and particularly with the results of 16.4.3.

i.e. with

$$\mathrm{Max}_{\tau_1} \sum_{\tau_2=1}^{\beta_2} \mathcal{H}(\tau_1, \tau_2)\eta_{\tau_2} \leqq 0.$$

Therefore the formula (17:5:b) of 17.5.2. gives

$$v'_2 \leqq 0.$$

So we see: Either $v'_1 \geqq 0$ or $v'_2 \leqq 0$, i.e.

(17:7) $$\text{Never } v'_1 < 0 < v'_2.$$

Now choose an arbitrary number w and replace the function $\mathcal{H}(\tau_1, \tau_2)$ by $\mathcal{H}(\tau_1, \tau_2) - w$.[1]

This replaces $K(\vec{\xi}, \vec{\eta})$ by $K(\vec{\xi}, \vec{\eta}) - w \sum_{\tau_1=1}^{\beta_1} \sum_{\tau_2=1}^{\beta_2} \xi_{\tau_1}\eta_{\tau_2}$, that is—as $\vec{\xi}, \vec{\eta}$ lie in S_{β_1}, S_{β_2} and so $\sum_{\tau_1=1}^{\beta_1} \xi_{\tau_1} = \sum_{\tau_2=1}^{\beta_2} \eta_{\tau_2} = 1$—by $K(\vec{\xi}, \vec{\eta}) - w$. Consequently v'_1, v'_2 are replaced by $v'_1 - w, v'_2 - w$.[2] Therefore application of (17:7) to these $v'_1 - w, v'_2 - w$ gives

(17:8) $$\text{Never } v'_1 < w < v'_2.$$

Now w was perfectly arbitrary. Hence for $v'_1 < v'_2$ it would be possible to choose w with $v'_1 < w < v'_2$ thus contradicting (17:8). So $v'_1 < v'_2$ is impossible, and we have proved that $v'_1 = v'_2$ as desired. This completes the proof.

17.7. Comparison of the Treatments by Pure and by Mixed Strategies

17.7.1. Before going further let us once more consider the meaning of the result of

$$v'_1 = v'_2.$$

The essential feature of this is that we have always $v'_1 = v'_2$ but not always $v_1 = v_2$,—i.e. always general strict determinateness, but not always special strict determinateness (cf. the beginning of 17.6.).

Or, to express it mathematically:

We have always

(17:9) $$\mathrm{Max}_{\vec{\xi}} \mathrm{Min}_{\vec{\eta}} K(\vec{\xi}, \vec{\eta}) = \mathrm{Min}_{\vec{\eta}} \mathrm{Max}_{\vec{\xi}} K(\vec{\xi}, \vec{\eta}),$$

[1] I.e. the game Γ is replaced by a new one which is played in precisely the same way as Γ except that at the end player 1 gets less (and player 2 gets more) by the fixed amount w than in Γ.

[2] This is immediately clear if we remember the interpretation of the preceding footnote.

i.e.

$$(17{:}10)\quad \text{Max}_{\vec{\xi}}\, \text{Min}_{\vec{\eta}} \sum_{\tau_1=1}^{\beta_1} \sum_{\tau_2=1}^{\beta_2} \mathcal{H}(\tau_1, \tau_2)\xi_{\tau_1}\eta_{\tau_2} = \text{Min}_{\vec{\eta}}\, \text{Max}_{\vec{\xi}} \sum_{\tau_1=1}^{\beta_1} \sum_{\tau_2=1}^{\beta_2} \mathcal{H}(\tau_1, \tau_2)\xi_{\tau_1}\eta_{\tau_2}.$$

Using (17:A) we may even write for this

$$(17{:}11)\quad \text{Max}_{\vec{\xi}}\, \text{Min}_{\tau_2} \sum_{\tau_1=1}^{\beta_1} \mathcal{H}(\tau_1, \tau_2)\xi_{\tau_1} = \text{Min}_{\vec{\eta}}\, \text{Max}_{\tau_1} \sum_{\tau_2=1}^{\beta_2} \mathcal{H}(\tau_1, \tau_2)\eta_{\tau_2}.$$

But we do not always have

$$(17{:}12)\qquad \text{Max}_{\tau_1}\, \text{Min}_{\tau_2}\, \mathcal{H}(\tau_1, \tau_2) = \text{Min}_{\tau_2}\, \text{Max}_{\tau_1}\, \mathcal{H}(\tau_1, \tau_2).$$

Let us compare (17:9) and (17:12): (17:9) is always true and (17:12) is not. Yet the difference between these is merely that of $\vec{\xi}$, $\vec{\eta}$, K and τ_1, τ_2, $\mathcal{H}$. Why does the substitution of the former for the latter convert the untrue assertion (17:12) into the true assertion (17:9)?

The reason is that the $\mathcal{H}(\tau_1, \tau_2)$ of (17:12) is a perfectly arbitrary function of its variables τ_1, τ_2 (cf. 14.1.3.), while the $K(\vec{\xi}, \vec{\eta})$ of (17:9) is an extremely special function of its variables $\vec{\xi}$, $\vec{\eta}$—i.e. of the $\xi_1, \cdots, \xi_{\beta_1}$, $\eta_1, \cdots, \eta_{\beta_2}$,—namely a bilinear form. (Cf. the first part of 17.6.) Thus the absolute generality of $\mathcal{H}(\tau_1, \tau_2)$ renders any proof of (17:12) impossible, while the special—bilinear form—nature of $K(\vec{\xi}, \vec{\eta})$ provides the basis for the proof of (17:9), as given in 17.6.[1]

17.7.2. While this is plausible it may seem paradoxical that $K(\vec{\xi}, \vec{\eta})$ should be more special than $\mathcal{H}(\tau_1, \tau_2)$, although the former obtained from the latter by a process which bore all the marks of a generalization: We obtained it by the replacement of our original strict concept of a pure strategy by the mixed strategies, as described in 17.2.; i.e. by the replacement of τ_1, τ_2 by $\vec{\xi}$, $\vec{\eta}$.

But a closer inspection dispels this paradox. $K(\vec{\xi}, \vec{\eta})$ is a very special function when compared with $\mathcal{H}(\tau_1, \tau_2)$; but its variables have an enormously

[1] That the $K(\vec{\xi}, \vec{\eta})$ is a bilinear form is due to our use of the "mathematical expectation" wherever probabilities intervene. It seems significant that the linearity of this concept is connected with the existence of a solution, in the sense in which we found one. Mathematically this opens up a rather interesting perspective: One might investigate which other concepts, in place of "mathematical expectation," would not interfere with our solution,—i.e. with the result of 17.6. for zero-sum two-person games.

The concept of "mathematical expectation" is clearly a fundamental one in many ways. Its significance from the point of view of the theory of utility was brought forth particularly in 3.7.1.

wider domain than the previous variables τ_1, τ_2. Indeed τ_1 had the finite set $(1, \cdots, \beta_1)$ for its domain, while $\vec{\xi}$ varies over the set S_{β_1}, which is a $(\beta_1 - 1)$-dimensional surface in the β_1-dimensional linear space S_{β_1} (cf. the end of 16.2.2. and 17.2.). Similarly for τ_2 and $\vec{\eta}$.[1]

There are actually among the $\vec{\xi}$ in S_{β_1} special points which correspond to the various τ_1 in $(1, \cdots, \beta_1)$. Given such a τ_1 we can form (as in 16.1.3. and at the end of 17.2.) the coordinate vector $\vec{\xi} = \vec{\delta}^{\tau_1}$, expressing the choice of the strategy $\Sigma_1^{\tau_1}$ to the exclusion of all others. We can correlate special $\vec{\eta}$ in S_{β_2} with the τ_2 in $(1, \cdots, \beta_2)$ in the same way: Given such a τ_2 we can form the coordinate vector $\vec{\eta} = \vec{\delta}^{\tau_2}$, expressing the choice of the strategy $\Sigma_2^{\tau_2}$ to the exclusion of all others.

Now clearly:

$$\begin{aligned} K(\vec{\delta}^{\tau_1}, \vec{\delta}^{\tau_2}) &= \sum_{\tau_1'=1}^{\beta_1} \sum_{\tau_2'=1}^{\beta_2} \mathcal{H}(\tau_1', \tau_2') \delta_{\tau_1'\tau_1} \delta_{\tau_2'\tau_2} \\ &= \mathcal{H}(\tau_1, \tau_2).^2 \end{aligned}$$

Thus the function $K(\vec{\xi}, \vec{\eta})$ contains, in spite of its special character, the entire function $\mathcal{H}(\tau_1, \tau_2)$ and it is therefore really the more general concept of the two, as it ought to be. It is actually more than $\mathcal{H}(\tau_1, \tau_2)$ since not all $\vec{\xi}, \vec{\eta}$ are of the special form $\vec{\delta}^{\tau_1}, \vec{\delta}^{\tau_2}$,—not all mixed strategies are pure.[3] One could say that $K(\vec{\xi}, \vec{\eta})$ is the extension of $\mathcal{H}(\tau_1, \tau_2)$ from the narrower domain of the τ_1, τ_2—i.e. of the $\vec{\delta}^{\tau_1}, \vec{\delta}^{\tau_2}$—to the wider domain of the $\vec{\xi}, \vec{\eta}$—i.e. to all of S_{β_1}, S_{β_2}—from the pure strategies to the mixed strategies. The fact that $K(\vec{\xi}, \vec{\eta})$ is a *bilinear form* expresses merely that this extension is carried out by *linear interpolation*. That it is this process which must be used, is of course due to the linear character of "mathematical expectation."[4]

[1] Observe that $\vec{\xi} = \{\xi_1, \cdots, \xi_{\beta_1}\}$ with the components $\xi_{\tau_1}, \tau_1 = 1, \cdots, \beta_1$, also contains τ_1; but there is a fundamental difference. In $\mathcal{H}(\tau_1, \tau_2)$, τ_1 itself is a variable. In $K(\vec{\xi}, \vec{\eta})$, $\vec{\xi}$ is a variable, while τ_1 is, so to say, a variable within the variable. $\vec{\xi}$ is actually a function of τ_1 (cf. the end of 16.1.2.) and this function as such is the variable of $K(\vec{\xi}, \vec{\eta})$. Similarly for τ_2 and $\vec{\eta}$.

Or, in terms of τ_1, τ_2: $\mathcal{H}(\tau_1, \tau_2)$ is a function of τ_1, τ_2 while $K(\vec{\xi}, \vec{\eta})$ is a function of functions of τ_1, τ_2 (in the mathematical terminology: a functional).

[2] The meaning of this formula is apparent if we consider what choice of strategies $\vec{\delta}^{\tau_1}, \vec{\delta}^{\tau_2}$ represent.

[3] I.e. several strategies may be used effectively with positive probabilities.

[4] The fundamental connection between the concept of numerical utility and the linear "mathematical expectation" was pointed out at the end of 3.7.1.

17.7.3. Reverting to (17:9)-(17:12), we see now that we can express the truth of (17:9)-(17-11) and the untruth of (17:12) as follows:

(17:9), (17:10) express that each player is fully protected against having his strategy found out by his opponent if he can use the mixed strategies $\vec{\xi}$, $\vec{\eta}$ instead of the pure strategies τ_1, τ_2. (17:11) states that this remains true if the player who finds out his opponent's strategy uses the τ_1, τ_2 while only the player whose strategy is being found out enjoys the protection of the $\vec{\xi}$, $\vec{\eta}$. The falsity of (17:12), finally, shows that both players—and particularly the player whose strategy happens to be found out—may not forego with impunity the protection of the $\vec{\xi}$, $\vec{\eta}$.

17.8. Analysis of General Strict Determinateness

17.8.1. We shall now reformulate the contents of 14.5.—as mentioned at the end of 17.4.—with particular consideration of the fact established in 17.6. that every zero-sum two-person game Γ is generally strictly determined. Owing to this result we may define:

$$\mathrm{v}' = \mathrm{Max}_{\vec{\xi}} \mathrm{Min}_{\vec{\eta}} \mathrm{K}(\vec{\xi}, \vec{\eta}) = \mathrm{Min}_{\vec{\eta}} \mathrm{Max}_{\vec{\xi}} \mathrm{K}(\vec{\xi}, \vec{\eta})$$
$$= \mathrm{Sa}_{\vec{\xi}|\vec{\eta}} \mathrm{K}(\vec{\xi}, \vec{\eta}).$$

(Cf. also (13:C*) in 13.5.2. and the end of 13.4.3.)

Let us form two sets $\bar{A}$, $\bar{B}$—subsets of S_{β_1}, S_{β_2}, respectively—in analogy to the definition of the sets A, B in (14:D:a), (14:D:b) of 14.5.1. These are the sets A^ϕ, B^ϕ of 13.5.1. (the ϕ corresponding to our K). We define:

(17:B:a) $\bar{A}$ is the set of those $\vec{\xi}$ (in S_{β_1}) for which $\mathrm{Min}_{\vec{\eta}} \mathrm{K}(\vec{\xi}, \vec{\eta})$ assumes its maximum value, i.e. for which

$$\mathrm{Min}_{\vec{\eta}} \mathrm{K}(\vec{\xi}, \vec{\eta}) = \mathrm{Max}_{\vec{\xi}} \mathrm{Min}_{\vec{\eta}} \mathrm{K}(\vec{\xi}, \vec{\eta}) = \mathrm{v}'.$$

(17:B:b) $\bar{B}$ is the set of those $\vec{\eta}$ (in S_{β_2}) for which $\mathrm{Max}_{\vec{\xi}} \mathrm{K}(\vec{\xi}, \vec{\eta})$ assumes its minimum value, i.e. for which

$$\mathrm{Max}_{\vec{\xi}} \mathrm{K}(\vec{\xi}, \vec{\eta}) = \mathrm{Min}_{\vec{\eta}} \mathrm{Max}_{\vec{\xi}} \mathrm{K}(\vec{\xi}, \vec{\eta}) = \mathrm{v}'.$$

It is now possible to repeat the argumentation of 14.5.

In doing this we shall use the homologous enumeration for the assertions (14:C:a)-(14:C:f) as in 14.5.[1]

[1] (a)-(f) will therefore appear in an order different from the natural one. This was equally true in 14.5., since the enumeration there was based upon that of 14.3.1., 14.3.3., and the argumentation in those paragraphs followed a somewhat different route.

We observe first:

(17:C:d) Player 1 can, by playing appropriately, secure for himself a gain $\geqq$ v',—irrespective of what player 2 does.

Player 2 can, by playing appropriately, secure for himself a gain $\geqq -$v',—irrespective of what player 1 does.

Proof: Let player 1 choose $\vec{\xi}$ from $\bar{A}$; then irrespective of what player 2 does, i.e. for every $\vec{\eta}$ we have $K(\vec{\xi}, \vec{\eta}) \geqq \text{Min}_{\vec{\eta}} K(\vec{\xi}, \vec{\eta}) = v'$. Let player 2 choose $\vec{\eta}$ from $\bar{B}$. Then irrespective of what player 1 does, i.e. for every $\vec{\xi}$, we have $K(\vec{\xi}, \vec{\eta}) \leqq \text{Max}_{\vec{\xi}} K(\vec{\xi}, \vec{\eta}) = v'$. This completes the proof.

Second, (17:C:d) is clearly equivalent to this:

(17:C:e) Player 2 can, by playing appropriately, make it sure that the gain of player 1 is $\leqq$ v', i.e. prevent him from gaining $>$ v', irrespective of what player 1 does.

Player 1 can, by playing appropriately, make it sure that the gain of player 2 is $\leqq -$v', i.e. prevent him from gaining $> -$v', irrespective of what player 2 does.

17.8.2. Third, we may now assert—on the basis of (17:C:d) and (17:C:e) and of the considerations in the proof of (17:C:d)—that:

(17:C:a) The good way (combination of strategies) for 1 to play the game Γ is to choose any $\vec{\xi}$ belonging to $\bar{A}$,—$\bar{A}$ being the set of (17:B:a) above.

(17:C:b) The good way (combination of strategies) for 2 to play the game Γ is to choose any $\vec{\eta}$ belonging to $\bar{B}$,—$\bar{B}$ being the set of (17:B:b) above.

Fourth, combination of the assertions of (17:C:d)—or equally well of those of (17:C:e)—gives:

(17:C:c) If both players 1 and 2 play the game Γ well—i.e. if $\vec{\xi}$ belongs to $\bar{A}$ and $\vec{\eta}$ belongs to $\bar{B}$—then the value of $K(\vec{\xi}, \vec{\eta})$ will be equal to the value of a play (for 1),—i.e. to v'.

We add the observation that (13:D*) in 13.5.2. and the remark concerning the sets $\bar{A}$, $\bar{B}$ before (17:B:a), (17:B:b) above give together this:

(17:C:f) Both players 1 and 2 play the game Γ well—i.e. $\vec{\xi}$ belongs to $\bar{A}$ and $\vec{\eta}$ belongs to $\bar{B}$—if and only if $\vec{\xi}$, $\vec{\eta}$ is a saddle point of $K(\vec{\xi}, \vec{\eta})$.

All this should make it amply clear that v′ may indeed be interpreted as the value of a play of Γ (for 1), and that $\bar{A}$, $\bar{B}$ contain the good ways of playing Γ for 1, 2, respectively. There is nothing heuristic or uncertain about the entire argumentation (17:C:a)-(17:C:f). We have made no extra hypotheses about the "intelligence" of the players, about "who has found out whose strategy" etc. Nor are our results for one player based upon any belief in the rational conduct of the other,—a point the importance of which we have repeatedly stressed. (Cf. the end of 4.1.2.; also 15.8.3.)

17.9. Further Characteristics of Good Strategies

17.9.1. The last results—(17:C:c) and (17:C:f) in 17.8.2.—give also a simple explicit characterization of the elements of our present solution,—i.e. of the number v′ and of the vector sets $\bar{A}$ and $\bar{B}$.

By (17:C:c) loc. cit., $\bar{A}$, $\bar{B}$ determine v′; hence we need only study $\bar{A}$, $\bar{B}$, and we shall do this by means of (17:C:f) id.

According to that criterion, $\overrightarrow{\xi}$ belongs to $\bar{A}$, and $\overrightarrow{\eta}$ belongs to $\bar{B}$ if and only if $\overrightarrow{\xi}$, $\overrightarrow{\eta}$ is a saddle point of $K(\overrightarrow{\xi}, \overrightarrow{\eta})$. This means that

$$K(\overrightarrow{\xi}, \overrightarrow{\eta}) = \left\{ \begin{array}{l} \mathrm{Max}_{\overrightarrow{\xi}'} K(\overrightarrow{\xi}', \overrightarrow{\eta}) \\ \mathrm{Min}_{\overrightarrow{\eta}'} K(\overrightarrow{\xi}, \overrightarrow{\eta}') \end{array} \right|$$

We make this explicit by using the expression (17:2) of 17.4.1. and 17.6. for $K(\overrightarrow{\xi}, \overrightarrow{\eta})$, and the expressions of the lemma (17:A) of 17.5.2. for $\mathrm{Max}_{\overrightarrow{\xi}'} K(\overrightarrow{\xi}', \overrightarrow{\eta})$ and $\mathrm{Min}_{\overrightarrow{\eta}'} K(\overrightarrow{\xi}, \overrightarrow{\eta}')$. Then our equations become:

$$\sum_{\tau_1=1}^{\beta_1} \sum_{\tau_2=1}^{\beta_2} \mathcal{H}(\tau_1, \tau_2)\xi_{\tau_1}\eta_{\tau_2} = \left\{ \begin{array}{l} \mathrm{Max}_{\tau_1'} \sum_{\tau_2=1}^{\beta_2} \mathcal{H}(\tau_1', \tau_2)\eta_{\tau_2} \\ \mathrm{Min}_{\tau_2'} \sum_{\tau_1=1}^{\beta_1} \mathcal{H}(\tau_1, \tau_2')\xi_{\tau_1} \end{array} \right|$$

Considering that $\sum_{\tau_1=1}^{\beta_1} \xi_{\tau_1} = \sum_{\tau_2=1}^{\beta_2} \eta_{\tau_2} = 1$, we can also write for these

$$\sum_{\tau_1=1}^{\beta_1} \left[\mathrm{Max}_{\tau_1'} \left\{ \sum_{\tau_2=1}^{\beta_2} \mathcal{H}(\tau_1', \tau_2)\eta_{\tau_2} \right\} - \sum_{\tau_2=1}^{\beta_2} \mathcal{H}(\tau_1, \tau_2)\eta_{\tau_2} \right] \xi_{\tau_1} = 0,$$

$$\sum_{\tau_2=1}^{\beta_2} \left[-\mathrm{Min}_{\tau_2'} \left\{ \sum_{\tau_1=1}^{\beta_1} \mathcal{H}(\tau_1, \tau_2')\xi_{\tau_1} \right\} + \sum_{\tau_1=1}^{\beta_1} \mathcal{H}(\tau_1, \tau_2)\xi_{\tau_1} \right] \eta_{\tau_2} = 0.$$

Now on the left-hand side of these equations the ξ_{τ_1}, η_{τ_2} have coefficients which are all $\geqq 0$.[1] The ξ_{τ_1}, η_{τ_2} themselves are also $\geqq 0$. Hence these

[1] Observe how the Max and Min occur there!

equations hold only when all terms of their left hand sides vanish separately. I.e. when for each $\tau_1 = 1, \cdots, \beta_1$ for which the coefficient is not zero, we have $\xi_{\tau_1} = 0$; and for each $\tau_2 = 1, \cdots, \beta_2$ for which the coefficient is not zero, we have $\eta_{\tau_2} = 0$.

Summing up:

(17:D) $\vec{\xi}$ belongs to $\bar{A}$ and $\vec{\eta}$ belongs to $\bar{B}$ if and only if these are true:

For each $\tau_1 = 1, \cdots, \beta_1$, for which $\sum_{\tau_2=1}^{\beta_2} \mathcal{H}(\tau_1, \tau_2)\eta_{\tau_2}$ does not assume its maximum (in τ_1) we have $\xi_{\tau_1} = 0$.

For each $\tau_2 = 1, \cdots, \beta_2$, for which $\sum_{\tau_1=1}^{\beta_1} \mathcal{H}(\tau_1, \tau_2)\xi_{\tau_1}$ does not assume its minimum (in τ_2) we have $\eta_{\tau_2} = 0$.

It is easy to formulate these principles verbally. They express this: If $\vec{\xi}, \vec{\eta}$ are good mixed strategies, then $\vec{\xi}$ excludes all strategies τ_1 which are not optimal (for player 1) against $\vec{\eta}$, and $\vec{\eta}$ excludes all strategies τ_2 which are not optimal (for player 2) against $\vec{\xi}$; i.e. $\vec{\xi}, \vec{\eta}$ are—as was to be expected—optimal against each other.

17.9.2. Another remark which may be made at this point is this:

(17:E) The game is specially strictly determined if and only if there exists for each player a good strategy which is a pure strategy.

In view of our past discussions, and particularly of the process of generalization by which we passed from pure strategies to mixed strategies, this assertion may be intuitively convincing. But we shall also supply a mathematical proof, which is equally simple. This is it:

We saw in the last part of 17.5.2. that both v_1 and v'_1 obtain by applying $\text{Max}_{\vec{\xi}}$ to $\text{Min}_{\tau_2} \sum_{\tau_1=1}^{\beta_1} \mathcal{H}(\tau_1, \tau_2)\xi_{\tau_1}$, only with different domains for $\vec{\xi}$: The set of all $\vec{\delta}^{\tau_1}$ $(\tau_1 = 1, \cdots, \beta_1)$ for v_1, and all of S_{β_1} for v'_1; i.e. the pure strategies in the first case, and the mixed ones in the second. Hence $v_1 = v'_1$, i.e. the two maxima are equal if and only if the maximum of the second domain is assumed (at least once) within the first domain. This means by (17:D) above that (at least) one pure strategy must belong to $\bar{A}$, i.e. be a good one. I.e.

(17:F:a) $v_1 = v'_1$ if and only if there exists for player 1 a good strategy which is a pure strategy.

Similarly:

(17:F:b) $v_2 = v'_2$ if and only if there exists for player 2 a good strategy which is a pure strategy.

Now $v'_1 = v'_2 = v'$ and strict determinateness means $v_1 = v_2 = v'$, i.e. $v_1 = v'_1$ and $v_2 = v'_2$. So (17:F:a), (17:F:b) give together (17:E).

17.10. Mistakes and Their Consequences. Permanent Optimality

17.10.1. Our past discussions have made clear what a good mixed strategy is. Let us now say a few words about the other mixed strategies. We want to express the distance from "goodness" for those strategies (i.e. vectors $\vec{\xi}$, $\vec{\eta}$) which are not good; and to obtain some picture of the consequences of a mistake—i.e. of the use of a strategy which is not good. However, we shall not attempt to exhaust this subject, which has many intriguing ramifications.

For any $\vec{\xi}$ in S_{β_1} and any $\vec{\eta}$ in S_{β_2} we form the numerical functions

(17:13:a) $$\alpha(\vec{\xi}) = v' - \operatorname{Min}_{\vec{\eta}} K(\vec{\xi}, \vec{\eta}),$$

(17:13:b) $$\beta(\vec{\eta}) = \operatorname{Max}_{\vec{\xi}} K(\vec{\xi}, \vec{\eta}) - v'.$$

By the lemma (17:A) of 17.5.2. equally

(17:13:a*) $$\alpha(\vec{\xi}) = v' - \operatorname{Min}_{\tau_2} \sum_{\tau_1=1}^{\beta_1} \mathcal{H}(\tau_1, \tau_2)\xi_{\tau_1},$$

(17:13:b*) $$\beta(\vec{\eta}) = \operatorname{Max}_{\tau_1} \sum_{\tau_2=1}^{\beta_2} \mathcal{H}(\tau_1, \tau_2)\eta_{\tau_2} - v'.$$

The definition

$$v' = \operatorname{Max}_{\vec{\xi}} \operatorname{Min}_{\vec{\eta}} K(\vec{\xi}, \vec{\eta}) = \operatorname{Min}_{\vec{\eta}} \operatorname{Max}_{\vec{\xi}} K(\vec{\xi}, \vec{\eta})$$

guarantees that always

$$\alpha(\vec{\xi}) \geqq 0, \qquad \beta(\vec{\eta}) \geqq 0.$$

And now (17:B:a), (17:B:b) and (17:C:a), (17:C:b) in 17.8. imply that $\vec{\xi}$ is good if and only if $\alpha(\vec{\xi}) = 0$, and $\vec{\eta}$ is good if and only if $\beta(\vec{\eta}) = 0$.

Thus $\alpha(\vec{\xi})$, $\beta(\vec{\eta})$ are convenient numerical measures for the general $\vec{\xi}$, $\vec{\eta}$ expressing their distance from goodness. The explicit verbal formulation of what $\alpha(\vec{\xi})$, $\beta(\vec{\eta})$ are, makes this interpretation even more plausible: The formulae (17:13:a), (17:13:b) or (17:13:a*), (17:13:b*) above make clear how much of a loss the player risks—relative to the value of a play for him[1] —by using this particular strategy. We mean here "risk" in the sense of the worst that can happen under the given conditions.[2]

It must be understood, however, that $\alpha(\vec{\xi}), \beta(\vec{\eta})$ do not disclose which strategy of the opponent will inflict this (maximum) loss upon the player who is using $\vec{\xi}$ or $\vec{\eta}$. It is, in particular, not at all certain that if the opponent uses some particular good strategy, i.e. an $\vec{\eta}_0$ in $\bar{B}$ or a $\vec{\xi}_0$ in $\bar{A}$, this in itself implies the maximum loss in question. If a (not good) $\vec{\xi}$ or $\vec{\eta}$ is used by the player, then the maximum loss will occur for those $\vec{\eta}\,'$ or $\vec{\xi}\,'$ of the opponent, for which

$$(17:14:a) \qquad K(\vec{\xi}, \vec{\eta}\,') = \mathrm{Min}_{\vec{\eta}}\, K(\vec{\xi}, \vec{\eta}),$$

$$(17:14:b) \qquad K(\vec{\xi}\,', \vec{\eta}) = \mathrm{Max}_{\vec{\xi}}\, K(\vec{\xi}, \vec{\eta}),$$

i.e. if $\vec{\eta}\,'$ is optimal against the given $\vec{\xi}$, or $\vec{\xi}\,'$ optimal against the given $\vec{\eta}$. And we have never ascertained whether any fixed $\vec{\eta}_0$ or $\vec{\xi}_0$ can be optimal against all $\vec{\xi}$ or $\vec{\eta}$.

17.10.2. Let us therefore call an $\vec{\eta}\,'$ or a $\vec{\xi}\,'$ which is optimal against all $\vec{\xi}$ or $\vec{\eta}$—i.e. which fulfills (17:14:a), or (17:14:b) in 17.10.1. for all $\vec{\xi}, \vec{\eta}$—*permanently optimal.* Any permanently optimal $\vec{\eta}\,'$ or $\vec{\xi}\,'$ is necessarily good; this should be clear conceptually and an exact proof is easy.[3] But

[1] I.e. we mean by loss the value of the play minus the actual outcome: $v' - K(\vec{\xi}, \vec{\eta})$ for player 1 and $(-v') - (-K(\vec{\xi}, \vec{\eta})) = K(\vec{\xi}, \vec{\eta}) - v'$ for player 2.

[2] Indeed, using the previous footnote and (17:13:a), (17:13:b)

$$\alpha(\vec{\xi}) = v' - \mathrm{Min}_{\vec{\eta}}\, K(\vec{\xi}, \vec{\eta}) = \mathrm{Max}_{\vec{\eta}}\, \{v' - K(\vec{\xi}, \vec{\eta})\},$$

$$\beta(\vec{\eta}) = \mathrm{Max}_{\vec{\xi}}\, K(\vec{\xi}, \vec{\eta}) - v' = \mathrm{Max}_{\vec{\xi}}\, \{K(\vec{\xi}, \vec{\eta}) - v'\}.$$

I.e. each is a maximum loss.

[3] *Proof:* It suffices to show this for $\vec{\eta}\,'$; the proof for $\vec{\xi}\,'$ is analogous.

Let $\vec{\eta}\,'$ be permanently optimal. Choose a $\vec{\xi}^{\,*}$ which is optimal against $\vec{\eta}\,'$, i.e. with

$$K(\vec{\xi}^{\,*}, \vec{\eta}\,') = \mathrm{Max}_{\vec{\xi}}\, K(\vec{\xi}, \vec{\eta}\,')$$

By definition

$$K(\vec{\xi}^{\,*}, \vec{\eta}\,') = \mathrm{Min}_{\vec{\eta}}\, K(\vec{\xi}^{\,*}, \vec{\eta}).$$

Thus $\vec{\xi}^{\,*}, \vec{\eta}\,'$ is a saddle point of $K(\vec{\xi}, \vec{\eta})$ and therefore $\vec{\eta}\,'$ belongs to $\bar{B}$—i.e. it is good—by (17:C:f) in 17.8.2.

the question remains: Are all good strategies also permanently optimal? And even: Do any permanently optimal strategies exist?

In general the answer is no. Thus in Matching Pennies or in Stone, Paper, Scissors, the only good strategy (for player 1 as well as for player 2) is $\vec{\xi} = \vec{\eta} = \{\frac{1}{2}, \frac{1}{2}\}$ or $\{\frac{1}{3}, \frac{1}{3}, \frac{1}{3}\}$, respectively.[1] If player 1 played differently—e.g. always "heads"[2] or always "stone"[2]—then he would lose if the opponent countered by playing "tails"[3] or "paper."[3] But then the opponent's strategy is not good—i.e. $\{\frac{1}{2}, \frac{1}{2}\}$ or $\{\frac{1}{3}, \frac{1}{3}, \frac{1}{3}\}$, respectively—either. If the opponent played the good strategy, then the player's mistake would not matter.[4]

We shall get another example of this—in a more subtle and complicated way—in connection with Poker and the necessity of "bluffing," in 19.2 and 19.10.3.

All this may be summed up by saying that while our good strategies are perfect from the defensive point of view, they will (in general) not get the maximum out of the opponent's (possible) mistakes,—i.e. they are not calculated for the offensive.

It should be remembered, however, that our deductions of 17.8. are nevertheless cogent; i.e. a theory of the offensive, in this sense, is not possible without essentially new ideas. The reader who is reluctant to accept this, ought to visualize the situation in Matching Pennies or in Stone, Paper, Scissors once more; the extreme simplicity of these two games makes the decisive points particularly clear.

Another caveat against overemphasizing this point is: A great deal goes, in common parlance, under the name of "offensive," which is not at all "offensive" in the above sense,—i.e. which is fully covered by our present theory. This holds for all games in which perfect information prevails, as will be seen in 17.10.3.[5] Also such typically "aggressive" operations (and which are necessitated by imperfect information) as "bluffing" in Poker.[6]

17.10.3. We conclude by remarking that there is an important class of (zero-sum two-person) games in which permanently optimal strategies exist. These are the games in which perfect information prevails, which we analyzed in 15. and particularly in 15.3.2., 15.6., 15.7. Indeed, a small modification of the proof of special strict determinateness of these games, as given loc. cit., would suffice to establish this assertion too. It would give permanently optimal pure strategies. But we do not enter upon these considerations here.

[1] Cf. 17.1. Any other probabilities would lead to losses when "found out." Cf. below.

[2] This is $\vec{\xi} = \vec{\delta}' = \{1, 0\}$ or $\{1, 0, 0\}$, respectively.

[3] This is $\vec{\eta} = \vec{\delta}'' = \{0, 1\}$ or $\{0, 1, 0\}$, respectively.

[4] I.e. the bad strategy of "heads" (or "stone") can be defeated only by "tails" (or "paper"), which is just as bad in itself.

[5] Thus Chess and Backgammon are included.

[6] The preceding discussion applies rather to the failure to "bluff." Cf. 19.2. and 19.10.3.

Since the games in which perfect information prevails are always specially strictly determined (cf. above), one may suspect a more fundamental connection between specially strictly determined games and those in which permanently optimal strategies exist (for both players). We do not intend to discuss these things here any further, but mention the following facts which are relevant in this connection:

(17:G:a) It can be shown that if permanently optimal strategies exist (for both players) then the game must be specially strictly determined.

(17:G:b) It can be shown that the converse of (17:G:a) is not true.

(17:G:c) Certain refinements of the concept of special strict determinateness seem to bear a closer relationship to the existence of permanently optimal strategies.

17.11. The Interchange of Players. Symmetry

17.11.1. Let us consider the role of symmetry, or more generally the effects of interchanging the players 1 and 2 in the game Γ. This will naturally be a continuation of the analysis of 14.6.

As was pointed out there, this interchange of the players replaces the function $\mathcal{H}(\tau_1, \tau_2)$ by $-\mathcal{H}(\tau_2, \tau_1)$. The formula (17:2) of 17.4.1. and 17.6. shows that the effect of this for $K(\vec{\xi}, \vec{\eta})$ is to replace it by $-K(\vec{\eta}, \vec{\xi})$. In the terminology of 16.4.2., we replace the matrix (of $\mathcal{H}(\tau_1, \tau_2)$ cf. 14.1.3.) by its negative transposed matrix.

Thus the perfect analogy of the considerations in 14. continues; again we have the same formal results as there, provided that we replace $\tau_1, \tau_2, \mathcal{H}(\tau_1, \tau_2)$ by $\vec{\xi}, \vec{\eta}, K(\vec{\xi}, \vec{\eta})$. (Cf. the previous occurrence of this in 17.4. and 17.8.)

We saw in 14.6. that this replacement of $\mathcal{H}(\tau_1, \tau_2)$ by $-\mathcal{H}(\tau_2, \tau_1)$ carries v_1, v_2 into $-v_2, -v_1$. A literal repetition of those considerations shows now that the corresponding replacement of $K(\vec{\xi}, \vec{\eta})$ by $-K(\vec{\eta}, \vec{\xi})$ carries v'_1, v'_2 into $-v'_2, -v'_1$. Summing up: Interchanging the players 1, 2, carries v_1, v_2, v'_1, v'_2 into $-v_2, -v_1, -v'_2, -v'_1$.

The result of 14.6. established for (special) strict determinateness was that $v = v_1 = v_2$ is carried into $-v = -v_1 = -v_2$. In the absence of that property no such refinement of the assertion was possible.

At the present we know that we always have general strict determinateness, so that $v' = v'_1 = v'_2$. Consequently this is carried into $-v' = -v'_1 = -v'_2$.

Verbally the content of this result is clear: Since we succeeded in defining a satisfactory concept of the value of a play of Γ (for the player 1), v', it is only reasonable that this quantity should change its sign when the roles of the players are interchanged.

17.11.2. We can also state rigorously when the game Γ is *symmetric*. This is the case when the two players 1 and 2 have precisely the same role in it,—i.e. if the game Γ is identical with that game which obtains from it by interchanging the two players 1, 2. According to what was said above, this means that

$$\mathcal{H}(\tau_1, \tau_2) = -\mathcal{H}(\tau_2, \tau_1),$$

or equivalently that

$$K(\vec{\xi}, \vec{\eta}) = -K(\vec{\eta}, \vec{\xi}).$$

This property of the matrix $\mathcal{H}(\tau_1, \tau_2)$ or of the bilinear form $K(\vec{\xi}, \vec{\eta})$ was introduced in 16.4.4. and called *skew-symmetry*.[1,2]

In this case v_1, v_2 must coincide with $-v_2$, $-v_1$; hence $v_1 = -v_2$, and since $v_1 \geqq v_2$, so $v_1 \geqq 0$. But v' must coincide with $-v'$; therefore we can even assert that

$$v' = 0.^{3}$$

So we see: The value of each play of a symmetrical game is zero.

It should be noted that the value v' of each play of a game Γ could be zero without Γ being symmetric. A game in which $v' = 0$ will be called *fair*.

The examples of 14.7.2., 14.7.3. illustrate this: Stone, Paper, Scissors is symmetric (and hence fair); Matching Pennies is fair (cf. 17.1.) without being symmetric.[4]

[1] For a matrix $\mathcal{H}(\tau_1, \tau_2)$ or for the corresponding bilinear form $K(\vec{\xi}, \vec{\eta})$ *symmetry* is defined by

$$\mathcal{H}(\tau_1, \tau_2) = \mathcal{H}(\tau_2, \tau_1),$$

or equivalently by

$$K(\vec{\xi}, \vec{\eta}) = K(\vec{\eta}, \vec{\xi}).$$

It is remarkable that symmetry of the game Γ is equivalent to skew-symmetry, and not to symmetry, of its matrix or bilinear form.

[2] Thus skew-symmetry means that a reflection of the matrix scheme of Fig. 15 in 14.1.3. on its main diagonal (consisting of the fields (1, 1), (2, 2), etc.) carries it into its own negative. (Symmetry, in the sense of the preceding footnote, would mean that it carries it into itself.)

Now the matrix scheme of Fig. 15 is rectangular; it has β_2 columns and β_1 rows. In the case under consideration its shape must be unaltered by this reflection. Hence it must be quadratic,—i.e. $\beta_1 = \beta_2$. This is so, however, automatically, since the players 1, 2 are assumed to have the same role in Γ.

[3] This is, of course, due to our knowing that $v_1' = v_2'$. Without this—i.e. without the general theorem (16:F) of 16.4.3.—we should assert for the v_1', v_2' only the same which we obtained above for the v_1, v_2: $v_1' = -v_2'$ and since $v_1' \geqq v_2'$, so $v_1' \geqq 0$.

[4] The players 1 and 2 have different roles in Matching Pennies: 1 tries to match, and 2 tries to avoid matching. Of course, one has a feeling that this difference is inessential and that the fairness of Matching Pennies is due to this inessentiality of the assymetry. This could be elaborated upon, but we do not wish to do this on this occasion. A better example of fairness without symmetry would be given by a game which is grossly unsymmetric, but in which the advantages and disadvantages of each player are so judiciously adjusted that a fair game—i.e. value $v' = 0$—results.

A not altogether successful attempt at such a game is the ordinary way of "Rolling Dice." In this game player 1—the "player"—rolls two dice, each of which bear the numbers 1, · · · , 6. Thus at each roll any total 2, · · · , 12 may result. These totals

In a symmetrical game the sets $\bar{A}$, $\bar{B}$ of (17:B:a), (17:B:b) in 17.8. are obviously identical. Since $\bar{A} = \bar{B}$ we may put $\vec{\xi} = \vec{\eta}$ in the final criterion (17:D) of 17.9. We restate it for this case:

(17:H) In a symmetrical game, $\vec{\xi}$ belongs to $\bar{A}$ if and only if this is true: For each $\tau_2 = 1, \cdots \beta_2$ for which $\sum_{\tau_1=1}^{\beta_1} \mathcal{H}(\tau_1, \tau_2)\xi_{\tau_1}$ does not assume its minimum (in τ_2) we have $\xi_{\tau_2} = 0$.

Using the terminology of the concluding remark of 17.9., we see that the above condition expresses this: $\vec{\xi}$ is optimal against itself.

17.11.3. The results of 17.11.1., 17.11.2.—that in every symmetrical game $v' = 0$—can be combined with (17:C:d) in 17.8. Then we obtain this:

(17:I) In a symmetrical game each player can, by playing appropriately, avoid loss[1] irrespective of what the opponent does.

We can state this mathematically as follows:

If the matrix $\mathcal{H}(\tau_1, \tau_2)$ is skew-symmetric, then there exists a vector $\vec{\xi}$ in S_{β_1} with

$$\sum_{\tau_1=1}^{\beta_1} \mathcal{H}(\tau_1, \tau_2)\xi_{\tau_1} \geqq 0 \qquad \text{for} \qquad \tau_2 = 1, \cdots, \beta_2.$$

This could also have been obtained directly, because it coincides with the last result (16:G) in 16.4.4. To see this it suffices to introduce there our present notations: Replace the i, j, $a(i, j)$ there by our τ_1, τ_2, $\mathcal{H}(\tau_1, \tau_2)$ and the $\vec{w}$ there by our $\vec{\xi}$.

---

have the following probabilities:

Total	2	3	4	5	6	7	8	9	10	11	12
Chance out of 36	1	2	3	4	5	6	5	4	3	2	1
Probability	$\frac{1}{36}$	$\frac{2}{36}$	$\frac{3}{36}$	$\frac{4}{36}$	$\frac{5}{36}$	$\frac{6}{36}$	$\frac{5}{36}$	$\frac{4}{36}$	$\frac{3}{36}$	$\frac{2}{36}$	$\frac{1}{36}$

The rule is that if the "player" rolls 7 or 11 ("natural") he wins. If he rolls 2, 3, or 12 he loses. If he rolls anything else (4, 5, 6, or 8, 9, 10) then he rolls again until he rolls either a repeat of the original one (in which case he wins), or a 7 (in which case he loses). Player 2 (the "house") has no influence on the play.

In spite of the great differences of the rules as they affect players 1 and 2 (the "player" and the "house") their chances are nearly equal: A simple computation, which we do not detail, shows that the "player" has 244 chances against 251 for the "house," out of a total of 495; i.e. the value of a play—played for a unit stake—is

$$\frac{244 - 251}{495} = -\frac{7}{495} = -1.414\%.$$

Thus the approximation to fairness is reasonably good, and it may be questioned whether more was intended.

[1] I.e. secure himself a gain $\geqq 0$.

It is even possible to base our entire theory on this fact, i.e. to derive the theorem of 17.6. from the above result. In other words: The general strict determinateness of all Γ can be derived from that one of the symmetric ones. The proof has a certain interest of its own, but we shall not discuss it here since the derivation of 17.6. is more direct.

The possibility of protecting oneself against loss (in a symmetric game) exists only due to our use of the mixed strategies $\vec{\xi}$, $\vec{\eta}$ (cf. the end of 17.7.). If the players are restricted to pure strategies τ_1, τ_2 then the danger of having one's strategy found out, and consequently of sustaining losses, exists. To see this it suffices to recall what we found concerning Stone, Paper, Scissors (cf. 14.7. and 17.1.1.). We shall recognize the same fact in connection with Poker and the necessity of "bluffing" in 19.2.1.

CHAPTER IV

ZERO-SUM TWO-PERSON GAMES: EXAMPLES

18. Some Elementary Games

18.1. The Simplest Games

18.1.1. We have concluded our general discussion of the zero-sum two-person game. We shall now proceed to examine specific examples of such games. These examples will exhibit better than any general abstract discussions could, the true significance of the various components of our theory. They will show, in particular, how some formal steps which are dictated by our theory permit a direct common-sense interpretation. It will appear that we have here a rigorous formalization of the main aspects of such "practical" and "psychological" phenomena as those to be mentioned in 19.2., 19.10. and 19.16.[1]

18.1.2. The size of the numbers β_1, β_2—i.e. the number of alternatives confronting the two players in the normalized form of the game—gives a reasonable first estimate for the degree of complication of the game Γ. The case that either, or both, of these numbers is 1 may be disregarded: This would mean that the player in question has no choice at all by which he can influence the game.[2] Therefore the simplest games of the class which interests us are those with

$$\beta_1 = \beta_2 = 2. \tag{18:1}$$

We saw in 14.7. that Matching Pennies is such a game; its matrix scheme was given in Figure 12 in 13.4.1. Another instance of such a game is Figure 14, id.

	1	2
1	$\mathcal{H}(1, 1)$	$\mathcal{H}(1, 2)$
2	$\mathcal{H}(2, 1)$	$\mathcal{H}(2, 2)$

Figure 27.

Let us now consider the most general game falling under (18:1), i.e. under Figure 27. This applies, e.g., to Matching Pennies if the various ways of matching do not necessarily represent the same gain (or a gain at all),

[1] We stress this because of the widely held opinion that these things are congenitally unfit for rigorous (mathematical) treatment.

[2] Thus the game would really be one of one person; but then, of course, no longer of zero sum. Cf. 12.2.

nor the various ways of not matching the same loss (or a loss at all).[1] We propose to discuss for this case the results of 17.8.,—the value of the game Γ and the sets of good strategies $\bar{A}$, $\bar{B}$. These concepts have been established by the general existential proof of 17.8. (based on the theorem of 17.6.); but we wish to obtain them again by explicit computation in this special case, and thereby gain some further insight into their functioning and their possibilities.

18.1.3. There are certain trivial adjustments which can be made on a game given by Figure 27, and which simplify an exhaustive discussion considerably.

First it is quite arbitrary which of the two choices of player 1 we denote by $\tau_1 = 1$ and by $\tau_1 = 2$; we may interchange these,—i.e. the two rows of the matrix.

Second, it is equally arbitrary which of the two choices of player 2 we denote by $\tau_2 = 1$ and by $\tau_2 = 2$; we may interchange these,—i.e. the two columns of the matrix.

Finally, it is also arbitrary which of the two players we call 1 and which 2; we may interchange these,—i.e. replace $\mathcal{H}(\tau_1, \tau_2)$ by $-\mathcal{H}(\tau_1, \tau_2)$ (cf. 14.6. and 17.11.). This amounts to interchanging the rows and the columns of the matrix, and changing the sign of its elements besides.

Putting everything together, we have here $2 \times 2 \times 2 = 8$ possible adjustments, all of which describe essentially the same game.

18.2. Detailed Quantitative Discussion of These Games

18.2.1. We proceed now to the discussion proper. This will consist in the consideration of several alternative possibilities, the "Cases" to be enumerated below.

These Cases are distinguished by the various possibilities which exist for the positions of those fields of the matrix where $\mathcal{H}(\tau_1, \tau_2)$ assumes its maximum and its minimum for both variables τ_1, τ_2 together. Their delimitations may first appear to be arbitrary; but the fact that they lead to a quick cataloguing of all possibilities justifies them *ex post*.

Consider accordingly $\text{Max}_{\tau_1, \tau_2} \mathcal{H}(\tau_1, \tau_2)$ and $\text{Min}_{\tau_1, \tau_2} \mathcal{H}(\tau_1, \tau_2)$. Each one of these values will be assumed at least once and might be assumed more than once;[2] but this does not concern us at this juncture. We begin now with the definition of the various Cases:

18.2.2. Case (*A*): It is possible to choose a field where $\text{Max}_{\tau_1, \tau_2}$ is assumed and one where $\text{Min}_{\tau_1, \tau_2}$ is assumed, so that the two are neither in the same row nor in the same column.

By interchanging $\tau_1 = 1, 2$ as well as $\tau_2 = 1, 2$ we can make the first-mentioned field (of $\text{Max}_{\tau_1, \tau_2}$) to be (1, 1). The second-mentioned field

[1] Comparison of Figs. 12 and 27 shows that in Matching Pennies $\mathcal{H}(1, 1) = \mathcal{H}(2, 2) = 1$ (gain on matching); $\mathcal{H}(1, 2) = \mathcal{H}(2, 1) = -1$ (loss on not matching).

[2] In Matching Pennies (cf. footnote 1 above) the $\text{Max}_{\tau_1, \tau_2}$ is 1 and is assumed at (1, 1) and (2, 2), while the $\text{Min}_{\tau_1, \tau_2}$ is -1 and is assumed at (1, 2) and (2, 1).

(of $\mathrm{Min}_{\tau_1, \tau_2}$) must then be (2, 2). Consequently we have

(18:2) $$\mathcal{H}(1, 1) \begin{Bmatrix} \geqq \mathcal{H}(1, 2) \geqq \\ \geqq \mathcal{H}(2, 1) \geqq \end{Bmatrix} \mathcal{H}(2, 2).$$

Therefore (1, 2) is a saddle point.[1]

Thus the game is strictly determined in this case and

(18:3) $$v' = v = \mathcal{H}(1, 2).$$

18.2.3. Case (*B*): It is impossible to make the choices as prescribed above:

Choose the two fields in question (of $\mathrm{Max}_{\tau_1, \tau_2}$ and $\mathrm{Min}_{\tau_1, \tau_2}$); then they are in the same row or in the same column. If the former should be the case, then interchange the players 1, 2, so that these two fields are at any rate in the same column.[2]

By interchanging $\tau_1 = 1, 2$ as well as $\tau_2 = 1, 2$ if necessary, we can again make the first-mentioned field (of $\mathrm{Max}_{\tau_1, \tau_2}$) to be (1, 1). So the column in question is $\tau_2 = 1$. The second-mentioned field (of $\mathrm{Min}_{\tau_1, \tau_2}$) must then be (2, 1).[3] Consequently we have:

(18:4) $$\mathcal{H}(1, 1) \begin{Bmatrix} \geqq \mathcal{H}(1, 2) \geqq \\ \geqq \mathcal{H}(2, 2) \geqq \end{Bmatrix} \mathcal{H}(2, 1).$$

Actually $\mathcal{H}(1, 1) = \mathcal{H}(1, 2)$ or $\mathcal{H}(2, 2) = \mathcal{H}(2, 1)$ are excluded because for the $\mathrm{Max}_{\tau_1, \tau_2}$ and $\mathrm{Min}_{\tau_1, \tau_2}$ fields they would permit the alternative choices of (1, 2), (2, 1) or (1, 1), (2, 2), thus bringing about Case (*A*).[4]

So we can strengthen (18:4) to

(18:5) $$\mathcal{H}(1, 1) \begin{Bmatrix} > \mathcal{H}(1, 2) \geqq \\ \geqq \mathcal{H}(2, 2) > \end{Bmatrix} \mathcal{H}(2, 1).$$

We must now make a further disjunction:

18.2.4. Case (B_1):

(18:6) $$\mathcal{H}(1, 2) \geqq \mathcal{H}(2, 2)$$

Then (18:5) can be strengthened to

(18:7) $$\mathcal{H}(1, 1) > \mathcal{H}(1, 2) \geqq \mathcal{H}(2, 2) > \mathcal{H}(2, 1).$$

Therefore (1, 2) is again a saddle point.

Thus the game is strictly determined in this case too; and again

(18:8) $$v' = v = \mathcal{H}(1, 2).$$

[1] Recall 13.4.2. Observe that we had to take (1, 2) and not (2, 1).

[2] This interchange of the two players changes the sign of every matrix element (cf. above), hence it interchanges $\mathrm{Max}_{\tau_1, \tau_2}$ and $\mathrm{Min}_{\tau_1, \tau_2}$. But they will nevertheless be in the same column.

[3] To be precise: It might also be (1, 1). But then $\mathcal{H}(\tau_1, \tau_2)$ has the same $\mathrm{Max}_{\tau_1, \tau_2}$ and $\mathrm{Min}_{\tau_1, \tau_2}$, and so it is a constant. Then we can use (2, 1) also for $\mathrm{Min}_{\tau_1, \tau_2}$.

[4] $\mathcal{H}(1, 1) = \mathcal{H}(2, 2)$ and $\mathcal{H}(1, 2) = \mathcal{H}(2, 1)$ are perfectly possible, as the example of Matching Pennies shows. Cf. footnote 1 on p. 170 and footnote 1 on p. 172.

18.2.5. Case (B_2):

(18:9) $$\mathcal{H}(1, 2) < \mathcal{H}(2, 2)$$

Then (18:5) can be strengthened to

(18:10) $$\mathcal{H}(1, 1) \geqq \mathcal{H}(2, 2) > \mathcal{H}(1, 2) \geqq \mathcal{H}(2, 1).$$ [1]

The game is not strictly determined.[2]

It is easy however, to find good strategies, i.e. a $\vec{\xi}$ in $\bar{A}$ and an $\vec{\eta}$ in $\bar{B}$, by satisfying the characteristic condition (17:D) of 17.9. We can do even more: We can choose $\vec{\eta}$ so that $\sum_{\tau_2=1}^{2} \mathcal{H}(\tau_1, \tau_2)\eta_{\tau_2}$ is the same for all τ_1 and $\vec{\xi}$ so that $\sum_{\tau_1=1}^{2} \mathcal{H}(\tau_1, \tau_2)\xi_{\tau_1}$ is the same for all τ_2. For this purpose we need:

(18:11) $$\begin{cases} \mathcal{H}(1, 1)\eta_1 + \mathcal{H}(1, 2)\eta_2 = \mathcal{H}(2, 1)\eta_1 + \mathcal{H}(2, 2)\eta_2. \\ \mathcal{H}(1, 1)\xi_1 + \mathcal{H}(2, 1)\xi_2 = \mathcal{H}(1, 2)\xi_1 + \mathcal{H}(2, 2)\xi_2. \end{cases}$$

This means

(18:12) $$\begin{aligned} \xi_1 : \xi_2 &= \mathcal{H}(2, 2) - \mathcal{H}(2, 1) : \mathcal{H}(1, 1) - \mathcal{H}(1, 2), \\ \eta_1 : \eta_2 &= \mathcal{H}(2, 2) - \mathcal{H}(1, 2) : \mathcal{H}(1, 1) - \mathcal{H}(2, 1). \end{aligned}$$

We must satisfy these ratios, subject to the permanent requirements

$$\begin{matrix} \xi_1 \geqq 0, & \xi_2 \geqq 0 & \xi_1 + \xi_2 = 1 \\ \eta_1 \geqq 0, & \eta_2 \geqq 0 & \eta_1 + \eta_2 = 1 \end{matrix}$$

This is possible because the prescribed ratios (i.e. the right-hand sides in (18:12)) are positive by (18:10). We have

$$\xi_1 = \frac{\mathcal{H}(2, 2) - \mathcal{H}(2, 1)}{\mathcal{H}(1, 1) + \mathcal{H}(2, 2) - \mathcal{H}(1, 2) - \mathcal{H}(2, 1)},$$
$$\xi_2 = \frac{\mathcal{H}(1, 1) - \mathcal{H}(1, 2)}{\mathcal{H}(1, 1) + \mathcal{H}(2, 2) - \mathcal{H}(1, 2) - \mathcal{H}(2, 1)}.$$

and further

$$\eta_1 = \frac{\mathcal{H}(2, 2) - \mathcal{H}(1, 2)}{\mathcal{H}(1, 1) + \mathcal{H}(2, 2) - \mathcal{H}(1, 2) - \mathcal{H}(2, 1)},$$
$$\eta_2 = \frac{\mathcal{H}(1, 1) - \mathcal{H}(2, 1)}{\mathcal{H}(1, 1) + \mathcal{H}(2, 2) - \mathcal{H}(1, 2) - \mathcal{H}(2, 1)}.$$

We can even show that these $\vec{\xi}$, $\vec{\eta}$ are unique, i.e. that $\bar{A}$, $\bar{B}$ possess no other elements.

[1] This is actually the case for Matching Pennies. Cf. footnotes 1 on p. 170 and 4 on p. 171.

[2] Clearly $v_1 = \text{Max}_{\tau_1} \text{Min}_{\tau_2} \mathcal{H}(\tau_1, \tau_2) = \mathcal{H}(1, 2)$, $v_2 = \text{Min}_{\tau_2} \text{Max}_{\tau_1} \mathcal{H}(\tau_1, \tau_2) = \mathcal{H}(2, 2)$, so $v_1 < v_2$.

Proof: If either $\vec{\xi}$ or $\vec{\eta}$ were something else than we found above, then $\vec{\eta}$ or $\vec{\xi}$ respectively must have a component 0, owing to the characteristic condition (17:D) of 17.9. But then $\vec{\eta}$ or $\vec{\xi}$ would differ from the above values since in these both components are positive. So we see: If either $\vec{\xi}$ or $\vec{\eta}$ differs from the above values, then both do. And then both must have a component 0. For both the other component is then 1, i.e. both are coordinate vectors.[1] Hence the saddle point of $K(\vec{\xi}, \vec{\eta})$ which they represent is really one of $\mathcal{H}(\tau_1, \tau_2)$,—cf. (17:E) in 17.9. Thus the game would be strictly determined; but we know that it is not in this case.

This completes the proof.

All four expressions in (18:11) are now seen to have the same value, namely

$$\frac{\mathcal{H}(1, 1)\mathcal{H}(2, 2) - \mathcal{H}(1, 2)\mathcal{H}(2, 1)}{\mathcal{H}(1, 1) + \mathcal{H}(2, 2) - \mathcal{H}(1, 2) - \mathcal{H}(2, 1)}$$

and by (17:5:a), (17:5:b) in 17.5.2. this is the value of v′. So we have

$$\text{(18:13)} \qquad v' = \frac{\mathcal{H}(1, 1)\mathcal{H}(2, 2) - \mathcal{H}(1, 2)\mathcal{H}(2, 1)}{\mathcal{H}(1, 1) + \mathcal{H}(2, 2) - \mathcal{H}(1, 2) - \mathcal{H}(2, 1)}.$$

18.3. Qualitative Characterizations

18.3.1. The formal results in 18.2. can be summarized in various ways which make their meaning clearer. We begin with this criterion:

The fields (1, 1), (2, 2) form one *diagonal* of the matrix scheme of Fig. 27, the fields (1, 2), (2, 1) form the other *diagonal.*

We say that two sets of numbers E and F are *separated* either if every element of E is greater than every element of F, or if every element of E is smaller than every element of F.

Consider now the Cases (A), (B_1), (B_2) of 18.2. In the first two cases the game is strictly determined and the elements on one diagonal of the matrix are not separated from those on the other.[2] In the last case the game is not strictly determined, and the elements on one diagonal of the matrix are separated from those on the other.[3]

Thus separation of the diagonals is necessary and sufficient for the game not being strictly determined. This criterion was obtained subject to the use made in 18.2. of the adjustments of 18.1.3. But the three processes of adjustment described in 18.1.3. affect neither strict determinateness nor separation of the diagonals.[4] Hence our first criterion is always valid. We restate it:

[1] $\{1, 0\}$ or $\{0, 1\}$.

[2] Case (A): $\mathcal{H}(1, 1) \geqq \mathcal{H}(1, 2) \geqq \mathcal{H}(2, 2)$ by (18:2). Case (B_1): $\mathcal{H}(1, 1) > \mathcal{H}(1, 2) \geqq \mathcal{H}(2, 2)$ by (18:7).

[3] Case (B_2): $\mathcal{H}(1, 1) \geqq \mathcal{H}(2, 2) > \mathcal{H}(1, 2) \geqq \mathcal{H}(2, 1)$ by (18:10).

[4] The first is evident since these are only changes in notation, inessential for the game. The second is immediately verified.

(18:A) The game is not strictly determined if and only if the elements on one diagonal of the matrix are separated from those on the other.

18.3.2. In case (B_2), i.e. when the game is not strictly determined, both the (unique) $\vec{\xi}$ of $\bar{A}$ and the (unique) $\vec{\eta}$ of $\bar{B}$ which we found, have both components $\neq 0$. This, as well as the statement of uniqueness, is unaffected by adjustments described in 18.1.3.[1] So we have:

(18:B) If the game is not strictly determined, then there exists only one good strategy $\vec{\xi}$ (i.e. in $\bar{A}$) and only one good strategy $\vec{\eta}$ (i.e. in $\bar{B}$), and both have both their components positive.

I.e. both players must really resort to mixed strategies.

According to (18:B) no component of $\vec{\xi}$ or $\vec{\eta}$ ($\vec{\xi}$ in $\bar{A}$, $\vec{\eta}$ in $\bar{B}$) is zero. Hence the criterion of 17.9. shows that the argument which preceded (18:11)—which was then sufficient without being necessary—is now necessary (and sufficient). Hence (18:11) must be satisfied, and therefore all of its consequences are true. This applies in particular to the values ξ_1, ξ_2, η_1, η_2 given after (18:11), and to the value of v′ given in (18:13). All these formulae thus apply whenever the game is not strictly determined.

18.3.3. We now formulate another criterion:

In a general matrix $\mathcal{H}(\tau_1, \tau_2)$—cf. Fig. 15 on p. 99—(we allow for a moment any β_1, β_2) we say that a row (say τ_1') or a column (say τ_2') *majorizes* another row (say τ_1'') or column (say τ_2''), respectively, if this is true for their corresponding elements without exception. I.e. if $\mathcal{H}(\tau_1', \tau_2) \geqq \mathcal{H}(\tau_1'', \tau_2)$ for all τ_2, or if $\mathcal{H}(\tau_1, \tau_2') \geqq \mathcal{H}(\tau_1, \tau_2'')$ for all τ_1.

This concept has a simple meaning: It means that the choice of τ_1' is at least as good for player 1 as that of τ_1''—or that the choice of τ_2' is at most as good for player 2 as that of τ_2''—and that this is so in both cases irrespective of what the opponent does.[2]

Let us now return to our present problem ($\beta_1 = \beta_2 = 2$). Consider again the Cases (A), (B_1), (B_2) of 18.2. In the first two cases a row or a column majorizes the other.[3] In the last case neither is true.[4]

Thus the fact that a row or a column majorizes the other is necessary and sufficient for Γ being strictly determined. Like our first criterion this is subject to the use made in 18.2. of the adjustments made in 18.1.3. And, as there, those processes of adjustment affect neither strict determinateness nor majorization of rows or columns. Hence our present criterion too is always valid. We restate it:

(18:C) The game Γ is strictly determined if and only if a row or a column majorizes the other.

[1] These too are immediately verified.

[2] This is, of course, an exceptional occurrence: In general the relative merits of two alternative choices will depend on what the opponent does.

[3] Case (A): Column 1 majorizes column 2, by (18:2) Case (B_1): Row 1 majorizes row 2 by (18:7).

[4] Case (B_2): (18:10) excludes all four possibilities, as is easily verified.

18.3.4. That the condition of (18:C) is sufficient for strict determinateness is not surprising: It means that for one of the two players one of his possible choices is under all conditions at least as good as the other (cf. above). Thus he knows what to do and his opponent knows what to expect, which is likely to imply strict determinateness.

Of course these considerations imply a speculation on the rationality of the behavior of the other player, from which our original discussion is free. The remarks at the beginning and at the end of 15.8. apply to a certain extent to this, much simpler, situation.

What really matters in this result (18:C) however is that the necessity of the condition is also established; i.e. that nothing more subtle than outright majorization of rows or columns can cause strict determinateness.

It should be remembered that we are considering the simplest possible case: $\beta_1 = \beta_2 = 2$. We shall see in 18.5. how conditions get more involved in all respects when β_1, β_2 increase.

18.4. Discussion of Some Specific Games. (Generalized Forms of Matching Pennies)

18.4.1. The following are some applications of the results of 18.2. and 18.3.

(*a*) Matching Pennies in its ordinary form, where the $\mathcal{H}$ matrix of Figure 27 is given by Figure 12 on p. 94. We know that this game has the value

$$v' = 0$$

and the (unique) good strategies

$$\vec{\xi} = \vec{\eta} = \{\tfrac{1}{2}, \tfrac{1}{2}\}$$

(Cf. 17.1. The formulae of 18.2. will, of course, give this immediately.)

18.4.2. (b) Matching Pennies, where matching on heads gives a double premium. Thus the matrix of Figure 27 differs from that of Figure 12 by the doubling of its (1, 1) element:

	1	2
1	2	−1
2	−1	1

Figure 28*a*.

The diagonals are separated (1 and 2 are > than −1), hence the good strategies are unique and mixed (cf. (18:A), (18:B)). By using the pertinent formulae of case (B_2) in 18.2.5., we obtain the value

$$v' = \tfrac{1}{5}$$

and the good strategies

$$\vec{\xi} = \{\tfrac{2}{5}, \tfrac{3}{5}\}, \qquad \vec{\eta} = \{\tfrac{2}{5}, \tfrac{3}{5}\}.$$

It will be observed that the premium put on matching heads has increased the value of a play for player 1 who tries to match. It also causes him to choose heads less frequently, since the premium makes this choice plausible and therefore dangerous. The direct threat of extra loss by being matched on heads influences player 2 in the same way. This verbal argument has some plausibility but is certainly not stringent. Our formulae which yielded this result, however, were stringent.

18.4.3. (c) Matching Pennies, where matching on heads gives a double premium but failing to match on a choice (by player 1) of heads gives a triple penalty. Thus the matrix of Figure 27 is modified as follows:

	1	2
1	2	−3
2	−1	1

Figure 28*b*.

The diagonals are separated (1 and 2, are > than −1, −3), hence the good strategies are unique and mixed (cf. as before). The formulae used before give the value

$$v' = -\tfrac{1}{7},$$

and the good strategies

$$\vec{\xi} = \{\tfrac{2}{7}, \tfrac{5}{7}\}, \qquad \vec{\eta} = \{\tfrac{4}{7}, \tfrac{3}{7}\}.$$

We leave it to the reader to formulate a verbal interpretation of this result, in the same sense as before. The construction of other examples of this type is easy along the lines indicated.

18.4.4. (d) We saw in 18.1.2. that these variants of Matching Pennies are, in a way, the simplest forms of zero-sum two-person games. By this circumstance they acquire a certain general significance, which is further corroborated by the results of 18.2. and 18.3.: indeed we found there that this class of games exhibits in their simplest forms the conditions under which strictly and not-strictly determined cases alternate. As a further addendum in the same spirit we point out that the relatedness of these games to Matching Pennies stresses only one particular aspect. Other games which appear in an entirely different material garb may, in reality, well belong to this class. We shall give an example of this:

The game to be considered is an episode from the Adventures of Sherlock Holmes.[1,2]

[1] *Conan Doyle:* The Adventures of Sherlock Holmes, New York, 1938, pp. 550–551.

[2] The situation in question is of course again to be appraised as a paradigm of many possible conflicts in practical life. It was expounded as such by *O. Morgenstern:* Wirtschaftsprognose, Vienna, 1928, p. 98.

The author does not maintain, however, some pessimistic views expressed id. or in "Vollkommene Voraussicht und wirtschaftliches Gleichgewicht," Zeitschrift für Nationalökonomie, Vol. 6, 1934.

Accordingly our solution also answers doubts in the same vein expressed by *K. Menger:* Neuere Fortschritte in den exacten Wissenschaften, "Einige neuere Fortschritte in der exacten Behandlung Socialwissenschaftlicher Probleme," Vienna, 1936, pp. 117 and 131.

Sherlock Holmes desires to proceed from London to Dover and hence to the Continent in order to escape from Professor Moriarty who pursues him. Having boarded the train he observes, as the train pulls out, the appearance of Professor Moriarty on the platform. Sherlock Holmes takes it for granted—and in this he is assumed to be fully justified—that his adversary, who has seen him, might secure a special train and overtake him. Sherlock Holmes is faced with the alternative of going to Dover or of leaving the train at Canterbury, the only intermediate station. His adversary—whose intelligence is assumed to be fully adequate to visualize these possibilities—has the same choice. Both opponents must choose the place of their detrainment in ignorance of the other's corresponding decision. If, as a result of these measures, they should find themselves, *in fine*, on the same platform, Sherlock Holmes may with certainty expect to be killed by Moriarty. If Sherlock Holmes reaches Dover unharmed he can make good his escape.

What are the good strategies, particularly for Sherlock Holmes? This game has obviously a certain similarity to Matching Pennies, Professor Moriarty being the one who desires to match. Let him therefore be player 1, and Sherlock Holmes be player 2. Denote the choice to proceed to Dover by 1 and the choice to quit at the intermediate station by 2. (This applies to both τ_1 and τ_2.)

Let us now consider the $\mathcal{H}$ matrix of Figure 27. The fields (1, 1) and (2, 2) correspond to Professor Moriarty catching Sherlock Holmes, which it is reasonable to describe by a very high value of the corresponding matrix element,—say 100. The field (2, 1) signifies that Sherlock Holmes successfully escaped to Dover, while Moriarty stopped at Canterbury. This is Moriarty's defeat as far as the present action is concerned, and should be described by a big negative value of the matrix element—in the order of magnitude but smaller than the positive value mentioned above—say, -50. The field (1, 2) signifies that Sherlock Holmes escapes Moriarty at the intermediate station, but fails to reach the Continent. This is best viewed as a tie, and assigned the matrix element 0.

The $\mathcal{H}$ matrix is given by Figure 29:

	1	2
1	100	0
2	-50	100

Figure 29.

As in (b), (c) above, the diagonals are separated (100 is $>$ than 0, -50); hence the good strategies are again unique and mixed. The formulae used before give the value (for Moriarty)

$$v' = 40$$

and the good strategies ($\overrightarrow{\xi}$ for Moriarty, $\overrightarrow{\eta}$ for Sherlock Holmes):

$$\overrightarrow{\xi} = \{\tfrac{3}{5}, \tfrac{2}{5}\}, \qquad \overrightarrow{\eta} = \{\tfrac{2}{5}, \tfrac{3}{5}\}.$$

Thus Moriarty should go to Dover with a probability of 60%, while Sherlock Holmes should stop at the intermediate station with a probability of 60%,—the remaining 40% being left in each case for the other alternative.[1]

18.5. Discussion of Some Slightly More Complicated Games

18.5.1. The general solution of the zero-sum two-person game which we obtained in 17.8. brings certain alternatives and concepts particularly into the foreground: The presence or absence of strict determinateness, the value v′ of a play, and the sets $\bar{A}$, $\bar{B}$ of good strategies. For all these we obtained very simple explicit characterizations and determinations in 18.2. These became even more striking in the reformulation of those results in 18.3.

This simplicity may even lead to some misunderstandings. Indeed, the results of 18.2., 18.3. were obtained by explicit computations of the most elementary sort. The combinatorial criteria of (18:A), (18:C) in 18.3. for strict determinateness were—at least in their final form—also considerably more straightforward than anything we have experienced before. This may give occasion to doubts whether the somewhat involved considerations of 17.8. (and the corresponding considerations of 14.5. in the case of strict determinateness) were necessary,—particularly since they are based on the mathematical theorem of 17.6. which necessitates our analysis of linearity and convexity in 16. If all this could be replaced by discussions in the style of 18.2., 18.3. then our mode of discussion of 16. and 17. would be entirely unjustified.[2]

This is not so. As pointed out at the end of 18.3., the great simplicity of the procedures and results of 18.2. and 18.3. is due to the fact that they apply only to the simplest type of zero-sum two-person games: the Matching Pennies class of games, characterized by $\beta_1 = \beta_2 = 2$. For the general case the more abstract machinery of 16. and 17. seems so far indispensable.

[1] The narrative of *Conan Doyle*—excusably—disregards mixed strategies and states instead the actual developments. According to these Sherlock Holmes gets out at the intermediate station and triumphantly watches Moriarty's special train going on to Dover. *Conan Doyle's* solution is the best possible under his limitations (to pure strategies), insofar as he attributes to each opponent the course which we found to be the more probable one (i.e. he replaces 60% probability by certainty). It is, however, somewhat misleading that this procedure leads to Sherlock Holmes's complete victory, whereas, as we saw above, the odds (i.e. the value of a play) are definitely in favor of Moriarty. (Our result for $\overrightarrow{\xi}$, $\overrightarrow{\eta}$ yields that Sherlock Holmes is as good as 48% dead when his train pulls out from Victoria Station. Compare in this connection the suggestion in *Morgenstern*, loc. cit., p. 98, that the whole trip is unnecessary because the loser could be determined before the start.)

[2] Of course it would not lack rigor, but it would be an unnecessary use of heavy mathematical machinery on an elementary problem.

It may help to see these things in their right proportions if we show by some examples how the assertions of 18.2., 18.3. fail for greater values of β.

18.5.2. It will actually suffice to consider games with $\beta_1 = \beta_2 = 3$. In fact they will be somewhat related to Matching Pennies,—more general only by introduction of a third alternative.

Thus both players will have the alternative choices 1, 2, 3 (i.e. the values for τ_1, τ_2). The reader will best think of the choice 1 in terms of choosing "heads," the choice 2 of choosing "tails" and the choice 3 as something like "calling off." Player 1 again tries to match. If either player "calls off," then it will not matter whether the other player chooses "heads" or "tails,"—the only thing of importance is whether he chooses one of these two at all or whether he "calls off" too. Consequently the matrix has now the appearance of Figure 30:

τ_1 \ τ_2	1	2	3
1	1	-1	γ
2	-1	1	γ
3	α	α	β

Figure 30.

The four first elements—i.e. the first two elements of the first two rows—are the familiar pattern of Matching Pennies (cf. Fig. 12). The two fields with α are operative when player 1 "calls off" and player 2 does not. The two elements with γ are operative in the opposite case. The element with β refers to the case where both players "call off." By assigning appropriate values (positive, negative or zero) we can put a premium or a penalty on any one of these occurrences, or make it indifferent.

We shall obtain all the examples we need at this juncture by specializing this scheme,—i.e. by choosing the above α, β, γ appropriately.

18.5.3. Our purpose is to show that none of the results (18:A), (18:B), (18:C) of 18.3. is generally true.

Ad (18:A): This criterion of strict determinateness is clearly tied to the special case $\beta_1 = \beta_2 = 2$: For greater values of β_1, β_2 the two diagonals do not even exhaust the matrix rectangle, and therefore the occurrence on the diagonal alone cannot be characteristic as before.

Ad (18:B): We shall give an example of a game which is not strictly determined, but where nevertheless there exists a good strategy which is pure for one player (but of course not for the other). This example has the further peculiarity that one of the players has several good strategies, while the other has only one.

We choose in the game of Figure 30 α, β, γ as follows:

τ_1 \ τ_2	1	2	3
1	1	-1	0
2	-1	1	0
3	α	α	$-\delta$

Figure 31.

$\alpha > 0$, $\delta > 0$. The reader will determine for himself which combinations of "calling off" are at a premium or are penalized in the previously indicated sense.

This is a complete discussion of the game, using the criteria of 17.8.

For $\vec{\xi} = \{\frac{1}{2}, \frac{1}{2}, 0\}$ always $K(\vec{\xi}, \vec{\eta}) = 0$, i.e. with this strategy player 1 cannot lose. Hence $v' \geqq 0$. For $\vec{\eta} = \vec{\delta}^{\,3} = \{0, 0, 1\}$ always $K(\vec{\xi}, \vec{\eta}) \leqq 0$;[1] i.e. with this strategy player 2 cannot lose. Hence $v' \leqq 0$. Thus we have

$$v' = 0$$

Consequently $\vec{\xi}$ is a good strategy if and only if always $K(\vec{\xi}, \vec{\eta}) \geqq 0$ and $\vec{\eta}$ is a good strategy if and only if always $K(\vec{\xi}, \vec{\eta}) \leqq 0$.[2] The former is easily seen to be true if and only if

$$\xi_1 = \xi_2 = \tfrac{1}{2}, \qquad \xi_3 = 0,$$

and the latter if and only if

$$\eta_1 = \eta_2 \leqq \frac{\delta}{2(\alpha + \delta)}, \qquad \eta_3 = 1 - 2\eta_1.$$

Thus the set $\bar{A}$ of all good strategies $\vec{\xi}$ contains precisely one element, and this is not a pure strategy. The set $\bar{B}$ of all good strategies $\vec{\eta}$, on the other hand, contains infinitely many strategies, and one of them is pure: namely $\vec{\eta} = \vec{\delta}^{\,3} = \{0, 0, 1\}$.

The sets $\bar{A}$, $\bar{B}$ can be visualized by making use of the graphical representation of Figure 21 (cf. Figures 32, 33):

Ad (18:C): We shall give an example of a game which is strictly determined but in which no two rows and equally no two columns majorize each other. We shall actually do somewhat more.

18.5.4. Allow for a moment any β_1, β_2. The significance of the majorization of rows or of columns by each other was considered at the end of 18.3. It was seen to mean that one of the players had a simple direct motive

[1] It is actually equal to $-\delta\, \xi_3$.

[2] We leave to the reader the simple verbal interpretation of these statements.

for neglecting one of his possible choices in favor of another,—and this narrowed the possibilities in a way which could be ultimately connected with strict determinateness.

Specifically: If the row τ_1'' is majorized by the row τ_1'—i.e. if $\mathcal{H}(\tau_1'', \tau_2) \leqq \mathcal{H}(\tau_1', \tau_2)$ for all τ_2—then player 1 need never consider the choice τ_1'', since τ_1' is at least as good for him in every contingency. And: If the column τ_2'' majorizes the column τ_2'—i.e. if $\mathcal{H}(\tau_1, \tau_2'') \geqq \mathcal{H}(\tau_1, \tau_2')$ for all τ_1—then player 2 need never consider the choice τ_2'', since τ_2' is at least as good for him in every contingency. (Cf. loc. cit., particularly footnote 2 on p. 174. These are of course only heuristic considerations, cf. footnote 1, p. 182.)

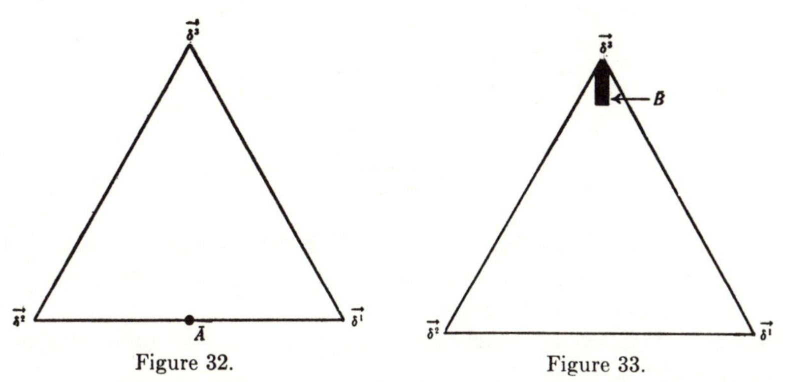

Figure 32. Figure 33.

Now we may use an even more general set-up: If the row τ_1'',—i.e. the player 1's pure strategy corresponding to τ_1''—is majorized by an average of all rows $\tau_1' \neq \tau_1''$—i.e. by a mixed strategy $\vec{\xi}$ with the component $\xi_{\tau_1''} = 0$—then it is still plausible to assume that player 1 need never consider the choice of τ_1'', since the other τ_1' are at least as good for him in every contingency. The mathematical expression of this situation is this:

$$\text{(18:14:a)} \qquad \begin{cases} \mathcal{H}(\tau_1'', \tau_2) \leqq \sum_{\tau_1=1}^{\beta_1} \mathcal{H}(\tau_1, \tau_2)\xi_{\tau_1} & \text{for all } \tau_2 \\ \vec{\xi} \text{ in } S_{\beta_1}, \qquad \xi_{\tau_1''} = 0. \end{cases}$$

The corresponding situation for player 2 arises if the column τ_2''—i.e. player 2's pure strategy corresponding to τ_2''—majorizes an average of all columns $\tau_2' \neq \tau_2''$,—i.e. a mixed strategy $\vec{\eta}$ with the component $\eta_{\tau_2''} = 0$. The mathematical expression of this situation is this:

$$\text{(18:14:b)} \qquad \begin{cases} \mathcal{H}(\tau_1, \tau_2'') \geqq \sum_{\tau_2=1}^{\beta_2} \mathcal{H}(\tau_1, \tau_2)\eta_{\tau_2} & \text{for all } \tau_1 \\ \vec{\eta} \text{ in } S_{\beta_2}, \qquad \eta_{\tau_2''} = 0. \end{cases}$$

The conclusions are the analogues of the above.

Thus a game in which (18:14:a) or (18:14:b) occurs, permits of an immediate and plausible narrowing of the possible choices for one of the players.[1]

18.5.5. We are now going to show that the applicability of (18:14:a), (18:14:b) is very limited: We shall specify a strictly determined game in which neither (18:14:a) nor (18:14:b) is ever valid.

Let us therefore return to the class of games of Figure 30. ($\beta_1 = \beta_2 = 3$). We choose $0 < \alpha < 1$, $\beta = 0$, $\gamma = -\alpha$:

τ_1 \ τ_2	1	2	3
1	1	-1	$-\alpha$
2	-1	1	$-\alpha$
3	α	α	0

Figure 34.

The reader will determine for himself which combinations of "calling off" are at a premium or are penalized in the previously indicated sense.

This is a discussion of the game:

The element (3, 3) is clearly a saddle point, so the game is strictly determined and

$$v = v' = 0.$$

It is not difficult to see now (with the aid of the method used in 18.5.3.), that the set $\bar{A}$ of all good strategies $\vec{\xi}$ as well as the set $\bar{B}$ of all good strategies $\vec{\eta}$, contains precisely one element: the pure strategy $\vec{\delta}^3 = \{0, 0, 1\}$.

On the other hand, the reader will experience little trouble in verifying that neither (18:14:a) nor (18:14:b) is ever valid here, i.e. that in Figure 34 no row is majorized by any average of the two other rows, and that no column majorizes any average of the other two columns.

18.6. Chance and Imperfect Information

18.6.1. The examples discussed in the preceding paragraphs make it clear that the role of chance—more precisely, of probability—in a game is not necessarily the obvious one, that which is directly provided for in the rules of the game. The games described in Figures 27 and 30 have rules

[1] This is of course a thoroughly heuristic argument. We do not need it, since we have the complete discussions of 14.5. and of 17.8. But one might suspect that it can be used to replace or at least to simplify those discussions. The example which we are going to give in the text seems to dispel any such hope.

There is another course which might produce results: If (18:14:a) or (18:14:b) holds, then a combination of it with 17.8. can be used to gain information about the sets of good strategies, $\bar{A}$ and $\bar{B}$. We do not propose to take up this subject here.

which do not provide for chance; the moves are personal without exception.[1] Nevertheless we found that most of them are not strictly determined, i.e. that their good strategies are mixed strategies involving the explicit use of probabilities.

On the other hand, our analysis of those games in which perfect information prevails showed that these are always strictly determined,—i.e. that they have good strategies which are pure strategies, involving no probabilities at all. (Cf. 15.)

Thus from the point of view of the players' behavior—i.e. of the strategies to be used—the important circumstance is whether the game is strictly determined or not, and not at all whether it contains any chance moves.

The results of 15. on games in which perfect information prevails indicate that there exists a close connection between strict determinateness and the rules which govern the players' state of information. To establish this point quite clearly, and in particular to show that the presence of chance moves is quite irrelevant, we shall now show this: In every (zero-sum two-person) game any chance move can be replaced by a combination of personal moves, so that the strategical possibilities of the game remain exactly the same. It will be necessary to allow for rules involving imperfect information of the players, but this is just what we want to demonstrate: That imperfect information comprises (among other things) all possible consequences of explicit chance moves.[2]

18.6.2. Let us consider, accordingly, a (zero-sum two-person) game Γ, and in it a chance move $\mathfrak{M}_\kappa$.[3] Enumerate the alternatives as usual by $\sigma_\kappa = 1, \cdots, \alpha_\kappa$ and assume that their probabilities $p_\kappa^{(1)}, \cdots, p_\kappa^{(\alpha_\kappa)}$ are all equal to $1/\alpha_\kappa$.[4] Now replace $\mathfrak{M}_\kappa$ by two personal moves $\mathfrak{M}'_\kappa$, $\mathfrak{M}''_\kappa$.

[1] The reduction of all games to the normalized form shows even more: It proves that every game is equivalent to one without chance moves, since the normalized form contains only personal moves.

[2] A direct way of removing chance moves exists of course after the introduction of the (pure) strategies and the umpire's choice, as described in 11.1. Indeed—as the last step in bringing a game into its normalized form—we eliminated the remaining chance moves by the explicit introduction of expectation values in 11.2.3.

But we now propose to eliminate the chance moves without upsetting the structure of the game so radically. We shall replace each chance move individually by personal moves (by two moves, as will be seen), so that their respective roles in determining the players' strategies will always remain differentiated and individually appraisable. This detailed treatment is likely to give a clearer idea of the structural questions involved than the summary procedure mentioned above.

[3] For our present purposes it is irrelevant whether the characteristics of $\mathfrak{M}_\kappa$ depend upon the previous course of the play or not.

[4] This is no real loss of generality. To see this, assume that the probabilities in question have arbitrary rational values,—say $r_1/t, \cdots, r_{\alpha_\kappa}/t$ ($r_1, \cdots, r_{\alpha_\kappa}$ and t integers). (Herein lies an actual restriction—but an arbitrary small one—since any probabilities can be approximated by rational ones to any desired degree.)

Now modify the chance move $\mathfrak{M}_\kappa$ so that it has $r_1 + \cdots + r_{\alpha_\kappa} = t$ alternatives (instead of α_κ), designated by $\sigma'_\kappa = 1, \cdots, t$ (instead of $\sigma_\kappa = 1, \cdots, \alpha_\kappa$); and so that each of the first r_1 values of σ'_κ has the same effect on the play as $\sigma_\kappa = 1$, each of the next r_2 values of σ'_κ the same as $\sigma_\kappa = 2$, etc., etc. Then giving all $\sigma'_\kappa = 1, \cdots, t$ the same probability $1/t$, has the same effect as giving $\sigma_\kappa = 1, \cdots, \alpha_\kappa$ the original probabilities $r_1/t, \cdots, r_{\alpha_\kappa}/t$.

$\mathfrak{M}'_\kappa$ and $\mathfrak{M}''_\kappa$ are personal moves of players 1 and 2 respectively. Both have α_κ alternatives; we denote the corresponding choices by $\sigma'_\kappa = 1, \cdots, \alpha_\kappa$ and $\sigma''_\kappa = 1, \cdots, \alpha_\kappa$. It is immaterial in which order these moves are made, but we prescribe that both moves must be made without any information concerning the outcome of any moves (including the other move $\mathfrak{M}'_\kappa$, $\mathfrak{M}''_\kappa$). We define a function $\delta(\sigma', \sigma'')$ by this matrix scheme. (Cf. Figure 35. The matrix element is $\delta(\sigma', \sigma'')$.[1]). The influence of $\mathfrak{M}'_\kappa$, $\mathfrak{M}''_\kappa$—i.e. of the corresponding (personal) choices σ'_κ, σ''_κ—on the outcome of the game is the same as that of $\mathfrak{M}_\kappa$ would have been with the corresponding (chance) choice $\sigma_\kappa = \delta(\sigma'_\kappa, \sigma''_\kappa)$. We denote this new game by Γ^*. We claim that the strategical possibilities of Γ^* are the same as those of Γ.

σ' \ σ''	1	2	....	$\alpha_\kappa - 1$	α_κ
1	1	α_κ	....	3	2
2	2	1	....	4	3
⋮	⋮	⋮	....	⋮	⋮
$\alpha_\kappa - 1$	$\alpha_\kappa - 1$	$\alpha_\kappa - 2$	....	1	α_κ
α_κ	α_κ	$\alpha_\kappa - 1$	....	2	1

Figure 35.

18.6.3. Indeed: Let player 1 use in Γ^* a given mixed strategy of Γ with the further specification concerning the move $\mathfrak{M}'_\kappa$,[2] to choose all $\sigma'_\kappa = 1, \cdots, \alpha_\kappa$ with the same probability $1/\alpha_\kappa$. Then the game Γ^*—with this strategy of player 1—will be from player 2's point of view the same as Γ. This is so because any choice of his at $\mathfrak{M}''_\kappa$ (i.e. any $\sigma''_\kappa = 1, \cdots, \alpha_\kappa$) produces the same result as the original chance move $\mathfrak{M}_\kappa$: One look at Figure 35 will show that the $\sigma'' = \sigma''_\kappa$ column of that matrix contains every number $\sigma = \delta(\sigma', \sigma'') = 1, \cdots, \alpha_\kappa$ precisely once,—i.e. that $\delta(\sigma', \sigma'')$ will assume every value $1, \cdots, \alpha_\kappa$ (owing to player 1's strategy) with the same probability $1/\alpha_\kappa$, just as $\mathfrak{M}_\kappa$ would have done. So from player 1's point of view, Γ^* is at least as good as Γ.

The same argument with players 1 and 2 interchanged—hence with the rows of the matrix in Figure 35 playing the roles which the column had above—shows that from player 2's point of view too Γ^* is at least as good as Γ.

[1] Arithmetically

$$\delta(\sigma', \sigma'') \begin{cases} = \sigma' - \sigma'' + 1 & \text{for} \quad \sigma' \geqq \sigma'' \\ = \sigma' - \sigma'' + 1 + \alpha_\kappa & \text{for} \quad \sigma' < \sigma'' \end{cases}$$

Hence $\delta(\sigma', \sigma'')$ is always one of the numbers $1, \cdots, \alpha_\kappa$.

[2] $\mathfrak{M}'_\kappa$ is his personal move, so his strategy must provide for it in Γ^*. There was no need for this in Γ, since $\mathfrak{M}_\kappa$ was a chance move.

Since the viewpoints of the two players are opposite, this means that Γ^* and Γ are equivalent.[1]

18.7. Interpretation of This Result

18.7.1. Repeated application to all chance moves of Γ, of the operation described in 18.6.2., 18.6.3., will remove them all,—thus proving the final contention of 18.6.1. The meaning of this result may be even better understood if we illustrate it by some practical instances of this manipulation.

(A) Consider the following quite elementary "game of chance." The two players decide, by a 50%–50% chance device, who pays the other one unit. The application of the device of 18.6.2., 18.6.3. transforms this game, which consists of precisely one chance move, into one of two personal moves. A look at the matrix of Figure 35 for $\alpha_\kappa = 2$ —with the $\delta(\sigma', \sigma'')$ values 1, 2 replaced by the actual payments 1, -1 —shows that it coincides with Figure 12. Remembering 14.7.2., 14.7.3. we see that this means—what is plain enough directly—that this is the game of Matching Pennies.

I.e.: Matching Pennies is the natural device to produce the probabilities $\frac{1}{2}$, $\frac{1}{2}$ by personal moves and imperfect information. (Recall 17.1.!)

(B) Modify (A) so as to allow for a "tie": The two players decide by a $33\frac{1}{3}$%, $33\frac{1}{3}$%, $33\frac{1}{3}$% chance device who pays the other one unit, or whether nobody pays anything at all. Apply again the device of 18.6.2., 18.6.3. Now the matrix of Figure 35 with $\alpha_\kappa = 3$—with the $\delta(\sigma', \sigma'')$ values 1, 2, 3 replaced by the actual payments 0, 1, -1—coincides with Figure 13. By 14.7.2., 14.7.3. we see that this is the game of Stone, Paper, Scissors.

I.e., Stone, Paper, Scissors is the natural device to produce the probabilities $\frac{1}{3}$, $\frac{1}{3}$, $\frac{1}{3}$ by personal moves and incomplete information. (Recall 17.1.!)

18.7.2. (C) The $\delta(\sigma', \sigma'')$ of Figure 35 can be replaced by another function, and even the domains $\sigma'_\kappa = 1, \cdots, \alpha_\kappa$ and $\sigma''_\kappa = 1, \cdots, \alpha_\kappa$ by other domains $\sigma'_\kappa = 1, \cdots, \alpha'_\kappa$ and $\sigma''_\kappa = 1, \cdots, \alpha''_\kappa$, provided that the following remains true: Every column of the matrix of Figure 35 contains each number $1, \cdots, \alpha_\kappa$ the same number of times,[2] and every row contains each number $1, \cdots, \alpha_\kappa$ the same number of times.[3] Indeed, the considerations of 18.6.2. made use of these two properties of $\delta(\sigma'_\kappa, \sigma''_\kappa)$ (and of $\alpha'_\kappa, \alpha''_\kappa$) only.

It is not difficult to see that the precaution of "cutting" the deck before dealing cards falls into this category. When one of the 52 cards has to be chosen by a chance move, with probability $\frac{1}{52}$, this is usually achieved by "mixing" the deck. This is meant to be a chance move, but if the player who mixes is dishonest, it may turn out to be a "personal" move of his.

[1] We leave it to the reader to cast these considerations into the precise formalism of 11. and 17.2., 17.8.: This presents absolutely no difficulties, but it is somewhat lengthy. The above verbal arguments convey the essential reason of the phenomenon under consideration in a clearer and simpler way—we hope.

[2] Hence $\alpha'_\kappa/\alpha_\kappa$ times; consequently α'_κ must be a multiple of α_κ.

[3] Hence $\alpha''_\kappa/\alpha_\kappa$ times; consequently α''_κ must be a multiple of α_κ.

As a protection against this, the other player is permitted to point out the place in the mixed deck, from which the card in question is to be taken, by "cutting" the deck at that point. This combination of two moves—even if they are personal—is equivalent to the originally intended chance move. The lack of information is, of course, the necessary condition for the effectiveness of this device.

Here $\alpha_\kappa = 52$, $\alpha'_\kappa = 52! =$ the number of possible arrangements of the deck, $\alpha''_\kappa = 52$ the number of ways of "cutting." We leave it to the reader to fill in the details and to choose the $\delta(\sigma'_\kappa, \sigma''_\kappa)$ for this set-up.[1]

19. Poker and Bluffing

19.1. Description of Poker

19.1.1. It has been stressed repeatedly that the case $\beta_1 = \beta_2 = 2$ as discussed in 18.3. and more specifically in 18.4., comprises only the very simplest zero-sum two-person games. We then gave in 18.5. some instances of the complications which can arise in the general zero-sum two-person game, but the understanding of the implications of our general result (i.e. of 17.8.) will probably gain more by the detailed discussion of a special game of the more complicated type. This is even more desirable because for the games with $\beta_1 = \beta_2 = 2$ the choices of the τ_1, τ_2, called (pure) strategies, scarcely deserve this name: just calling them "moves" would have been less exaggerated. Indeed, in these extremely simple games there could be hardly any difference between the extensive and the normalized form; and so the identity of moves and strategies, a characteristic of the normalized form, is inescapable in these games. We shall now consider a game in the extensive form in which the player has several moves, so that the passage to the normalized form and to strategies is no longer a vacuous operation.

19.1.2. The game of which we give an exact discussion is Poker.[2] However, actual Poker is really a much too complicated subject for an exhaustive discussion and so we shall have to subject it to some simplifying modifica-

[1] We assumed that the mixing is used to produce only one card. If whole "hands" are dealt, "cutting" is not an absolute safeguard. A dishonest mixer can produce correlations within the deck which one "cut" cannot destroy, and the knowledge of which gives this mixer an illegitimate advantage.

[2] The general considerations concerning Poker and the mathematical discussions of the variants referred to in the paragraphs which follow, were carried out by *J. von Neumann* in 1926–28, but not published before. (Cf. a general reference in "Zur Theorie der Gesellschaftsspiele," Math. Ann., Vol. 100 [1928]). This applies in particular to the symmetric variant of 19.4.-19.10., the variants (A), (B) of 19.11.-19.13., and to the entire interpretation of "Bluffing" which dominates all these discussions. The unsymmetric variant (C) of 19.14.-19.16. was considered in 1942 for the purposes of this publication.

The work of *E. Borel* and *J. Ville*, referred to in footnote 1 on p. 154, also contains considerations on Poker (Vol. IV, 2: "Applications aux Jeux de Hasard," Chap. V: "Le jeu de Poker"). They are very instructive, but mainly evaluations of probabilities applied to Poker in a more or less heuristic way, without a systematic use of any underlying general theory of games.

A definite strategical phase of Poker ("La Relance" = "The Overbid") is analyzed on pp. 91–97 loc. cit. This may be regarded also as a simplified variant of Poker,—

tions, some of which are, indeed, quite radical.[1] It seems to us, nevertheless, that the basic idea of Poker and its decisive properties will be conserved in our simplified form. Therefore it will be possible to base general conclusions and interpretations on the results which we are going to obtain by the application of the theory previously established.

To begin with, Poker is actually played by any number of persons,[2] but since we are now in the discussion of zero-sum two-person games, we shall set the number of players at two.

The game of Poker begins by dealing to each player 5 cards out of a deck.[3] The possible combinations of 5 which he may get in this way—there are 2,598,960 of them[4]—are called "hands" and arranged in a linear order, i.e. there is an exhaustive rule defining which hand is the strongest of all, which is the second, third, · · · strongest down to the weakest.[5] Poker is played in many variants which fall into two classes: "Stud" and "Draw" games. In a Stud game the player's hand is dealt to him in its entirety at the very beginning, and he has to keep it unchanged throughout the entire play. In "Draw" games there are various ways for a player to exchange all or part of his hand, and in some variants he may get his hand in several successive stages in the course of the play. Since we wish to discuss the simplest possible form, we shall examine the Stud game only.

In this case there is no point in discussing the hands as hands, i.e. as combinations of cards. Denoting the total number of hands by S—$S = 2,598,960$ for a full deck, as we saw—we might as well say that each

---

comparable to the two which we consider in 19.4.-19.10. and 19.14-19.16. It is actually closely related to the latter.

The reader who wishes to compare these two variants, may find the following indications helpful:

(I) Our bids a, b correspond to $1 + \alpha$, 1 loc. cit.

(II) The difference between our variant of 19.4.-19.10. and that in loc. cit. is this: If player 1 begins with a "low" bid, then our variant provides for a comparison of hands, while that in loc. cit. makes him lose the amount of the "low" bid unconditionally. I.e. we treated this initial "low" bid as "seeing"—cf. the discussion at the beginning of 19.14., particularly footnote 1 on p. 211—while in loc. cit. it is treated as "passing." We believe that our treatment is a better approximation to the phase in question in real Poker; and in particular that it is needed for a proper analysis and interpretation of "Bluffing." For technical details cf. footnote 1 on p. 219.

[1] Cf. however 19.11. and the end of 19.16.

[2] The "optimum"—in a sense which we do not undertake to interpret—is supposed to be 4 or 5.

[3] This is occasionally a full deck of 52 cards, but for smaller numbers of participants only parts of it—usually 32 or 28—are used. Sometimes one or two extra cards with special functions, "jokers," are added.

[4] This holds for a full deck. The reader who is conversant with the elements of combinatorics will note that this is the number of "combinations without repetitions of 5 out of 52":

$$\binom{52}{5} = \frac{52 \cdot 51 \cdot 50 \cdot 49 \cdot 48}{1 \cdot 2 \cdot 3 \cdot 4 \cdot 5} = 2,598,960.$$

[5] This description involves the well known technical terms "Royal Flush," "Straight Flush," "Four of a Kind," "Full House," etc. There is no need for us to discuss them here.

player draws a number $s = 1, \cdots, S$ instead. The idea is that $s = S$ corresponds to the strongest possible hand, $s = S - 1$ to the second strongest, etc., and finally $s = 1$ to the weakest. Since a "square deal" amounts to assuming that all possible hands are dealt with the same probability, we must interpret the drawing of the above number s as a chance move, each one of the possible values $s = 1, \cdots, S$ having the same probability $1/S$. Thus the game begins with two chance moves: The drawing of the number s for player 1 and for player 2,[1] which we denote by s_1 and s_2.

19.1.3. The next phase of the general game of Poker consists of the making of "Bids" by the players. The idea is that after one of the players has made a bid, which involves a smaller or greater amount of money, his opponent has the choice of "Passing," "Seeing," or "Overbidding." Passing means that he is willing to pay, without further argument, the amount of his last preceding bid (which is necessarily lower than the present bid). In this case it is irrelevant what hands the two players hold. The hands are not disclosed at all. "Seeing" means that the bid is accepted: the hands will be compared and the player with the stronger hand receives the amount of the present bid. "Seeing" terminates the play. "Overbidding" means that the opponent counters the present bid by a higher one, in which the roles of the players are reversed and the previous bidder has the choice of Passing, Seeing or Overbidding, etc.[2]

19.2. Bluffing

19.2.1. The point in all this is that a player with a strong hand is likely to make high bids—and numerous overbids—since he has good reason to expect that he will win. Consequently a player who has made a high bid, or overbid, may be assumed by his opponent—*a posteriori!*—to have a strong hand. This may provide the opponent with a motive for "Passing." However, since in the case of "Passing" the hands are not compared, even a player with a weak hand may occasionally obtain a gain against a stronger opponent by creating the (false) impression of strength by a high bid, or by overbid,—thus conceivably inducing his opponent to pass.

This maneuver is known as "Bluffing." It is unquestionably practiced by all experienced players. Whether the above is its real motivation may be doubted; actually a second interpretation is conceivable. That is if a player is known to bid high only when his hand is strong, his opponent is likely to pass in such cases. The player will, therefore, not be able to collect on high bids, or on numerous overbids, in just those cases where his actual strength gives him the opportunity. Hence it is desirable for him to create

[1] In actual Poker the second player draws from a deck from which the first player's hand has already been removed. We disregard this as we disregard some other minor complications of Poker.

[2] This scheme is usually complicated by the necessity of making unconditional payments, the "ante," at the start,—in some variants for the first bidder, in others for all those who wish to participate, again in others extra payments are required for the privilege of drawing, etc. We disregard all this.

uncertainty in his opponent's mind as to this correlation,—i.e. to make it known that he does occasionally bid high on a weak hand.

To sum up: Of the two possible motives for Bluffing, the first is the desire to give a (false) impression of strength in (real) weakness; the second is the desire to give a (false) impression of weakness in (real) strength. Both are instances of inverted signaling (cf. 6.4.3.),—i.e. of misleading the opponent. It should be observed however that the first type of Bluffing is most successful when it "succeeds," i.e. when the opponent actually "passes," since this secures the desired gain; while the second is most successful when it "fails," i.e. when the opponent "sees," since this will convey to him the desired confusing information.[1]

19.2.2. The possibility of such indirectly motivated—hence apparently irrational—bids has also another consequence. Such bids are necessarily risky, and therefore it can conceivably be worth while to make them riskier by appropriate counter measures,—thus restricting their use by the opponent. But such counter measures are *ipso facto* also indirectly motivated moves.

We have expounded these heuristic considerations at such length because our exact theory makes a disentanglement of all these mixed motives possible. It will be seen in 19.10. and in 19.15.3., 19.16.2. how the phenomena which surround Bluffing can be understood quantitatively, and how the motives are connected with the main strategic features of the game, like possession of the initiative, etc.

19.3. Description of Poker (Continued)

19.3.1. Let us now return to the technical rules of Poker. In order to avoid endless overbidding the number of bids is usually limited.[2] In order to avoid unrealistically high bids—with hardly foreseeable irrational effects upon the opponent—there are also maxima for each bid and overbid. It is also customary to prohibit too small overbids; we shall subsequently indicate what appears to be a good reason for this (cf. the end of 19.13.). We shall express these restrictions on the size of bids and overbids in the simplest possible form: We shall assume that two numbers a, b

$$a > b > 0$$

[1] At this point we might be accused once more of disregarding our previously stated guiding principle; the above discussion obviously assumes a series of plays (so that statistical observation of the opponent's habits is possible) and it has a definitely "dynamical" character. And yet we have repeatedly professed that our considerations must be applicable to one isolated play and also that they are strictly statical.

We refer the reader to 17.3., where this apparent contradiction has been carefully examined. Those considerations are fully valid in this case too, and should justify our procedure. We shall add now only that our inconsistency—the use of many plays and of a dynamical terminology—is a merely verbal one. In this way we were able to make our discussions more succinct and more akin to the way that these things are talked about in everyday language. But in 17.3. it was elaborated how all these questionable pictures can be replaced by the strictly static problem of finding a good strategy.

[2] This is the stop rule of 7.2.3.

are given *ab initio*, and that for every bid there are only two possibilities: the bid may be "high," in which case it is a; or "low," in which case it is b. By varying the ratio a/b—which is clearly the only thing that matters—we can make the game risky when a/b is much greater than 1, or relatively safe when a/b is only a little greater than 1.

The limitation of the number of bids and overbids will now be used for a simplification of the entire scheme. In the actual play one of the players begins with the initial bid; after that the players alternate.

The advantage or disadvantage contained in the possession of the initiative by one player—but concurrent with the necessity of acting first!—constitutes an interesting problem in itself. We shall discuss an (unsymmetric) form of Poker where this plays a role in 19.14., 19.15. But we wish at first to avoid being saddled with this problem too. In other words, we wish to avoid for the moment all deviations from symmetry, so as to obtain the other essential features of Poker in their purest and simplest form. We shall therefore assume that the two players both make initial bids, each one ignorant of the other's choice. Only after both have made this bid is each one informed of what the other did, i.e. whether his bid was "high" or "low."

19.3.2. We simplify further by giving to the players only the choice of "Passing" or "Seeing," i.e. by excluding "Overbidding." Indeed, "Overbidding" is only a more elaborate and intensive expression of the tendency which is already contained in a high initial bid. Since we wish to do things as simply as possible, we shall avoid providing several channels for the same tendency. (Cf. however (C) in 19.11. and 19.14., 19.15.).

Accordingly we prescribe these conditions: Consider the moment when both players are informed of each other's bids. If it then develops that both bid "high" or that both bid "low," then the hands are compared and the player with the stronger hand receives the amount a or b respectively from his opponent. If their hands are equal, no payment is made. If on the other hand one bids "high" and one bids "low," then the player with the low bid has the choice of "Passing" or "Seeing." "Passing" means that he pays to the opponent the amount of the low bid (without any consideration of their hands). "Seeing" means that he changes over from his "low" bid to the "high" one, and the situation is treated as if they both had bid "high" in the first place.

19.4. Exact Formulation of the Rules

19.4. We can now sum up the preceding description of our simplified Poker by giving an exact statement of the rules agreed upon:

First, by a chance move each player obtains his "hand," a number $s = 1, \cdots S$, each one of these numbers having the same probability $1/S$. We denote the hands of players 1, 2, by s_1, s_2 respectively.

After this each player will, by a personal move, choose either a or b, the "high" or "low" bid. Each player makes his choice (bid) informed about his own hand, but not about his opponent's hand or choice (bid). Lastly, each player is informed about the other's choice but not about his hand. (Each still knows his own hand and choice.) If it turns out that one bids "high" and the other "low," then the latter has the choice of "Seeing" or "Passing."

This is the play. When it is concluded the payments are made as follows: If both players bid "high," or if one bids "high," and the other bids "low" but subsequently "Sees," then for $s_1 \begin{matrix} > \\ = \\ < \end{matrix} s_2$ player 1 obtains from player 2 the amount $\begin{matrix} a \\ 0 \\ -a \end{matrix}$ respectively. If both players bid "low," then for $s_1 \begin{matrix} > \\ = \\ < \end{matrix} s_2$ player 1 obtains from player 2 the amount $\begin{matrix} b \\ 0 \\ -b \end{matrix}$ respectively. If one player bids "high," and the other bids "low" and subsequently "Passes," then the "high bidder" being $\begin{matrix} 1 \\ 2 \end{matrix}$, player 1 obtains from player 2 the amount $\begin{matrix} b \\ -b \end{matrix}$.[1]

19.5. Description of the Strategies

19.5.1. A (pure) strategy in this game consists clearly of the following specifications: To state for every "hand" $s = 1, \cdots, S$ whether a "high" or a "low" bid will be made, and in the latter case the further statement whether, if this "low" bid runs into a "high" bid of the opponent, the player will "See" or "Pass." It is simpler to describe this by a numerical index $i_s = 1, 2, 3$; $i_s = 1$ meaning a "high" bid; $i_s = 2$ meaning a "low" bid with subsequent "Seeing" (if the occasion arises); $i_s = 3$ meaning a "low" bid with subsequent "Passing" (if the occasion arises). Thus the strategy is a specification of such an index i_s for every $s = 1, \cdots, S$,—i.e. of the sequence $i_1, \cdots, i_S$.

This applies to both players 1 and 2. Accordingly we shall denote the above strategy by $\Sigma_1(i_1, \cdots i_S)$ or $\Sigma_2(j_1, \cdots, j_S)$.

Thus each player has the same number of strategies: as many as there are sequences $i_1, \cdots, i_S$,—i.e. precisely 3^S. With the notations of 11.2.2.

$$\beta_1 = \beta_2 = \beta = 3^S.$$

[1] For the sake of absolute formal correctness this should still be arranged according to the patterns of 6. and 7. in Chapter II. Thus the two first-mentioned chance moves (the dealing of hands) should be called moves 1 and 2; the two subsequent personal moves (the bids), moves 3 and 4; and the final personal move ("Passing" or "Seeing"), move 5.

In the case of move 5, both the player whose personal move it is, and the number of alternatives, depend on the previous course of the play as described in 7.1.2. and 9.1.5. (If both players bid "high" or both bid "low," then the number of alternatives is 1, and it does not matter to which player we ascribe this vacuous personal move. If one bids "high" and the other bids "low," then the personal move is the "low" bidder's).

A consistent use of the notations loc. cit. would also necessitate writing σ_1, σ_2 for s_1, s_2; σ_3, σ_4 for the "high" or "low" bid; σ_5 for "Passing" or "Seeing."

We leave it to the reader to iron out all these differences.

If we wanted to adhere rigorously to the notations of loc. cit., we should now enumerate the sequences $i_1, \cdots, i_S$ with a $\tau_1 = 1, \cdots, \beta$ and then denote the (pure) strategies of the players 1, 2 by $\Sigma_1^{\tau_1}, \Sigma_2^{\tau_2}$. But we prefer to continue with our present notations.

We must now express the payment which player 1 receives if the strategies $\Sigma_1(i_1, \cdots, i_S)$, $\Sigma_2(j_1, \cdots, j_S)$ are used by the two players. This is the matrix element $\mathcal{H}(i_1, \cdots, i_S|j_1, \cdots, j_S)$.[1]

If the players have actually the "hands" s_1, s_2 then the payment received by player 1 can be expressed in this way (using the rules stated above): It is $\mathcal{L}_{sgn(s_1-s_2)}(i_{s_1}, j_{s_2})$ where $sgn(s_1 - s_2)$ is the sign of $s_1 - s_2$,[2] and where the three functions

$$\mathcal{L}_+(i, j), \qquad \mathcal{L}_0(i, j), \qquad \mathcal{L}_-(i, j) \qquad i, j = 1, 2, 3.$$

can be represented by the following matrix schemes:[3]

$i \backslash j$	1	2	3
1	a	a	b
2	a	b	b
3	$-b$	b	b

$\mathcal{L}_+(i, j)$

Figure 36.

$i \backslash j$	1	2	3
1	0	0	b
2	0	0	0
3	$-b$	0	0

$\mathcal{L}_0(i, j)$

Figure 37.

$i \backslash j$	1	2	3
1	$-a$	$-a$	b
2	$-a$	$-b$	$-b$
3	$-b$	$-b$	$-b$

$\mathcal{L}_-(i, j)$

Figure 38.

Now s_1, s_2 originate from chance moves, as described above. Hence

$$\mathcal{H}(i_1, \cdots, i_S|j_1, \cdots, j_S) = \frac{1}{S^2} \sum_{s_1, s_2 = 1}^{S} \mathcal{L}_{sgn(s_1-s_2)}(i_{s_1}, j_{s_2}).^4$$

19.5.2. We now pass to the (mixed) strategies in the sense of 17.2. These are the vectors $\vec{\xi}, \vec{\eta}$ belonging to S_β. Considering the notations

[1] The entire sequence $i_1, \cdots, i_S$ is the row index, and the entire sequence $j_1, \cdots, j_S$ is the column index. In our original notations the strategies were $\Sigma_1^{\tau_1}, \Sigma_2^{\tau_2}$ and the matrix element $\mathcal{H}(\tau_1, \tau_2)$.

[2] I.e. $\begin{matrix} + \\ 0 \\ - \end{matrix}$ for $s_1 \begin{matrix} > \\ = \\ < \end{matrix} s_2$ respectively. It expresses in an arithmetical form which hand is stronger.

[3] The reader will do well to compare these matrix schemes with our verbal statements of the rules, and to verify their appropriateness.

Another circumstance which is worth observing is that the symmetry of the game corresponds to the identities

$$\mathcal{L}_+(i, j) \equiv -\mathcal{L}_-(j, i), \qquad \mathcal{L}_0(i, j) \equiv -\mathcal{L}_0(j, i)$$

[4] The reader may verify

$$\mathcal{H}(i_1, \cdots, i_S|j_1, \cdots, j_S) = -\mathcal{H}(j_1, \cdots, j_S|i_1, \cdots, i_S)$$

as a consequence of the relations at the end of footnote 3 above. I.e.

$$\mathcal{H}(i_1, \cdots, i_S|j_1, \cdots, j_S)$$

is skew-symmetric, expressing once more the symmetry of the game.

which we are now using, we must index the components of these vectors also in the new way: We must write $\xi_{i_1,\ldots,i_S}$, $\eta_{j_1,\ldots,j_S}$ instead of ξ_{τ_1}, η_{τ_2}.

We express (17:2) of 17.4.1., which evaluates the expectation value of player 1's gain

$$K(\vec{\xi}, \vec{\eta}) = \sum_{i_1,\ldots,i_S,j_1,\ldots,j_S} \mathcal{H}(i_1, \cdots i_S | j_1, \cdots, j_S) \xi_{i_1,\ldots,i_S} \eta_{j_1,\ldots,j_S}$$

$$= \frac{1}{S^2} \sum_{i_1,\ldots,i_S,j_1,\ldots,j_S} \sum_{s_1,s_2} \mathcal{L}_{sgn(s_1-s_2)}(i_{s_1}, j_{s_2}) \xi_{i_1,\ldots,i_S} \eta_{j_1,\ldots,j_S}.$$

There is an advantage in interchanging the two Σ and writing

$$K(\vec{\xi}, \vec{\eta}) = \frac{1}{S^2} \sum_{s_1,s_2} \sum_{i_1,\ldots,i_S,j_1,\ldots,j_S} \mathcal{L}_{sgn(s_1-s_2)}(i_{s_1}, j_{s_2}) \xi_{i_1,\ldots,i_S} \eta_{j_1,\ldots,j_S}.$$

If we now put

$$\rho_i^{s_1} = \sum_{\substack{i_1,\ldots,i_S \text{ excluding } i_{s_1} \\ i_{s_1} = i}} \xi_{i_1,\ldots,i_S} \tag{19:1}$$

$$\sigma_j^{s_2} = \sum_{\substack{j_1,\ldots,j_S \text{ excluding } j_{s_2} \\ j_{s_2} = j}} \eta_{j_1,\ldots,j_S} \tag{19:2}$$

then the above equation becomes

$$K(\vec{\xi}, \vec{\eta}) = \frac{1}{S^2} \sum_{s_1,s_2} \sum_{i,j} \mathcal{L}_{sgn(s_1-s_2)}(i, j) \rho_i^{s_1} \sigma_j^{s_2}. \tag{19:3}$$

It is worth while to expound the meaning of (19:1)-(19:3) verbally.

(19:1) shows that $\rho_i^{s_1}$ is the probability that player 1, using the mixed strategy $\vec{\xi}$, will choose i when his "hand" is s_1; (19:2) shows that $\sigma_j^{s_2}$ is the probability that player 2, using the mixed strategy $\vec{\eta}$, will choose j when his "hand" is s_2.[1] Now it is intuitively clear that the expectation value $K(\vec{\xi}, \vec{\eta})$ depends on these probabilities $\rho_i^{s_1}$, $\sigma_j^{s_2}$ only, and not on the underlying probabilities $\xi_{i_1,\ldots,i_S}$, $\eta_{j_1,\ldots,j_S}$ themselves.[2] The formula (19:3) can

[1] We know from 19.4. that i or $j = 1$ means a "high" bid, $i = 2, 3$ a "low" bid with (the intention of) a subsequent "Seeing" or "Passing" respectively.

[2] This means that two different mixtures of (pure) strategies may in actual effect be the same thing.

Let us illustrate this by a simple example. Put $S = 2$, i.e. let there be only a "high" and a "low" hand. Consider $i = 2, 3$ as one thing, i.e. let there be only a "high" and a

easily be seen to be correct in this direct way: It suffices to remember the meaning of the $\mathcal{L}_{sgn(s_1-s_2)}(i, j)$ and the interpretation of the $\rho_i^{s_1}$, $\sigma_j^{s_2}$.

19.5.3. It is clear, both from the meaning of the $\rho_i^{s_1}$, $\sigma_j^{s_2}$ and from their formal definition (19:1), (19:2), that they fulfill the conditions

$$\text{(19:4)} \qquad \text{all } \rho_i^{s_1} \geqq 0, \qquad \sum_{i=1}^{3} \rho_i^{s_1} = 1$$

$$\text{(19:5)} \qquad \text{all } \sigma_j^{s_2} \geqq 0, \qquad \sum_{j=1}^{3} \sigma_j^{s_2} = 1$$

On the other hand, any $\rho_i^{s_1}$, $\sigma_j^{s_2}$ which fulfill these conditions can be obtained from suitable $\vec{\xi}$, $\vec{\eta}$ by (19:1), (19:2). This is clear mathematically,[1] and intuitively as well. Any such system of $\rho_i^{s_1}$, $\sigma_j^{s_2}$ is one of probabilities which define a possible *modus procedendi*,—so they must correspond to some mixed strategy.

(19:4), (19:5) make it opportune to form the 3-dimensional vectors

$$\vec{\rho}^{\,s_1} = \{\rho_1^{s_1}, \rho_2^{s_1}, \rho_3^{s_1}\}, \qquad \vec{\sigma}^{\,s_2} = \{\sigma_1^{s_2}, \sigma_2^{s_2}, \sigma_3^{s_2}\}.$$

Then (19:4), (19:5) state precisely that all $\vec{\rho}^{\,s_1}$, $\vec{\sigma}^{\,s_2}$ belong to S_3.

This shows how much of a simplification the introduction of these vectors is: $\vec{\xi}$ (or $\vec{\eta}$) was a vector in S_β, i.e. depending on $\beta - 1 = 3^S - 1$ numerical constants; the $\vec{\rho}^{\,s_1}$ (or the $\vec{\sigma}^{\,s_2}$) are S vectors in S_3, i.e. each one depends on 2 numerical constants; hence they amount together to $2S$ numerical constants. And $3^S - 1$ is much greater than $2S$, even for moderate S.[2]

"low" bid. Then there are four possible (pure) strategies, to which we shall give names:

"Bold": Bid "high" on every hand.

"Cautious": Bid "low" on every hand.

"Normal": Bid "high" on a "high" hand, "low" on a "low" hand.

"Bluff": Bid "high" on a "low" hand, "low" on a "high" hand.

Then a 50–50 mixture of "Bold" and "Cautious" is in effect the same thing as a 50–50 mixture of "Normal" and "Bluff": both mean that the player will—according to chance—bid 50–50 "high" or "low" on any hand.

Nevertheless these are, in our present notations, two different "mixed" strategies,—i.e. vectors $\vec{\xi}$.

This means, of course, that our notations, which were perfectly suited to the general case, are redundant for many particular games. This is a frequent occurrence in mathematical discussions with general aims.

There was no reason to take account of this redundance as long as we were working out the general theory. But we shall remove it now for the particular game under consideration.

[1] Put e.g. $\xi_{i_1, \ldots, i_S} = \rho_{i_1}^{1} \cdot \ldots \cdot \rho_{i_S}^{S}$, $\eta_{j_1, \ldots, j_S} = \sigma_{j_1}^{1} \cdot \ldots \cdot \sigma_{j_S}^{S}$ and verify the (17:1:a), (17:1:b) of 17.2.1. as consequences of the above (19:4), (19:5).

[2] Actually S is about $2\frac{1}{2}$ millions (cf. footnote 4 on p. 187); so $3^S - 1$ and $2S$ are both great, but the former is quite exorbitantly the greater.

19.6. Statement of the Problem

19.6. Since we are dealing with a symmetric game, we can use the characterization of the good (mixed) strategies—i.e. of the $\vec{\xi}$ in $\bar{A}$—given in (17:H) of 17.11.2. It stated this: $\vec{\xi}$ must be optimal against itself,—i.e. $\mathrm{Min}_{\vec{\eta}} \mathrm{K}(\vec{\xi}, \vec{\eta})$ must be assumed for $\vec{\eta} = \vec{\xi}$.

Now we saw in 19.5. that $\mathrm{K}(\vec{\xi}, \vec{\eta})$ depends actually on the $\vec{\rho}^{s_1}, \vec{\sigma}^{s_2}$. So we may write for it, $\mathrm{K}(\vec{\rho}^1, \cdots, \vec{\rho}^S | \vec{\sigma}^1, \cdots \vec{\sigma}^S)$. Then (19:3) in 19.5.2. states (we rearrange the Σ somewhat)

$$(19:6) \quad \mathrm{K}(\vec{\rho}^1, \cdots, \vec{\rho}^S | \vec{\sigma}^1, \cdots, \vec{\sigma}^S) = \frac{1}{S^2} \sum_{s_1, i} \sum_{s_2, j} \mathcal{L}_{sgn(s_1 - s_2)}(i, j) \rho_i^{s_1} \sigma_j^{s_2}.$$

And the characteristic of the $\vec{\rho}^1, \cdots, \vec{\rho}^S$ of a good strategy is that

$$\mathrm{Min}_{\vec{\sigma}^1, \ldots, \vec{\sigma}^S} \mathrm{K}(\vec{\rho}^1, \cdots, \vec{\rho}^S | \vec{\sigma}^1, \cdots, \vec{\sigma}^S)$$

is assumed at $\vec{\sigma}^1 = \vec{\rho}^1, \cdots, \vec{\sigma}^S = \vec{\rho}^S$. The explicit conditions for this can be found in essentially the same way as in the similar problem of 17.9.1.; we will give a brief alternative discussion.

The $\mathrm{Min}_{\vec{\sigma}^1, \ldots, \vec{\sigma}^S}$ of (19:6) amounts to a minimum with respect to each $\vec{\sigma}^1, \cdots, \vec{\sigma}^S$ separately. Consider therefore such a $\vec{\sigma}^{s_2}$. It is restricted only by the requirement to belong to S_3,—i.e. by

$$\text{all } \sigma_j^{s_2} \geqq 0, \qquad \sum_{j=1}^{3} \sigma_j^{s_2} = 1.$$

(19:6) is a linear expression in these three components $\sigma_1^{s_1}, \sigma_2^{s_2}, \sigma_3^{s_3}$. Hence it assumes its minimum with respect to $\vec{\sigma}^{s_2}$ there where all those components $\sigma_j^{s_2}$ which do not have the smallest possible coefficient (with respect to j, cf. below), vanish.

The coefficient of $\sigma_j^{s_2}$ is

$$\frac{1}{S^2} \sum_{s_1, i} \mathcal{L}_{sgn(s_1 - s_2)}(i, j) \rho_i^{s_1} \qquad \text{to be denoted by } \frac{1}{S} \gamma_j^{s_2}.$$

Thus (19:6) becomes

$$(19:7) \qquad \mathrm{K}(\vec{\rho}^1, \cdots, \vec{\rho}^S | \vec{\sigma}^1, \cdots, \vec{\sigma}^S) = \frac{1}{S} \sum_{s_2, j} \gamma_j^{s_2} \sigma_j^{s_2}.$$

And the condition for the minimum (with respect to $\overrightarrow{\sigma}^{s_2}$) is this:

(19:8) For each pair s_2, j, for which $\gamma_j^{s_2}$ does not assume its minimum (in j [1]), we have $\sigma_j^{s_2} = 0$.

Hence the characteristic of a good strategy—minimization at $\overrightarrow{\sigma}^1 = \overrightarrow{\rho}^1, \cdots, \overrightarrow{\sigma}^S = \overrightarrow{\rho}^S$—is this:

(19:A) $\overrightarrow{\rho}^1, \cdots, \overrightarrow{\rho}^S$ describe a good strategy, i.e. a $\overrightarrow{\xi}$ in $\bar{A}$, if and only if this is true:

For each pair s_2, j for which $\gamma_j^{s_2}$ does not assume its minimum (in j [1]), we have $\rho_j^{s_2} = 0$.

We finally state the explicit expressions for the $\gamma_j^{s_2}$, of course by using the matrix schemes of Figures 36–38. They are

(19:9:a) $$\gamma_1^{s_2} = \frac{1}{S}\left\{\sum_{s_1=1}^{s_2-1}(-a\rho_1^{s_1} - a\rho_2^{s_1} - b\rho_3^{s_1}) - b\rho_3^{s_2} + \sum_{s_1=s_2+1}^{S}(a\rho_1^{s_1} + a\rho_2^{s_1} - b\rho_3^{s_1})\right\},$$

(19:9:b) $$\gamma_2^{s_2} = \frac{1}{S}\left\{\sum_{s_1=1}^{s_2-1}(-a\rho_1^{s_1} - b\rho_2^{s_1} - b\rho_3^{s_1}) + \sum_{s_1=s_2+1}^{S}(a\rho_1^{s_1} + b\rho_2^{s_1} + b\rho_3^{s_1})\right\},$$

(19:9:c) $$\gamma_3^{s_2} = \frac{1}{S}\left\{\sum_{s_1=1}^{s_2-1}(b\rho_1^{s_1} - b\rho_2^{s_1} - b\rho_3^{s_1}) + b\rho_1^{s_2} + \sum_{s_1=s_2+1}^{S}(b\rho_1^{s_1} + b\rho_2^{s_1} + b\rho_3^{s_1})\right\}$$

19.7. Passage from the Discrete to the Continuous Problem

19.7.1. The criterion (19:A) of 19.6., together with the formulae (19:7), (19:9:a), (19:9:b), (19:9:c), can be used to determine all good strategies.[2] This discussion is of a rather tiresome combinatorial character, involving the analysis of a number of alternatives. The results which are obtained

[1] We mean in j and *not* in s_2, j!

[2] This determination has been carried out by one of us and will be published elsewhere.

are qualitatively similar to those which we shall derive below under somewhat modified assumptions, except for certain differences in very delicate detail which may be called the "fine structure" of the strategy. We shall say more about this in 19.12.

For the moment we are chiefly interested in the main features of the solution and not in the question of "fine structure." We begin by turning our attention to the "granular" structure of the sequence of possible hands $s = 1, \cdots, S$.

If we try to picture the strength of all possible "hands" on a scale from 0% to 100%, or rather of fractions from 0 to 1, then the weakest possible hand, 1, will correspond to 0, and the strongest possible hand, S, to 1. Hence the "hand" $s(= 1, \cdots, S)$ should be placed at $z = \frac{s-1}{S-1}$ on this scale. I.e. we have this correspondence:

Possible "hands"								
	Old scale:	$s =$	1	2	3	.....	$S-1$	S
	New scale:	$z =$	0	$\frac{1}{S-1}$	$\frac{2}{S-1}$	.....	$\frac{S-2}{S-1}$	1

Figure 39.

Thus the values of z fill the interval

$$(19{:}10) \qquad 0 \leqq z \leqq 1$$

very densely,[1] but they form nevertheless a *discrete* sequence. This is the "granular" structure referred to above. We will now replace it by a continuous one.

I.e. we assume that the chance move which chooses the hand s—i.e. z—may produce any z of the interval (19:10). We assume that the probability of any part of (19:10) is the length of that part, i.e. that z is *equidistributed* over (19:10).[2] We denote the "hands" of the two players 1, 2 by z_1, z_2 respectively.

19.7.2. This change entails that we replace the vectors $\vec{\rho}^{\,s_1}$, $\vec{\sigma}^{\,s_2}$ ($s_1, s_2, = 1, \cdots, S$) by vectors $\vec{\rho}^{\,z_1}$, $\vec{\sigma}^{\,z_2}$ ($0 \leqq z_1, z_2 \leqq 1$); but they are, of course, still probability vectors of the same nature as before, i.e. belonging to S_3. In consequence, the components (probabilities) $\rho_i^{s_1}$, $\sigma_j^{s_2}$ ($s_1, s_2 = 1, \cdots, S$; $i, j = 1, 2, 3$) give way to the components $\rho_i^{z_1}$, $\sigma_j^{z_2}$ ($0 \leqq z_1, z_2 \leqq 1; i, j = 1, 2, 3$). Similarly the $\gamma_j^{s_2}$ (in (19:9:a), (19:9:b), (19:9:c) of 19.6.) become $\gamma_j^{z_2}$.

We now rewrite the expressions for K and the γ_j^s in the formulae (19:7), (19:9:a), (19:9:b), (19:9:c) in 19.6. Clearly all sums

[1] It will be remembered (cf. footnote 4 on p. 187) that S is about $2\frac{1}{2}$ millions.
[2] This is the so-called geometrical probability.

$$\frac{1}{S}\sum_{s_1=1}^{S}, \frac{1}{S}\sum_{s_2=1}^{S}$$

must be replaced by integrals

$$\int_0^1 \cdots dz_1, \int_0^1 \cdots dz_2,$$

sums

$$\frac{1}{S}\sum_{s_1=1}^{s_2-1}, \quad \frac{1}{S}\sum_{s_1=s_2+1}^{S}$$

by integrals

$$\int_0^{z_2} \cdots dz_1, \int_{z_2}^1 \cdots dz_1,$$

while isolated terms behind a factor $1/S$ may be neglected.[1,2] These being understood, the formulae for K and the γ_j^s (i.e. γ_j^z) become:

(19:7*) $$\mathrm{K} = \sum_j \int_0^1 \gamma_j^{z_2}\sigma_j^{z_2} dz_2$$

(19:9:a*) $$\gamma_1^{z_2} = \int_0^{z_2} (-a\rho_1^{z_1} - a\rho_2^{z_1} - b\rho_3^{z_1})dz_1 + \int_{z_2}^1 (a\rho_1^{z_1} + a\rho_2^{z_1} - b\rho_3^{z_1})dz_1,$$

(19:9:b*) $$\gamma_2^{z_2} = \int_0^{z_2} (-a\rho_1^{z_1} - b\rho_2^{z_1} - b\rho_3^{z_1})dz_1 + \int_{z_2}^1 (a\rho_1^{z_1} + b\rho_2^{z_1} + b\rho_3^{z_1})dz_1,$$

(19:9:c*) $$\gamma_3^{z_2} = \int_0^{z_2} (b\rho_1^{z_1} - b\rho_2^{z_1} - b\rho_3^{z_1})dz_1 + \int_{z_2}^1 (b\rho_1^{z_1} + b\rho_2^{z_1} + b\rho_3^{z_1})dz_1.$$

And the characterization (19:A) of 19.6. goes over into this:

(19:B) The $\overrightarrow{\rho^z}$ $(0 \leqq z \leqq 1)$ (they all belong to S_3) describe a good strategy if and only if this is true:

For each z, j for which γ_j^z does not assume its minimum (in j [3]), we have $\rho_j^z = 0$.[4]

[1] Specifically we mean the middle terms $-b\rho_3^{s_2}$ and $b\rho_1^{s_2}$ in (19:9:a) and (19:9:c).

[2] These terms correspond to $s_1 = s_2$, in our present set-up to $z_1 = z_2$, and since the z_1, z_2 are continuous variables, the probability of their (fortuitous) coincidence is indeed 0.

Mathematically one may describe these operations by saying that we are now carrying out the limiting process $S \to \infty$.

[3] We mean in j and *not* in z, j!

[4] The formulae (19:7*), (19:9:a*), (19:9:6*), (19:9:c*) and this criterion could also have been derived directly by discussing this "continuous" arrangement, with the $\overrightarrow{\rho^{z_1}}$, $\overrightarrow{\rho^{z_2}}$ in place of the $\overrightarrow{\xi}$, $\overrightarrow{\eta}$ from the start. We preferred the lengthier and more explicit procedure followed in 19.4.-19.7. in order to make the rigor and the completeness of our procedure apparent. The reader will find it a good exercise to carry out the shorter direct discussion, mentioned above.

It would be tempting to build up a theory of games, into which such continuous parameters enter, systematically and directly; i.e. in sufficient generality for applications like the present one, and without the necessity for a limiting process from discrete games.

An interesting step in this direction was taken by *J. Ville* in the work referred to in footnote 1 on p. 154: pp. 110–113 loc. cit. The continuity assumptions made there seem, however, to be too restrictive for many applications,—in particular for the present one.

19.8. Mathematical Determination of the Solution

19.8.1. We now proceed to the determination of the good strategies $\overrightarrow{\rho}^{\,s}$, i.e. of the solution of the implicit condition (19:B) of 19.7.

Assume first that $\rho_2^{z'} > 0$ ever happens.[1] For such a z necessarily $\text{Min}_j \gamma_j^z = \gamma_2^z$ hence $\gamma_1^z \geqq \gamma_2^z$ i.e.

$$\gamma_2^z - \gamma_1^z \leqq 0.$$

Substituting (19:9:a*), (19:9:b*) into this gives

$$(19{:}11) \qquad (a - b)\left(\int_0^z \rho_2^{z_1} dz_1 - \int_z^1 \rho_2^{z_1} dz_1\right) + 2b \int_z^1 \rho_3^{z_1} dz_1 \leqq 0.$$

Now let z^0 be the upper limit of these z with $\rho_2^z > 0$.[2] Then (19:11) holds by continuity for $z = z^0$ too. As $\rho_2^{z_1} > 0$ does not occur for $z_1 > z^0$—by hypothesis—so the $\int_{z_0}^1 \rho_2^{z_1}\, dz_1$ term in (19:11) is now 0. So we may write it with $+$ instead of $-$, and (19:11) becomes:

$$(a - b) \int_0^1 \rho_2^{z_1} dz_1 + 2b \int_{z^0}^1 \rho_3^{z_1} dz_1 \leqq 0.$$

But $\rho_2^{z_1}$ is always $\geqq 0$ and sometimes > 0, by hypothesis; hence the first term is > 0.[3,4] The second term is clearly $\geqq 0$. So we have derived a contradiction. I.e. we have shown

$$(19{:}12) \qquad \rho_2^z \equiv 0.$$ [5]

19.8.2. Having eliminated $j = 2$ we now analyze the relationship of $j = 1$ and $j = 3$. Since $\rho_2^z = 0$ so $\rho_1^z + \rho_3^z \equiv 1$ i.e.:

$$(19{:}13) \qquad \rho_3^z = 1 - \rho_1^z,$$

and consequently

$$(19{:}14) \qquad 0 \leqq \rho_1^z \leqq 1.$$

Now there may exist in the interval $0 \leqq z \leqq 1$ subintervals in which always $\rho_1^z \equiv 0$ or always $\rho_1^z \equiv 1$.[6] A z which is not inside any interval of

[1] I.e. that the good strategy under consideration provides for $j = 2$, i.e. "low" bidding with (the intention of) subsequent "Seeing," under certain conditions.

[2] I.e. the greatest z^0 for which $\rho_2^z > 0$ occurs arbitrarily near to z^0. (But we do not require $\rho_2^z > 0$ for all $z < z^0$.) This z^0 exists certainly if the z with $\rho_2^z > 0$ exist.

[3] Of course $a - b > 0$.

[4] It does not seem necessary to go into the detailed fine points of the theory of integration, measure, etc. We assume that our functions are smooth enough so that a positive function has a positive integral etc. An exact treatment could be given with ease if we made use of the pertinent mathematical theories mentioned above.

[5] The reader should reformulate this verbally: We excluded "low" bids with (the intention of) subsequent "Seeing" by analysing conditions for the (hypothetical) upper limit of the hands for which this would be done; and showed that near there, at least, an outright "high" bid would be preferable.

This is, of course, conditioned by our simplification which forbids "overbidding."

[6] I.e. where the strategy directs the player to bid always "high," or where it directs him to bid always "low" (with subsequent "Passing").

either kind—i.e. arbitrarily near to which both $\rho_1^{z'} \neq 0$ and $\rho_1^{z'} \neq 1$ occur—will be called *intermediate*. Since $\rho_1^{z'} \neq 0$ or $\rho_1^{z'} \neq 1$ (i.e. $\rho_3^{z'} \neq 0$) imply $\text{Min}_j\ \gamma_j^{z'} = \gamma_1^{z'}$ or $\gamma_3^{z'}$ respectively, therefore we see: Both $\gamma_1^{z'} \leqq \gamma_3^{z'}$ and $\gamma_1^{z'} \geqq \gamma_3^{z'}$ occur arbitrarily near to an intermediate z. Hence for such a z, $\gamma_1^z = \gamma_3^z$ by continuity,[1] i.e.

$$\gamma_3^z - \gamma_1^z = 0.$$

Substituting (19:9:a*), (19:9:c*) and recalling (19:12), (19:13), gives

$$(a + b) \int_0^z \rho_1^{z_1} dz_1 - (a - b) \int_z^1 \rho_1^{z_1} dz_1 + 2b \int_z^1 (1 - \rho_1^{z_1}) dz_1 = 0$$

i.e.

(19:15) $$(a + b) \left(\int_0^z \rho_1^{z_1} dz_1 - \int_z^1 \rho_1^{z_1} dz_1 \right) + 2b(1 - z) = 0.$$

Consider next two intermediate z', z''. Apply (19:15) to $z = z'$ and $z = z''$ and subtract. Then

$$2(a + b) \int_{z'}^{z''} \rho_1^{z_1} dz_1 - 2b(z'' - z') = 0$$

obtains, i.e.

(19:16) $$\frac{1}{z'' - z'} \int_{z'}^{z''} \rho_1^{z_1} dz_1 = \frac{b}{a + b}.$$

Verbally: Between two intermediate z', z'' the average of ρ_1^z is $\frac{b}{a + b}$.

So neither $\rho_1^z \equiv 0$ nor $\rho_1^z \equiv 1$ can hold throughout the interval

$$z' \leqq z \leqq z''$$

since that would yield the average 0 or 1. Hence this interval must contain (at least) a further intermediate z, i.e. between any two intermediate places there lies (at least) a third one. Iteration of this result shows that between two intermediate $\bar{z}'$, $\bar{z}''$ the further intermediate z lie everywhere dense. Hence the $\bar{z}'$, $\bar{z}''$ for which (19:16) holds lie everywhere dense between z', z''. But then (19:16) must hold for all $\bar{z}'$, $\bar{z}''$ between z', z'', by continuity.[2] This leaves no alternative but that $\rho_1^z = \frac{b}{a + b}$ everywhere between z', z''.[3]

[1] The γ_j^z are defined by integrals (19:9:a*), (19:9:b*), (19:9:c*), hence they are certainly continuous.

[2] The integral in (19:16) is certainly continuous.

[3] Clearly isolated exceptions covering a z area 0—i.e. of total probability zero (e.g. a finite number of fixed z's)—could be permitted. They alter no integrals. An exact mathematical treatment would be easy but does not seem to be called for in this context (cf. footnote 4 on p. 199). So it seems simplest to assume $\rho_1^z = \frac{b}{a + b}$ in $\bar{z}' \leqq z \leqq \bar{z}''$ without any exceptions.

This ought to be kept in mind when appraising the formulae of the next pages which deal with the interval $\bar{z}' \leqq z \leqq \bar{z}''$ on one hand, and with the intervals $0 \leqq z < \bar{z}'$ and $\bar{z}'' < z' \leqq 1$ on the other; i.e. which count the points $\bar{z}'$, $\bar{z}''$ to the first-mentioned interval. This is, of course, irrelevant: two fixed isolated points—$\bar{z}'$ and $\bar{z}''$ in this case—could be disposed of in any way (cf. above).

The reader must observe, however, that while there is no significant difference

19.8.3. Now if intermediate z exist at all, then there exists a smallest one and a largest one; choose $\bar{z}'$, $\bar{z}''$ as these. We have

(19:17) $$\rho_1^z = \frac{b}{a+b} \quad \text{throughout} \quad \bar{z}' \leqq z \leqq \bar{z}''.$$

If no intermediate z exist, then we must have $\rho_1^z \equiv 0$ (for all z) or $\rho_1^z \equiv 1$ (for all z). It is easy to see that neither is a solution.[1] Thus intermediate z do exist and with them $\bar{z}'$, $\bar{z}''$ exist and (19:17) is valid.

19.8.4. The left hand side of (19:15) is $\gamma_3^z - \gamma_1^z$ for all z; hence for $z = 1$

$$\gamma_3^1 - \gamma_1^1 = (a+b)\int_0^1 \rho_1^{z_1} dz_1 > 0,$$

(since $\rho_1^{z_1} \equiv 0$ is excluded). By continuity $\gamma_3^z - \gamma_1^z > 0$, i.e. $\gamma_1^z < \gamma_3^z$ remains true even when z is merely near enough to 1. Hence $\rho_3^z = 0$, i.e. $\rho_1^z = 1$ for these z. Thus (19:17) necessitates $\bar{z}'' < 1$. Now no intermediate z exists in $\bar{z}'' \leqq z \leqq 1$; hence we have $\rho_1^z \equiv 0$ or $\rho_1^z \equiv 1$ throughout this interval. Our preceding result excludes the former. Hence

(19:18) $$\rho_1^z \equiv 1 \quad \text{throughout} \quad \bar{z}'' \leqq z \leqq 1.$$

19.8.5. Consider finally the lower end of (19:17), $\bar{z}'$. If $\bar{z}' > 0$ then we have an interval $0 \leqq z \leqq \bar{z}'$. This interval contains no intermediate z; hence we have $\rho_1^z \equiv 0$ or $\rho_1^z \equiv 1$ throughout $0 \leqq z \leqq \bar{z}'$. The first derivative of $\gamma_3^z - \gamma_1^z$, i.e. of the left side of (19:15), is clearly $2(a+b)\rho_1^z - 2b$. Hence in $0 \leqq z < \bar{z}'$ this derivative is $2(a+b)\cdot 0 - 2b = -2b < 0$ if $\rho_1^z \equiv 0$ there, $2(a+b)\cdot 1 - 2b = 2a > 0$ if $\rho_1^z \equiv 1$ there, i.e. $\gamma_3^z - \gamma_1^z$ is monotone decreasing or increasing respectively, throughout $0 \leqq z < \bar{z}'$. Since its value is 0 at the upper end (the intermediate point $\bar{z}'$), we have $\gamma_3^z - \gamma_1^z > 0$ or < 0 respectively, i.e. $\gamma_1^z < \gamma_3^z$ or $\gamma_3^z < \gamma_1^z$ respectively, throughout $0 \leqq z < \bar{z}'$. The former necessitates $\rho_3^z \equiv 0$, $\rho_1^z \equiv 1$ the latter $\rho_1^z \equiv 0$ in $0 \leqq z < \bar{z}'$; but the hypotheses with which we started were $\rho_1^z \equiv 0$ or $\rho_1^z \equiv 1$ respectively, there. So there is a contradiction in each case.

Consequently

(19:19) $$\bar{z}' = 0.$$

19.8.6. And now we determine $\bar{z}''$ by expressing the validity of (19:15) for the intermediate $z = \bar{z}' = 0$. Then (19:15) becomes

$$-(a+b)\int_0^1 \rho_1^{z_1} dz_1 + 2b = 0$$

$$\int_0^1 \rho_1^{z_1} dz_1 = \frac{2b}{a+b}$$

between a $<$ and a $\leqq$ when the z's themselves are compared, this is not so for the γ_j^z. Thus we saw that $\gamma_1^z > \gamma_3^z$ implies $\rho_1^z = 0$, while $\gamma_1^z \geqq \gamma_3^z$ has no such consequence. (Cf. also the discussion of Fig. 41 and of Figs. 47, 48.)

[1] I.e. bidding "low" (with subsequent "Pass") under all conditions is not a good strategy; nor is bidding "high" under all conditions.

Mathematical proof: For $\rho_1^z \equiv 0$: Compute $\gamma_1^0 = -b$, $\gamma_3^0 = b$ hence $\gamma_1^0 < \gamma_3^0$ contradicting $\rho_3^0 = 1 \neq 0$. For $\rho_1^z \equiv 1$: Compute $\gamma_1^0 = a$, $\gamma_3^0 = b$ hence $\gamma_3^0 < \gamma_1^0$ contradicting $\rho_1^0 = 1 \neq 0$.

But (19:17), (19:18), (19:19) give

$$\int_0^1 \rho_1^{z_1} dz_1 = \bar{z}'' \cdot \frac{b}{a+b} + (1 - \bar{z}'') \cdot 1$$
$$= 1 - \frac{a}{a+b} \cdot \bar{z}''.$$

So we have

$$1 - \frac{a}{a+b} \bar{z}'' = \frac{2b}{a+b}$$
$$\frac{a}{a+b} \bar{z}'' = 1 - \frac{2b}{a+b} = \frac{a-b}{a+b},$$

i.e.

(19:20) $$\bar{z}'' = \frac{a-b}{a}$$

Combining (19:17), (19:18), (19:19), (19:20) gives:

(19:21) $$\rho_1^z \begin{cases} = \dfrac{b}{a+b} & \text{for} \quad 0 \leqq z \leqq \dfrac{a-b}{a} \\ = 1 & \text{for} \quad \dfrac{a-b}{a} < z \leqq 1. \end{cases}$$

Together with (19:12), (19:13) this characterizes the strategy completely.

19.9. Detailed Analysis of the Solution

19.9.1. The results of 19.8. ascertain that there exists one and only one good strategy in the form of Poker under consideration.[1] It is described by (19:21), (19:12), (19:13) in 19.8. We shall give a graphical picture of this strategy which will make it easier to discuss it verbally in what follows. (Cf. Figure 40. The actual proportions of this figure correspond to $a/b \sim 3$.)

The line —— plots the curve $\rho = \rho_1^z$. Thus the height of —— above the line $\rho = 0$ is the probability of a "high" bid: ρ_1^z; the height of the line $\rho = 1$ above —— is the probability of a "low" bid (necessarily with subsequent "pass"): $\rho_3^z = 1 - \rho_1^z$.

19.9.2. The formulae (19:9:a*), (19:9:b*), (19:9:c*) of 19.7. permit us now to compute the coefficients γ_j^z. We give the graphical representations instead of the formulae, leaving the elementary verification to the reader. (Cf. Figure 41. The actual proportions are those of Figure 40, i.e. $a/b \sim 3$—cf. there.) The line —— plots the curve $\gamma = \gamma_1^z$; the line $\cdots\cdots$ plots the curve $\gamma = \gamma_2^z$; the line ----- plots the curve $\gamma = \gamma_3^z$. The figure shows that

[1] We have actually shown only that nothing else but the strategy determined in 19.8. can be good. That this strategy is indeed good, could be concluded from the established existence of (at least) a good strategy, although our passage to the "continuous" case may there create some doubt. But we shall verify below that the strategy in question is good, i.e. that it fulfills (19:B) of 19.7.

—— and ----- (i.e. γ_1^z and γ_3^z) coincide in $0 \leqq z \leqq \frac{a-b}{a}$ and that · · · · · and ----- (i.e. γ_2^z and γ_3^z) coincide in $\frac{a-b}{a} \leqq z \leqq 1$. All three curves are made

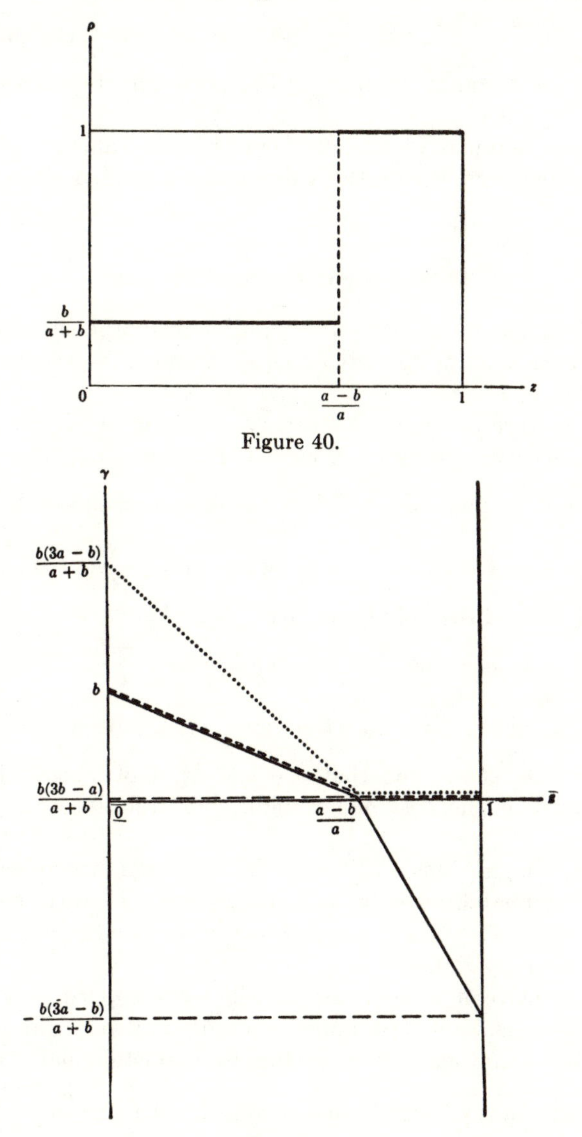

Figure 40.

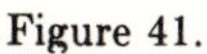

Figure 41.

of two linear pieces each, joining at $z = \frac{a-b}{a}$. The actual values of the γ_j^z at the critical points $z = 0, \frac{a-b}{a}, 1$ are given in the figure.[1]

[1] The simple computational verification of these results is left to the reader.

19.9.3. Comparison of Figures 40 and 41 shows that our strategy is indeed good, i.e. that it fulfills (19:B) of 19.7. Indeed: In $0 \leqq z \leqq \frac{a-b}{a}$ where both $\rho_1^z \neq 0$, $\rho_3^z \neq 0$ both γ_1^z and γ_3^z are the lowest curves, i.e. equal to $\mathrm{Min}_j\ \gamma_j^z$. In $\frac{a-b}{a} < z \leqq 1$ where only $\rho_1^z \neq 0$ there only γ_1^z is the lowest curve, i.e. equal to $\mathrm{Min}_j\ \gamma_j^z$. (The behavior of γ_2^z does not matter, since always $\rho_2^z = 0$.)

We can also compute K from (19:7*) in 19.7., the value of a play. $K = 0$ is easily obtained; and this is the value to be expected since the game is symmetric.

19.10. Interpretation of the Solution

19.10.1. The results of 19.8., 19.9., although mathematically complete, call for a certain amount of verbal comment and interpretation, which we now proceed to give.

First the picture of the good strategy, as given in Figure 40, indicates that for a sufficiently strong hand $\rho_1^z = 1$; i.e. that the player should then bid "high," and nothing else. This is the case for hands $z > \frac{a-b}{a}$. For weaker hands, however, $\rho_1^z = \frac{b}{a+b}$, $\rho_3^z = 1 - \rho_1^z = \frac{a}{a+b}$, so both $\rho_1^z, \rho_2^z \neq 0$; i.e. the player should then bid irregularly "high" and "low" (with specified probabilities). This is the case for hands $z \leqq \frac{a-b}{a}$. The "high" bids (in this case) should be rarer than the "low" ones, indeed $\frac{\rho_3^z}{\rho_1^z} = \frac{a}{b}$ and $a > b$. This last formula shows too that the last kind of "high" bids become increasingly rare if the cost of a "high" bid (relative to a "low" one) increases.

Now these "high" bids on "weak" hands—made irregularly, governed by (specified) probabilities only, and getting rarer when the cost of "high" bidding is increased—invite an obvious interpretation: These are the "Bluffs" of ordinary Poker.

Due to the extreme simplifications which we applied to Poker for the purpose of this discussion, "Bluffing" comes up in a very rudimentary form only; but the symptoms are nevertheless unmistakable: The player is advised to bid always "high" on a strong hand $\left(z > \frac{a-b}{a}\right)$ and to bid mostly "low" $\left(\text{with the probability } \frac{a}{a+b}\right)$ on a "low" one $\left(z < \frac{a-b}{a}\right)$ but with occasional, irregularly distributed "Bluffs" (with the probability $\left.\frac{b}{a+b}\right)$.

19.10.2. Second, the conditions in the zone of "Bluffing," $0 \leqq z \leqq \frac{a-b}{a}$, throw some light on another matter too,—the consequences of deviating from the good strategy, "permanent optimality," "defensive," "offensive," as discussed in 17.10.1., 17.10.2.

Assume that player 2 deviates from the good strategy, i.e. uses probabilities σ_j^z which may differ from the ρ_j^z obtained above. Assume, furthermore, that player 1 still uses those ρ_j^z, i.e. the good strategy. Then we can use for the γ_j^z of (19:9a*), (19:9:b*), (19:9:c*) in 19.7., the graphical representation of Figure 41, and express the outcome of the play—for player 1—by (19:7*) in 19.7.

(19:22) $$\mathrm{K} = \sum_j \int_0^1 \gamma_j^z \sigma_j^z dz.$$

Consequently player 2's σ_j^z are optimal against player 1's ρ_j^z if the analogue of the condition (19:8) in 19.6. is fulfilled:

(19:C) For each pair z, j for which γ_j^z does not assume its minimum (in j [1]) we have $\sigma_j^z = 0$.

I.e. (19:C) is necessary and sufficient for σ_j^z being just as good against ρ_j^z as ρ_j^z itself,—that is, giving a $\mathrm{K} = 0$. Otherwise σ_j^z is worse,—that is, giving a $\mathrm{K} > 0$. In other words:

(19:D) A *mistake*, i.e. a strategy σ_j^z which deviates from the good strategy ρ_j^z will cause *no* losses when the opponent sticks to the good strategy if and only if the σ_j^z fulfill (19:C) above.

Now one glance at Figure 41 suffices to make it clear that (19:C) means $\sigma_2^z = \sigma_3^z = 0$ for $z > \frac{a-b}{a}$ but merely $\sigma_2^z = 0$ for $z \leqq \frac{a-b}{a}$.[2] I.e.: (19:C) prescribes "high" bidding, and nothing else, for strong hands $\left(z > \frac{a-b}{a}\right)$; it forbids "low" bidding with subsequent "Seeing" for all hands, but it fails to prescribe the probability ratio of "high" bidding and of "low" bidding (with subsequent "Passing") for weak hands, i.e. in the zone of "Bluffing" $\left(z \leqq \frac{a-b}{a}\right)$.

19.10.3. Thus any deviation from the good strategy which involves more than just incorrect "Bluffing," leads to immediate losses. It suffices for the opponent to stick to the good strategy. Incorrect "Bluffing" causes no losses against an opponent playing the good strategy; but the

[1] We mean in j, and not in z, j!

[2] Actually even $\sigma_2 \neq 0$ would be permitted at the one place $z = \frac{a-b}{a}$. But this isolated value of z has probability 0 and so it can be disregarded. Cf. footnote 3 on p. 200.

opponent could inflict losses by deviating appropriately from the good strategy. I.e. the importance of "Bluffing" lies not in the actual play, played against a good player, but in the protection which it provides against the opponent's potential deviations from the good strategy. This is in agreement with the remarks made at the end of 19.2., particularly with the second interpretation which we proposed there for "Bluffing."[1] Indeed, the element of uncertainty created by "Bluffing" is just that type of constraint on the opponent's strategy to which we referred there, and which was analyzed at the end of 19.2.

Our results on "bluffing" fit in also with the conclusions of 17.10.2. We see that the unique good strategy of this variant of Poker is not permanently optimal; hence no permanently optimal strategy exists there. (Cf. the first remarks in 17.10.2., particularly footnote 3 on p. 163.) And "Bluffing" is a defensive measure in the sense discussed in the second half of 17.10.2.

19.10.4. Third and last, let us take a look at the offensive steps indicated loc. cit., i.e. the deviations from good strategy by which a player can profit from his opponent's failure to "Bluff" correctly.

We reverse the roles: Let player 1 "Bluff" incorrectly, i.e. use ρ_j^z different from those of Figure 40. Since only incorrect "Bluffing" is involved, we still assume

$$\rho_2^z = 0 \qquad \text{for all} \qquad z$$
$$\left.\begin{matrix}\rho_1^z = 1 \\ \rho_3^z = 0\end{matrix}\right\} \qquad \text{for all} \qquad z > \frac{a-b}{a}.$$

So we are interested only in the consequences of

$$(19{:}23) \qquad \rho_1^z \gtrless \frac{b}{a+b} \qquad \text{for some} \qquad z = z_0 < \frac{a-b}{a}.^{2}$$

The left hand side of (19:15) in 19.8. is still a valid expression for $\gamma_3^z - \gamma_1^z$. Consider now a $z < z_0$. Then $\gtrless$ in (19:23) leaves $\int_0^z \rho_1^{z_1} dz_1$ unaffected, but it $\begin{matrix}\text{increases}\\ \text{decreases}\end{matrix}$ $\int_z^1 \rho_1^{z_1} dz_1$ hence it $\begin{matrix}\text{decreases}\\ \text{increases}\end{matrix}$ the left hand side of (19:15), i.e. $\gamma_3^z - \gamma_1^z$. Since $\gamma_3^z - \gamma_1^z$ would be 0 without the change (19:23) (cf. Figure 41), so it will now be $\lessgtr 0$. I.e. $\gamma_3^z \lessgtr \gamma_1^z$. Consider next a z in

$$z_0 < z \leqq \frac{a-b}{a}.$$

[1] All this holds for the form of Poker now under consideration. For further viewpoints cf. 19.16.

[2] We need this really for more than one z, cf. footnote 3 on p. 200. The simplest assumption is that these inequalities hold in a small neighborhood of the z_0 in question.

It would be easy to treat this matter rigorously in the sense of footnote 4 on p. 199 and of footnote 3 on p. 200. We refrain from doing it for the reason stated there.

Then $\gtrless$ in (19:23) $\begin{matrix}\text{increases}\\ \text{decreases}\end{matrix}$ $\int_0^z \rho_{1'}^{z_1} dz_1$ while it leaves $\int_z^1 \rho_{1'}^{z_1} dz_1$ unaffected; hence it $\begin{matrix}\text{increases}\\ \text{decreases}\end{matrix}$ the left hand side of (19:15), i.e. $\gamma_3^z - \gamma_1^z$. Since $\gamma_3^z - \gamma_1^z$ would be 0 without the change (19:23) (cf. Fig. 41), so it will now be $\gtrless 0$. I.e. $\gamma_3^z \gtrless \gamma_1^z$. Summing up:

(19:E) The change (19:23) with $\gtrless$ causes

$$\gamma_3^z \lesseqgtr \gamma_1^z \quad \text{for} \quad z < z_0,$$
$$\gamma_3^z \gtrless \gamma_1^z \quad \text{for} \quad z_0 < z \leqq \frac{a-b}{a}.$$

Hence the opponent can gain, i.e. decrease the K of (19:22), by using σ_j^z which differ from the present ρ_j^z: For $z < z_0$ by increasing $\begin{matrix}\sigma_3^z\\ \sigma_1^z\end{matrix}$ at the expense of $\begin{matrix}\sigma_1^z\\ \sigma_3^z\end{matrix}$, i.e. by $\begin{matrix}\text{decreasing}\\ \text{increasing}\end{matrix}$ σ_1^z from the value of ρ_1^z, $\frac{b}{a+b}$, to the extreme value $\begin{matrix}0\\ 1\end{matrix}$. And for $z_0 < z \leqq \frac{a-b}{a}$ by increasing $\begin{matrix}\sigma_1^z\\ \sigma_3^z\end{matrix}$ at the expense of $\begin{matrix}\sigma_3^z\\ \sigma_1^z\end{matrix}$ i.e. by $\begin{matrix}\text{increasing}\\ \text{decreasing}\end{matrix}$ σ_1^z from the value of ρ_1^z, $\frac{b}{a+b}$ to the extreme value $\begin{matrix}1\\ 0\end{matrix}$. In other words:

(19:F) If the opponent "Bluffs" too $\begin{matrix}\text{much}\\ \text{little}\end{matrix}$ for a certain hand z_0, then he can be punished by the following deviations from the good strategy: "Bluffing" $\begin{matrix}\text{less}\\ \text{more}\end{matrix}$ for hands weaker than z_0, and "Bluffing" $\begin{matrix}\text{more}\\ \text{less}\end{matrix}$ for hands stronger than z_0.

I.e. by imitating his mistake for hands which are stronger than z_0 and by doing the opposite for weaker ones.

These are the precise details of how correct "Bluffing" protects against too much or too little "Bluffing" of the opponent, and its immediate consequences. Reflections in this direction could be carried even beyond this point, but we do not propose to pursue this subject any further.

19.11. More General Forms of Poker

19.11. While the discussions which we have now concluded throw a good deal of light on the strategical structure and the possibilities of Poker, they succeeded only due to our far reaching simplification of the rules of the game. These simplifications were formulated and imposed in 19.1., 19.3. and 19.7. For a real understanding of the game we should now make an effort to remove them.

By this we do not mean that all the fanciful complications of the game which we have eliminated (cf. 19.1.) must necessarily be reinstated,[1]

[1] Nor do we wish, yet to consider anything but a two-person game!

but some simple and important features of the game were equally lost and their reconsideration would be of great advantage. We mean in particular:

(A) The "hands" should be discrete, and not continuous. (Cf. 19.7.)

(B) There should be more than two possible ways to bid. (Cf. 19.3.)

(C) There should be more than one opportunity for each player to bid, and alternating bids, instead of simultaneous ones, should also be considered. (Cf. 19.3.)

The problem of meeting these desiderata (A), (B), (C) simultaneously—and finding the good strategies—is unsolved. Therefore we must be satisfied for the moment to add (A), (B), (C) separately.

The complete solutions for (A) and for (B) are known, while for (C) only a very limited amount of progress has been made. It would lead too far to give all these mathematical deductions in detail, but we shall report briefly the results concerning (A), (B), (C).

19.12. Discrete Hands

19.12.1. Consider first (A). I.e. let us return to the discrete scale of hands $s = 1, \cdots, S$ as introduced at the end of 19.1.2., and used in 19.4-19.7. In this case the solution is in many ways similar to that of Figure 40. Generally $\rho_2^s = 0$ and there exists a certain s^0 such that $\rho_1^s = 1$ for $s > s^0$, while $\rho_1^s \neq 0, 1$ for $s < s^0$. Also, if we change to the z scale (cf. Fig. 39), then $\frac{s^0 - 1}{S - 1}$ is very nearly $\frac{a - b}{a}$.[1] So we have a zone of "Bluffing" and above it a zone of "high" bids,—just as in Fig. 40.

But the ρ_1^s for $s < s^0$, i.e. in the zone of "Bluffing," are not at all equal to or near to the $\frac{b}{a + b}$ of Fig. 40.[2] They oscillate around this value by amounts which depend on certain arithmetical peculiarities of S but do not tend to disappear for $S \to \infty$. The averages of the ρ_1^s however, tend to $\frac{b}{a + b}$.[3] In other words:

The good strategy of the discrete game is very much like the good strategy of the continuous game: this is true for all details as far as the division into two zones (of "Bluffing" and of "high" bids) is concerned; also for the positions and sizes of these zones, and for the events in the zone of "high" bids. But in the zone of "Bluffing" it applies only to average statements (concerning several hands of approximately equal strength). The precise procedures for individual hands may differ widely from those

[1] Precisely: $\frac{s^0 - 1}{S - 1} \to \frac{a - b}{a}$ for $S \to \infty$.

[2] I.e. not $\rho_1^s \to \frac{b}{a + b}$ for $S \to \infty$ whatever the variability of s.

[3] Actually $\frac{1}{2}(\rho_1^s + \rho_1^{s+1}) = \frac{b}{a + b}$ for most $s < s^0$.

given in Figure 40, and depend on arithmetical peculiarities of s and S (with respect to a/b).[1]

19.12.2. Thus the strategy which corresponds more precisely to Figure 40 —i.e. where $\rho_1^s \equiv \frac{b}{a+b}$ for all $s < s^0$—is not good, and it differs quite considerably from the good one. Nevertheless it can be shown that the maximal loss which can be incurred by playing this "average" strategy is not great. More precisely, it tends to 0 for $S \to \infty$.[2]

So we see: In the discrete game the correct way of "Bluffing" has a very complicated "fine structure," which however secures only an extremely small advantage to the player who uses it.

This phenomenon is possibly typical, and recurs in much more complicated real games. It shows how extremely careful one must be in asserting or expecting continuity in this theory.[3] But the practical importance —i.e. the gains and losses caused—seems to be small, and the whole thing is probably *terra incognita* to even the most experienced players.

19.13. m possible Bids

19.13.1. Consider, second, (B): I.e. let us keep the hands continuous, but permit bidding in more than two ways. I.e. we replace the two bids

$$a > b\ (> 0)$$

by a greater number, say m, ordered:

$$a_1 > a_2 > \cdots > a_{m-1} > a_m\ (> 0).$$

In this case too the solution bears a certain similarity to that of Figure 40.[4] There exists a certain z^0 [5] such that for $z > z^0$ the player should make the highest bid, and nothing else, while for $z < z^0$ he should make irregularly various bids (always including the highest bid a, but also others), with specified probabilities. Which bids he must make and with what proba-

[1] Thus in the equivalent of Figure 40, the left part of the figure will not be a straight line $\left(\rho = \frac{b}{a+b} \text{ in } 0 \leqq z \leqq \frac{a-b}{a}\right)$, but one which oscillates violently around this average.

[2] It is actually of the order $1/S$. Remember that in real Poker S is about $2\frac{1}{2}$ millions. (Cf. footnote 4 on p. 187.)

[3] Recall in this connection the remarks made in the second part of footnote 4 on p. 198.

[4] It has actually been determined only under the further restriction of the rules which forbids "Seeing" a higher bid. I.e. each player is expected to make his final, highest bid at once, and to "Pass" (and accept the consequences) if the opponent's bid should turn out higher than his.

[5] Analogue of the $z = \frac{a-b}{a}$ in Figure 40.

bilities, is determined by the value of z.[1] So we have a zone of "Bluffing" and above it a zone of "high" bids—actually of the highest bid and nothing else—just as in Figure 40. But the "Bluffing"—in its own zone $z \leqq z^0$—has a much more complicated and varying structure than in Figure 40.

We shall not go into a detailed analysis of this structure, although it offers some quite interesting aspects. We shall, however, mention one of its peculiarities.

19.13.2. Let two values

$$a > b > 0$$

be given, and use them as highest and lowest bids:

$$a_1 = a, \qquad a_m = b.$$

Now let $m \to \infty$ and choose the remaining bids $a_2, \cdots, a_{m-1}$ so that they fill the interval

$$b \leqq x \leqq a \tag{19:24}$$

with unlimited increasing density. (Cf. the two examples to be given in footnote 2 below.) If the good strategy described above now tends to a limit—i.e. to an asymptotic strategy for $m \to \infty$—then one could interpret this as a good strategy for the game in which only upper and lower bounds are set for the bids (a and b), and the bids can be anything in between (i.e. in (19:24)). I.e. the requirement of a minimum interval between bids mentioned at the beginning of 19.3. is removed.

Now this is not the case. E.g. we can interpolate the $a_2, \cdots, a_{m-1}$ between $a_1 = a$ and $a_m = b$ both in arithmetic and in geometric sequence.[2] In both cases an asymptotic strategy obtains for $m \to \infty$ but the two strategies differ in many essential details.

If we consider the game in which all bids (19:24) are permitted, as one in its own right, then a direct determination of its good strategies is possible.

[1] If the bids which he must make are $a_1, a_p, a_q, \cdots, a_n (1 < p < q < \cdots < n)$, then it can be shown that their probabilities must be

$$\frac{1}{ca_1}, \frac{1}{ca_p}, \frac{1}{ca_q}, \cdots \frac{1}{ca_n}, \left(c = \frac{1}{a_1} + \frac{1}{a_p} + \cdots + \frac{1}{a_n}\right)$$

respectively. I.e. if a certain bid is to be made at all, then its probability must be inversely proportional to the cost.

Which $a_p, a_q, \cdots a_m$ actually occur for a given z is determined by a more complicated criterion, which we shall not discuss here.

Observe that the c above was needed only to make the sum of all probabilities equal to 1. The reader may verify for himself that the probabilities in Figure 40 have the above values.

[2] The first one is defined by

$$a_p = \frac{1}{m-1}((m-p)a + (p-1)b) \qquad \text{for} \qquad p = 1, 2, \cdots, m-1, m$$

the second one is defined by

$$a_p = \sqrt[m-1]{a^{m-p}b^{p-1}} \qquad \text{for} \qquad p = 1, 2, \cdots, m-1, m.$$

It turns out that both strategies mentioned above are good, together with many others.

This shows to what complications the abandonment of a minimum interval between bids can lead: a good strategy of the limiting case cannot be an approximation for the good strategies of all nearby cases with a finite number of bids. The concluding remarks of 19.12. are thus re-emphasized.

19.14. Alternate Bidding

19.14.1. Third and last, consider (C): The only progress so far made in this direction is that we can replace the simultaneous bids of the two players by two successive ones; i.e. by an arrangement in which player 1 bids first and player 2 bids afterwards.

Thus the rules stated in 19.4. are modified as follows:

First each player obtains, by a chance move, his hand $s = 1, \cdots, S$, each one of these numbers having the same probability $1/S$. We denote the hands of players 1, 2, by s_1, s_2 respectively.

After this[1] player 1 will, by a personal move, choose either a or b,—the "high" or the "low" bid.[2] He does this informed about his own hand but not about the opponent's hand. If his bid is "low," then the play is concluded. If his bid is "high," then player 2 will, by a personal move, choose either a or b,—the "high" or the "low" bid.[3] He does this informed about his own hand, and about the opponent's choice, but not his hand.

This is the play. When it is concluded, the payments are made as follows: If player 1 bids "low," then for $s_1 \begin{matrix} > \\ = \\ < \end{matrix} s_2$ player 1 obtains from player 2 the amount $\begin{matrix} b \\ 0 \\ -b \end{matrix}$ respectively. If both players bid "high," then for $s_1 \begin{matrix} > \\ = \\ < \end{matrix} s_2$ player 1 obtains from player 2 the amount $\begin{matrix} a \\ 0 \\ -a \end{matrix}$ respectively. If player 1 bids "high" and player 2 bids "low," then player 1 obtains from player 2 the amount b.[4]

19.14.2. The discussion of the pure and mixed strategies can now be carried out, essentially as we did for our original variant of Poker in 19.5.

We give the main lines of this discussion in a way which will be perfectly clear for the reader who remembers the procedure of 19.4.-19.7.

A pure strategy in this game consists clearly of the following specifications: to state for every hand $s = 1, \cdots, S$ whether a "high" or a "low" bid will be made. It is simpler to describe this by a numerical index $i_s = 1, 2$; $i_s = 1$ meaning a "high" bid, $i_s = 2$ meaning a "low" bid. Thus

[1] We continue from here on as if player 2 had already made the "low" bid, and this were player 1's turn to "See" or to "Overbid." We disregard "Passing" at this stage.

[2] I.e. "Overbid" or "See," cf. footnote 1 above.

[3] I.e. "See" or "Pass." Observe the shift of meaning since footnote 2 above.

[4] In interpreting these rules, recall the above footnotes. From the formalistic point of view, footnote 1 on p. 191 should be recalled, *mutatis mutandis.*

the strategy is a specification of such an index i_s for every $s = 1, \cdots, S$ i.e. of a sequence $i_1, \cdots, i_S$.

This applies to both players 1 and 2; accordingly we shall denote the above strategy by $\Sigma_1(i_1, \cdots, i_S)$ or $\Sigma_2(j_1, \cdots, j_S)$. Thus each player has the same number of strategies,—as many as there are sequences $i_1, \cdots, i_S$; i.e. precisely 2^S. With the notations of 11.2.2.

$$\beta_1 = \beta_2 = \beta = 2^S.$$

(But the game is not symmetrical!)

We must now express the payment which player 1 receives if the strategies $\Sigma_1(i_1, \cdots, i_S)$, $\Sigma_2(j_1, \cdots, j_S)$ are used by the two players. This is the matrix element $\mathcal{H}(i_1, \cdots, i_S|j_1, \cdots, j_S)$. If the players have actually hands s_1, s_2 then the payment received by player 1 can be expressed in this way (using the rules stated above): It is $\mathcal{L}_{sgn(s_1-s_2)}(i_{s_1}, j_{s_2})$ where $sgn(s_1 - s_2)$ is the sign of $s_1 - s_2$ and where the three functions

$$\mathcal{L}_+(i, j), \qquad \mathcal{L}_0(i, j), \qquad \mathcal{L}_-(i, j)$$

can be represented by the following matrix schemes:

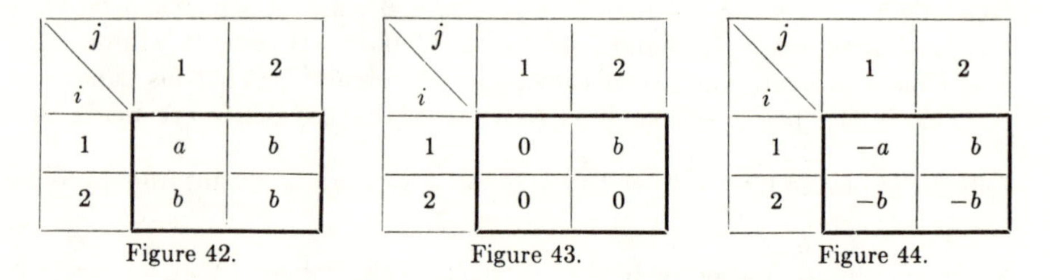

i \ j	1	2
1	a	b
2	b	b

Figure 42.

i \ j	1	2
1	0	b
2	0	0

Figure 43.

i \ j	1	2
1	$-a$	b
2	$-b$	$-b$

Figure 44.

Now s_1, s_2 originate from chance moves, as described above. Hence:

$$\mathcal{H}(i_1, \cdots, i_S|j_1, \cdots, j_S) = \frac{1}{S^2} \sum_{s_1, s_2 = 1}^{S} \mathcal{L}_{sgn(s_1-s_2)}(i_{s_1}, j_{s_2}).$$

19.14.3. We now pass to the mixed strategies in the sense of 17.2. These are vectors $\vec{\xi}, \vec{\eta}$ belonging to S_β. We must index the components of these vectors like the (pure) strategies: we must write $\xi_{i_1, \ldots, i_S}, \eta_{i_1, \ldots, i_S}$ instead of $\xi_{\tau_1}, \eta_{\tau_2}$.

We express (17:2) of 17.4.1. which evaluates the expectation of player 1's gain

$$\begin{aligned} K(\vec{\xi}, \vec{\eta}) &= \sum_{i_1, \cdots, i_S, j_1, \cdots, j_S} \mathcal{H}(i_1, \cdots, i_S|j_1, \cdots, j_S)\xi_{i_1, \ldots, i_S}\eta_{j_1, \ldots, j_S} \\ &= \frac{1}{S^2} \sum_{i_1, \cdots, i_S, j_1, \cdots, j_S} \sum_{s_1, s_2} \mathcal{L}_{sgn(s_1-s_2)}(i_{s_1}, j_{s_2})\xi_{i_1, \ldots, i_S}\eta_{j_1, \ldots, j_S}. \end{aligned}$$

There is an advantage in interchanging the two Σ and writing

$$K(\vec{\xi}, \vec{\eta}) = \frac{1}{S^2} \sum_{s_1, s_2} \sum_{i_1, \cdots, i_S, j_1, \cdots, j_S} \mathcal{L}_{sgn(s_1 - s_2)}(i_{s_1}, j_{s_2}) \xi_{i_1, \ldots, i_S} \eta_{j_1, \ldots, j_S}.$$

If we now put

$$(19{:}25) \qquad \rho_i^{s_1} = \sum_{\substack{i_1, \cdots, i_S \text{ excluding } i_{s_1} \\ i_{s_1} = i}} \xi_{i_1, \ldots, i_S},$$

$$(19{:}26) \qquad \sigma_j^{s_2} = \sum_{\substack{j_1, \cdots, j_S \text{ excluding } j_{s_2} \\ j_{s_2} = j}} \eta_{j_1, \ldots, j_S},$$

then the above equation becomes

$$(19{:}27) \qquad K(\vec{\xi}, \vec{\eta}) = \frac{1}{S^2} \sum_{s_1, s_2} \sum_{i, j} \mathcal{L}_{sgn(s_1 - s_2)}(i, j) \rho_i^{s_1} \sigma_j^{s_2}.$$

19.14.4. All this is precisely as in 19.5.2. As there, (19:25) shows that $\rho_i^{s_1}$ is the probability that player 1, using the mixed strategy $\vec{\xi}$ will choose i when his hand is s_1. (19:26) shows that $\sigma_j^{s_2}$ is the probability that player 2, using the mixed strategy $\vec{\eta}$ will choose j when his hand is s_2. It is again clear intuitively that the expectation value $K(\vec{\xi}, \vec{\eta})$ depends upon these probabilities only, and not on the underlying probabilities $\xi_{i_1, \ldots, i_S}$, $\eta_{j_1, \ldots, j_S}$ themselves. (19:27) expresses this and could have easily been derived directly, on this basis.

It is also clear, both from the meaning of the $\rho_i^{s_1}$, $\sigma_j^{s_2}$ and from their formal definitions (19:25), (19:26), that they fulfill the conditions:

$$(19{:}28) \qquad \text{all } \rho_i^{s_1} \geqq 0 \qquad \sum_{i=1}^{2} \rho_i^{s_1} = 1$$

$$(19{:}29) \qquad \text{all } \sigma_j^{s_2} \geqq 0 \qquad \sum_{j=1}^{2} \sigma_j^{s_2} = 1,$$

and that any $\rho_i^{s_1}$, $\sigma_j^{s_2}$ which fulfill these conditions can be obtained from suitable $\vec{\xi}$, $\vec{\eta}$ by (19:25), (19:26). (Cf. the corresponding step in 19.5.2. particularly footnote 1 on p. 194.) It is therefore opportune to form the 2-dimensional vectors

$$\vec{\rho}^{s_1} = \{\rho_1^{s_1}, \rho_2^{s_1}\}, \qquad \vec{\sigma}^{s_2} = \{\sigma_1^{s_2}, \sigma_2^{s_2}\}.$$

Then (19:28), (19:29) state precisely that all $\vec{\rho}^{s_1}$, $\vec{\sigma}^{s_2}$ belong to S_2.

Thus $\overrightarrow{\xi}$ (or $\overrightarrow{\eta}$) was a vector in S_β i.e. depending on $\beta - 1 = 2^S - 1$ constants; the $\overrightarrow{\rho}^{s_1}$ (or $\overrightarrow{\sigma}^{s_2}$) are S vectors in S_2 i.e. each one depends on one numerical constant, hence they amount together to S numerical constants. So we have reduced $2^S - 1$ to S. (Cf. the end of 19.5.3.)

19.14.5. We now rewrite (19:27) as in 19.6.

$$\text{(19:30)} \qquad \mathrm{K}(\overrightarrow{\rho}^1, \cdots, \overrightarrow{\rho}^S | \overrightarrow{\sigma}^1, \cdots, \overrightarrow{\sigma}^S) = \frac{1}{S} \sum_{s_2, j} \gamma_j^{s_2} \sigma_j^{s_2},$$

with the coefficients

$$\frac{1}{S} \gamma_j^{s_2} = \frac{1}{S^2} \sum_{s_1, i} \mathfrak{L}_{sgn(s_1 - s_2)}(i, j) \rho_i^{s_1},$$

i.e. using the matrix schemes of Figures 42–44,

$$\text{(19:31:a)} \quad \gamma_1^{s_2} = \frac{1}{S} \left\{ \sum_{s_1=1}^{s_2-1} (-a\rho_1^{s_1} - b\rho_2^{s_1}) + \sum_{s_1=s_2+1}^{S} (a\rho_1^{s_1} + b\rho_2^{s_1}) \right\}$$

$$\text{(19:31:b)} \quad \gamma_2^{s_2} = \frac{1}{S} \left\{ \sum_{s_1=1}^{s_2-1} (b\rho_1^{s_1} - b\rho_2^{s_1}) + b\rho_1^{s_2} + \sum_{s_1=s_2+1}^{S} (b\rho_1^{s_1} + b\rho_2^{s_1}) \right\}.$$

Since the game is no longer symmetric, we need also the corresponding formulae in which the roles of the two players are interchanged. This is:

$$\text{(19:32)} \qquad \mathrm{K}(\overrightarrow{\rho}^1, \cdots, \overrightarrow{\rho}^S | \overrightarrow{\sigma}^1, \cdots, \overrightarrow{\sigma}^S) = \frac{1}{S} \sum_{s_1, i} \delta_i^{s_1} \rho_i^{s_1},$$

with the coefficients

$$\frac{1}{S} \delta_i^{s_1} = \frac{1}{S^2} \sum_{s_2, j} \mathfrak{L}_{sgn(s_1 - s_2)}(i, j) \sigma_j^{s_2}$$

i.e. using the matrix schemes of Figures 42–44,

$$\text{(19:33:a)} \quad \delta_1^{s_1} = \frac{1}{S} \left\{ \sum_{s_2=1}^{s_1-1} (a\sigma_1^{s_2} + b\sigma_2^{s_2}) + b\sigma_2^{s_1} + \sum_{s_2=s_1+1}^{S} (-a\sigma_1^{s_2} + b\sigma_2^{s_2}) \right\}$$

$$\text{(19:33:b)} \quad \delta_2^{s_1} = \frac{1}{S} \left\{ \sum_{s_2=1}^{s_1-1} (b\sigma_1^{s_2} + b\sigma_2^{s_2}) + \sum_{s_2=s_1+1}^{S} (-b\sigma_1^{s_2} - b\sigma_2^{s_2}) \right\}.$$

The criteria for good strategies are now essentially repetitions of those in 19.6. I.e. due to the asymmetry of the variant now under consideration our present criterion will be obtained from the general criterion (17:D) of 17.9. in the same way as that of 19.6. could be obtained from the symmetrical criterion at the end of 17.11.2. I.e.:

(19:G) The $\vec{\rho}^{\,1}, \cdots, \vec{\rho}^{\,S}$ and the $\vec{\sigma}^{\,1}, \cdots, \vec{\sigma}^{\,S}$—they all belong to S_2—describe good strategies if and only if this is true:

For each s_2, j, for which $\gamma_j^{s_2}$ does not assume its minimum (in j [1]) we have $\sigma_j^{s_2} = 0$. For each s_1, i for which $\delta_i^{s_1}$ does not assume its maximum (in i [1]) we have $\rho_i^{s_1} = 0$.

19.14.6. Now we replace the discrete hands s_1, s_2 by continuous ones, in the sense of 19.7. (Cf. in particular Figure 39 there.) As described in 19.7. this replaces the vectors $\vec{\rho}^{\,s_1}$, $\vec{\sigma}^{\,s_2}$ $(s_1, s_2 = 1, \cdots, S)$ by vectors $\vec{\rho}^{\,z_1}$, $\vec{\sigma}^{\,z_2}$ $(0 \leqq z_1, z_2 \leqq 1)$, which are still probability vectors of the same nature as before, i.e. belonging to S_2. So the components $\rho_i^{s_1}$, $\sigma_j^{s_2}$ make place for the components $\rho_i^{z_1}$, $\sigma_j^{z_2}$. Similarly the $\delta_i^{s_1}$, $\gamma_j^{s_2}$ become $\delta_i^{z_1}$, $\gamma_j^{z_2}$. The sums in our formulae (19:30), (19:31:a), (19:31:b), and (19:32), (19:33:a), (19:33:b) go over into integrals, just as in (19:7*), (19:9:a*), (19:9:b*), (19:9:c*) in 19.7. So we obtain:

(19:30*) $$\mathrm{K} = \sum_j \int_0^1 \gamma_j^{z_2}\sigma_j^{z_2}dz_2,$$

(19:31:a*) $$\gamma_1^{z_2} = \int_0^{z_2} (-a\rho_1^{z_1} - b\rho_2^{z_1})dz_1 + \int_{z_2}^1 (a\rho_1^{z_1} + b\rho_2^{z_1})dz_1,$$

(19:31:b*) $$\gamma_2^{z_2} = \int_0^{z_2} (b\rho_1^{z_1} - b\rho_2^{z_1})dz_1 + \int_{z_2}^1 (b\rho_1^{z_1} + b\rho_2^{z_1})dz_1,$$

and

(19:32*) $$\mathrm{K} = \sum_i \int_0^1 \delta_i^{z_1}\rho_i^{z_1}dz_1,$$

(19:33:a*) $$\delta_1^{z_1} = \int_0^{z_1} (a\sigma_1^{z_2} + b\sigma_2^{z_2})dz_2 + \int_{z_1}^1 (-a\sigma_1^{z_2} + b\sigma_2^{z_2})dz_2,$$

(19:33:b*) $$\delta_2^{z_1} = \int_0^{z_1} (b\sigma_1^{z_2} + b\sigma_2^{z_2})dz_2 + \int_{z_2}^1 (-b\sigma_1^{z_2} - b\sigma_2^{z_2})dz_2.$$

Our criterion for good strategies is now equally transformed. (This is the same transition as that from the discrete criterion of 19.6. to the continuous criterion of 19.7.) We obtain

(19:H) The $\vec{\rho}^{\,z_1}$ and the $\vec{\sigma}^{\,z_2}$—$(0 \leqq z_1, z_2 \leqq 1)$ they all belong to S_2—describe good strategies if and only if this is true:

For each z_2, j for which $\gamma_j^{z_2}$ does not assume its minimum (in j [2]) we have $\sigma_j^{z_2} = 0$. For each z_1, i for which $\delta_i^{z_1}$ does not assume its maximum (in i [2]) we have $\rho_i^{z_1} = 0$.

[1] We mean in j (i) and not in s_2, j (s_1, i)!

[2] We mean in j (i) and not in z_2, j (z_1, j)!

19.15. Mathematical Description of All Solutions

19.15.1. The determination of the good strategies $\vec{\rho}^{\,z}$ and $\vec{\sigma}^{\,z}$, i.e. of the solutions of the implicit condition stated at the end of 19.14., can be carried out completely. The mathematical methods which achieve this are similar to those with which we determined in 19.8. the good strategies of our original variant of Poker,—i.e. the solutions of the implicit condition stated at the end of 19.7.

We shall not give the mathematical discussion here, but we shall describe the good strategies $\vec{\rho}^{\,z}$ and $\vec{\sigma}^{\,z}$ which it produces.

There exists one and only one good strategy $\vec{\rho}^{\,z}$ while the good strategies $\vec{\sigma}^{\,z}$ form an extensive family. (Cf. Figures 45–46. The actual proportions of these figures correspond to $a/b \sim 3$.)

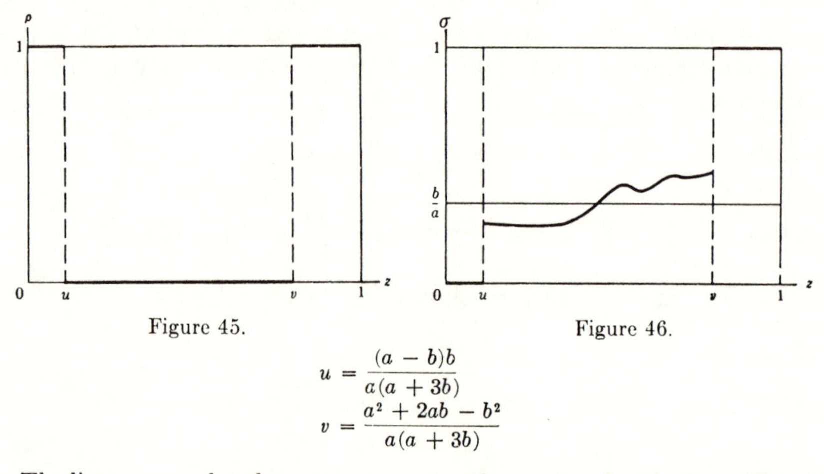

Figure 45. Figure 46.

$$u = \frac{(a - b)b}{a(a + 3b)}$$
$$v = \frac{a^2 + 2ab - b^2}{a(a + 3b)}$$

The lines —— plot the curves $\rho = \rho_1^z$ and $\sigma = \sigma_1^z$ respectively. Thus the height of —— above the line $\rho = 0$ $(\sigma = 0)$ is the probability of a "high" bid, ρ_1^z (σ_1^z); the height of the line $\rho = 1$ $(\sigma = 1)$ above —— is the probability of a "low" bid, $\rho_2^z = 1 - \rho_1^z$ $(\sigma_2^z = 1 - \sigma_1^z)$. The irregular ⁄ part of the $\sigma = \sigma_1^z$ curve (in Figure 46) in the interval $u \leqq z \leqq v$ represents the multiplicity of the good strategies $\vec{\sigma}^{\,z}$: Indeed, this part of the $\sigma = \sigma_1^z$ curve is subject to the following (necessary and sufficient) conditions:

$$\frac{1}{v - z_0} \int_{z_0}^{v} \sigma_1^z dz \begin{cases} = \dfrac{b}{a} & \text{when} \quad z_0 = u \\ \geqq \dfrac{b}{a} & \text{when} \quad u < z_0 < v. \end{cases}$$

Verbally: Between u and v the average of σ_1^z is b/a, and on any right end of this interval the average of σ_1^z is $\geqq b/a$.

Thus both $\overrightarrow{\rho}^{z}$ and $\overrightarrow{\sigma}^{z}$ exhibit three different types of behavior on these three intervals:[1]

First: $0 \leqq z < u$. Second: $u \leqq z \leqq v$. Third: $v < z \leqq 1$. The lengths of these three intervals are u, $v - u$, $1 - v$, and the somewhat complicated expressions for u, v can be best remembered with the help of these easily verified ratios:

$$u:1 - v = a - b:a + b$$
$$v - u:1 - v = a:b.$$

19.15.2. The formulae (19:31:a*), (19:31:b*) and (19:33:a*), (19:33:b*) of 19.14.6. permit us now to compute the coefficients γ_j^z, δ_i^z. We give (as in 19.9. in Figure 41) the graphical representations, instead of the formulae, leaving the elementary verification to the reader. For identification of the $\overrightarrow{\rho}^{z}$, $\overrightarrow{\sigma}^{z}$ as good strategies only the differences, $\delta_1^z - \delta_2^z$, $\gamma_2^z - \gamma_1^z$ matter: Indeed, the criterion at the end of 19.14. can be formulated as stating that whenever this difference is > 0 then $\rho_2^z = 0$ or $\sigma_2^z = 0$ respectively, and that whenever

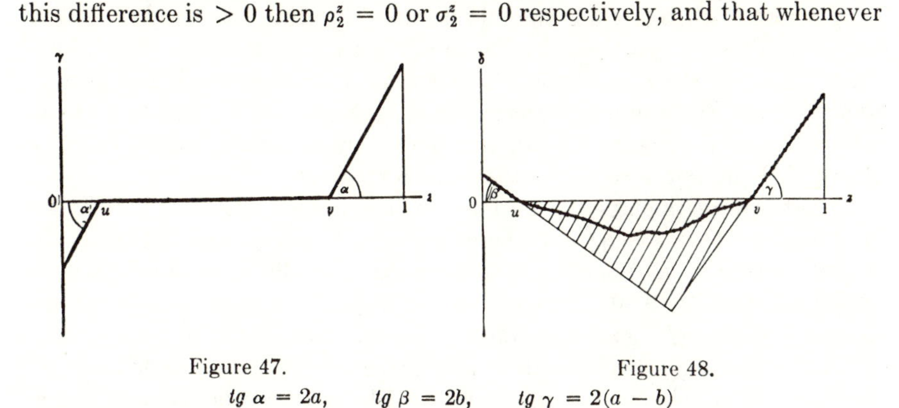

Figure 47. Figure 48.

$tg\ \alpha = 2a$, $tg\ \beta = 2b$, $tg\ \gamma = 2(a - b)$

this difference is < 0 then $\rho_1^z = 0$ or $\sigma_1^z = 0$ respectively. We give therefore the graphs of these differences. (Cf. Figures 47, 48. The actual proportions are those of Figures 45, 46; i.e. $a/b \sim 3$,—cf. there.)

The line ———— plots the curve $\gamma = \gamma_2^z - \gamma_1^z$; the line ▬▬▬ plots the curve $\delta = \delta_1^z - \delta_2^z$. The irregular part of the $\delta = \delta_1^z - \delta_2^z$ curve (in Figure 48) in the interval $u \leqq z \leqq v$ corresponds to the similarly irregular part of the $\sigma = \sigma_1^z$ curve (in Figure 46) in the same interval,—i.e. it also represents the multiplicity of the good strategies $\overrightarrow{\sigma}^{z}$. The restriction to which that part of the $\sigma = \sigma_1^z$ curve is subjected (cf. the discussion after Figure 46) means that this part of the $\delta = \delta_1^z - \delta_2^z$ curve must lie within the shaded triangle ///////// (cf. Figure 48).

19.15.3. Comparison of Figure 45 with Figure 47, and of Figure 46 with Figure 48 shows that our strategies are indeed good, i.e. that they fulfill (19:H). We leave it to the reader to verify this, in analogy with the comparison of Figure 40 and Figure 41 in 19.9.

[1] Concerning the endpoints of these intervals, etc., cf. footnote 3 on p. 200.

The value of K can also be obtained from (19:30*) or (19:32*) in 19.14.6. The result is:

$$\mathrm{K} = bu = \frac{(a - b)b^2}{a(a + 3b)}.$$[1]

Thus player 1 has a positive expectation value for the play,—i.e. an advantage[2] which is plausibly imputable to his possessing the initiative.

19.16. Interpretation of the Solutions. Conclusions

19.16.1. The results of 19.15. should now be discussed in the same way as those of 19.8., 19.9. were in 19.10. We do not wish to do this at full length, but just to make a few remarks on this subject.

We see that instead of the two zones of Figure 40 three zones appear in Figures 45, 46. The highest one (farthest to the right) corresponds to "high" bids, and nothing else, in all these figures (i.e. for both players). The behavior of the other zones, however, is not so uniform.

For player 2 (Figure 46) the middle zone describes that kind of "Bluffing" which we had on the lowest zone in Figure 40,—irregular "high" and "low" bids on the same hand. But the probabilities, while not entirely arbitrary, are not uniquely determined as in Figure 40.[3] And there exists a lowest zone (in Figure 46) where player 2 must always bid "low,"—i.e. where his hand is too weak for that mixed conduct.

Furthermore, in player 2's middle zone the γ_j^z show the same indifference as in Figure 41— $\gamma_2^z - \gamma_1^z = 0$ there,—both in Figure 41 and in Figure 47—so the motives for his conduct in this zone are as indirect as those discussed in the last part of 19.10. Indeed, these "high" bids are more of a defense against "Bluffing," than "Bluffing" proper. Since this bid of player 2 concludes the play, there is indeed no motive for the latter, while there is a need to put a rein on the opponent's "Bluffing" by occasional "high" bids,—by "Seeing" him.

For player 1 (Figure 45) the situation is different. He must bid "high," and nothing else, in the lowest zone; and bid "low," and nothing else, in the middle zone. These "high" bids on the very weakest hands—while the bid on the medium hands is "low"—are aggressive "Bluffing" in its

[1] For numerical orientation: If $a/b = 3$, which is the ratio on which all our figures are based, then $u = \frac{1}{9}$, $v = \frac{7}{9}$ and $\mathrm{K} = \frac{b}{9}$.

[2] For $a/b \sim 3$ this is about $b/9$ (cf. footnote 1 above), i.e. about 11 per cent. of the "low" bid.

[3] Cf. the discussion after Figure 46. Indeed, it is even possible to meet those requirements with $\sigma_1^z = 0$ and 1 only; e.g. $\sigma_1^z = 0$ in the lower $\frac{a-b}{a}$ fraction and $\sigma_1^z = 1$ in the upper $\frac{b}{a}$ fraction of the middle interval.

The existence of such a solution (i.e. never $\sigma_1^z \neq 0, 1$,—by Figure 45 never $\rho_1^z \neq 0, 1$ either) means, of course, that this variant is strictly determined. But a discussion on that basis (i.e. with pure strategies) will not disclose solutions like the one actually drawn in Figure 46.

purest form. The δ_i^z are not at all indifferent in this zone of "Bluffing" (i.e. the lowest zone): $\delta_1^z - \delta_2^z > 0$ there in Figure 48.—i.e. any failure to "Bluff" under these conditions leads to instant losses.

19.16.2. Summing up: Our new variant of Poker distinguishes two varieties of "Bluffing": the purely aggressive one practiced by the player who has the initiative; and a defensive one—"Seeing" irregularly, even with a medium hand, the opponent who is suspected of "Bluffing"—practiced by the player who bids last. Our original variant where the initiative was split between the two players—because they bid simultaneously—contained a procedure which we can now recognize as a mixture of these two things.[1]

All this gives valuable heuristic hints how real Poker—with longer sequences of (alternating) bids and overbids—ought to be approached. The mathematical problem is difficult, but probably not beyond the reach of the techniques that are available. It will be considered in other publications.

[1] The variant of *E. Borel*, referred to in footnote 2 on p. 186, is treated loc. cit. in a way which bears a certain resemblance to our procedure. Using our terminology, the course of *E. Borel* can be described as follows:

The Max-Min (Max for player 1, Min for player 2) is determined both for pure and for mixed strategies. The two are identical,—i.e. this variant is strictly determined. The good strategies which are obtained in this way are rather similar to those of our Figure 46. Accordingly the characteristics of "Bluffing" do not appear as clearly as in our Figures 40 and 45. Cf. the analogous considerations in the text above.

CHAPTER V

ZERO-SUM THREE-PERSON GAMES

20. Preliminary Survey

20.1. General Viewpoints

20.1.1. The theory of the zero-sum two-person game having been completed, we take the next step in the sense of 12.4.: We shall establish the theory of the zero-sum three-person game. This will bring entirely new viewpoints into play. The types of games discussed thus far have had also their own characteristic problems. We saw that the zero-sum one-person game was characterized by the emergence of a maximum problem and the zero-sum two-person game by the clear cut opposition of interest which could no longer be described as a maximum problem. And just as the transition from the one-person to the zero-sum two-person game removed the pure maximum character of the problem, so the passage from the zero-sum two-person game to the zero-sum three-person game obliterates the pure opposition of interest.

20.1.2. Indeed, it is apparent that the relationships between two players in a zero-sum three-person game can be manifold. In a zero-sum two-person game anything one player wins is necessarily lost by the other and *vice versa*,—so there is always an absolute antagonism of interests. In a zero-sum three-person game a particular move of a player—which, for the sake of simplicity, we assume to be clearly advantageous to him—may be disadvantageous to both other players, but it may also be advantageous to one and (*a fortiori*) disadvantageous to the other opponent.[1] Thus some players may occasionally experience a parallelism of interests and it may be imagined that a more elaborate theory will have to decide even whether this parallelism is total, or partial, etc. On the other hand, opposition of interest must also exist in the game (it is zero-sum)—and so the theory will have to disentangle the complicated situations which may ensue.

It may happen, in particular, that a player has a choice among various policies: That he can adjust his conduct so as to get into parallelism of interest with another player, or the opposite; that he can choose with which of the other two players he wishes to establish such a parallelism, and (possibly) to what extent.

[1] All this, of course, is subject to all the complications and difficulties which we have already recognized and overcome in the zero-sum two-person game: whether a particular move is advantageous or disadvantageous to a certain player may not depend on that move alone, but also on what other players do. However, we are trying first to isolate the new difficulties and to analyze them in their purest form. Afterward we shall discuss the interrelation with the old difficulties.

20.1.3. As soon as there is a possibility of choosing with whom to establish parallel interests, this becomes a case of choosing an ally. When alliances are formed, it is to be expected that some kind of a mutual understanding between the two players involved will be necessary. One can also state it this way: A parallelism of interests makes a cooperation desirable, and therefore will probably lead to an agreement between the players involved. An opposition of interests, on the other hand, requires presumably no more than that a player who has elected this alternative act independently in his own interest.

Of all this there can be no vestige in the zero-sum two-person game. Between two players, where neither can win except (precisely) the other's loss, agreements or understandings are pointless.[1] This should be clear by common sense. If a formal corroboration (proof) be needed, one can find it in our ability to complete the theory of the zero-sum two-person game without ever mentioning agreements or understandings between players.

20.2. Coalitions

20.2.1. We have thus recognized a qualitatively different feature of the zero-sum three-person game (as against the zero-sum two-person game). Whether it is the only one is a question which can be decided only later. If we succeed in completing the theory of the zero-sum three-person game without bringing in any further new concepts, then we can claim to have established this uniqueness. This will be the case essentially when we reach 23.1. For the moment we simply observe that this is a new major element in the situation, and we propose to discuss it fully before taking up anything else.

Thus we wish to concentrate on the alternatives for acting in cooperation with, or in opposition to, others, among which a player can choose. I.e. we want to analyze the possibility of *coalitions*—the question between which players, and against which player, coalitions will form.[2]

[1] This is, of course, different in a general two-person game (i.e. one with variable sum): there the two players may conceivably cooperate to produce a greater gain. Thus there is a certain similarity between the general two-person game and the zero-sum three-person game.

We shall see in Chap. XI, particularly in 56.2.2., that there is a general connection behind this: the general n-person game is closely related to the zero-sum $n + 1$-person game.

[2] The following seems worth noting: coalitions occur first in a zero-sum game when the number of participants in the game reaches three. In a two-person game there are not enough players to go around: a coalition absorbs at least two players, and then nobody is left to oppose. But while the three-person game of itself implies coalitions, the scarcity of players is still such as to circumscribe these coalitions in a definite way: a coalition must consist of precisely two players and be directed against precisely one (the remaining) player.

If there are four or more players, then the situation becomes considerably more involved,—several coalitions may form, and these may merge or oppose each other, etc. Some instances of this appear at the end of 36.1.2., et seq., the end of 37.1.2., et seq.; another allied phenomenon at the end of 38.3.2.

Consequently it is desirable to form an example of a zero-sum three-person game in which this aspect is foremost and all others are suppressed; i.e., a game in which the coalitions are the only thing that matters, and the only conceivable aim of all players.[1]

20.2.2. At this point we may mention also the following circumstance: A player can at best choose between two possible coalitions, since there are two other players either of whom he may try to induce to cooperate with him against the third. We shall have to elucidate by the study of the zero-sum three-person game just how this choice operates, and whether any particular player has such a choice at all. If, however, a player has only one possibility of forming a coalition (in whatever way we shall *in fine* interpret this operation) then it is not quite clear in what sense there is a coalition at all: moves forced upon a player in a unique way by the necessities of the rules of the game are more in the nature of a (one sided) strategy than of a (cooperative) coalition. Of course these considerations are rather vague and uncertain at the present stage of our analysis. We bring them up nevertheless, because these distinctions will turn out to be decisive.

It may also seem uncertain, at this stage at least, how the possible choices of coalitions which confront one player are related to those open to another; indeed, whether the existence of several alternatives for one player implies the same for another.

21. The Simple Majority Game of Three Persons

21.1. Description of the Game

21.1. We now formulate the example mentioned above: a simple zero-sum three-person game in which the possibilities of understandings—i.e. coalitions—between the players are the only considerations which matter.

This is the game in question:

Each player, by a personal move, chooses the number of one of the two other players.[2] Each one makes his choice uninformed about the choices of the two other players.

After this the payments will be made as follows: If two players have chosen each other's numbers we say that they form a *couple*.[3] Clearly

[1] This is methodically the same device as our consideration of Matching Pennies in the theory of the zero-sum two-person game. We had recognized in 14.7.1. that the decisive new feature of the zero-sum two-person game was the difficulty of deciding which player "finds out" his opponent. Matching Pennies was the game in which this "finding out" dominated the picture completely, where this mattered and nothing else.

[2] Player 1 chooses 2 or 3, player 2 chooses 1 or 3, player 3 chooses 1 or 2.

[3] It will be seen that the formation of a couple is in the interest of the players who create it. Accordingly our discussion of understandings and coalitions in the paragraphs which follow will show that the players combine into a coalition in order to be able to form a couple. The difference between the concepts of a couple and a coalition nevertheless should not be overlooked: A couple is a formal concept which figures in the set of rules of the game which we define now; a coalition is a notion belonging to the theory concerning this game (and, as will be seen, many other games).

there will be precisely one couple, or none at all.[1,2] If there is precisely one couple, then the two players who belong to it get one-half unit each, while the third (excluded) player correspondingly loses one unit. If there is no couple, then no one gets anything.[3]

The reader will have no difficulty in recognizing the actual social processes for which this game is a highly schematized model. We shall call it the *simple majority game* (*of three players*).

21.2. Analysis of the Game. Necessity of "Understandings"

21.2.1. Let us try to understand the situation which exists when the game is played.

To begin with, it is clear that there is absolutely nothing for a player to do in this game but to look for a partner,—i.e. for another player who is prepared to form a couple with him. The game is so simple and absolutely devoid of any other strategic possibilities that there just is no occasion for any other reasoned procedure. Since each player makes his personal move in ignorance of those of the others, no collaboration of the players can be established during the course of the play. Two players who wish to collaborate must get together on this subject before the play,—i.e. outside the game. The player who (in making his personal move) lives up to his agreement (by choosing the partner's number) must possess the conviction that the partner too will do likewise. As long as we are concerned only with the rules of the game, as stated above, we are in no position to judge what the basis for such a conviction may be. In other words what, if anything, *enforces* the "sanctity" of such agreements? There may be games which themselves—by virtue of the rules of the game as defined in 6.1. and 10.1.—provide the mechanism for agreements and for their enforcement.[4] But we cannot base our considerations on this possibility, since a game *need* not provide this mechanism; the simple majority game described above certainly does not. Thus there seems to be no escape from the necessity of considering agreements concluded outside the game. If we do not allow for them, then it is hard to see what, if anything, will govern the conduct of a player in a simple majority game. Or, to put this in a somewhat different form:

[1] I.e. there cannot be simultaneously two different couples. Indeed, two couples must have one player on common (since there are only three players), and the number chosen by this player must be that of the other player in both couples,—i.e. the two couples are identical.

[2] It may happen that no couples exist: e.g., if 1 chooses 2, 2 chooses 3, and 3 chooses 1.

[3] For the sake of absolute formal correctness this should still be arranged according to the patterns of 6. and 7. in Chap. II. We leave this to the reader, as in the analogous situation discussed in footnote 1 on p. 191.

[4] By providing personal moves of one player, about which only one other player is informed and which contain (possibly conditional) statements of the first player's future policy; and by prescribing for him to adhere subsequently to these statements, or by providing (in the functions which determine the outcome of a game) penalties for the non-adherence.

We are trying to establish a theory of the rational conduct of the participants in a given game. In our consideration of the simple majority game we have reached the point beyond which it is difficult to go in formulating such a theory without auxiliary concepts such as "agreements," "understandings," etc. On a later occasion we propose to investigate what theoretical structures are required in order to eliminate these concepts. For this purpose the entire theory of this book will be required as a foundation, and the investigation will proceed along the lines indicated in Chapter XII, and particularly in 66. At any rate, at present our position is too weak and our theory not sufficiently advanced to permit this "self-denial." We shall therefore, in the discussions which follow, make use of the possibility of the establishment of coalitions outside the game; this will include the hypothesis that they are respected by the contracting parties.

21.2.2. These agreements have a certain amount of similarity with "conventions" in some games like Bridge—with the fundamental difference, however, that those affected only *one* "organization" (i.e. one player split into two "persons") while we are now confronted with the relationship of two players. At this point the reader may reread with advantage our discussion of "conventions" and related topics in the last part of 6.4.2. and 6.4.3., especially footnote 2 on p. 53.

21.2.3. If our theory were applied as a statistical analysis of a long series of plays of the same game—and not as the analysis of one isolated play—an alternative interpretation would suggest itself. We should then view agreements and all forms of cooperation as establishing themselves by repetition in such a long series of plays.

It would not be impossible to derive a mechanism of enforcement from the player's desire to maintain his record and to be able to rely on the record of his partner. However, we prefer to view our theory as applying to an individual play. But these considerations, nevertheless, possess a certain significance in a virtual sense. The situation is similar to the one which we encountered in the analysis of the (mixed) strategies of a zero-sum two-person game. The reader should apply the discussions of 17.3. *mutatis mutandis* to the present situation.

21.3. Analysis of the Game: Coalitions. The Role of Symmetry

21.3. Once it is conceded that agreements may exist between the players in the simple majority game, the path is clear. This game offers to players who collaborate an infallible opportunity to win—and the game does not offer to anybody opportunities for rational action of any other kind. The rules are so elementary that this point ought to be fully convincing.

Again the game is wholly symmetric with respect to the three players. That is true as far as the rules of the game are concerned: they do not offer to any player any possibility which is not equally open to any other player. What the players do within these possibilities is, of course, another matter. Their conduct may be unsymmetric; indeed, since understandings, i.e. coalitions, are sure to arise, it will of necessity be unsymmetric. Among the three players there is room for only one coalition (of two players) and one player will necessarily be left out. It is quite instructive to observe how the rules of the game are absolutely fair (in this case, symmetric), but the conduct of the players will necessarily not be.[1,2]

Thus the only significant strategic feature of this game is the possibility of coalitions between two players.[3] And since the rules of the game are perfectly symmetrical, all three possible coalitions[4] must be considered on the same footing. If a coalition is formed, then the rules of the game provide that the two allies get one unit from the third (excluded) player—each one getting one-half unit.

Which of these three possible coalitions will form, is beyond the scope of the theory,—at least at the present stage of its development. (Cf. the end of 4.3.2.) We can say only that it would be irrational if no coalitions were formed at all, but as to which particular coalition will be formed must depend on conditions which we have not yet attempted to analyze.

22. Further Examples

22.1. Unsymmetric Distribution. Necessity of Compensations

22.1.1. The remarks of the preceding paragraphs exhaust, at least for the time being, the subject of the simple majority game. We must now begin to remove, one by one, the extremely specializing assumptions which characterized this game: its very special nature was essential for us in order to observe the role of coalitions in a pure and isolated form—*in vitro*—

[1] We saw in 17.11.2. that no such thing occurs in the zero-sum two-person games. There, if the rules of the game are symmetric, both players get the same amount (i.e. the value of the game is zero), and both have the same good strategies. I.e. there is no reason to expect a difference in their conduct or in the results which they ultimately obtain.

It is on emergence of coalitions—when more than two players are present—and of the "squeeze" which they produce among the players, that the peculiar situation described above arises. (In our present case of three players the "squeeze" is due to the fact that each coalition can consist of only two players, i.e. less than the total number of players but more than one-half of it. It would be erroneous, however, to assume that no such "squeeze" obtains for a greater number of players.)

[2] This is, of course, a very essential feature of the most familiar forms of social organizations. It is also an argument which occurs again and again in the criticism directed against these institutions, most of all against the hypothetical order based upon "*laisser faire.*" It is the argument that even an absolute, formal fairness—symmetry of the rules of the game—does not guarantee that the use of these rules by the participants will be fair and symmetrical. Indeed, this "does not guarantee" is an understatement: it is to be expected that any exhaustive theory of rational behavior will show that the participants are driven to form coalitions in unsymmetric arrangements.

To the extent to which an exact theory of these coalitions is developed, a real understanding of this classical criticism is achieved. It seems worth emphasizing that this characteristically "social" phenomenon occurs only in the case of three or more participants.

[3] Such a coalition is in this game, of course, simply an agreement to choose each other's numbers, so as to form a couple in the sense of the rules. This situation was foreseen already at the beginning of 4.3.2.

[4] Between players 1,2; 1,3; 2,3.

but now this step is completed. We must begin to adjust our ideas to more general situations.

22.1.2. The specialization which we propose to remove first is this: In the simple majority game any coalition can get one unit from the opponent; the rules of the game provide that this unit must be divided evenly among the partners. Let us now consider a game in which each coalition offers the same total return, but where the rules of the game provide for a different distribution. For the sake of simplicity let this be the case only in the coalition of players 1 and 2, where player 1, say, is favored by an amount ϵ. The rules of the modified game are therefore as follows:

The moves are the same as in the simple majority game described in 21.1. The definition of a couple is the same too. If the couple 1,2 forms, then player 1 gets the amount $\frac{1}{2} + \epsilon$ [1], player 2 gets the amount $\frac{1}{2} - \epsilon$, and player 3 loses one unit. If any other couple forms (i.e. 1,3 or 2,3) then the two players which belong to it get one-half unit each while the third (excluded) player loses one unit.

What will happen in this game?

To begin with, it is still characterized by the possibility of three coalitions —corresponding to the three possible couples—which may arise in it. *Prima facie* it may seem that player 1 has an advantage, since at least in his couple with player 2 he gets more by ϵ than in the original, simple majority game.

However, this advantage is quite illusory. If player 1 would really insist on getting the extra ϵ in the couple with player 2, then this would have the following consequence: The couple 1,3 would never form, because the couple 1,2 is more desirable from 1's point of view; the couple 1,2 would never form, because the couple 2,3 is more desirable from 2's point of view; but the couple 2,3 is entirely unobstructed, since it can be brought about by a coalition of 2,3 who then need pay no attention to 1 and his special desires. Thus the couple 2,3 and no other will form; and player 1 will not get $\frac{1}{2} + \epsilon$ nor even one-half unit, but he will certainly be the excluded player and lose one unit.

So any attempt of player 1 to keep his privileged position in the couple 1,2 is bound to lead to disaster for him. The best he can do is to take steps which make the couple 1,2 just as attractive for 2 as the competing couple 2,3. That is to say, he acts wisely if, in case of the formation of a couple with 2, he returns the extra ϵ to his partner. It should be noted that he cannot keep any fraction of ϵ; i.e., if he should try to keep an extra amount ϵ' for himself,[2] then the above arguments could be repeated literally with ϵ' in place of ϵ.[3]

[1] It seems natural to assume $0 < \epsilon < \frac{1}{2}$.

[2] We mean of course $0 < \epsilon' < \epsilon$.

[3] So the motives for player 1's ultimate disaster—the certain formation of couple 2,3—would be weaker, but the disaster the same and just as certain as before. Cf. in this connection footnote 1 on p. 228.

22.1.3. One could try some other variations of the original, simple, majority game, still always maintaining that the total value of each coalition is one unit. E.g. we could consider rules where player 1 gets the amount $\frac{1}{2} + \epsilon$ in each couple 1,2, 1,3; while players 2 and 3 split even in the couple 2,3. In this case neither 2 nor 3 would care to cooperate with 1 if 1 should try to keep his extra ϵ or any fraction thereof. Hence any such attempt of player 1 would again lead with certainty to a coalition of 2,3 against him and to a loss of one unit.

Another possibility would be that two players are favored in all couples with the third: e.g. in the couples 1,3 and 2,3, players 1 and 2 respectively get $\frac{1}{2} + \epsilon$ while 3 gets only $\frac{1}{2} - \epsilon$; and in the couple 1,2 both get one-half unit each. In this case both players 1 and 2 would lose interest in a coalition with each other, and player 3 will become the desirable partner for each of them. One must expect that this will lead to a competitive bidding for his cooperation. This must ultimately lead to a refund to player 3 of the extra advantage ϵ. Only this will bring the couple 1,2 back into the field of competition and thereby restore equilibrium.

22.1.4. We leave to the reader the consideration of further variants, where all three players fare differently in all three couples. Furthermore we shall not push the above analysis further, although this could be done and would even be desirable in order to answer some plausible objections. We are satisfied with having established some kind of a general plausibility for our present approach which can be summarized as follows: It seems that what a player can get in a definite coalition depends not only on what the rules of the game provide for that eventuality, but also on the other (competing) possibilities of coalitions for himself and for his partner. Since the rules of the game are absolute and inviolable, this means that under certain conditions *compensations* must be paid among coalition partners; i.e. that a player must have to pay a well-defined price to a prospective coalition partner. The amount of the compensations will depend on what other alternatives are open to each of the players.

Our examples above have served as a first illustration for these principles. This being understood, we shall now take up the subject *de novo* and in more generality, and handle it in a more precise manner.[1]

22.2. Coalitions of Different Strength. Discussion

22.2.1. In accordance with the above we now take a far reaching step towards generality. We consider a game in which this is the case:

If players 1,2 cooperate, then they can get the amount c, and no more, from player 3; if players 1,3 cooperate, they can get the amount b, and no more, from player 2; if players 2,3 cooperate, they can get the amount a, and no more, from player 1.

[1] This is why we need not analyze any further the heuristic arguments of this paragraph—the discussion of the next paragraphs takes care of everything.

All these possibilities were anticipated at the beginning of 4.3.2. and in 4.3.3.

We make no assumptions whatsoever concerning further particulars about the rules of this game. So we need not describe by what steps—of what order of complication—the above amounts are secured. Nor do we state how these amounts are divided between the partners, whether and how either partner can influence or modify this distribution, etc.

We shall nevertheless be able to discuss this game completely. But it will be necessary to remember that a coalition is probably connected with compensations passing between the partners. The argument is as follows:

22.2.2. Consider the situation of player 1. He can enter two alternative coalitions: with player 2 or with player 3. Assume that he attempts to retain an amount x under all conditions. In this case player 2 cannot count upon obtaining more than the amount $c - x$ in a coalition with player 1. Similarly player 3 cannot count on getting more than the amount $b - x$ in a coalition with player 1. Now if the sum of these upper bounds—i.e. the amount $(c - x) + (b - x)$—is less than what players 2 and 3 can get by combining with each other in a coalition, then we may safely assume that player 1 will find no partner.[1] A coalition of 2 and 3 can obtain the amount a. So we see: If player 1 desires to get an amount x under all conditions, then he is disqualified from any possibility of finding a partner if his x fulfills

$$(c - x) + (b - x) < a.$$

I.e. the desire to get x is unrealistic and absurd unless

$$(c - x) + (b - x) \geqq a.$$

This inequality may be written equivalently as

$$x \leqq \frac{-a + b + c}{2}.$$

We restate this:

(22:1:a) Player 1 cannot reasonably maintain a claim to get under all conditions more than the amount $\alpha = \dfrac{-a + b + c}{2}$.

The same considerations may be repeated for players 2 and 3, and they give:

(22:1:b) Player 2 cannot reasonably maintain a claim to get under all conditions more than the amount $\beta = \dfrac{a - b + c}{2}$.

(22:1:c) Player 3 cannot reasonably maintain a claim to get under all conditions more than the amount $\gamma = \dfrac{a + b - c}{2}$.

[1] We assume, of course, that a player is not indifferent to any possible profit, however small. This was implicit in our discussion of the zero-sum two-person game as well.

The traditional idea of the "*homo oeconomicus*," to the extent to which it is clearly conceived at all, also contains this assumption.

22.2.3. Now the criteria (22:1:a)-(22:1:c) were only necessary ones, and one could imagine *a priori* that further considerations could further lower their upper bounds, α, β, γ—or lead to some other restrictions of what the players can aim for. This is not so, as the following simple consideration shows.

One verifies immediately that

$$\alpha + \beta = c, \qquad \alpha + \gamma = b, \qquad \beta + \gamma = a.$$

In other words: If the players 1,2,3 do not aim at more than permitted by (22:1:a), (22:1:b), (22:1:c), i.e. than α, β, γ respectively, then any two players who combine can actually obtain these amounts in a coalition. Thus these claims are fully justified. Of course only two players—the two who form a coalition—can actually obtain their "justified" dues. The third player, who is excluded from the coalition, will not get α, β, γ respectively, but $-a$, $-b$, $-c$ instead.[1]

22.3. An Inequality. Formulae

22.3.1. At this point an obvious question presents itself: Any player 1,2,3 can get the amount α, β, γ respectively if he succeeds in entering a coalition; if he does not succeed, he gets instead only $-a$, $-b$, $-c$. This makes sense only if α, β, γ are greater than the corresponding $-a$, $-b$, $-c$, since otherwise a player might not want to enter a coalition at all, but might find it more advantageous to play for himself. So the question is whether the three differences

$$\begin{aligned} p &= \alpha - (-a) = \alpha + a, \\ q &= \beta - (-b) = \beta + b, \\ r &= \gamma - (-c) = \gamma + c, \end{aligned}$$

are all $\geqq 0$.

It is immediately seen that they are all equal to each other. Indeed:

$$p = q = r = \frac{a + b + c}{2}.$$

We denote this quantity by $\Delta/2$. Then our question is whether

$$\Delta = a + b + c \geqq 0.$$

This inequality can be demonstrated as follows:

22.3.2. A coalition of the players 1,2 can obtain (from player 3) the amount c and no more. If player 1 plays alone, then he can prevent players 2,3 from reducing him to a result worse than $-a$ since even a coalition of players 2,3 can obtain (from player 1) the amount $+a$ and no more; i.e. player 1 can get the amount $-a$ for himself without any outside help. Similarly, player 2 can get the amount $-b$ for himself without any outside help. Consequently the two players 1,2 between them can get the amount

[1] These are indeed the amounts which a coalition of the other players can wrest from players 1,2,3 respectively. The coalition cannot take more.

$-(a+b)$ even if they fail to cooperate with each other. Since the maximum they can obtain together under any conditions is c, this implies $c \geqq -a-b$ i.e. $\Delta = a+b+c \geqq 0$.

22.3.3. This proof suggests the following remarks:

First: We have based our argument on player 1. Owing to the symmetry of the result $\Delta = a+b+c \geqq 0$ with respect to the three players, the same inequality would have obtained if we had analyzed the situation of player 2 or player 3. This indicates that there exists a certain symmetry in the role of the three players.

Second: $\Delta = 0$ means $c = -a-b$ or just as well $\alpha = -a$, and the two corresponding pairs of equations which obtain by the cyclic permutation of the three players. So in this case no coalition has a *raison d'être:* Any two players can obtain, without cooperating, the same amount which they can produce in perfect cooperation (e.g. for players 1 and 2 this amount is $-a-b=c$). Also, after all is said and done, each player who succeeds in joining a coalition gets no more than he could get for himself without outside help (e.g. for player 1 this amount is $\alpha = -a$).

If, on the other hand, $\Delta > 0$ then every player has a definite interest in joining a coalition. The advantage contained in this is the same for all three players: $\Delta/2$.

Here we have again an indication of the symmetry of certain aspects of the situation for all players: $\Delta/2$ is the inducement to seek a coalition; it is the same for all players.

22.3.4. Our result can be expressed by the following table:

Player		1	2	3
Value of a play	With coalition	α	β	γ
	Without coalition	$-a$	$-b$	$-c$

Figure 49.

If we put

$$a' = -a + \frac{1}{3}\Delta = \alpha - \frac{1}{6}\Delta = \frac{-2a+b+c}{3},$$

$$b' = -b + \frac{1}{3}\Delta = \beta - \frac{1}{6}\Delta = \frac{a-2b+c}{3},$$

$$c' = -c + \frac{1}{3}\Delta = \gamma - \frac{1}{6}\Delta = \frac{a+b-2c}{3},$$

then we have

$$a' + b' + c' = 0,$$

and we can express the above table equivalently in the following manner:

(22:A) A play has for the players 1,2,3 the *basic values* a', b', c' respectively. (This is a possible valuation, since the sum of these values is zero, cf. above). The play will, however, certainly be attended by the formation of a coalition. Those two players who form it get (beyond their basic values) a *premium* of $\Delta/6$ and the excluded player sustains a loss of $-\Delta/3$.

Thus the inducement to form a coalition is $\Delta/2$ for each player, and always $\Delta/2 \geqq 0$.

23. The General Case

23.1. Exhaustive Discussion. Inessential and Essential Games

23.1.1. We can now remove all restrictions.

Let Γ be a perfectly arbitrary zero-sum three-person game. A simple consideration suffices to bring it within the reach of the analysis of 22.2., 22.3. We argue as follows:

If two players, say 1 and 2, decide to cooperate completely—postponing temporarily, for a later settlement, the question of distribution, i.e. of the compensations to be paid between partners—then Γ becomes a zero-sum two-person game. The two players in this new game are: the coalition 1,2 (which is now a composite player consisting of two "natural persons"), and the player 3. Viewed in this manner Γ falls under the theory of the zero-sum two-person game of Chapter III. Each play of this game has a well defined value (we mean the v' defined in 17.4.2.). Let us denote by c the value of a play for the coalition 1,2 (which in our present interpretation is one of the players).

Similarly we can assume an absolute coalition between players 1,3 and view Γ as a zero-sum two-person game between this coalition and the player 2. We then denote by b the value of a play for the coalition 1,3.

Finally we can assume an absolute coalition between players 2,3, and view Γ as a zero-sum two-person game between this coalition and the player 1. We then denote by a the value of a play for the coalition 2,3.

It ought to be understood that we do not—yet!—assume that any such coalition will necessarily arise. The quantities a, b, c are merely computationally defined; we have formed them on the basis of the main (mathematical) theorem of 17.6. (For explicit expressions of a, b, c cf. below.)

23.1.2. Now it is clear that the zero-sum three-person game Γ falls entirely within the domain of validity of 22.2., 22.3.: a coalition of the players 1,2 or 1,3 or 2,3 can obtain (from the excluded players 3 or 2 or 1) the amounts c, b, a respectively, and no more. Consequently all results of 22.2., 22.3. hold, in particular the one formulated at the end which describes every player's situation with and without a coalition.

23.1.3. These results show that the zero-sum three-person game falls into two quantitatively different categories, corresponding to the possibilities $\Delta = 0$ and $\Delta > 0$. Indeed:

$\Delta = 0$: We have seen that in this case coalitions have no *raison d'être*, and each player can get the same amount for himself, by playing a "lone hand" against all others, as he could obtain by any coalition. In this case, and in this case alone, it is possible to assume a unique value of each play for each player,—the sum of these values being zero. These are the basic values a', b', c' mentioned at the end of 22.3. In this case the formulae of 22.3. show that $a' = \alpha = -a$, $b' = \beta = -b$, $c' = \gamma = -c$. We shall call a game in this case, in which it is inessential to consider coalitions, an *inessential* game.

$\Delta > 0$: In this case there is a definite inducement to form coalitions, as discussed at the end of 22.3. There is no need to repeat the description given there; we only mention that now $\alpha > a' > -a$, $\beta > b' > -b$, $\gamma > c' > -c$. We shall call a game in this case, in which coalitions are essential, an *essential* game.

Our above classification, inessential and essential, applies at present only to zero-sum three-person games. But we shall see subsequently that it can be extended to all games and that it is a differentiation of central importance.

23.2. Complete Formulae

23.2. Before we analyze this result any further, let us make a few purely mathematical remarks about the quantities a, b, c—and the $\alpha, \beta, \gamma, a', b', c', \Delta$ based upon them—in terms of which our solution was expressed.

Assume the zero-sum three-person game Γ in the normalized form of 11.2.3. There the players 1,2,3 choose the variables τ_1, τ_2, τ_3 respectively (each one uninformed about the two other choices) and get the amounts $\mathcal{H}_1(\tau_1, \tau_2, \tau_3), \mathcal{H}_2(\tau_1, \tau_2, \tau_3), \mathcal{H}_3(\tau_1, \tau_2, \tau_3)$ respectively. Of course (the game is zero-sum):

$$\mathcal{H}_1(\tau_1, \tau_2, \tau_3) + \mathcal{H}_2(\tau_1, \tau_2, \tau_3) + \mathcal{H}_3(\tau_1, \tau_2, \tau_3) \equiv 0.$$

The domains of the variables are:

$$\tau_1 = 1, 2, \cdots, \beta_1,$$
$$\tau_2 = 1, 2, \cdots, \beta_2,$$
$$\tau_3 = 1, 2, \cdots, \beta_3.$$

Now in the two-person game which arises between an absolute coalition of players 1,2, and the player 3, we have the following situation:

The composite player 1,2 has the variables τ_1, τ_2; the other player 3 has the variable τ_3. The former gets the amount

$$\mathcal{H}_1(\tau_1, \tau_2, \tau_3) + \mathcal{H}_2(\tau_1, \tau_2, \tau_3) \equiv -\mathcal{H}_3(\tau_1, \tau_2, \tau_3),$$

the latter the negative of this amount.

A mixed strategy of the composite player 1,2 is a vector $\overrightarrow{\xi}$ of $S_{\beta_1\beta_2}$, the components of which we may denote by ξ_{τ_1,τ_2}.[1] Thus the $\overrightarrow{\xi}$ of $S_{\beta_1\beta_2}$ are characterized by

$$\xi_{\tau_1,\tau_2} \geqq 0, \qquad \sum_{\tau_1,\tau_2} \xi_{\tau_1,\tau_2} = 1.$$

[1] The number of pairs τ_1, τ_2 is, of course, $\beta_1\beta_2$.

A mixed strategy of the player 3 is a vector $\overrightarrow{\eta}$ of S_{β_3} the components of which we denote by η_{τ_3}. The $\overrightarrow{\eta}$ of S_{β_3} are characterized by

$$\eta_{\tau_3} \geqq 0, \qquad \sum_{\tau_3} \eta_{\tau_3} = 1.$$

The bilinear form $K(\overrightarrow{\xi}, \overrightarrow{\eta})$ of (17:2) in 17.4.1. is therefore

$$K(\overrightarrow{\xi}, \overrightarrow{\eta}) \equiv \sum_{\tau_1, \tau_2, \tau_3} \{\mathcal{H}_1(\tau_1, \tau_2, \tau_3) + \mathcal{H}_2(\tau_1, \tau_2, \tau_3)\} \xi_{\tau_1, \tau_2} \eta_{\tau_3}$$
$$\equiv - \sum_{\tau_1, \tau_2, \tau_3} \mathcal{H}_3(\tau_1, \tau_2, \tau_3) \xi_{\tau_1, \tau_2} \eta_{\tau_3},$$

and finally

$$c = \operatorname{Max}_{\overrightarrow{\xi}} \operatorname{Min}_{\overrightarrow{\eta}} K(\overrightarrow{\xi}, \overrightarrow{\eta}) = \operatorname{Min}_{\overrightarrow{\eta}} \operatorname{Max}_{\overrightarrow{\xi}} K(\overrightarrow{\xi}, \overrightarrow{\eta}).$$

The expressions for b, a obtain from this by cyclical permutations of the players 1,2,3 in all details of this representation.

We repeat the formulae expressing $\alpha, \beta, \gamma, a', b', c'$ and Δ:

$$\Delta = a + b + c \qquad \text{necessarily} \quad \geqq 0,$$
$$\alpha = \frac{-a + b + c}{2}, \qquad a' = \frac{-2a + b + c}{3},$$
$$\beta = \frac{a - b + c}{2}, \qquad b' = \frac{a - 2b + c}{3},$$
$$\gamma = \frac{a + b - c}{2}, \qquad c' = \frac{a + b - 2c}{3},$$

and we have

$$\Delta \geqq 0,$$
$$a' + b' + c' = 0,$$
$$\alpha = a' + \frac{\Delta}{6}, \qquad \beta = b' + \frac{\Delta}{6}, \qquad \gamma = c' + \frac{\Delta}{6},$$
$$-a = a' - \frac{\Delta}{3}, \qquad -b = b' - \frac{\Delta}{3}, \qquad -c = c' - \frac{\Delta}{3}.$$

24. Discussion of an Objection

24.1. The Case of Perfect Information and Its Significance

24.1.1. We have obtained a solution for the zero-sum three-person game which accounts for all possibilities and which indicates the direction that the search for the solutions of the n-person game must take: the analysis of all possible coalitions, and the competitive relationship which they bear to each other,—which should determine the compensations that players who want to form a coalition will pay to each other.

We have noticed already that this will be a much more difficult problem for $n \geqq 4$ players than it was for $n = 3$ (cf. footnote 2, p. 221).

Before we attack this question, it is well to pause for a moment to reconsider our position. In the discussions which follow we shall put the main stress on the formation of coalitions and the compensations between the participants in those coalitions, using the theory of the zero-sum two-person game to determine the values of the ultimate coalitions which oppose each other after all players have "taken sides" (cf. 25.1.1., 25.2.). But is this aspect of the matter really as universal as we propose to claim?

We have adduced already some strong positive arguments for it, in our discussion of the zero-sum three-person game. Our ability to build the theory of the n-person game (for all n) on this foundation will, *in fine*, be the decisive positive argument. But there is a negative argument—an objection—to be considered, which arises in connection with those games where perfect information prevails.

The objection which we shall now discuss applies only to the above mentioned special category of games. Thus it would not, if found valid, provide us with an alternative theory that applies to all games. But since we claim a general validity for our proposed stand, we must invalidate all objections, even those which apply only to some special case.[1]

24.1.2. Games with perfect information have already been discussed in 15. We saw there that they have important peculiarities and that their nature can be understood fully only when they are considered in the extensive form—and not merely in the normalized one on which our discussion chiefly relied (cf. also 14.8.).

The analysis of 15. began by considering n-person games (for all n), but in its later parts we had to narrow it to the zero-sum two-person game. At the end, in particular, we found a verbal method of discussing it (cf. 15.8.) which had some remarkable features: First, while not entirely free from objections, it seemed worth considering. Second, the argumentation used was rather different from that by which we had resolved the general case of the zero-sum two-person game—and while applicable only to this special case, it was more straight forward than the other argumentation. Third, it led—for the zero-sum two-person games with perfect information—to the same result as our general theory.

Now one might be tempted to use this argumentation for $n \geqq 3$ players too; indeed a superficial inspection of the pertinent paragraph 15.8.2. does not immediately disclose any reason why it should be restricted (as there) to $n = 2$ players (cf., however, 15.8.3.). But this procedure makes no mention of coalitions or understandings between players, etc.; so if it is usable for $n = 3$ players, then our present approach is open to grave doubts.[2] We

[1] In other words: in claiming general validity for a theory one necessarily assumes the burden of proof against all objectors.

[2] One might hope to evade this issue, by expecting to find $\Delta = 0$ for all zero-sum three-person games with perfect information. This would make coalitions unnecessary. Cf. the end of 23.1.

Just as games with perfect information avoided the difficulties of the theory of zero-sum two-person games by being strictly determined (cf. 15.6.1.), they would now avoid those of the zero-sum three-person games by being inessential.

propose to show therefore why the procedure of 15.8. is inconclusive when the number of players is three or more.

To do this, let us repeat some characteristic steps of the argumentation in question (cf. 15.8.2., the notations of which we are also using).

24.2. Detailed Discussion. Necessity of Compensations between Three or More Players

24.2.1. Consider accordingly a game Γ in which perfect information prevails. Let $\mathfrak{M}_1, \mathfrak{M}_2, \cdots, \mathfrak{M}_\nu$ be its moves, $\sigma_1, \sigma_2, \cdots, \sigma_\nu$ the choices connected with these moves, $\pi(\sigma_1, \cdots, \sigma_\nu)$ the play characterized by these choices, and $\mathfrak{F}_j(\pi(\sigma_1, \cdots, \sigma_\nu))$ the outcome of this play for the player $j(= 1, 2, \cdots, n)$.

Assume that the moves $\mathfrak{M}_1, \mathfrak{M}_2, \cdots, \mathfrak{M}_{\nu-1}$ have already been made, the outcome of their choices being $\sigma_1, \sigma_2, \cdots, \sigma_{\nu-1}$ and consider the last move $\mathfrak{M}_\nu$ and its σ_ν. If this is a chance move—i.e. $k_\nu(\sigma_1, \cdots, \sigma_{\nu-1}) = 0$,—then the various possible values $\sigma_\nu = 1, 2, \cdots, \alpha_\nu(\sigma_1, \cdots, \sigma_{\nu-1})$ have the probabilities $p_\nu(1), p_\nu(2), \cdots, p_\nu(\alpha_\nu(\sigma_1, \cdots, \sigma_{\nu-1}))$, respectively. If this is a personal move of player k—i.e. $k_\nu(\sigma_1, \cdots, \sigma_{\nu-1}) = k = 1, 2, \cdots, n$,—then player k will choose σ_ν so as to make $\mathfrak{F}_k(\pi(\sigma_1, \cdots, \sigma_{\nu-1}, \sigma_\nu))$ a maximum. Denote this σ_ν by $\sigma_\nu(\sigma_1, \cdots, \sigma_{\nu-1})$. Thus one can argue that the value of the play is already known (for each player $j = 1, \cdots n$) after the moves $\mathfrak{M}_1, \mathfrak{M}_2, \cdots, \mathfrak{M}_{\nu-1}$ (and before $\mathfrak{M}_\nu$!),—i.e. as a function of $\sigma_1, \sigma_2, \cdots, \sigma_{\nu-1}$ alone. Indeed: by the above it is

$$\mathfrak{F}'_j(\pi'(\sigma_1, \cdots, \sigma_{\nu-1})) \begin{cases} = \displaystyle\sum_{\sigma_\nu=1}^{\alpha_\nu(\sigma_1, \cdots, \sigma_{\nu-1})} p_\nu(\sigma_\nu)\mathfrak{F}_j(\pi(\sigma_1, \cdots, \sigma_{\nu-1}, \sigma_\nu)) \\ \qquad \text{for} \quad k_\nu(\sigma_1, \cdots, \sigma_{\nu-1}) = 0, \\ = \mathfrak{F}_j(\pi(\sigma_1, \cdots, \sigma_{\nu-1}, \sigma_\nu(\sigma_1, \cdots, \sigma_{\nu-1}))), \\ \qquad \text{where } \sigma_\nu = \sigma_\nu(\sigma_1, \cdots, \sigma_{\nu-1}) \text{ maximizes} \\ \qquad \mathfrak{F}_k(\pi(\sigma_1, \cdots, \sigma_{\nu-1}, \sigma_\nu)) \text{ for} \\ \qquad k_\nu(\sigma_1, \cdots, \sigma_{\nu-1}) = k = 1, \cdots, n. \end{cases}$$

Consequently we can treat the game Γ as if it consisted of the moves $\mathfrak{M}_1, \mathfrak{M}_2, \cdots, \mathfrak{M}_{\nu-1}$ only (without $\mathfrak{M}_\nu$).

By this device we have removed the last move $\mathfrak{M}_\nu$. Repeating it, we can similarly remove successively the moves $\mathfrak{M}_{\nu-1}, \mathfrak{M}_{\nu-2}, \cdots, \mathfrak{M}_2, \mathfrak{M}_1$ and finally obtain a definite value of the play (for each player $j = 1, 2, \cdots n$).

24.2.2. For a critical appraisal of this procedure consider the last two steps $\mathfrak{M}_{\nu-1}, \mathfrak{M}_\nu$ and assume that they are personal moves of two different

---

This, however, is not the case. To see that, it suffices to modify the rules of the simple majority game (cf. 21.1.) as follows: Let the players 1,2,3 make their personal moves (i.e. the choices of τ_1, τ_2, τ_3 respectively, cf. loc. cit.) in this order, each one being informed about the anterior moves. It is easy to verify that the values c, b, a of the three coalitions 1,2, 1,3, 2,3 are the same as before

$$c = b = a = 1, \qquad \Delta = a + b + c = 3 > 0.$$

A detailed discussion of this game, with particular respect to the considerations of 21.2., would be of a certain interest, but we do not propose to continue this subject further at present.

players, say 1,2 respectively. In this situation we have assumed that player 2 will certainly choose σ_ν so as to maximize $\mathfrak{F}_2(\sigma_1, \cdots, \sigma_{\nu-1}, \sigma_\nu)$. This gives a $\sigma_\nu = \sigma_\nu(\sigma_1, \cdots, \sigma_{\nu-1})$. Now we have also assumed that player 1, in choosing $\sigma_{\nu-1}$ can rely on this; i.e. that he may safely replace the $\mathfrak{F}_1(\sigma_1, \cdots, \sigma_{\nu-1}, \sigma_\nu)$, (which is what he will really obtain), by $\mathfrak{F}_1(\sigma_1, \cdots, \sigma_{\nu-1}, \sigma_\nu(\sigma_1, \cdots, \sigma_{\nu-1}))$ and maximize this latter quantity.[1] But can he rely on this assumption?

To begin with, $\sigma_\nu(\sigma_1, \cdots, \sigma_{\nu-1})$ may not even be uniquely determined: $\mathfrak{F}_2(\sigma_1, \cdots, \sigma_{\nu-1}, \sigma_\nu)$ may assume its maximum (for given $\sigma_1, \cdots, \sigma_{\nu-1}$) at several places σ_ν. In the zero-sum two-person game this was irrelevant: there $\mathfrak{F}_1 \equiv -\mathfrak{F}_2$, hence two σ_ν which give the same value to $\mathfrak{F}_2$, also give the same value to $\mathfrak{F}_1$.[2] But even in the zero-sum three-person game, $\mathfrak{F}_2$ does not determine $\mathfrak{F}_1$, due to the existence of the third player and his $\mathfrak{F}_3$! So it happens here for the first time that a difference which is unimportant for one player may be significant for another player. This was impossible in the zero-sum two-person game, where each player won (precisely) what the other lost.

What then must player 1 expect if two σ_ν are of the same importance for player 2, but not for player 1? One must expect that he will try to induce player 2 to choose the σ_ν which is more favorable to him. He could offer to pay to player 2 any amount up to the difference this makes for him.

This being conceded, one must envisage that player 1 may even try to induce player 2 to choose a σ_ν which does not maximize $\mathfrak{F}_2(\sigma_1, \cdots, \sigma_{\nu-1}, \sigma_\nu)$. As long as this change causes player 2 less of a loss than it causes player 1 a gain,[3] player 1 can compensate player 2 for his loss, and possibly even give up to him some part of his profit.

24.2.3. But if player 1 can offer this to player 2, then he must also count on similar offers coming from player 3 to player 2. I.e. there is no certainty at all that player 2 will, by his choice of σ_ν, maximize $\mathfrak{F}_2(\sigma_1, \cdots, \sigma_{\nu-1}, \sigma_\nu)$. In comparing two σ_ν one must consider whether player 2's loss is overcompensated by player 1's or player 3's gain, since this could lead to understandings and compensations. I.e. one must analyze whether a coalition 1,2 or 2,3 would gain by any modification of σ_ν.

24.2.4. This brings the coalitions back into the picture. A closer analysis would lead us to the considerations and results of 22.2., 22.3., 23. in every detail. But it does not seem necessary to carry this out here in complete detail: after all, this is just a special case, and the discussion of 22.2., 22.3., 23. was of absolutely general validity (for the zero-sum three-person game)

[1] Since this is a function of $\sigma_1, \cdots, \sigma_{\nu-2}, \sigma_{\nu-1}$ only, of which $\sigma_1, \cdots, \sigma_{\nu-2}$ are known at $\mathfrak{M}_{\nu-1}$, and $\sigma_{\nu-1}$ is controlled by player 1, he is able to maximize it.

He cannot in any sense maximize $\mathfrak{F}_1(\sigma_1, \cdots, \sigma_{\nu-1}, \sigma_\nu)$ since that also depends on σ_ν which he neither knows nor controls.

[2] Indeed, we refrained in 15.8.2. from mentioning $\mathfrak{F}_2$ at all: instead of maximizing $\mathfrak{F}_2$, we talked of minimizing $\mathfrak{F}_1$. There was no need even to introduce $\sigma_\nu(\sigma_1, \cdots, \sigma_{\nu-1})$ and everything was described by Max and Min operations on $\mathfrak{F}_1$.

[3] I.e. when it happens at the expense of player 3.

provided that the consideration of understandings and compensations, i.e. of coalitions, is permitted.

We wanted to show that the weakness of the argument of 15.8.2., already recognized in 15.8.3., becomes destructive exactly when we go beyond the zero-sum two-person games, and that it leads precisely to the mechanism of coalitions etc. foreseen in the earlier paragraphs of this chapter. This should be clear from the above analysis, and so we can return to our original method in dealing with zero-sum three-person games,—i.e. claim full validity for the results of 22.2., 22.3., 23.

CHAPTER VI

FORMULATION OF THE GENERAL THEORY: ZERO-SUM n-PERSON GAMES

25. The Characteristic Function

25.1. Motivation and Definition

25.1.1. We now turn to the zero-sum n-person game for general n. The experience gained in Chapter V concerning the case $n = 3$ suggests that the possibilities of coalitions between players will play a decisive role in the general theory which we are developing. It is therefore important to evolve a mathematical tool which expresses these "possibilities" in a quantitative way.

Since we have an exact concept of "value" (of a play) for the zero-sum two-person game, we can also attribute a "value" to any given group of players, provided that it is opposed by the coalition of all the other players. We shall give these rather heuristic indications an exact meaning in what follows. The important thing is, at any rate, that we shall thus reach a mathematical concept on which one can try to base a general theory—and that the attempt will, *in fine*, prove successful.

Let us now state the exact mathematical definitions which carry out this program.

25.1.2. Suppose then that we have a game Γ of n players who, for the sake of brevity, will be denoted by $1, 2, \cdots, n$. It is convenient to introduce the set $I = (1, 2, \cdots, n)$ of all these players. Without yet making any predictions or assumptions about the course a play of this game is likely to take, we observe this: if we group the players into two parties, and treat each party as an absolute coalition—i.e. if we assume full cooperation within each party—then a zero-sum two-person game results.[1] Precisely: Let S be any given subset of I, $-S$ its complement in I. We consider the zero-sum two-person game which results when all players k belonging to S cooperate with each other on the one hand, and all players k belonging to $-S$ cooperate with each other on the other hand.

Viewed in this manner Γ falls under the theory of the zero-sum two-person game of Chapter III. Each play of this game has a well defined value (we mean the v' defined in 17.8.1.). Let us denote by $\mathrm{v}(S)$ the value of a play for the coalition of all k belonging to S (which, in our present interpretation, is one of the players).

[1] This is exactly what we did in the case $n = 3$ in 23.1.1. The general possibility was already alluded to at the beginning of 24.1.

Mathematical expressions for $\mathrm{v}(S)$ obtain as follows:[1]

25.1.3. Assume the zero-sum n-person game Γ in the normalized form of 11.2.3. There each player $k = 1, 2, \cdots, n$ chooses a variable τ_k (each one uninformed about the $n - 1$ other choices) and gets the amount

$$\mathcal{H}_k(\tau_1, \tau_2, \cdots, \tau_n).$$

Of course (the game is zero-sum):

$$\sum_{k=1}^{n} \mathcal{H}_k(\tau_1, \cdots, \tau_n) \equiv 0. \tag{25:1}$$

The domains of the variables are:

$$\tau_k = 1, \cdots, \beta_k \qquad \text{for} \qquad k = 1, 2, \cdots, n.$$

Now in the two-person game which arises between an absolute coalition of all players k belonging to S (player 1′) and that one of all players k belonging to $-S$ (player 2′), we have the following situation:

The composite player 1′ has the aggregate of variables τ_k where k runs over all elements of S. It is necessary to treat this aggregate as one variable and we shall therefore designate it by one symbol τ^S. The composite player 2′ has the aggregate of variables τ_k where k runs over all elements of $-S$. This aggregate too is one variable, which we designate by the symbol τ^{-S}. The player 1′ gets the amount

$$\bar{\mathcal{H}}(\tau^S, \tau^{-S}) = \sum_{k \text{ in } S} \mathcal{H}_k(\tau_1, \cdots, \tau_n) = -\sum_{k \text{ in } -S} \mathcal{H}_k(\tau_1, \cdots, \tau_n);\text{[2]} \tag{25:2}$$

the player 2′ gets the negative of this amount.

A mixed strategy of the player 1′ is a vector $\overrightarrow{\xi}$ of S_{β^S},[3] the components of which we denote by ξ_{τ^S}. Thus the $\overrightarrow{\xi}$ of S_{β^S} are characterized by

$$\xi_{\tau^S} \geqq 0, \qquad \sum_{\tau^S} \xi_{\tau^S} = 1.$$

A mixed strategy of the player 2′ is a vector $\overrightarrow{\eta}$ of $S_{\beta^{-S}}$,[4] the components of which we denote by $\eta_{\tau^{-S}}$. Thus the $\overrightarrow{\eta}$ of $S_{\beta^{-S}}$ are characterized by

$$\eta_{\tau^{-S}} \geqq 0, \qquad \sum_{\tau^{-S}} \eta_{\tau^{-S}} = 1.$$

[1] This is a repetition of the construction of 23.2., which applied only to the special case $n = 3$.

[2] The τ^S, τ^{-S} of the first expression form together the aggregate of the $\tau_1, \cdots, \tau_n$ of the two other expressions; so τ^S, τ^{-S} determine those $\tau_1, \cdots, \tau_n$.

The equality of the two last expressions is, of course, only a restatement of the zero-sum property.

[3] β^S is the number of possible aggregates τ^S, i.e. the product of all β_k where k runs over all elements of S.

[4] β^{-S} is the number of possible aggregates τ^{-S}, i.e. the product of all β_k where k runs over all elements of $-S$.

The bilinear form $K(\vec{\xi}, \vec{\eta})$ of (17:2) in 17.4.1. is therefore

$$K(\vec{\xi}, \vec{\eta}) = \sum_{\tau^S, \tau^{-S}} \bar{\mathcal{H}}(\tau^S, \tau^{-S}) \xi_{\tau^S} \eta_{\tau^{-S}},$$

and finally

$$v(S) = \mathrm{Max}_{\vec{\xi}} \; \mathrm{Min}_{\vec{\eta}} \; K(\vec{\xi}, \vec{\eta}) = \mathrm{Min}_{\vec{\eta}} \; \mathrm{Max}_{\vec{\xi}} \; K(\vec{\xi}, \vec{\eta}).$$

25.2. Discussion of the Concept

25.2.1. The above function $v(S)$ is defined for all subsets S of I and has real numbers as values. Thus it is, in the sense of 13.1.3., a *numerical set function*. We call it the *characteristic function of the game* Γ. As we have repeatedly indicated, we expect to base the entire theory of the zero-sum n-person game on this function.

It is well to visualize what this claim involves. We propose to determine everything that can be said about coalitions between players, compensations between partners in every coalition, mergers or fights between coalitions, etc., in terms of the characteristic function $v(S)$ alone. *Prima facie*, this program may seem unreasonable, particularly in view of these two facts:

(a) An altogether fictitious two-person game, which is related to the real n-person game only by a theoretical construction, was used to define $v(S)$. Thus $v(S)$ is based on a hypothetical situation, and not strictly on the n-person game itself.

(b) $v(S)$ describes what a given coalition of players (specifically, the set S) can obtain from their opponents (the set $-S$)—but it fails to describe how the proceeds of the enterprise are to be divided among the partners k belonging to S. This division, the "imputation," is indeed directly determined by the individual functions $\mathcal{H}_k(\tau_1, \cdots, \tau_n)$, k belonging to S, while $v(S)$ depends on much less. Indeed, $v(S)$ is determined by their partial sum $\bar{\mathcal{H}}(\tau^S, \tau^{-S})$ alone, and even by less than that since it is the saddle value of the bilinear form $K(\vec{\xi}, \vec{\eta})$ based on $\bar{\mathcal{H}}(\tau^S, \tau^{-S})$ (cf. the formulae of 25.1.3.).

25.2.2. In spite of these considerations we expect to find that the characteristic function $v(S)$ determines everything, including the "imputation" (cf. (b) above). The analysis of the zero-sum three-person game in Chapter V indicates that the direct distribution (i.e., "imputation") by means of the $\mathcal{H}_k(\tau_1, \cdots, \tau_n)$ is necessarily offset by some system of "compensations" which the players must make to each other before coalitions can be formed. The "compensations" should be determined essentially by the possibilities which exist for each partner in the coalition S (i.e. for each k belonging to S), to forsake it and to join some other coalition T. (One may have to consider also the influence of possible simultaneous and concerted desertions by sets of several partners in S etc.) I.e. the "imputation" of $v(S)$ to the players k belonging to S should be determined by the other $v(T)$[1]—and not by the $\mathcal{H}_k(\tau_1, \cdots, \tau_n)$. We have demonstrated this for the zero-sum three-person game in Chapter V. One of the main objectives of the theory we are trying to build up is to establish the same thing for the general n-person game.

25.3. Fundamental Properties

25.3.1. Before we undertake to elucidate the importance of the characteristic function $v(S)$ for the general theory of games, we shall investigate this function as a mathematical entity in itself. We know that it is a numerical set function, defined for all subsets S of $I = (1, 2, \cdots, n)$ and we now propose to determine its essential properties.

It will turn out that they are the following:

(25:3:a) $$v(\ominus) = 0,$$

(25:3:b) $$v(-S) = -v(S),$$

(25:3:c) $$v(S \cup T) \geqq v(S) + v(T), \quad \text{if} \quad S \cap T = \ominus.$$

We prove first that the characteristic set function $v(S)$ of every game fulfills (25:3:a)-(25:3:c).

25.3.2. The simplest proof is a conceptual one, which can be carried out with practically no mathematical formulae. However, since we gave exact mathematical expressions for $v(S)$ in 25.1.3., one might desire a strictly mathematical, formalistic proof—in terms of the operations Max and Min and the appropriate vectorial variables. We emphasize therefore that our conceptual proof is strictly equivalent to the desired formalistic, mathematical one, and that the translation can be carried out without any real difficulty. But since the conceptual proof makes the essential ideas clearer, and in a briefer and simpler way, while the formalistic proof would involve a certain amount of cumbersome notations, we prefer to give the former. The reader who is interested may find it a good exercise to construct the formalistic proof by translating our conceptual one.

25.3.3. *Proof* of (25:3:a):[2] The coalition $\ominus$ has no members, so it always gets the amount zero, therefore $v(\ominus) = 0$.

Proof of (25:3:b): $v(S)$ and $v(-S)$ originate from the same (fictitious) zero-sum two-person game,—the one played by the coalition S against

[1] All this is very much in the sense of the remarks in 4.3.3. on the role of "virtual" existence.

[2] Observe that we are treating even the empty set $\ominus$ as a coalition. The reader should think this over carefully. In spite of its strange appearance, the step is harmless—and quite in the spirit of general set theory. Indeed, it would be technically quite a nuisance to exclude the empty set from consideration.

Of course this empty coalition has no moves, no variables, no influence, no gains, and no losses. But this is immaterial.

The complementary set of $\ominus$, the set of all players I, will also be treated as a possible coalition. This too is the convenient procedure from the set-theoretical point of view. To a lesser extent this coalition also may appear to be strange, since it has no opponents. Although it has an abundance of members—and hence of moves and variables—it will (in a zero-sum game) equally have nothing to influence, and no gains or losses. But this too is immaterial.

the coalition $-S$. The value of a play of this game for its two composite players is indeed $v(S)$ and $v(-S)$ respectively. Therefore $v(-S) = -v(S)$.

Proof of (25:3:c): The coalition S can obtain from its opponents (by using an appropriate mixed strategy) the amount $v(S)$ and no more. The coalition T can obtain similarly the amount $v(T)$ and no more. Hence the coalition $S \cup T$ can obtain from its opponents the amount $v(S) + v(T)$, even if the subcoalitions S and T fail to cooperate with each other.[1] Since the maximum which the coalition $S \cup T$ can obtain under any condition is $v(S \cup T)$ this implies $v(S \cup T) \geqq v(S) + v(T)$.[2]

25.4. Immediate Mathematical Consequences

25.4.1. Before we go further let us draw some conclusions from the above (25:3:a)-(25:3:c). These will be derived in the sense that they hold for *any* numerical set function $v(S)$ which fulfills (25:3:a)-(25:3:c) irrespective of whether or not it is the characteristic function of a zero-sum n-person game Γ.

(25:4) $$v(I) = 0.$$

Proof:[3] By (25:3:a), (25:3:b), $v(I) = v(-\ominus) = -v(\ominus) = 0$.

(25:5) $$v(S_1 \cup \cdots \cup S_p) \geqq v(S_1) + \cdots + v(S_p)$$

if $S_1, \cdots, S_p$ are pairwise disjunct subsets of I.

Proof: Immediately by repeated application of (25:3:c).

(25:6) $$v(S_1) + \cdots + v(S_p) \leqq 0$$

if $S_1, \cdots, S_p$ are a decomposition of I, i.e. pairwise disjunct subsets of I with the sum I.

Proof: We have $S_1 \cup \cdots \cup S_p = I$, hence $v(S_1 \cup \cdots \cup S_p) = 0$ by (25:4). Therefore (25:6) follows from (25:5).

25.4.2. While (25:4)-(25:6) are consequences of (25:3:a)-(25:3:c), they—and even somewhat less—can replace (25:3:a)-(25:3:c) equivalently. Precisely:

(25:A) The conditions (25:3:a)-(25:3:c) are equivalent to the assertion of (25:6) for the values $p = 1, 2, 3$ only; but (25:6) must then be stated for $p = 1, 2$ with an $=$ sign, and for $p = 3$ with a $\leqq$ sign.

[1] Observe that we are now using $S \cap T = \ominus$. If S and T had common elements, we could not break up the coalition $S \cup T$ into the subcoalitions S and T.

[2] This proof is very nearly a repetition of the proof of $a + b + c \geqq 0$ in 22.3.2. One could even deduce our (25:3:c) from that relation: Consider the decomposition of I into the three disjunct subsets S, T, $-(S \cup T)$. Treat the three corresponding (hypothetical) absolute coalitions as the three players of the zero-sum three-person game into which this transforms Γ. Then $v(S)$, $v(T)$, $v(S \cup T)$ correspond to the $-a$, $-b$, c loc. cit.; hence $a + b + c \geqq 0$ means $-v(S) - v(T) + v(S \cup T) \geqq 0$; i.e. $v(S \cup T) \geqq v(S) + v(T)$.

[3] For a $v(S)$ originating from a game, both (25:3:a) and (25:4) are conceptually contained in the remark of footnote 2 on p. 241.

Proof: (25:6) for $p = 2$ with an $=$ sign states $v(S) + v(-S) = 0$ (we write S for S_1, hence S_2 is $-S$); i.e. $v(-S) = -v(S)$ which is exactly (25:3:b).

(25:6) for $p = 1$ with an $=$ sign states $v(I) = 0$ (in this case S_1 must be I)—which is exactly (25:4). Owing to (25:3:b), this is exactly the same as (25:3:a). (Cf. the above proof of (25:4).)

(25:6) for $p = 3$ with an $\leqq$ sign states $v(S) + v(T) + v(-(S \cup T)) \leqq 0$ (we write S, T for S_1, S_2; hence S_3 is $-(S \cup T)$), i.e.

$$-v(-(S \cup T)) \geqq v(S) + v(T).$$

By (25:3:b) this becomes $v(S \cup T) \geqq v(S) + v(T)$ which is exactly (25:3:c).

So our assertions are equivalent precisely to the conjunction of (25:3:a)-(25:3:c).

26. Construction of a Game with a Given Characteristic Function

26.1. The Construction

26.1.1. We now prove the converse of 25.3.1.: That for any numerical set function $v(S)$ which fulfills the conditions (25:3:a)-(25:3:c) there exists a zero-sum n-person game Γ of which this $v(S)$ is the characteristic function.

In order to avoid confusion it is better to denote the given numerical set function which fulfills (25:3:a)-(25:3:c) by $v_0(S)$. We shall define with its help a certain zero-sum n-person game Γ, and denote the characteristic function of this Γ by $v(S)$. It will then be necessary to prove that $v(S) \equiv v_0(S)$.

Let therefore a numerical set function $v_0(S)$ which fulfills (25:3:a)-(25:3:c) be given. We define the zero-sum n-person game Γ as follows:[1]

Each player $k = 1, 2, \cdots, n$ will, by a personal move, choose a subset S_k of I which contains k. Each one makes his choice independently of the choice of the other players.[2]

After this the payments to be made are determined as follows:

Any set S of players, for which

(26:1) $\qquad S_k = S \qquad$ for every k belonging to S

is called a *ring*.[3,4] Any two rings with a common element are identical.[5]

[1] This game Γ is essentially a more general analogue of the simple majority game of three persons, defined in 21.1. We shall accompany the text which follows with footnotes pointing out the details of this analogy.

[2] The n-element set I has 2^{n-1} subsets S containing k, which we can enumerate by an index $\tau_k(S) = 1, 2, \cdots, 2^{n-1}$. If we now let the player k choose, instead of S_k, its index $\tau_k = \tau_k(S_k) = 1, 2, \cdots, 2^{n-1}$, then the game is already in the normalized form of 11.2.3. Clearly all $\beta_k = 2^{n-1}$.

[3] The rings are the analogues of the couples in 21.1. The contents of footnote 3 on p. 222 apply accordingly; in particular the rings are the formal concept in the set of rules of the game which induces the coalitions which influence the actual course of each play.

[4] Verbally: A ring is a set of players, in which every one has chosen just this set.

The analogy with the definition of a couple in 21.1. is clear. The differences are due to formal convenience: in 21.1. we made each player designate the other element of the couple which he desires; now we expect him to indicate the entire ring. A closer analysis of this divergence would be easy enough, but it does not seem necessary.

[5] *Proof:* Let S and T be two rings with a common element k; then by (26:1) $S_k = S$ and $S_k = T$, and so $S = T$.

In other words: The totality of all rings (which have actually formed in a play) is a system of pairwise disjunct subsets of I.

Each player who is contained in none of the rings thus defined forms by himself a (one-element) set which is called a *solo set*. Thus the totality of all rings and solo sets (which have actually formed in a play) is a decomposition of I; i.e. a system of pairwise disjunct subsets of I with the sum I. Denote these sets by $C_1, \cdots, C_p$ and the respective numbers of their elements by $n_1, \cdots, n_p$.

Consider now a player k. He belongs to precisely one of these sets $C_1, \cdots, C_p$ say to C_q. Then player k gets the amount

$$(26:2) \qquad \frac{1}{n_q} \mathrm{v}_0(C_q) - \frac{1}{n} \sum_{r=1}^{p} \mathrm{v}_0(C_r).^1$$

This completes the description of the game Γ. We shall now show that this Γ is a zero-sum n-person game and that it has the desired characteristic function $\mathrm{v}_0(S)$.

26.1.2. *Proof* of the zero-sum character: Consider one of the sets C_q. Each one of the n_q players belonging to it gets the same amount, stated in (26:2). Hence the players of C_q together get the amount

$$(26:3) \qquad \mathrm{v}_0(C_q) - \frac{n_q}{n} \sum_{r=1}^{p} \mathrm{v}_0(C_r).$$

In order to obtain the total amount which all players $1, \cdots, n$ get, we must sum the expression (26:3) over all sets C_q, i.e. over all $q = 1, \cdots, p$. This sum is clearly

$$\sum_{q=1}^{p} \mathrm{v}_0(C_q) - \sum_{r=1}^{p} \mathrm{v}_0(C_r),$$

i.e. zero.[2]

Proof that the characteristic function is $\mathrm{v}_0(S)$: Denote the characteristic function of Γ by $\mathrm{v}(S)$. Remember that (25:3:a)-(25:3:c) hold for $\mathrm{v}(S)$ because it is a characteristic function, and for $\mathrm{v}_0(S)$ by hypothesis. Consequently (25:4)-(25:6) also hold for both $\mathrm{v}(S)$ and $\mathrm{v}_0(S)$.

We prove first that

$$(26:4) \qquad \mathrm{v}(S) \geqq \mathrm{v}_0(S) \qquad \text{for all subsets } S \text{ of } I.$$

If S is empty, then both sides are zero by (25:3:a). So we may assume that S is not empty. In this case a coalition of all players k belonging to S can

[1] The course of the play, that is the choices $S_1, \cdots, S_n$—or, in the sense of footnote 2 on p. 243, the choices $\tau_1, \cdots, \tau_n$—determine the $C_1, \cdots, C_p$, and thus the expression (26:2). Of course (26:2) is the $\mathcal{H}_k(\tau_1, \cdots, \tau_n)$ of the general theory.

[2] Obviously $\sum_{q=1}^{p} n_q = n$.

govern the choices of its S_k so as with certainty to make S a ring. It suffices for every k in S to choose his $S_k = S$. Whatever the other players (in $-S$) do, S will thus be one of the sets (rings or solo sets) $C_1, \cdots, C_p$, say C_q. Thus each k in $C_q = S$ gets the amount (26:2); hence the entire coalition S gets the amount (26:3). Now we know that the system

$$C_1, \cdots, C_p$$

is a decomposition of I; hence by (25:6) $\sum_{r=1}^{p} v_0(C_r) \leqq 0$. That is, the expression (26:3) is $\geqq v_0(C_q) = v_0(S)$.[1] In other words, the players belonging to the coalition S can secure for themselves at least the amount $v_0(S)$ irrespective of what the players in $-S$ do. This means that $v(S) \geqq v_0(S)$; i.e. (26:4).

Now we can establish the desired formula

$$(26:5) \qquad v(S) = v_0(S).$$

Apply (26:4) to $-S$. Owing to (25:3:b) this means $-v(S) \geqq -v_0(S)$, i.e.

$$(26:6) \qquad v(S) \leqq v_0(S).$$

(26:4), (26:6) give together (26:5).[2]

26.2. Summary

26.2. To sum up: in paragraphs 25.3.-26.1. we have obtained a complete mathematical characterization of the characteristic functions $v(S)$ of all possible zero-sum n-person games Γ. If the surmise which we expressed in 25.2.1. proves to be true, i.e. if we shall be able to base the entire theory of the game on the global properties of the coalitions as expressed by $v(S)$, then our characterization of $v(S)$ has revealed the exact mathematical substratum of the theory. Thus the characterization of $v(S)$ and the functional relations (25:3:a)-(25:3:c) are of fundamental importance.

We shall therefore undertake a first mathematical analysis of the meaning and of the immediate properties of these relations. We call the functions which satisfy them *characteristic functions*—even when they are viewed in themselves, without reference to any game.

27. Strategic Equivalence. Inessential and Essential Games

27.1. Strategic Equivalence. The Reduced Form

27.1.1. Consider a zero-sum n-person game Γ with the characteristic function $v(S)$. Let also a system of numbers $\alpha_1^0, \cdots, \alpha_n^0$ be given. We

[1] Observe that the expression (26:3), i.e. the total amount obtained by the coalition S, is not determined by the choices of the players in S alone. But we derived for it a lower bound $v_0(S)$, which is determined.

[2] Observe that in our discussion of the good strategies of the (fictitious) two-person game between the coalitions S and $-S$ (our above proof really amounted to that), we considered only pure strategies, and no mixed ones. In other words, all these two-person games happened to be strictly determined.

This, however, is irrelevant for the end which we are now pursuing.

now form a new game Γ' which agrees with Γ in all details except for this: Γ' is played in exactly the same way as Γ, but when all is over, player k gets in Γ' the amount which he would have got in Γ (after the same play), plus α_k^0. (Observe that the $\alpha_1^0, \cdots, \alpha_n^0$ are absolute constants!) Thus if Γ is brought into the normalized form of 11.2.3. with the functions

$$\mathcal{H}_k(\tau_1, \cdots, \tau_n),$$

then Γ' is also in this normalized form, with the corresponding functions $\mathcal{H}'_k(\tau_1, \cdots, \tau_n) \equiv \mathcal{H}_k(\tau_1, \cdots, \tau_n) + \alpha_k^0$. Clearly Γ' will be a zero-sum n-person game (along with Γ) if and only if

$$(27:1) \qquad \sum_{k=1}^{n} \alpha_k^0 = 0,$$

which we assume.

Denote the characteristic function of Γ' by $\mathrm{v}'(S)$, then clearly

$$(27:2) \qquad \mathrm{v}'(S) \equiv \mathrm{v}(S) + \sum_{k \text{ in } S} \alpha_k^0.$$[1]

Now it is apparent that the strategic possibilities of the two games Γ and Γ' are exactly the same. The only difference between these two games consists of the fixed payments α_k^0 after each play. And these payments are absolutely fixed; nothing that any or all of the players can do will modify them. One could also say that the position of each player has been shifted by a fixed amount, but that the strategic possibilities, the inducements and possibilities to form coalitions etc., are entirely unaffected. In other words: If two characteristic functions $\mathrm{v}(S)$ and $\mathrm{v}'(S)$ are related to each other by (27:2)[2], then every game with the characteristic function $\mathrm{v}(S)$ is fully equivalent from all strategic points of view to some game with the characteristic function $\mathrm{v}'(S)$, and conversely. I.e. $\mathrm{v}(S)$ and $\mathrm{v}'(S)$ describe two strategically equivalent families of games. In this sense $\mathrm{v}(S)$ and $\mathrm{v}'(S)$ may themselves be considered equivalent.

Observe that all this is independent of the surmise restated in 26.2., according to which all games with the same $\mathrm{v}(S)$ have the same strategic characteristics.

27.1.2. The transformation (27:2) (we need pay no attention to (27:1), cf. footnote 2 above) replaces, as we have seen, the set functions $\mathrm{v}(S)$ by a

[1] The truth of this relation becomes apparent if one recalls how $\mathrm{v}(S)$, $\mathrm{v}'(S)$ were defined with the help of the coalition S. It is also easy to prove (27:2) formalistically with the help of the $\mathcal{H}_k(\tau_1, \cdots, \tau_n)$, $\mathcal{H}'_k(\tau_1, \cdots, \tau_n)$.

[2] Under these conditions (27:1) follows and need not be postulated separately. Indeed, by (25:4) in 25.4.1., $\mathrm{v}(I) = \mathrm{v}'(I) = 0$, hence (27:2) gives

$$\sum_{k \text{ in } I} \alpha_k^0 = 0; \qquad \text{i.e. } \sum_{k=1}^{n} \alpha_k^0 = 0.$$

strategically fully equivalent set-function $v'(S)$. We therefore call this relationship *strategic equivalence*.

We now turn to a mathematical property of this concept of strategic equivalence of characteristic functions.

It is desirable to pick from each family of characteristic functions $v(S)$ in strategic equivalence a particularly simple representative $\bar{v}(S)$. The idea is that, given $v(S)$, this representative $\bar{v}(S)$ should be easy to determine, and that on the other hand two $v(S)$ and $v'(S)$ would be in strategic equivalence if and only if their representatives $\bar{v}(S)$ and $\bar{v}'(S)$ are identical. Besides, we may try to choose these representatives $\bar{v}(S)$ in such a fashion that their analysis is simpler than that of the original $v(S)$.

27.1.3. When we started from characteristic functions $v(S)$ and $v'(S)$, then the concept of strategic equivalence could be based upon (27:2) alone; (27:1) ensued (cf. footnote 2, p. 246). However, we propose to start now from one characteristic function $v(S)$ alone, and to survey all possible $v'(S)$ which are in strategic equivalence with it—in order to choose the representative $\bar{v}(S)$ from among them. Therefore the question arises which systems $\alpha_1^0, \cdots, \alpha_n^0$ we may use, i.e. for which of these systems (using (27:2)) the fact that $v(S)$ is a characteristic function entails the same for $v'(S)$. The answer is immediate, both by what we have said so far, and by direct verification: The condition (27:1) is necessary and sufficient.[1]

Thus we have the n indeterminate quantities $\alpha_1^0, \cdots, \alpha_n^0$ at our disposal in the search for a representative $\bar{v}(S)$; but the $\alpha_1^0, \cdots, \alpha_n^0$ are subject to one restriction: (27:1). So we have $n - 1$ free parameters at our disposal.

27.1.4. We may therefore expect that we can subject the desired representative $\bar{v}(S)$ to $n - 1$ requirements. As such we choose the equations

(27:3) $$\bar{v}((1)) = \bar{v}((2)) = \cdots = \bar{v}((n)).$$[2]

I.e. we require that every one-man coalition—every player left to himself—should have the same value.

We may substitute (27:2) into (27:3) and state this together with (27:1), and so formulate all our requirements concerning the $\alpha_1^0, \cdots, \alpha_n^0$. So we obtain:

(27:1*) $$\sum_{k=1}^{n} \alpha_k^0 = 0,$$

(27:2*) $$v((1)) + \alpha_1^0 = v((2)) + \alpha_2^0 = \cdots = v((n)) + \alpha_n^0.$$

It is easy to verify that these equations are solved by precisely one system of $\alpha_1^0, \cdots, \alpha_n^0$:

[1] This detailed discussion may seem pedantic. We gave it only to make clear that when we start with two characteristic functions $v(S)$ and $v'(S)$ then (27:1) is superfluous, but when we start with one characteristic function only, then (27:1) is needed.

[2] Observe that these are $n - 1$ and not n equations.

(27:4) $$\alpha_k^0 = -\mathrm{v}((k)) + \frac{1}{n}\sum_{j=1}^{n} \mathrm{v}((j)).$$[1]

So we can say:

(27:A) We call a characteristic function $\bar{\mathrm{v}}(S)$ *reduced* if and only if it satisfies (27:3). Then every characteristic function $\mathrm{v}(S)$ is in strategic equivalence with precisely one reduced $\bar{\mathrm{v}}(S)$. This $\bar{\mathrm{v}}(S)$ is given by the formulae (27:2) and (27:4), and we call it the *reduced form* of $\mathrm{v}(S)$.

The reduced functions will be the representatives for which we have been looking.

27.2. Inequalities. The Quantity γ

27.2. Let us consider a reduced characteristic function $\bar{\mathrm{v}}(S)$. We denote the joint value of the n terms in (27:3) by $-\gamma$, i.e.

(27:5) $$-\gamma = \bar{\mathrm{v}}((1)) = \bar{\mathrm{v}}((2)) = \cdots \bar{\mathrm{v}}((n)).$$

We can state (27:5) also this way:

(27:5*) $\bar{\mathrm{v}}(S) = -\gamma$ for every one-element set S.

Combination with (25:3:b) in 25.3.1. transforms (27:5*) into

(27:5**) $\bar{\mathrm{v}}(S) = \gamma$ for every $(n - 1)$-element set S.

We re-emphasize that any one of (27:5), (27:5*), (27:5**) is—besides defining γ—just a restatement of (27:3), i.e. a characterization of the reduced nature of $\bar{\mathrm{v}}(S)$.

Now apply (25:6) in 25.4.1. to the one-element sets $S_1 = (1), \cdots, S_n = (n)$. (So $p = n$). Then (27:5) gives $-n\gamma \leqq 0$, i.e.:

(27:6) $$\gamma \geqq 0.$$

Consider next an arbitrary subset S of I. Let p be the number of its elements: $S = (k_1, \cdots, k_p)$. Now apply (25:5) in 25.4.1. to the one-element sets $S_1 = (k_1), \cdots, S_p = (k_p)$. Then (27:5) gives

$$\bar{\mathrm{v}}(S) \geqq -p\gamma.$$

Apply this also to $-S$ which has $n - p$ elements. Owing to (25:3:b) in 25.3.1., the above inequality now becomes

$$-\bar{\mathrm{v}}(S) \geqq -(n-p)\gamma; \quad \text{i.e.} \quad \bar{\mathrm{v}}(S) \leqq (n-p)\gamma.$$

[1] *Proof:* Denote the joint value of the n terms in (27:2*) by β. Then (27:2*) amounts to $\alpha_k^0 = -\mathrm{v}((k)) + \beta$, and so (27:1*) becomes

$$n\beta - \sum_{k=1}^{n} \mathrm{v}((k)) = 0; \quad \text{i.e. } \beta = \frac{1}{n}\sum_{k=1}^{n} \mathrm{v}((k)).$$

Combining these two inequalities gives:

(27:7) $\quad -p\gamma \leqq \bar{v}(S) \leqq (n-p)\gamma \quad$ for every p-element set S.

(27:5*) and $\bar{v}(\ominus) = 0$ (i.e. (25:3:a) in 25.3.1.) can also be formulated this way:

(27:7*) For $p = 0, 1$ we have $=$ in the first relation of (27:7).

(27:5**) and $\bar{v}(I) = 0$ (i.e. (25:4) in 25.4.1.) can also be formulated this way:

(27:7**) For $p = n - 1, n$ we have $=$ in the second relation of (27:7).

27.3. Inessentiality and Essentiality

27.3.1. In analyzing these inequalities it is best now to distinguish two alternatives.

This distinction is based on (27:6):

First case: $\gamma = 0$. Then (27:7) gives $\bar{v}(S) = 0$ for all S. This is a perfectly trivial case, in which the game is manifestly devoid of further possibilities. There is no occasion for any strategy of coalitions, no element of struggle or competition: each player may play a lone hand, since there is no advantage in any coalition. Indeed, every player can get the amount zero for himself irrespective of what the others are doing. And in no coalition can all its members together get more than zero. Thus the value of a play of this game is zero for every player, in an absolutely unequivocal way.

If a general characteristic function $v(S)$ is in strategic equivalence with such a $\bar{v}(S)$—i.e. if its reduced form is $\bar{v}(S) \equiv 0$—then we have the same conditions, only shifted by α_k^0 for the player k. A play of a game Γ with this characteristic function $v(S)$ has unequivocally the value α_k^0 for the player k: he can get this amount even alone, irrespective of what the others are doing. No coalition could do better *in toto*.

We call a game Γ, the characteristic function $v(S)$ of which has such a reduced form $\bar{v}(S) \equiv 0$, *inessential.*[1]

27.3.2. Second case: $\gamma > 0$. By a change in unit[2] we could make $\gamma = 1$.[3] This obviously affects none of the strategically significant aspects of the game, and it is occasionally quite convenient to do. At this moment, however, we do not propose to do this.

In the present case, at any rate, the players will have good reasons to want to form coalitions. Any player who is left to himself loses the amount γ (i.e. he gets $-\gamma$, cf. (27:5*) or (27:7*)), while any $n - 1$ players who

[1] That this coincides with the meaning given to the word inessential in 23.1.3. (in the special case of a zero-sum three-person game) will be seen at the end of 27.4.1.

[2] Since payments are made, we mean the monetary unit. In a wider sense it might be the unit of utility. Cf. 2.1.1.

[3] This would not have been possible in the first case, where $\gamma = 0$.

cooperate win together the amount γ (i.e. their coalition gets γ, cf. (27:5**) or (27:7**)).[1]

Hence an appropriate strategy of coalitions is now of great importance.

We call a game Γ *essential* when its characteristic function $v(S)$ has a reduced form $\bar{v}(S)$ not $\equiv 0$.[2]

27.4. Various Criteria. Non-additive Utilities

27.4.1. Given a characteristic function $v(S)$, we wish to have an explicit expression for the γ of its reduced form $\bar{v}(S)$. (Cf. above.)

Now $-\gamma$ is the joint value of the $\bar{v}((k))$, i.e. of the $v((k)) + \alpha_k^0$, and this is by (27:4) $\frac{1}{n}\sum_{j=1}^{n} v((j))$.[3] Hence

$$\gamma = -\frac{1}{n}\sum_{j=1}^{n} v((j)). \tag{27:8}$$

Consequently we have:

(27:B) The game Γ is *inessential* if and only if

$$\sum_{j=1}^{n} v((j)) = 0 \qquad \text{(i.e. } \gamma = 0\text{)},$$

and it is *essential* if and only if

$$\sum_{j=1}^{n} v((j)) < 0 \qquad \text{(i.e. } \gamma > 0\text{)}.^4$$

For a zero-sum three-person game we have, with the notations of 23.1., $v((1)) = -a$, $v((2)) = -b$, $v((3)) = -c$; so $\gamma = \frac{1}{3}\Delta$. Therefore our concepts of essential and inessential specialize to those of 23.1.3. in the case of a zero-sum three-person game. Considering the interpretation of these concepts in both cases, this was to be expected.

[1] This is, of course, not the whole story There may be other coalitions—of > 1 but $< n - 1$ players—which are worth aspiring to. (If this is to happen, $n - 1$ must exceed 1 by more than 1,—i.e. $n \geqq 4$.) This depends upon the $\bar{v}(S)$ of the sets S with > 1 but $< n - 1$ elements. But only a complete and detailed theory of games can appraise the role of these coalitions correctly.

Our above comparison of isolated players and $n - 1$ player coalitions (the biggest coalitions which have anybody to oppose!) suffices only for our present purpose: to establish the importance of coalitions in this situation.

[2] Cf. again footnote 1 on p. 249.

[3] So $-\gamma$ is the β of footnote 1 on p. 248.

[4] We have seen already that one or the other must be the case, since $\sum_{j=1}^{n} v((j)) \leqq 0$ as well as $\gamma \geqq 0$.

27.4.2. We can formulate some other criteria of inessentiality:

(27:C) The game Γ is inessential if and only if its characteristic function $v(S)$ can be given this form:

$$v(S) \equiv \sum_{k \text{ in } S} \alpha_k^0$$

for a suitable system $\alpha_1^0, \cdots, \alpha_n^0$.

Proof: Indeed, this expresses by (27:2) precisely that $v(S)$ is in strategic equivalence with $\bar{v}(S) \equiv 0$. As this $\bar{v}(S)$ is reduced, it is then the reduced form of $v(S)$—and this is the meaning of inessentiality.

(27:D) The game Γ is inessential if and only if its characteristic function $v(S)$ has always $=$ in (25:3:c) of 25.3.1.; i.e. when

$$v(S \cup T) = v(S) + v(T) \quad \text{if} \quad S \cap T = \ominus.$$

Proof: Necessity: A $v(S)$ of the form given in (27:C) above obviously possesses this property.

Sufficiency: Repeated application of this equation gives $=$ in (25:5) of 25.4.1.; i.e.

$$v(S_1 \cup \cdots \cup S_p) = v(S_1) + \cdots + v(S_p)$$
$$\text{if } S_1, \cdots, S_p \text{ are pairwise disjunct.}$$

Consider an arbitrary S, say $S = (k_1, \cdots, k_p)$. Then $S_1 = (k_1), \cdots, S_p = (k_p)$ give

$$v(S) = v((k_1)) + \cdots + v((k_p)).$$

So we have

$$v(S) = \sum_{k \text{ in } S} \alpha_k^0$$

with $\alpha_1^0 = v((1)), \cdots, \alpha_n^0 = v((n))$ and so Γ is inessential by (27:C).

27.4.3. Both criteria (27:C) and (27:D) express that the values of all coalitions arise additively from those of their constituents.[1] It will be remembered what role the additivity of value, or rather its frequent absence, has played in economic literature. The cases in which value is not generally additive were among the most important, but they offered significant difficulties to every theoretical approach; and one cannot say that these difficulties have ever been really overcome. In this connection one should recall the discussions of concepts like complementarity, total value, imputation, etc. We are now getting into the corresponding phase of our theory; and it is significant that we find additivity only in the unin-

[1] The reader will understand that we are using the word "value" (of the coalition S) for the quantity $v(S)$.

teresting (inessential) case, while the really significant (essential) games have a non-additive characteristic function.[1]

Those readers who are familiar with the mathematical theory of measure will make this further observation: the additive $\mathrm{v}(S)$—i.e. the inessential games—are exactly the measure functions of I, which give I the total measure zero. Thus the general characteristic functions $\mathrm{v}(S)$ are a new generalization of the concept of measure. These remarks are in a deeper sense connected with the preceding ones concerning economic value. However, it would lead too far to pursue this subject further.[2]

27.5. The Inequalities in the Essential Case

27.5.1. Let us return to the inequalities of 27.2., in particular to (27:7), (27:7*), (27:7**). For $\gamma = 0$ (inessential case) everything is trivially clear. Assume therefore that $\gamma > 0$ (essential case).

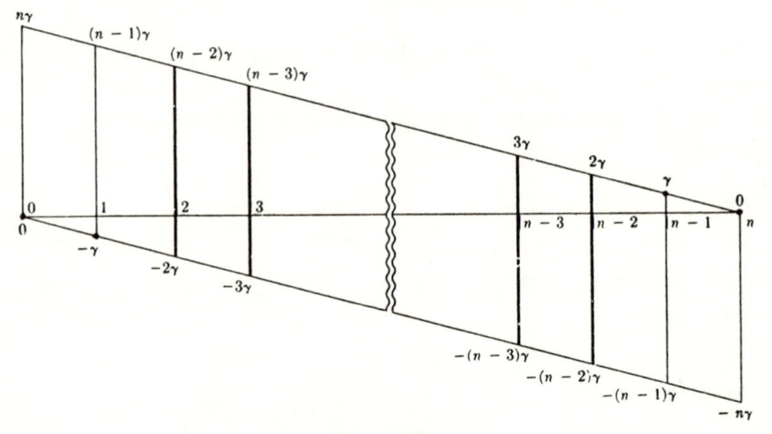

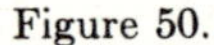
Figure 50.

Abscissa: p, number of elements of S. Dot at 0, $-\gamma$, γ, 0 or heavy line: Range of possible values $\bar{\mathrm{v}}(S)$ for the S with the corresponding p.

Now (27:7), (27:7*), (27:7**) set a range of possible values for $\bar{\mathrm{v}}(S)$ for every number p of elements in S. This range is pictured for each $p = 0, 1, 2, \cdots, n-2, n-1, n$ in Figure 50.

We can add the following remarks:

27.5.2. First: It will be observed that in an essential game—i.e. when $\gamma > 0$—necessarily $n \geqq 3$. Otherwise the formulae (27:7), (27:7*), (27:7**)—or Figure 50, which expresses their content—lead to a conflict: For $n = 1$ or 2 an $(n-1)$-element set S has 0 or 1 elements, hence its

[1] We are, of course, concerned at this moment only with a particular aspect of the subject: we are considering values of coalitions only—i.e. of concerted acts of behavior—and not of economic goods or services. The reader will observe, however, that the specialization is not as far reaching as it may seem: goods and services stand really for the economic act of their exchange—i.e. for a concerted act of behavior.

[2] The theory of measure reappears in another connection. Cf. 41.3.3.

$\bar{v}(S)$ must on the one hand be γ, and on the other hand 0 or $-\gamma$, which is impossible.[1]

Second: For the smallest possible number of participants in an essential game, i.e. for $n = 3$, the formulae (27:7), (27:7*), (27:7**)—or Figure 50—determine everything: they state the values of $\bar{v}(S)$ for $0, 1, n - 1, n$-element sets S; and for $n = 3$ the following are all possible element numbers: 0, 1, 2, 3. (Cf. also a remark in footnote 1 on p. 250.) This is in harmony with the fact which we found in 23.1.3., according to which there exists only one type of essential zero-sum three-person games.

Third: For greater numbers of participants, i.e. for $n \geqq 4$, the problem assumes a new complexion. As formulae (27:7), (27:7*), (27:7**)—or Figure 50—show, the element number of p of the set S can now have other values than $0, 1, n - 1, n$. I.e. the interval

$$2 \leqq p \leqq n - 2 \tag{27:9}$$

now becomes available.[2] It is in this interval that the above formulae no longer determine a unique value of $\bar{v}(S)$; they set for it only the interval

$$-p\gamma \leqq \bar{v}(S) \leqq (n - p)\gamma, \tag{27:7}$$

the length of which is $n\gamma$ for every p (cf. again Figure 50).

27.5.3. In this connection the question may be asked whether really the entire interval (27:7) is available,—i.e. whether it cannot be narrowed further by some new, more elaborate considerations concerning $\bar{v}(S)$. The answer is: No. It is actually possible to define for every $n \geqq 4$ a single game Γ_p in which, for each p of (27:9), $\bar{v}(S)$ assumes both values $-p\gamma$ and $(n - p)\gamma$ for suitable p-element sets S. It may suffice to mention the subject here without further elaboration.

To sum up: The real ramifications of the theory of games appear only when $n \geqq 4$ is reached. (Cf. footnote 1 on p. 250, where the same idea was expounded.)

27.6. Vector Operations on Characteristic Functions

27.6.1. In concluding this section some remarks of a more formal nature seem appropriate.

The conditions (25:3:a)-(25:3:c) in 25.3.1., which describe the characteristic function $v(S)$, have a certain vectorial character: they allow analogues of the vector operations, defined in 16.2.1., of *scalar multiplication*, and of *vector addition*. More precisely:

Scalar multiplication: Given a constant $t \geqq 0$ and a characteristic function $v(S)$, then $tv(S) \equiv u(S)$ is also a characteristic function. Vector addition: Given two characteristic functions $v(S)$, $w(S)$;[3] then

[1] Of course, in a zero-sum one-person game nothing happens at all, and for the zero-sum two-person games we have a theory in which no coalitions appear. Hence the inessentiality of all these cases is to be expected.

[2] It has $n - 3$ elements; and this number is positive as soon as $n \geqq 4$.

[3] Everything here must refer to the same n and to the same set of players

$$I = (1, 2, \cdots, n).$$

$v(S) + w(S) \equiv z(S)$ is also a characteristic function. The only difference from the corresponding definitions of 16.2. is that we had to require $t \geqq 0$.[1,2]

27.6.2. The two operations defined above allow immediate practical interpretation:

Scalar multiplication: If $t = 0$, then this produces $u(S) \equiv 0$, i.e. the eventless game considered in 27.3.1. So we may assume $t > 0$. In this case our operation amounts to a change of the unit of utility, namely to its multiplication by the factor t.

Vector addition: This corresponds to the *superposition* of the games corresponding to $v(S)$ and to $w(S)$. One would imagine that the same players $1, 2, \cdots, n$ are playing these two games simultaneously, but independently. I.e., no move made in one game is supposed to influence the other game, as far as the rules are concerned. In this case the characteristic function of the combined game is clearly the sum of those of the two constituent games.[3]

27.6.3. We do not propose to enter upon a systematic investigation of these operations, i.e. of their influence upon the strategic situations in the games which they affect. It may be useful, however, to make some remarks on this subject—without attempting in any way to be exhaustive.

We observe first that combinations of the operations of scalar multiplication and vector addition also can now be interpreted directly. Thus the characteristic function

$$z(S) \equiv tv(S) + sw(S) \tag{27:10}$$

belongs to the game which arises by superposition of the games of $v(S)$ and $w(S)$ if their units of utility are first multiplied by t and s respectively.

If $s = 1 - t$, then (27:10) corresponds to the formation of the center of gravity in the sense of (16:A:c) in 16.2.1.

It will appear from the discussion in 35.3.4. (cf. in particular footnote 1 on p. 304 below) that even this seemingly elementary operation can have very involved consequences as regards strategy.

We observe next that there are some cases where our operations have no consequences in strategy.

First, the scalar multiplication by a $t > 0$ alone, being a mere change in unit, has no such consequences.

[1] Indeed, $t < 0$ would upset (25:3:c) in 25.3.1. Note that a multiplication of the original $\mathcal{H}_k(\tau_1, \cdots, \tau_n)$ with a $t < 0$ would be perfectly feasible. It is simplest to consider a multiplication by $t = -1$, i.e. a change in sign. But a change of sign of the $\mathcal{H}_k(\tau_1, \cdots, \tau_n)$ does not at all correspond to a change of sign of the $v(S)$. This should be clear by common sense, as a reversal of gains and losses modifies all strategic considerations in a very involved way. (This reversal and some of its consequences are familiar to chess players.) A formal corroboration of our assertion may be found by inspecting the definitions of 25.1.3.

[2] Vector spaces with this restriction of scalar multiplication are sometimes called *positive vector spaces*. We do not need to enter upon their systematic theory.

[3] This should be intuitively obvious. An exact verification with the help of 25.1.3. involves a somewhat cumbersome notation, but no real difficulties.

Second—and this is of greater significance—the strategic equivalence discussed in 27.1. is a superposition: we pass from the game of $v(S)$ to the strategically equivalent game of $v'(S)$ by superposing on the former an inessential game.[1] (Cf. (27:1) and (27:2) in 27.1.1. and, concerning inessentiality, 27.3.1. and (27:C) in 27.4.2.) We may express this in the following way: we know that an inessential game is one in which coalitions play no role. The superposition of such a game on another one does not disturb strategic equivalence, i.e. it leaves the strategic structure of that game unaffected.

28. Groups, Symmetry and Fairness

28.1. Permutations, Their Groups, and Their Effect on a Game

28.1.1. Let us now consider the role of symmetry, or more generally, the effects of interchanging the players $1, \cdots, n$—or their numbers—in an n-person game Γ. This will naturally be an extension of the corresponding study made in 17.11. for the zero-sum two-person game.

This analysis begins with what is in the main a repetition of the steps taken in 17.11. for $n = 2$. But since the interchanges of the symbols $1, \cdots, n$ offer for a general n many more possibilities than for $n = 2$, it is indicated that we should go about it somewhat more systematically.

Consider the n symbols $1, \cdots, n$. Form any *permutation* P of these symbols. P is described by stating for every $i = 1, \cdots, n$, into which i^P (also $= 1, \cdots, n$), P carries it. So we write:

(28:1) $$P: i \to i^P,$$

or by way of complete enumeration:

(28:2) $$P: \begin{pmatrix} 1, & 2, & \cdots, & n \\ 1^P, & 2^P, & \cdots, & n^P \end{pmatrix}.$$ [2]

Among the permutations some deserve special mention:

(28:A:a) The *identity* I_n which leaves every $i(= 1, \cdots, n)$ unchanged:

$$i \to i^{I_n} = i.$$

(28:A:b) Given two permutations P, Q, their *product* PQ, which consists in carrying out first P and then Q:

$$i \to i^{PQ} = (i^P)^Q.$$

[1] With the characteristic function $w(S) \equiv \sum_{k \text{ in } S} \alpha_k^0$, then in our above notations

$$v'(S) \equiv v(S) + w(S)$$

[2] Thus for $n = 2$, the interchange of the two elements 1, 2 is the permutation $\begin{pmatrix} 1, 2 \\ 2, 1 \end{pmatrix}$. The identity (cf. below) is $I_n = \begin{pmatrix} 1, 2, \cdots, n \\ 1, 2, \cdots, n \end{pmatrix}$.

The number of all possible permutations is the *factorial* of n,

$$n! = 1 \cdot 2 \cdot \ . \ . \ . \ \cdot n,$$

and they form together the *symmetric group* of permutations Σ_n. Any subsystem G of Σ_n which fulfills these two conditions:

(28:A:a*) I_n belongs to G,

(28:A:b*) PQ belongs to G if P and Q do,

is a *group* of permutations.[1]

A permutation P carries every subset S of $I = (1, \cdots, n)$ into another subset S^P.[2]

28.1.2. After these general and preparatory remarks we now proceed to apply their concepts to an arbitrary n-person game Γ.

Perform a permutation P on the symbols $1, \cdots, n$ denoting the players of Γ. I.e. denote the player $k = 1, \cdots, n$ by k^P instead of k; this transforms the game Γ into another game Γ^P. The replacement of Γ by Γ^P must make its influence felt in two respects: in the influence which each player exercises on the course of the play,—i.e. in the index k of the variable τ_k which each player chooses; and in the outcome of the play for him,—i.e. in the index k of the function $\mathcal{H}_k$ which expresses this.[3] So Γ^P is again in the normalized form, with functions $\mathcal{H}_k^P(\tau_1, \cdots, \tau_n)$, $k = 1, \cdots, n$. In expressing $\mathcal{H}_k^P(\tau_1, \cdots, \tau_n)$ by means of $\mathcal{H}_k(\tau_1, \cdots, \tau_n)$, we must remember: the player k in Γ had $\mathcal{H}_k$; now he is k^P in Γ^P, so he has $\mathcal{H}_{k^P}^P$. If we form $\mathcal{H}_{k^P}^P$ with the variables $\tau_1, \cdots, \tau_n$, then we express the outcome of the game Γ^P when the player whose designation in Γ^P is k chooses τ_k. So the player k in Γ who is k^P in Γ^P chooses τ_{k^P}. So the variables in $\mathcal{H}_k$ must be $\tau_{1^P}, \cdots, \tau_{n^P}$. We have therefore:

(28:3) $$\mathcal{H}_{k^P}^P(\tau_1, \cdots, \tau_n) \equiv \mathcal{H}_k(\tau_{1^P}, \cdots, \tau_{n^P}).$$ [4,5]

[1] For the important and extensive theory of groups compare *L. C. Mathewson:* Elementary Theory of Finite Groups, Boston 1930; *W. Burnside:* Theory of Groups of Finite Order, 2nd Ed. Cambridge 1911; *A. Speiser:* Theorie der Gruppen von endlicher Ordnung, 3rd Edit. Berlin 1937.

We shall not need any particular results or concepts of group theory, and mention the above literature only for the use of the reader who may want to acquire a deeper insight into that subject.

Although we do not wish to tie up our exposition with the intricacies of group theory, we nevertheless introduced some of its basic *termini* for this reason: a real understanding of the nature and structure of symmetry is not possible without some familiarity with (at least) the elements of group theory. We want to prepare the reader who may want to proceed in this direction, by using the correct terminology.

For a fuller exposition of the relationship between symmetry and group theory, cf. *H. Weyl:* Symmetry, Journ. Washington Acad. of Sciences, Vol. XXVIII (1938), pp. 253ff.

[2] If $S = (k_1, \cdots, k_p)$, then $S^P = (k_1^P, \cdots, k_p^P)$.

[3] Cf. the similar situation for $n = 2$ in footnote 1 on p. 109.

[4] The reader will observe that the superscript P for the index k of the functions $\mathcal{H}$ themselves appears on the left-hand side, while the superscript P for the indices k of the variables τ_k appear on the right-hand side. This is the correct arrangement; and the argument preceding (28:3) was needed to establish it.

The importance of getting this point faultless and clear lies in the fact that we could

Denote the characteristic functions of Γ and Γ^P by $v(S)$ and $v^P(S)$ respectively. Since the players, who form in Γ^P the set S^P, are the same ones who form in Γ the set S, we have

(28:4) $$v^P(S^P) \equiv v(S) \qquad \text{for every } S.$$[1]

28.1.3. If (for a particular P) Γ coincides with Γ^P, then we say that Γ is *invariant* or *symmetric* with respect to P. By virtue of (28:3) this is expressed by

(28:5) $$\mathcal{H}_{k^P}(\tau_1, \cdots, \tau_n) \equiv \mathcal{H}_k(\tau_{1^P}, \cdots, \tau_{n^P}).$$

When this is the case, then (28:4) becomes

(28:6) $$v(S^P) \equiv v(S) \qquad \text{for every } S.$$

Given any Γ, we can form the system G_Γ of all P with respect to which Γ is symmetric. It is clear from (28:A:a), (28:A:b) above, that the identity I_n belongs to G_Γ, and that if P, Q belong to G_Γ, then their product PQ does too. So G_Γ is a group by (28:A:a*), (28:A:b*) above. We call G_Γ the *invariance group* of Γ.

Observe that (28:6) can now be stated in this form:

(28:7) $v(S) = v(T)$ if there exists a P in G_Γ with $S^P = T$, i.e. which carries S into T.

The size of G_Γ—i.e. the number of its elements—gives some sort of a measure of "how symmetric" Γ is. If every permutation P (other than identity I_n) changes Γ, then G_Γ consists of I_n alone,—Γ is *totally unsymmetric.* If no permutation P changes Γ, then G_Γ contains all P, i.e. it is the symmetric group Σ_n,—Γ is *totally symmetric.* There are, of course, numerous intermediate cases between these two extremes, and the precise structure of Γ's symmetry (or lack of it) is disclosed by the group G_Γ.

28.1.4. The condition after (28:7) implies that S and T have the same number of elements. The converse implication, however, need not be true if G_Γ is small enough, i.e. if Γ is unsymmetric enough. It is therefore

---

not otherwise be sure that successive applications of the superscripts P and Q (in this order) to Γ will give the same result as a (single) application of the superscript PQ to Γ. The reader may find the verification of this a good exercise in handling the calculus of permutations.

For $n = 2$ and $P = \begin{pmatrix} 1, 2 \\ 2, 1 \end{pmatrix}$, application of P on either side had the same effect, so it is not necessary to be exhaustive on this point. Cf. footnote 1 on p. 109.

[5] In the zero-sum two-person game, $\mathcal{H} \equiv \mathcal{H}_1 \equiv -\mathcal{H}_2$, and similarly $\mathcal{H}^P \equiv \mathcal{H}_1^P \equiv -\mathcal{H}_2^P$. Hence in this case $\left(\text{cf. above, } n = 2 \text{ and } P = \begin{pmatrix} 1, 2 \\ 2, 1 \end{pmatrix}\right)$ (28:3) becomes $\mathcal{H}^P(\tau_1, \tau_2) \equiv -\mathcal{H}(\tau_2, \tau_1)$. This is in accord with the formulae of 14.6. and 17.11.2.

But this simplification is possible only in the zero-sum two-person game; in all other cases we must rely upon the general formula (28:3) alone.

[1] This conceptual proof is clearer and simpler than a computational one, which could be based on the formulae of 25.1.3. The latter, however, would cause no difficulties either, only more extensive notations.

of interest to consider those groups $G = G_\Gamma$ which permit this converse implication, i.e. for which the following is true:

(28:8) If S, T have the same number of elements, then there exists a P in G with $S^P = T$,—i.e. which carries S into T.

This condition (28:8) is obviously satisfied when G is the symmetric group Σ_n, i.e. for the $G = G_\Gamma = \Sigma_n$ of a totally symmetric Γ. It is also satisfied for certain smaller groups,—i.e. for certain Γ of less than total symmetry.[1]

28.2. Symmetry and Fairness

28.2.1. At any rate, whenever (28:8) holds for $G = G_\Gamma$, we can conclude from (28:7):

(28:9) $v(S)$ depends only upon the number of elements in S.

That is:

(28:10) $$v(S) = v_p$$

where p is the number of elements in S, $(p = 0, 1, \cdots, n)$.

Consider the conditions (25:3:a)-(25:3:c) in 25.3.1., which give an exhaustive description of all characteristic functions $v(S)$. It is easy to rewrite them for v_p when (28:10) holds. They become:

(28:11:a) $$v_0 = 0,$$

(28:11:b) $$v_{n-p} = -v_p,$$

(28:11:c) $$v_{p+q} \geqq v_p + v_q \quad \text{for} \quad p + q \leqq n.$$

(27:3) in 27.1.4. is clearly a consequence of (28:10) (i.e. of (28:9)), so that such a $v(S)$ is automatically reduced,—with $\gamma = -v_1$. We have therefore, in particular, (27:7), (27:7*), (27:7**) in 27.2., i.e. the conditions of Figure 50.

Condition (28:11:c) can be rewritten, by a procedure which is parallel to that of (25:A) in 25.4.2.

[1] For $n = 2$, Σ_n contains, besides the identity, only one more permutation ($p = \begin{pmatrix} 1, 2 \\ 2, 1 \end{pmatrix}$ cf. several preceding references); so $G = \Sigma_n$ is the only possibility of any symmetry.

Consider therefore $n \geqq 3$, and call G *set-transitive* if it fulfills (28:8). The question, which $G \neq \Sigma_n$ are then set-transitive, is of a certain group-theoretical interest, but we need not concern ourselves with it in this work.

For the reader who is interested in group theory we nevertheless mention:

There exists a subgroup of Σ_n which contains half of its elements (i.e. $\frac{1}{2}n!$), known as the *alternating group* $\mathfrak{A}_n$. This group is of great importance in group theory and has been extensively discussed there. For $n \geqq 3$ it is easily seen to be set-transitive too.

So the real question is this: for which $n \geqq 3$ do there exist set-transitive groups $G \neq \Sigma_n; \mathfrak{A}_n$?

It is easy to show that for $n = 3, 4$ none exist. For $n = 5, 6$ such groups do exist. (For $n = 5$ a set-transitive group G with 20 elements exists, while Σ_5, $\mathfrak{A}_5$ have 120, 60 elements respectively. For $n = 6$ a set-transitive group G with 120 elements exists, while Σ_6, $\mathfrak{A}_6$ have 720, 360 elements respectively.) For $n = 7, 8$ rather elaborate group-theoretical arguments show that no such groups exist. For $n = 9$ the question is still open. It seems probable that no such groups exist for any $n > 9$, but this assertion has not yet been established for all these n.

Put $r = n - p - q$; then (28:11:b) permits us to state (28:11:c) as follows:

(28:11:c*) $\quad v_p + v_q + v_r \leqq 0 \quad$ if $\quad p + q + r = n.$

Now (28:11:c*) is symmetric with respect to p, q, r;[1] hence we may make $p \leqq q \leqq r$ by an appropriate permutation. Furthermore, when $p = 0$ (hence $r = n - q$), then (28:11:c*) follows from (28:11:a), (28:11:b) (even with $=$). Thus we may assume $p \neq 0$. So we need to require (28:11:c*) only for $1 \leqq p \leqq q \leqq r$, and therefore the same is true for (28:11:c). Observe finally that, as $r = n - p - q$, the inequality $q \leqq r$ means $p + 2q \leqq n$. We restate this:

(28:12) It suffices to require (28:11:c) only when

$$1 \leqq p \leqq q, \qquad p + 2q \leqq n.$$[2]

28.2.2. The property (28:10) of the characteristic function is a consequence of symmetry, but this property is also important in its own right. This becomes clear when we consider it in the simplest possible special case: for $n = 2$.

Indeed, for $n = 2$ (28:10) simply means that the v' of 17.8.1. vanishes.[3] This means in the terminology of 17.11.2., that the game Γ is fair. We extend this concept: The n-person game Γ is *fair* when its characteristic function $v(S)$ fulfills (28:9), i.e. when it is a v_p of (28:10). Now, as in 17.11.2., this notion of fairness of the game embodies what is really essential in the concept of symmetry. It must be remembered, however, that the concept of fairness—and similarly that of total symmetry—of the game may or may not imply that all individual players can expect the same fate in an individual play (provided that they play well). For $n = 2$ this implication did hold, but not for $n \geqq 3$! (Cf. 17.11.2. for the former, and footnotes 1 and 2 on p. 225 for the latter.)

28.2.3. We observe, finally, that by (27:7), (27:7*), (27:7**) in 27.2., or by Figure 50, all reduced games are symmetric and hence fair, when $n = 3$, but not when $n \geqq 4$. (Cf. the discussion in 27.5. 2.) Now the unrestricted zero-sum n-person game is brought into its reduced form by the fixed extra payments $\alpha_1, \cdots, \alpha_n$ (to the players $1, \cdots, n$, respectively), as described in 27.1. Thus the unfairness of a zero-sum three-person game—i.e. what is really effective in its asymmetry—is exhaustively expressed by these $\alpha_1, \alpha_2, \alpha_3$; that is, by fixed, definite payments. (Cf. also the "basic values," a', b', c' of 22.3.4.) In a zero-sum n-person game with

[1] Both in its assertion and in its hypothesis!

[2] These inequalities replace the original $p + q \leqq n$; they are obviously much stronger. As they imply $3p \leqq p + 2q \leqq n$ and $1 + 2q \leqq p + 2q \leqq n$, we have

$$p \leqq \frac{n}{3}, \qquad q \leqq \frac{n-1}{2}.$$

[3] By definition $v' = v((1)) = -v((2))$. For $n = 2$ the only essential assertion of (28:9) (which is equivalent to (28:10)) is $v((1)) = v((2))$. Due to the above, this means precisely that $v' = -v'$, i.e. that $v' = 0$.

$n \geqq 4$, this is no longer always possible, since the reduced form need not be fair. That is, there may exist, in such a game, much more fundamental differences between the strategic positions of the players, which cannot be expressed by the $\alpha_1, \cdots, \alpha_n$,—i.e. by fixed, definite payments. This will become amply clear in the course of Chapter VII. In the same connection it is also useful to recall footnote 1 on p. 250.

29. Reconsideration of the Zero-sum Three-person Game

29.1. Qualitative Discussion

29.1.1. We are now prepared for the main undertaking: To formulate the principles of the theory of the zero-sum n-person game.[1] The characteristic function $v(S)$, which we have defined in the preceding sections, provides the necessary tool for this operation.

Our procedure will be the same as before: We must select a special case to serve as a basis for further investigation. This shall be one which we have already settled and which we nevertheless deem sufficiently characteristic for the general case. By analyzing the (partial) solution found in this special case, we shall then try to crystallize the rules which should govern the general case. After what we said in 4.3.3. and in 25.2.2., it ought to be plausible that the zero-sum three-person game will be the special case in question.

29.1.2. Let us therefore reconsider the argument by which our present solution of the zero-sum three-person game was obtained. Clearly the essential case will be the one of interest. We know now that we may as well consider it in its reduced form, and that we may also choose $\gamma = 1$.[2] The characteristic function in this case is completely determined, as discussed in the second case of 27.5.2.:

$$(29{:}1) \qquad v(S) = \begin{cases} 0 \\ -1 \\ 1 \\ 0 \end{cases} \quad \text{when } S \text{ has} \quad \begin{cases} 0 \\ 1 \\ 2 \\ 3 \end{cases} \text{elements.}^{3}$$

We saw that in this game everything is decided by the (two-person) coalitions which form, and our discussions[4] produced the following main conclusions:

Three coalitions may form, and accordingly the three players will finish the play with the following results:

[1] Of course the general n-person game will still remain, but we shall be able to solve it with the help of the zero-sum games. The greatest step is the present one: the passage to the zero-sum n-person games.

[2] Cf. 27.1.4. and 27.3.2.

[3] In the notation of 23.1.1. this means $a = b = c = 1$. The general parts of the discussions referred to were those in 22.2., 22.3., 23. The above specialization takes us actually back to the earlier (more special) case of 22.1. So our considerations of 27.1. (on strategic equivalence and reduction) have actually this effect in the zero-sum three-person games: they carry the general case back into the preceding special one, as stated above.

[4] In 22.2.2., 22.2.3.; but these are really just elaborations of those in 22.1.2., 22.1.3.

Coalition \ Player	1	2	3
(1, 2)	$\frac{1}{2}$	$\frac{1}{2}$	-1
(1, 3)	$\frac{1}{2}$	-1	$\frac{1}{2}$
(2, 3)	-1	$\frac{1}{2}$	$\frac{1}{2}$

Figure 51.

This "solution" calls for interpretation, and the following remarks suggest themselves in particular:[1]

29.1.3.

(29:A:a) The three distributions specified above correspond to all strategic possibilities of the game.

(29:A:b) None of them can be considered a solution in itself; it is the system of all three and their relationship to each other which really constitute the solution.

(29:A:c) The three distributions possess together, in particular, a "stability" to which we have referred thus far only very sketchily. Indeed no equilibrium can be found outside of these three distributions; and so one should expect that any kind of negotiation between the players must always *in fine* lead to one of these distributions.

(29:A:d) Again it is conspicuous that this "stability" is only a characteristic of all three distributions viewed together. No one of them possesses it alone; each one, taken by itself, could be circumvented if a different coalition pattern should spread to the necessary majority of players.

29.1.4. We now proceed to search for an exact formulation of the heuristic principles which lead us to the solutions of Figure 51, always keeping in mind the remarks (29:A:a)-(29:A:d).

A more precise statement of the intuitively recognizable "stability" of the system of three distributions in Figure 51—which should be a concise summary of the discussions referred to in footnote 4 on p. 260—leads us back to a position already taken in the earlier, qualitative discussions.[2] It can be put as follows:

(29:B:a) If any other scheme of distribution should be offered for consideration to the three players, then it will meet with

[1] These remarks take up again the considerations of 4.3.3. In connection with (29:A:d) the second half of 4.6.2. may also be recalled.

[2] These viewpoints permeate all of 4.4.-4.6., but they appear more specifically in 4.4.1. and 4.6.2.

rejection for the following reason: a sufficient number of players[1] prefer, in their own interest, at least one of the distributions of the solution (i.e. of Figure 51), and are convinced or can be convinced[2] of the possibility of obtaining the advantages of that distribution.

(29:B:b) If, however, one of the distributions of the solution is offered, then no such group of players can be found.

We proceed to discuss the merits of this heuristic principle in a more exact way.

29.2. Quantitative Discussion

29.2.1. Suppose that β_1, β_2, β_3 is a possible method of distribution between the players 1,2,3. I.e.

$$\beta_1 + \beta_2 + \beta_3 = 0.$$

Then, since by definition $v((i))(= -1)$ is the amount that player i can get for himself (irrespective of what all others do), he will certainly block any distribution with $\beta_i < v((i))$. We assume accordingly that

$$\beta_i \geqq v((i)) = -1.$$

We may permute the players 1,2,3 so that

$$\beta_1 \geqq \beta_2 \geqq \beta_3.$$

Now assume $\beta_2 < \frac{1}{2}$. Then *a fortiori* $\beta_3 < \frac{1}{2}$. Consequently the players 2,3 will both prefer the last distribution of Figure 51,[3] where they both get the higher amount $\frac{1}{2}$.[4] Besides, it is clear that they can get the advantage of that distribution (irrespective of what the third player does), since the amounts $\frac{1}{2}$, $\frac{1}{2}$ which it assigns to them do not exceed together $v((2, 3)) = 1$.

If, on the other hand, $\beta_2 \geqq \frac{1}{2}$, then *a fortiori* $\beta_1 \geqq \frac{1}{2}$. Since $\beta_3 \geqq -1$, this is possible only when $\beta_1 = \beta_2 = \frac{1}{2}$, $\beta_3 = -1$, i.e. when we have the first distribution of Figure 51. (Cf. footnote 3 above.)

[1] Of course, in this case, two.

[2] What this "convincing" means was discussed in 4.4.3. Our discussion which follows will make it perfectly clear.

[3] Since we made an unspecified permutation of the players 1,2,3 the last distribution of Fig. 51 really stands for all three.

[4] Observe that each one of these two players profits by such a change separately and individually. It would not suffice to have only the totality (of these two) profit. Cf., e.g., the first distribution of Fig. 51 with the second; the players 1,3 as a totality would profit by the change from the former to the latter,—and nevertheless the first distribution is just as good a constituent of the solution as any other.

In this particular change, player 3 would actually profit (getting $\frac{1}{2}$ instead of -1), and for player 1 the change is indifferent (getting $\frac{1}{2}$ in both cases). Nevertheless player 1 will not act unless further compensations are made—and these can be disregarded in this connection. For a more careful discussion of this point, cf. the last part of this section.

This establishes (29:B:a) at the end of 29.1.4. (29:B:b) loc. cit. is immediate: in each of the three distributions of Figure 51 there is, to be sure, one player who is desirous of improving his standing,[1] but since there is only one, he is not able to do so. Neither of his two possible partners gains anything by forsaking his present ally and joining the dissatisfied player: already each gets $\frac{1}{2}$, and they can get no more in any alternative distribution of Figure 51.[2]

29.2.2. This point may be clarified further by some heuristic elaboration.

We see that the dissatisfied player finds no one who desires spontaneously to be his partner, and he can offer no positive inducement to anyone to join him; certainly none by offering to concede more than $\frac{1}{2}$ from the proceeds of their future coalition. The reason for regarding such an offer as ineffective can be expressed in two ways: on purely formal grounds this offer may be excluded because it corresponds to a distribution which is outside the scheme of Figure 51; the real subjective motive for which any prospective partner would consider it unwise[3] to accept a coalition under such conditions is most likely the fear of subsequent disadvantage,—there may be further negotiations preceding the formation of a coalition, in which he would be found in a particularly vulnerable position. (Cf. the analysis in 22.1.2., 22.1.3.)

So there is no way for the dissatisfied player to overcome the indifference of the two possible partners. We stress: there is, on the side of the two possible partners no positive motive against a change into another distribution of Figure 51, but just the indifference characteristic of certain types of stability.[4]

30. The Exact Form of the General Definitions

30.1. The Definitions

30.1.1. We return to the case of the zero-sum n-person game Γ with general n. Let the characteristic function of Γ be $v(S)$.

We proceed to give the decisive definitions.

In accordance with the suggestions of the preceding paragraphs we mean by a *distribution* or *imputation* a set of n numbers $\alpha_1, \cdots, \alpha_n$ with the following properties

$$\text{(30:1)} \qquad \alpha_i \geqq v((i)) \quad \text{for} \quad i = 1, \cdots, n,$$

$$\text{(30:2)} \qquad \sum_{i=1}^{n} \alpha_i = 0.$$

[1] The one who gets -1.

[2] The reader may find it a good exercise to repeat this discussion with a general (not reduced) $v(S)$,—i.e. with general a, b, c, and the quantities of 22.3.4. The result is the same; it cannot be otherwise, since our theory of strategic equivalence and reduction is correct. (Cf. footnote 3 on p. 260.)

[3] Or unsound, or unethical.

[4] At every change from one distribution of Fig. 51 to another, one player is definitely against, one definitely for it; and so the remaining player blocks the change by his indifference.

It may be convenient to view these systems $\alpha_1, \cdots, \alpha_n$ as vectors in the n-dimensional linear space L_n in the sense of 16.1.2.:

$$\overrightarrow{\alpha} = \{\alpha_1, \cdots, \alpha_n\}$$

A set S (i.e. a subset of $I = 1, \cdots, n$) is called *effective* for the imputation $\overrightarrow{\alpha}$, if

$$\sum_{i \text{ in } S} \alpha_i \leqq \mathrm{v}(S). \tag{30:3}$$

An imputation $\overrightarrow{\alpha}$ *dominates* another imputation $\overrightarrow{\beta}$, in symbols

$$\overrightarrow{\alpha} \succ \overrightarrow{\beta},$$

if there exists a set S with the following properties:

(30:4:a) S is not empty,

(30:4:b) S is effective for $\overrightarrow{\alpha}$,

(30:4:c) $\alpha_i > \beta_i$ for all i in S.

A set V of imputations is a *solution* if it possesses the following properties:

(30:5:a) No $\overrightarrow{\beta}$ in V is dominated by an $\overrightarrow{\alpha}$ in V,

(30:5:b) Every $\overrightarrow{\beta}$ not in V is dominated by some $\overrightarrow{\alpha}$ in V.

(30:5:a) and (30:5:b) can be stated as a single condition:

(30:5:c) The elements of V are precisely those imputations which are undominated by any element of V.

(Cf. footnote 1 on p. 40.)

30.1.2. The meaning of these definitions can, of course, be visualized when we recall the considerations of the preceding paragraphs and also of the earlier discussions of 4.4.3.

To begin with, our distributions or imputations correspond to the more intuitive notions of the same name in the two places referred to. What we call an effective set is nothing but the players who "are convinced or can be convinced" of the possibility of obtaining what they are offered by $\overrightarrow{\alpha}$; cf. again 4.4.3 and (29:B:a) in 29.1.4. The condition (30:4:c) in the definition of domination expresses that all these players have a positive motive for preferring $\overrightarrow{\alpha}$ to $\overrightarrow{\beta}$. It is therefore apparent that we have defined domination entirely in the spirit of 4.4.1., and of the preference described by (29:B:a) in 29.1.4.

The definition of a solution agrees completely with that given in 4.5.3., as well as with (29:B:a), (29:B:b) in 29.1.4.

30.2. Discussion and Recapitulation

30.2.1. The motivation for all these definitions has been given at the places to which we referred in detail in the course of the last paragraph. We shall nevertheless re-emphasize some of their main features—particularly the concept of a solution.

We have already seen in 4.6. that our concept of a solution of a game corresponds precisely to that of a "standard of behavior" of everyday parlance. Our conditions (30:5:a), (30:5:b), which correspond to the conditions (4:A:a), (4:A:b) of 4.5.3., express just the kind of "inner stability" which is expected of workable standards of behavior. This was elaborated further in 4.6. on a qualitative basis. We can now reformulate those ideas in a rigorous way, considering the exact character which the discussion has now assumed. The remarks we wish to make are these:[1]

30.2.2.

(30:A:a) Consider a solution V. We have *not* excluded for an imputation $\overrightarrow{\beta}$ in V the existence of an outside imputation $\overrightarrow{\alpha}'$ (not in V) with $\overrightarrow{\alpha}' \succ \overrightarrow{\beta}$.[2] If such an $\overrightarrow{\alpha}'$ exists, the attitude of the players must be imagined like this: If the solution V (i.e. this system of imputations) is "accepted" by the players 1, $\cdots$, n, then it must impress upon their minds the idea that only the imputations $\overrightarrow{\beta}$ in V are "sound" ways of distribution. An $\overrightarrow{\alpha}'$ not in V with $\overrightarrow{\alpha}' \succ \overrightarrow{\beta}$, although preferable to an effective set of players, will fail to attract them, because it is "unsound." (Cf. our detailed discussion of the zero-sum three-person game, especially as to the reason for the aversion of each player to accept more than the determined amount in a coalition. Cf. the end of 29.2. and its references.) The view of the "unsoundness" of $\overrightarrow{\alpha}'$ may also be supported by the existence of an $\overrightarrow{\alpha}$ in V with $\overrightarrow{\alpha} \succ \overrightarrow{\alpha}'$ (cf. (30:A:b) below). All these arguments are, of course, circular in a sense and again depend on the selection of V as a "standard of behavior," i.e. as a criterion of "soundness." But this sort of circularity is not unfamiliar in everyday considerations dealing with "soundness."

[1] The remarks (30:A:a)-(30:A:d) which follow are a more elaborate and precise presentation of the ideas of 4.6.2. Remark (30:A:e) bears the same relationship to 4.6.3.

[2] Indeed, we shall see in (31:M) of 31.2.3. that an imputation $\overrightarrow{\beta}$, for which *never* $\overrightarrow{\alpha}' \succ \overrightarrow{\beta}$, exists only in inessential games.

(30:A:b) If the players $1, \cdots, n$ have accepted the solution V as a "standard of behavior," then the ability to discredit with the help of V (i.e. of its elements) any imputation not in V, is necessary in order to maintain their faith in V. Indeed, for every outside $\overrightarrow{\alpha}'$ (not in V) there must exist an $\overrightarrow{\alpha}$ in V with $\overrightarrow{\alpha} \succ \overrightarrow{\alpha}'$. (This was our postulate (30:5:b).)

(30:A:c) Finally there must be no inner contradiction in V, i.e. for $\overrightarrow{\alpha}, \overrightarrow{\beta}$ in V, never $\overrightarrow{\alpha} \succ \overrightarrow{\beta}$. (This was our other postulate (30:5:a).)

(30:A:d) Observe that if domination, i.e. the relation $\succ$, were transitive, then the requirements (30:A:b) and (30:A:c) (i.e. our postulates (30:5:a) and (30:5:b)) would exclude the rather delicate situation in (30:A:a). Specifically: In the situation of (30:A:a), $\overrightarrow{\beta}$ belongs to V, $\overrightarrow{\alpha}'$ does not, and $\overrightarrow{\alpha}' \succ \overrightarrow{\beta}$. By (30:A:b) there exists an $\overrightarrow{\alpha}$ in V so that $\overrightarrow{\alpha} \succ \overrightarrow{\alpha}'$. Now if domination were transitive we could conclude that $\overrightarrow{\alpha} \succ \overrightarrow{\beta}$, which contradicts (30:A:c) since $\overrightarrow{\alpha}, \overrightarrow{\beta}$ both belong to V.

(30:A:e) The above considerations make it even more clear that only V in its entirety is a solution and possesses any kind of stability —but none of its elements individually. The circular character stressed in (30:A:a) makes it plausible also that several solutions V may exist for the same game. I.e. several stable standards of behavior may exist for the same factual situation. Each of these would, of course, be stable and consistent in itself, but in conflict with all others. (Cf. also the end of 4.6.3. and the end of 4.7.)

In many subsequent discussions we shall see that this multiplicity of solutions is, indeed, a very general phenomenon.

30.3. The Concept of Saturation

30.3.1. It seems appropriate to insert at this point some remarks of a more formalistic nature. So far we have paid attention mainly to the meaning and motivation of the concepts which we have introduced, but the notion of solution, as defined above, possesses some formal features which deserve attention.

The formal—logical—considerations which follow will be of no immediate use, and we shall not dwell upon them extensively, continuing afterwards more in the vein of the preceding treatment. Nevertheless we deem that these remarks are useful here for a more complete understanding of the structure of our theory. Furthermore, the procedures to be used here will have an important technical application in an entirely different connection in 51.1.-51.4.

30.3.2. Consider a domain (set) D for the elements x, y of which a certain relation $x\mathcal{R}y$ exists. The validity of $\mathcal{R}$ between two elements x, y of D is expressed by the formula $x\mathcal{R}y$.[1] $\mathcal{R}$ is defined by a statement specifying unambiguously for which pairs x, y of D, $x\mathcal{R}y$ is true, and for which it is not. If $x\mathcal{R}y$ is equivalent to $y\mathcal{R}x$, then we say that $x\mathcal{R}y$ is *symmetric*. For any relation $\mathcal{R}$ we can define a new relation $\mathcal{R}^s$ by specifying $x\mathcal{R}^s y$ to mean the conjunction of $x\mathcal{R}y$ and $y\mathcal{R}x$. Clearly $\mathcal{R}^s$ is always symmetric and coincides with $\mathcal{R}$ if and only if $\mathcal{R}$ is symmetric. We call $\mathcal{R}^s$ the *symmetrized* form of $\mathcal{R}$.[2]

We now define:

(30:B:a) A subset A of D is $\mathcal{R}$-*satisfactory* if and only if $x\mathcal{R}y$ holds for all x, y of A.

(30:B:b) A subset A of D and an element y of D are $\mathcal{R}$-*compatible* if and only if $x\mathcal{R}y$ holds for all x of A.

From these one concludes immediately:

(30:C:a) A subset A of D is $\mathcal{R}$-*satisfactory* if and only if this is true: The y which are $\mathcal{R}$-compatible with A form a superset of A.

We define next:

(30:C:b) A subset A of D is $\mathcal{R}$-*saturated* if and only if this is true: The y which are $\mathcal{R}$-compatible with A form precisely the set A.

Thus the requirement which must be added to (30:C:a) in order to secure (30:C:b) is this:

(30:D) If y is not in A, then it is not $\mathcal{R}$-compatible with A; i.e. there exists an x in A such that not $x\mathcal{R}y$.

Consequently $\mathcal{R}$-saturation may be equivalently defined by (30:B:a) and (30:D).

30.3.3. Before we investigate these concepts any further, we give some examples. The verification of the assertions made in them is easy and will be left to the reader.

First: Let D be any set and $x\mathcal{R}y$ the relation $x = y$. Then $\mathcal{R}$-satisfactoriness of A means that A is either empty or a one-element set, while $\mathcal{R}$-saturation of A means that A is a one-element set.

Second: Let D be a set of real numbers and $x\mathcal{R}y$ the relation $x \leqq y$.[3] Then $\mathcal{R}$-satisfactoriness of A means the same thing as above,[4] while $\mathcal{R}$-saturation of A means that A is a one-element set, consisting of the greatest element of D. Thus there exists no such A if D has no greatest element

[1] It is sometimes more convenient to use a formula of the form $\mathcal{R}(x, y)$, but for our purposes $x\mathcal{R}y$ is preferable.

[2] Some examples: Let D consist of all real numbers. The relations $x = y$ and $x \neq y$ are symmetric. None of the four relations $x \leqq y$, $x \geqq y$, $x < y$, $x > y$ is symmetric. The symmetrized form of the two former is $x = y$ (conjunction of $x \leqq y$ and $x \geqq y$), the symmetrized form of the two latter is an absurdity (conjunction of $x < y$ and $x > y$).

[3] D could be any other set in which such a relation is defined, cf. the second example in 65.4.1.

[4] Cf. footnote 1 on p. 268.

(e.g. for the set of all real numbers) and A is unique if D has a greatest element (e.g. when it is finite).

Third: Let D be the plane and $x\mathfrak{R}y$ express that the points x, y have the same height (ordinate). Then $\mathfrak{R}$-satisfactoriness of A means that all points of A have the same height, i.e. lie on one parallel to the axis of abscissae. $\mathfrak{R}$-saturation means that A is precisely a line parallel to the axis of abscissae.

Fourth: Let D be the set of all imputations, and $x\mathfrak{R}y$ the negation of the domination $x \succ y$. Then comparison of our (30:B:a), (30:D) with (30:5:a), (30:5:b) in 30.1.1., or equally of (30:C:b) with (30:5:c) id. shows: $\mathfrak{R}$-saturation of A means that A is a solution.

30.3.4. One look at the condition (30:B:a) suffices to see that satisfactoriness for the relation $x\mathfrak{R}y$ is the same as for the relation $y\mathfrak{R}x$ and so also for their conjunction $x\mathfrak{R}^s y$. In other words: $\mathfrak{R}$-satisfactoriness is the same thing as $\mathfrak{R}^s$-satisfactoriness.

Thus satisfactoriness is a concept which need be studied only on symmetric relations.

This is due to the x, y symmetric form of the definitory condition (30:B:a). The equivalent condition (30:C:a) does not exhibit this symmetry, but of course this does not invalidate the proof.

Now the definitory condition (30:C:b) for $\mathfrak{R}$-saturation is very similar in structure to (30:C:a). It is equally asymmetric. However, while (30:C:a) possesses an equivalent symmetric form (30:B:a), this is not the case for (30:C:b). The corresponding equivalent form for (30:C:b) is, as we know, the conjunction of (30:B:a) and (30:D)—and (30:D) is not at all symmetric. I.e. (30:D) is essentially altered if $x\mathfrak{R}y$ is replaced by $y\mathfrak{R}x$. So we see:

(30:E) While $\mathfrak{R}$-satisfactoriness in unaffected by the replacement of $\mathfrak{R}$ by $\mathfrak{R}^s$, it does not appear that this is the case for $\mathfrak{R}$-saturation.

Condition (30:B:a) (amounting to $\mathfrak{R}$-satisfactoriness) is the same for $\mathfrak{R}$ and $\mathfrak{R}^s$. Condition (30:D) for $\mathfrak{R}^s$ is implied by the same for $\mathfrak{R}$ since $\mathfrak{R}^s$ implies $\mathfrak{R}$. So we see:

(30:F) $\mathfrak{R}^s$-saturation is implied by $\mathfrak{R}$-saturation.

The difference between these two types of saturation referred to above is a real one: it is easy to give an explicit example of a set which is $\mathfrak{R}^s$-saturated without being $\mathfrak{R}$-saturated.[1]

Thus the study of saturation cannot be restricted to symmetric relations.

30.3.5. For symmetric relations $\mathfrak{R}$ the nature of saturation is simple enough. In order to avoid extraneous complications we assume for this section that $x\mathfrak{R}x$ is always true.[2]

[1] E.g.: The first two examples of 30.3.3. are in the relation of $\mathfrak{R}^s$ and $\mathfrak{R}$ to each other (cf. footnote 2 on p. 267); their concepts of satisfactoriness are identical, but those of saturation differ.

[2] This is clearly the case for our decisive example of 30.3.3.: $x\mathfrak{R}y$ the negation of $x \succ y$ —since never $x \succ x$.

Now we prove:

(30:G) Let $\mathcal{R}$ be symmetric. Then the $\mathcal{R}$-saturation of A is equivalent to its being maximal $\mathcal{R}$-satisfactory. I.e. it is equivalent to: A is $\mathcal{R}$-satisfactory, but no proper superset of A is.

Proof: $\mathcal{R}$-saturation means $\mathcal{R}$-satisfactoriness (i.e. condition (30:B:a)) together with condition (30:D). So we need only prove: If A is $\mathcal{R}$-satisfactory, then (30:D) is equivalent to the non-$\mathcal{R}$-satisfactoriness of all proper supersets of A.

Sufficiency of (30:D): If $B \supset A$ is $\mathcal{R}$-satisfactory, then any y in B, but not in A, violates (30:D).[1]

Necessity of (30:D): Consider a y which violates (30:D). Then

$$B = A \cup (y) \supset A.$$

Now B is $\mathcal{R}$-satisfactory, i.e. for x', y' in B, always $x'\mathcal{R}y'$. Indeed, when x', y' are both in A, this follows from the $\mathcal{R}$-satisfactoriness of A. If x', y' are both $= y$, we are merely asserting $y\mathcal{R}y$. If one of x', y' is in A, and the other $= y$, then the symmetry of $\mathcal{R}$ allows us to assume x' in A, $y' = y$. Now our assertion coincides with the negation of (30:D).

If $\mathcal{R}$ is not symmetric, we can only assert this:

(30:H) $\mathcal{R}$-saturation of A implies its being maximal $\mathcal{R}$-satisfactory.

Proof: Maximal $\mathcal{R}$-satisfactoriness is the same as maximal $\mathcal{R}^S$-satisfactoriness, cf. (30:E). As $\mathcal{R}^S$ is symmetric, this amounts to $\mathcal{R}^S$-saturation by (30:G). And this is a consequence of $\mathcal{R}$-saturation by (30:F).

The meaning of the result concerning a symmetric $\mathcal{R}$ is the following: Starting with any $\mathcal{R}$-satisfactory set, this set must be increased as long as possible,—i.e. until any further increase would involve the loss of $\mathcal{R}$-satisfactoriness. In this way *in fine* a maximal $\mathcal{R}$-satisfactory set is obtained, —i.e. an $\mathcal{R}$-saturated one by (30:G).[2] This argument secures not only the existence of $\mathcal{R}$-saturated sets, but it also permits us to infer that every $\mathcal{R}$-satisfactory set can be extended to an $\mathcal{R}$-saturated one.

[1] Note that none of the extra restrictions on $\mathcal{R}$ has been used so far.

[2] This process of exhaustion is elementary—i.e. it is over after a finite number of steps—when D is finite.

However, since the set of all imputations is usually infinite, the case of an infinite D is important. When D is infinite, it is still heuristically plausible that the process of exhaustion referred to can be carried out by making an infinite number of steps. This process, known as *transfinite induction,* has been the object of extensive set-theoretical studies. It can be performed in a rigorous way which is dependent upon the so-called *axiom of choice.*

The reader who is interested will find literature in *F. Hausdorff*, footnote 1, on p. 61. Cf. also *E. Zermelo*, Beweis dass jede Menge wohlgeordnet werden kann. Math. Ann. Vol. 59 (1904) p. 514ff. and Math. Ann. Vol. 65 (1908) p. 107ff.

These matters carry far from our subject and are not strictly necessary for our purposes. We do not therefore dwell further upon them.

It should be noted that every subset of an $\mathcal{R}$-saturated set is necessarily $\mathcal{R}$-satisfactory.[1] The above assertion means therefore that the converse statement is also true.

30.3.6. It would be very convenient if the existence of solutions in our theory could be established by such methods. The *prima facie* evidence, however, is against this: the relation which we must use, $x\mathcal{R}y$—negation of the domination $x \succ y$, cf. 30.3.3.—is clearly asymmetrical. Hence we cannot apply (30:G), but only (30:H): maximal satisfactoriness is only necessary, but may not be sufficient for saturation, i.e. for being a solution.

That this difficulty is really deep seated can be seen as follows: If we could replace the above $\mathcal{R}$ by a symmetric one, this could not only be used to prove the existence of solutions, but it would also prove in the same operation the possibility of extending any $\mathcal{R}$-satisfactory set of imputations to a solution (cf. above). Now it is probable that every game possesses a solution, but we shall see that there exist games in which certain satisfactory sets are subsets of no solutions.[2] Thus the device of replacing $\mathcal{R}$ by something symmetric cannot work because this would be equally instrumental in proving the first assertion, which is presumably true, and the second one, which is certainly not true.[3]

The reader may feel that this discussion is futile, since the relation $x\mathcal{R}y$ which we must use ("not $x \succ y$") is *de facto* asymmetric. From a technical point of view, however, it is conceivable that another relation $x\mathcal{S}y$ may be found with the following properties: $x\mathcal{S}y$ is not equivalent to $x\mathcal{R}y$; indeed, $\mathcal{S}$ is symmetric, while $\mathcal{R}$ is not, but $\mathcal{S}$-saturation is equivalent to $\mathcal{R}$-saturation. In this case $\mathcal{R}$-saturated sets would have to exist because they are the $\mathcal{S}$-saturated ones, and the $\mathcal{S}$-satisfactory—but not necessarily the $\mathcal{R}$-satisfactory—sets would always be extensible to $\mathcal{S}$-saturated, i.e. $\mathcal{R}$-saturated ones.[4] This program of attack on the existence problem of solutions is not as arbitrary as it may seem. Indeed, we shall see later a similar problem which is solved in precisely this way (cf. 51.4.3.). All this is, however, for the time being just a hope and a possibility.

30.3.7. In the last section we considered the question whether every $\mathcal{R}$-satisfactory set is a subset of an $\mathcal{R}$-saturated set. We noted that for the relation $x\mathcal{R}y$ which we must use ("not $x \succ y$," asymmetric) the answer is in the negative. A brief comment upon this fact seems to be worth while.

If the answer had been in the affirmative it would have meant that any set fulfilling (30:B:a) can be extended to one fulfilling (30:B:a) and (30:D); or, in the notations of 30.1.1., that any set of imputations fulfilling (30:5:a) can be extended to one fulfilling (30:5:a) and (30:5:b).

[1] Clearly property (30:B:a) is not lost when passing to a subset.

[2] Cf. footnote 2 on p. 285.

[3] This is a rather useful principle of the technical side of mathematics. The inappropriateness of a method can be inferred from the fact that if it were workable at all it would prove too much.

[4] The point is that $\mathcal{R}$- and $\mathcal{S}$-saturation are assumed to be equivalent to each other, but $\mathcal{R}$- and $\mathcal{S}$-satisfactoriness are not expected to be equivalent.

It is instructive to restate this in the terminology of 4.6.2. Then the statement becomes: Any standard of behavior which is free from inner contradiction can be extended to one which is stable,—i.e. not only free from inner contradictions, but also able to upset all imputations outside of it.

The observation in 30.3.6., according to which the above is not true in general, is of some significance: in order that a set of rules of behavior should be the nucleus (i.e. a subset) of a stable standard of behavior, it may have to possess deeper structural properties than mere freedom from inner contradictions.[1]

30.4. Three Immediate Objectives

30.4.1. We have formulated the characteristics of a solution of an unrestricted zero-sum n-person game and can therefore begin the systematic investigation of the properties of this concept. In conjunction with the early stages of this investigation it seems appropriate to carry out three special enquiries. These deal with the following special cases:

First: Throughout the discussions of 4., the idea recurred that the unsophisticated concept of a solution would be that of an imputation,—i.e. in our present terminology, of a one-element set V. In 4.4.2. we saw specifically that this would amount to finding a "first" element with respect to domination. We saw in the subsequent parts of 4., as well as in our exact discussions of 30.2., that it is mainly the intransitivity of our concept of domination which defeats this endeavor and forces us to introduce sets of imputations V as solutions.

It is, therefore, of interest—now that we are in a position to do it—to give an exact answer to the following question: For which games do one-element solutions V exist? What else can be said about the solutions of such games?

Second: The postulates of 30.1.1. were extracted from our experiences with the zero-sum three-person game, in its essential case. It is, therefore, of interest to reconsider this case in the light of the present, exact theory. Of course, we know—indeed this was a guiding principle throughout our discussions—that the solution which we obtained by the preliminary methods of 22., 23., are solutions in the sense of our present postulates too. Nevertheless it is desirable to verify this explicitly. The real point, however, is to ascertain whether the present postulates do not ascribe to those games further solutions as well. (We have already seen that it is not inconceivable that there should exist several solutions for the same game.)

We shall therefore determine all solutions for the essential zero-sum three-person games—with results which are rather surprising, but, as we shall see, not unreasonable.

[1] If the relation $\mathcal{S}$ referred to at the end of 30.3.6. could be found, then this $\mathcal{S}$—and not $\mathcal{R}$—would disclose which standards of behavior are such nuclei (i.e. subsets): the $\mathcal{S}$-satisfactory ones.

Cf. the similar situation in 51.4., where the corresponding operation is performed successfully.

30.4.2. These two items exhaust actually all zero-sum games with $n \leqq 3$. We observed in the first remark of 27.5.2. that for $n = 1, 2$, these games are inessential; so this, together with the inessential and the essential cases of $n = 3$, takes care of everything in $n \leqq 3$.

When this program is fulfilled we are left with the games $n \geqq 4$—and we know already that difficulties of a new kind begin with them (cf. the allusions of footnote 1, p. 250, and the end of 27.5.3.).

30.4.3. Third: We introduced in 27.1. the concept of strategic equivalence. It appeared plausible that this relationship acts as its name expresses: two games which are linked by it offer the same strategical possibilities and inducements to form coalitions, etc. Now that we have put the concept of a solution on an exact basis, this heuristic expectation demands a rigorous proof.

These three questions will be answered in (31:P) in 31.2.3.; in 32.2.; and in (31:Q) in 31.3.3., respectively.

31. First Consequences

31.1. Convexity, Flatness, and Some Criteria for Domination

31.1.1. This section is devoted to proving various auxiliary results concerning solutions, and the other concepts which surround them, like inessentiality, essentiality, domination, effectivity. Since we have now put all these notions on an exact basis, the possibility as well as the obligation arises to be absolutely rigorous in establishing their properties. Some of the deductions which follow may seem pedantic, and it may appear occasionally that a verbal explanation could have replaced the mathematical proof. Such an approach, however, would be possible for only part of the results of this section and, taking everything into account, the best plan seems to be to go about the whole matter systematically with full mathematical rigor.

Some principles which play a considerable part in finding solutions are (31:A), (31:B), (31:C), (31:F), (31:G), (31:H), which for certain coalitions decide *a priori* that they must always, or never, be taken into consideration. It seemed appropriate to accompany these principles with verbal explanations (in the sense indicated above) in addition to their formal proofs.

The other results possess varying interest of their own in different directions. Together they give a first orientation of the circumstances which surround our newly won concepts. The answers to the first and third questions in 30.4. are given in (31:P) and (31:Q). Another question which arose previously is settled in (31:M).

31.1.2. Consider two imputations $\overrightarrow{\alpha}$, $\overrightarrow{\beta}$ and assume that it has become necessary to decide whether $\overrightarrow{\alpha} \succ \overrightarrow{\beta}$ or not. This amounts to deciding whether or not there exists a set S with the properties (30:4:a)-(30:4:c) in 30.1.1. One of these, (30:4:c) is

$$\alpha_i > \beta_i \qquad \text{for all } i \text{ in } S.$$

We call this the *main condition*. The two others, (30:4:a), (30:4:b), are the *preliminary conditions*.

Now one of the major technical difficulties in handling this concept of domination—i.e. in finding solutions V in the sense of 30.1.1.—is the presence of these preliminary conditions. It is highly desirable to be able, so to say, to short circuit them, i.e. to discover criteria under which they are certainly satisfied, and others under which they are certainly not satisfied. In looking for criteria of the latter type, it is by no means necessary that they should involve non-fulfillment of the preliminary conditions for all imputations $\overrightarrow{\alpha}$—it suffices if they involve it for all those imputations $\overrightarrow{\alpha}$ which fulfill the main condition for some other imputation $\overrightarrow{\beta}$. (Cf. the proofs of (31:A) or (31:F), where exactly this is utilized.)

We are interested in criteria of this nature in connection with the question of determining whether a given set V of imputations is a solution or not; i.e. whether it fulfills the conditions (30:5:a), (30:5:b)—the condition (30:5:c)—in 30.1.1. This amounts to determining which imputations $\overrightarrow{\beta}$ are dominated by elements of V.

Criteria which dispose of the preliminary conditions summarily, in the situation described above, are most desirable if they contain no reference at all to $\overrightarrow{\alpha}$,[1,2] i.e. if they refer to S alone. (Cf. (31:F), (31:G), (31:H).) But even criteria which involve $\overrightarrow{\alpha}$ may be desirable. (Cf. (31:A).) We shall consider even a criterion which deals with S and $\overrightarrow{\alpha}$ by referring to the behavior of another $\overrightarrow{\alpha}'$. (Of course, both in V. Cf. (31:B).)

In order to cover all these possibilities, we introduce the following terminology:

We consider proofs which aim at the determination of all imputations $\overrightarrow{\beta}$, which are dominated by elements of a given set of imputations V. We are thus concerned with the relations $\overrightarrow{\alpha} \succ \overrightarrow{\beta}$ ($\overrightarrow{\alpha}$ in V), and the question whether a certain set S meets our preliminary requirements for such a relation. We call S *certainly necessary* if we know (owing to the fulfillment by S of some appropriate criterion) that S and $\overrightarrow{\alpha}$ always meet the preliminary conditions. We call a set S *certainly unnecessary*, if we know (again owing to the fulfillment by S of some appropriate criterion, but which may now involve other things too, cf. above) that the possibility that S and $\overrightarrow{\alpha}$ meet

[1] The point being that in our original definition of $\overrightarrow{\alpha} \succ \overrightarrow{\beta}$ the preliminary conditions refer to S and to $\overrightarrow{\alpha}$ (but not to $\overrightarrow{\beta}$). Specifically (30:4:b) does.

[2] The hypothetical element of V, which should dominate $\overrightarrow{\beta}$.

the preliminary conditions can be disregarded (because this never happens, or for any other reason. Cf. also the qualifications made above).

These considerations may seem complicated, but they express a quite natural technical standpoint.[1]

We shall now give certain criteria of the certainly necessary and of the certainly unnecessary characters. After each criterion we shall give a verbal explanation of its content, which, it is hoped, will make our technique clearer to the reader.

31.1.3. First, three elementary criteria:

(31:A) S is certainly unnecessary for a given $\vec{\alpha}$ (in V) if there exists an i in S with $\alpha_i = v((i))$.

Explanation: A coalition need never be considered if it does not promise to every participant (individually) definitely more than he can get for himself.

Proof: If $\vec{\alpha}$ fulfills the main condition for some imputation, then $\alpha_i > \beta_i$. Since $\vec{\beta}$ is an imputation, so $\beta_i \geqq v((i))$. Hence $\alpha_i > v((i))$. This contradicts $\alpha_i = v((i))$.

(31:B) S is certainly unnecessary for a given $\vec{\alpha}$ (in V) if it is certainly necessary (and being considered) for another $\vec{\alpha}'$ (in V), such that

$$\alpha_i' \geqq \alpha_i \qquad \text{for all } i \text{ in } S. \tag{31:1}$$

Explanation: A coalition need not be considered if another one, which has the same participants and promises every one (individually) at least as much, is certain to receive consideration.

Proof: Let $\vec{\alpha}$ and $\vec{\beta}$ fulfill the main condition: $\alpha_i > \beta_i$ for all i in S. Then $\vec{\alpha}'$ and $\vec{\beta}$ fulfill it also, by (31:1), $\alpha_i' > \beta_i$ for all i in S. Since S and $\vec{\alpha}'$ are

[1] For the reader who is familiar with formal logic we observe the following:

The attributes "certainly necessary" and "certainly unnecessary" are of a logical nature. They are characterized by our ability to show (by any means whatever) that a certain logical omission will invalidate no proof (of a certain kind). Specifically: Let a proof be concerned with the domination of a $\vec{\beta}$ by an element $\vec{\alpha}$ of V. Assume that this domination $\vec{\alpha} \succ \vec{\beta}$ occurring with the help of the set S ($\vec{\alpha}$ in V) be under consideration. Then this proof remains correct if we treat S and $\vec{\alpha}$ (when they possess the attribute in question) as if they always (or never) fulfilled the preliminary conditions,—without our actually investigating these conditions. In the mathematical proofs which we shall carry out, this procedure will be applied frequently.

It can even happen that the same S will turn out (by the use of two different criteria) to be both certainly necessary and certainly unnecessary (for the same $\vec{\alpha}$,—e.g. for all of them). This means merely that neither of the two omissions mentioned above spoils any proof. This can happen, for instance, when $\vec{\alpha}$ fulfills the main condition for no imputation. (An example is obtained by combining (31:F) and (31:G) in the case described in (31:E:b). Another is pointed out in footnote 1 on p. 310, and in footnote 1 on p. 431.)

being considered, they thus establish that $\overrightarrow{\beta}$ is dominated by an element of V, and it is unnecessary to consider S and $\overrightarrow{\alpha}$.

(31:C) S is certainly unnecessary if another set $T \subseteq S$ is certainly necessary (and is being considered).

Explanation: A coalition need not be considered if a part of it is already certain to receive consideration.

Proof: Let $\overrightarrow{\alpha}$ (in V) and $\overrightarrow{\beta}$ fulfill the main condition for S; then they will obviously fulfill it *a fortiori* for $T \subseteq S$. Since T and $\overrightarrow{\alpha}$ are being considered, they thus establish that $\overrightarrow{\beta}$ is dominated by an element of V and it is unnecessary to consider S and $\overrightarrow{\alpha}$.

31.1.4. We now introduce some further criteria, and on a somewhat broader basis than immediately needed. For this purpose we begin with the following consideration:

For an arbitrary set $S = (k_1, \cdots, k_p)$ apply (25:5) in 25.4.1., with $S_1 = (k_1), \cdots, S_p = (k_p)$. Then

$$\mathrm{v}(S) \geqq \mathrm{v}((k_1)) + \cdots + \mathrm{v}((k_p))$$

obtains, i.e.

$$\mathrm{v}(S) \geqq \sum_{k \text{ in } S} \mathrm{v}((k)). \tag{31:2}$$

The excess of the left-hand side of (31:2) over the right hand side expresses the total advantage (for all participants together) inherent in the formation of the coalition S. We call this the *convexity* of S. If this advantage vanishes, i.e. if

$$\mathrm{v}(S) = \sum_{k \text{ in } S} \mathrm{v}((k)), \tag{31:3}$$

then we call S *flat*.

The following observations are immediate:

(31:D) The following sets are always flat:
(31:D:a) The empty set,
(31:D:b) Every one-element set,
(31:D:c) Every subset of a flat set.
(31:E) Any one of the following assertions is equivalent to the inessentiality of the game:
(31:E:a) $I = (1, \cdots, n)$ is flat,
(31:E:b) There exists an S such that both S and $-S$ are flat,
(31:E:c) Every S is flat.

Proof: Ad (31:D:a), (31:D:b): For these sets (31:3) is obvious.

Ad (31:D:c): Assume $S \subseteq T$, T flat. Put $R = T - S$. Then by (31:2)

(31:4) $$\mathrm{v}(S) \geqq \sum_{k \text{ in } S} \mathrm{v}((k)),$$

(31:5) $$\mathrm{v}(R) \geqq \sum_{k \text{ in } R} \mathrm{v}((k)).$$

Since T is flat, so by (30:3)

(31:6) $$\mathrm{v}(T) = \sum_{k \text{ in } T} \mathrm{v}((k)).$$

As $S \cap R = \ominus$, $S \cup R = T$; therefore

$$\mathrm{v}(S) + \mathrm{v}(R) \leqq \mathrm{v}(T),$$
$$\sum_{k \text{ in } S} \mathrm{v}((k)) + \sum_{k \text{ in } R} \mathrm{v}((k)) = \sum_{k \text{ in } T} \mathrm{v}((k)).$$

Hence (31:6) implies

(31:7) $$\mathrm{v}(S) + \mathrm{v}(R) \leqq \sum_{k \text{ in } S} \mathrm{v}((k)) + \sum_{k \text{ in } R} \mathrm{v}((k)).$$

Now comparison of (31:4), (31:5) and (31:7) shows that we must have equality in all of them. But equality in (31:4) expresses just the flatness of S.

Ad (31:E:a): The assertion coincides with (27:B) in 27.4.1.

Ad (31:E:c): The assertion coincides with (27:C) in 27.4.2.

Ad (31:E:b): For an inessential game this is true for any S owing to (31:E:c). Conversely, if this is true for (at least one) S, then

$$\mathrm{v}(S) = \sum_{k \text{ in } S} \mathrm{v}((k)), \qquad \mathrm{v}(-S) = \sum_{k \text{ not in } S} \mathrm{v}((k)),$$

hence by addition (use (25:3:b) in 25.3.1.),

$$0 = \sum_{k=1}^{n} \mathrm{v}((k)),$$

i.e. the game is inessential by (31:E:a) or by (27:B) in 27.4.1.

31.1.5. We are now in a position to prove:

(31:F) S is certainly unnecessary if it is flat.

Explanation: A coalition need not be considered if the game allows no total advantage (for all its participants together) over what they would get for themselves as independent players.[1]

[1] Observe that this is related to (31:A), but not at all identical with it! Indeed: (31:A) deals with the α_i, i.e. with the promises made to each participant individually. (31:F) deals with $\mathrm{v}(S)$ (which determines flatness), i.e. with the possibilities of the game for all participants together. But both criteria correlate these with the $\mathrm{v}((i))$, i.e. with what each player individually can get for himself.

Proof: If $\overrightarrow{\alpha} \succ \overrightarrow{\beta}$ with the help of this S then we have: Necessarily $S \neq \ominus$. $\alpha_i > \beta_i$ for all i in S and $\beta_i \geqq v((i))$, hence $\alpha_i > v((i))$. So $\sum_{i \text{ in } S} \alpha_i > \sum_{i \text{ in } S} v((i))$. As S is flat, this means $\sum_{i \text{ in } S} \alpha_i > v(S)$. But S must be effective, $\sum_{i \text{ in } S} \alpha_i \leqq v(S)$, which is a contradiction.

(31:G) S is certainly necessary if $-S$ is flat and $S \neq \ominus$.

Explanation: A coalition must be considered if it (is not empty and) opposes one of the kind described in (31:F).

Proof: The preliminary conditions are fulfilled for all imputations $\overrightarrow{\alpha}$.

Ad (30:4:a): $S \neq \ominus$ was postulated.

Ad (30:4:b): Always $\alpha_i \geqq v((i))$, so $\sum_{i \text{ not in } S} \alpha_i \geqq \sum_{i \text{ not in } S} v((i))$. Since $\sum_{i=1}^{n} \alpha_i = 0$, the left-hand side is equal to $-\sum_{i \text{ in } S} \alpha_i$. Since $-S$ is flat, the right-hand side is equal to $v(-S)$, i.e. (use (25:3:b) in 25.3.1.) to $-v(S)$. So $-\sum_{i \text{ in } S} \alpha_i \geqq -v(S)$, $\sum_{i \text{ in } S} \alpha_i \leqq v(S)$, i.e. S is effective.

From (31:F), (31:G) we obtain by specialization:

(31:H) A p-element set is certainly necessary if $p = n - 1$, and certainly unnecessary if $p = 0, 1, n$.

Explanation: A coalition must be considered if it has only one opponent. A coalition need not be considered, if it is empty or consists of one player only (!), or if it has no opponents.

Proof: $p = n - 1$: $-S$ has only one element, hence it is flat by (31:D) above. The assertion now follows from (31:G).

$p = 0, 1$: Immediate by (31:D) and (31:F).

$p = n$: In this case necessarily $S = I = (1, \cdots, n)$ rendering the main condition unfulfillable. Indeed, that now requires $\alpha_i > \beta_i$ for all $i = 1, \cdots, n$, hence $\sum_{i=1}^{n} \alpha_i > \sum_{i=1}^{n} \beta_i$. But as $\overrightarrow{\alpha}, \overrightarrow{\beta}$ are imputations, both sides vanish,—and this is a contradiction.

Thus those p for which the necessity of S is in doubt, are restricted to $p \neq 0, 1, n - 1, n$, i.e. to the interval

$$2 \leqq p \leqq n - 2. \tag{31:8}$$

This interval plays a role only when $n \geqq 4$. The situation discussed is similar to that at the end of 27.5.2. and in 27.5.3., and the case $n = 3$ appears once more as one of particular simplicity.

31.2. The System of All Imputations. One-element Solutions

31.2.1. We now discuss the structure of the system of all imputations.

(31:I) For an inessential game there exists precisely one imputation:

(31:9) $\vec{\alpha} = \{\alpha_1, \cdots ; \alpha_n\}, \quad \alpha_i = v((i)) \quad \text{for} \quad i = 1, \cdots, n.$

For an essential game there exist infinitely many imputations —an $(n-1)$-dimensional continuum—but (31:9) is not one of them.

Proof: Consider an imputation

$$\vec{\beta} = \{\beta_1, \cdots, \beta_n\},$$

and put

$$\beta_i = v((i)) + \epsilon_i \quad \text{for} \quad i = 1, \cdots, n.$$

Then the characteristic conditions (30:1), (30:2) of 30.1.1. become

(31:10) $\epsilon_i \geqq 0 \quad \text{for} \quad i = 1, \cdots, n.$

(31:11) $\sum_{i=1}^{n} \epsilon_i = - \sum_{i=1}^{n} v((i)).$

If Γ is inessential, then (27:B) in 27.4.1. gives $- \sum_{i=1}^{n} v((i)) = 0$; so (31:10), (31:11) amount to $\epsilon_1 = \cdots = \epsilon_n = 0$, i.e. (31:9) is the unique imputation.

If Γ is essential, the (27:B) in 27.4.1. gives $- \sum_{i=1}^{n} v((i)) > 0$, so (31:10), (31:11) possess infinitely many solutions, which form an $(n-1)$-dimensional continuum;[1] so the same is true for the imputations $\vec{\beta}$. But the $\vec{\alpha}$ of (31:9) is not one of them, because $\epsilon_1 = \cdots = \epsilon_n = 0$ now violate (31:11).

An immediate consequence:

(31:J) A solution V is never empty.

Proof: I.e. the empty set $\ominus$ is not a solution. Indeed: Consider any imputation $\vec{\beta}$,—there exists at least one by (31:I). $\vec{\beta}$ is not in $\ominus$ and no $\vec{\alpha}$ in $\ominus$ has $\vec{\alpha} \succ \vec{\beta}$. So $\ominus$ violates (30:5:b) in 30.1.1.[2]

31.2.2. We have pointed out before that the simultaneous validity of

(31:12) $\vec{\alpha} \succ \vec{\beta}, \quad \vec{\beta} \succ \vec{\alpha}$

is by no means impossible.[3] However:

(31:K) Never $\vec{\alpha} \succ \vec{\alpha}$.

[1] There is only one equation: (31:11).

[2] This argument may seem pedantic; but if the conditions for the imputations conflicted (i.e. without (31:I)), then V $= \ominus$ would be a solution.

[3] The sets S of these two dominations would have to be disjunct. By (31:H) these S must have $\geqq 2$ elements each. Hence (31:12) can occur only when $n \geqq 4$.

By a more detailed consideration even $n = 4$ can be excluded; but for every $n \geqq 5$ (31:12) is really possible.

Proof: (30:4:a), (30:4:c) in 30.1.1. conflict for $\vec{\alpha} = \vec{\beta}$.

(31:L) Given an essential game and an imputation $\vec{\alpha}$, there exists an imputation $\vec{\beta}$ such that $\vec{\beta} \succ \vec{\alpha}$ but not $\vec{\alpha} \succ \vec{\beta}$.[1]

Proof: Put

$$\vec{\alpha} = \{\alpha_1, \cdots, \alpha_n\}.$$

Consider the equation

$$\alpha_i = \mathrm{v}((i)). \tag{31:13}$$

Since the game is essential, (31:I) excludes the proposition that (31:13) be valid for all $i = 1, \cdots, n$. Let (31:13) fail, say for $i = i_0$. Since $\vec{\alpha}$ is an imputation, so $\alpha_{i_0} \geqq \mathrm{v}((i_0))$, hence the failure of (31:13) means $\alpha_{i_0} > \mathrm{v}((i_0))$, i.e.

$$\alpha_{i_0} = \mathrm{v}((i_0)) + \epsilon, \qquad \epsilon > 0. \tag{31:14}$$

Now define a vector

$$\vec{\beta} = \{\beta_1, \cdots, \beta_n\},$$

by

$$\beta_{i_0} = \alpha_{i_0} - \epsilon = \mathrm{v}((i_0)),$$

$$\beta_i = \alpha_i + \frac{\epsilon}{n-1} \qquad \text{for} \qquad i \neq i_0.$$

These equations make it clear that $\beta_i \geqq \mathrm{v}((i))$[2] and that $\sum_{i=1}^{n} \beta_i = \sum_{i=1}^{n} \alpha_i = 0$.[3] So $\vec{\beta}$ is an imputation along with $\vec{\alpha}$.

We now prove the two assertions concerning $\vec{\alpha}, \vec{\beta}$.

$\vec{\beta} \succ \vec{\alpha}$: We have $\beta_i > \alpha_i$ for all $i \neq i_0$, i.e. for all i in the set $S = -(i_0)$. This set has $n - 1$ elements and it fulfills the main condition (for $\vec{\beta}, \vec{\alpha}$), hence (31:H) gives $\vec{\beta} \succ \vec{\alpha}$.

Not $\vec{\alpha} \succ \vec{\beta}$: Assume that $\vec{\alpha} \succ \vec{\beta}$. Then a set S fulfilling the main condition must exist, which is not excluded by (31:H). So S must have $\geqq 2$ elements. So an $i \neq i_0$ in S must exist. The former implies $\beta_i > \alpha_i$

[1] Hence $\vec{\alpha} \neq \vec{\beta}$.

[2] For $i = i_0$, we have actually $\beta_{i_0} = \mathrm{v}((i_0))$. For $i \neq i_0$, we have $\beta_i > \alpha_i \geqq \mathrm{v}((i))$.

[3] $\sum_{i=1}^{n} \beta_i = \sum_{i=1}^{n} \alpha_i$ because the difference of β_i and α_i is ϵ for one value of i ($i = i_0$) and $-\frac{\epsilon}{n-1}$ for $n - 1$ values of i (all $i \neq i_0$).

(by the construction of $\vec{\beta}$); the latter implies $\alpha_i > \beta_i$ (owing to the main condition)—and this is a contradiction.

31.2.3. We can draw the conclusions in which we were primarily interested:

(31:M) An imputation $\vec{\alpha}$, for which never $\vec{\alpha}' \succ \vec{\alpha}$, exists if and only if the game is inessential.[1]

Proof: Sufficiency: If the game is inessential, then it possesses by (31:I) precisely one imputation $\vec{\alpha}$, and this has the desired property by (31:K).

Necessity: If the game is essential, and $\vec{\alpha}$ is an imputation, then $\vec{\alpha}' = \vec{\beta}$ of (31:L) gives $\vec{\alpha}' = \vec{\beta} \succ \vec{\alpha}$.

(31:N) A game which possesses a one-element solution[2] is necessarily inessential.

Proof: Denote the one-element solution in question by $V = (\vec{\alpha})$. This V must satisfy (30:5:b) in 30.1.1. This means under our present circumstances: Every $\vec{\beta}$ other than $\vec{\alpha}$ is dominated by $\vec{\alpha}$. I.e.:

$$\vec{\beta} \neq \vec{\alpha} \quad \text{implies} \quad \vec{\alpha} \succ \vec{\beta}.$$

Now if the game is essential, then (31:L) provides a $\vec{\beta}$ which violates this condition.

(31:O) An inessential game possesses precisely one solution V. This is the one-element set $V = (\vec{\alpha})$ with the $\vec{\alpha}$ of (31:I).

Proof: By (31:I) there exists precisely one imputation, the $\vec{\alpha}$ of (31:I). A solution V cannot be empty by (31:J); hence the only possibility is $V = (\vec{\alpha})$. Now $V = (\vec{\alpha})$ is indeed a solution, i.e. it fulfills (30:5:a), (30:5:b) in 30.1.1.: the former by (31:K), the latter because $\vec{\alpha}$ is the only imputation by (31:I).

We can now answer completely the first question of 30.4.1.:

(31:P) A game possesses a one-element solution (cf. footnote 2 above) if and only if it is inessential; and then it possesses no other solutions.

Proof: This is just a combination of the results of (31:N) and (31:O).

[1] Cf. the considerations of (30:A:a) in 30.2.2., and particularly footnote 2 on p. 265.

[2] We do not exclude the possibility that this game may possess other solutions as well, which may or may not be one-element sets. Actually this never happens (under our present hypothesis), as the combination of the result of (31:N) with that of (31:O)—or the result of (31:P)—shows. But the present consideration is independent of all this.

31.3. The Isomorphism Which Corresponds to Strategic Equivalence

31.3.1. Consider two games Γ and Γ' with the characteristic functions $v(S)$ and $v'(S)$ which are strategically equivalent in the sense of 27.1. We propose to prove that they are really equivalent from the point of view of the concepts defined in 30.1.1. This will be done by establishing an *isomorphic correspondence* between the entities which form the substratum of the definitions of 30.1.1., i.e. the imputations. That is, we wish to establish a one-to-one correspondence, between the imputations of Γ and those of Γ', which is isomorphic with respect to those concepts, i.e. which carries effective sets, domination, and solutions for Γ into those for Γ'.

The considerations are merely an exact elaboration of the heuristic indications of 27.1.1., hence the reader may find them unnecessary. However, they give quite an instructive instance of an "isomorphism proof," and, besides, our previous remarks on the relationship of verbal and of exact proofs may be applied again.

31.3.2. Let the strategic equivalence be given by $\alpha_1^0, \cdots, \alpha_n^0$ in the sense of (27:1), (27:2) in 27.1.1. Consider all imputations $\overrightarrow{\alpha} = \{\alpha_1, \cdots, \alpha_n\}$ of Γ and all imputations $\overrightarrow{\alpha}' = \{\alpha_1', \cdots, \alpha_n'\}$ of Γ'. We look for a one-to-one correspondence

$$(31:15) \qquad \overrightarrow{\alpha} \rightleftarrows \overrightarrow{\alpha}'$$

with the specified properties.

What (31:15) ought to be is easily guessed from the motivation at the beginning of 27.1.1. We described there the passage from Γ to Γ' by adding to the game a fixed payment of α_k^0 to the player k. Applying this principle to the imputations means

$$(31:16) \qquad \alpha_k' = \alpha_k + \alpha_k^0 \quad \text{for} \quad k = 1, \cdots, n.$$[1]

Accordingly we define the correspondence (31:15) by the equations (31:16).

31.3.3. We now verify the asserted properties of (31:15), (31:16).

The imputations of Γ are mapped on the imputations of Γ': This means by (30:1), (30:2) in 30.1.1., that

$$(31:17) \qquad \alpha_i \geqq v((i)) \quad \text{for} \quad i = 1, \cdots. n.$$

$$(31:18) \qquad \sum_{i=1}^{n} \alpha_i = 0,$$

go over into

$$(31:17^*) \qquad \alpha_i' \geqq v'((i)) \quad \text{for} \quad i = 1, \cdots, n,$$

$$(31:18^*) \qquad \sum_{i=1}^{n} \alpha_i' = 0.$$

[1] If we introduce the (fixed) vector $\overrightarrow{\alpha}^0 = \{\alpha_1^0, \cdots, \alpha_n^0\}$ then (31:16) may be written vectorially $\overrightarrow{\alpha}' = \overrightarrow{\alpha} + \overrightarrow{\alpha}^0$. I.e. it is a translation (by $\overrightarrow{\alpha}$) in the vector space of the imputations.

This is so for (31:17), (31:17*) because $v'((i)) = v((i)) + \alpha_i^0$ (by (27:2) in 27.1.1.), and for (31:18), (31:18*) because $\sum_{i=1}^{n} \alpha_i^0 = 0$ (by (27:1) id.).

Effectivity for Γ goes over into effectivity for Γ': This means by (30:3) in 30.1.1., that

$$\sum_{i \text{ in } S} \alpha_i \leqq v(S)$$

goes over into

$$\sum_{i \text{ in } S} \alpha'_i \leqq v'(S).$$

This becomes evident by comparison of (31:16) with (27:2).

Domination for Γ goes over into domination for Γ': This means the same thing for (30:4:a)-(30:4:c) in 30.1.1. (30:4:a) is trivial; (30:4:b) is effectivity, which we settled: (30:4:c) asserts that $\alpha_i > \beta_i$ goes over into $\alpha'_i > \beta'_i$ which is obvious. The solutions of Γ are mapped on the solutions of Γ': This means the same for (30:5:a), (30:5:b) (or (30:5:c)) in 30.1.1. These conditions involve only domination, which we settled.

We restate these results:

(31:Q) If two zero-sum games Γ and Γ' are strategically equivalent, then there exists an isomorphism between their imputations—i.e. a one-to-one mapping of those of Γ on those of Γ' which leaves the concepts defined in 30.1.1. invariant.

32. Determination of all Solutions of the Essential Zero-sum Three-person Game

32.1. Formulation of the Mathematical Problem. The Graphical Method

32.1.1. We now turn to the second problem formulated in 30.4.1.: The determination of all solutions for the essential zero-sum three-person games.

We know that we may consider this game in the reduced form and that we can choose $\gamma = 1$.[1] The characteristic function in this case is completely determined as we have discussed before:[2]

$$\text{(32:1)} \qquad v(S) = \begin{cases} 0 \\ -1 \\ 1 \\ 0 \end{cases} \quad \text{when } S \text{ has } \begin{cases} 0 \\ 1 \\ 2 \\ 3 \end{cases} \text{elements.}$$

An imputation is a vector

$$\overrightarrow{\alpha} = \{\alpha_1, \alpha_2, \alpha_3\}$$

[1] Cf. the discussion at the beginning of 29.1., or the references there given: the end of 27.1. and the second remark in 27.3.

[2] Cf. the discussion at the beginning of 29.1., or the second case of 27.5.

whose three components must fulfill (30:1), (30:2) in 30.1.1.* These conditions now become (considering (30:1))

$$\text{(32:2)} \qquad \alpha_1 \geqq -1, \qquad \alpha_2 \geqq -1, \qquad \alpha_3 \geqq -1,$$

$$\text{(32:3)} \qquad \alpha_1 + \alpha_2 + \alpha_3 = 0.$$

We know, from (31:I) in 31.2.1., that these α_1, α_2, α_3 form only a two-dimensional continuum—i.e. that they should be representable in the plane. Indeed, (32:3) makes a very simple plane representation possible.

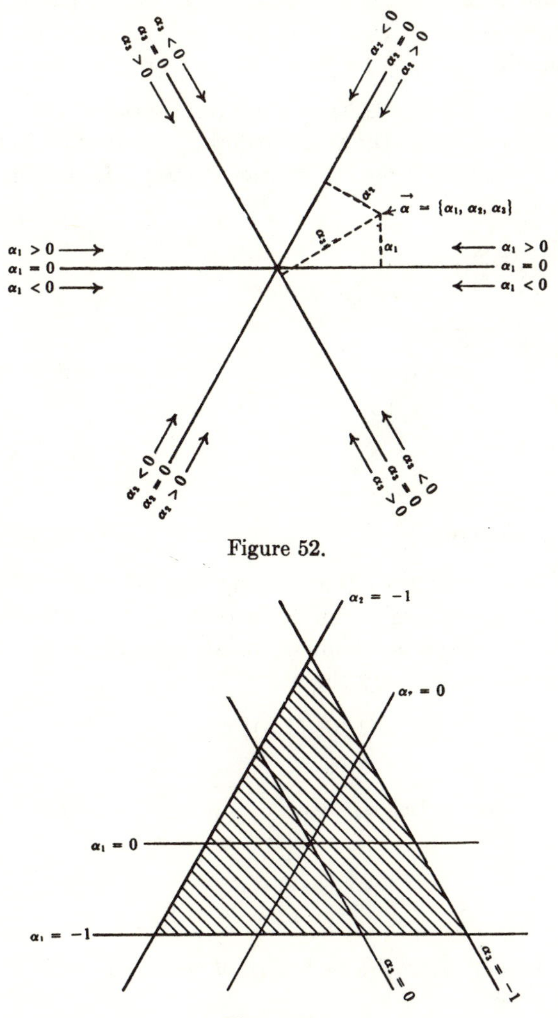

Figure 52.

Figure 53.

32.1.2. For this purpose we take *three axes* in the plane, making angles of 60° with each other. For any point of the plane we define α_1, α_2, α_3 by directed perpendicular distances from these three axes. The whole arrangement, and in particular the signs to be ascribed to the α_1, α_2, α_3

are given in Figure 52. It is easy to verify that for any point the algebraic sum of these three perpendicular distances vanishes and that conversely any triplet $\overrightarrow{\alpha} = \{\alpha_1, \alpha_2, \alpha_3\}$ for which the sum vanishes, corresponds to a point.

So the plane representation of Figure 52 expresses precisely the condition (32:3). The remaining condition (32:2) is therefore the equivalent of a restriction imposed upon the point $\overrightarrow{\alpha}$ within the plane of Figure 52. This restriction is obviously that it must lie on or within the triangle formed by the three lines $\alpha_1 = -1$, $\alpha_2 = -1$, $\alpha_3 = -1$. Figure 53 illustrates this.

Thus the shaded area, to be called the *fundamental triangle*, represents the $\overrightarrow{\alpha}$ which fulfill (32:2), (32:3)—i.e. all imputations.

32.1.3. We next express the relationship of domination in this graphical representation. As $n = 3$, we know from (31:H) (cf. also the discussion of

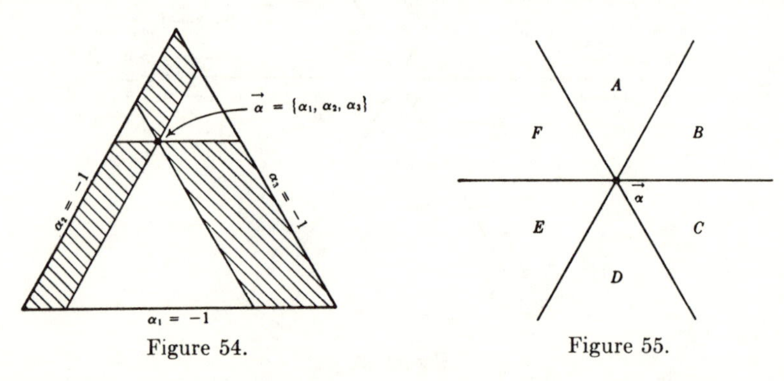

Figure 54. Figure 55.

(31:8) at the end of 31.1.5.) that among the subsets S of $I = (1, 2, 3)$ those of two elements are certainly necessary, and all others certainly unnecessary. I.e., the sets which we must consider in our determination of all solutions V are precisely these:

$$(1,2);\ (1,3);\ (2,3).$$

Thus for

$$\overrightarrow{\alpha} = \{\alpha_1, \alpha_2, \alpha_3\}, \qquad \overrightarrow{\beta} = \{\beta_1, \beta_2, \beta_3\}$$

domination

$$\overrightarrow{\alpha} \succ \overrightarrow{\beta}$$

means that

(32:4) Either $\alpha_1 > \beta_1$, $\alpha_2 > \beta_2$; or $\alpha_1 > \beta_1$, $\alpha_3 > \beta_3$; or $\alpha_2 > \beta_2$, $\alpha_3 > \beta_3$.

Diagrammatically: $\overrightarrow{\alpha}$ dominates the points in the shaded areas, and no others,[1] in Figure 54.

[1] In particular, no points on the boundary lines of these areas.

Thus the point $\vec{\alpha}$ dominates three of the six sextants indicated in Figure 55 (namely A, C, E). From this one concludes easily that $\vec{\alpha}$ is dominated by the three other sextants (namely B, D, F). So the only points which do not dominate $\vec{\alpha}$ and are not dominated by it, lie on the three lines (i.e. six half-lines) which separate these sextants. I.e.:

(32:5) If neither of $\vec{\alpha}$, $\vec{\beta}$ dominates the other, then the direction from $\vec{\alpha}$ to $\vec{\beta}$ is parallel to one of the sides of the fundamental triangle.

32.1.4. Now the systematic search for all solutions can begin.

Consider a solution V, i.e. a set in the fundamental triangle which fulfills the conditions (30:5:a), (30:5:b) of 30.1.1. In what follows we shall use these conditions currently, without referring to them explicitly on each occasion.

Since the game is essential, V must contain at least two points[1] say $\vec{\alpha}$ and $\vec{\beta}$. By (32:5) the direction from $\vec{\alpha}$ to $\vec{\beta}$ is parallel to one of the sides of the fundamental triangle; and by a permutation of the numbers of the players 1,2,3 we can arrange this to be the side $\alpha_1 = -1$, i.e. the horizontal. So $\vec{\alpha}$, $\vec{\beta}$ lie on a horizontal line l. Now two possibilities arise and we treat them separately:

(a) Every point of V lies on l.

(b) Some points of V do not lie on l.

32.2. Determination of All Solutions

32.2.1. We consider (b) first. Any point not on l must fulfill (32:5) with respect to both $\vec{\alpha}$ and $\vec{\beta}$, i.e. it must be the third vertex of one of the two equilateral triangles with the base $\vec{\alpha}$, $\vec{\beta}$: one of the two points $\vec{\alpha}\,'$, $\vec{\alpha}\,''$ of Figure 56. So either $\vec{\alpha}\,'$ or $\vec{\alpha}\,''$ belongs to V. Any point of V which differs from $\vec{\alpha}$, $\vec{\beta}$ and $\vec{\alpha}\,'$ or $\vec{\alpha}\,''$ must again fulfill (32:5), but now with respect to all three points $\vec{\alpha}$, $\vec{\beta}$ and $\vec{\alpha}\,'$ or $\vec{\alpha}\,''$. This, however, is impossible, as an inspection of Figure 56 immediately reveals. So V consists of precisely these three points,—i.e. of the three vertices of a triangle which is in the position of triangle I or triangle II of Figure 57. Comparison of Figure 57 with Figures 54 or 55 shows that the vertices of triangle I leave the interior of this triangle undominated. This rules out I.[2]

[1] This is also directly observable in Fig. 54.

[2] This provides the example referred to in 30.3.6.: The three vertices of triangle I do not dominate each other, i.e. they form a satisfactory set in the sense of loc. cit. They are nevertheless unsuitable as a subset of a solution.

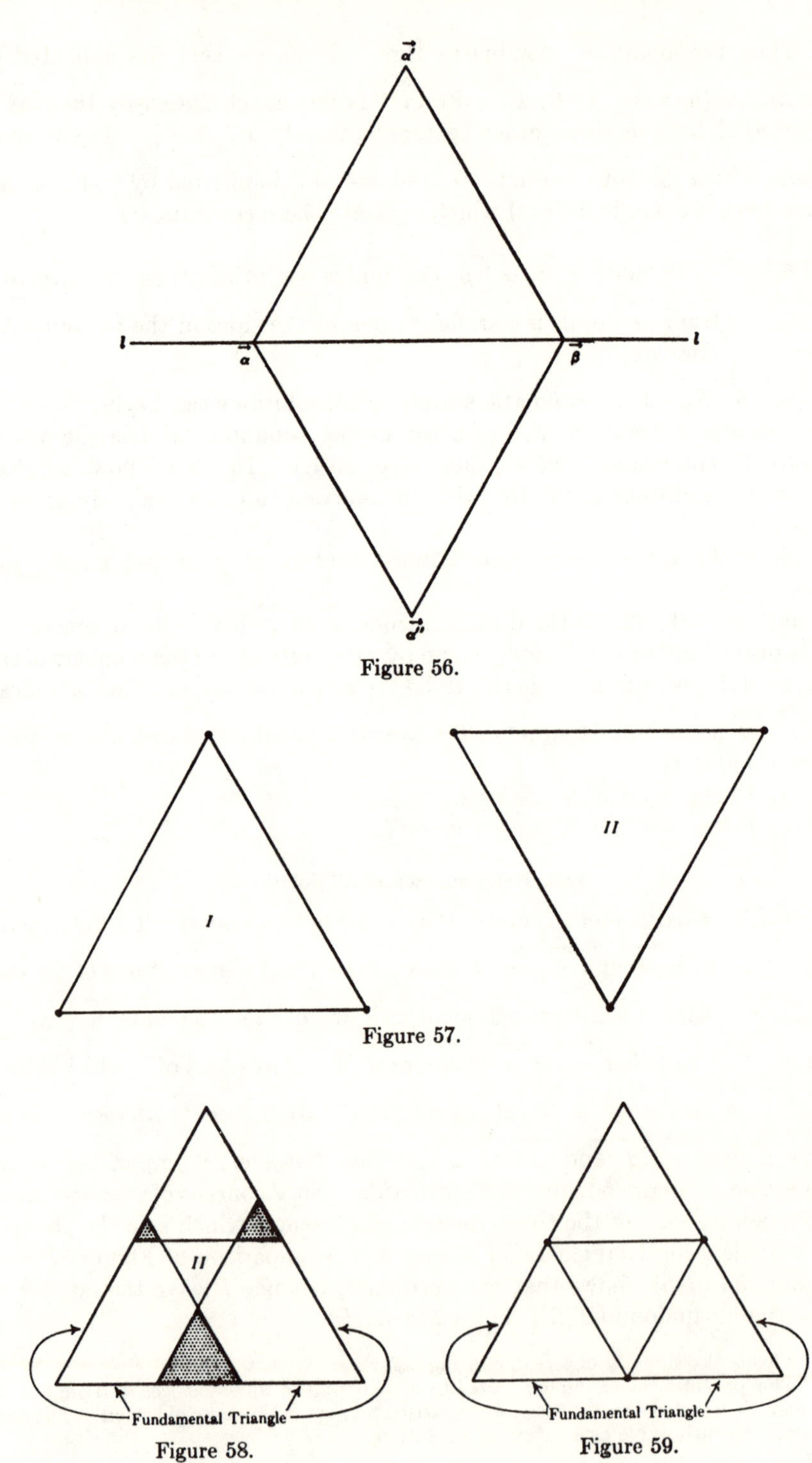

Figure 56.

Figure 57.

Figure 58.

Figure 59.

The same comparison shows that the vertices of triangle *II* leave undominated the dotted areas indicated in Figure 58. Hence triangle *II* must be placed in such a manner in the fundamental triangle that these dotted areas fall entirely outside the fundamental triangle. This means that the three vertices of *II* must lie on the three sides of the fundamental triangle, as indicated in Figure 59. Thus these three vertices are the middle points of the three sides of the fundamental triangle.

Comparison of Figure 59 with Figure 54 or Figure 55 shows that this set V is indeed a solution. One verifies immediately that these three middle points are the points (vectors)

$$(32:6) \qquad \{-1, \tfrac{1}{2}, \tfrac{1}{2}\}, \{\tfrac{1}{2}, -1, \tfrac{1}{2}\}, \{\tfrac{1}{2}, \tfrac{1}{2}, -1\},$$

i.e. that this solution V is the set of Figure 51.

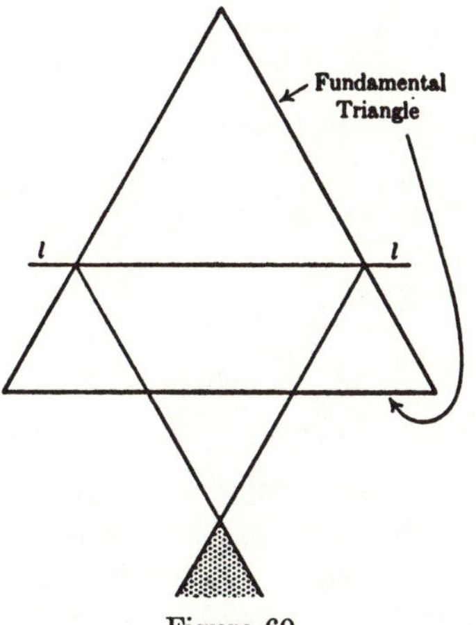

Figure 60.

32.2.2. Let us now consider (a) in 32.1.4. In this case all of V lies on the horizontal line l. By (32:5) no two points of l dominate each other, so that every point of l is undominated by V. Hence every point of l (in the fundamental triangle) must belong to V. I.e., V is precisely that part of l which is in the fundamental triangle. So the elements $\overrightarrow{\alpha} = \{\alpha_1, \alpha_2, \alpha_3\}$ of V are characterized by an equation

$$(32:7) \qquad \alpha_1 = c.$$

Diagrammatically: Figure 60.

Comparison of Figure 60 with Figures 54 or 55 shows that the line l leaves the dotted area indicated on Figure 60 undominated. Hence the line l must be placed in such a manner in the fundamental triangle that the dotted area falls entirely outside the fundamental triangle. This means that l must lie below the middle points of those two sides of the fundamental triangle which it intersects.[1] In the terminology of (32:7): $c < \frac{1}{2}$. On the other hand, $c \geqq -1$ is necessary to make l intersect the fundamental triangle at all. So we have:

(32:8) $$-1 \leqq c < \tfrac{1}{2}.$$

Comparison of Figure 60 with Figures 54 or 55 shows that under these conditions[2] the set V—i.e. l—is indeed a solution.

But the form (32:7) of this solution was brought about by a suitable permutation of the numbers 1,2,3. Hence we have two further solutions, characterized by

(32:7*) $$\alpha_2 = c,$$

and characterized by

(32:7**) $$\alpha_3 = c,$$

always with (32:8)

32.2.3. Summing up:

This is a complete list of solutions:

(32:A) For every c which fulfills (32:8): The three sets (32:7) (32:7*), (32:7**).

(32:B) The set (32:6).

33. Conclusions

33.1. The Multiplicity of Solutions. Discrimination and Its Meaning

33.1.1. The result of 32. calls for careful consideration and comment. We have determined all solutions of the essential zero-sum three-person game. In 29.1., before the rigorous definitions of 30.1. were formulated, we had already decided which solution we wanted; and this solution reappears now as (32:B). But we found other solutions besides: the (32:A), which are infinitely many sets, each one of them an infinite set of imputations itself. What do these supernumerary solutions stand for?

Consider, e.g., the form (32:7) of (32:A). This gives a solution for every c of (32:8) consisting of all imputations $\vec{\alpha} = \{\alpha_1, \alpha_2, \alpha_3\}$ which fulfill (32:7), i.e. $\alpha_1 = c$. Besides this, they must fulfill only the requirements,

[1] The limiting position of l, going through the middle points themselves, must be excluded. The reason is that in this position the vertex of the dotted area would lie on the fundamental triangle,—and this is inadmissable since that point too is undominated by V, i.e. by l.

Observe that the corresponding prohibition did not occur in case (b), i.e. for the dotted areas of Figure 58. Their vertices too were undominated by V, but they belong to V. In our present position of V, on the other hand, the vertex under consideration does not belong to V, i.e. to l.

This exclusion of the limiting position causes the $<$—and not the $\leqq$—in the inequality which follows.

[2] (32:8), i.e. l intersects the fundamental triangle, but below its middle.

(30:1), (30:2) of 30.1.1.,—i.e. (32:2), (32:3) of 32.1.1. In other words: Our solution consists of all

$$(33{:}1) \qquad \overrightarrow{\alpha} = \{c, a, -c-a\}, \qquad -1 \leqq a \leqq 1 - c.$$

The interpretation of this solution consists manifestly of this: One of the players (in this case 1) is being discriminated against by the two others (in this case 2,3). They *assign* to him the amount which he gets, c. This amount is the same for all imputations of the solution, i.e. of the accepted standard of behavior. The place in society of player 1 is prescribed by the two other players; he is excluded from all negotiations that may lead to coalitions. Such negotiations do go on, however, between the two other players: the distribution of their share, $-c$, depends entirely upon their bargaining abilities. The solution, i.e. the accepted standard of behavior, imposes absolutely no restrictions upon the way in which this share is divided between them,—expressed by a, $-c-a$.[1] This is not surprising. Since the excluded player is absolutely "tabu," the threat of the partner's desertion is removed from each participant of the coalition. There is no way of determining any definite division of the spoils.[2,3]

Incidentally: It is quite instructive to see how our concept of a solution as a set of imputations is able to take care of this situation also.

33.1.2. There is more that should be said about this "discrimination" against a player.

First, it is not done in an entirely arbitrary manner. The c, in which discrimination finds its quantitative expression, is restricted by (32:8) in 32.2.2. Now part of (32:8), $c \geqq -1$, is clear enough in its meaning, but the significance of the other part $c < \frac{1}{2}$[4] is considerably more recondite (cf. however, below). It all comes back to this: Even an arbitrary system of discriminations can be compatible with a stable standard of behavior—i.e. order of society—but it may have to fulfill certain quantitative conditions, in order that it may not impair that stability.

Second, the discrimination need not be clearly disadvantageous to the player who is affected. It cannot be clearly advantageous,—i.e. his fixed value c cannot be equal to or better than the best the others may expect. This would mean, by (33:1), that $c \geqq 1 - c$, i.e. $c \geqq \frac{1}{2}$,—which is exactly what (32:8) forbids. But it would be clearly disadvantageous only for $c = -1$; and this is a possible value for c (by (32:8)), but not the only one. $c = -1$ means that the player is not only excluded, but also exploited to

[1] Except that both must be $\geqq -1$—i.e. what the player can get for himself, without any outside help.

a, $-c-a \geqq -1$ is, of course, the $-1 \leqq a \leqq 1 - c$ of (33:1).

[2] Cf. the discussions at the end of 25.2. Observe that the arguments which we adduced there to motivate the primate of $v(S)$ have ceased to operate in this particular case—and $v(S)$ nevertheless determines the solutions!

[3] Observe that due to (32:8) in 32.2.2., the "spoils", i.e. the amount $-c$, can be both positive and negative.

[4] And that $=$ is excluded in $c < \frac{1}{2}$, but not in $c \geqq -1$.

100 per cent. The remaining c (of (32:8)) with $-1 < c < \frac{1}{2}$ correspond to gradually less and less disadvantageous forms of segregation.

33.1.3. It seems remarkable that our concept of a solution is able to express all these nuances of non-discriminatory (32:B), and discriminatory (32:A), standards of behavior—the latter both in their 100 per cent injurious form, $c = -1$, and in a continuous family of less and less injurious ones $-1 < c < \frac{1}{2}$. It is particularly significant that we did not look for any such thing—the heuristic discussions of 29.1 were certainly not in this spirit—but we were nevertheless forced to these conclusions by the rigorous theory itself. And these situations arose even in the extremely simple framework of the zero-sum three-person game!

For $n \geqq 4$ we must expect a much greater wealth of possibilities for all sorts of schemes of discrimination, prejudices, privileges, etc. Besides these, we must always look out for the analogues of the solution (32:B), i.e. the nondiscriminating "objective" solutions. But we shall see that the conditions are far from simple. And we shall also see that it is precisely the investigation of the discriminatory "inobjective" solutions which leads to a proper understanding of the general non-zero-sum games—and thence to application to economics.

33.2. Statics and Dynamics

33.2. At this point it may be advantageous to recall the discussions of 4.8.2. concerning statics and dynamics. What we said then applies now; indeed it was really meant for the phase which our theory has now reached.

In 29.2. and in the places referred to there, we considered the negotiations, expectations and fears which precede the formation of a coalition and which determine its conditions. These were all of the quasi-dynamic type described in 4.8.2. The same applies to our discussion in 4.6. and again in 30.2., of how various imputations may or may not dominate each other depending on their relationship to a solution; i.e., how the conducts approved by an established standard of behavior do not conflict with each other, but can be used to discredit the non-approved varieties.

The excuse, and the necessity, for using such considerations in a static theory were set forth on that occasion. Thus it is not necessary to repeat them now.

CHAPTER VII

ZERO-SUM FOUR-PERSON GAMES

34. Preliminary Survey

34.1. General Viewpoints

34.1. We are now in possession of a general theory of the zero-sum n-person game, but the state of our information is still far from satisfactory. Save for the formal statement of the definitions we have penetrated but little below the surface. The applications which we have made—i.e. the special cases in which we have succeeded in determining our solutions—can be rated only as providing a preliminary orientation. As pointed out in 30.4.2., these applications cover all cases $n \leqq 3$, but we know from our past discussions how little this is in comparison with the general problem. Thus we must turn to games with $n \geqq 4$ and it is only here that the full complexity of the interplay of coalitions can be expected to emerge. A deeper insight into the nature of our problems will be achieved only when we have mastered the mechanisms which govern these phenomena.

The present chapter is devoted to zero-sum four-person games. Our information about these still presents many lacunae. This compels an inexhaustive and chiefly casuistic treatment, with its obvious shortcomings.[1] But even this imperfect exposition will disclose various essential qualitative properties of the general theory which could not be encountered previously, (for $n \leqq 3$). Indeed, it will be noted that the interpretation of the mathematical results of this phase leads quite naturally to specific "social" concepts and formulations.

34.2. Formalism of the Essential Zero-sum Four-person Game

34.2.1. In order to acquire an idea of the nature of the zero-sum four-person games we begin with a purely descriptive classification.

Let therefore an arbitrary zero-sum four-person game Γ be given, which we may as well consider in its reduced form: and also let us choose $\gamma = 1$.[2] These assertions correspond, as we know from (27:7*) and (27:7**) in 27.2., to the following statements concerning the characteristic functions:

$$(34:1) \qquad \mathrm{v}(S) = \begin{cases} 0 \\ -1 \\ 1 \\ 0 \end{cases} \quad \text{when } S \text{ has } \begin{cases} 0 \\ 1 \\ 3 \\ 4 \end{cases} \text{ elements.}$$

[1] E.g., a considerable emphasis on heuristic devices.

[2] Cf. 27.1.4. and 27.3.2. The reader will note the analogy between this discussion and that of 29.1.2. concerning the zero-sum three-person game. About this more will be said later.

Thus only the v(S) of the two-element sets S remain undetermined by these normalizations. We therefore direct our attention to these sets.

The set $I = (1,2,3,4)$ of all players possesses six two-element subsets S:

$$(1,2),\ (1,3),\ (1,4),\ (2,3),\ (2,4),\ (3,4).$$

Now the v(S) of these sets cannot be treated as independent variables, because each one of these S has another one of the same sequence as its complement. Specifically: the first and the last, the second and the fifth, the third and the fourth, are complements of each other respectively. Hence their v(S) are the negatives of each other. It is also to be remembered that by the inequality (27:7) in 27.2. (with $n = 4$, $p = 2$) all these v(S) are $\leqq 2$, $\geqq -2$. Hence if we put

$$(34:2) \qquad \begin{cases} \mathrm{v}((1,4)) = 2x_1, \\ \mathrm{v}((2,4)) = 2x_2, \\ \mathrm{v}((3,4)) = 2x_3, \end{cases}$$

then we have

$$(34:3) \qquad \begin{cases} \mathrm{v}((2,3)) = -2x_1, \\ \mathrm{v}((1,3)) = -2x_2, \\ \mathrm{v}((1,2)) = -2x_3, \end{cases}$$

and in addition

$$(34:4) \qquad -1 \leqq x_1,\ x_2,\ x_3 \leqq 1.$$

Conversely: If any three numbers x_1, x_2, x_3 fulfilling (34:4) are given, then we can define a function v(S) (for all subsets S of $I = (1,2,3,4)$) by (34:1)-(34:3), but we must show that this v(S) is the characteristic function of a game. By 26.1. this means that our present v(S) fulfills the conditions (25:3:a)-(25:3:c) of 25.3.1. Of these, (25:3:a) and (25:3:b) are obviously fulfilled, so only (25:3:c) remains. By 25.4.2. this means showing that

$$\mathrm{v}(S_1) + \mathrm{v}(S_2) + \mathrm{v}(S_3) \leqq 0 \qquad \text{if } S_1,\ S_2,\ S_3 \text{ are a decomposition of } I.$$

(Cf. also (25:6) in 25.4.1.) If any of the sets S_1, S_2, S_3 is empty, the two others are complements and so we even have equality by (25:3:a), (25:3:b) in 25.3.1. So we may assume that none of the sets S_1, S_2, S_3 is empty. Since four elements are available altogether, one of these sets, say $S_1 = S$, must have two elements, while the two others are one-element sets. Thus our inequality becomes

$$\mathrm{v}(S) - 2 \leqq 0, \qquad \text{i.e.} \qquad \mathrm{v}(S) \leqq 2.$$

If we express this for all two-element sets S, then (34:2), (34:3) transform the inequality into

$$\begin{array}{ccc} 2x_1 \leqq 2, & 2x_2 \leqq 2, & 2x_3 \leqq 2, \\ -2x_1 \leqq 2, & -2x_2 \leqq 2, & -2x_3 \leqq 2, \end{array}$$

which is equivalent to the assumed (34:4). Thus we have demonstrated:

(34:A) The essential zero-sum four-person games (in their reduced form with the choice $\gamma = 1$) correspond exactly to the triplets of numbers x_1, x_2, x_3 fulfilling the inequalities (34:4). The correspondence between such a game, i.e. its characteristic function, and its x_1, x_2, x_3 is given by the equations (34:1)-(34:3).[1]

34.2.2. The above representation of the essential zero-sum four-person games by triplets of numbers x_1, x_2, x_3 can be illustrated by a simple geometrical picture. We can view the numbers x_1, x_2, x_3 as the Cartesian coordinates of a point.[2] In this case the inequalities (34:4) describe a part

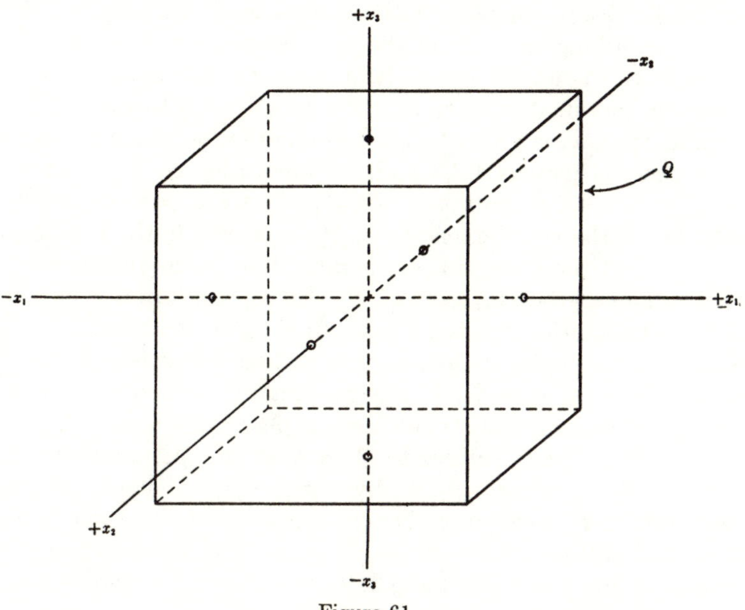

Figure 61.

of space which exactly fills a cube Q. This cube is centered at the origin of the coordinates, and its edges are of length 2 because its six faces are the six planes

$$x_1 = \pm 1, \qquad x_2 = \pm 1, \qquad x_3 = \pm 1,$$

as shown in Figure 61.

Thus each essential zero-sum four-person game Γ is represented by precisely one point in the interior or on the surface of this cube, and *vice versa.* It is quite useful to view these games in this manner and to try to correlate their peculiarities with the geometrical conditions in Q. It will be particularly instructive to identify those games which correspond to definite significant points of Q.

[1] The reader may now compare our result with that of 29.1.2. concerning the zero-sum three-person games. It will be noted how the variety of possibilities has increased.

[2] We may also consider these numbers as the components of a vector in L_3 in the sense of 16.1.2. et seq. This aspect will sometimes be the more convenient, as in footnote 1 on p. 304.

But even before carrying out this program, we propose to consider certain questions of symmetry. We want to uncover the connections between the permutations of the players 1,2,3,4, and the geometrical transformations (motions) of Q. Indeed: by 28.1. the former correspond to the symmetries of the game Γ, while the latter obviously express the symmetries of the geometrical object.

34.3. Permutations of the Players

34.3.1. In evolving the geometrical representation of the essential zero-sum four-person game we had to perform an arbitrary operation, i.e. one which destroyed part of the symmetry of the original situation. Indeed, in describing the $\text{v}(S)$ of the two-element sets S, we had to single out three from among these sets (which are six in number), in order to introduce the coordinates x_1, x_2, x_3. We actually did this in (34:2), (34:3) by assigning the player 4 a particular role and then setting up a correspondence between the players 1,2,3 and the quantities x_1, x_2, x_3 respectively (cf. (34:2)). Thus a permutation of the players 1,2,3 will induce the same permutation of the coordinates x_1, x_2, x_3—and so far the arrangement is symmetric. But these are only six permutations from among the total of 24 permutations of the players 1,2,3,4.[1] So a permutation which replaces the player 4 by another one is not accounted for in this way.

34.3.2. Let us consider such a permutation. For reasons which will appear immediately, consider the permutation A, which interchanges the players 1 and 4 with each other and also the players 2 and 3.[2] A look at the equations (34:2), (34:3) suffices to show that this permutation leaves x_1 invariant, while it replaces x_2, x_3 by $-x_2$, $-x_3$. Similarly one verifies: The permutation B, which interchanges 2 and 4, and also 1 and 3, leaves x_2 invariant and replaces x_1, x_3 by $-x_1$, $-x_3$. The permutation C, which interchanges 3 and 4 and also 1 and 2, leaves x_3 invariant and replaces x_1, x_2 by $-x_1$, $-x_2$.

Thus each one of the three permutations A, B, C affects the variables x_1, x_2, x_3 only as far as their signs are concerned, each changing two signs and leaving the third invariant.

Since they also carry 4 into 1,2,3, respectively, they produce all permutations of the players 1,2,3,4, if combined with the six permutations of the players 1,2,3. Now we have seen that the latter correspond to the six permutations of x_1, x_2, x_3 (without changes in sign). Consequently the 24 permutations of 1,2,3,4 correspond to the six permutations of x_1, x_2, x_3, each one in conjunction with no change of sign or with two changes of sign.[3]

[1] Cf. 28.1.1., following the definitions (28:A:a), (28:A:b).

[2] With the notations of 29.1.:

$$A = \begin{pmatrix} 1,2,3,4 \\ 4,3,2,1 \end{pmatrix}, \quad B = \begin{pmatrix} 1,2,3,4 \\ 3,4,1,2 \end{pmatrix}, \quad C = \begin{pmatrix} 1,2,3,4 \\ 2,1,4,3 \end{pmatrix}.$$

[3] These sign changes are $1 + 3 = 4$ possibilities in each case, so we have $6 \times 4 = 24$ operations on x_1, x_2, x_3 to represent the 24 permutations of 1,2,3,4,—as it should be.

34.3.3. We may also state this as follows: If we consider all movements in space which carry the cube Q into itself, it is easily verified that they consist of the permutations of the coordinate axes x_1, x_2, x_3 in combination with any reflections on the coordinate planes (i.e. the planes x_2, x_3; x_1, x_3; x_1, x_2). Mathematically these are the permutations of x_1, x_2, x_3 in combination with any changes of sign among the x_1, x_2, x_3. These are 48 possibilities.[1] Only half of these, the 24 for which the number of sign changes is even (i.e. 0 or 2), correspond to the permutations of the players.

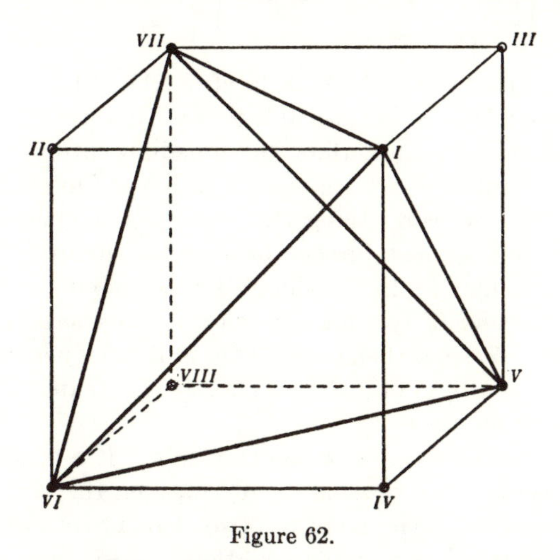

Figure 62.

It is easily verified that these are precisely the 24 which not only carry the cube Q into itself, but also the tetrahedron I, V, VI, VII, as indicated in Figure 62. One may also characterize such a movement by observing that it always carries a vertex • of Q into a vertex •; and equally a vertex ○ into a vertex ○, but never a • into a ○.[2]

We shall now obtain a much more immediate interpretation of these statements by describing directly the games which correspond to specific points of the cube Q: to the vertices • or ○, to the center (the origin in Figure 61), and to the main diagonals of Q.

35. Discussion of Some Special Points in the Cube Q

35.1. The Corner I (and V, VI, VII)

35.1.1. We begin by determining the games which correspond to the four corners •: I, V, VI, VII. We have seen that they arise from each other by suitable permutations of the players 1,2,3,4. Therefore it suffices to consider one of them, say I.

[1] For each variable x_1, x_2, x_3 there are two possibilities: change or no change. This gives altogether $2^3 = 8$ possibilities. Combination with the six permutations of x_1, x_2, x_3 yields $8 \times 6 = 48$ operations.

[2] This group of motions is well known in group theory and particularly in crystallography, but we shall not elaborate the point further.

The point I corresponds to the values 1,1,1 of the coordinates x_1, x_2, x_3. Thus the characteristic function $v(S)$ of this game is:

$$(35:1)\quad v(S) = \begin{cases} 0 \\ -1 \\ 2 \\ \\ -2 \\ 1 \\ 0 \end{cases} \text{when } S \text{ has } \begin{cases} 0 \\ 1 \\ 2 \quad \text{(and 4 belongs to } S) \\ \quad \text{elements} \\ 2 \quad \text{(and 4 does not belong to } S) \\ 3 \\ 4 \end{cases}$$

(Verification is immediate with the help of (34:1), (34:2), (34:3) in 34.2.1.) Instead of applying the mathematical theory of Chapter VI to this game, let us first see whether it does not allow an immediate intuitive interpretation.

Observe first that a player who is left to himself loses the amount -1. This is manifestly the worst thing that can ever happen to him since he can protect himself against further losses without anybody else's help.[1] Thus we may consider a player who gets this amount -1 as completely defeated. A coalition of two players may be considered as defeated if it gets the amount -2, since then each player in it must necessarily get -1.[2,3] In this game the coalition of any two players is defeated in this sense if it does not comprise player 4.

Let us now pass to the complementary sets. If a coalition is defeated in the above sense, it is reasonable to consider the complementary set as a winning coalition. Therefore the two-element sets which contain the player 4 must be rated as winning coalitions. Also since any player who remains isolated must be rated as defeated, three-person coalitions always win. This is immaterial for those three-element coalitions which contain the player 4, since in these coalitions two members are winning already if the player 4 is among them. But it is essential that 1,2,3 be a winning coalition, since all its proper subsets are defeated.[4]

[1] This view of the matter is corroborated by our results concerning the three-person game in 23. and 32.2., and more fundamentally by our definition of the imputation in 30.1.1., particularly condition (30:1).

[2] Since neither he nor his partner need accept less than -1, and they have together -2, this is the only way in which they can split.

[3] In the terminology of 31.1.4.: this coalition is flat. There is of course no gain, and therefore no possible motive for two players to form such a coalition. But if it happens that the two other players have combined and show no desire to acquire a third ally, we may treat the remaining two as a coalition even in this case.

[4] We warn the reader that, although we have used the words "defeated" and "winning" almost as *termini technici*, this is not our intention. These concepts are, indeed, very well suited for an exact treatment. The "defeated" and "winning" coalitions actually coincide with the sets S considered in (31:F) and in (31:G) in 31.1.5.; those for which S is flat or $-S$ is flat, respectively. But we shall consider this question in such a way only in Chap. X.

For the moment our considerations are absolutely heuristic and ought to be taken in the same spirit as the heuristic discussions of the zero-sum three-person game in 21., 22. The only difference is that we shall be considerably briefer now, since our experience and routine have grown substantially in the discussion.

As we now possess an exact theory of solutions for games already, we are under

35.1.2. So it is plausible to view this as a struggle for participation in any one of the various possible coalitions:

(35:2) $$(1,4),\ (2,4),\ (3,4),\ (1,2,3),$$

where the amounts obtainable for these coalitions are:

(35:3) $$\mathrm{v}((1,4)) = \mathrm{v}((2,4)) = \mathrm{v}((3,4)) = 2, \qquad \mathrm{v}((1,2,3)) = 1.$$

Observe that this is very similar to the situation which we found in the essential zero-sum three-person game, where the winning coalitions were:

(35:2*) $$(1,2),\ (1,3),\ (2,3),$$

and the amounts obtainable for these coalitions:

(35:3*) $$\mathrm{v}((1,2)) = \mathrm{v}((1,3)) = \mathrm{v}((2,3)) = 1.$$

In the three-person game we determined the distribution of the proceeds (35:3*) among the winners by assuming: A player in a winning coalition should get the same amount no matter which is the winning coalition. Denoting these amounts for the players 1,2,3 by α, β, γ respectively, (35:3*) gives

(35:4*) $$\alpha + \beta = \alpha + \gamma = \beta + \gamma = 1$$

from which follows

(35:5*) $$\alpha = \beta = \gamma = \tfrac{1}{2}.$$

These were indeed the values which those considerations yielded.

Let us assume the same principle in our present four-person game. Denote by α, β, γ, δ, respectively, the amount that each player 1,2,3,4 gets if he succeeds in participating in a winning coalition. Then (35:3) gives

(35:4) $$\alpha + \delta = \beta + \delta = \gamma + \delta = 2, \qquad \alpha + \beta + \gamma = 1,$$

from which follows

(35:5) $$\alpha = \beta = \gamma = \tfrac{1}{3}, \qquad \delta = \tfrac{5}{3}.$$

All the heuristic arguments used in 21., 22., for the three-person game could be repeated.[1]

35.1.3. Summing up:

(35:A) This is a game in which the player 4 is in a specially favored position to win: any one ally suffices for him to form a winning coalition. Without his cooperation, on the other hand, three players must combine. This advantage also expresses itself in

obligation to follow up this preliminary heuristic analysis by an exact analysis which is based rigorously on the mathematical theory. We shall come to this. (Cf. loc. cit. above, and also the beginning of 36.2.3.)

[1] Of course, without making this thereby a rigorous discussion on the basis of 30.1.

the amounts which each player 1,2,3,4 should get when he is among the winners—if our above heuristic deduction can be trusted. These amounts are $\frac{1}{3}$, $\frac{1}{3}$, $\frac{1}{3}$, $\frac{5}{3}$ respectively. It is to be noted that the advantage of player 4 refers to the case of victory only; when defeated, all players are in the same position (i.e. get -1).

The last mentioned circumstance is, of course, due to our normalization by reduction. Independently of any normalization, however, this game exhibits the following trait: One player's quantitative advantage over another may, when both win, differ from what it is when both lose.

This cannot happen in a three-person game, as is apparent from the formulation which concludes 22.3.4. Thus we get a first indication of an important new factor that emerges when the number of participants reaches four.

35.1.4. One last remark seems appropriate. In this game player 4's strategic advantage consisted in the fact that he needed only one ally for victory, whereas without him a total of three partners was necessary. One might try to pass to an even more extreme form by constructing a game in which every coalition that does not contain player 4 is defeated. It is essential to visualize that this is not so, or rather that such an advantage is no longer of a strategic nature. Indeed in such a game

$$\mathrm{v}(S) = \begin{cases} 0 \\ -1 \\ -2 \\ -3 \end{cases} \quad \text{if } S \text{ has } \begin{cases} 0 \\ 1 \\ 2 \\ 3 \end{cases} \text{elements and 4 does not belong to } S,$$

hence

$$\mathrm{v}(S) = \begin{cases} 3 \\ 2 \\ 1 \\ 0 \end{cases} \quad \text{if } S \text{ has } \begin{cases} 1 \\ 2 \\ 3 \\ 4 \end{cases} \text{elements and 4 belongs to } S.$$

This is not reduced, as

$$\mathrm{v}((1)) = \mathrm{v}((2)) = \mathrm{v}((3)) = -1, \qquad \mathrm{v}((4)) = 3.$$

If we apply the reduction process of 27.1.4. to this $\mathrm{v}(S)$ we find that its reduced form is

$$\bar{\mathrm{v}}(S) \equiv 0.$$

i.e. the game is inessential. (This could have been shown directly by (27:B) in 27.4.) Thus this game has a uniquely determined value for each player 1,2,3,4: -1, -1, -1, 3, respectively.

In other words: Player 4's advantage in this game is one of a fixed payment (i.e. of cash), and not one of strategic possibilities. The former is, of course, more definite and tangible than the latter, but of less theoretical interest since it can be removed by our process of reduction.

35.1.5. We observed at the beginning of this section that the corners V, VI, VII differ from I only by permutations of the players. It is easily verified that the special role of player 4 in I is enjoyed by the players 1,2,3, in V, VI, VII, respectively.

35.2. The Corner *VIII* (and *II*, *III*, *IV*). The Three-person Game and a "Dummy"

35.2.1. We next consider the games which correspond to the four corners $\circ$: $II, III, IV, VIII$. As they arise from each other by suitable permutations of the players 1,2,3,4, it suffices to consider one of them, say VIII.

The point VIII corresponds to the values -1, -1, -1 of the coordinates x_1, x_2, x_3. Thus the characteristic function $v(S)$ of this game is:

$$(35:6)\quad v(S) = \begin{cases} 0 \\ -1 \\ -2 \\ \\ 2 \\ 1 \\ 0 \end{cases} \quad \text{when } S \text{ has} \quad \begin{cases} 0 \\ 1 \\ 2 \qquad \text{(and 4 belongs to } S) \\ \quad \text{elements} \\ 2 \quad \text{(and 4 does not belong to } S) \\ 3 \\ 4 \end{cases}$$

(Verification is immediate with the help of (34:1), (34:2), (34:3) in 34.2.1.) Again, instead of applying to this game the mathematical theory of Chapter VI, let us first see whether it does not allow an immediate intuitive interpretation.

The important feature of this game is that the inequality (25:3:c) in 25.3. becomes an equality, i.e.:

$$(35:7)\qquad v(S \cup T) = v(S) + v(T) \qquad \text{if} \qquad S \cap T = \ominus,$$

when $T = (4)$. That is: If S represents a coalition which does not contain the player 4, then the addition of 4 to this coalition is of no advantage; i.e. it does not affect the strategic situation of this coalition nor of its opponents in any way. This is clearly the meaning of the additivity expressed by (35:7).[1]

35.2.2. This circumstance suggests the following conclusion,—which is of course purely heuristic.[2] Since the accession of player 4 to any

[1] Note that the indifference in acquiring the cooperation of 4 is expressed by (35:7), and not by

$$v(S \cup T) = v(S).$$

That is, a player is "indifferent" as a partner, not if his accession does not alter the value of a coalition but if he brings into the coalition exactly the amount which—and no more than—he is worth outside.

This remark may seem trivial; but there exists a certain danger of misunderstanding, particularly in non-reduced games where $v((4)) > 0$,—i.e. where the accession of 4 (although strategically indifferent!) actually increases the value of a coalition.

Observe also that the indifference of S and $T = (4)$ to each other is a strictly reciprocal relationship.

[2] We shall later undertake exact discussion on the basis of 30.1. At that time it will be found also that all these games are special cases of more general classes of some importance. (Cf. Chap. IX, particularly 41.2.)

coalition appears to be a matter of complete indifference to both sides, it seems plausible to assume that player 4 has no part in the transactions that constitute the strategy of the game. He is isolated from the others and the amount which he can get for himself— $v(S) = -1$ —is the actual value of the game for him. The other players 1,2,3, on the other hand, play the game strictly among themselves; hence they are playing a three-person game. The values of the original characteristic function $v(S)$ which describes the original three-person game are:

$$\left.\begin{aligned} v(\ominus) &= 0, \\ v((1)) &= v((2)) = v((3)) = -1, \\ v((1,2)) &= v((1,3)) = v((2,3)) = 2, \\ v((1,2,3)) &= 1, \end{aligned}\right\} \begin{array}{l} I' = (1,2,3) \text{ is now the set} \\ \text{of all players.} \end{array} \tag{35:6*}$$

(Verify this from (35:6).)

At first sight this three-person game represents the oddity that $v(I')$ (I' is now the set of all players!) is not zero. This, however, is perfectly reasonable: by eliminating player 4 we transform the game into one which is not of zero sum; since we assessed player 4 a value -1, the others retain together a value 1. We do not yet propose to deal with this situation systematically. (Cf. footnote 2 on p. 299.) It is obvious, however, that this condition can be remedied by a slight generalization of the transformation used in 27.1. We modify the game of 1,2,3 by assuming that each one got the amount $\frac{1}{3}$ in cash in advance, and then compensating for this by deducting equivalent amounts from the $v(S)$ values in (35:6*). Just as in 27.1., this cannot affect the strategy of the game, i.e. it produces a strategically equivalent game.[1]

After consideration of the compensations mentioned above[2] we obtain the new characteristic function:

$$\begin{aligned} v'(\ominus) &= 0, \\ v'((1)) &= v'((2)) = v'((3)) = -\tfrac{4}{3}, \\ v'((1,2)) &= v'((1,3)) = v'((2,3)) = \tfrac{4}{3}, \\ v'((1,2,3)) &= 0. \end{aligned} \tag{35:6**}$$

This is the reduced form of the essential zero-sum three-person game discussed in 32.—except for a difference in unit: We have now $\gamma = \frac{4}{3}$ instead

[1] In the terminology of 27.1.1.: $\alpha_1^0 = \alpha_2^0 = \alpha_3^0 = -\frac{1}{3}$. The condition there which we have infringed is (27:1): $\sum_i \alpha_i^0 = 0$. This is necessary since we started with a non-zero-sum game.

Even $\sum_i \alpha_i^0 = 0$ could be safeguarded if we included player 4 in our considerations, putting $\alpha_4^0 = 1$. This would leave him just as isolated as before, but the necessary compensation would make $v((4)) = 0$, with results which are obvious.

One can sum this up by saying that in the present situation it is not the reduced form of the game which provides the best basis of discussion among all strategically equivalent forms.

[2] I.e. deduction of as many times $\frac{1}{3}$ from $v(S)$ as S has elements.

of the $\gamma = 1$ of (32:1) in 32.1.1. Thus we can apply the heuristic results of 23.1.3., or the exact results of 32.[1] Let us restrict ourselves, at any rate, to the solution which appears in both cases and which is the simplest one: (32:B) of 32.2.3. This is the set of imputations (32:6) in 32.2.1., which we must multiply by the present value of $\gamma = \frac{4}{3}$; i.e.:

$$\{-\tfrac{4}{3}, \tfrac{2}{3}, \tfrac{2}{3}\}, \{\tfrac{2}{3}, -\tfrac{4}{3}, \tfrac{2}{3}\}, \{\tfrac{2}{3}, \tfrac{2}{3}, -\tfrac{4}{3}\}.$$

(The players are, of course, 1,2,3.) In other words: The aim of the strategy of the players 1,2,3 is to form any coalition of two; a player who succeeds in this, i.e. who is victorious, gets $\frac{2}{3}$, and a player who is defeated gets $-\frac{4}{3}$. Now each of the players 1,2,3 of our original game gets the extra amount $\frac{1}{3}$ beyond this,—hence the above amounts $\frac{2}{3}$, $-\frac{4}{3}$ must be replaced by $1, -1$.

35.2.3. Summing up:

(35:B) This is a game in which the player 4 is excluded from all coalitions. The strategic aim of the other players 1,2,3 is to form any coalition of two. Player 4 gets -1 at any rate. Any other player 1,2,3 gets the amount 1 when he is among the winners, and the amount -1 when he is defeated. All this is based on heuristic considerations.

One might say more concisely that this four-person game is only an "inflated" three-person game: the essential three-person game of the players 1,2,3, inflated by the addition of a "dummy" player 4. We shall see later that this concept is of a more general significance. (Cf. footnote 2 on p. 299.)

35.2.4. One might compare the dummy role of player 4 in this game with the exclusion a player undergoes in the discriminatory solution (32:A) in 32.2.3., as discussed in 33.1.2. There is, however, an important difference between these two phenomena. In our present set-up, player 4 has really no contribution to make to any coalition at all; he stands apart by virtue of the characteristic function $v(S)$. Our heuristic considerations indicate that he should be excluded from all coalitions in all acceptable solutions. We shall see in 46.9. that the exact theory establishes just this. The excluded player of a discriminatory solution in the sense of 33.1.2. is excluded only in the particular situation under consideration. As far as the characteristic function of that game is concerned, his role is the same as that of all other players. In other words: The "dummy" in our present game is excluded by virtue of the objective facts of the situation (the characteristic function $v(S)$).[2] The excluded player in a discriminatory solution is excluded solely by the arbitrary (though stable) "prejudices" that the particular standard of behavior (solution) expresses.

[1] Of course the present discussion is heuristic in any event. As to the exact treatment, cf. footnote 2 on p. 299.

[2] This is the "physical background," in the sense of 4.6.3.

We observed at the beginning of this section that the corners *II*, *III*, *IV* differ from *VIII* only by permutations of the players. It is easily verified that the special role of player 4 in *VIII* is enjoyed by the players 1,2,3 in *II*, *III*, *IV*, respectively.

35.3. Some Remarks Concerning the Interior of Q

35.3.1. Let us now consider the game which corresponds to the center of Q, i.e. to the values 0,0,0 of the coordinates x_1, x_2, x_3. This game is clearly unaffected by any permutation of the players 1,2,3,4, i.e. it is symmetric. Observe that it is the only such game in Q, since total symmetry means invariance under all permutations of x_1, x_2, x_3 and sign changes of any two of them (cf. 34.3.); hence $x_1 = x_2 = x_3 = 0$.

The characteristic function $\mathrm{v}(S)$ of this game is:

$$(35:8) \qquad \mathrm{v}(S) = \begin{cases} 0 \\ -1 \\ 0 \\ 1 \\ 0 \end{cases} \text{ when } S \text{ has } \begin{cases} 0 \\ 1 \\ 2 \\ 3 \\ 4 \end{cases} \text{elements.}^{1}$$

(Verification is immediate with the help of (34:1), (34:2), (34:3) in 34.2.1.) The exact solutions of this game are numerous; indeed, one must say that they are of a rather bewildering variety. It has not been possible yet to order them and to systematize them by a consistent application of the exact theory, to such an extent as one would desire. Nevertheless the known specimens give some instructive insight into the ramifications of the theory. We shall consider them in somewhat more detail in 37. and 38.

At present we make only this (heuristic) remark: The idea of this (totally) symmetric game is clearly that any majority of the players (i.e. any coalition of three) wins, whereas in case of a tie (i.e. when two coalitions form, each consisting of two players) no payments are made.

35.3.2. The center of Q represented the only (totally) symmetric game in our set-up: with respect to all permutations of the players 1,2,3,4. The geometrical picture suggests consideration of another symmetry as well: with respect to all permutations of the coordinates x_1, x_2, x_3. In this way we select the points of Q with

$$(35:9) \qquad x_1 = x_2 = x_3,$$

which form a *main diagonal* of Q, the line

$$(35:10) \qquad I\text{-center-}VIII.$$

We saw at the beginning of 34.3.1. that this symmetry means precisely that the game is invariant with respect to all permutations of the players 1,2,3. In other words:

[1] This representation shows once more that the game is symmetric, and uniquely characterized by this property. Cf. the analysis of 28.2.1.

The main diagonal (35:9), (35:10) represents all those games which are symmetric with respect to the players 1,2,3, i.e. where only player 4 may have a special role.

Q has three other main diagonals (II-center-V, III-center-VI, IV-center VII), and they obviously correspond to those games where another player (players 1,2,3, respectively) alone may have a special role.

Let us return to the main diagonal (35:9), (35:10). The three games which we have previously considered (I, VIII, Center) lie on it; indeed in all these games only player 4 had a special role.[1] Observe that the entire category of games is a one-parameter variety. Owing to (35:9), such a game is characterized by the value x_1 in

$$-1 \leqq x_1 \leqq 1. \tag{35:11}$$

The three games mentioned above correspond to the extreme values $x_1 = 1$, $x_1 = -1$ and to the middle value $x_1 = 0$. In order to get more insight into the working of the exact theory, it would be desirable to determine exact solutions for all these values of x_1, and then to see how these solutions shift as x_1 varies continuously along (35:10). It would be particularly instructive to find out how the qualitatively different kinds of solutions recognized for the special values $x_1 = -1, 0, 1$ go over into each other. In 36.3.2. we shall give indications about the information that is now available in this regard.

35.3.3. Another question of interest is this: Consider a game, i.e. a point in Q, where we can form some intuitive picture of what solutions to expect, e.g. the corner VIII. Then consider a game in the immediate neighborhood of VIII, i.e. one with only slightly changed values of x_1, x_2, x_3. Now it would be desirable to find exact solutions for these neighboring games, and to see in what details they differ from the solutions of the original game,—i.e. how a small distortion of x_1, x_2, x_3 distorts the solutions.[2] Special cases of this problem will be considered in 36.1.2., and at the end of 37.1.1., as well as in 38.2.7.

35.3.4. So far we have considered games that are represented by points of Q in more or less special positions.[3] A more general, and possibly more typical problem arises when the representative point X is somewhere in the interior of Q, in "general" position,—i.e. in a position with no particular distinguishing properties.

Now one might think that a good heuristic lead for the treatment of the problem in such points is provided by the following consideration. We have some heuristic insight into the conditions at the corners I-VIII (cf. 35.1. and 35.2.). Any point X of Q is somehow "surrounded" by these corners; more precisely, it is their center of gravity, if appropriate weights

[1] In the center not even he.

[2] This procedure is familiar in mathematical physics, where it is used in attacking problems which cannot be solved in their general form for the time being: it is the analysis of *perturbations*.

[3] Corners, the center, and entire main diagonals.

are used. Hence one might suspect that the strategy of the games, represented by X, is in some way a combination of the strategies of the (more familiar) strategies of the games represented by *I-VIII*. One might even hope that this "combination" will in some sense be similar to the formation of the center of gravity which related X to *I-VIII*.[1]

We shall see in 36.3.2. and in 38.2.5.7. that this is true in limited parts of Q, but certainly not over all of Q. In fact, in certain interior areas of Q phenomena occur which are qualitatively different from anything exhibited by *I-VIII*. All this goes to show what extreme care must be exercised in dealing with notions involving strategy, or in making guesses about them. The mathematical approach is in such an early stage at present that much more experience will be needed before one can feel any self-assurance in this respect.

36. Discussion of the Main Diagonals

36.1. The Part Adjacent to the Corner *VIII.*: Heuristic Discussion

36.1.1. The systematic theory of the four-person game has not yet advanced so far as to furnish a complete list of solutions for all the games represented by all points of Q. We are not able to specify even one solution for every such game. Investigations thus far have succeeded only in determining solutions (sometimes one, sometimes more) in certain parts of Q. It is only for the eight corners *I-VIII* that a demonstrably complete list of solutions has been established. At the present the parts of Q in which solutions are known at all form a rather haphazard array of linear, plane and spatial areas. They are distributed all over Q but do not fill it out completely.

The exhaustive list of solutions which are known for the corners *I-VIII* can easily be established with help of the results of Chapters IX and X, where these games will be fitted into certain larger divisions of the general theory. At present we shall restrict ourselves to the casuistic approach

[1] Consider two points $X = \{x_1, x_2, x_3\}$ and $Y = \{y_1, y_2, y_3\}$ in Q. We may view these as vectors in L_3 and it is indeed in this sense that the formation of a center of gravity

$$tX + (1 - t)Y = \{tx_1 + (1 - t)y_1, tx_2 + (1 - t)y_2, tx_3 + (1 - t)y_3\}$$

is understood. (Cf. (16:A:c) in 16.2.1.)

Now if $X = \{x_1, x_2, x_3\}$ and $Y = \{y_1, y_2, y_3\}$ define the characteristic functions $\mathrm{v}(S)$ and $\mathrm{w}(S)$ in the sense of (34:1)-(34:3) in 34.2.1., then $tX + (1 - t)Y$ will give, by the same algorithm, a characteristic function

$$\mathrm{u}(S) \equiv t\mathrm{v}(S) + (1 - t)\mathrm{w}(S).$$

(It is easy to verify this relationship by inspection of the formulae which we quoted.) And this same $\mathrm{u}(S)$ was introduced as center of gravity of $\mathrm{v}(S)$ and $\mathrm{w}(S)$ by (27:10) in 27.6.3.

Thus the considerations of the text are in harmony with those of 27.6. That we are dealing with centers of gravity of more than two points (eight: I-VIII) instead of only two, is not essential: the former operation can be obtained by iteration of the latter.

It follows from these remarks that the difficulties which are pointed out in the text below have a direct bearing on 27.6.3., as was indicated there.

which consists in describing particular solutions in cases where such are known. It would scarcely serve the purpose of this exposition to give a precise account of the momentary state of these investigations[1] and it would take up an excessive amount of space. We shall only give some instances which, it is hoped, are reasonably illustrative.

36.1.2. We consider first conditions on the main diagonal *I*-Center-*VIII* in Q near its end at *VIII*, $x_1 = x_2 = x_3 = -1$ (cf. 35.3.3.), and we shall try

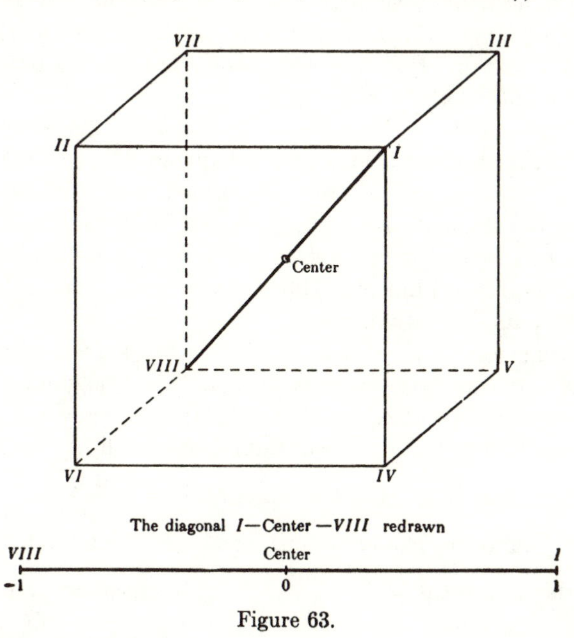

Figure 63.

to extend over the $x_1 = x_2 = x_3 > -1$ as far as possible. (Cf. Figure 63.) On this diagonal

$$(36{:}1)\quad \mathrm{v}(S) = \begin{cases} 0 & \\ -1 & \\ 2x_1 & \\ & \text{when } S \text{ has} \\ -2x_1 & \\ 1 & \\ 0 & \end{cases} \begin{cases} 0 & \\ 1 & \\ 2 & \text{(and 4 belongs to } S) \\ \text{elements} & \\ 2 & \text{(and 4 does not belong to } S) \\ 3 & \\ 4 & \end{cases}$$

(Observe that this gives (35:1) in 35.1.1. for $x_1 = 1$ and (35:6) in 35.2.1. for $x_1 = -1$.) We assume that $x_1 > -1$ but not by too much,—just how much excess is to be permitted will emerge later. Let us first consider this situation heuristically.

Since x_1 is supposed to be not very far from -1, the discussion of 35.2. may still give some guidance. A coalition of two players from among

[1] This will be done by one of us in subsequent mathematical publications.

1,2,3 may even now be the most important strategic aim, but it is no longer the only one: the formula (35:7) of 35.2.1. is not true, but instead

$$\text{(36:2)} \qquad \mathrm{v}(S \cup T) > \mathrm{v}(S) + \mathrm{v}(T) \qquad \text{if} \qquad S \cap T = \ominus$$

when $T = (4)$.[1] Indeed, it is easily verified from (36:1) that this excess is always[2] $2(1 + x_1)$. For $x_1 = -1$ this vanishes, but we have x_1 slightly > -1, so the expression is slightly > 0. Observe that for the preceding coalition of two players other than player 4, the excess in (36:2)[3] is by (36:1) always $2(1 - x_1)$. For $x_1 = -1$ this is 4, and as we have x_1 slightly > -1, it will be only slightly < 4.

Thus the first coalition (between two players, other than player 4), is of a much stronger texture than any other (where player 4 enters into the picture),—but the latter cannot be disregarded nevertheless. Since the first coalition is the stronger one, it may be suspected that it will form first and that once it is formed it will act as one player in its dealings with the two others. Then some kind of a three-person game may be expected to take place for the final crystallization.

36.1.3. Taking, e.g. (1,2) for this "first" coalition, the surmised three-person game is between the players (1,2), 3,4.[4] In this game the a, b, c of 23.1. are $a = \mathrm{v}((3,4)) = 2x_1$, $b = \mathrm{v}((1,2,4)) = 1$, $c = \mathrm{v}((1,2,3)) = 1$.[5] Hence, if we may apply the results obtained there (all of this is extremely heuristic!) the player (1,2) gets the amount $\alpha = \dfrac{-a + b + c}{2} = 1 - x_1$, if successful (in joining the last coalition), and $-a = -2x_1$ if defeated. The player 3 gets the amount $\beta = \dfrac{a - b + c}{2} = x_1$ if successful, and $-b = -1$ if defeated. The player 4 gets the amount $\gamma = \dfrac{a + b - c}{2} = x_1$ if successful, and $-c = -1$ if defeated.

Since "first" coalitions (1,3), (2,3) may form, just as well as (1,2), there are the same heuristic reasons as in the first discussion of the three-person game (in 21.-22.) to expect that the partners of these coalitions will split even. Thus, when such a coalition is successful (cf. above), its members may be expected to get $\dfrac{1 - x_1}{2}$ each, and when it is defeated x_1 each.

36.1.4. Summing up: If these surmises prove correct, the situation is as follows:

[1] Unless $S = \ominus$ or $-T$, in which case there is always $=$ in (36:2). I.e. in the present situation S must have one or two elements.

[2] By footnote 1 above, S has one or two elements and it does not contain 4.

[3] I.e., now S, T are two one-element sets, not containing player 4.

[4] One might say that (1,2) is a juridical person, while 3,4 are, in our picture, natural persons.

[5] In all the formulae which follow, remember that x_1 is near to -1,—i.e. presumably negative; hence $-x_1$ is a gain, and x_1 is a loss.

If the "first" coalition is (1,2), and if it is successful in finding an ally, and if the player who joins it in the final coalition is player 3, then the players 1,2,3,4 get the amounts $\frac{1-x_1}{2}$, $\frac{1-x_1}{2}$, x_1, -1 respectively. If the player who joins the final coalition is player 4, then these amounts are replaced by $\frac{1-x_1}{2}$, $\frac{1-x_1}{2}$, -1, x_1. If the "first" coalition (1,2) is not successful, i.e. if the players 3,4 combine against it, then the players get the amounts $-x_1$, $-x_1$, x_1, x_1 respectively.

If the "first" coalition is (1,3) or (2,3), then the corresponding permutation of the players 1,2,3 must be applied to the above.

36.2. The Part Adjacent to the Corner *VIII.*: Exact Discussion

36.2.1. It is now necessary to submit all this to an exact check. The heuristic suggestion manifestly corresponds to the following surmise:

Let V be the set of these imputations:

$$(36:3)\quad \begin{aligned} \overrightarrow{\alpha}' &= \left\{\frac{1-x_1}{2}, \frac{1-x_1}{2}, x_1, -1\right\} \\ \overrightarrow{\alpha}'' &= \left\{\frac{1-x_1}{2}, \frac{1-x_1}{2}, -1, x_1\right\} \\ \overrightarrow{\alpha}''' &= \{-x_1, -x_1, x_1, x_1\} \end{aligned}$$

and the imputations which originate from these by permuting the players, (i.e. the components) 1,2,3.

(Cf. footnote 5, p. 306.) We expect that this V is a solution in the rigorous sense of 30.1. if x_1 is near to -1 and we must determine whether this is so, and precisely in what interval of the x_1.

This determination, if carried out, yields the following result:

(36:A) The set V of (36:3) is a solution if and only if

$$-1 \leqq x_1 \leqq -\tfrac{1}{5}.$$

This then is the answer to the question, how far (from the starting point $x_1 = -1$, i.e. the corner *VIII*) the above heuristic consideration guides to a correct result.[1]

36.2.2. The proof of (36:A) can be carried out rigorously without any significant technical difficulty. It consists of a rather mechanical disposal

[1] We wish to emphasize that (36:A) does not assert that V is (in the specified range of x_1) the only solution of the game in question. However, attempts with numerous similarly built sets failed to disclose further solutions for $x_1 \leqq -\frac{1}{5}$ (i.e. in the range of (36:A)). For x_1 slightly $> -\frac{1}{5}$ (i.e. slightly outside the range of (36:A)), where the V of (36:A) is no longer a solution, the same is true for the solution which replaces it. Cf. (36:B) in 36.3.1.

We do not question, of course, that other solutions of the "discriminatory" type, as repeatedly discussed before, always exist. But they are fundamentally different from the finite solutions V which are now under consideration.

These are the arguments which seem to justify our view that some qualitative change in the nature of the solutions occurs at

$$x_1 = -\tfrac{1}{5} \text{ (on the diagonal } I\text{-center-}VIII).$$

of a series of special cases, and does not contribute anything to the clarification of any question of principle.[1] The reader may therefore omit reading it if he feels so disposed, without losing the connection with the main course of the exposition. He should remember only the statement of the results in (36:A).

Nevertheless we give the proof in full for the following reason: The set V of (36:3) was found by heuristic considerations, i.e. without using the exact theory of 30.1. at all. The rigorous proof to be given is based on 30.1. alone, and thereby brings us back to the only ultimately satisfactory standpoint, that of the exact theory. The heuristic considerations were only a device to guess the solution, for want of any better technique; and it is a fortunate feature of the exact theory that its solutions can occasionally be guessed in this way. But such a guess must afterwards be justified by the exact method, or rather that method must be used to determine in what domain (of the parameters involved) the guess was admissible.

We give the exact proof in order to enable the reader to contrast and to compare explicitly these two procedures,—the heuristic and the rigorous.

36.2.3. The *proof* is as follows:

If $x_1 = -1$, then we are in the corner VIII, and the V of (36:3) coincides with the set which we introduced heuristically (as a solution) in 35.2.3., and which can easily be justified rigorously (cf. also footnote 2 on p. 299). Therefore we disregard this case now, and assume that

(36:4) $$x_1 > -1.$$

We must first establish which sets $S \subseteq I = (1,2,3,4)$ are certainly necessary or certainly unnecessary in the sense of 31.1.2.—since we are carrying out a proof which is precisely of the type considered there.

The following observations are immediate:

(36:5) By virtue of (31:H) in 31.1.5., three-element sets S are certainly necessary, two-element sets are dubious, and all other sets are certainly unnecessary.[2]

(36:6) Whenever a two-element set turns out to be certainly necessary, we may disregard all those three-element sets of which it is a subset, owing to (31:C) in 31.1.3.

Consequently we shall now examine the two-element sets. This of course must be done for all the $\overrightarrow{\alpha}$ in the set V of (36:3).

[1] The reader may contrast this proof with some given in connection with the theory of the zero-sum two-person game, e.g. the combination of 16.4. with 17.6. Such a proof is more transparent, it usually covers more ground, and gives some qualitative elucidation of the subject and its relation to other parts of mathematics. In some later parts of this theory such proofs have been found, e.g. in 46. But much of it is still in the primitive and technically unsatisfactory state of which the considerations which follow are typical.

[2] This is due to $n = 4$.

Consider first those two-element sets S which occur in conjunction with $\overrightarrow{\alpha}'$.[1] As $\alpha'_4 = -1$ we may exclude by (31:A) in 31.1.3. the possibility that S contains 4. $S = (1,2)$ would be effective if $\alpha'_1 + \alpha'_2 \leqq \mathrm{v}((1,2))$, i.e. $1 - x_1 \leqq -2x_1$, $x_1 \leqq -1$ which is not the case by (36:4). $S = (1,3)$ is effective if $\alpha'_1 + \alpha'_3 \leqq \mathrm{v}((1,3))$, i.e. $\frac{1 + x_1}{2} \leqq -2x_1$, $x_1 \leqq -\frac{1}{5}$. Thus the condition

$$(36:7) \qquad x_1 \leqq -\tfrac{1}{5}$$

which we assume to be satisfied makes its first appearance. $S = (2,3)$ we do not need, since 1 and 2 play the same role in $\overrightarrow{\alpha}'$ (cf. footnote 1 above).

We now pass to $\overrightarrow{\alpha}''$. As $\alpha''_3 = -1$ we now exclude the S which contains 3 (cf. above). $S = (1,2)$ is disposed of as before, since $\overrightarrow{\alpha}'$ and $\overrightarrow{\alpha}''$ agree in these components. $S = (1,4)$ would be effective if $\alpha''_1 + \alpha''_4 \leqq \mathrm{v}((1,4))$, i.e. $\frac{1 + x_1}{2} \leqq 2x_1$, $x_1 \geqq \frac{1}{3}$ which, by (36:7), is not the case. $S = (2,4)$ is discarded in the same way.

Finally we take $\overrightarrow{\alpha}'''$. $S = (1,2)$ is effective: $\alpha'''_1 + \alpha'''_2 = \mathrm{v}((1,2))$ i.e. $-2x_1 = -2x_1$. $S = (1,3)$ need not be considered for the following reason: We are already considering $S = (1,2)$ for $\overrightarrow{\alpha}'''$, if we interchange 2 and 3 (cf. footnote 1 above) this goes over into $(1,3)$, with the components $-x_1, -x_1$. Our original $S = (1,3)$ for $\overrightarrow{\alpha}'''$ with the components $-x_1, x_1$ is thus rendered unnecessary by (31:B) in 31.1.3., as $-x_1 \geqq x_1$ owing to (36:7). $S = (2,3)$ is discarded in the same way. $S = (1,4)$ would be effective if $\alpha'''_1 + \alpha'''_4 \leqq \mathrm{v}((1,4))$ i.e. $0 \leqq 2x_1$, $x_1 \geqq 0$, which, by (36:7), is not the case. $S = (2,4)$ is discarded in the same way. $S = (3,4)$ is effective: $\alpha'''_3 + \alpha'''_4 = \mathrm{v}((3,4))$, i.e. $2x_1 = 2x_1$.

Summing up:

(36:8) Among the two-element sets S the three given below are certainly necessary, and all others are certainly unnecessary:

$$(1,3) \text{ for } \overrightarrow{\alpha}', \ (1,2) \text{ and } (3,4) \text{ for } \overrightarrow{\alpha}'''.$$

Concerning three-element sets S: By (31:A) in 31.1.3. we may exclude those containing 4 for $\overrightarrow{\alpha}'$ and 3 for $\overrightarrow{\alpha}''$. Consequently only $(1,2,3)$ is left for $\overrightarrow{\alpha}'$ and $(1,2,4)$ for $\overrightarrow{\alpha}''$. Of these the former is excluded by (36:6), as it contains the set $(1,3)$ of (36:8). For $\overrightarrow{\alpha}'''$ every three-element set

[1] Here, and in the entire discussion which follows, we shall make use of the freedom to apply permutations of 1,2,3 as stated in (36:3), in order to abbreviate the argumentation. Hence the reader must afterwards apply these permutations of 1,2,3 to our results.

contains the set (1,2) or the set (3,4) of (36:8); hence we may exclude it by (36:6).

Summing up:

(36:9) Among the three-element sets S, the one given below is certainly necessary, and all others are certainly unnecessary:[1]

$$(1,2,4) \text{ for } \overrightarrow{\alpha}''.$$

36.2.4. We now verify (30:5:a) in 30.1.1., i.e. that no $\overrightarrow{\alpha}$ of V dominates any $\overrightarrow{\beta}$ of V.

$\overrightarrow{\alpha} = \overrightarrow{\alpha}'$: By (36:8), (36:9) we must use $S = (1,3)$. Can $\overrightarrow{\alpha}'$ dominate with this S any 1,2,3 permutation of $\overrightarrow{\alpha}'$ or $\overrightarrow{\alpha}''$ or $\overrightarrow{\alpha}'''$? This requires first the existence of a component $< x_1$ (this is the 3 component of $\overrightarrow{\alpha}'$) among the 1,2,3 components of the imputation in question. Thus $\overrightarrow{\alpha}'$ and $\overrightarrow{\alpha}'''$ are excluded.[2] Even in $\overrightarrow{\alpha}''$ the 1,2 components are excluded (cf. footnote 2) but the 3 component will do. But now another one of the 1,2,3 components of this imputation $\overrightarrow{\alpha}''$ must be $< \frac{1 - x_1}{2}$ (this is the 1 component of $\overrightarrow{\alpha}'$), and this is not the case; the 1,2 components of $\overrightarrow{\alpha}''$ are both $= \frac{1 - x_1}{2}$.

$\overrightarrow{\alpha} = \overrightarrow{\alpha}''$: By (36:8), (36:9) we must use $S = (1,2,4)$. Can $\overrightarrow{\alpha}''$ dominate with this S any 1,2,3 permutation of $\overrightarrow{\alpha}'$ or $\overrightarrow{\alpha}''$ or $\overrightarrow{\alpha}'''$? This requires first that the 4 component of the imputation in question be $< x_1$ (this is the 4 component of $\overrightarrow{\alpha}''$). Thus $\overrightarrow{\alpha}''$ and $\overrightarrow{\alpha}'''$ are excluded. For $\overrightarrow{\alpha}'$ we must require further that two of its 1,2,3 components be $< \frac{1 - x_1}{2}$ (this is the 1 as well as the 2 component of $\overrightarrow{\alpha}''$), and this is not the case; only one of these components is $\neq \frac{1 - x_1}{2}$.

$\overrightarrow{\alpha} = \overrightarrow{\alpha}'''$: By (36:8), (36:9) we must use $S = (1,2)$ and then $S = (3,4)$. $S = (1,2)$: Can $\overrightarrow{\alpha}'''$ dominate with this S as described above? This requires the existence of two components $< -x_1$ (this is the 1 as well as the 2 component of $\overrightarrow{\alpha}'''$) among the 1,2,3 components of the imputation in question. This is not the case for $\overrightarrow{\alpha}'''$, as only one of these components is $\neq -x_1$

[1] As every three-element set is certainly necessary by (36:5) above, this is another instance of the phenomenon mentioned at the end of footnote 1 on p. 274.

[2] Indeed $\frac{1 - x_1}{2} \geqq x_1$, i.e. $x_1 \leqq \frac{1}{3}$ and $-x_1 \geqq x_1$, i.e. $x_1 \leqq 0$—both by (36:7).

there. Nor is it the case for $\overrightarrow{\alpha}'$ or $\overrightarrow{\alpha}''$, as only one of those components is $\neq \frac{1-x_1}{2}$ there.[1] $S = (3,4)$: Can $\overrightarrow{\alpha}'''$ dominate with this S as described above? This requires first that the 4 component of the imputation in question be $< x_1$ (this is the 4 component of $\overrightarrow{\alpha}'''$). Thus $\overrightarrow{\alpha}''$ and $\overrightarrow{\alpha}'''$ are excluded. For $\overrightarrow{\alpha}'$ we must require further the existence of a component $< x_1$ (this is the 3 component of $\overrightarrow{\alpha}'''$) among its 1,2,3 components, and this is not the case; all these components are $\geqq x_1$ (cf. footnote 2 on p. 310).

This completes the verification of (30:5:a).

36.2.5. We verify next (30:5:b) in 30.1.1., i.e. that an imputation $\overrightarrow{\beta}$ which is undominated by the elements of V must belong to V.

Consider a $\overrightarrow{\beta}$ undominated by the elements of V. Assume first that $\beta_4 < x_1$. If any one of $\beta_1, \beta_2, \beta_3$ were $< x_1$, we could make (by permuting 1,2,3) $\beta_3 < x_1$. This gives $\overrightarrow{\alpha}''' \succ \overrightarrow{\beta}$ with $S = (3,4)$ of (36:8). Hence

$$\beta_1, \beta_2, \beta_3 \geqq x_1.$$

If any two of $\beta_1, \beta_2, \beta_3$ were $< \frac{1-x_1}{2}$, we could make (by permuting 1,2.3) $\beta_1, \beta_2 < \frac{1-x_1}{2}$. This gives $\overrightarrow{\alpha}'' \succ \overrightarrow{\beta}$ with $S = (1,2,4)$ of (36:9). Hence, at most one of $\beta_1, \beta_2, \beta_3$ is $< \frac{1-x_1}{2}$ i.e. two are $\geqq \frac{1-x_1}{2}$. By permuting 1,2,3, we can thus make

$$\beta_1, \beta_2 \geqq \frac{1-x_1}{2}.$$

Clearly $\beta_4 \geqq -1$. Thus each component of $\overrightarrow{\beta}$ is $\geqq$ the corresponding component of $\overrightarrow{\alpha}'$, and since both are imputations[2] it follows that they coincide: $\overrightarrow{\beta} = \overrightarrow{\alpha}'$, and so it is in V.

Assume next that $\beta_4 \geqq x_1$. If any two of $\beta_1, \beta_2, \beta_3$ were $< -x_1$, we could make (by permuting 1,2,3) $\beta_1, \beta_2 < -x_1$. This gives $\overrightarrow{\alpha}''' \succ \overrightarrow{\beta}$ with $S = (1,2)$ of (36:8). Hence, at most one of $\beta_1, \beta_2, \beta_3$ is $< -x_1$, i.e. two are $\geqq -x_1$. By permuting 1,2,3 we can make

$$\beta_1, \beta_2 \geqq -x_1.$$

If $\beta_3 \geqq x_1$, then all this implies that each component of $\overrightarrow{\beta}$ is $\geqq$ the cor-

[1] And $\frac{1-x_1}{2} \geqq -x_1$, i.e. $x_1 \geqq -1$.

[2] Consequently for both the sum of all components is the same: zero.

responding component of $\overrightarrow{\alpha}'''$, and since both are imputations (cf. footnote 2, on p. 311) it follows that they coincide: $\overrightarrow{\beta} = \overrightarrow{\alpha}'''$, and so it is in V.

Assume therefore that $\beta_3 < x_1$. If any one of β_1, β_2 were $< \frac{1 - x_1}{2}$, we could make (by permuting 1,2) $\beta_1 < \frac{1 - x_1}{2}$. This gives $\overrightarrow{\alpha}' \succ \overrightarrow{\beta}$ with $S = (1,3)$ of (36:8). Hence

$$\beta_1, \beta_2 \geqq \frac{1 - x_1}{2}.$$

Clearly $\beta_3 \geqq -1$. Thus each component of $\overrightarrow{\beta}$ is $\geqq$ the corresponding component of $\overrightarrow{\alpha}''$, and since both are imputations (cf. footnote 2, p. 311), it follows that they coincide: $\overrightarrow{\beta} = \overrightarrow{\alpha}''$, and so it is in V.

This completes the verification of (30:5:b).[1]

So we have established the criterion (36:A).[2]

36.3. Other Parts of the Main Diagonals

36.3.1. When x_1 passes outside the domain (36:A) of 36.2.1., i.e. when it crosses its border at $x_1 = -\frac{1}{5}$, then the V of (36:3) id. ceases to be a solution. It is actually possible to find a solution which is valid for a certain domain in $x_1 > -\frac{1}{5}$ (adjoining $x_1 = -\frac{1}{5}$), which obtains by adding to the V of (36:3) the further imputations

(36:10) $\overrightarrow{\alpha}^{IV} = \left\{\frac{1 - x_1}{2}, -x_1, \frac{-1 + x_1}{2}, x_1\right\}$ and permutations as in (36:3).[3]

The exact statement is actually this:

(36:B) The set V of (36:3) and (36:10) is a solution if and only if

$$-\frac{1}{5} < x_1 \leqq 0.$$[4]

[1] The reader will observe that in the course of this analysis all sets of (36:8), (36:9) have been used for dominations, and $\overrightarrow{\beta}$ had to be equated successively to all three $\overrightarrow{\alpha}'$, $\overrightarrow{\alpha}''$, $\overrightarrow{\alpha}'''$ of (36:3).

[2] Concerning $x_1 = -1$, cf. the remarks made at the beginning of this proof.

[3] An inspection of the above proof shows that when x_1 becomes $> -\frac{1}{5}$, this goes wrong: The set $S = (1,3)$ (and with it (2,3)) is no longer effective for $\overrightarrow{\alpha}'$. Of course this rehabilitates the three-element set $S = (1,2,3)$ which was excluded solely because (1,3) (and (2,3)) is contained in it.

Thus domination by this element of V, $\overrightarrow{\alpha}'$, now becomes more difficult, and it is therefore not surprising that an increase of the set V must be considered in the search for a solution.

[4] Observe the discontinuity at $x_1 = -\frac{1}{5}$ which belongs to (36:A) and not to (36:B)! The exact theory is quite unambiguous, even in such matters.

The proof of (36:B) is of the same type as that of (36:A) given above, and we do not propose to discuss it here.

The domains (36:A) and (36:B) exhaust the part $x_1 \leqq 0$ of the entire available interval $-1 \leqq x_1 \leqq 1$—i.e. the half *VIII*-Center of the diagonal *VIII*-Center-*I*.

36.3.2. Solutions of a nature similar to V described in (36:A) of 36.2.1. and in (36:B) of 36.3.1., have been found on the other side $x_1 > 0$—i.e. the half Center-*I* of the diagonal—as well. It turns out that on this half, qualitative changes occur of the same sort as in the first half covered by (36:A) and (36:B). Actually three such intervals exist, namely:

(36:C) $\quad 0 \leqq x_1 < \frac{1}{9},$

(36:D) $\quad \frac{1}{9} < x_1 \leqq \frac{1}{3},$

(36:E) $\quad \frac{1}{3} \leqq x_1 \leqq 1.$

(Cf. Figure 64, which is to be compared with Figure 63.)

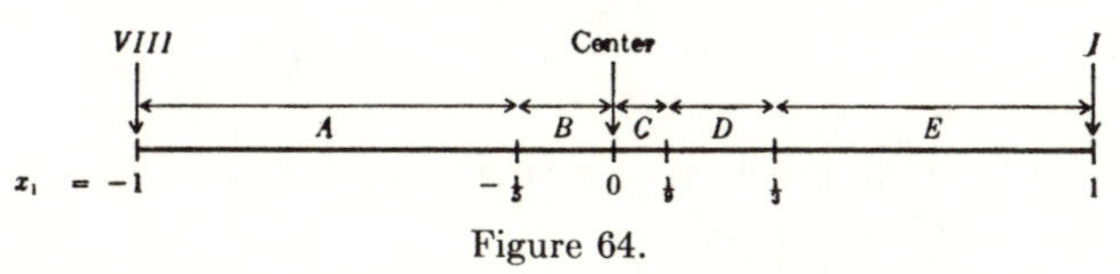

Figure 64.

We shall not discuss the solutions pertaining to (36:C), (36:D), (36:E).[1]

The reader may however observe this: $x_1 = 0$ appears as belonging to both (neighboring) domains (36:B) and (36:C), and similarly $x_1 = \frac{1}{3}$ to both domains (36:D) and (36:E). This is so because, as a close inspection of the corresponding solutions V shows that while qualitative changes in the nature of V occur at $x_1 = 0$ and $\frac{1}{3}$, these changes are not discontinuous.

The point $x_1 = \frac{1}{9}$, on the other hand, belongs to neither neighboring domain (36:C) or (36:D). It turns out that the types of solutions V which are valid in these two domains are both unusable at $x_1 = \frac{1}{9}$. Indeed, the conditions at this point have not been sufficiently clarified thus far.

37. The Center and Its Environs

37.1. First Orientation Concerning the Conditions around the Center

37.1.1. The considerations of the last section were restricted to a one-dimensional subset of the cube Q: The diagonal *VIII*-center-*I*. By using the permutations of the players 1,2,3,4, as described in 34.3., this can be made to dispose of all four main diagonals of Q. By techniques that are similar to those of the last section, solutions can also be found along some other one-dimensional lines in Q. Thus there is quite an extensive net of lines in Q on which solutions are known. We do not propose to enumerate them, particularly because the information that is available now corresponds probably to only a transient state of affairs.

[1] Another family of solutions, which also cover part of the same territory, will be discussed in 38.2. Cf. in particular 38.2.7., and footnote 2 on p. 328.

This, however, should be said: such a search for solutions along isolated one-dimensional lines, when the whole three-dimensional body of the cube Q waits for elucidation, cannot be more than a first approach to the problem. If we can find a three-dimensional part of the cube—even if it is a small one—for all points of which the same qualitative type of solutions can be used, we shall have some idea of the conditions which are to be expected. Now such a three-dimensional part exists around the center of Q. For this reason we shall discuss the conditions at the center.

37.1.2. The center corresponds to the values 0,0,0 of the coordinates x_1, x_2, x_3 and represents, as pointed out in 35.3.1., the only (totally) symmetric game in our set-up. The characteristic function of this game is:

$$(37:1) \qquad \mathrm{v}(S) = \begin{cases} 0 \\ -1 \\ 0 \\ 1 \\ 0 \end{cases} \quad \text{when } S \text{ has } \begin{cases} 0 \\ 1 \\ 2 \\ 3 \\ 4 \end{cases} \text{elements.}$$

(Cf. (35:8) id.) As in the corresponding cases in 35.1., 35.2., 36.1., we begin again with a heuristic analysis.

This game is obviously one in which the purpose of all strategic efforts is to form a three-person coalition. A player who is left alone is clearly a loser, any coalition of 3 in the same sense a winner, and if the game should terminate with two coalitions of two players each facing each other, then this must obviously be interpreted as a tie.

The qualitative question which arises here is this: The aim in this game is to form a coalition of three. It is probable that in the negotiations which precede the play a coalition of two will be formed first. This coalition will then negotiate with the two remaining players, trying to secure the cooperation of one of them against the other. In securing the adherence of this third player, it seems questionable whether he will be admitted into the final coalition on the same conditions as the two original members. If the answer is affirmative, then the total proceeds of the final coalition, 1, will be divided equally among the three participants: $\frac{1}{3}, \frac{1}{3}, \frac{1}{3}$. If it is negative, then the two original members (belonging to the first coalition of two) will probably both get the same amount, but more than $\frac{1}{3}$. Thus 1 will be divided somewhat like this: $\frac{1}{3} + \epsilon, \frac{1}{3} + \epsilon, \frac{1}{3} - 2\epsilon$ with an $\epsilon > 0$.

37.1.3. The first alternative would be similar to the one which we encountered in the analysis of the point I in 35.1. Here the coalition (1,2,3), if it forms at all, contains its three participants on equal terms. The second alternative corresponds to the situation in the interval analyzed in 36.1.-2. Here any two players (neither of them being player 4) combined first, and this coalition then admitted either one of the two remaining players on less favorable terms.

37.1.4. The present situation is not a perfect analogue of either of these.

In the first case the coalition (1,2) could not make stiff terms to player 3 because they absolutely needed him: if 3 combined with 4, then 1 and 2 would be completely defeated; and (1,2) could not, as a coalition, combine with 4 against 3, since 4 needed only one of them to be victorious (cf. the description in 35.1.3.). In our present game this is not so: the coalition (1,2) can use 3 as well as 4, and even if 3 and 4 combine against it, only a tie results.

In the second case the discrimination against the member who joins the coalition of three participants last is plausible, since the original coalition of two is of a much stronger texture than the final coalition of three. Indeed, as x_1 tends to -1, the latter coalition tends to become worthless; cf. the remarks at the end of 36.1.2. In our present game no such qualitative difference can be recognized: the first coalition (of two) accounts for the difference between defeat and tie, while formation of the final coalition (of three) decides between tie and victory.

We have no satisfactory basis for a decision except to try both alternatives. Before we do this, however, an important limitation of our considerations deserves attention.

37.2. The Two Alternatives and the Role of Symmetry

37.2.1. It will be noted that we assume that the same one of the two alternatives above holds for all four coalitions of three players. Indeed, we are now looking for symmetric solutions only, i.e. solutions which contain, along with an imputation $\overrightarrow{\alpha} = \{\alpha_1, \alpha_2, \alpha_3, \alpha_4\}$, all its permutations.

Now a symmetry of the game by no means implies in general the corresponding symmetry in each one of its solutions. The discriminatory solutions discussed in 33.1.1. make this clear already for the three-person game. We shall find in 37.6. further instances of this for the symmetric four-person game now under consideration.

It must be expected, however, that asymmetric solutions for a symmetric game are of too recondite a character to be discovered by a first heuristic survey like the present one. (Cf. the analogous occurrence in the three-person game, referred to above.) This then is our excuse for looking at present only for symmetric solutions.

37.2.2. One more thing ought to be said: it is not inconceivable that, while asymmetric solutions exist, general organizational principles, like those corresponding to our above two alternatives, are valid either for the totality of all participants or not at all. This surmise gains some strength from the consideration that the number of participants is still very low, and may actually be too low to permit the formation of several groups of participants with different principles of organization. Indeed, we have only four participants, and ample evidence that three is the minimum number for any kind of organization. These somewhat vague considerations will find exact corroboration in at least one special instance in (43:L) et seq. of 43.4.2. For the present case, however, we are not able to support them by any rigorous proof.

37.3. The First Alternative at the Center

37.3.1. Let us now consider the two alternatives of 37.1.2. We take them up in reverse order.

Assume first that the two original participants admit the third one under much less favorable conditions. Then the first coalition (of two) must be considered as the core on which the final coalition (of three) crystallizes. In this last phase the first coalition must therefore be expected to act as one player in its dealings with the two others, thus bringing about something like a three-person game. If this view is sound, then we may repeat the corresponding considerations of 36.1.3.

Taking, e.g. (1,2), as the "first" coalition, the surmised three-person game is between the players (1, 2), 3,4. The considerations referred to above therefore apply literally, only with changed numerical values: $a = 0$, $b = c = 1$ and so $\alpha = 1$, $\beta = \gamma = 0$.[1]

Since the "first" coalition may consist of any two players, there are heuristic reasons similar to those in the discussion of the three-person game (in 21.-22.) to expect that the partners in it will split even: when an ally is found, as well as when a tie results, the amount to be divided being 1 or 0 respectively.[2]

37.3.2. Summing up: if the above surmises prove correct, the situation is as follows:

If the "first" coalition is (1,2) and if it is successful in finding an ally, and if the player who joins it in the final coalition is 3, then the players 1,2,3,4 get the amounts $\frac{1}{2}$, $\frac{1}{2}$, 0, -1 respectively. If the "first" coalition is not successful, i.e. if a tie results, then these amounts are replaced by 0,0,0,0.

If the distribution of the players is different, then the corresponding permutation of the players 1,2,3,4 must be applied to the above.

It is now necessary to submit all this to an exact check. The heuristic suggestion manifestly corresponds to the following surmise:

Let V be the set of these following imputations

(37:2) $\overrightarrow{\alpha}' = \{\frac{1}{2}, \frac{1}{2}, 0, -1\}$, $\overrightarrow{\alpha}'' = \{0, 0, 0, 0\}$ and the imputations which originate from these by permuting the players (i.e. the components) 1,2,3,4.

We expect that this V is a solution.

[1] The essential difference between this discussion and that referred to, is that player 4 is no longer excluded from the "first" coalition.

[2] The argument in this case is considerably weaker than in the case referred to (or in the corresponding application in 36.1.3.) since every "first" coalition may now wind up in two different ways (tie or victory). The only satisfactory decision as to the value of the argumentation obtains when the exact theory is applied. The desired justification is actually contained in the proof of 38.2.1.-3.; indeed, it is the special case

$$y_1 = y_2 = y_3 = y_4 = 1$$

of (38:D) in 38.2.3.

A rigorous consideration, of the same type as that which constitutes 36.2., shows that this V is indeed a solution in the sense of 30.1. We do not give it here, particularly because it is contained in a more general proof which will be given later. (Cf. the reference of footnote 2 on p. 316.)

37.4. The Second Alternative at the Center

37.4.1. Assume next that the final coalition of three contains all its participants on equal terms. Then if this coalition is, say (1,2,3), the players 1,2,3,4 get the amounts $\frac{1}{3}$, $\frac{1}{3}$; $\frac{1}{3}$, -1 respectively.

It would be rash to conclude from this that we expect the set of imputations V to which this leads, to be a solution; i.e. the set of these imputations $\overrightarrow{\alpha} = \{\alpha_1, \alpha_2, \alpha_3, \alpha_4\}$:

$$(37:3) \qquad \overrightarrow{\alpha}''' = \{\tfrac{1}{3}, \tfrac{1}{3}, \tfrac{1}{3}, -1\} \qquad \text{and permutations as in (37:2).}$$

We have made no attempt as yet to understand how this formation of the final coalition in "one piece" comes about, without assuming the previous existence of a favored two-person core.

37.4.2. In the previous solution of (37:2) such an explanation is discernable. The stratified form of the final coalition is expressed by the imputation $\overrightarrow{\alpha}'$ and the motive for just this scheme of distribution lies in the threat of a tie, expressed by the imputation $\overrightarrow{\alpha}''$. To put it exactly: the $\overrightarrow{\alpha}'$ form a solution only in conjunction with the $\overrightarrow{\alpha}''$, and not by themselves.

In (37:3) this second element is lacking. A direct check in the sense of 30.1. discloses that the $\overrightarrow{\alpha}'''$ fulfill condition (30:5:a) there, but not (30:5:b). I.e. they do not dominate each other, but they leave certain other imputations undominated. Hence further elements must be added to V.[1]

This addition can certainly not be the $\overrightarrow{\alpha}'' = \{0,0,0,0\}$ of (37:2) since that imputation happens to be dominated by $\overrightarrow{\alpha}'''$.[2] In other words the extension (i.e. stabilization, in the sense of 4.3.3.) of $\overrightarrow{\alpha}'''$ to a solution must be achieved by entirely different imputations (i.e. threats) in the case of the $\overrightarrow{\alpha}'''$ of (37:3) as in the case of the $\overrightarrow{\alpha}'$ of (37:2).

It seems very difficult to find a heuristic motivation for the steps which are now necessary. Luckily, however, a rigorous procedure is possible from here on, thus rendering further heuristic considerations unnecessary.

[1] To avoid misunderstandings: It is by no means generally true that any set of imputations which do not dominate each other can be extended to a solution. Indeed, the problem of recognizing a given set of imputations as being a subset of some (unknown) solution is still unsolved. Cf. 30.3.7.

In the present case we are just expressing the hope that such an extension will prove possible for the V of (37:3), and this hope will be further justified below.

[2] With $S = (1,2,3)$.

Indeed, one can prove rigorously that there exists one and only one symmetric extension of the V of (37:3) to a solution. This is the addition of these imputations $\overrightarrow{\alpha} = \{\alpha_1, \alpha_2, \alpha_3, \alpha_4\}$:

(37:4) $\overrightarrow{\alpha}^{IV} = \{\frac{1}{3}, \frac{1}{3}, -\frac{1}{3}, -\frac{1}{3}\}$ and permutations as in (37:2).

37.4.3. If a common-sense interpretation of this solution, i.e. of its constituent $\overrightarrow{\alpha}^{IV}$ of (37:4), is wanted, it must be said that it does not seem to be a tie at all (like the corresponding $\overrightarrow{\alpha}''$ of (37:2))—rather, it seems to be some kind of compromise between a part (two members) of a possible victorious coalition and the other two players. However, as stated above, we do not attempt to find a full heuristic interpretation for the V of (37:3) and (37:4); indeed it may well be that this part of the exact theory is already beyond such possibilities.[1] Besides, some subsequent examples will illustrate the peculiarities of this solution on a much wider basis. Again we refrain from giving the exact proof referred to above.

37.5. Comparison of the Two Central Solutions

37.5.1. The two solutions (37:2) and (37:3), (37:4), which we found for the game representing the center, present a new instance of a possible multiplicity of solutions. Of course we had observed this phenomenon before, namely in the case of the essential three-person game in 33.1.1. But there all solutions but one were in some way abnormal (we described this by terming them "discriminatory"). Only one solution in that case was a finite set of imputations; that solution alone possessed the same symmetry as the game itself (i.e. was symmetric with respect to all players). This time conditions are quite different. We have found two solutions which are both finite sets of imputations,[2] and which possess the full symmetry of the game. The discussion of 37.1.2. shows that it is difficult to consider either solution as "abnormal" or "discriminatory" in any sense; they are distinguished essentially by the way in which the accession of the last participant to the coalition of three is treated, and therefore seem to correspond to two perfectly normal principles of social organization.

37.5.2. If anything, the solution (37:3), (37:4) may seem the less normal one. Both in (37:2) and in (37:3), (37:4) the character of the solution was determined by those imputations which described a complete decision, $\overrightarrow{\alpha}'$ and $\overrightarrow{\alpha}'''$ respectively. To these the extra "stabilizing" imputations, $\overrightarrow{\alpha}''$ and $\overrightarrow{\alpha}^{IV}$, had to be added. Now in the first solution this extra $\overrightarrow{\alpha}''$ had an

[1] This is, of course, a well known occurrence in mathematical-physical theories, even if they originate in heuristic considerations.

[2] An easy count of the imputations given and of their different permutations shows that the solution (37:2) consists of 13 elements, and the solution (37:3), (37:4) of 10.

obvious heuristic interpretation as a tie, while in the second solution the nature of the extra $\overrightarrow{\alpha}^{IV}$ appeared to be more complex.

A more thorough analysis discloses, however, that the first solution is surrounded by some peculiar phenomena which can neither be explained nor foreseen by the heuristic procedure which provided easy access to this solution.

These phenomena are quite instructive from a general point of view too, because they illustrate in a rather striking way some possibilities and interpretations of our theory. We shall therefore analyze them in some detail in what follows. We add that a similar expansion of the second solution has not been found up to now.

37.6. Unsymmetrical Central Solutions

37.6.1. To begin with, there exist some finite but asymmetrical solutions which are closely related to (37:2) in 37.3.2. because they contain some of the imputations $\{\frac{1}{2}, \frac{1}{2}, 0, -1\}$.[1] One of these solutions is the one which obtains when we approach the center along the diagonal I-Center-VIII from either side, and use there the solutions referred to in 36.3. I.e.: it obtains by continuous fit to the domains (36:B) and (36:C) there mentioned. (It will be remembered that the point $x_1 = 0$, i.e. the center, belongs to both these domains, cf. 36.3.2.) Since this solution can be taken also to express a *sui generis* principle of social organization, we shall describe it briefly.

This solution possesses the same symmetry as those which belong to the games on the diagonal I-Center-VIII, as it is actually one of them: symmetric with respect to players 1,2,3, while player 4 occupies a special position.[2] We shall therefore describe it in the same way we did the solutions on the diagonal, e.g. in (36:3) in 36.2.1. Here only permutations of the players 1,2,3 are suppressed, while in the descriptions of (37:3) and (37:4) we suppressed all permutations of the players 1,2,3,4.

37.6.2. For the sake of a better comparison, we restate with this notation (i.e. allowing for permutations of 1,2,3 only) the definition of our first fully-symmetric solution (37:2) in 37.3.2. It consists of these imputations:[3]

(37:2*)
$$\begin{aligned}\overrightarrow{\beta}' &= \{\tfrac{1}{2}, \tfrac{1}{2}, 0, -1\}\\ \overrightarrow{\beta}'' &= \{\tfrac{1}{2}, \tfrac{1}{2}, -1, 0\}\\ \overrightarrow{\beta}''' &= \{\tfrac{1}{2}, 0, -1, \tfrac{1}{2}\}\\ \overrightarrow{\beta}^{IV} &= \{0, 0, 0, 0\}\end{aligned}$$
and the imputations which originate from these by permuting the players 1,2,3.

[1] I.e. some but not all of the 12 permutations of this imputation.

[2] That the position of player 4 in the solution is really different from that of the others, is what distinguishes this solution from the two symmetric ones mentioned before.

[3] Our $\overrightarrow{\beta}'$, $\overrightarrow{\beta}''$, $\overrightarrow{\beta}'''$ exhaust the α' of (37:2) in 37.3.2., while $\overrightarrow{\beta}^{IV}$ is $\overrightarrow{\alpha}''$, id. $\overrightarrow{\alpha}'$ had to be represented by the three imputations $\overrightarrow{\beta}'$, $\overrightarrow{\beta}''$, $\overrightarrow{\beta}'''$ because this system of representation makes it necessary to state in which one of the three possible positions of that imputation (i.e. the values $\frac{1}{2}$, 0, -1) the player 4 is found.

Now the (asymmetric) solution to which we refer consists of these imputations:

(37:5) $\overrightarrow{\beta}', \overrightarrow{\beta}'', \overrightarrow{\beta}^{IV}$ as in (37:2*); $\overrightarrow{\beta}^{V} = \{\frac{1}{2}, 0, -\frac{1}{2}, 0\}$[1] — and the imputations which originate from these by permuting the players 1,2,3.

Once more we omit giving the proof that (37:5) is a solution. Instead we shall suggest an interpretation of the difference between this solution and that of (37:2)—i.e. of the first (symmetric) solution in 37.3.2.

37.6.3. This difference consists in replacing

$$\overrightarrow{\beta}''' = \{\tfrac{1}{2}, 0, -1, \tfrac{1}{2}\}$$

by

$$\overrightarrow{\beta}^{V} = \{\tfrac{1}{2}, 0, -\tfrac{1}{2}, 0\}.$$

That is: the imputation $\overrightarrow{\beta}'''$—in which the player 4 would belong to the "first" coalition (cf. 37.3.1.), i.e. to the group which wins the maximum amount $\frac{1}{2}$—is removed, and replaced by another imputation $\overrightarrow{\beta}^{V}$. Player 4 now gets somewhat less and the losing player among 1,2,3 (in this arrangement player 3) gets somewhat more than in $\overrightarrow{\beta}'''$. This difference is precisely $\frac{1}{2}$, so that player 4 is reduced to the tie position 0, and player 3 moves from the completely defeated position -1 to an intermediate position $-\frac{1}{2}$.

Thus players 1,2,3 form a "privileged" group and no one from the outside will be admitted to the "first" coalition. But even among the three members of the privileged group the wrangle for coalition goes on, since the "first" coalition has room for two participants only. It is worth noting that a member of the privileged group may even be completely defeated, as in $\overrightarrow{\beta}''$,—but only by a majority of his "class" who form the "first" coalition and who may admit the "unprivileged" player 4 to the third membership of the "final" coalition, to which he is eligible.

37.6.4. The reader will note that this describes a perfectly possible form of social organization. This form is discriminatory to be sure, although not in the simple way of the "discriminatory" solutions of the three-person game. It describes a more complex and a more delicate type of social inter-relation, due to the solution rather than to the game itself.[2] One may think it somewhat arbitrary, but since we are considering a "society" of very small size, all possible standards of behavior must be adjusted rather precisely and delicately to the narrowness of its possibilities.

We scarcely need to elaborate the fact that similar discrimination against any other player (1,2,3 instead of 4) could be expressed by suitable

[1] This imputation $\overrightarrow{\beta}^{V}$ is reminiscent in its arrangement of $\overrightarrow{\alpha}^{IV}$ in (37:4) of 37.4.2., but it has not been possible to make anything of that analogy.

[2] As to this feature, cf. the discussion of 35.2.4.

solutions, which could then be associated with the three other diagonals of the cube Q.

38. A Family of Solutions for a Neighborhood of the Center

38.1. Transformation of the Solution Belonging to the First Alternative at the Center

38.1.1. We continue the analysis of the ramifications of solution (37:2) in 37.3.2. It will appear that it can be subjected to a peculiar transformation without losing its character as a solution.

This transformation consists in multiplying the imputations (37:2) of 37.3.2. by a common (positive) numerical factor z. In this way the following set of imputations obtains:

(38:1) $$\begin{aligned} \vec{\gamma}' &= \left\{\frac{z}{2}, \frac{z}{2}, 0, -z\right\} \\ \vec{\gamma}'' &= \{0, 0, 0, 0\} \end{aligned}$$ and the imputations which originate from these by permuting the players 1,2,3,4.

In order that these be imputations, all their components must be $\geqq -1$ (i.e. the common value of the $v((i))$. As $z > 0$ this means only that $-z \geqq -1$, i.e. we must have

(38:2) $$0 < z \leqq 1.$$

For $z = 1$ our (38:1) coincides with (37:2) of 37.3.2. It would not seem likely *a priori* that (38:1) should be a solution for the same game for any other z of (38:2). And yet a simple discussion shows that it is a solution if and only if $z > \frac{2}{3}$—i.e. when (38:2) is replaced by

(38:3) $$\tfrac{2}{3} < z \leqq 1.$$

The importance of this family of solutions is further increased by the fact that it can be extended to a certain three-dimensional piece surrounding the center of the cube Q. We shall give the necessary discussion in full, because it offers an opportunity to demonstrate a technique that may be of wider applicability in these investigations.

The interpretation of these results will be attempted afterwards.

38.1.2. We begin by observing that consideration of the set V defined by the above (38:1) for the game described by (37:1) in 37.1.2. (i.e. the center of Q), could be replaced by consideration of the original set V of (37:2) in 37.3.2. in another game. Indeed, our (38:1) was obtained from (37:2) by multiplying by z. Instead of this we could keep (37:2) and multiply the characteristic function (37:1) by $1/z$; this would destroy the normalization $\gamma = 1$ which was necessary for the geometrical representation by Q (cf. 34.2.2.)—but we propose to accept that.

What we are now undertaking can be formulated therefore as follows:

So far we have started with a given game, and have looked for solutions. Now we propose to reverse this process, starting with a solution and looking for the game. Precisely: we start with a given set of imputations V, and ask for which characteristic function $\mathrm{v}(S)$ (i.e. games) this V is a solution.[1]

Multiplication of the $\mathrm{v}(S)$ of (37:1) in 37.1.2. by a common factor means that we still demand

(38:4) $\qquad \mathrm{v}(S) = 0 \qquad$ when S is a two-element set,

but beyond this only the reduced character of the game (cf. 27.1.4.), i.e.

(38:5) $$\mathrm{v}((1)) = \mathrm{v}((2)) = \mathrm{v}((3)) = \mathrm{v}((4)).$$

Indeed, this joint value of (38:5) is $-1/z$ and therefore (38:4), (38:5) and (25:3:a) (25:3:b) in 25.3.1. yield that this $\mathrm{v}(S)$ is just (37:1) multiplied by $1/z$. Our assertion (38:3) above means that the V of (37:2) in 37.3.2. is a solution for (38:4), (38:5) if and only if the joint value of (38:5) (i.e. $-1/z$) is $\leqq -1$ and $> -\frac{3}{2}$.

38.1.3. Now we shall go one step further and drop the requirement of reduction, i.e. (38:5). So we demand of $\mathrm{v}(S)$ only (38:4), restricting its values for two-element sets S. We restate the final form of our question:

(38:A) Consider all zero-sum four-person games where

(38:6) $\qquad \mathrm{v}(S) = 0 \qquad$ for all two-element sets S.

For which among these is the set V of (37:2) in 37.3.2. a solution?

It will be noted that since we have dropped the requirements of normalization and reduction of $\mathrm{v}(S)$ all connections with the geometrical representation in Q are severed. A special manipulation will be necessary therefore at the end, in order to put the results which we shall obtain back into the framework of Q.

38.2. Exact Discussion

38.2.1. The unknowns of the problem (38:A) are clearly the values

(38:7) $$\mathrm{v}((1)) = -y_1, \quad \mathrm{v}((2)) = -y_2, \quad \mathrm{v}((3)) = -y_3, \quad \mathrm{v}((4)) = -y_4.$$

We propose to determine what restrictions the condition in (38:A) actually places on these numbers y_1, y_2, y_3, y_4.

[1] This reversed procedure is quite characteristic of the elasticity of the mathematical method—for the kind and degree of freedom which exists there. Although initially it deflects the inquiry into a direction which must be considered unnatural from any but the strictest mathematical point of view, it is nevertheless effective; by an appropriate technical manipulation it finally discloses solutions which have not been found in any other way.

After our previous examples where the guidance came from heuristic considerations, it is quite instructive to study this case where no heuristic help is relied on and solutions are found by a purely mathematical trick,—the reversal referred to above.

For the reader who might be dissatisfied with the use of such devices (i.e. exclusively technical and non-conceptual ones), we submit that they are freely and legitimately used in mathematical analysis.

We have repeatedly found the heuristic procedure easier to handle than the rigorous one. The present case offers an example of the opposite.

This game is no longer symmetric.[1] Hence the permutations of the players 1,2,3,4 are now legitimate only if accompanied by the corresponding permutations of y_1, y_2, y_3, y_4.[2]

To begin with, the smallest component with which a given player k is ever associated in the vectors of (37:2) in 37.3.2., is -1. Hence the vectors will be imputations if and only if $-1 \geqq v((k))$ i.e.

(38:8) $$y_k \geqq 1 \quad \text{for} \quad k = 1,2,3,4.$$

The character of V as a set of imputations is thus established; let us now see whether it is a solution. This investigation is similar to the proof given in 36.2.3-5.

38.2.2. The observations (36:5), (36:6), of 36.2.3., apply again. A two-element set $S = (i, j)$ is effective for $\overrightarrow{\alpha} = \{\alpha_1, \alpha_2, \alpha_3, \alpha_4\}$ when $\alpha_i + \alpha_j \leqq 0$ (cf. (38:A)). Hence we have for the $\overrightarrow{\alpha}'$, $\overrightarrow{\alpha}''$ of (37:2): In $\overrightarrow{\alpha}''$ every two-element set S is effective. In $\overrightarrow{\alpha}'$: No two-element set S which does not contain the player 4 is effective, while those which do contain him, $S = (1,4)$, (2,4), (3,4) clearly are. However, if we consider $S = (1,4)$, we may discard the two others; $S = (2,4)$ arises from it by interchanging 1 and 2, which does not affect $\overrightarrow{\alpha}'$;[3] $S = (3,4)$ is actually inferior to it after 1 and 3 are interchanged, since $\frac{1}{2} \geqq 0$.[4]

Summing up:

(38:B) Among the two-element sets S, those given below are certainly necessary, and all others are certainly unnecessary:

$$(1,4) \text{ for } \overrightarrow{\alpha}',^{5} \text{ all for } \overrightarrow{\alpha}''.$$

Concerning three-element sets: Owing to the above we may exclude by (36:6) all three-element sets for $\overrightarrow{\alpha}''$, and for $\overrightarrow{\alpha}'$ those which contain (1,4) or (2,4).[6] This leaves only $S = (1,2,3)$ for $\overrightarrow{\alpha}'$.

Summing up:

(38:C) Among the three-element sets S, the one given below is certainly necessary, and all others are certainly unnecessary:

$$(1,2,3) \text{ for } \overrightarrow{\alpha}'.$$

[1] Unless $y_1 = y_2 = y_3 = y_4$.

[2] But there is nothing objectionable in such uses of the permutations of 1,2,3,4 as we made in the formulation of (37:2) in 37.3.2.

[3] This permutation and similar ones later are clearly legitimate devices in spite of footnote 1 above. Observe footnote 1 on p. 309 and footnote 2 above.

[4] As $\alpha_4' = -1$ we could discard all these sets, including $S = (1,4)$, when $v((4)) = -1$; i.e. when $y_4 = 1$, which is a possibility. But we are under no obligation to do this. We prefer not to do it, in order to be able to treat $y_4 = 1$ and $y_4 > 1$ together.

[5] And all permutations of 1,2,3,4; these modify $\overrightarrow{\alpha}'$ too.

[6] The latter obtains from the former by interchanging 1 and 2, which does not affect $\overrightarrow{\alpha}'$.

We leave to the reader the verification of (30:5:a) in 30.1.1., i.e. that no $\overrightarrow{\alpha}'$ of V dominates any $\overrightarrow{\beta}$ of V. (Cf. the corresponding part of the proof in 36.2.4. Actually the proof of (30:5:b), which follows, also contains the necessary steps.)

38.2.3. We next verify (30:5:b) in 30.1.1., i.e. that an imputation $\overrightarrow{\beta}$ which is undominated by the elements of V must belong to V.

Consider a $\overrightarrow{\beta}$ undominated by the elements of V. If any two of β_1, β_2, β_3, β_4 were < 0, we could make these (by permuting 1,2,3,4) $\beta_1, \beta_2 < 0$. This gives $\overrightarrow{\alpha}'' \succ \overrightarrow{\beta}$ with $S = (1,2)$ of (38:B). Hence at most one of β_1, β_2, β_3, β_4 is < 0. If none is < 0, then all are $\geqq 0$. So each component of $\overrightarrow{\beta}$ is $\geqq$ the corresponding component of $\overrightarrow{\alpha}''$, and since both are imputations (cf. footnote 2 on p. 311), it follows that they coincide, $\overrightarrow{\beta} = \overrightarrow{\alpha}''$; and so it is in V.

Hence precisely one of β_1, β_2, β_3, β_4 is < 0. By permuting 1,2,3,4 we can make it β_4.

If any two of β_1, β_2, β_3 were $< \frac{1}{2}$, we could make these (by permuting 1,2,3) $\beta_1, \beta_2 < \frac{1}{2}$. Besides, $\beta_4 < 0$. So the interchange of 3 and 4 gives $\overrightarrow{\alpha}' \succ \overrightarrow{\beta}$ with $S = (1,2,3)$ of (38:C). Hence at most one of β_1, β_2, β_3 is $< \frac{1}{2}$. If none is $< \frac{1}{2}$, then $\beta_1, \beta_2, \beta_3 \geqq \frac{1}{2}$. Hence $\beta_4 \leqq -\frac{3}{2}$. But $\beta_4 \geqq v((4)) = -y_4$, so this necessitates $-y_4 \leqq -\frac{3}{2}$, i.e. $y_4 \geqq \frac{3}{2}$. Thus we need $y_4 < \frac{3}{2}$ to exclude this possibility, and as we are permuting freely 1,2,3,4, we even need

$$(38:9) \qquad y_k < \tfrac{3}{2} \quad \text{for} \quad k = 1,2,3,4.$$

If this condition is satisfied, then we can conclude that precisely one of β_1, β_2, β_3 is $< \frac{1}{2}$. By permuting 1,2,3, we can make it β_3.

So $\beta_1, \beta_2 \geqq \frac{1}{2}$, $\beta_3 \geqq 0$. If $\beta_4 \geqq -1$,[1] then each component of $\overrightarrow{\beta}$ is $\geqq$ the corresponding component of $\overrightarrow{\alpha}'$, and since both are imputations (cf. footnote 2 on p. 311) it follows that they coincide: $\overrightarrow{\beta} = \overrightarrow{\alpha}'$ and so it is in V.

Hence $\beta_4 < -1$. Also $\beta_3 < \frac{1}{2}$. So interchange of 1 and 3 gives $\overrightarrow{\alpha}' \succ \overrightarrow{\beta}$ with $S = (1,4)$ of (38:B).

This, at last, is a contradiction, and thereby completes the verification of (30:5:b) in 30.1.1.

The condition (38:9), which we needed for this proof, is really necessary: it is easy to verify that

$$\overrightarrow{\beta}' = \{\tfrac{1}{2}, \tfrac{1}{2}, \tfrac{1}{2}, -\tfrac{3}{2}\}$$

[1] If $v((4)) = -1$, i.e. if $y_4 = 1$, then this is certainly the case; but we do not wish to assume it. (Cf. footnote 4 on p. 323.)

is undominated by our V, and the only way to prevent it from being an imputation is to have $-\frac{3}{2} < v((4)) = -y_4$, i.e. $y_4 < \frac{3}{2}$.[1] Permuting 1,2,3,4 then gives (38:9).

Thus we need precisely (38:8) and (38:9). Summing up:

(38:D) The V of (37:2) in 37.3.2. is a solution for a game of (38:A) (with (38:6), (38:7) there) if and only if

(38:10) $$1 \leqq y_k < \frac{3}{2} \quad \text{for} \quad k = 1,2,3,4.$$

38.2.4. Let us now reintroduce the normalization and reduction which we abandoned temporarily, but which are necessary in order to refer these results to Q, as pointed out immediately after (38:A).

The reduction formulae of 27.1.4. show that the share of the player k must be altered by the amount α_k^0 where

$$\begin{aligned} \alpha_k^0 &= -v((k)) + \tfrac{1}{4}\{v((1)) + v((2)) + v((3)) + v((4))\} \\ &= y_k - \tfrac{1}{4}(y_1 + y_2 + y_3 + y_4) \end{aligned}$$

and

$$\begin{aligned} \gamma &= -\tfrac{1}{4}\{v((1)) + v((2)) + v((3)) + v((4))\} \\ &= \tfrac{1}{4}(y_1 + y_2 + y_3 + y_4). \end{aligned}$$

For a two-element set $S = (i, j)$, $v(S)$ is increased from its original value 0 to

$$\begin{aligned} \alpha_i^0 + \alpha_j^0 &= y_i + y_j - \tfrac{1}{2}(y_1 + y_2 + y_3 + y_4) \\ &= \tfrac{1}{2}(y_i + y_j - y_k - y_l) \end{aligned}$$

(k, l are the two players other than i, j).

The above γ is clearly $\geqq 1 > 0$ (by (38:10)), hence the game is essential. The normalization is now carried out by dividing the characteristic function, as well as every player's share, by γ. Thus for $S = (i, j)$, $v(S)$ is now modified further to

$$\frac{\alpha_i^0 + \alpha_j^0}{\gamma} = 2\,\frac{y_i + y_j - y_k - y_l}{y_1 + y_2 + y_3 + y_4}.$$

This then is the normalized and reduced form of the characteristic function, as used in 34.2.1. for the representation by Q. (34:2) id. gives, together with the above expression, the formulae

[1] Observe that the failure of V to dominate this $\overrightarrow{\beta}'$ could not be corrected by adding $\overrightarrow{\beta}'$ to V (when $y_4 \leqq \frac{3}{2}$). Indeed, $\overrightarrow{\beta}'$ dominates $\overrightarrow{\alpha}'' = \{0,0,0,0\}$ with $S = (1,2,3)$, so it would be necessary to remove $\overrightarrow{\alpha}''$ from V, thereby creating new undominated imputations, etc.

If $y_1 = y_2 = y_3 = y_4 = \frac{3}{2}$, then a change of unit by $\frac{2}{3}$ brings our game back to the form (37:1) of 37.1.2., and it carries the above $\overrightarrow{\beta}'$ into the $\overrightarrow{\alpha}^{IV} = \{\frac{1}{3}, \frac{1}{3}, \frac{1}{3}, -1\}$ of (37:3) in 37.4.1. Thus further attempts to make our V over into a solution would probably transform it gradually into (37:3), (37:4) in 37.4.1-2. This is noteworthy, since we started with (37:2) in 37.3.2.

These connections between the two solutions (37:2) and (37:3), (37:4) should be investigated further.

(38:11)
$$\begin{cases} x_1 = \dfrac{y_1 - y_2 - y_3 + y_4}{y_1 + y_2 + y_3 + y_4}, \\ x_2 = \dfrac{-y_1 + y_2 - y_3 + y_4}{y_1 + y_2 + y_3 + y_4}, \\ x_3 = \dfrac{-y_1 - y_2 + y_3 + y_4}{y_1 + y_2 + y_3 + y_4}, \end{cases}$$

for the coordinates x_1, x_2, x_3 in Q.

38.2.5. Thus (38:10) and (38:11) together define the part of Q in which these solutions—i.e. the solution (37:2) in 37.3.2., transformed as indicated above—can be used. This definition is exhaustive, but implicit. Let us make it explicit. I.e., given a point of Q with the coordinates x_1, x_2, x_3, let us decide whether (38:10) and (38:11) can then be satisfied together (by appropriate y_1, y_2, y_3, y_4).

We put for the hypothetical y_1, y_2, y_3, y_4

(38:12)
$$y_1 + y_2 + y_3 + y_4 = \frac{4}{z}$$

with z indefinite. Then the equations (38:11) become

(38:12*)
$$\begin{cases} y_1 - y_2 - y_3 + y_4 = \dfrac{4x_1}{z}, \\ -y_1 + y_2 - y_3 + y_4 = \dfrac{4x_2}{z}, \\ -y_1 - y_2 + y_3 + y_4 = \dfrac{4x_3}{z}. \end{cases}$$

(38:12) and (38:12*) can be solved with respect to y_1, y_2, y_3, y_4:

(38:13)
$$\begin{cases} y_1 = \dfrac{1 + x_1 - x_2 - x_3}{z}, & y_2 = \dfrac{1 - x_1 + x_2 - x_3}{z}, \\ y_3 = \dfrac{1 - x_1 - x_2 + x_3}{z}, & y_4 = \dfrac{1 + x_1 + x_2 + x_3}{z}. \end{cases}$$

Now (38:11) is satisfied, and we must use our freedom in choosing z to satisfy (38:10).

Let w be the greatest and v the smallest of the four numbers

(38:14)
$$\begin{matrix} u_1 = 1 + x_1 - x_2 - x_3, & u_2 = 1 - x_1 + x_2 - x_3, \\ u_3 = 1 - x_1 - x_2 + x_3, & u_4 = 1 + x_1 + x_2 + x_3. \end{matrix}$$

These are known quantities, since x_1, x_2, x_3 are assumed to be given.

Now (38:10) clearly means that $1 \leqq v/z$ and that $w/z < \frac{3}{2}$, i.e. it means that

(38:15)
$$\tfrac{2}{3}w < z \leqq v.$$

Obviously this condition can be fulfilled (for z) if and only if

(38:16)
$$\tfrac{2}{3}w < v.$$

And if (38:16) is satisfied, then condition (38:15) allows infinitely many values—an entire interval—for z.

38.2.6. Before we draw any conclusions from (38:15), (38:16), we give the explicit formulae which express what has become of the solution (37:2) of 37.3.2. owing to our transformations. We must take the imputations $\overrightarrow{\alpha}'$, $\overrightarrow{\alpha}''$, loc. cit., add the amount α_k to the component k (i.e. to the player k's share), and divide this by γ.

These manipulations transform the possible values of the component k —which are $\frac{1}{2}$, 0, -1 in (37:2)—as follows. We consider first $k = 1$, and use the above expressions for α_k and γ as well as (38:13). Then:

$$\frac{1}{2} \text{ goes into } \frac{\frac{1}{2} + \alpha_1}{\gamma} = \frac{2 + 4y_1 - (y_1 + y_2 + y_3 + y_4)}{y_1 + y_2 + y_3 + y_4} = \frac{z}{2} + x_1 - x_2 - x_3,$$

$$0 \text{ goes into } \frac{\alpha_1}{\gamma} = \frac{4y_1 - (y_1 + y_2 + y_3 + y_4)}{y_1 + y_2 + y_3 + y_4} = x_1 - x_2 - x_3,$$

$$-1 \text{ goes into } \frac{-1 + \alpha_1}{\gamma} = \frac{-4 + 4y_1 - (y_1 + y_2 + y_3 + y_4)}{y_1 + y_2 + y_3 + y_4} = -z + x_1 - x_2 - x_3.$$

For the other $k = 2,3,4$ these expressions are changed only in so far that their $x_1 - x_2 - x_3$ is replaced by $-x_1 + x_2 - x_3$, $-x_1 - x_2 + x_3$, $x_1 + x_2 + x_3$, respectively.[1]

Summing up (and recalling (38:14)):

(38:E) The component k is transformed as follows:
$\frac{1}{2}$ goes into $z/2 + u_k - 1$,
0 goes into $u_k - 1$,
-1 goes into $-z + u_k - 1$,
with the u_1, u_2, u_3, u_4 of (38:14).

We leave it to the reader to restate (37:2) with the modification (38:E), paying due attention to carrying out correctly the permutations 1,2,3,4 which are required there.

It will be noted that for the center—i.e. $x_1 = x_2 = x_3 = 0$—(38:E) reproduces the formulae (38:1) of 38.1.1., as it should.

38.2.7. We now return to the discussion of (38:15), (38:16).

Condition (38:16) expresses that the four numbers u_1, u_2, u_3, u_4 of (38:14) are not too far apart—that their minimum is more than $\frac{2}{3}$ of their maximum —i.e. that on a relative scale their sizes vary by less than 2:3.

This is certainly true at the center, where $x_1 = x_2 = x_3 = 0$; there u_1, u_2, u_3, u_4 are all $= 1$. Hence in this case $v = w = 1$, and (38:15)

[1] This is immediate, owing to the form of the equations (38:13), and equally by considering the influence of the permutations of the players 1,2,3,4 on the coordinates x_1, x_2, x_3 as described in 34.3.2.

becomes $\frac{2}{3} < z \leqq 1$ proving the assertions made earlier in this discussion (cf. (38:3) in 38.1.1.).

Denote the part of Q in which (38:16) is true by Z. Then even a sufficiently small neighborhood of the center belongs to Z.[1] So Z is a three-dimensional piece in the interior of Q, containing the center in its own interior.

We can also express the relationship of Z to the diagonals of Q, say to I-Center-VIII. Z contains the following parts of that diagonal. (Use Figure 64): on one side precisely C, on the other a little less than half of B.[2] We add that these solutions are different from the family of solutions valid in (36:B) and (36:C) which were referred to in 36.3.

38.3. Interpretation of The Solutions

38.3.1. The family of solutions which we have thus determined possesses several remarkable features.

We note first that for every game for which this family is a solution at all (i.e. in every point of Z) it gives infinitely many solutions.[3] And all we said in 37.5.1. applies again: these solutions are finite sets of imputations[4] and possess the full symmetry of the game.[5] Thus there is no "discrimination" in any one of these solutions. Nor can the differences in "organizational principles," which we discussed loc. cit., be ascribed to them. There is nevertheless a simple "organizational principle" that can be stated in a qualitative verbal form, to distinguish these solutions. We proceed to formulate it.

38.3.2. Consider (38:E), which expresses the changes to which (37:2) in 37.3.2. is to be subjected. It is clear that the worst possible outcome for the player k in this solution is the last expression (since this corresponds to -1),—i.e. $-z + u_k - 1$. This expression is $>$ or $= -1$, according to whether z is $<$ or $= u_k$. Now u_1, u_2, u_3, u_4 are the four numbers of (38:14), the smallest of which is v. By (38:15) $z \leqq v$, i.e. always $-z + u_k - 1 \geqq -1$, and $=$ occurs only for the greatest possible value of z, $z = v$,—and then only for those k for which u_k attains its minimum, v.

[1] If x_1, x_2, x_3 differ from 0 by $< \frac{1}{15}$ then each of the four numbers u_1, u_2, u_3, u_4 of (38:14) is $< 1 + \frac{3}{15} = \frac{6}{5}$ and $> 1 - \frac{3}{15} = \frac{4}{5}$; hence on a relative size they vary by $< \frac{6}{5} : \frac{4}{5} = \frac{3}{2}$. So we are still in Z. In other words: Z contains a cube with the same center as Q, but with $\frac{1}{15}$ of Q's (linear) size.

Actually Z is somewhat bigger than this, its volume is about $\frac{1}{1000}$ of the volume of Q.

[2] On that diagonal $x_1 = x_2 = x_3$, so the u_1, u_2, u_3, u_4 are: (three times) $1 - x_1$ and $1 + 3x_1$. So for $x_1 \geqq 0$, $v = 1 - x_1$, $w = 1 + 3x_1$, hence (38:16) becomes $x_1 < \frac{1}{9}$. And for $x_1 \leqq 0$, $v = 1 + 3x_1$, $w = 1 - x_1$, hence (38:16) becomes $> -\frac{1}{11}$. So the intersection is this:

$$0 \leqq x_1 < \tfrac{1}{9} \qquad \text{(this is precisely } C\text{)}$$
$$0 \geqq x_1 > -\tfrac{1}{11} \qquad (B \text{ is } 0 \geqq x_1 > -\tfrac{1}{5}).$$

[3] The solution which we found contained four parameters: y_1, y_2, y_3, y_4 while the games for which they are valid had only three parameters: x_1, x_2, x_3.

[4] Each one has 13 elements, like (37:2) in 37.3.2.

[5] In the center $x_1 = x_2 = x_3 = 0$ we have $y_1 = y_2 = y_3 = y_4$ (cf. (38:13)), i.e. symmetry in 1,2,3,4. On the diagonal $x_1 = x_2 = x_3$ we have $y_1 = y_2 = y_3$ (cf. (38:13)), i.e. symmetry in 1,2,3.

We restate this:

(38:F) In this family of solutions, even as the worst possible outcome, a player k faces, in general, something that is definitely better than what he could get for himself alone,—i.e. $v((k)) = -1$. This advantage disappears only when z has its greatest possible value, $z = v$, and then only for those k for which the corresponding number u_1, u_2, u_3, u_4 in (38:14) attains the minimum in (38:14).

In other words: In these solutions a defeated player is in general not "exploited" completely, not reduced to the lowest possible level—the level which he could maintain even alone, i.e. $v((k)) = -1$. We observed before such restraint on the part of a victorious coalition, in the "milder" kind of "discriminatory" solutions of the three-person game discussed in 33.1. (i.e. when $c > -1$, cf. the end of 33.1.2.). But there only one player could be the object of this restraint in any one solution, and this phenomenon went with his exclusion from the competition for coalitions. Now there is no discrimination or segregation; instead this restraint applies to all players in general, and in the center of Q ((38:1) in 38.1.1., with $z < 1$) the solution is even symmetric![1]

38.3.3. Even when z assumes its maximum value v, in general only one player will lose this advantage, since in general the four numbers u_1, u_2, u_3, u_4 of (38:14) are different from each other and only one is equal to their minimum v. All four players will lose it simultaneously only if u_1, u_2, u_3, u_4 are all equal to their minimum v—i.e. to each other—and one look at (38:14) suffices to show that this happens only when $x_1 = x_2 = x_3 = 0$, i.e. at the center.

This phenomenon of not "exploiting" a defeated player completely is a very important possible (but by no means necessary) feature of our solutions,—i.e. of social organizations. It is likely to play a greater role in the general theory also.

We conclude by stating that some of the solutions which we mentioned, but failed to describe in 36.3.2., also possess this feature. These are the solutions in C of Figure 64. But nevertheless they differ from the solutions which we have considered here.

[1] There is a quantitative difference of some significance as well. Both in our present set-up (four-person game, center of Q) and in the one referred to (three-person game in the sense of 33.1.), the best a player can do (in the solutions which we found) is $\frac{1}{2}$, and the worst is -1.

The upper limit of what he may get in case of defeat, in those of our solutions where he is not completely "exploited," is now $-\frac{2}{3}$ (i.e. $-z$ with $\frac{2}{3} < z \leqq 1$) and it was $\frac{1}{2}$ then (i.e. c with $-1 \leqq c < \frac{1}{2}$). So this zone now covers the fraction $\frac{(-\frac{2}{3}) - (-1)}{\frac{1}{2} - (-1)} = \frac{\frac{1}{3}}{\frac{3}{2}} = \frac{2}{9}$, i.e. $22\frac{2}{9}$ % of the significant interval, while it then covered 100%.

CHAPTER VIII

SOME REMARKS CONCERNING $n \geqq 5$ PARTICIPANTS

39. The Number of Parameters in Various Classes of Games

39.1. The Situation for $n = 3,4$

39.1. We know that the essential games constitute our real problem and that they may always be assumed in the reduced form and with $\gamma = 1$. In this representation there exists precisely one zero-sum three-person game, while the zero-sum four-person games form a three-dimensional manifold.[1] We have seen further that the (unique) zero-sum three-person game is automatically symmetric, while the three-dimensional manifold of all zero-sum four-person games contains precisely one symmetric game.

Let us express this by stating, for each one of the above varieties of games, how many dimensions it possesses,—i.e. how many indefinite parameters must be assigned specific (numerical) values in order to characterize a game of that class. This is best done in the form of a table, given in Figure 65 in a form extended to all $n \geqq 3$.[2] Our statements above reappear in the entries $n = 3,4$ of that table.

39.2. The Situation for All $n \geqq 3$

39.2.1. We now complete the table by determining the number of parameters of the zero-sum n-person game, both for the class of all these games, and for that of the symmetric ones.

The characteristic function is an aggregate of as many numbers $\mathrm{v}(S)$ as there are subsets S in $I = (1, \cdots, n)$,—i.e. of 2^n. These numbers are subject to the restrictions (25:3:a)-(25:3:c) of 25.3.1., and to those due to the reduced character and the normalization $\gamma = 1$, expressed by (27:5) in 27.2. Of these (25:3:b) fixes $\mathrm{v}(-S)$ whenever $\mathrm{v}(S)$ is given, hence it halves the number of parameters:[3] so we have 2^{n-1} instead of 2^n. Next (25:3:a) fixes one of the remaining $\mathrm{v}(S)$: $\mathrm{v}(\ominus)$; (27:5) fixes n of the remaining $\mathrm{v}(S)$: $\mathrm{v}((1)), \cdots, \mathrm{v}((n))$; hence they reduce the number of parameters by $n + 1$.[4] So we have $2^{n-1} - n - 1$ parameters. Finally (25:3:c) need not be considered, since it contains only inequalities.

39.2.2. If the game is symmetric, then $\mathrm{v}(S)$ depends only on the number of elements p of S: $\mathrm{v}(S) = \mathrm{v}_p$, cf. 28.2.1. Thus it is an aggregate of as many numbers v_p as there are $p = 0, 1, \cdots, n$,—i.e. $n + 1$. These

[1] Concerning the general remarks, cf. 27.1.4. and 27.3.2.; concerning the zero-sum three-person game cf. 29.1.2.; concerning the zero-sum four-person game cf. 34.2.1.

[2] There are no essential zero-sum games for $n = 1,2$!

[3] S and $-S$ are never the same set!

[4] $S = \ominus, (1), \cdots, (n)$ differ from each other and from each other's complements.

numbers are subject to the restrictions (28:11:a)-(28:11:c) of 28.2.1.; the reduced character is automatic, and we demand also $v_1 = -\gamma = -1$. (28:11:b) fixes v_{n-p} when v_p is given; hence it halves the numbers of those parameters for which $n - p \neq p$. When $n - p = p$[1]—i.e. $n = 2p$, which happens only when n is even, and then $p = n/2$ —(28:11:b) shows that this v_p must vanish. So we have $\frac{n+1}{2}$ parameters if n is odd and $\frac{n}{2}$ if n is even, instead of the original $n + 1$. Next (28:11:a) fixes one of the remaining v_p: v_0; and $v_1 = -\gamma = -1$ fixes another one of the remaining v_p: v_1; hence they reduce the number of parameters by 2:[2] so we have $\frac{n+1}{2} - 2$ or $\frac{n}{2} - 2$ parameters. Finally (28:11:c) need not be considered since it contains only inequalities.

39.2.3. We collect all this information in the table of Figure 65. We also state explicitly the values arising by specialization to $n = 3,4,5,6,7,8,$— the first two of which were referred to previously.

Number of players	All games	Symmetric games
3	0*	0*
4	3	0*
5	10	1
6	25	1
7	56	2
8	119	2
. . .	. . .	. . .
n	$2^{n-1} - n - 1$	$\frac{n+1}{2} - 2$ for n odd $\frac{n}{2} - 2$ for n even

* Denotes the game is unique.

Figure 65.—Essential games. (Reduced and $\gamma = 1$.)

The rapid increase of the entries in the left-hand column of Figure 65 may serve as another indication, if one be needed, how the complexity of a game increases with the number of its participants. It seems noteworthy

[1] Contrast this with footnote 3 on p. 330!

[2] $p = 0, 1$ differ from each other and from each other's $n - p$. (The latter only because of $n \geq 3$.)

that there is an increase in the right-hand column too, i.e. for the symmetric games, but a much slower one.

40. The Symmetric Five-person Game

40.1. Formalism of the Symmetric Five-person Game

40.1.1. We shall not attempt a direct attack on the zero-sum five-person game. The systematic theory is not far enough advanced to allow it; and for a descriptive and casuistic approach (as used for the zero-sum, four-person game) the number of its parameters, 10, is rather forbidding.

It is possible however to examine the symmetric zero-sum five-person games in the latter sense. The number of parameters, 1, is small but not zero, and this is a qualitatively new phenomenon deserving consideration. For $n = 3,4$ there existed only one symmetric game, so it is for $n = 5$ that it happens for the first time that the structure of the symmetric game shows any variety.

40.1.2. The symmetric zero-sum five-person game is characterized by the v_p, $p = 0,1,2,3,4,5$ of 28.2.1., subject to the restrictions (28:11:a)-(28:11:c) formulated there. (28:11:a), (28:11:b) state (with $\gamma = 1$)

(40:1) $$v_0 = 0, \qquad v_1 = -1, \qquad v_4 = 1, \qquad v_5 = 0$$

and $v_2 = -v_3$, i.e.

(40:2) $$v_2 = -\eta, \qquad v_3 = \eta$$

Now (28:11:c) states $v_{p+q} \geqq v_p + v_q$ for $p + q \leqq 5$ and we can subject p, q to the further restrictions of (28:12) id. Therefore $p = 1$, $q = 1,2$,[1] and so these two inequalities obtain (using (40:1), (40:2)):

$$p = 1,\ q = 1: \quad -2 \leqq -\eta; \qquad p = 1,\ q = 2: \quad -1 - \eta \leqq \eta;$$

i.e.

(40:3) $$-\tfrac{1}{2} \leqq \eta \leqq 2.$$

Summing up:

(40:A) The symmetric zero-sum five-person game is characterized by one parameter η with the help of (40:1), (40:2). The domain of variability of η is (40:3).

40.2. The Two Extreme Cases

40.2.1. It may be useful to give a direct picture of the symmetric games described above. Let us first consider the two extremes of the interval (40:3):

$$\eta = 2, \ -\tfrac{1}{2}.$$

[1] This is easily verified by inspection of (28:12), or by using the inequalities of footnote 2 on p. 259. These give $1 \leqq p \leqq \frac{5}{3}$, $1 \leqq q \leqq 2$; hence as p, q are integers, $p = 1$, $q = 1,2$.

Consider first $\eta = 2$: In this case $v(S) = -2$ for every two-element set S; i.e. every coalition of two players is defeated.[1] Thus a coalition of three (being the set complementary to the former) is a winning coalition. This tells the whole story: In the gradual crystallization of coalitions, the point at which the passage from defeat to victory occurs is when the size increases from two to three, and at this point the transition is 100%.[2]

Summing up:

(40:B) $\eta = 2$ describes a game in which the only objective of all players is to form coalitions of three players.

40.2.2. Consider next $\eta = -\frac{1}{2}$. In this case we argue as follows:

$$v(S) = \begin{cases} 1 \\ \\ \frac{1}{2} \end{cases} \quad \text{when } S \text{ has } \begin{cases} 4 \\ \\ 2 \end{cases} \text{ elements.}$$

A coalition of four always wins.[3]

Now the above formula shows that a coalition of two is doing just as well, *pro rata*, as a coalition of four; hence it is reasonable to consider the former just as much winning coalitions as the latter. If we take this broader view of what constitutes winning, we may again affirm that the whole story of the game has been told: In the formation of coalitions, the point at which the passage from defeat to victory occurs is when the size increases from one to two; at this point the transition is 100%.[4]

Summing up:

(40:C) $\eta = -\frac{1}{2}$ describes a game in which the only objective of all players is to form coalitions of two players.

40.2.3. On the basis of (40:B) and (40:C) it would be quite easy to guess heuristically solutions for their respective games. This, as well as the exact proof that those sets of imputations are really solutions, is easy; but we shall not consider this matter further.

Before we pass to the consideration of the other η of (40:3) let us remark that (40:B) and (40:C) are obviously the simplest instances of a general

[1] Cf. the discussion in 35.1.1., particularly footnote 4 on p. 296.

[2] One player is just as much defeated as two, four are no more victorious than three. Of course a coalition of three has no motive to take in a fourth partner; it seems (heuristically) plausible that if they do they will accept him only on the worst possible terms. But nevertheless such a coalition of four wins if viewed as a unit, since the remaining isolated player is defeated.

[3] In any zero-sum n-person game any coalition of $n - 1$ wins, since an isolated player is always defeated. Cf. loc. cit. above.

[4] One player is defeated, two or four players are victorious. A coalition of three players is a composite case deserving some attention: $v(S)$ is $-\frac{1}{2}$ for a three-element set S, i.e. it obtains from the $\frac{1}{2}$ of a two-element set by addition of -1. Thus a coalition of three is no better than a winning coalition of two (which it contains) plus the remaining isolated and defeated player separately. This coalition is just a combination of a winning and a defeated group whose situation is entirely unaltered by this operation.

method of defining games. This procedure (which is more general than that of Chapter X, referred to in footnote 4 on p. 296) will be considered exhaustively elsewhere (for asymmetric games also). It is subject to some restrictions of an arithmetical nature; thus it is clear that there can be no (essential symmetric zero-sum) n-person game in which every coalition of p is winning if p is a divisor of n, since then n/p such coalitions could be formed and everybody would win with no loser left. On the other hand the same requirement with $p = n - 1$ does not restrict the game at all (cf. footnote 3, p. 333).

40.3. Connection between the Symmetric Five-person Game and the 1,2,3-symmetric Four-person Game

40.3.1. Consider now the η in the interior of (40:3). The situation is somewhat similar to that discussed at the end of 35.3. We have some heuristic insight into the conditions at the two ends of (40:3) (cf. above). Any point η of (40:3) is somehow "surrounded" by these end-points. More precisely, it is their center of gravity if appropriate weights are used.[1] The remarks made loc. cit. apply again: while this construction represents all games of (40:3) as combinations of the extreme cases (40:B), (40:C), it is nevertheless not justified to expect that the strategies of the former can be obtained by any direct process from those of the latter. Our experiences in the case of the zero-sum four-person game speak for themselves.

There is, however, another analogy with the four-person game which gives some heuristic guidance. The number of parameters in our case is the same as for those zero-sum four-person games which are symmetric with respect to the players 1,2,3; we have now the parameter η which runs over

$$-\tfrac{1}{2} \leqq \eta \leqq 2, \qquad (40:3)$$

while the games referred to had the parameter x_1 which varies over

$$-1 \leqq x_1 \leqq 1.^{2} \qquad (40:4)$$

This analogy between the (totally) symmetric five-person game and the 1,2,3-symmetric four-person game is so far entirely formal. There is, however, a deeper significance behind it. To see this we proceed as follows:

40.3.2. Consider a symmetric five-person game Γ with its η in (40:3). Let us now modify this game by combining the players 4 and 5 into one person, i.e. one player 4'. Denote the new game by Γ'. It is important to realize that Γ' is an entirely new game: we have not asserted that in Γ players 4 and 5 will necessarily act together, form a coalition, etc., or that there are any generally valid strategical considerations which would motivate just this coalition.[3] We have forced 4 and 5 to combine; we did this by modifying the rules of the game and thereby replacing Γ by Γ'.

[1] The reader can easily carry out this composition in the sense of footnote 1 on p. 304, relying on our equations (40:1), (40:2) in 40.1.2.

[2] Cf. 35.3.2. In the representation in Q used there, $x_1 = x_2 = x_3$.

[3] This ought to be contrasted with the discussion in 36.1.2., where a similar combination of two players was formed under such conditions that this merger seemed strategically justified.

Now Γ is a symmetric five-person game, while Γ' is a 1,2,3-symmetric four-person game.[1] Given the η of Γ we shall want to determine the x_1 of Γ' in order to see what correspondence of (40:3) and (40:4) this defines. Afterwards we shall investigate whether there are not, in spite of what was said above, some connections between the strategies—i.e. the solutions—of Γ and Γ'.

The characteristic function $v'(S)$ of Γ' is immediately expressible in terms of the characteristic function $v(S)$ of Γ. Indeed:

$$\begin{array}{ll} v'((1)) = v((1)) = -1, & v'((2)) = v((2)) = -1, \\ v'((3)) = v((3)) = -1, & v'((4')) = v((4,5)) = -\eta; \\ v'((1,2)) = v((1,2)) = -\eta, & v'((1,3)) = v((1,3)) = -\eta, \\ v'((2,3)) = v((2,3)) = -\eta, & v'((1,4')) = v((1,4,5)) = \eta, \\ v'((2,4')) = v((2,4,5)) = \eta, & v'((3,4')) = v((3,4,5)) = \eta; \\ v'((1,2,3)) = v((1,2,3)) = \eta, & v'((1,2,4')) = v((1,2,4,5)) = 1, \\ v'((1,3,4')) = v((1,3,4,5)) = 1, & v'((2,3,4')) = v((2,3,4,5)) = 1; \end{array}$$

and of course

$$v'(\ominus) = v'((1,2,3,4')) = 0.$$

While Γ was normalized and reduced, Γ' is neither; and we must bring Γ' into that form since we want to compute its x_1, x_2, x_3, i.e. refer it to the Q of 34.2.2.

Let us therefore apply first the normalization formulae of 27.1.4. They show that the share of the player $k = 1,2,3,4'$ must be altered by the amount α_k^0 where

$$\alpha_k^0 = -v'((k)) + \tfrac{1}{4}\{v'((1)) + v'((2)) + v'((3)) + v'((4'))\},$$

and

$$\gamma = -\tfrac{1}{4}\{v'((1)) + v'((2)) + v'((3)) + v'((4'))\}.$$

Hence

$$\alpha_1^0 = \alpha_2^0 = \alpha_3^0 = \frac{1-\eta}{4}, \qquad \alpha_4^0 = -\frac{3(1-\eta)}{4}, \qquad \gamma = \frac{3+\eta}{4}.$$

This γ is clearly $\geqq \dfrac{3-\frac{1}{2}}{4} = \dfrac{5}{8} > 0$ (by (40:3)); hence the game is essential. The normalization is now carried out by dividing every player's share by γ.

Thus for a two-element set $S = (i, j)$, $v'(S)$ is replaced by

$$v''(S) = \frac{v'(S) + \alpha_i^0 + \alpha_j^0}{\gamma}.$$

Consequently a simple computation yields

$$v''((1,2)) = v''((1,3)) = v''((2,3)) = -\frac{2(3\eta - 1)}{3+\eta},$$

$$v''((1,4')) = v''((2,4')) = v''((3,4')) = \frac{2(3\eta - 1)}{3+\eta}.$$

[1] The participants in Γ are players 1,2,3,4,5, who all have the same role in the original Γ. The participants in Γ' are players 1,2,3 and the composite player (4,5): 4'. Clearly 1,2,3 have still the same role, but 4' is different.

This then is the normalized and reduced form of the characteristic function, as used in 34.2. for the representation by Q. (34:2) in 34.2.1 gives, together with the above expression, the formula

$$x_1 = x_2 = x_3 = \frac{3\eta - 1}{3 + \eta}.$$

Taking $x_1 = x_2 = x_3$ for granted, this relation can also be written as follows:

$$(3 - x_1)(3 + \eta) = 10. \tag{40:5}$$

Now it is easy to verify that (40:5) maps the η-domain (40:3) on the x_1-domain (40:4). The mapping is obviously monotone. Its details are shown in Figure 66 and in the adjoining table of corresponding values of x_1 and η. The curve in this figure represents the relation (40:5) in the x_1, η-plane. This curve is clearly (an arc of) a hyperbola.

40.3.3. Our analysis of the 1,2,3-symmetric four-person game has culminated in the result stated in 36.3.2.: The game, i.e. the diagonal I-Center-VIII in Q which represents them, is divided into five classes A–E, each of which is characterized by a certain qualitative type of solutions. The positions of the zones A–E on the diagonal I-Center-VIII, i.e. the interval $-1 \leqq x_1 \leqq 1$, are shown in Figure 64.

The present results suggest therefore the consideration of the corresponding classes of symmetric five-person games Γ in the hope that some heuristic lead for the detection of their solutions may emerge from their comparison with the 1,2,3-symmetric four-person games Γ, class by class.

Using the table in Figure 66 we obtain the zones $\bar{A}$–$\bar{E}$ in $-\frac{1}{2} \leqq \eta \leqq 2$, which are the images of the zones A–E in $-1 \leqq x_1 \leqq 1$. The details appear in Figure 67.

A detailed analysis of the symmetric five-person games can be carried out on this basis. It discloses that the zones $\bar{A}$, $\bar{B}$ do indeed play the role which we expect, but that the zones $\bar{C}$, $\bar{D}$, $\bar{E}$ must be replaced by others, $\bar{C}'$, $\bar{D}'$. These zones $\bar{A}$–$\bar{D}'$ in $-\frac{1}{2} \leqq \eta \leqq 2$ and their inverse images A–D' in $-1 \leqq x_1 \leqq 1$ (again obtained with the help of the table of Figure 66) are shown in Figure 68.

It is remarkable that the x_1-diagram of Figure 68 shows more symmetry than that of Figure 67, although it is the latter which is significant for the 1,2,3-symmetric four-person games.

40.3.4. The analysis of symmetric five-person games has also some heuristic value beyond the immediate information it gives. Indeed, by comparing the symmetric five-person game Γ and the 1,2,3-symmetric four-person game Γ' which corresponds to it, and by studying the differences between their solutions, one observes the strategic effects of the merger of players 4 and 5 in one (composite) player 4'. To the extent to which the solutions present no essential differences (which is the case in the zones $\bar{A}$, $\bar{B}$, as indicated above) one may say that this merger did not affect the really

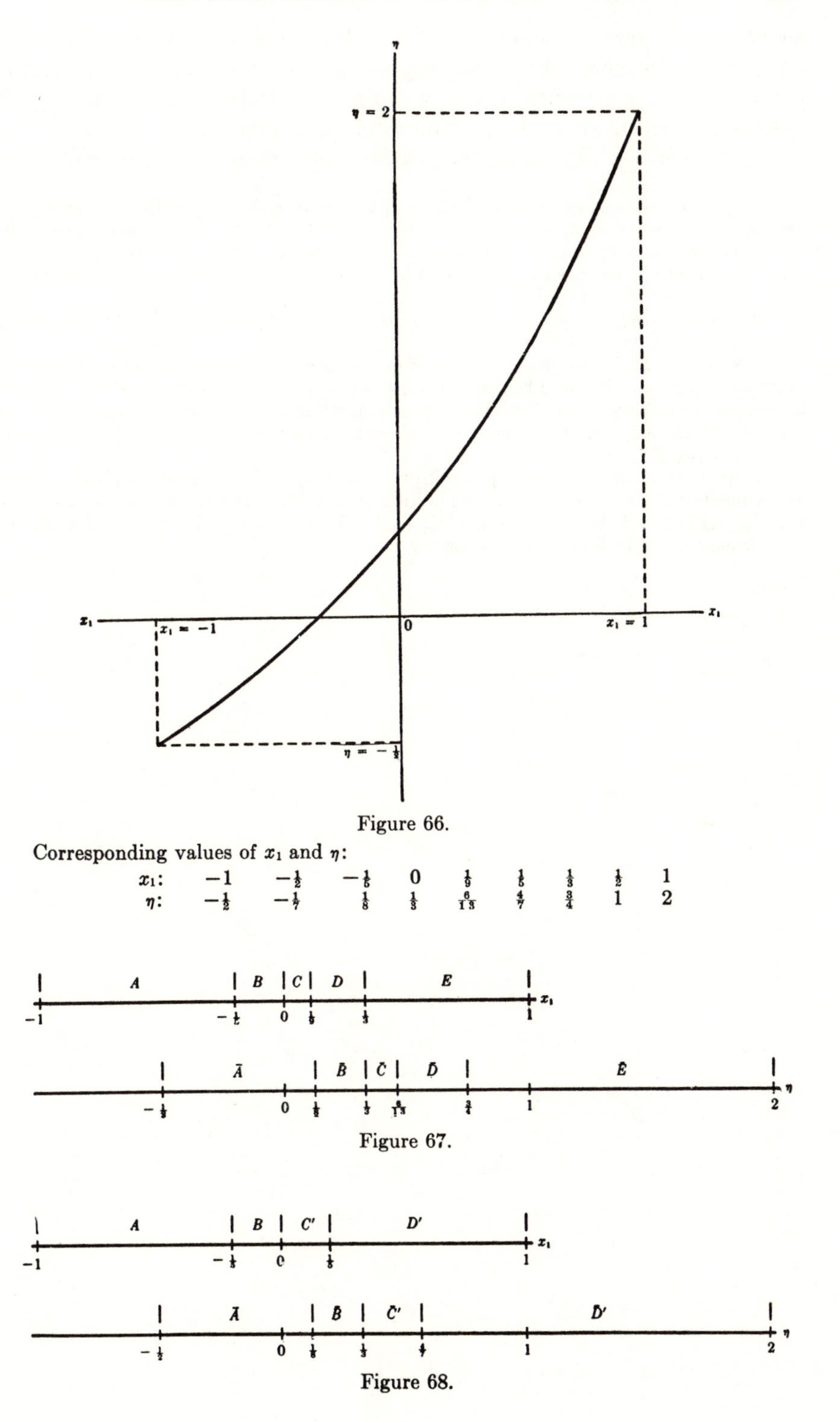

Figure 66.

Corresponding values of x_1 and η:

x_1:	-1	$-\frac{1}{2}$	$-\frac{1}{5}$	0	$\frac{1}{9}$	$\frac{1}{5}$	$\frac{1}{3}$	$\frac{1}{2}$	1
η:	$-\frac{1}{2}$	$-\frac{1}{7}$	$\frac{1}{8}$	$\frac{1}{3}$	$\frac{6}{13}$	$\frac{4}{7}$	$\frac{3}{4}$	1	2

Figure 67.

Figure 68.

significant strategic considerations.[1] On the other hand, when such differences arise (this happens in the remaining zones) we face the interesting situation that even when 4 and 5 happen to cooperate in Γ, their joint position is dislocated by the possibility of their separation.[2]

Space forbids a fuller discussion based on the rigorous concept of solutions.

[1] Of course one must expect, in solutions of Γ, arrangements where the players 4 and 5 are ultimately found in opposing coalitions. It is clear that this can have no parallel in Γ'. All we mean by the absence of essential differences is that those imputations in a solution of Γ which indicate a coalition of 4 and 5 should correspond to equivalent imputations in the solution of Γ'.

These ideas require further elaboration, which is possible, but it would lead too far to undertake it now.

[2] Already in 22.2., our first discussion of the three-person game disclosed that the division of proceeds within a coalition is determined by the possibilities of each partner in case of separation. But this situation which we now visualize is different. In our present Γ it can happen that even the total share of player 4 plus player 5 is influenced by this "virtual" fact.

A qualitative idea of such a possibility is best obtained by considering this: When a preliminary coalition of 4 and 5 is bargaining with prospective further allies, their bargaining position will be different if their coalition is known to be indissoluble (in Γ') than when the opposite is known to be a possibility (in Γ).

CHAPTER IX

COMPOSITION AND DECOMPOSITION OF GAMES

41. Composition and Decomposition

41.1. Search for n-person Games for Which All Solutions Can Be Determined

41.1.1. The last two chapters will have conveyed a specific idea of the rapidity with which the complexity of our problem increases as the number n of participants goes to $4,5, \cdots$ etc. In spite of their incompleteness, those considerations tended to be so voluminous that it must seem completely hopeless to push this—casuistic—approach beyond five participants.[1] Besides, the fragmentary character of the results gained in this manner very seriously limits their usefulness in informing us about the general possibilities of the theory.

On the other hand, it is absolutely vital to get some insight into the conditions which prevail for the greater values of n. Quite apart from the fact that these are most important for the hoped for economic and sociological applications, there is also this to consider: With every increase of n, qualitatively new phenomena appeared. This was clear for each of $n = 2,3,4$ (cf. 20.1.1., 20.2., 35.1.3., and also the remarks of footnote 2 on p. 221), and if we did not observe it for $n = 5$ this may be due to our lack of detailed information about this case. It will develop later, (cf. the end of 46.12.), that very important qualitative phenomena make their first appearance for $n = 6$.

41.1.2. For these reasons it is imperative that we find some technique for the attack on games with higher n. In the present state of things we cannot hope for anything systematic or exhaustive. Consequently the natural procedure is to find some special classes of games involving many participants[2] that can be decisively dealt with. It is a general experience in many parts of the exact and natural sciences that a complete understanding of suitable special cases—which are technically manageable, but which embody the essential principles—has a good chance to be the pacemaker for the advance of the systematic and exhaustive theory.

We will formulate and discuss two such families of special cases. They can be viewed as extensive generalizations of two four-person games—so that each one of these will be the prototype of one of the two families. These two four-person games correspond to the 8 corners of the cube Q, introduced in 34.2.2.: Indeed, we saw that those corners presented only

[1] As was seen in Chapter VIII, it was already necessary for five participants to restrict ourselves to the symmetric case.

[2] And in such a manner that each one plays an essential role!

two strategically different types of games—the corners *I*, *V*, *VI*, *VII*, discussed in 35.1. and the corners *II*, *III*, *IV*, *VIII*, discussed in 35.2. Thus the corners *I* and *VIII* of *Q* are the prototypes of those generalizations to which this chapter and the following one will be devoted.

41.2. The First Type. Composition and Decomposition

41.2.1. We first consider the corner *VIII* of *Q*, discussed in 35.2. As was brought out in 35.2.2. this game has the following conspicuous feature: The four participants fall into two separate sets (one of three elements and the other of one) which have no dealings with each other. I.e. the players of each set may be considered as playing a separate game, strictly among themselves and entirely unrelated to the other set.

The natural generalization of this is to a game Γ of $n = k + l$ participants, with the following property: The participants fall into two sets of k and l elements, respectively, which have no dealings with each other. I.e. the players of each set may be considered playing a separate game, say Δ and H respectively, strictly among themselves and entirely unrelated to the other set.[1]

We will describe this relationship between the games Γ, Δ, H by the following nomenclature: *Composition* of Δ, H produces Γ, and conversely Γ can be *decomposed* into the *constituents* Δ, H.[2]

41.2.2. Before we deal with the above verbal definitions in an exact way, some qualitative remarks may be appropriate:

First, it should be noted that our procedure of composition and decomposition is closely analogous to one which has been successfully applied in many parts of modern mathematics.[3] As these matters are of a highly technical mathematical nature, we will not say more about them here. Suffice it to state that our present procedure was partly motivated by those analogies. The exhaustive but not trivial results, which we shall obtain

[1] In the original game of 35.2. the second set consisted of one isolated player, who was also termed a "dummy." This suggests an alternative generalization to the above one: A game in which the participants fall into two sets such that those of the first set play a game strictly among themselves etc., while those of the second set have no influence upon the game, neither regarding their own fate, nor that of the others. (These are then the "dummies.")

This is, however, a special case of the generalization in the text. It is subsumed in it by taking the game H of the second set as an inessential one, i.e. one which has a definite value for each one of its participants that cannot be influenced by anybody. (Cf. 27.3.1. and the end of 43.4.2. A player in an inessential game could conceivably deteriorate his position by playing inappropriately. We ought to exclude this possibility for a "dummy"—but this point is of little importance.)

The general discussion, which we are going to carry out (both games Δ and H essential) will actually disclose a phenomenon which does not arise in the special case to which the corner *VIII* of 35.2. belongs—i.e. the case of "dummies" (H inessential). The new phenomenon will be discussed in 46.7., 46.8., and the case of "dummies"—where nothing new happens—in 46.9.

[2] It would seem natural to extend the concepts of composition and decomposition to more than 2 constituents. This will be carried out in 43.2., 43.3.

[3] Cf. *G. Birkhoff & S. MacLane:* A Survey of Modern Algebra. New York, 1941, Chapt. XIII.

and also be able to use for further interpretations are a rather encouraging symptom from a technical point of view.

41.2.3. Second, the reader may feel that the operation of composition is of an entirely formal and fictitious nature. Why should two games, Δ and H, played by two distinct sets of players and having absolutely no influence upon each other, be considered as one game Γ?

Our result will disclose that the complete separation of the games Δ and H, as far as the rules are concerned, does not necessarily imply the same for their solutions. I.e.: Although the two sets of players cannot influence each other directly, nevertheless when they are regarded as one set,—one society—there may be stable standards of behaviour which establish correlations between them.[1] The significance of this circumstance will be elaborated more fully when we reach it loc. cit.

41.2.4. Besides, it should be noted that this procedure of composition is quite customary in the natural sciences as well as in economic theory. Thus it is perfectly legitimate to consider two separate mechanical systems—situated, to take an extreme case, say one on Jupiter and one on Uranus—as one. It is equally feasible to consider the internal economies of two separate countries—the connections between which are disregarded—as one. This is, of course, the preliminary step before introducing the interacting forces between those systems. Thus we could choose in our first example as those two systems the two planets Jupiter and Uranus themselves (both in the gravitational field of the Sun), and then introduce as interaction the gravitational forces which the planets exert on each other. In our second example, the interaction enters with the consideration of international trade, international capital movements, migrations, etc.

We could equally use the decomposable game Γ as a stepping stone to other games in its neighborhood, which, in their turn, permit no decomposition.[2]

In our present considerations, however, these latter modifications will not be considered. Our interest is in the correlations introduced by the solutions referred to at the beginning of this paragraph.

41.3. Exact Definitions

41.3.1. Let us now proceed to the strictly mathematical description of the composition and decomposition of games.

Let k players $1', \cdots, k'$, forming the set $J = (1', \cdots, k')$ play the game Δ; and l players $1'', \cdots, l''$, forming the set $K = (1'', \cdots, l'')$ play the game H. We re-emphasize that Δ and H are disjoint sets of players and[3] that the games Δ and H are without any influence upon each

[1] There is some analogy between this and the phenomenon noted before (cf. 21.3., 37.2.1.) that a symmetry of the game need not imply the same symmetry in all solutions.

[2] Cf. 35.3.3., applied to the neighborhood of corner *I*., which according to 35.2. is a decomposable game. The remark of footnote 2 on p. 303 on perturbations is also pertinent.

[3] If the same players $1, \cdots, n$ are playing simultaneously two games, then an entirely different situation prevails. That is the superposition of games referred to in

other. Denote the characteristic functions of these two games by $v_\Delta(S)$ and $v_H(T)$ respectively, where $S \subseteq J$ and $T \subseteq K$.

In forming the composite game Γ, it is convenient to use the same symbols $1', \cdots, k', 1'', \cdots, l''$ for its $n = k + l$ players.[1] They form the set $I = J \cup K = (1', \cdots, k', 1'', \cdots, l'')$.

Clearly every set $R \subseteq I$ permits a unique representation

(41:1) $$R = S \cup T, \qquad S \subseteq J, \qquad T \subseteq K;$$

the inverse of this formula being

(41:2) $$S = R \cap J, \qquad T = R \cap K.$$[2]

Denote the characteristic function of the game Γ by $v_\Gamma(R)$ with $R \subseteq I$. The intuitive fact that the games Δ and H combine without influencing each other to Γ has this quantitative expression: The value in Γ of a coalition $R \subseteq I$ obtains by addition of the value in Δ of its part S ($\subseteq J$) in J and of the value in H of its part T ($\subseteq K$) in K. Expressed by a formula:

(41:3) $v_\Gamma(R) = v_\Delta(S) + v_H(T)$ where R, S, T are linked by (41:1), i.e. (41:2).[3]

41.3.2. The form (41:3) expressed the composite $v_\Gamma(R)$ by means of its constituents $v_\Delta(S)$, $v_H(T)$. However, it also contains the answer to the inverse problem: To express $v_\Delta(S)$, $v_H(T)$ by $v_\Gamma(R)$.

Indeed $v_\Delta(\ominus) = v_H(\ominus) = 0$.[4] Hence putting alternately $T = \ominus$ and $S = \ominus$ in (41:3) gives:

(41:4) $$v_\Delta(S) = v_\Gamma(S) \quad \text{for} \quad S \subseteq J,$$
(41:5) $$v_H(T) = v_\Gamma(T) \quad \text{for} \quad T \subseteq K.$$[5]

We are now in a position to express the fact of the *decomposability* of the game Γ with respect to the two sets J and K. I.e.: the given game Γ (among the elements of $I = J \cup K$) is such that it can be decomposed into two suitable games Δ (among the elements J) and H (among the elements of K). As stated, this is an implicit property of Γ involving the existence of the unknown Δ, H. But it will be expressed as an explicit property of Γ.

Indeed: If two such Δ, H exist, then they cannot be anything but those described by (41:4), (41:5). Hence the property of Γ in question is, that the

---

27.6.2. and also in 35.3.4. Its influences on the strategy are much more complex and scarcely describable by general rules, as was pointed out at the latter loc. cit.

[1] Instead of the usual $1, \cdots, n$.

[2] These formulae (41:1), (41:2) have an immediate verbal meaning. The reader may find it profitable to formulate it.

[3] Of course, a rigorous deduction on the basis of 25.1.3. is feasible without difficulty. All of 25.3.2. applies in this case.

[4] Note that the empty set $\ominus$ is a subset of both J and K; since J and K are disjoint, it is their only common subset.

[5] This is an instance of the technical usefulness of our treating the empty set $\ominus$ as a coalition. Cf. footnote 2 on p. 241.

Δ, H of (41:4), (41:5) fulfill (41:3). Substituting, therefore, (41:4), (41:5) into (41:3), using (41:1) to express R in terms of S, T gives this:

(41:6) $\quad \mathrm{v}_\Gamma(S \cup T) = \mathrm{v}_\Gamma(S) + \mathrm{v}_\Gamma(T) \quad$ for $\quad S \subseteq J, \quad T \subseteq K.$

Or, if we use (41:2) (expressing S, T in terms of R) in place of (41:1)

(41:7) $\quad \mathrm{v}_\Gamma(R) = \mathrm{v}_\Gamma(R \cap J) + \mathrm{v}_\Gamma(R \cap K) \quad$ for $\quad R \subseteq I.$

41.3.3. In order to see the role of the equations (41:6), (41:7) in the proper light, a detailed reconsideration of the basic principles upon which they rest, is necessary. This will be done in sections 41.4.-42.5.2. which follow. However, two remarks concerning the interpretation of these equations can be made immediately.

First: (41:6) expresses that a coalition between a set $S \subseteq J$ and a set $T \subseteq K$ has no attraction—that while there may be motives for players within J to combine with each other, and similarly for players within K, there are no forces acting across the boundaries of J and K.

Second: To those readers who are familiar with the mathematical theory of measure, we make this further observation in continuation of that made at the end of 27.4.3.: (41:7) is exactly Carathéodory's definition of measurability. This concept is quite fundamental for the theory of additive measure and Carathéodory's approach to it appears to be the technically superior one to date.[1] Its emergence in the present context is a remarkable fact which seems to deserve further study.

41.4. Analysis of Decomposability

41.4.1. We obtained the criteria (41:6), (41:7) of Γ's decomposability by substituting the $\mathrm{v}_\Delta(S)$, $\mathrm{v}_H(T)$ obtained from (41:4), (41:5) into the fundamental condition (41:3). However, this deduction contains a lacuna: We did not verify whether it is possible to find two games Δ, H which produce the $\mathrm{v}_\Delta(S)$, $\mathrm{v}_H(T)$ formally defined by (41:4), (41:5).

There is no difficulty in formalizing these extra requirements. As we know from 25.3.1. they mean that $\mathrm{v}_\Delta(S)$ and $\mathrm{v}_H(T)$ fulfill the conditions (25:3:a)-(25:3:c) eod. It must be understood that we assume the given $\mathrm{v}_\Gamma(R)$ to originate from a game Γ, i.e. that $\mathrm{v}_\Gamma(R)$ fulfills these conditions. Hence the following question presents itself:

(41:A) $\mathrm{v}_\Gamma(R)$ fulfills (25:3:a)-(25:3:c) in 25.3.1. together with the above (41:6), i.e. (41:7). Will then the $\mathrm{v}_\Delta(S)$ and $\mathrm{v}_H(T)$ of (41:4), (41:5) also fulfill (25:3:a)-(25:3:c) in 25.3.1.? Or, if this is not the case, which further postulate must be imposed upon $\mathrm{v}_\Gamma(R)$?

To decide this, we check (25:3:a)-(25:3:c) of 25.3.1. separately for $\mathrm{v}_\Delta(S)$ and $\mathrm{v}_H(T)$. It is convenient to take them up in a different order.

41.4.2. Ad (25:3:a): By virtue of (41:4), (41:5), this is the same statement for $\mathrm{v}_\Delta(S)$ and $\mathrm{v}_H(T)$ as for $\mathrm{v}_\Gamma(R)$.

[1] Cf. *C. Carathéodory:* Vorlesungen über Reelle Funktionen, Berlin, 1918, Chapt. V.

Ad (25:3:c): By virtue of (41:4), (41:5), this carries over to $v_\Delta(S)$ and $v_H(T)$ from $v_\Gamma(R)$—it amounts only to a restriction from the $R \subseteq I$ to $S \subseteq J$ and $T \subseteq K$.

Before discussing the remaining (25:3:b), we insert a remark concerning (25:4) of 25.4.1. Since this is a consequence of (25:3:a)-(25:3:c), it is legitimate for us to draw conclusions from it—and it will be seen that this anticipation simplifies the analysis of (25:3:b).

From here on we will have to use promiscuously complementary sets in I, J, K. It is, therefore, necessary to avoid the notation $-S$, and to write instead $I - S$, $J - S$, $K - S$, respectively.

Ad (25:4): For $v_\Delta(S)$ and $v_H(T)$ the role of the set I is taken over by the sets J and K, respectively. Hence this condition becomes:

$$v_\Delta(J) = 0,$$
$$v_H(K) = 0.$$

Owing to (41:4), (41:5), this means

(41:8) $$v_\Gamma(J) = 0,$$

(41:9) $$v_\Gamma(K) = 0.$$

Since $K = I - J$, therefore (25:3:b) (applied to $v_\Gamma(S)$ for which it was assumed to hold) gives

(41:10) $$v_\Gamma(J) + v_\Gamma(K) = 0.$$

Thus (41:8) and (41:9) imply each other by virtue of the identity (41:10).

In (41:8) or (41:9) we have actually a new condition, which does not follow from (41:6) or (41:7).

Ad (25:3:b): We will derive its validity for $v_\Delta(S)$ and $v_H(T)$ from the assumed one for $v_\Gamma(R)$. By symmetry it suffices to consider $v_\Delta(S)$.

The relation to be proven is

(41:11) $$v_\Delta(S) + v_\Delta(J - S) = 0.$$

By (41:4) this means

(41:12) $$v_\Gamma(S) + v_\Gamma(J - S) = 0.$$

Owing to (41:8), which we must require anyhow, this may be written

(41:13) $$v_\Gamma(S) + v_\Gamma(J - S) = v_\Gamma(J)$$

(Of course, $S \subseteq J$.)

To prove (41:13), apply (25:3:b) for $v_\Gamma(R)$ to $R = J - S$ and $R = J$. For these sets $I - R = S \cup K$ and $I - R = K$, respectively. So (41:13) becomes

$$v_\Gamma(S) - v_\Gamma(S \cup K) = -v_\Gamma(K),$$

i.e.

$$v_\Gamma(S \cup K) = v_\Gamma(S) + v_\Gamma(K),$$

and this is the special case of (41:6) with $T = K$.

Thus we have filled in the lacuna mentioned at the beginning of this paragraph and answered the questions of (41:A).

(41:B) The further postulate which must be imposed upon $v_\Gamma(R)$ is this: (41:8), i.e. (41:9).

All these put together answer the question of 41.3.2. concerning decomposability:

(41:C) The game Γ is decomposable with respect to the sets J and K (cf. 41.3.2.) if and only if it fulfills these conditions: (41:6), i.e. (41:7) and (41:8), i.e. (41:9).

41.5. Desirability of a Modification

41.5.1. The two conditions which we proved equivalent to decomposability in (41:C) are of very different character. (41:6) (i.e. (41:7)) is the really essential one, while (41:8) (i.e. (41.9)) expresses only a rather incidental circumstance. We will justify this rigorously below, but first a qualitative remark will be useful. The prototype of our concept of decomposition was the game referred to at the beginning of 41.2.1.: the game represented by the corner *VIII* of 35.2. Now this game fulfilled (41:6), but not (41:8). (The former follows from (35:7) in 35.2.1., the latter from $v(J) = v((1,2,3)) = 1 \neq 0$.) We nevertheless considered that game as decomposable (with $J = (1,2,3)$, $K = (4)$)—how is it then possible, that it violates the condition (41:8) which we found to be necessary for the decomposability?

41.5.2. The answer is simple: For the above game the constituents Δ (in $J = (1,2,3)$) and H (in $K = (4)$) do not completely satisfy (25:3:a)-(25:3:c) in 25.3.1. To be precise, they do not fulfill the consequence (25:4) in 25.4.1.: $v_\Delta(J) = v_H(K) = 0$ is not true (and it was from this condition that we derived (41:8)). In other words: the constituents of Γ are not zero-sum games. This point, of course, was perfectly clear in 35.2.2., where it received due consideration.

Consequently we must endeavor to get rid of the condition (41:8), recognizing that this may force us to consider other than zero-sum games.

42. Modification of the Theory

42.1. No Complete Abandoning of the Zero-sum Condition

42.1. Complete abandonment of the zero-sum condition for our games[1] would mean that the functions $\mathcal{H}_k(\tau_1, \cdots, \tau_n)$ which characterized it in the sense of 11.2.3. are entirely unrestricted. I.e. that the requirement

(42:1) $$\sum_{k=1}^{n} \mathcal{H}_k(\tau_1, \cdots, \tau_n) \equiv 0$$

[1] We again denote the players by $1, \cdots, n$.

of 11.4. and 25.1.3. is dropped, with nothing else to take its place. This would necessitate a revision of considerable significance, since the construction of the characteristic function in 25. depended upon (25:1), i.e. (42:1), and would therefore have to be taken up *de novo*.

Ultimately this revision will become necessary (cf. Chapter XI) but not yet at the present stage.

In order to get a precise idea of just what is necessary now, let us make the auxiliary considerations contained in 42.2.1., 42.2.2. below.

42.2. Strategic Equivalence. Constant-sum Games

42.2.1. Consider a zero-sum game Γ which may or may not fulfill conditions (41:6) and (41:8). Pass from Γ to a strategically equivalent game Γ' in the sense of 27.1.1., 27.1.2., with the $\alpha_1^0, \cdots, \alpha_n^0$ described there. It is evident, that (41:6) for Γ is equivalent to the same for Γ'.[1]

The situation is altogether different for (41:8). Passage from Γ to Γ' changes the left hand side of (41:8) by $\sum_{k \text{ in } J} \alpha_k^0$, hence the validity of (41:8) in one case is by no means implied by that in the other. Indeed this is true:

(42:A) For every Γ it is possible to choose a strategically equivalent game Γ' so that the latter fulfills (41:8).

Proof: The assertion is[1] that we can choose $\alpha_1^0, \cdots, \alpha_n^0$ with $\sum_{k=1}^{n} \alpha_k^0 = 0$ (this is (27:1) in 27.1.1.) so that

$$v(J) + \sum_{k \text{ in } J} \alpha_k^0 = 0$$

Now this is obviously possible if $J \neq \ominus$ or I, since then $\sum_{k \text{ in } J} \alpha_k^0$ can be given any assigned value. For $J = \ominus$ or I, there is nothing to prove, as then $v(J) = 0$ by (25:3:a) in 25.3.1. and (25:4) in 25.4.1.

This result can be interpreted as follows: If we refrain from considering other than zero-sum games,[2] then condition (41:6) expresses that while the game Γ may not be decomposable itself, it is strategically equivalent to some decomposable game Γ'.[3]

42.2.2. The above rigorous result makes it clear where the weakness of our present arrangement lies. Decomposability is an important strategic property and it is therefore inconvenient that of two strategically equivalent games one may be termed decomposable without the other. It is, therefore,

[1] By (27:2) in 27.1.1. Observe that the $v_{\Gamma}(S)$, $v_{\Gamma'}(S)$ of (42:A) are the $v(S)$, $v'(S)$ of (27:2) loc. cit.

[2] I.e. we require this not only for Γ, but also for its constituents Δ, H.

[3] The treatment of the constituents in 35.2.2. amounts to exactly this, as an inspection of footnote 1 on p. 300 shows explicitly.

desirable to widen these concepts so that decomposability becomes an invariant under strategic equivalence.

In other words: We want to modify our concept so that the transformation (27:2) of 27.1.1., which defines strategic equivalence, does not interfere with the relationship between a decomposable game Γ and its constituents Δ and H. This relationship is expressed by (41:3):

(42:2) $$v_\Gamma(S \cup T) = v_\Delta(S) + v_H(T) \quad \text{for} \quad S \subseteq J, \quad T \subseteq K.$$

Now if we use (27:2) with the same α_k^0 for all three games Γ, Δ, H then (42:2) is manifestly undisturbed. The only trouble is with the preliminary condition (27:1). This states for Γ, Δ, H that

$$\sum_{k \text{ in } I} \alpha_k^0 = 0, \qquad \sum_{k \text{ in } J} \alpha_k^0 = 0, \qquad \sum_{k \text{ in } K} \alpha_k^0 = 0,$$

respectively—and while we now assume the first relation true, the two others may fail.

Hence the natural way out is to discard (27:1) of 27.1.1. altogether. I.e. to widen the domain of games, which we consider, by including all those games which are strategically equivalent to zero-sum ones by virtue of the transformation formula (27:2) alone—without demanding (27:1).

As was seen in 27.1.1. this amounts to replacing the functions

$$\mathcal{H}_k(\tau_1, \cdots, \tau_n)$$

of the latter by new functions

$$\mathcal{H}'_k(\tau_1, \cdots, \tau_n) \equiv \mathcal{H}_k(\tau_1, \cdots, \tau_n) + \alpha_k^0.$$

(The $\alpha_1^0, \cdots, \alpha_n^0$ are no longer subject to (27:1)). The systems of functions $\mathcal{H}'_k(\tau_1, \cdots, \tau_n)$ which are obtained in this way from the system of functions $\mathcal{H}_k(\tau_1, \cdots, \tau_n)$ which fulfill (42:1) in 42.1. are easy to characterize. The characteristic is (in place of (42:1) loc. cit.) the property

(42:3) $$\sum_{k=1}^{n} \mathcal{H}'_k(\tau_1, \cdots, \tau_n) \equiv s.$$[1]

Summing up:

(42:B) We are widening the domain of games which we consider, by passing from the *zero-sum games* to the *constant-sum games*.[2] At the same time, we widen the concept of strategic equivalence,

[1] s is an arbitrary constant $\gtreqless 0$. In the transformation (27:2) which produces this game from a zero-sum one, there is obviously

$$\sum_{k=1}^{n} \alpha_k^0 = s.$$

[2] This gives a precise meaning to the statement at the beginning of 42.1. according to which we are not yet prepared to consider all games unrestrictedly.

introduced in 27.1.1., by defining it again by transformation (27:2) loc. cit., but dropping the condition (27:1) eod.

42.2.3. It is essential to recognize that our above generalizations do not alter our main ideas on strategic equivalence. This is best done by considering the following two points.

First, we stated in 25.2.2. that we proposed to understand all quantitative properties of a game by means of its characteristic function alone. One must realize that the reasons for this are just as good in our present domain of constant-sum games as in the original (and narrower) one of zero-sum games. The reason is this:

(42:C) Every constant-sum game is strategically equivalent to a zero-sum game.

Proof: The transformation (27:2) obviously replaces the s of (42:3) above by $s + \sum_{k=1}^{n} \alpha_k^0$. Now it is possible to choose the $\alpha_1^0, \cdots, \alpha_n^0$ so as to make this $s + \sum_{k=1}^{n} \alpha_k^0 = 0$, i.e. to carry the given constant-sum game into a (strategically equivalent) zero-sum game.

Second, our new concept of strategic equivalence was only necessary for the sake of the new (non-zero-sum) games that we introduced. For the old (zero-sum) games it is equivalent to the old concept. In other words: If two zero-sum games obtain from each other by means of the transformation (27:2) in 27.1.1., then (27:1) is automatically fulfilled. Indeed, this was already observed in footnote 2 on p. 246.

42.3. The Characteristic Function in the New Theory

42.3.1. Given a constant-sum game Γ' (with the $\mathcal{H}_k'(\tau_1, \cdots, \tau_n)$ fulfilling (42:3)), we could introduce its characteristic function $v'(S)$ by repeating the definitions of 25.1.3.[1] On the other hand, we may follow the procedure suggested by the argumentation of 42.2.2., 42.2.3.: We can obtain Γ' with the functions $\mathcal{H}_k'(\tau_1, \cdots, \tau_n)$ from a zero-sum game Γ with the functions $\mathcal{H}_k(\tau_1, \cdots, \tau_n)$ as in 42.2.2., i.e. by

(42:4) $$\mathcal{H}_k'(\tau_1, \cdots, \tau_n) \equiv \mathcal{H}_k(\tau_1, \cdots, \tau_n) + \alpha_k^0$$

with appropriate $\alpha_1^0, \cdots, \alpha_n^0$ (cf. footnote 1 on p. 246), and then define the characteristic function $v'(S)$ of Γ' by means of (27:2) in 27.1.1., i.e. by

(42:5) $$v'(S) \equiv v(S) + \sum_{k \text{ in } S} \alpha_k^0.$$

[1] The whole arrangement of 25.1.3. can be repeated literally, although Γ' is no longer zero-sum, with two exceptions. In (25:1) and (25:2) of 25.1.3. we must add s to the extreme right hand term. (This is so, because we now have (42:3) in place of (42.1).) This difference is entirely immaterial.

Now the two procedures are equivalent, i.e. the $v'(S)$ of (42:4), (42:5) coincides with the one obtained by the reapplication of 25.1.3. Indeed, an inspection of the formulae of 25.1.3. shows immediately, that the substitution of (42:4) there produces the result (42:5).[1,2]

42.3.2. $v(S)$ is a characteristic function of a zero-sum game, if and only if it fulfills the conditions (25:3:a)-(25:3:c) of 25.3.1., as was pointed out there and in 26.2. (The proof was given in 25.3.3. and 26.1.) What do these conditions become in the case of a constant-sum game?

In order to answer this question, let us remember, that (25:3:a)-(25:3:c) loc. cit. imply (25:4) in 25.4.1. Hence, we can add (25:4) to them, and modify (25:3:b) by adding $v(I)$ to its right hand side (this is no change owing to (25:4)). Thus the characterization of the $v(S)$ of all zero-sum games becomes this:

(42:6:a) $$v(\ominus) = 0,$$

(42:6:b) $$v(S) + v(-S) = v(I),$$

(42:6:c) $$v(S) + v(T) \leqq v(S \cup T) \quad \text{if} \quad S \cap T = \ominus,$$

and

(42:6:d) $$v(I) = 0.$$

Now the $v'(S)$ of all constant-sum games obtain from these $v(S)$ by subjecting them to the transformation (42:5) of 42.3.1. How does this transformation affect (42:6:a)-(42:6:d)?

One verifies immediately, that (42:6:a)-(42:6:c) are entirely unaffected, while (42:6:d) is completely obliterated.[3] So we see:

(42:D) $v(S)$ is the characteristic function of a constant-sum game if and only if it satisfies the conditions (42:6:a)-(42:6:c).

(We write from now on $v(S)$ for $v'(S)$).

As mentioned above, (42:6:d) is no longer valid. However, we have

(42:6:d*) $$v(I) = s.$$

Indeed, this is clear from (42:3), considering the procedure of 25.1.3. It can also be deduced by comparing footnote 1 on p. 347 and footnote 3 above (our $v(S)$ is the $v'(S)$ there). Besides (42:6:d*) is intuitively clear: A coalition of all players obtains the fixed sum s of the game.

[1] The verbal equivalent of this consideration is easily found.

[2] Had we decided to define $v'(S)$ by means of (42:2), (42:5) only, a question of ambiguity would have arisen. Indeed: A given constant-sum game Γ' can obviously be obtained from many different zero-sum games Γ by (42:4), will then (42:5) always yield the same $v'(S)$?

It would be easy to prove directly that this is the case. This is unnecessary, however, because we have shown that the $v'(S)$ of (42:5) is always equal to that one of 25.1.3.—and that $v'(S)$ is defined unambiguously, with the help of Γ' alone.

[3] According to (42:5), the right hand side of (42:6:d) goes over into $\sum_{i \text{ in } I} \alpha_i^0$, i.e. $\sum_{i=1}^{n} \alpha_i^0$, and this sum is completely arbitrary.

42.4. Imputations, Domination, Solutions in the New Theory

42.4.1. From now on, we are considering characteristic functions of any constant-sum game, i.e. functions $v(S)$ subject to (42:6:a)-(42:6:c) only.

Our first task in this wider domain, is naturally that of extending to it the concepts of imputations, dominations, and solutions as defined in 30.1.1.

Let us begin with the *distributions* or *imputations*. We can take over from 30.1.1. their interpretation as vectors

$$\vec{\alpha} = \{\alpha_1, \cdots, \alpha_n\}.$$

Of the conditions (30:1), (30:2) eod. we may conserve (30:1):

$$\text{(42:7)} \qquad \alpha_i \geqq v((i))$$

unchanged—the reasons referred to there[1] are just as valid now as then. (30:2) eod., however, must be modified. The constant-sum of the game being s (cf. (42:3) and (42:6:d*) above), each imputation should distribute this amount—i.e. it is natural to postulate

$$\text{(42:8)} \qquad \sum_{i=1}^{n} \alpha_i = s.$$

By (42:6:d*) this is equivalent to

$$\text{(42:8*)} \qquad \sum_{i=1}^{n} \alpha_i = v(I).$$[2]

The definitions of *effectivity, domination, solution* we take over unchanged from 30.1.1.[3] the supporting arguments brought forward in the discussions which led up to those definitions, appear to lose no strength by our present generalization.

42.4.2. These considerations receive their final corroborations by observing this:

(42:E) For our new concept of strategic equivalence of constant-sum games Γ, Γ',[4] there exists an isomorphism of their imputations, i.e. a one-to-one mapping of those of Γ on those of Γ', which leaves the concepts of 30.1.1.[5] invariant.

This is an analogue of (31:Q) in 31.3.3. and it can be demonstrated in the same way. As there, we define the correspondence

$$\text{(42:9)} \qquad \vec{\alpha} \rightleftarrows \vec{\alpha}'$$

[1] $\alpha_i < v((i))$ would be unacceptable, cf. e.g. the beginning of 29.2.1.

[2] For the special case of a zero-sum game $s = v(I) = 0$ so (42:8), (42:8*) coincide—as they must—with (30:2) loc. cit.

[3] I.e. (30:3); (30:4:a)–(30:4:c); (30:5:a), (30:5:b), or (30:5:c) loc. cit., respectively.

[4] As defined at the end of 42.2.2., i.e. by (27:2) in 27.1.1., without (27:1) eod.

[5] As redefined in 42.4.1.

between the imputations $\overrightarrow{\alpha} = \{\alpha_1, \cdots, \alpha_n\}$ of Γ and the imputations $\overrightarrow{\alpha}' = \{\alpha'_1, \cdots, \alpha'_n\}$ of Γ' by

(42:10) $$\alpha'_k = \alpha_k + \alpha^0_k$$

where the $\alpha^0_1, \cdots, \alpha^0_n$ are those of (27:2) in 27.1.1.

Now the proof of (31:Q) in 31.3.3. carries over almost literally. The one difference is that (30:2) of 30.1.1. is replaced by our (42:8) but since (27:2) in 27.1.1. gives $v'(I) = v(I) + \sum_{i=1}^{n} \alpha^0_i$, this too takes care of itself.[1] The reader who goes over 31.3. again, will see that everything else said there applies equally to the present case.

42.5. Essentiality, Inessentiality, and Decomposability in the New Theory

42.5.1. We know from (42:C) in 42.2.3. that every constant-sum game is strategically equivalent to a zero-sum game. Hence (42:E) allows us to carry over the general results of 31. from the zero-sum games to the constant-sum ones always passing from the latter class to the former one by strategic equivalence.

This forces us to define *inessentiality* for a constant-sum game by strategic equivalence to an inessential zero-sum game. We may state therefore:

(42:F) A zero-sum game is inessential if and only if it is strategically equivalent to the game with $\bar{v}(S) \equiv 0$. (Cf. 23.1.3. or (27:C) in 27.4.2.) By the above, the same is true for a constant-sum game. (But we must use our new definitions of inessentiality and of strategic equivalence.)

Essentiality is, of course, defined as negation of inessentiality.

Application of the transformation formula (42:5) of 42.3.1. to the criteria of 27.4. shows, that there are only minor changes.

(27:8) in 27.4.1. must be replaced by

(42:11) $$\gamma = \frac{1}{n}\left\{v(I) - \sum_{j=1}^{n} v((j))\right\}$$

since the right hand side of this formula is invariant under (42:5) and it goes over into (27:8) loc. cit. for $v(I) = 0$ (i.e. the zero-sum case).

The substitution of (42:11) for (27:8) necessitates replacement of the 0 on the right hand side of both formulae in the criterion (27:B) of 27.4.1. by

[1] And this was the only point in the proof referred to, at which $\sum_{i=1}^{n} \alpha^0_i = 0$ (i.e. (27:1) in 27.1.1., which we no longer require) is used.

$v(I)$. The criteria (27:C), (27:D) of 27.4.2. are invariant under (42:5), and hence unaffected.

42.5.2. We can now return to the discussion of composition and decomposition in 41.3.-41.4., in the wider domain of all constant-sum games.

All of 41.3. can be repeated literally.

When we come to 41.4., the question (41:A) formulated there again presents itself. In order to determine whether any postulates beyond (41:6), i.e. (41:7) of 41.3.2. are now needed, we must investigate (42:6:a)-(42:6:c) in 42.3.2., instead of (25:3:a)-(25:3:c) in 25.3.1. (for all three of $v_\Gamma(R)$, $v_\Delta(S)$, $v_H(T)$).

(42:6:a), (42:6:c) are immediately disposed of, exactly as (25:3:a), (25:3:c) in 41.4. As to (42:6:b), the proof of (25:3:b) in 41.4. is essentially applicable, but this time no extra condition arises (like (41:8) or (41:9) loc. cit.). To simplify matters, we give this proof in full.

Ad (42:6:b): We will derive its validity for $v_\Delta(S)$ and $v_H(T)$ from the assumed one for $v_\Gamma(R)$. By symmetry it suffices to consider $v_\Delta(S)$.

The relation to be proven is

(42:12) $$v_\Delta(S) + v_\Delta(J - S) = v_\Delta(J).$$

By (41:4) this means

(42:12*) $$v_\Gamma(S) + v_\Gamma(J - S) = v_\Gamma(J).$$

To prove (42:12*) apply (42:6:b) for $v_\Gamma(R)$ to $R = J - S$ and $R = J$. For these $I - R = S \cup K$ and $I - R = K$, respectively. So (42:12*) becomes

$$v_\Gamma(S) + v_\Gamma(I) - v_\Gamma(S \cup K) = v_\Gamma(I) - v_\Gamma(K),$$

i.e.

$$v_\Gamma(S \cup K) = v_\Gamma(S) + v_\Gamma(K),$$

and this is the special case of (41:6) with $T = K$.

Thus we have improved upon the result (41:C) of 41.4. as follows:

(42:G) In the domain of all constant-sum games the game Γ is decomposable with respect to the sets J and K (cf. 41.3.2.) if and only if it fulfills the condition (41:6), i.e. (41:7).

42.5.3. Comparison of (41:C) in 41.4. and of (42:G) in 42.5.2. shows that the passage from zero-sum to constant-sum games rids us of the unwanted condition (41:8), i.e. (41:9) for decomposability.

Decomposability is now defined by (41:6), i.e. (41:7) alone, and it is invariant under strategic equivalence—as it should be.

We also know that when a game Γ is decomposed into two (constituent) games Δ and H (all of them constant-sum only!), we can make all these games zero-sum by strategic equivalence. (Cf. (42:C) in 42.2.3. for Γ, and then (42:A) in 42.2.1. et sequ. for Δ, H.)

Thus we can always use one of the two domains of games—zero-sum or constant-sum—whichever is more convenient for the problem just under consideration.

In the remainder of this chapter we will continue to consider constant-sum games, unless the opposite is explicitly stated.

43. The Decomposition Partition

43.1. Splitting Sets. Constituents

43.1. We defined the decomposability of a game Γ not *per se*, but with respect to a decomposition of the set I of all players into two complementary sets, J, K.

Therefore it is feasible to take this attitude: Consider the game Γ as given, and the sets J, K as variable. Since J determines K (indeed $K = I - J$), it suffices to treat J as the variable. Then we have this question:

Given a game Γ (with the set of players I) for which sets $J \subseteq I$ (and the corresponding $K = I - J$) is Γ decomposable?

We call those $J(\subseteq I)$ for which this is the case the *splitting sets of* Γ. The constituent game Δ which obtains in this decomposition (cf. 41.2.1. and (41:4) of 41.3.2.) is the *J-constituent of* Γ.[1]

A splitting set J is thus defined by (41:6), i.e. (41:7) in 41.3.2., where $K = I - J$ must be substituted.

The reader will note that this concept has a very simple intuitive meaning: A splitting set is a self contained group of players, who neither influence, nor are influenced by, the others as far as the rules of the game are concerned.

43.2. Properties of the System of All Splitting Sets

43.2.1. The totality of all splitting sets of a given game is characterized by an aggregate of simple properties. Most of these have an intuitive immediate meaning, which may make mathematical proof seem unnecessary. We will nevertheless proceed systematically and give proofs, stating the intuitive interpretations in footnotes. Throughout what follows we write $v(S)$ for $v_\Gamma(S)$ (the characteristic function of Γ).

(43:A) J is a splitting set if and only if its complement $K = I - J$ is one.[2]

Proof: The decomposability of Γ involves J and K symmetrically.

(43:B) $\ominus$ and I are splitting sets.[3]

Proof: (41:6) or (41:7) with $J = \ominus$, $K = I$ are obviously true, as $v(\ominus) = 0$.

[1] By the same definition the game H (cf. 41.2.1. and (41:5) in 41.3.2.) is then the K-constituent ($K = I - J$) of Γ.

[2] That a set of players is self-contained in the sense of 43.1., is clearly the same statement, as that the complement is self-contained.

[3] That these are self-contained is tautological.

43.2.2.

(43:C) $J' \cap J''$ and $J' \cup J''$ are splitting sets if J', J'' are.[1]

Proof: Ad $J' \cup J''$: As J', J'' are splitting sets, we have (41:6) for J, K equal to J', $I - J'$ and J'', $I - J''$. We wish to prove it for $J' \cup J''$, $I - (J' \cup J'')$. Consider therefore two $S \subseteq J' \cup J''$, $T \subseteq I - (J' \cup J'')$. Let S' be the part of S in J', then $S'' = S - S'$ lies in the complement of J', and as $S \subseteq J' \cup J''$, S'' also lies in J''. So $S = S' + S''$, $S' \subseteq J'$, $S'' \subseteq J''$. Now $S' \subseteq J'$, $S'' \subseteq I - J'$ and (41:6) for J', $I - J'$ give

(43:1) $$v(S) = v(S') + v(S'').$$

Next $S'' \subseteq I - J'$ and $T \subseteq I - (J' \cup J'') \subseteq I - J'$ so $S'' \cup T \subseteq I - J'$. Also $S' \subseteq J'$. Clearly $S' \cup (S'' \cup T) = S \cup T$. Hence (41:6) for J', $I - J'$ also gives

(43:2) $$v(S \cup T) = v(S') + v(S'' \cup T).$$

Finally $S'' \subseteq J''$ and $T \subseteq I - (J' \cup J'') \subseteq I - J''$. Hence (41:6) for J'', $I - J''$ gives

(43:3) $$v(S'' \cup T) = v(S'') + v(T).$$

Now substitute (43:3) into (43:2) and then contract the right hand side by (43:1). This gives

$$v(S \cup T) = v(S) + v(T),$$

which is (41:6), as desired.

Ad $J' \cap J''$: Use (43:A) and the above result. As J', J'' are splitting sets, the same obtains successively for $I - J'$, $I - J''$, $(I - J') \cup (I - J'')$ which is clearly $I - (J' \cap J'')$[2], and $J' \cap J''$—the last one being the desired expression.

43.3. Characterization of the System of All Splitting Sets. The Decomposition Partition

43.3.1. It may be that there exist no other splitting sets than the trivial ones $\ominus$, I (cf. (43:B) above). In that case, we call the game Γ *indecomposable.*[3] Without studying this question any further,[4] we continue to investigate the splitting sets of Γ.

[1] The intersection $J' \cap J''$: It may strike the reader as odd, that two self-contained sets J', J'' should have a non-empty intersection at all. This is possible, however, as the example $J' = J''$ shows. The deeper reason is that a self-contained set may well be the sum of smaller self-contained sets (proper subsets). (Cf. (43:H) in 43.3.) Our present assertion is that if two self-contained sets J', J'' have a non-empty intersection $J' \cap J''$, then this intersection is such a self-contained subset. In this form it will probably appear plausible.

The sum $J' \cup J''$: That the sum of two self-contained sets will again be self-contained stands to reason. This may be somewhat obscured when a non-empty intersection $J' \cap J''$ exists, but this case is really harmless as discussed above. The proof which follows is actually primarily an exact account of the ramifications of just this case.

[2] The complement of the intersection is the sum of the complements.

[3] Actually most games are indecomposable; otherwise the criterion (42:G) in 42.5.2. requires the restrictive equations (41:6), (41:7) in 41.3.2.

[4] Yet! Cf. footnote 3 and its references.

(43:D) Consider a splitting set J of Γ and the J-constituent Δ of Γ. Then a $J' \subseteq J$ is a splitting set of Δ if and only if it is one of Γ.[1]

Proof: Considering (41:4), J' is a splitting set of Δ by virtue of (41:6) when

(43:4) $\quad v(S \cup T) = v(S) + v(T) \quad$ for $\quad S \subseteq J', \quad T \subseteq J - J'.$

(We write $v(S)$ for $v_\Gamma(S)$). Again by (41:6) J' is a splitting set of Γ when

(43:5) $\quad v(S \cup T) = v(S) + v(T) \quad$ for $\quad S \subseteq J', \quad T \subseteq I - J'.$

We must prove the equivalence of (43:4) and (43:5). As $J \subseteq I$, so (43:4) is clearly a special case of (43:5)—hence we need only prove that (43:4) implies (43:5).

Assume, therefore, (43:4). We may use (41:6) for Γ with J, $K = I - J$.

Consider two $S \subseteq J'$, $T \subseteq I - J'$. Let T' be the part of T in J, then $T'' = T - T'$ lies in $I - J$. So $T = T' \cup T''$, $T' \subseteq J$, $T'' \subseteq I - J$ and (41:6) for Γ with J, $I - J$ give

(43:6)
$$v(T) = v(T') + v(T'').$$

Next $S \subseteq J' \subseteq J$ and $T' \subseteq J$ so $S \cup T' \subseteq J$. Also $T'' \subseteq I - J$. Clearly $(S \cup T') \cup T'' = S \cup T$. Hence (41:6) for Γ with J, $I - J$ also gives

(43:7)
$$v(S \cup T) = v(S \cup T') + v(T'').$$

Finally $S \subseteq J'$ and $T' \subseteq I - J'$ and $T' \subseteq J$, so $T' \subseteq J - J'$. Hence (43:4) gives

(43:8)
$$v(S \cup T') = v(S) + v(T').$$

Now substitute (43:8) into (43:7) and then contract the right hand side by (43:6). This gives precisely the desired (43:5).

43.3.2. (43:D) makes it worth while to consider those splitting sets J, for which $J \neq \ominus$, but no proper subset $J' \neq \ominus$ of J is a splitting set. We call such a set J, for obvious reasons, a *minimal splitting set.*

Consider our definitions of indecomposability and of minimality. (43:D) implies immediately:

(43:E) The J-component Δ (of Γ) is indecomposable if and only if J is a minimal splitting set.

The minimal splitting sets form an arrangement with very simple properties, and they determine the totality of all splitting sets. The statement follows:

(43:F) Any two different minimal splitting sets are disjunct.

(43:G) The sum of all minimal splitting sets is I.

[1] To be self-contained within a self-contained set, is the same thing as to be such in the original (total) set. The statement may seem obvious; that it is not so, will appear from the proof.

(43:H) By forming all sums of all possible aggregates of minimal splitting sets, we obtain precisely the totality of all splitting sets.[1]

Proof: Ad (43:F): Let J', J'' be two minimal splitting sets which are not disjunct. Then $J' \cap J'' \neq \ominus$ is splitting by (43:C), as it is $\subseteq J'$ and $\subseteq J''$. So the minimality of J' and J'' implies that $J' \cap J''$ is equal to both J' and J''. Hence $J' = J''$.

Ad (43:G): It suffices to show that every k in I belongs to some minimal splitting set.

There exist splitting sets which contain the player k (i.e. I); let J be the intersection of all of them. J is splitting by (43:C). If J were not minimal, then there would exist a splitting set $J' \neq \ominus, J$, which is $\subseteq J$. Now $J'' = J - J' = J \cap (I - J')$ is also a splitting set by (43:A), (43:C), and clearly also $J'' \neq \ominus, J$. Either J' or $J'' = J - J'$ must contain k—say that J' does. Then J' is among the sets of which J is the intersection. Hence $J' \supseteq J$. But as $J' \subseteq J$ and $J' \neq J$, this is impossible.

Ad (43:H): Every sum of minimal splitting sets is splitting by (43:C), so we need only prove the converse.

Let K be a splitting set. If J is minimal splitting, then $J \cap K$ is splitting by (43:C), also $J \cap K \subseteq J$—hence either $J \cap K = \ominus$ or $J \cap K = J$. In the first case J, K are disjunct, in the second $J \subseteq K$. So we see:

(43:I) Every minimal splitting set J is either disjunct with K or $\subseteq K$.

Let K' be the sum of the former J, and K'' the sum of the latter. $K' \cup K''$ is the sum of all minimal splitting sets, hence by (43:G)

$$K' \cup K'' = I. \tag{43:9}$$

By their origin K' is disjunct with K, and K'' is $\subseteq K$. I.e.

$$K' \subseteq I - K, \qquad K'' \subseteq K. \tag{43:10}$$

Now (43:9), (43:10) together necessitate $K'' = K$; hence K is a sum of a suitable aggregate of minimal sets, as desired.

43.3.3. (43:F), (43:G) make it clear that the minimal splitting sets form a partition in the sense of 8.3.1., with the sum I. We call this the *decomposition partition of* Γ, and denote it by Π_Γ. Now (43:H) can be expressed as follows:

(43:H*) A splitting set $K \subseteq I$ is characterized by the following property: The points of each element of Π_Γ go together as far as K is concerned—i.e. each element of Π_Γ lies completely inside or completely outside of K.

[1] The intuitive meaning of these assertions should be quite clear. They characterize the structure of the maximum possibilities of decomposition of Γ in a plausible way.

Thus Π_Γ expresses how far the decomposition of Γ in I can be pushed, without destroying those ties which the rules of Γ establish between players.[1] By virtue of (43:E) the elements of Π_Γ are also characterized by the fact that they decompose Γ into indecomposable constituents.

43.4. Properties of the Decomposition Partition

43.4.1. The nature of the decomposition partition Π_Γ being established, it is natural to study the effect of the fineness of this partition. We wish to analyze only the two extreme possibilities: When Π_Γ is as fine as possible, i.e. when it dissects I down to the one-element sets—and when Π_Γ is as coarse as possible, i.e. when it does not dissect I at all. In other words: In the first case Π_Γ is the system of all one-element sets (in I)—in the second case Π_Γ consists of I alone.

The meaning of these two extreme cases is easily established:

(43:J) Π_Γ is the system of all one-element sets (in I) if and only if the game is inessential.

Proof: It is clear from (43:H) or (43:H*) that the stated property of Π_Γ is equivalent of saying that all sets $J(\subseteq I)$ are splitting. I.e. (by 43.1.) that for any two complementary sets J and $K(= I - J)$ the game Γ is decomposable. This means that (41:6) holds in all those cases. This implies, however, that the condition imposed by (41:6) on S, T (i.e. $S \subseteq J$, $T \subseteq K$) means merely that S, T are disjunct. Thus our statement becomes

$$v(S \cup T) = v(S) + v(T) \qquad \text{for} \qquad S \cap T = \ominus$$

Now this is precisely the condition of inessentiality by (27:D) in 27.4.2.

(43:K) Π_Γ consists of I if and only if the game Γ is indecomposable.

Proof: It is clear from (43:H) (or (43:H*)), that the stated property of Π_Γ is equivalent to saying that $\ominus$, I are the only splitting sets. But this is exactly the definition of indecomposability at the beginning of 43.3.

These results show that indecomposability and inessentiality are two opposite extremes for a game. In particular, inessentiality means that the decomposition of Γ, described at the end of 43.3., can be pushed through to the individual players, without ever severing any tie that the rules of the game Γ establish.[2] The reader should compare this statement with our original definition of inessentiality in 27.3.1.

43.4.2. The connection between inessentiality, decomposability, and the number n of players is as follows:

$n = 1$: This case is scarcely of practical importance. Such a game is clearly indecomposable,[3] and it is at the same time inessential by the first remark in 27.5.2.

[1] I.e. without impairing the self-containedness of the resulting sets.
[2] I.e. that every player is self-contained in this game.
[3] As I is a one-element set $\ominus$, I are its only subsets.

It should then be noted that indecomposability and inessentiality are by (43:J), (43:K) incompatible when $n \geqq 2$, but not when $n = 1$.

$n = 2$: Such a game, too, is necessarily inessential by the first remark of 27.5.2. Hence it is decomposable.

$n \geqq 3$: For these games decomposability is an exceptional occurrence. Indeed, decomposability implies (41:6) with some $J \neq \ominus, I$; hence $K = I - J \neq \ominus, I$. So we can choose j in J, k in K. Then (41:6) with $S = (j)$, $T = (k)$ gives

(43:11) $$\mathrm{v}((j, k)) = \mathrm{v}((j)) + \mathrm{v}((k)).$$

Now the only equations which the values of $\mathrm{v}(S)$ must satisfy, are (25:3:a), (25:3:b) of 25.3.1. (if zero-sum games are considered) or (42:6:a), (42:6:b) of 42.3.2. (43:11) is neither of these, since only the sets (j), (k), (j, k) occur in (43:11) and these are none of the sets occurring in those equations—i.e. $\ominus$ or I or complements—as $n \geqq 3$.[1] Thus (43:11) is an extra condition which is not fulfilled in general.

By the above an indecomposable game cannot have $n = 2$ hence it has $n = 1$ or $n \geqq 3$. Combining this with (43:E), we obtain the following peculiar result:

(43:L) Every element of the decomposition partition Π_Γ is either a one-element set, or else it has $n \geqq 3$ elements.

Note that the one-element sets in Π_Γ are the one-element splitting sets[2] —i.e. they correspond to those players who are self-contained, separated from the remainder of the game (from the point of view of the strategy of coalitions). They are the "dummies" in the sense of 35.2.3. and footnote 1 on p. 340. Consequently, our result (43:L) expresses this fact: Those players who are not "dummies," are grouped in indecomposable constituent games of $n \geqq 3$ players each.

This appears to be a general principle of social organization.

44. Decomposable Games. Further Extension of the Theory

44.1. Solutions of a (Decomposable) Game and Solutions of Its Constituents

44.1. We have completed the descriptive part of our study of composition and decomposition. Let us now pass to the central part of the problem: The investigation of the solutions in a decomposable game.

Consider a game Γ which is decomposable for J and $I - J = K$, with the J- and K-constituents Δ and H. We use strategic equivalence, as explained at the beginning of 42.5.3., to make all three games zero-sum.

Assume that the solutions for Δ as well as those for H are known; does this then determine the solution for Γ? In other words: How do the solutions for a decomposable game obtain from those for its constituents?

Now there exists a surmise in this respect which appears to be the *prima facie* plausible one, and we proceed to formulate it.

[1] For $n = 2$ it is otherwise; $(j, k) = I$, (j) and (k) are complements.
[2] Such a splitting set is, of course, automatically minimal.

44.2. Composition and Decomposition of Imputations and of Sets of Imputations

44.2.1. Let us use the notations of 41.3.1. But as we write $\mathrm{v}(S)$ for $\mathrm{v}_\Gamma(S)$ this also replaces by (41:4), (41:5), $\mathrm{v}_\Delta(S)$, $\mathrm{v}_\mathrm{H}(S)$.

On the other hand, we must distinguish between imputations for Γ, Δ, H.[1] In expressing this distinction, it is better to indicate the set of players to whom an imputation refers, instead of the game in which they are engaged. I.e. we will affix to them the symbols I, J, K rather than Γ, Δ, H. In this sense we denote the imputations for I (i.e. Γ) by

$$(44:1)\qquad \overrightarrow{\alpha}_I = \{\alpha_{1'}, \cdots, \alpha_{k'}, \alpha_{1''}, \cdots, \alpha_{l''}\},$$

and those for J, K (i.e. Δ, H) by

$$(44:2)\qquad \overrightarrow{\beta}_J = \{\beta_{1'}, \cdots, \beta_{k'}\},$$

$$(44:3)\qquad \overrightarrow{\gamma}_K = \{\gamma_{1''}, \cdots, \gamma_{l''}\}.$$

If three such imputations are linked by the relationship

$$(44:4)\qquad \begin{array}{lll} \alpha_{i'} = \beta_{i'} & \text{for} & i' = 1', \cdots, k', \\ \alpha_{j''} = \gamma_{j''} & \text{for} & j'' = 1'', \cdots, l'', \end{array}$$

then we say that $\overrightarrow{\alpha}_I$ obtains by *composition* for $\overrightarrow{\beta}_J$, $\overrightarrow{\gamma}_K$, that $\overrightarrow{\beta}_J$, $\overrightarrow{\gamma}_K$ obtain by *decomposition* from $\overrightarrow{\alpha}_I$ (for J, K), and that $\overrightarrow{\beta}_J$, $\overrightarrow{\gamma}_K$ are the (J-, K-) *constituents* of $\overrightarrow{\alpha}_I$.

Since we are now dealing with zero-sum games, all these imputations must fulfill the conditions (30:1), (30:2) of 30.1.1. Now one verifies immediately for $\overrightarrow{\alpha}_I$, $\overrightarrow{\beta}_J$, $\overrightarrow{\gamma}_K$ linked by (44:4).

Ad (30:1) of 30.1.1.: The validity of this for $\overrightarrow{\beta}_J$, $\overrightarrow{\gamma}_K$ is clearly equivalent to its validity for $\overrightarrow{\alpha}_I$.

Ad (30:2) of 30.1.1.: For $\overrightarrow{\beta}_J$, $\overrightarrow{\gamma}_K$ this states (using (44:4))

$$(44:5)\qquad \sum_{i'=1'}^{k'} \alpha_{i'} = 0,$$

$$(44:6)\qquad \sum_{j''=1''}^{l''} \alpha_{j''} = 0.$$

For $\overrightarrow{\alpha}_I$ it amounts to

$$(44:7)\qquad \sum_{i'=1'}^{k'} \alpha_{i'} + \sum_{j''=1''}^{l''} \alpha_{j''} = 0.$$

[1] It is now convenient to re-introduce the notations of 41.3.1. for the players.

Thus its validity for $\overrightarrow{\beta}_J$, $\overrightarrow{\gamma}_K$ implies the same for $\overrightarrow{\alpha}_I$, while its validity for $\overrightarrow{\alpha}_I$ does not imply the same for $\overrightarrow{\beta}_J$, $\overrightarrow{\gamma}_K$—indeed (44:7) does imply the equivalence of (44:5) and (44:6), but it fails to imply the validity of either one.

So we have:

(44:A) Any two imputations $\overrightarrow{\beta}_J$, $\overrightarrow{\gamma}_K$ can be composed to an $\overrightarrow{\alpha}_I$, while an imputation $\overrightarrow{\alpha}_I$ can be decomposed of two $\overrightarrow{\beta}_J$, $\overrightarrow{\gamma}_K$ if and only if it fulfills (44:5), i.e. (44:6).

We call such an $\overrightarrow{\alpha}_I$ decomposable (for J, K).

44.2.2. This situation is similar to that which prevails for the games themselves: Composition is always possible, while decomposition is not. Decomposability is again an exceptional occurrence.[1]

It ought to be noted, finally, that the concept of composition of imputations has a simple intuitive meaning. It corresponds to the same operation of "viewing as one" two separate occurrences, which played the corresponding role for games in 41.2.1., 41.2.3., 41.2.4. Decomposition of an $\overrightarrow{\alpha}_I$ (into $\overrightarrow{\beta}_J$, $\overrightarrow{\gamma}_K$) is possible if and only if the two self-contained sets of players J, K are given by the sets of imputations $\overrightarrow{\alpha}_I$ precisely their "just dues"—which are zero. This is the meaning of the condition (44:A) (i.e. of (44:5), (44:6)).

44.2.3. Consider a set V_J of imputations $\overrightarrow{\beta}_J$ and a set W_K of imputations $\overrightarrow{\gamma}_K$. Let U_I be the set of those imputations $\overrightarrow{\alpha}_I$ which obtain by composition of all $\overrightarrow{\beta}_J$ in V_J with all $\overrightarrow{\gamma}_K$ in W_K. We then say that U_I obtains by *composition* from V_J, W_K, that V_J, W_K obtain by *decomposition* from U_I (for J, K), and that V_J, W_K are the (J-, K-) *constituents* of U_I.

Clearly the operation of composition can always be carried out, whatever V_J, W_K—whereas a given U_I need not allow decomposition (for J, K). If U_I can be decomposed, we call it *decomposable* (for J, K).

Note that this decomposability of U_I restricts it very strongly; it implies, among other things that all elements $\overrightarrow{\alpha}_I$ of U_I must be decomposable (cf. the interpretation at the end of 44.2.2.).

In order to interpret these concepts for the sets of imputations U_I, V_J, W_K more thoroughly, it is convenient to restrict ourselves to solutions of the games Γ, Δ, H.

[1] There are great technical differences between the concepts of decomposability etc., for games and for imputations. Observe, however, the analogy between (41:4), (41:5) in 41.3.2.; (41:8), (41:9), (41:10) in 41.4.2.; and our (44:4), (44:5), (44:6), (44:7).

44.3. Composition and Decomposition of Solutions. The Main Possibilities and Surmises

44.3.1. Let V_J, W_K be two solutions for the games Δ, H respectively. Their composition yields an imputation set U_I which one might expect to be a solution for the game Γ. Indeed, U_I is the expression of a standard of behavior which can be formulated as follows. We give the verbal formulation in the text under (44:B:a)-(44:B:c), stating the mathematical equivalents in footnotes, which, as the reader will verify, add up precisely to our definition of composition.

(44:B:a) The players of J always obtain together exactly their "just dues" (zero), and the same is true for the players of K.[1]

(44:B:b) There is no connection whatever between the fate of players in the set J and in the set K.[2]

(44:B:c) The fate of the players in J is governed by the standard of behavior V_J,[3] the fate of the players in K is governed by the standard of behavior W_K.[4]

If the two constituent games are imagined to occur absolutely separate from each other, then this is the plausible way of viewing their separate solutions V_J, W_K as one solution U_I of the composite game Γ.

However, since a solution is an exact concept, this assertion needs a proof. I.e. we must demonstrate this:

(44:C) If V_J, W_K are solutions of Δ, H, then their composition U_I is a solution of Γ.

44.3.2. This, by the way, is another instance of the characteristic relationship between common sense and mathematical rigour. Although an assertion (in the present case that U_I is a solution whenever V_J, W_K are) is required by common sense, it has no validity within the theory (in this case based on the definitions of 30.1.1.) unless proved mathematically. To this extent it might seem that rigour is more important than common sense. This, however, is limited by the further consideration that if the mathematical proof fails to establish the common sense result, then there is a strong case for rejecting the theory altogether. Thus the primate of the mathematical procedure extends only to establish checks on the theories—in a way which would not be open to common sense alone.

[1] Every element $\overrightarrow{\alpha}_I$ of U_I is decomposable.

[2] Any $\overrightarrow{\beta}_J$ which is used in forming U_I and any $\overrightarrow{\gamma}_K$, which is used in forming U_I, give by composition an element $\overrightarrow{\alpha}_I$ of U_I.

[3] The above mentioned $\overrightarrow{\beta}_J$ are precisely the elements of V_J.

[4] The above mentioned $\overrightarrow{\gamma}_K$ are precisely the elements of W_K.

It will be seen that (44:C) is true, although not trivial.

One might be tempted to expect that the converse of (44:C) is also true, i.e. to demand a proof of this:

(44:D) If U_I is a solution of Γ, then it can be decomposed into solutions V_J, W_K of Δ, H.

This is *prima facie* quite plausible: Since Γ is the composition of what are for all intents and purposes two entirely separate games, how could any solution of Γ fail to exhibit this composite structure?

The surprising fact is, however, that (44:D) is not true in general. The reader might think that this should induce us to abandon—or at least to modify materially—our theory (i.e. 30.1.1.) if we take the above methodological statement seriously. Yet we will show, that the "common sense" basis for (44:D) is quite questionable. Indeed, our result, contradicting (44:D) will provide a very plausible interpretation—which connects it successfully with well known phenomena in social organizations.

44.3.3. The proper understanding of the failure of (44:D) and of the validity of the theory which replaces it, necessitates rather detailed considerations. Before we enter upon these, it might be useful to make, in anticipation, some indications as to how the failure of (44:D) occurs.

It is natural, to split (44:D) into two assertions:

(44:D:a) If U_I is a solution of Γ, then it is decomposable (for J, K).

(44:D:b) If a solution U_I of Γ is decomposable (for J, K), then its constituents V_J, W_K are solutions for Δ, H.

Now it will appear that (44:D:b) is true, and (44:D:a) is false. I.e. it can happen that a decomposable game Γ possesses an indecomposable solution.[1]

However, the decomposability of a solution (or of any set of imputations) is expressed by (44:B:a)-(44:B:c) in 44.3.1. So one or more of these conditions must fail for the indecomposable solution referred to above. Now it will be seen (cf. 46.11.) that the condition which is not satisfied is (44:B:a). This may seem to be very grave, because (44:B:a) is the primary condition in the sense that when it fails, the conditions (44:B:b), (44:B:c) cannot even be formulated.

The concept of decomposition possesses a certain elasticity. This appeared in 42.2.1., 42.2.2. and 42.5.2., where we succeeded in ridding ourselves of an inconvenient auxiliary condition connected with the decomposability of a game by modifying that concept. It will be seen that our difficulties will again be met by this procedure—so that (44:D) will be replaced by a correct and satisfactory theorem. Hence we must aim at modifying our arrangements, so that the condition (44:B:a) can be discarded.

We will succeed in doing this, and then it will appear that conditions (44:B:b), (44:B:c) make no difficulties and that a complete result can be obtained.

[1] This is similar to the phenomenon that a symmetric game may possess an asymmetric solution. Cf. 37.2.1.

44.4. Extension of the Theory. Outside Sources

44.4.1. It is now time to discard the normalization which we introduced (temporarily) in 44.1.: That the games under consideration are zero-sum. We return to the standpoint of 42.2.2. according to which the games are constant-sum.

These being understood, consider a game Γ which is decomposable (for J, K) with J-, K-constituents Δ, H.

The theory of composability and decomposability of imputations, as given in 44.2.1., 44.2.2. could now be repeated with insignificant changes. (44:1)-(44:4) may be taken over literally, while (44:5)-(44:7) are only modified in their right hand sides. Since (30:2) of 30.1.1. has been replaced by (42:8*) of 42.4.1. those formulae (44:5)-(44:7) now become:

(44:5*) $$\sum_{i'=1'}^{k'} \alpha_{i'} = \mathrm{v}(J),$$

(44:6*) $$\sum_{j''=1''}^{l''} \alpha_{j''} = \mathrm{v}(K),$$

and

(44:7*) $$\sum_{i'=1'}^{k'} \alpha_{i'} + \sum_{j''=1''}^{l''} \alpha_{j''} = \mathrm{v}(I) = \mathrm{v}(J) + \mathrm{v}(K).$$

(The last equation on the right hand side by (42:6:b) in 42.3.2., or equally by (41:6) in 41.3.2. with $S = J$, $T = K$.) The situation is exactly as in 44.2.1., indeed, it really arises from that one by the isomorphism of 42.4.2. Thus $\overrightarrow{\alpha}_I$ fulfills (44:7*), but for its decomposability (44:5*), (44:6*) are needed—and (44:7*) does imply the equivalence of (44:5*) and (44:6*), but it fails to imply the validity of either.

So the criterion of decomposability (44:A) in 44.2.1. is again true, only with our (44:5*), (44:6*) in place of its (44:5), (44:6). And the final conclusion of 44.2.2. may be repeated: Decomposition of an $\overrightarrow{\alpha}_I$ (into $\overrightarrow{\beta}_J$, $\overrightarrow{\gamma}_K$) is possible if and only if the two self contained sets of players J, K are given by this imputation $\overrightarrow{\alpha}_I$ precisely their just dues—which are now $\mathrm{v}(J)$, $\mathrm{v}(K)$.[1]

Since we know that this limitation of the decomposability of imputations—the reason for (44:B:a) in 44.3.1.—is a source of difficulties, we have to remove it. This means removal of the conditions (44:5*), (44:6*), i.e. of the condition (42:8*) in 42.4.1. from which they originate.

44.4.2. According to the above, we will attempt to work the theory of a constant-sum game Γ with a new concept of imputations, which is based on (42:7) of 42.4.1. (i.e. on (30:1) of 30.1.1.) alone, without (42:8*) in 42.4.1. In other words[2]

[1] Instead of zero, as loc. cit.

[2] We again denote the players by $1, \cdots, n$.

An *extended imputation* is a system of numbers $\alpha_1, \cdots, \alpha_n$ with this property:

(44:8) $$\alpha_i \geqq v((i)) \quad \text{for} \quad i = 1, \cdots, n.$$

We impose no conditions upon $\sum_{i=1}^{n} \alpha_i$. We view these extended imputations, too, as vectors

$$\overrightarrow{\alpha} = \{\alpha_1, \cdots, \alpha_n\}.$$

44.4.3. It will now be necessary to reconsider all our definitions which are rooted in the concepts of imputation—i.e. those of 30.1.1. and 44.2.1. But, before we do this, it is well to interpret this notion of extended imputations.

The essence of this concept is that it represents a distribution of certain amounts between the players, without demanding that they should total up to the constant sum of the game Γ.

Such an arrangement would be extraneous to the picture that the players are only dealing with each other. However, we have always conceived of imputations as a distributive scheme proposed to the totality of all players. (This idea pervades, e.g. all of 4.4., 4.5.; it is quite explicit in 4.4.1.) Such a proposal may come from one of the players,[1] but this is immaterial. We can equally imagine, that outside sources submit varying imputations to the consideration of the players of Γ. All this harmonizes with our past considerations, but in all this, those "outside sources" manifested themselves only by making suggestions—without contributing to, or withdrawing from, the proceeds of the game.

44.5. The Excess

44.5.1. Now our present concept of extended imputations may be taken to express that the "outside sources" can make suggestions which actually involve contributions or withdrawals, i.e. transfers. For the extended imputation $\overrightarrow{\alpha} = \{\alpha_1, \cdots, \alpha_n\}$ the amount of this transfer is

(44:9) $$e = \sum_{i=1}^{n} \alpha_i - v(I)$$

and will be called the *excess of* $\overrightarrow{\alpha}$. Thus

(44:10)
$$\begin{aligned} e &> 0 && \text{for a contribution,} \\ e &= 0 && \text{if no transfer takes place,} \\ e &< 0 && \text{for a withdrawal.} \end{aligned}$$

[1] Who tries to form a coalition. Since we consider the entire imputation as his proposal, this necessitates our assuming that he is even making propositions to those players, who will not be included in the coalition. To these he may offer their respective minima $v((i))$ (possibly more, cf. 38.3.2. and 38.3.3.). There may also be players in intermediate positions "between included and excluded" (cf. the second alternative in 37.1.3.). Of course, those less favored players may make their dissatisfaction effective, this leads to the concept of domination, etc.

It will be necessary to subject this to certain suitable limitations, in order to obtain realistic problems; and we will take due account of this.

It is important to realize how these transfers interact with the game. The transfers are part of the suggestions made from outside, which are accepted or rejected by the players, weighed against each other, according to the principles of domination, etc.[1] In the course of this process, any dissatisfied set of players may fall back upon the game Γ, which is the sole criterion of the effectivity of their preference of their situation in one (extended) imputation against another.[2] Thus the game, the physical background of the social process under consideration, determines the stability of all details of the organization—but the initiative comes through the outside suggestions, circumscribed by the limitations of the excess referred to above.

44.5.2. The simplest form that this "limitation" of the excess can take, consists in prescribing its value e explicitly. In interpreting this prescription, (44:10) should be remembered.

The situation which exists when $e \gtrless 0$ may at first seem paradoxical.

This is particularly true when $e < 0$, i.e. when a withdrawal from outside is attempted. Why should the players, who could fall back on a game of constant sum $v(I)$ accept an inferior total? I.e. how can a "standard of behavior," a "social order," based on such a principle be stable? There is, nevertheless an answer: The game is only worth $v(I)$ if all players form a coalition, act in concert. If they are split into hostile groups, then each group may have to estimate its chances more pessimistically and such a division may stabilize totals that are inferior to $v(I)$.[3]

The alternative $e > 0$, i.e. when the outside interference consists of a free gift, may seem less difficult to accept. But in this case too, it will be

[1] This is, of course, a narrow and possibly even somewhat arbitrary description of the social process. It should be remembered, however, that we use it only for a definite and limited purpose: To determine stable equilibria, i.e. solutions. The concluding remarks of 4.6.3. should make this amply clear.

[2] We are, of course, alluding to the definitions of effectivity and domination, cf. 4.4.1. and the beginning of 4.4.3.—given in exact form in 30.1.1. We will extend the exact definitions to our present concepts in 44.7.1.

[3] For a first quantitative orientation, in the heuristic manner: If the players are grouped into disjunct sets (coalitions) $S_1, \cdots, S_p$, then the total of their own valuations is $v(S_1) + \cdots + v(S_p)$. This is $\leqq v(I)$ by (42:6:c) in 42.3.2.

Oddly enough, this sum is actually $= v(I)$ when $p = 2$ by (42:6:b) in 42.3.2.—i.e. in this model the disagreements between three or more groups are the effective sources of damage.

Clearly by (42:6:c) in 42.3.2. the above sums $v(S_1) + \cdots + v(S_p)$ are all $\geqq \sum_{i=1}^{n} v((i))$. On the other hand, this latter expression is one of them (put $p = n$, $S_i = (i)$). So the damage is greatest when each player is isolated from all others.

The whole phenomenon disappears, therefore, when $\sum_{i=1}^{n} v((i)) = v(I)$, i.e. when the game is inessential. (Cf. (42:11) in 42.5.1.)

necessary to study the game in order to see how the distribution of this gift among the players can be governed by stable arrangements. It has to be expected, that the optimistic appraisal of their own chances, derived from the possibilities of the various coalitions in which they might participate will determine the players in making their claims. The theory must then provide their adjustment to the available total.

44.6. Limitations of the Excess. The Non-isolated Character of a Game in the New Setup

44.6.1. These considerations indicate that the excess e must be neither too small (when $e < 0$), nor too large (when $e > 0$). In the former case a situation would arise where each player would prefer to fall back on the game, even if the worst should happen, i.e. if he has to play it isolated.[1] In the latter case it will happen that the "free gift" is "too large," i.e. that no player in any imagined coalition can make such claims as to exhaust the available total. Then the very magnitude of the gift will act as a dissolvent on the existing mechanisms of organizations.

We will see in 45. that these qualitative considerations are correct and we will get from rigorous deductions the details of their operation and the precise value of the excess at which they become effective.

44.6.2. In all these considerations the game Γ can no longer be considered as an isolated occurrence, since the excess is a contribution or a withdrawal by an outside source. This makes it intelligible that this whole train of ideas should come up in connection with the decomposition theory of the game Γ. The constituent games Δ, H are indeed no longer entirely isolated, but coexistent with each other.[2] Thus, there is a good reason to look at Δ, H in this way—whether the composite game Γ should be treated in the old manner (i.e. as isolated), or in the new one, may be debatable. We shall see, however, that this ambiguity for Γ does not influence the result essentially, whereas the broader attitude concerning Δ, H proves to be absolutely necessary (cf. 46.8.3. and also 46.10.).

When a game Γ is considered in the above sense, as a non-isolated occurrence, with contributions or withdrawals by an outside source, one might be tempted to do this: Treat this outside source also as a player, including him together with the other players into a larger game Γ'. The rules of Γ' (which includes Γ) must then be devised in such a manner as to provide a mechanism for the desired transfers. We shall be able to meet this demand with the help of our final results, but the problem has some intricacies that are better considered only at that stage.

[1] This happens when the proposed total $\mathrm{v}(I) + e$ is $< \sum_{i=1}^{n} \mathrm{v}((i))$. As the last expression is equal to $\mathrm{v}(I) - n\gamma$ (by (42:11) in 42.5.1.) this means $e < -n\gamma$.

We will see in 45.1. that this is precisely the criterion for e being "too small."

[2] This in spite of the absence of "interactions," as far as the rules of the game are concerned; cf. 41.2.3., 41.2.4.

44.7. Discussion of the New Setup $E(e_0)$, $F(e_0)$

44.7.1. The reconsideration of our old definitions mentioned at the beginning of 44.4.3. is a very simple matter.

For the *extended imputations* we have the new definitions of 44.4.2. The definitions of *effectivity* and *domination* we take over unchanged from 30.1.1.[1]—the supporting arguments brought forward in the discussion which led up to those definitions appear to lose no strength by our present generalizations. The same applies to our definition of *solutions* eod.[2] with one caution: The definition of a solution referred to makes the concept of a solution dependent upon the set of all imputations in which it is formed. Now in our present setup of extended imputations we shall have to consider limitations concerning them—notably concerning their excesses—as indicated in 44.5.1. These restrictions will determine the set of all extended imputations to be considered and thereby the concept of a solution.

44.7.2. Specifically we shall consider two types of limitations.

First, we shall consider the case where the value of the excess is prescribed. Then we have an equation

$$e = e_0 \tag{44:11}$$

with a given e_0. The meaning of this restriction is that the transfer from outside is prescribed, in the sense of the discussion of 44.5.2.

Second, we shall consider the case where only an upper limit of the excess is prescribed. Then we have an inequality

$$e \leqq e_0 \tag{44:12}$$

with a given e_0. The meaning of this restriction is that the transfer from outside is assigned a maximum (from the point of view of the players who receive it).

The case in which we are really interested is the first one, i.e. that one of 44.5.2. The second case will prove technically useful for the clarification of the first one—although its introduction may at first seem artificial. We refrain from considering further alternatives because we will be able to complete the indicated discussion with these two cases alone.

Denote the set of all extended imputations fulfilling (44:11) (first case) by $E(e_0)$. Considering (44:9) in 44.5.1., we can write (44:11) as

$$\sum_{i=1}^{n} \alpha_i = \mathrm{v}(I) + e_0. \tag{44:11*}$$

Denote the set of all extended imputations fulfilling (44:12) (second case) by $F(e_0)$. Considering (44:9) in 44.5.1., we can write (44:12) as

$$\sum_{i=1}^{n} \alpha_i \leqq \mathrm{v}(I) + e_0. \tag{44:12*}$$

[1] I.e. (30:3); (30:4:a)-(30:4:c) loc. cit., respectively.

[2] I.e. (30:5:a), (30:5:b) or (30:5:c) eod.

For the sake of completeness, we repeat the characterization of an extended imputation which must be added to (44:11*), as well as to (44:12*):

(44:13) $\alpha_i \geqq \mathrm{v}((i)), \quad \text{for} \quad i = 1, \cdots, n.$

Note that the definitions of (44:9) as well as (44:11*), (44:12*) and (44:13) are invariant under the isomorphism of 42.4.2.

44.7.3. Now the definition of a solution can be taken over from 30.1.1. Because of the central role of this concept we restate that definition, adjusted to the present conditions. Throughout the definition which follows, $E(e_0)$ can be replaced by $F(e_0)$, as indicated by [ ].

A set $\mathsf{V} \subseteq E(e_0)$ $[F(e_0)]$ is a *solution* for $E(e_0)$ $[F(e_0)]$ if it possesses the following properties:

(44:E:a) No $\overrightarrow{\beta}$ in V is dominated by an $\overrightarrow{\alpha}$ in V.

(44:E:b) Every $\overrightarrow{\beta}$ of $E(e_0)[F(e_0)]$ not in V is dominated by some $\overrightarrow{\alpha}$ in V.

(44:E:a) and (44:E:b) can be stated as a single condition:

(44:E:c) The elements of V are those elements of $E(e_0)$ $[F(e_0)]$ which are undominated by any element of V.

It will be noted that $E(0)$ takes us back to the original 30.1.1. (zero-sum game) and 42.4.1. (constant-sum game).

44.7.4. The concepts of *composition*, *decomposition* and *constituents* of extended imputations can again be defined by (44:1)-(44:4) of 44.2.1. As pointed out in 44.4.2. the technical purpose of our extending the concept of imputation is now fulfilled. Decomposition as well as composition can now always be carried out.

The connection of these concepts with the sets $E(e_0)$ and $F(e_0)$ is not so simple; we will deal with it as the necessity arises.

For the *composition*, *decomposition* and *constituents* of sets of extended imputations the definitions of 44.2.3. can now be repeated literally.

45. Limitations of the Excess. Structure of the Extended Theory

45.1. The Lower Limit of the Excess

45.1. In the setups of 30.1.1. and of 42.4.1. imputations always existed. It is now different: Either set $E(e_0)$, $F(e_0)$ may be empty for certain e_0. Obviously this happens when (44:11*) or (44:12*) of 44.7.2. conflict with (44:13) eodem—and this is clearly the case for

$$\mathrm{v}(I) + e_0 < \sum_{i=1}^{n} \mathrm{v}((i))$$

in both alternatives. As the right hand side is equal to $\mathrm{v}(I) - n\gamma$ by (42:11) in 42.5.1., this means

(45:1) $$e_0 < -n\gamma$$

If $E(e_0)$ [$F(e_0)$] is empty, then the empty set is clearly a solution for it—and since it is its only subset, it is also its only solution.[1] If, on the other hand, $E(e_0)$ [$F(e_0)$] is not empty, then none of its solutions can be empty. This follows by literal repetition of the proof of (31:J) in 31.2.1.

The right hand side of the inequality (45:1) is determined by the game Γ; we introduce this notation for it (with the opposite sign, and using (42:11) in 42.5.1.):

(45:2) $$|\Gamma|_1 = n\gamma = \mathrm{v}(I) - \sum_{i=1}^{n} \mathrm{v}((i)).$$

Now we can sum up our observations as follows:

(45:A) If

$$e_0 < -|\Gamma|_1,$$

then $E(e_0)$, $F(e_0)$ are empty and the empty set is their only solution. Otherwise neither $E(e_0)$ nor $F(e_0)$ nor any solution of either can be empty.

This result gives the first indication, that "too small" values of e_0 (i.e. e) in the sense of 44.6.1. exist. Actually, it corroborates the quantitative estimate of footnote 1 on p. 366.

45.2. The Upper Limit of the Excess. Detached and Fully Detached Imputations

45.2.1. Let us now turn to those values of e_0 (i.e. e), which are "too large" in the sense of 44.6.1. When does the disorganizing influence of the magnitude of e, which we there foresaw, manifest itself?

As indicated in 44.6.1., the critical phenomenon is this: The excess may be too large to be exhausted by the claims which any player in any imagined coalition can possibly make. We proceed to formulate this idea in a quantitative way.

It is best to consider the extended imputations $\overrightarrow{\alpha}$ themselves, instead of their excesses e. Such an $\overrightarrow{\alpha}$ is past any claims which may be made in any coalition, if it assigns to the players of each (non-empty) set $S \subseteq I$ more than those players could get by forming a coalition in Γ, i.e. if

(45:3) $$\sum_{i \text{ in } S} \alpha_i > \mathrm{v}(S) \qquad \text{for every non-empty set } S \subseteq I.$$

Comparing this with (30:3) in 30.1.1. shows that our criterion amounts to demanding that every non-empty set S be ineffective for $\overrightarrow{\alpha}$.

In our actual deductions it will prove advantageous to widen (45:3) somewhat by including the limiting case of equality. The condition then becomes

[1] In spite of its triviality, this circumstance should not be overlooked. The text actually repeats footnote 2 on p. 278.

(45:4) $$\sum_{i \text{ in } S} \alpha_i \geqq v(S) \quad \text{for every} \quad S \subseteq I.$$[1]

It is convenient to give these $\overrightarrow{\alpha}$ a name. We call the $\overrightarrow{\alpha}$ of (45:3) *fully detached*, and those of (45:4) *detached*. As indicated, the latter concept will be really needed in our proofs—both *termini* are meant to express that the extended imputation is detached from the game, i.e. that it cannot be effectively supported within the game by any coalition.

45.2.2. One more remark is useful:

The only restriction imposed upon extended imputations is (44:13) of 44.7.2.:

(45:5) $$\alpha_i \geqq v((i)) \quad \text{for} \quad i = 1, \cdots, n.$$

Now if the requirement (45:4) of detachedness is fulfilled—and hence *a fortiori* if the requirement (45:3) of full detachedness is fulfilled—then it is unnecessary to postulate the condition (45:5) as well. Indeed, (45:5) is the special case of (45:4) for $S = (i)$.

This remark will be made use of implicitly in the proofs which follow.

45.2.3. Now we can revert to the excesses, i.e. characterize those which belong to detached (or fully detached) imputations. This is the formal characterization:

(45:B) The game Γ determines a number $|\Gamma|_2$ with the following properties:

(45:B:a) A fully detached extended imputation with the excess e exists if and only if

$$e > |\Gamma|_2.$$

(45:B:b) A detached extended imputation with the excess e exists if and only if

$$e \geqq |\Gamma|_2.$$[2]

Proof: Existence of a detached $\overrightarrow{\alpha}$[3]: Let α^0 be the maximum of all $v(S)$, $S \subseteq I$ (so $\alpha^0 \geqq v(\ominus) = 0$). Put $\overrightarrow{\alpha}^0 = \{\alpha_1^0, \cdots, \alpha_n^0\} = \{\alpha^0, \cdots, \alpha^0\}$. Then for every non-empty $S \subseteq I$ we have $\sum_{i \text{ in } S} \alpha_i^0 \geqq \alpha^0 \geqq v(S)$. This is (45:4), so $\overrightarrow{\alpha}^0$ is detached.

[1] It is no longer necessary to exclude $S = \ominus$, since (45:4) unlike (45:3) is true when $S = \ominus$. Indeed, then both sides vanish.

[2] The intuitive meaning of these statements is quite simple: It is plausible that in order to produce a detached or a fully detached imputation, a certain (positive) minimum excess is required. $|\Gamma|_2$ is this minimum, or rather lower limit. Since the notions "detached" and "fully detached" differ only in a limiting case (the = sign in (45:4)), it stands to reason that their lower limits be the same. These things find an exact expression in (45:B).

[3] Note that it is necessary to prove this! The evaluation which we give here is crude, for more precise ones cf. (45:F) below.

Properties of the detached $\overrightarrow{\alpha}$: According to the above, detached

$$\overrightarrow{\alpha} = \{\alpha_1, \cdots, \alpha_n\}$$

exist, and with them their excesses $e = \sum_{i=1}^{n} \alpha_i - v(I)$. By (45:4) (with $S = I$) all these e are $\geqq 0$. Hence it follows by continuity, that these e have a minimum e^*. Choose a detached $\overrightarrow{\alpha}^* = \{\alpha_1^*, \cdots, \alpha_n^*\}$ with this excess e^*.[1]

We now put

$$(45:6) \qquad |\Gamma|_2 = e^*.$$

Proof of (45:B:a), (45:B:b): If $\overrightarrow{\alpha} = \{\alpha_1, \cdots, \alpha_n\}$ is detached, then by definition $e = \sum_{i=1}^{n} \alpha_i - v(I) \geqq e^*$. If $\overrightarrow{\alpha} = \{\alpha_1, \cdots, \alpha_n\}$ is fully detached, then (45:3) remains true if we subtract a sufficiently small $\delta > 0$ from each α_i. So $\overrightarrow{\alpha}' = \{\alpha_1 - \delta, \cdots, \alpha_n - \delta\}$ is detached. Hence by definition $e - n\delta = \sum_{i=1}^{n} (\alpha_i - \delta) - v(I) \geqq e^*$, $e > e^*$.

Consider now the detached $\overrightarrow{\alpha}^* = \{\alpha_1^*, \cdots, \alpha_n^*\}$ with

$$\sum_{i=1}^{n} \alpha_i^* - v(I) = e^*.$$

Then (45:4) holds for $\overrightarrow{\alpha}^*$; hence (45:3) holds if we increase each α_i^* by a $\delta > 0$. So $\overrightarrow{\alpha}'' = \{\alpha_1^* + \delta, \cdots, \alpha_n^* + \delta\}$ is fully detached. Its excess is $e = \sum_{i=1}^{n} (\alpha_i^* + \delta) - v(I) = e^* + n\delta$. So every $e = e^* + n\delta$, $\delta > 0$, i.e. every $e > e^*$, is the excess of a fully detached imputation—hence *a fortiori* of a detached one; and e^* is, of course, the excess of a detached imputation $\overrightarrow{\alpha}^*$.

Thus all parts of (45:B:a), (45:B:b) hold for (45:6).

45.2.4. The fully detached and the detached extended imputations are also closely connected with the concept of domination. The properties involved are given in (45:C) and (45:D) below. They form a peculiar antithesis to each other. This is remarkable, since our two concepts are strongly analogous to each other—indeed, the second one arises from the first one by the inclusion of its limiting cases.

[1] This continuity argument is valid because the = sign is included in (45:4).

(45:C) A fully detached extended imputation $\overrightarrow{\alpha}$ dominates no other extended imputation $\overrightarrow{\beta}$.

Proof: If $\overrightarrow{\alpha} \succ \overrightarrow{\beta}$, then $\overrightarrow{\alpha}$ must possess a non-empty effective set.

(45:D) An extended imputation $\overrightarrow{\alpha}$ is detached if and only if it is dominated by no other extended imputation $\overrightarrow{\beta}$.

Proof: Sufficiency of being detached: Let $\overrightarrow{\alpha} = \{\alpha_1, \cdots, \alpha_n\}$ be detached. Assume *a contrario* $\overrightarrow{\beta} \succ \overrightarrow{\alpha}$, with the effective set S. Then S is not empty; $\alpha_i < \beta_i$ for i in S. So $\sum_{i \text{ in } S} \alpha_i < \sum_{i \text{ in } S} \beta_i \leqq \mathrm{v}(S)$ contradicting (45:4).

Necessity of being detached: Assume that $\overrightarrow{\alpha} = \{\alpha_1, \cdots, \alpha_n\}$ is not detached. Let S be a (necessarily non-empty) set for which (45:4) fails, i.e. $\sum_{i \text{ in } S} \alpha_i < \mathrm{v}(S)$. Then for a sufficiently small $\delta > 0$, even

$$\sum_{i \text{ in } S} (\alpha_i + \delta) \leqq \mathrm{v}(S).$$

Put $\overrightarrow{\beta} = \{\beta_1, \cdots, \beta_n\} = \{\alpha_1 + \delta, \cdots, \alpha_n + \delta\}$, then always $\alpha_i < \beta_i$ and S is effective for $\overrightarrow{\beta}$: $\sum_{i \text{ in } S} \beta_i \leqq \mathrm{v}(S)$. Thus $\overrightarrow{\beta} \succ \overrightarrow{\alpha}$.

45.3. Discussion of the Two Limits $|\Gamma|_1$, $|\Gamma|_2$. Their Ratio

45.3.1. The two numbers $|\Gamma|_1$ and $|\Gamma|_2$, as defined in (45:2) of 45.1. and in (45:B) of 45.2.3. are both in a way quantitative measures of the essentiality of Γ. More precisely:

(45:E) If Γ is inessential, then $|\Gamma|_1 = 0$, $|\Gamma|_2 = 0$.
If Γ is essential, then $|\Gamma|_1 > 0$, $|\Gamma|_2 > 0$.

Proof: The statements concerning $|\Gamma|_1$, which is $= n\gamma$ by (45:2) of 45.1., coincide with the definitions of inessentiality and essentiality of 27.3., as reasserted in 42.5.1.

The statements concerning $|\Gamma|_2$ follow from those concerning $|\Gamma|_1$, by means of the inequalities of (45:F), which we can use here.

45.3.2. The quantitative relationship of $|\Gamma|_1$ and $|\Gamma|_2$ is characterized as follows:

Always

(45:F) $$\frac{1}{n-1}|\Gamma|_1 \leqq |\Gamma|_2 \leqq \frac{n-2}{2}|\Gamma|_1.$$

Proof: As we know, $|\Gamma|_1$ and $|\Gamma|_2$ are invariant under strategic equivalence, hence we may assume the game Γ to be zero-sum, and even reduced in the sense of 27.1.4. We can now use the notations and relations of 27.2.

Since $|\Gamma|_1 = n\gamma$, we want to prove that

$$\frac{n}{n-1}\gamma \leqq |\Gamma|_2 \leqq \frac{n(n-2)}{2}\gamma. \tag{45:7}$$

Proof of the first inequality of (45:7): Let $\overrightarrow{\alpha} = \{\alpha_1, \cdots, \alpha_n\}$ be detached. Then (45:4) gives for the $(n-1)$-element set $S = I - (k)$, $\sum_{i=1}^{n} \alpha_i - \alpha_k = \sum_{i \text{ in } S} \alpha_i \geqq \mathrm{v}(S) = \gamma$, i.e.

$$\sum_{i=1}^{n} \alpha_i - \alpha_k \geqq \gamma. \tag{45:8}$$

Summing (45:8) over $k = 1, \cdots, n$, gives $n\sum_{i=1}^{n} \alpha_i - \sum_{k=1}^{n} \alpha_k \geqq n\gamma$, i.e. $(n-1)\sum_{i=1}^{n} \alpha_i \geqq n\gamma$, $\sum_{i=1}^{n} \alpha_i \geqq \frac{n}{n-1}\gamma$. Now $\mathrm{v}(I) = 0$, so $e = \sum_{i=1}^{n} \alpha_i$. Thus $e \geqq \frac{n}{n-1}\gamma$ for all detached imputations; hence $|\Gamma|_2 \geqq \frac{n}{n-1}\gamma$.

Proof of the second inequality of (45:7):

Put $\alpha^{00} = \frac{n-2}{2}\gamma$, and $\overrightarrow{\alpha}^{00} = \{\alpha_1^{00}, \cdots, \alpha_n^{00}\} = \{\alpha^{00}, \cdots, \alpha^{00}\}$.

This $\overrightarrow{\alpha}^{00}$ is detached, i.e. it fulfills (45:4) for all $S \subseteq I$. Indeed: Let p be the number of elements of S. Now we have:

$p = 0$: $S = \ominus$, (45:4) is trivial.

$p = 1$: $S = (i)$, (45:4) becomes $\alpha^{00} \geqq \mathrm{v}((i))$, i.e. $\frac{n-2}{2}\gamma \geqq -\gamma$ which is obvious.

$p \geqq 2$: (45:4) becomes $p\alpha^{00} \geqq \mathrm{v}(S)$, but by (27:7) in 27.2.

$$\mathrm{v}(S) \leqq (n-p)\gamma,$$

so it suffices to prove $p\alpha^{00} \geqq (n-p)\gamma$ i.e. $p\,\frac{n-2}{2}\gamma \geqq (n-p)\gamma$. This amounts to $p\,\frac{n}{2}\gamma \geqq n\gamma$, which follows from $p \geqq 2$.

Thus $\overrightarrow{\alpha}^{00}$ is indeed detached. As $\mathrm{v}(I) = 0$, the excess is

$$e^{00} = n\alpha^{00} = \frac{n(n-2)}{2}\gamma.$$

Hence $|\Gamma|_2 \leqq \frac{n(n-2)}{2}\gamma$.

45.3.3. It is worth while to consider the inequalities of (45:F) for $n = 1,2,3,4, \cdots$ successively:

$n = 1,2$: In these cases the coefficient $\frac{1}{n-1}$ of the lower bound of the inequality is greater than the coefficient $\frac{n-2}{2}$ of the upper bound.[1] This may seem absurd. But since Γ is necessarily inessential for $n = 1,2$ (cf. the first remark in 27.5.2.), we have in these cases $|\Gamma|_1 = 0$, $|\Gamma|_2 = 0$, and so the contradictions disappear.

$n = 3$: In this case the two coefficients $\frac{1}{n-1}$ and $\frac{n-2}{2}$ coincide: Both are equal to $\frac{1}{2}$. So the inequalities merge to an equation:

$$|\Gamma|_2 = \tfrac{1}{2}|\Gamma|_1. \tag{45:9}$$

$n \geqq 4$: In these cases the coefficient $\frac{1}{n-1}$ of the lower bound is definitely smaller than the coefficient $\frac{n-2}{2}$ of the upper bound.[2] So now the inequalities leave a non-vanishing interval open for $|\Gamma_2|$.

The lower bound $|\Gamma|_2 = \frac{1}{n-1}|\Gamma|_1$ is precise, i.e. there exists for each $n \geqq 4$ an essential game for which it is assumed. There also exist for each $n \geqq 4$ essential games with $|\Gamma|_2 > \frac{1}{n-1}|\Gamma|_1$, but it is probably not possible to reach the upper bound of our inequality, $|\Gamma|_2 = \frac{n-2}{2}|\Gamma|_1$. The precise value of the upper bound has not yet been determined. We do not need to discuss these things here any further.[3]

45.3.4. In a more qualitative way, we may therefore say that $|\Gamma|_1, |\Gamma|_2$ are both quantitative measures of the essentiality of the game Γ. They measure it in two different, and to a certain extent, independent ways. Indeed, the ratio $|\Gamma|_2/|\Gamma|_1$, which never occurs for $n = 1,2$ (no essential games!), and is a constant for $n = 3$ (its value is $\frac{1}{2}$), is variable with Γ for each $n \geqq 4$.

We saw in 45.1., 45.2., that these two quantities actually measure the limits, within which a dictated excess will not "disorganize" the players, in the sense of 44.6.1. Judging from our results, an excess $e < -|\Gamma|_1$ is "too small" and an excess $e > |\Gamma|_2$ is "too great" in that sense. This view will be corroborated in a much more precise sense in 46.8.

[1] They are $\infty, -\frac{1}{2}$ for $n = 1$; $1, 0$ for $n = 2$. Note also the paradoxical values ∞ and $-\frac{1}{2}$!

[2] $\frac{1}{n-1} < \frac{n-2}{2}$ means $2 < (n-1)(n-2)$ which is clearly the case for all $n \geqq 4$.

[3] For $n = 4$ our inequality is $\frac{1}{3}|\Gamma|_1 \leqq |\Gamma|_2 \leqq |\Gamma|_1$. As mentioned above, we know an essential game with $|\Gamma|_2 = \frac{1}{3}|\Gamma|_1$ and also one with $|\Gamma|_2 = \frac{1}{2}|\Gamma|_1$.

45.4. Detached Imputations and Various Solutions. The Theorem Connecting $E(e_0)$, $F(e_0)$

45.4.1. (44:E:c) in the definition of a solution in 44.7.3. and our result (45:D) in 45.2.4. give immediately:

(45:G) A solution V for $E(e_0)$ [$F(e_0)$] must contain every detached extended imputation of $E(e_0)$ [$F(e_0)$].

The importance of this result is due to its role in the following consideration.

After what was said at the beginning of 44.7.2. about the roles of $E(e_0)$ and $F(e_0)$, the importance of establishing the complete inter-relationship between these two cases will be obvious. I.e. we must determine the connection between the solutions for $E(e_0)$ and $F(e_0)$.

Now the whole difference between $E(e_0)$ and $F(e_0)$ and their solutions is not easy to appraise in an intuitive way. It is difficult to see *a priori* why there should be any difference at all: In the first case the "gift," made to the players from the outside, has the prescribed value e_0, in the second case it has the prescribed maximum value e_0. It is difficult to see how the "outside source," which is willing to contribute up to e_0 can ever be allowed to contribute less than e_0 in a "stable" standard of behavior (i.e. solution). However, our past experience will caution us against rash conclusions in this respect. Thus we saw in 33.1. and 38.3. that already three and four-person games possess solutions in which an isolated and defeated player is not "exploited" up to the limit of the physical possibilities—and the present case bears some analogy to that.

45.4.2. (45:G) permits us to make a more specific statement:

A detached extended imputation $\overrightarrow{\alpha}$ belongs by (45:G) to every solution for $F(e_0)$, if it belongs to $F(e_0)$. On the other hand, $\overrightarrow{\alpha}$ clearly cannot belong to any solution for $E(e_0)$ if it does not belong to $E(e_0)$. We now define:

(45:10) $D^*(e_0)$ is the set of all detached extended imputations $\overrightarrow{\alpha}$ in $F(e_0)$, but not in $E(e_0)$.

So we see: Any solution of $F(e_0)$ contains all elements of $D^*(e_0)$; any solution of $E(e_0)$ contains no element of $D^*(e_0)$. Consequently $F(e_0)$ and $E(e_0)$ have certainly no solution in common if $D^*(e_0)$ is not empty.

Now the detached $\overrightarrow{\alpha}$ of $D^*(e_0)$ are characterized by having an excess $e \leqq e_0$, but not $e = e_0$—i.e. by

(45:11) $$e < e_0.$$

From this we conclude:

(45:H) $D^*(e_0)$ is empty if and only if

$$e_0 \leqq |\Gamma|_2.$$

Proof: Owing to (45:B) and to (45:11) above, the non-emptiness of $D^*(e_0)$ is equivalent to the existence of an e with $|\Gamma|_2 \leqq e < e_0$—i.e. to $e_0 > |\Gamma|_2$. Hence the emptiness of $D^*(e_0)$ amounts to $e_0 \leqq |\Gamma|_2$.

Thus the solutions for $F(e_0)$ and for $E(e_0)$ are sure to differ, when $e_0 > |\Gamma|_2$. This is further evidence that e_0 is "too large" for normal behavior when it is $> |\Gamma|_2$.

45.4.3. Now we can prove that the difference indicated above is the only one between the solutions for $E(e_0)$ and for $F(e_0)$. More precisely:

(45:I) The relationship

(45:12) $$\mathsf{V} \rightleftarrows \mathsf{W} = \mathsf{V} \cup D^*(e_0)$$

establishes a one-to-one relationship between all solutions V for $E(e_0)$ and all solutions W for $F(e_0)$.

This will be demonstrated in the next section.

45.5. Proof of the Theorem

45.5.1. We begin by proving some auxiliary lemmas.

The first one consists of a perfectly obvious observation, but of wide applicability:

(45:J) Let the two extended imputations $\overrightarrow{\gamma} = \{\gamma_1, \cdots, \gamma_n\}$ and $\overrightarrow{\delta} = \{\delta_1, \cdots, \delta_n\}$ bear the relationship

(45:13) $$\gamma_i \geqq \delta_i \quad \text{for all} \quad i = 1, \cdots, n;$$

then for every $\overrightarrow{\alpha}$, $\overrightarrow{\alpha} \succ \overrightarrow{\gamma}$ implies $\overrightarrow{\alpha} \succ \overrightarrow{\delta}$.

The meaning of this result is, of course, that (45:13) expresses some kind of inferiority of $\overrightarrow{\delta}$ to $\overrightarrow{\gamma}$—in spite of the intransitivity of domination. This inferiority is, however, not as complete as one might expect. Thus one cannot make the plausible inference of $\overrightarrow{\gamma} \succ \overrightarrow{\beta}$ from $\overrightarrow{\delta} \succ \overrightarrow{\beta}$, because the effectivity of a set S for $\overrightarrow{\delta}$ may not imply the same for $\overrightarrow{\gamma}$. (The reader should recall the basic definitions of 30.1.1.)

It should also be observed, that (45:J) emerges only because we have extended the concept of imputations. For our older definitions (cf. 42.4.1.) we would have had $\sum_{i=1}^{n} \gamma_i = \sum_{i=1}^{n} \delta_i$; hence $\gamma_i \geqq \delta_i$ for all $i = 1, \cdots, n$ necessitates $\gamma_i = \delta_i$ for all $i = 1, \cdots, n$, i.e. $\overrightarrow{\gamma} = \overrightarrow{\delta}$.

45.5.2. Now four lemmas leading directly to the desired proof of (45:I).

(45:K) If $\overrightarrow{\alpha} \succ \overrightarrow{\beta}$ with $\overrightarrow{\alpha}$ detached and in $F(e_0)$ and $\overrightarrow{\beta}$ in $E(e_0)$, then there exists an $\overrightarrow{\alpha}' \succ \overrightarrow{\beta}$ with $\overrightarrow{\alpha}'$ detached and in $E(e_0)$.

Proof: Let S be the set of (30:4:a)-(30:4:c) in 30.1.1. for the domination $\overrightarrow{\alpha} \succ \overrightarrow{\beta}$. $S = I$ would imply $\alpha_i > \beta_i$ for all $i = 1, \cdots, n$ so

$$\sum_{i=1}^{n} \alpha_i - v(I) > \sum_{i=1}^{n} \beta_i - v(I).$$

But as $\overrightarrow{\alpha}$ is in $F(e_0)$ and $\overrightarrow{\beta}$ in $E(e_0)$, so $\sum_{i=1}^{n} \alpha_i - v(I) \leqq e_0 = \sum_{i=1}^{n} \beta_i - v(I)$, contradicting the above.

So $S \neq I$. Choose, therefore, an $i_0 = 1, \cdots, n$, not in S. Define $\overrightarrow{\alpha}' = \{\alpha'_1, \cdots, \alpha'_n\}$ with

$$\alpha'_{i_0} = \alpha_{i_0} + \epsilon,$$
$$\alpha'_i = \alpha_i \quad \text{for} \quad i \neq i_0,$$

choosing $\epsilon \geqq 0$ so that $\sum_{i=1}^{n} \alpha'_i - v(I) = e_0$. Thus all $\alpha'_i \geqq \alpha_i$; hence $\overrightarrow{\alpha}'$ is detached and it is clearly in $E(e_0)$. Again, as $\alpha'_i = \alpha_i$ for $i \neq i_0$, hence for all i in S, so our $\overrightarrow{\alpha} \succ \overrightarrow{\beta}$ implies $\overrightarrow{\alpha}' \succ \overrightarrow{\beta}$.

(45:L) Every solution W for $F(e_0)$ has the form (45:12) of (45:I) for a unique $\mathsf{V} \subseteq E(e_0)$.[1]

Proof: Obviously the V in question—if it exists at all—is the intersection $\mathsf{W} \cap E(e_0)$, so it is unique. In order that (45:12) should hold for

$$\mathsf{V} = \mathsf{W} \cap E(e_0),$$

we need only that the remainder of W be equal to $D^*(e_0)$, i.e.

$$\mathsf{W} - E(e_0) = D^*(e_0). \tag{45:14}$$

Let us therefore prove (45:14).

Every element of $D^*(e_0)$ is detached and in $F(e_0)$—so it is in W by (45:G). Again, it is not in $E(e_0)$, so it is in $\mathsf{W} - E(e_0)$. Thus

$$\mathsf{W} - E(e_0) \supseteq D^*(e_0). \tag{45:15}$$

If also

$$\mathsf{W} - E(e_0) \subseteq D^*(e_0), \tag{45:16}$$

then (45:15), (45:16) together give (45:14), as desired. Assume therefore, that (45:16) is not true.

Accordingly, consider an $\overrightarrow{\alpha} = \{\alpha_1, \cdots, \alpha_n\}$ in $\mathsf{W} - E(e_0)$ and not in $D^*(e_0)$. Then $\overrightarrow{\alpha}$ is in $F(e_0)$, but not in $E(e_0)$, so $\sum_{i=1}^{n} \alpha_i - v(I) < e_0$. As

[1] We do not yet assert that this V is a solution for $E(e_0)$—that will come in (45:M).

$\overrightarrow{\alpha}$ is not in $D^*(e_0)$, this excludes its being detached. Hence there exists a non-empty set S with $\sum_{i \text{ in } S} \alpha_i < \mathrm{v}(S)$.

Now form $\overrightarrow{\alpha}' = \{\alpha'_1, \cdots, \alpha'_n\}$ with

$$\begin{aligned} \alpha'_i &= \alpha_i + \epsilon && \text{for } i \text{ in } S, \\ \alpha'_i &= \alpha_i && \text{for } i \text{ not in } S, \end{aligned}$$

choosing $\epsilon > 0$ so that still $\sum_{i=1}^{n} \alpha'_i - \mathrm{v}(I) \leqq e_0$ and $\sum_{i \text{ in } S} \alpha'_i \leqq \mathrm{v}(S)$. So $\overrightarrow{\alpha}'$ is in $F(e_0)$. If it is not in W, then (as W is a solution for $F(e_0)$) there exists a $\overrightarrow{\beta}$ in W with $\overrightarrow{\beta} \succ \overrightarrow{\alpha}'$. As all $\alpha'_i \geqq \alpha_i$, this implies $\overrightarrow{\beta} \succ \overrightarrow{\alpha}$ by (45:J). This is impossible, since both $\overrightarrow{\beta}, \overrightarrow{\alpha}$ belong to (the solution) W. Hence $\overrightarrow{\alpha}'$ must be in W. Now $\alpha'_i > \alpha_i$ for all i in S, and $\sum_{i \text{ in } S} \alpha'_i \leqq \mathrm{v}(S)$. So $\overrightarrow{\alpha}' \succ \overrightarrow{\alpha}$. But as both $\overrightarrow{\alpha}', \overrightarrow{\alpha}$ belong to (the solution) W, this is a contradiction.

(45:M) The V of (45:L) is a solution for $E(e_0)$.

Proof: $\mathsf{V} \subseteq E(e_0)$ is clear, and V fulfills (44:E:a) of 44.7.3. along with W (which is a solution for $F(e_0)$), since $\mathsf{V} \subseteq \mathsf{W}$. So we need only verify (44:E:b) of 44.7.3.

Consider a $\overrightarrow{\beta}$ in $E(e_0)$, but not in V. Then $\overrightarrow{\beta}$ is also in $F(e_0)$ but not in W, hence there exists an $\overrightarrow{\alpha}$ in W with $\overrightarrow{\alpha} \succ \overrightarrow{\beta}$ (W is a solution for $F(e_0)$!). If this $\overrightarrow{\alpha}$ belongs to $E(e_0)$, then it belongs to $\mathsf{W} \cap E(e_0) = \mathsf{V}$, i.e. we have an $\overrightarrow{\alpha}$ in $E(e_0)$ with $\overrightarrow{\alpha} \succ \overrightarrow{\beta}$.

If $\overrightarrow{\alpha}$ does not belong to $E(e_0)$, then it belongs to $\mathsf{W} - E(e_0) = D^*(e_0)$, and so it is detached. Thus $\overrightarrow{\alpha} \succ \overrightarrow{\beta}$, $\overrightarrow{\alpha}$ detached and in $F(e_0)$. Hence there exists by (45:K) an $\overrightarrow{\alpha}' \succ \overrightarrow{\beta}$, $\overrightarrow{\alpha}'$ detached and in $E(e_0)$. By (45:G) this $\overrightarrow{\alpha}'$ belongs to W, ($E(e_0) \subseteq F(e_0)$, W is a solution for $F(e_0)$!); hence it belongs to $\mathsf{W} \cap E(e_0) = \mathsf{V}$. So we have an $\overrightarrow{\alpha}'$ in $E(e_0)$ with $\overrightarrow{\alpha}' \succ \overrightarrow{\beta}$.]

Thus (44:E:b) of 44.7.3. holds at any rate.

(45:N) If V is a solution for $E(e_0)$, then the W of (45:12) in (45:I) is a solution for $F(e_0)$.

Proof: $\mathsf{W} \subseteq F(e_0)$ is clear, so we must prove (44:E:a), (44:E:b) of 44.7.3.

Ad (44:E:a): Assume $\overrightarrow{\alpha} \succ \overrightarrow{\beta}$ for two $\overrightarrow{\alpha}, \overrightarrow{\beta}$ in W. $\overrightarrow{\alpha} \succ \overrightarrow{\beta}$ and (45:D) exclude that $\overrightarrow{\beta}$ be detached. So $\overrightarrow{\beta}$ is not in $D^*(e_0)$, hence it is in

$$\mathsf{W} - D^*(e_0) = \mathsf{V}.$$

Hence $\overrightarrow{\alpha} \succ \overrightarrow{\beta}$ excludes that $\overrightarrow{\alpha}$ too be in (the solution) V. So $\overrightarrow{\alpha}$ is in

$$\mathsf{W} - \mathsf{V} = D^*(e_0).$$

Consequently $\overrightarrow{\alpha}$ is detached.

Now (45:K) produces an $\overrightarrow{\alpha}' \succ \overrightarrow{\beta}$ which is detached and in $E(e_0)$. Being detached, $\overrightarrow{\alpha}'$ belongs by (45:G) to (the solution for $E(e_0)$) V. As $\overrightarrow{\alpha}', \overrightarrow{\beta}$ both belong to (the solution) V and $\overrightarrow{\alpha}' \succ \overrightarrow{\beta}$, this is a contradiction.

Ad (44:E:b): Consider a $\overrightarrow{\beta} = \{\beta_1, \cdots, \beta_n\}$ in $F(e_0)$, but not in W. Now form $\overrightarrow{\beta}(\epsilon) = \{\beta_1(\epsilon), \cdots, \beta_n(\epsilon)\} = \{\beta_1 + \epsilon, \cdots, \beta_n + \epsilon\}$ for every $\epsilon \geqq 0$. Let ϵ increase from 0 until one of these two things occurs for the first time:

(45:17) $\qquad \overrightarrow{\beta}(\epsilon)$ is in $E(e_0)$,[1]

(45:18) $\qquad \overrightarrow{\beta}(\epsilon)$ is detached.[2]

We distinguish these two possibilities:

(45:17) happens first, say for $\epsilon = \epsilon_1 \geqq 0$: $\overrightarrow{\beta}(\epsilon_1)$ is in $E(e_0)$, but it is not detached.

If $\epsilon_1 = 0$, then $\overrightarrow{\beta} = \overrightarrow{\beta}(0)$ is in $E(e_0)$. As $\overrightarrow{\beta}$ is not in $\mathsf{V} \subseteq \mathsf{W}$, there exists an $\overrightarrow{\alpha} \succ \overrightarrow{\beta}$ in (the solution for $E(e_0)$) V. *A fortiori* $\overrightarrow{\alpha}$ in W.

Assume next $\epsilon_1 > 0$, and $\overrightarrow{\beta}(\epsilon_1)$ in V. As $\overrightarrow{\beta}(\epsilon_1)$ is not detached, there exists a (non-empty) $S \subseteq I$ with $\sum_{i \text{ in } S} \beta_i(\epsilon_1) < \mathrm{v}(S)$. Besides, always $\beta_i(\epsilon_1) > \beta_i$. So $\overrightarrow{\beta}(\epsilon_1) \succ \overrightarrow{\beta}$. And $\overrightarrow{\beta}(\epsilon_1)$ is in V, hence *a fortiori* in W.

Assume, finally, $\epsilon_1 > 0$ and $\overrightarrow{\beta}(\epsilon_1)$ not in V. As $\overrightarrow{\beta}(\epsilon_1)$ is in $E(e_0)$, there exists an $\overrightarrow{\alpha} \succ \overrightarrow{\beta}(\epsilon_1)$ in (the solution for $E(e_0)$) V. Since always $\overrightarrow{\beta}_i(\epsilon_1) > \beta_i$, $\overrightarrow{\alpha} \succ \overrightarrow{\beta}(\epsilon_1)$ implies $\overrightarrow{\alpha} \succ \overrightarrow{\beta}$ by (45:J). And $\overrightarrow{\alpha}$ is in V, hence *a fortiori* in W.

(45:18) happens first, or simultaneously with (45:17), say for $\epsilon = \epsilon_2 \geqq 0$: $\overrightarrow{\beta}(\epsilon_2)$ is still in $F(e_0)$, and it is detached.

If $\overrightarrow{\beta}(\epsilon_2)$ is in $E(e_0)$, then it is by (45:G) in (the solution for $E(e_0)$) V. If $\overrightarrow{\beta}(\epsilon_2)$ is not in $E(e_0)$, then it is in $D^*(e_0)$. So $\overrightarrow{\beta}(\epsilon_2)$ is at any rate in W.

[1] I.e. the excess of $\overrightarrow{\beta}(\epsilon)$ is $= e_0$. For $\overrightarrow{\beta}(0) = \overrightarrow{\beta}$ is in $F(e_0)$, i.e. its excess is $\leqq e_0$, and the excess of $\overrightarrow{\beta}(\epsilon)$ increases with ϵ.

[2] I.e. $\sum_{i \text{ in } S} \beta_i(\epsilon) \geqq \mathrm{v}(S)$ for all $S \subseteq I$. Each $\sum_{i \text{ in } S} \beta_i(\epsilon)$ increases with ϵ.

This excludes $\epsilon_2 = 0$, since $\overrightarrow{\beta} = \overrightarrow{\beta}(0)$ is not in $\mathbb{W}$. So $\epsilon_2 > 0$.

For $0 < \epsilon < \epsilon_2$, $\overrightarrow{\beta}(\epsilon)$ is not detached, so there exists a non-empty $S \subseteq I$ with $\sum_{i \text{ in } S} \beta_i(\epsilon) < v(S)$. Hence there exists by continuity a non-empty $S \subseteq I$ even with $\sum_{i \text{ in } S} \beta_i(\epsilon_2) \leqq v(S)$. Besides, always $\beta_i(\epsilon_2) > \beta_i$, hence $\overrightarrow{\beta}(\epsilon_2) \succ \overrightarrow{\beta}$ And $\overrightarrow{\beta}(\epsilon_2)$ belongs to $\mathbb{W}$.

Summing up: In every case there exists an $\overrightarrow{\alpha} \succ \overrightarrow{\beta}$ in $\mathbb{W}$. (This $\overrightarrow{\alpha}$ was $\overrightarrow{\alpha}$, $\overrightarrow{\beta}(\epsilon_1)$, $\overrightarrow{\alpha}$, $\overrightarrow{\beta}(\epsilon_2)$ above, respectively.) So (44:E:b) is fulfilled.

We can now give the promised proof:

Proof of (45:I): Immediate, by combining (45:L), (45:M), (45:N).

45.6. Summary and Conclusions

45.6.1. Our main results, obtained so far, can be summarized as follows:

(45:O) If

(45:O:a) $$e_0 < -|\Gamma|_1,$$

then $E(e_0)$, $F(e_0)$ are empty and the empty set is their only solution.

If

(45:O:b) $$-|\Gamma|_1 \leqq e_0 \leqq |\Gamma|_2,$$

then $E(e_0)$, $F(e_0)$ are not empty, both have the same solutions, which are all not empty.

If

(45:O:c) $$e_0 > |\Gamma|_2,$$

then $E(e_0)$, $F(e_0)$ are not empty, they have no solution in common, all their solutions are not empty.

Proof: Immediate by combining (45:A), (45:I) and (45:H).

This result makes the critical character of the points $e_0 = -|\Gamma|_1$, $|\Gamma|_2$ quite clear and it further strengthens the views expressed at the end of 45.1. and following (45:H) in 45.4.2. concerning these points: That it is here where e_0 becomes "too small" or "too large" in the sense of 44.6.1.

45.6.2. We are also able now to prove some relations which will be useful later (in 46.5.).

(45:P) Let $\mathbb{W}$ be a non-empty solution for $F(e_0)$, i.e. assume that $e_0 \geqq -|\Gamma|_1$. Then

(45:P:a) $$\text{Max}_{\overrightarrow{\alpha} \text{ in } \mathbb{W}} e(\overrightarrow{\alpha}) = e_0$$

(45:P:b) $$\text{Min}_{\vec{\alpha} \text{ in } \mathbb{W}} e(\vec{\alpha}) = \text{Min}\,(e_0, |\Gamma|_2).$$ [1]

Also

(45:P:c) $$\text{Max}_{\vec{\alpha} \text{ in } \mathbb{W}} e(\vec{\alpha}) - \text{Min}_{\vec{\alpha} \text{ in } \mathbb{W}} e(\vec{\alpha}) = \text{Max}\,(0, e_0 - |\Gamma|_2).$$ [2]

Proof: (45:P:c) follows from (45:P:a), (45:P:b) since

$$e_0 - \text{Min}\,(e_0, |\Gamma|_2) = \text{Max}\,(e_0 - e_0, e_0 - |\Gamma|_2) = \text{Max}\,(0, e_0 - |\Gamma|_2).$$

We now prove (45:P:a), (45:P:b).

Write $\mathbb{W} = \mathbb{V} \cup D^*(e_0)$, $\mathbb{V}$ a solution for $E(e_0)$, following (45:I). As $e_0 \geqq -|\Gamma|_1$, so $\mathbb{V}$ is not empty (by (45:A) or (45:O)). As we know $e(\vec{\alpha}) = e_0$ throughout $\mathbb{V}$ and $e(\vec{\alpha}) < e_0$ throughout $D^*(e_0)$.

Now for $e_0 \leqq |\Gamma|_2$, $D^*(e_0)$ is empty (by (45:H)), so

(45:19) $$\text{Max}_{\vec{\alpha} \text{ in } \mathbb{W}} e(\vec{\alpha}) = \text{Max}_{\vec{\alpha} \text{ in } \mathbb{V}} e(\vec{\alpha}) = e_0,$$

(45:20) $$\text{Min}_{\vec{\alpha} \text{ in } \mathbb{W}} e(\vec{\alpha}) = \text{Min}_{\vec{\alpha} \text{ in } \mathbb{V}} e(\vec{\alpha}) = e_0.$$

And for $e_0 > |\Gamma|_2$, $D^*(e_0)$ is not empty (again by (45:H)), it is the set of all detached $\vec{\alpha}$ with $e(\vec{\alpha}) < e_0$. Hence by (45:B:b) in 45.2.3. these $e(\vec{\alpha})$ have a minimum, $|\Gamma|_2$. So we have in this case:

(45:19*) $$\text{Max}_{\vec{\alpha} \text{ in } \mathbb{W}} e(\vec{\alpha}) = \text{Max}_{\vec{\alpha} \text{ in } \mathbb{V}} e(\vec{\alpha}) = e_0,$$

(45:20*) $$\text{Min}_{\vec{\alpha} \text{ in } \mathbb{W}} e(\vec{\alpha}) = \text{Min}_{\vec{\alpha} \text{ in } D^*(e_0)} e(\vec{\alpha}) = |\Gamma|_2.$$

(45:19), (45:19*) together give our (45:P:a), and (45:20), (45:20*) give together our (45:P:b).

46. Determination of All Solutions in a Decomposable Game

46.1. Elementary Properties of Decompositions

46.1.1. Let us now return to the decomposition of a game Γ.

Let Γ be decomposable for J, $K(= I - J)$ with Δ, H as its J-, K-constituents.

Given any extended imputation $\vec{\alpha} = \{\alpha_1, \cdots, \alpha_n\}$ for I, we form its J-, K-constituents $\vec{\beta}$, $\vec{\gamma}$ ($\beta_i = \alpha_i$ for i in J, $\gamma_i = \alpha_i$ for i in K), and their excesses

[1] Our assertion includes the claim that these $\text{Max}_{\vec{\alpha} \text{ in } \mathbb{W}}$ and $\text{Min}_{\vec{\alpha} \text{ in } \mathbb{W}}$ exist.

[2] Verbally: The maximum excess in the solution $\mathbb{W}$ is the maximum excess allowed in $F(e_0)$: e_0. The minimum excess in the solution $\mathbb{W}$ is again e_0, unless $e_0 > |\Gamma|_2$, in which case it is only $|\Gamma|_2$. I.e. the minimum is as nearly e_0 as possible, considering that it must never exceed $|\Gamma|_2$.

The "width" of the interval of excesses in $\mathbb{W}$ is the excess of e_0 over $|\Gamma|_2$, if any.

$$(46:1)\quad \begin{cases} \text{Excess of } \overrightarrow{\alpha} \text{ in } I\colon e = e(\overrightarrow{\alpha}) = \sum_{i=1}^{n} \alpha_i - \mathrm{v}(I), \\ \text{Excess of } \overrightarrow{\beta} \text{ in } J\colon f = f(\overrightarrow{\alpha}) = \sum_{i \text{ in } J} \alpha_i - \mathrm{v}(J), \\ \text{Excess of } \overrightarrow{\gamma} \text{ in } K\colon g = g(\overrightarrow{\alpha}) = \sum_{i \text{ in } K} \alpha_i - \mathrm{v}(K).^{1} \end{cases}$$

Since

$$(46:2)\qquad \mathrm{v}(J) + \mathrm{v}(K) = \mathrm{v}(I)$$

(by (42:6:b) in 42.3.2., or equally by (41:6) in 41.3.2. with $S = J$, $T = K$), therefore

$$(46:3)\qquad e = f + g$$

(46:A) We have

(46:A:a) $$|\Gamma|_1 = |\Delta|_1 + |\mathrm{H}|_1,$$

(46:A:b) $$|\Gamma|_2 = |\Delta|_2 + |\mathrm{H}|_2.$$

(46:A:c) Γ is inessential if and only if Δ, H are both inessential.

Proof: Ad (46:A:a): Apply the definition (45:2) in 45.1. to Γ, Δ, H in turn.

$$(46:4)\qquad |\Gamma|_1 = \mathrm{v}(I) - \sum_{i \text{ in } I} \mathrm{v}((i)),$$

$$(46:5)\qquad |\Delta|_1 = \mathrm{v}(J) - \sum_{i \text{ in } J} \mathrm{v}((i)),$$

$$(46:6)\qquad |\mathrm{H}|_1 = \mathrm{v}(K) - \sum_{i \text{ in } K} \mathrm{v}((i)).$$

Comparing (46:4) with the sum of (46:5) and (46:6) gives (46:A:a), owing to (46:2).

Ad (46:A:b): Let $\overrightarrow{\alpha}$, $\overrightarrow{\beta}$, $\overrightarrow{\gamma}$ be as above (before (46:1)). Then $\overrightarrow{\alpha}$ is detached (in I) if

$$\sum_{i \text{ in } R} \alpha_i \geqq \mathrm{v}(R) \qquad \text{for all} \qquad R \subseteq I.$$

Recalling (41:6) in 41.3.2. we may write for this

$$(46:7)\qquad \sum_{i \text{ in } S} \alpha_i + \sum_{i \text{ in } T} \alpha_i \geqq \mathrm{v}(S) + \mathrm{v}(T) \qquad \text{for all} \qquad S \subseteq J, \qquad T \subseteq K.$$

Again $\overrightarrow{\beta}$, $\overrightarrow{\gamma}$ are detached (in J, K) if

$$(46:8)\qquad \sum_{i \text{ in } S} \alpha_i \geqq \mathrm{v}(S) \qquad \text{for all} \qquad S \subseteq J,$$

$$(46:9)\qquad \sum_{i \text{ in } T} \alpha_i \geqq \mathrm{v}(T) \qquad \text{for all} \qquad T \subseteq K.$$

[1] Up to this point it was not necessary to give explicit expression to the dependence of $\overrightarrow{\alpha}$'s excess e upon $\overrightarrow{\alpha}$. We do this now for e as well as for f, g.

Now (46:7) is equivalent to (46:8), (46:9). Indeed: (46:7) obtains by adding (46:8) and (46:9); and (46:7) specializes for $T = \ominus$ to (46:8) and for $S = \ominus$ to (46:9).

Thus $\vec{\alpha}$ is detached, if and only if its (J-, K-) constituents $\vec{\beta}$, $\vec{\gamma}$ are both detached. As their excesses e and f, g are correlated by (46:3), this gives for their minima (cf. (45:B:b))

$$|\Gamma|_2 = |\Delta|_2 + |\mathrm{H}|_2,$$

i.e. our formula (46:A:b).

Ad (46:A:c): Immediate by combining (46:A:a) or (46:A:b) with (45:E) as applied to Γ, Δ, H.

The quantities $|\Gamma|_1$, $|\Gamma|_2$ are both quantitative measures of the essentiality of the game Γ, in the sense of 45.3.1. Our above result states that both are additive for the composition of games.

46.1.2. Another lemma which will be useful in our further discussions:

(46:B) If $\vec{\alpha} \succ \vec{\beta}$ (for Γ), then the set S of 30.1.1. for this domination can be chosen with $S \subseteq J$ or $S \subseteq K$ without any loss of generality.[1]

Proof: Consider the set S of 30.1.1. for the domination $\vec{\alpha} \succ \vec{\beta}$. If accidentally $S \subseteq J$ or $S \subseteq K$, then there is nothing to prove, so we may assume that neither $S \subseteq J$ nor $S \subseteq K$. Consequently $S = S_1 \cup T_1$, where $S_1 \subseteq J$, $T_1 \subseteq K$, and neither S_1 nor T_1 is empty.

We have $\alpha_i > \beta_i$ for all i in S, i.e. for all i in S_1, as well as for all i in T_1. Finally

$$\sum_{i \text{ in } S} \alpha_i \leqq \mathrm{v}(S).$$

The left hand side is clearly equal to $\sum_{i \text{ in } S_1} \alpha_i + \sum_{i \text{ in } T_1} \alpha_i$, while the right hand side is equal to $\mathrm{v}(S_1) + \mathrm{v}(T_1)$ by (41:6) in 41.3.2. Thus

$$\sum_{i \text{ in } S_1} \alpha_i + \sum_{i \text{ in } T_1} \alpha_i \leqq \mathrm{v}(S_1) + \mathrm{v}(T_1),$$

hence at least one of

$$\sum_{i \text{ in } S_1} \alpha_i \leqq \mathrm{v}(S_1), \qquad \sum_{i \text{ in } T_1} \alpha_i \leqq \mathrm{v}(T_1)$$

must be true.

Thus of the three conditions of domination in 30.1.1. (for $\vec{\alpha} \succ \vec{\beta}$) (30:4:a), (30:4:c) holds for both of S_1, T_1 and (30:4:b) for at least one of them. Hence, we may replace our original S by either S_1 ($\subseteq J$) or T_1($\subseteq K$).

This completes the proof.

[1] I.e. this extra restriction on S does not (in this case!) modify the concept of domination.

46.2. Decomposition and Its Relation to the Solutions: First Results Concerning $F(e_0)$

46.2.1. We now direct our course towards the main objective of this part of the theory: The determination of all solutions U_I of the decomposable game Γ. This will be achieved in 46.6., concluding a chain of seven lemmas.

We begin with some purely descriptive observations.

Consider a solution U_I for $F(e_0)$ of Γ. If U_I is empty, there is nothing more to say. Let us assume, therefore, that U_I is not empty—owing to (45:A) (or equally to (45:O)) this is equivalent to

$$e_0 \geqq -|\Gamma|_1 = -|\Delta|_1 - |\mathrm{H}|_1.$$

Using the notations of (46:1) in 46.1.1. we form:

$$(46:10)\quad \begin{cases} \operatorname{Max}_{\vec{\alpha} \text{ in } \mathsf{U}_I} f(\vec{\alpha}) = \bar{\varphi}, \\ \operatorname{Min}_{\vec{\alpha} \text{ in } \mathsf{U}_I} f(\vec{\alpha}) = \underline{\varphi}, \\ \operatorname{Max}_{\vec{\alpha} \text{ in } \mathsf{U}_I} g(\vec{\alpha}) = \bar{\psi}, \\ \operatorname{Min}_{\vec{\alpha} \text{ in } \mathsf{U}_I} g(\vec{\alpha}) = \underline{\psi}.^{1} \end{cases}$$

[1] That all these quantities can be formed, i.e. that the maxima and the minima in question exist and are assumed, can be ascertained by a simple continuity consideration. Indeed $f(\vec{\alpha}) = \sum_{i \text{ in } J} \alpha_i - \mathrm{v}(J)$ and $g(\vec{\alpha}) = \sum_{i \text{ in } K} \alpha_i - \mathrm{v}(K)$ are both continuous functions of $\vec{\alpha}$, i.e. of its components $\alpha_1, \cdots, \alpha_n$. The existence of their maxima and minima is therefore a well known consequence of the continuity properties of the domain of $\vec{\alpha}$—the set U_I.

For the reader who is acquainted with the necessary mathematical background—topology—we give the precise statement and its proof. (The underlying mathematical facts are discussed e.g. by *C. Carathéodory*, loc. cit., footnote 1 on p. 343. Cf. there pp. 136–140, particularly theorem 5).

U_I is a set in the n-dimensional linear space L_n (Cf. 30.1.1.). In order to be sure that every continuous function has a maximum and a minimum in U_I, we must know that U_I is *bounded* and *closed*.

Now we prove:

(*) Any solution U for $F(e_0)$ $[E(e_0)]$ of an n-person game Γ is a bounded and closed set in L_n.

Proof: Boundedness: If $\vec{\alpha} = \{\alpha_1, \cdots, \alpha_n\}$ belongs to U, then every $\alpha_i \geqq \mathrm{v}((i))$ and $\sum_{i=1}^{n} \alpha_i - \mathrm{v}(I) \leqq e_0$, hence $\alpha_i \leqq \mathrm{v}(I) + e_0 - \sum_{j \neq i} \alpha_j \leqq \mathrm{v}(I) + e_0 - \sum_{j \neq i} \mathrm{v}((i))$. So each α_i is restricted to the fixed interval

$$\mathrm{v}((i)) \leqq \alpha_i \leqq \mathrm{v}(I) + e_0 - \sum_{j \neq i} \mathrm{v}((i)),$$

and so these $\vec{\alpha}$ form a bounded set.

Closedness: This is equivalent to the openness of the complement of U. That set is, by (30:5:c) in 30.1.1., the set of all $\vec{\beta}$ which are dominated by any $\vec{\alpha}$ of U. (Observe

Given two $\overrightarrow{\alpha} = \{\alpha_1, \cdots, \alpha_n\}$, $\overrightarrow{\beta} = \{\beta_1, \cdots, \beta_n\}$ there exists a unique $\overrightarrow{\gamma} = \{\gamma_1, \cdots, \gamma_n\}$ which has the same J-component as $\overrightarrow{\alpha}$, and the same K-component as $\overrightarrow{\beta}$:

(46:11)
$$\begin{aligned} \gamma_i &= \alpha_i \quad \text{for } i \text{ in } J, \\ \gamma_i &= \beta_i \quad \text{for } i \text{ in } K. \end{aligned}$$

46.2.2. We now prove:

(46:C) If $\overrightarrow{\alpha}$, $\overrightarrow{\beta}$ belong to U_I, then the $\overrightarrow{\gamma}$ of (46:11) belongs to U_I if and only if

(46:C:a)
$$f(\overrightarrow{\alpha}) + g(\overrightarrow{\beta}) \leqq e_0.$$

Incidentally

(46:C:b)
$$e(\overrightarrow{\gamma}) = f(\overrightarrow{\alpha}) + g(\overrightarrow{\beta}).$$

Proof: Formula (46:C:b): By (46:3) in 46.1.1. $e(\overrightarrow{\gamma}) = f(\overrightarrow{\gamma}) + g(\overrightarrow{\gamma})$, and clearly $f(\overrightarrow{\gamma}) = f(\overrightarrow{\alpha})$, $g(\overrightarrow{\gamma}) = g(\overrightarrow{\beta})$.

Necessity of (46:C:a): Since $\mathsf{U}_I \subseteq F(e_0)$, therefore $e(\overrightarrow{\gamma}) \leqq e_0$ is necessary and by (46:C:b) this coincides with (46:C:a).

Sufficiency of (46:C:a): $\overrightarrow{\gamma}$ is clearly an extended imputation, along with $\overrightarrow{\alpha}$, $\overrightarrow{\beta}$, and (46:C:a), (46:C:b) guarantee that $\overrightarrow{\gamma}$ belongs to $F(e_0)$.[1]

Now assume that $\overrightarrow{\gamma}$ is not in U_I. Then there exists a $\overrightarrow{\delta} \succ \overrightarrow{\gamma}$ in U_I. The set S of 30.1.1. for this domination may be chosen by (46:B) with $S \subseteq J$ or $S \subseteq K$. Now clearly $\overrightarrow{\delta} \succ \overrightarrow{\gamma}$ implies, when $S \subseteq J$ that $\overrightarrow{\delta} \succ \overrightarrow{\alpha}$,

---

that we are introducing the solution character of U at this point!)

For any $\overrightarrow{\alpha}$ denote the set of all $\overrightarrow{\beta} \succ \overrightarrow{\alpha}$ by $D_{\overrightarrow{\alpha}}$. Then the complement of U is the sum of all $D_{\overrightarrow{\alpha}}$, $\overrightarrow{\alpha}$ of U.

Since the sum of any number (even of infinitely many) open sets is again open, it suffices to prove the openness of each $D_{\overrightarrow{\alpha}}$, i.e. this: If $\overrightarrow{\beta} \succ \overrightarrow{\alpha}$, then for every $\overrightarrow{\beta}'$ which is sufficiently near to $\overrightarrow{\beta}$, we have also $\overrightarrow{\beta}' \succ \overrightarrow{\alpha}$. Now in the definition of domination, $\overrightarrow{\beta} \succ \overrightarrow{\alpha}$ by (30:4:a)-(30:4:c) in 30.1.1., $\overrightarrow{\beta}$ appears in the condition (30:4:c) only. And the validity of (30:4:c) is clearly not impaired by a sufficiently small change of β_i, since (30:4:c) is a $<$ relation.

(Note that the same is not true for $\overrightarrow{\alpha}$, because $\overrightarrow{\alpha}$ appears in (30:4:b) also, and (30:4:b) might be destroyed by arbitrary small changes, since (30:4:b) is a $\leqq$ relation. But we needed this property for $\overrightarrow{\beta}$, and not for $\overrightarrow{\alpha}$!)

[1] This is the only use of (46:C:a).

and when $S \subseteq K$ that $\overrightarrow{\delta} \succ \overrightarrow{\beta}$. As $\overrightarrow{\delta}, \overrightarrow{\alpha}, \overrightarrow{\beta}$ belong to U_I, both alternatives are impossible.

Hence $\overrightarrow{\gamma}$ must belong to U_I, as asserted.

We restate (46:C) in an obviously equivalent form:

(46:D) Let V_J be the set of all J-constituents and W_K the set of all K-constituents of U_I.

Then U_I obtains from these V_J and W_K as follows:

U_I is the set of all those $\overrightarrow{\gamma}$, which have a J-constituent $\overrightarrow{\alpha}'$ in V_J and a K-constituent $\overrightarrow{\beta}'$ in W_K such that

(46:12) $$e(\overrightarrow{\alpha}') + e(\overrightarrow{\beta}') \leqq e_0.$$ [1]

46.3. Continuation

46.3. Recalling the definition of U_I's decomposability (for J, K) in (44:B) in 44.3.1., one sees with little difficulty, that it is equivalent to this:

U_I obtains from the V_J, W_K of (46:D) as outlined there, but without the condition (46:12).

Thus (46:12) may be interpreted as expressing just to what extent U_I is *not* decomposable. This is of some interest in the light of what was said in 44.3.3. about (44:D:a) there.

One may even go a step further: The necessity of (46:12) in (46:D) is easy to establish. (It corresponds to (46:C:a), i.e. to the very simple first two steps in the proof of (46:C)). Hence (46:D) expresses that U_I is no further from decomposability, than unavoidable.

All this, in conjunction with (44:D:b) in 44.3.3., suggests strongly that V_J, W_K ought to be solutions of Δ, H. With our present extensions of all concepts it is necessary, however, to decide which $F(f_0)$, $F(g_0)$ to take; f_0 being the excess we propose to use in J, and g_0 the one in K.[2] It will appear that the $\bar{\varphi}$, $\bar{\psi}$ of 46.2.1. are these f_0, g_0.

Indeed, we can prove:

(46:E)

(46:E:a) V_J is a solution of Δ for $F(\bar{\varphi})$,

(46:E:b) W_K is a solution of H for $F(\bar{\psi})$.

It is convenient, however, to derive first another result:

[1] Note that these $\overrightarrow{\alpha}'$, $\overrightarrow{\beta}'$ are not the $\overrightarrow{\alpha}$, $\overrightarrow{\beta}$ of (46:C)—they are their J-, K-constituents as well as those of $\overrightarrow{\gamma}$. $e(\overrightarrow{\alpha}')$, $e(\overrightarrow{\beta}')$ are the excesses of $\overrightarrow{\alpha}'$, $\overrightarrow{\beta}'$ formed in J, K. But they are equal to $f(\overrightarrow{\alpha})$, $g(\overrightarrow{\beta})$ as well as to $f(\overrightarrow{\gamma})$, $g(\overrightarrow{\gamma})$. (All of this is related to (46:C)).

[2] The reader will note that this is something like a question of distributing the given excess e_0 in I between J and K.

(46:F)

(46:F:a) $$\bar{\varphi} + \underline{\psi} = e_0,$$

(46:F:b) $$\underline{\varphi} + \bar{\psi} = e_0.$$

Note that in (46:E), as well as in (46:F), the parts (a), (b) obtain from each other by interchanging J, Δ, $\bar{\varphi}$, $\underline{\varphi}$ with K, H, $\bar{\psi}$, $\underline{\psi}$. Hence it suffices to prove in each case only one of (a), (b)—we chose (a).

Proof of (46:F:a): Choose an $\vec{\alpha}$ in U_I for which $f(\vec{\alpha})$ assumes its maximum $\bar{\varphi}$. Since necessarily $e(\vec{\alpha}) \leqq e_0$, and since by definition $g(\vec{\alpha}) \geqq \underline{\psi}$, therefore (46:3) in 46.1.1. gives

(46:13) $$\bar{\varphi} + \underline{\psi} \leqq e_0.$$

Assume now that (46:F:a) is not true. Then (46:13) would imply further

(46:14) $$\bar{\varphi} + \underline{\psi} < e_0.$$

Use the above $\vec{\alpha}$ in U_I with $f(\vec{\alpha}) = \bar{\varphi}$, and chose also a $\vec{\beta}$ in U_I for which $g(\vec{\beta})$ assumes its minimum $\underline{\psi}$. Then $f(\vec{\alpha}) + g(\vec{\beta}) = \bar{\varphi} + \underline{\psi} \leqq e_0$ (by (46:13) or (46:14)). Thus the $\vec{\gamma}$ of (46:C) belongs to U_I, too. Again (46:C) together with (46:14) gives

$$e(\vec{\gamma}) = f(\vec{\alpha}) + g(\vec{\beta}) = \bar{\varphi} + \underline{\psi} < e_0,$$

i.e. $\sum_{i=1}^{n} \gamma_i < \mathrm{v}(I) + e_0$. Now define

$$\vec{\delta} = \{\delta_1, \cdots, \delta_n\} = \{\gamma_1 + \epsilon, \cdots, \gamma_n + \epsilon\},$$

choosing $\epsilon > 0$ so that $\sum_{i=1}^{n} \delta_i = \mathrm{v}(I) + e_0$. Thus $\vec{\delta}$ belongs to $F(e_0)$.

If $\vec{\delta}$ did not belong to U_I, then an $\vec{\eta} \succ \vec{\delta}$ would exist in U_I. By (45:J) $\vec{\eta} \succ \vec{\gamma}$, which is impossible, since $\vec{\eta}$, $\vec{\gamma}$ are both in U_I. Hence $\vec{\delta}$ belongs to U_I. Now $\sum_{i \text{ in } J} \delta_i - \mathrm{v}(J) > \sum_{i \text{ in } J} \gamma_i - \mathrm{v}(J) = \sum_{i \text{ in } J} \alpha_i - \mathrm{v}(J)$, i.e. $f(\vec{\delta}) > f(\vec{\alpha}) = \bar{\varphi}$, contradicting the definition of $\bar{\varphi}$.

Consequently (46:F:a) must be true and the proof is completed.

Proof of (46:E:a): If $\vec{\alpha}\,'$ belongs to V_J, then it is the J-constituent of an $\vec{\alpha}$ of U_I. Hence (cf. footnote 1 on p. 386) $e(\vec{\alpha}\,') = f(\vec{\alpha}) \leqq \bar{\varphi}$, so that $\vec{\alpha}\,'$ belongs to $F(\bar{\varphi})$. Thus $\mathsf{V}_J \subseteq F(\bar{\varphi})$.

So our task is to prove (44:E:a), (44:E:b) of 44.7.3.

Ad (44:E:a): Assume, that $\overrightarrow{\alpha}' \succ \overrightarrow{\beta}'$ happened for two $\overrightarrow{\alpha}', \overrightarrow{\beta}'$ in V_J. Then $\overrightarrow{\alpha}', \overrightarrow{\beta}'$ are the J-constituents of two $\overrightarrow{\gamma}, \overrightarrow{\delta}$ in U_I. But $\overrightarrow{\alpha}' \succ \overrightarrow{\beta}'$ clearly implies $\overrightarrow{\gamma} \succ \overrightarrow{\delta}$, which is impossible.

Ad (44:E:b): Consider an $\overrightarrow{\alpha}'$ in $F(\bar{\varphi})$ but not in V_J. Then by definition $e(\overrightarrow{\alpha}') \leqq \bar{\varphi}$. Use the $\overrightarrow{\beta}$ in U_I mentioned in the above proof of (46:F:a), for which $g(\overrightarrow{\beta}) = \underline{\psi}$. Let $\overrightarrow{\beta}'$ be the K-constituent of this $\overrightarrow{\beta}$, so that $\overrightarrow{\beta}'$ is in W_K and $e(\overrightarrow{\beta}') = g(\overrightarrow{\beta}) = \underline{\psi}$. Thus $e(\overrightarrow{\alpha}') + e(\overrightarrow{\beta}') \leqq \bar{\varphi} + \underline{\psi} = e_0$ (use (46:F:a)). Form the $\overrightarrow{\gamma}$ (for I), which has the J-, K-constituents $\overrightarrow{\alpha}', \overrightarrow{\beta}'$. Then $e(\overrightarrow{\gamma}) = e(\overrightarrow{\alpha}') + e(\overrightarrow{\beta}') \leqq e_0$ i.e. $\overrightarrow{\gamma}$ belongs to $F(e_0)$.

$\overrightarrow{\gamma}$ does not belong to U_I because its J-constituent $\overrightarrow{\alpha}'$ does not belong to V_J. Hence, there exists a $\overrightarrow{\delta} \succ \overrightarrow{\gamma}$ in (the solution for $F(e_0)$) U_I.

Let S be the set of 30.1.1. for the domination $\overrightarrow{\delta} \succ \overrightarrow{\gamma}$. By (46:B) we may assume that $S \subseteq J$ or $S \subseteq K$.

Assume first that $S \subseteq K$. As $\overrightarrow{\gamma}$ has the same K-constituents $\overrightarrow{\beta}'$ as $\overrightarrow{\beta}$, we can conclude from $\overrightarrow{\delta} \succ \overrightarrow{\gamma}$ that $\overrightarrow{\delta} \succ \overrightarrow{\beta}$. Since both $\overrightarrow{\delta}, \overrightarrow{\beta}$ belong to U_I, this is impossible.

Consequently $S \subseteq J$. Denote the J-constituent of $\overrightarrow{\delta}$ by $\overrightarrow{\delta}'$; as $\overrightarrow{\delta}$ belongs to U_I, therefore $\overrightarrow{\delta}'$ belongs to V_J. $\overrightarrow{\gamma}$ has the J-constituent $\overrightarrow{\alpha}'$. Hence we can conclude from $\overrightarrow{\delta} \succ \overrightarrow{\gamma}$ that $\overrightarrow{\delta}' \succ \overrightarrow{\alpha}'$.

Thus we have the desired $\overrightarrow{\delta}'$ from V_J with $\overrightarrow{\delta}' \succ \overrightarrow{\alpha}'$.

46.4. Continuation

46.4.1. (46:D), (46:E) expressed the general solution U_I of Γ in terms of appropriate solutions of V_J, W_K of Δ, H. It is natural, therefore, to try to reverse this procedure: To start with the V_J, W_K and to obtain U_I.

It must be remembered, however, that the V_J, W_K of (46:D) are not entirely arbitrary. If we reconsider the definitions (46:10) of 46.2.1. in this light of (46:D), then we see that they can also be stated in this form:

$$(46:15)\quad \begin{cases} \operatorname{Max}_{\overrightarrow{\alpha}' \text{ in } V_J} e(\overrightarrow{\alpha}') = \bar{\varphi}, \\ \operatorname{Min}_{\overrightarrow{\alpha}' \text{ in } V_J} e(\overrightarrow{\alpha}') = \underline{\varphi}, \\ \operatorname{Max}_{\overrightarrow{\beta}' \text{ in } W_K} e(\overrightarrow{\beta}') = \bar{\psi}, \\ \operatorname{Min}_{\overrightarrow{\beta}' \text{ in } W_K} e(\overrightarrow{\beta}') = \underline{\psi}. \end{cases}$$

And (46:F) expresses a relationship of these $\bar{\varphi}$, $\underline{\varphi}$, $\bar{\psi}$, $\underline{\psi}$ which are determined by V_J, W_K—with each other and with e_0.

46.4.2. We will show that this is the only restraint that must be imposed upon the V_J, W_K. To do this, we start with two arbitrary non-empty solutions V_J, W_K of Δ, H (which need not have been obtained from any solution U_I of Γ), and assert as follows:

(46:G) Let V_J be a non-empty solution of Δ for $F(\bar{\varphi})$ and W_K a non-empty solution of H for $F(\bar{\psi})$. Assume that $\bar{\varphi}$, $\bar{\psi}$ fulfill (46:15) above, and also that with the $\underline{\varphi}$, $\underline{\psi}$ of (46:15)

(46:16)
$$\bar{\varphi} + \underline{\psi} = \underline{\varphi} + \bar{\psi} = e_0.$$

For any $\overrightarrow{\alpha}'$ of V_J and any $\overrightarrow{\beta}'$ of W_K with

(46:17)
$$e(\overrightarrow{\alpha}') + e(\overrightarrow{\beta}') \leqq e_0,$$

form the $\overrightarrow{\gamma}$ (for I) which has the J-, K-components $\overrightarrow{\alpha}'$, $\overrightarrow{\beta}'$.

Denote the set of all these $\overrightarrow{\gamma}$ by U_I.

The U_I which are obtained in this way are precisely all solutions of Γ for $F(e_0)$.

Proof: All U_I of the stated character are obtained in this way: Apply (46:D) to U_I forming its V_J, W_K. Then all our assertions are contained in (46:D), (46:E), (46:F) together with (46:15).

All U_I obtained in this way have the stated character: Consider an U_I constructed with the help of V_J, W_K as described above. We have to prove that this U_I is a solution Γ for $F(e_0)$.

For every $\overrightarrow{\gamma}$ of U_I our (46:17) gives $e(\overrightarrow{\gamma}) = e(\overrightarrow{\alpha}') + e(\overrightarrow{\beta}') \leqq e_0$, so that $\overrightarrow{\gamma}$ belongs to $F(e_0)$. Thus $U_I \subseteq F(e_0)$.

So our task is to prove (44:E:a), (44:E:b) of 44.7.3.

Ad (44:E:a): Assume that $\overrightarrow{\eta} \succ \overrightarrow{\gamma}$ happened for two $\overrightarrow{\eta}, \overrightarrow{\gamma}$ in U_I. Let $\overrightarrow{\alpha}'$, $\overrightarrow{\beta}'$ be the J-, K-constituents of $\overrightarrow{\gamma}$ and $\overrightarrow{\delta}'$, $\overrightarrow{\epsilon}'$ the J-, K-constituents of $\overrightarrow{\eta}$ from which they obtain as described above. Let S be the set of 30.1.1. for the domination $\overrightarrow{\eta} \succ \overrightarrow{\gamma}$. By (46:B) we may assume that $S \subseteq J$ or $S \subseteq K$. Now $S \subseteq J$ would cause $\overrightarrow{\eta} \succ \overrightarrow{\gamma}$ to imply $\overrightarrow{\delta}' \succ \overrightarrow{\alpha}'$, which is impossible, since $\overrightarrow{\delta}'$, $\overrightarrow{\alpha}'$ both belong to V_J; and $S \subseteq K$ would cause $\overrightarrow{\eta} \succ \overrightarrow{\gamma}$ to imply $\overrightarrow{\epsilon}' \succ \overrightarrow{\beta}'$ which is impossible, since $\overrightarrow{\epsilon}'$, $\overrightarrow{\beta}'$ both belong to W_K.

Ad (44:E:b): Assume *per absurdum*, the existence of a $\overrightarrow{\gamma}$ in $F(e_0)$ but not in U_I, such that there is no $\overrightarrow{\eta}$ of U_I with $\overrightarrow{\eta} \succ \overrightarrow{\gamma}$. Let $\overrightarrow{\alpha}'$, $\overrightarrow{\beta}'$ be the J-, K-constituents of $\overrightarrow{\gamma}$.

Assume first $e(\overrightarrow{\alpha}') \leqq \bar{\varphi}$. Then $\overrightarrow{\alpha}'$ belongs to $F(\bar{\varphi})$. Consequently either $\overrightarrow{\alpha}'$ belongs to V_J or there exists a $\overrightarrow{\delta}'$ in V_J with $\overrightarrow{\delta}' \succ \overrightarrow{\alpha}'$. In the latter case choose an $\overrightarrow{\epsilon}'$ in W_K for which $e(\overrightarrow{\epsilon}')$ assumes its minimum value $\underline{\psi}$. Form the $\overrightarrow{\eta}$ with the J-, K-constituents $\overrightarrow{\delta}'$, $\overrightarrow{\epsilon}'$. As $\overrightarrow{\delta}'$, $\overrightarrow{\epsilon}'$ belong to V_J, W_K, respectively, and as $e(\overrightarrow{\delta}') + e(\overrightarrow{\epsilon}') \leqq \bar{\varphi} + \underline{\psi} = e_0$, therefore $\overrightarrow{\eta}$ belongs to U_I. Besides $\overrightarrow{\eta} \succ \overrightarrow{\gamma}$ owing to $\overrightarrow{\delta}' \succ \overrightarrow{\alpha}'$ (these being their J-constituents). Thus $\overrightarrow{\eta}$ contradicts our original assumption concerning $\overrightarrow{\gamma}$. Hence we have demonstrated, for the case under consideration, that $\overrightarrow{\alpha}'$ must belong to V_J.

In other words:

(46:18) Either $\overrightarrow{\alpha}'$ belongs to V_J, or $e(\overrightarrow{\alpha}') > \bar{\varphi}$.

Observe that in the first case necessarily $e(\overrightarrow{\alpha}') \geqq \underline{\varphi}$, and of course in the second case $e(\overrightarrow{\alpha}') > \bar{\varphi} \geqq \underline{\varphi}$. Consequently:

(46:19) At any rate $e(\overrightarrow{\alpha}') \geqq \underline{\varphi}$.

Interchanging J and K carries (46:18), (46:19) into these:

(46:20) Either $\overrightarrow{\beta}'$ belongs to W_K or $e(\overrightarrow{\beta}') > \bar{\psi}$.

(46:21) At any rate $e(\overrightarrow{\beta}') \geqq \underline{\psi}$.

Now if we had the second alternative of (46:18), then this gives in conjunction with (46:21)

$$e(\overrightarrow{\gamma}) = e(\overrightarrow{\alpha}') + e(\overrightarrow{\beta}') > \bar{\varphi} + \underline{\psi} = e_0,$$

which is impossible, as $\overrightarrow{\gamma}$ belongs $F(e_0)$. The second alternative of (46:20) is equally impossible.

Thus we have the first alternatives in both (46:18) and (46:20), i.e. $\overrightarrow{\alpha}'$, $\overrightarrow{\beta}'$ belong to V_J, W_K. As $\overrightarrow{\gamma}$ belongs to $F(e_0)$, therefore

$$e(\overrightarrow{\alpha}') + e(\overrightarrow{\beta}') = e(\overrightarrow{\gamma}) \leqq e_0.$$

Consequently $\overrightarrow{\gamma}$ must belong to U_I—contradicting our original assumption.

46.5. The Complete Result in $F(e_0)$

46.5.1. The result (46:G) is, in spite of its completeness, unsatisfactory in one respect: The conditions (46:16) and (46:17) on which it depends are altogether implicit. We will, therefore, replace them by equivalent, but much more transparent conditions.

To do this, we begin with the numbers $\bar{\varphi}$, $\bar{\psi}$ which we assume to be given first. Which solutions V_J, W_K of Δ, H for $F(\bar{\varphi})$, $F(\bar{\psi})$ can we then use in the sense of (46:G)?

First of all, V_J, W_K must be non-empty; application of (45:A) or (45:O) to Δ, H (instead of Γ) shows that this means

$$\bar{\varphi} \geqq -|\Delta|_1, \qquad \bar{\psi} \geqq -|\mathrm{H}|_1. \tag{46:22}$$

Consider next (46:15). Apply (45:P) of 45.6.1. to Δ, H (instead of Γ). Then (45:P:a) secures the two Max-equations of (46:15), while (45:P:b) transforms the two Min-equations of (46:15) into

$$\underline{\varphi} = \text{Min}\ (\bar{\varphi}, |\Delta|_2), \qquad \underline{\psi} = \text{Min}\ (\bar{\psi}, |\mathrm{H}|_2). \tag{46:23}$$

Let us, therefore, define $\underline{\varphi}$, $\underline{\psi}$ by (46:23).

Now we express (46:16), i.e.

$$\bar{\varphi} + \underline{\psi} = \underline{\varphi} + \bar{\psi} = e_0. \tag{46:16}$$

The first equation of (46:16) may also be written as

$$\bar{\varphi} - \underline{\varphi} = \bar{\psi} - \underline{\psi},$$

i.e. by (46:23)

$$\text{Max}\ (0, \bar{\varphi} - |\Delta|_2) = \text{Max}\ (0, \bar{\psi} - |\mathrm{H}|_2).\text{[1]} \tag{46:24}$$

46.5.2. Now two cases are possible:

Case (a): Both sides of (46:24) are zero. Then in each Max of (46:24) the 0-term is $\geqq$ than the other term, i.e. $\bar{\varphi} - |\Delta|_2 \leqq 0$, $\bar{\psi} - |\mathrm{H}|_2 \leqq 0$, i.e.,

$$\bar{\varphi} \leqq |\Delta|_2, \qquad \bar{\psi} \leqq |\mathrm{H}|_2. \tag{46:25}$$

Conversely: If (46:25) holds, then (46:24) becomes $0 = 0$, i.e. it is automatically satisfied. Now the definition (46:23) becomes

$$\underline{\varphi} = \bar{\varphi}, \qquad \underline{\psi} = \bar{\psi}, \tag{46:26}$$

and so the full condition (46:16) becomes[2]

$$\bar{\varphi} + \bar{\psi} = e_0. \tag{46:27}$$

(46:25) and (46:27) give also

$$e_0 \leqq |\Delta|_2 + |\mathrm{H}|_2 = |\Gamma|_2. \tag{46:28}$$

Case (b): Both sides of (46:24) are not zero. Then in each Max of (46:24) the 0-term is $<$ than the other term—i.e. $\bar{\varphi} - |\Delta|_2 > 0$, $\bar{\psi} - |\mathrm{H}|_2 > 0$, i.e.

$$\bar{\varphi} > |\Delta|_2, \qquad \bar{\psi} > |\mathrm{H}|_2.\text{[3]} \tag{46:29}$$

[1] Cf. (45:P:c) and its proof.

[2] Of which we used only the first part to obtain (46:24), on which this discussion is based.

[3] Note that the important point is that (46:25), (46:29) exhaust all possibilities—i.e. that we cannot have $\bar{\varphi} \leqq |\Delta|_2$, $\bar{\psi} > |\mathrm{H}|_2$, or $\bar{\varphi} > |\Delta|_2$, $\bar{\psi} \leqq |\mathrm{H}|_2$. This is, of course, due to the equation (46:24), which forces that both sides vanish or neither.

The meaning of this will appear in the lemmas which follow.

Conversely: If (46:29) holds, then (46:24) becomes $\bar{\varphi} - |\Delta|_2 = \bar{\psi} - |\mathrm{H}|_2$, i.e. it is not automatically satisfied. We can express (46:24) by writing

(46:30) $$\bar{\varphi} = |\Delta|_2 + \omega, \qquad \bar{\psi} = |\mathrm{H}|_2 + \omega,$$

and then (46:29) becomes simply

(46:31) $$\omega > 0.$$

Now the definition (46:23) becomes

(46:32) $$\underline{\varphi} = |\Delta|_2, \qquad \underline{\psi} = |\mathrm{H}|_2,$$

and so the full condition (46:16)[1] becomes

$$|\Delta|_2 + |\mathrm{H}|_2 + \omega = e_0,$$

i.e.

(46:33) $$e_0 = |\Gamma|_2 + \omega.$$

(46:31) and (46:33) give also

(46:34) $$e_0 > |\Gamma|_2.$$

46.5.3. Summing up:

(46:H) The conditions (46:16), (46:17) of (46:G) amount to this: One of the two following cases must hold:

Case (a): (1) $$-|\Gamma|_1 \leqq e_0 \leqq |\Gamma|_2,$$

together with

(2) $$-|\Delta|_1 \leqq \bar{\varphi} \leqq |\Delta|_2,$$

(3) $$-|\mathrm{H}|_1 \leqq \bar{\psi} \leqq |\mathrm{H}|_2,$$

and

(4) $$\bar{\varphi} + \bar{\psi} = e_0.$$

Case (b): (1) $$e_0 > |\Gamma|_2,$$

together with

(2) $$\bar{\varphi} > |\Delta|_2,$$

(3) $$\bar{\psi} > |\mathrm{H}|_2,$$

and

(4) $$e_0 - |\Gamma|_2 = \bar{\varphi} - |\Delta|_2 = \bar{\psi} - |\mathrm{H}|_2.$$[2]

Proof: Case (a): We knew all along, that $e_0 \geqq -|\Gamma|_1$ and $\bar{\varphi} \geqq -|\Delta|_1$, $\bar{\psi} \geqq -|\mathrm{H}|_1$. The other conditions coincide with (46:28), (46:25), (46:27) which contain the complete description of this case.

Case (b): These conditions coincide with (46:34), (46:29), (46:30), (46:33) which contain the complete description of the case (after elimination of ω which subsumes (46:31) under (1)-(3)).

[1] Cf. footnote 2 on p. 391.

[2] The reader will note that while (1)-(3) for (a) and for (b) show a strong analogy, the final condition (4) is entirely different for (a) and for (b). Nevertheless, all this was obtained by the rigorous discussion of one consistent theory!

More will be said about this later.

46.6. The Complete Result in $E(e_0)$

46.6. (46:G) and (46:H) characterize the solutions of Γ for $F(e_0)$ in a complete and explicit way. It is now apparent, too, that the cases (a), (b) of (46:H) coincide with (45:O:b), (45:O:c) in 45.6.1.: Indeed (a), (b) of (46:H) are distinguished by their conditions (1), and these are precisely (45:O:b), (45:O:c).

We now combine the results of (46:G), (46:H) with those of (45:I), (45:O). This will give us a comprehensive picture of the situation, utilizing all our information.

(46:I) If

(46:I:a) (1)
$$e_0 < -|\Gamma|_1,$$

then the empty set is the only solution of Γ, for $E(e_0)$ as well as for $F(e_0)$.

If

(46:I:b) (1)
$$-|\Gamma|_1 \leqq e_0 \leqq |\Gamma|_2,$$

then Γ has the same solutions $\bar{\mathsf{U}}_I$ for $E(e_0)$ and for $F(e_0)$. These $\bar{\mathsf{U}}_I$ are precisely those sets, which obtain in the following manner:

Choose any two $\bar{\varphi}$, $\bar{\psi}$ so that

(2)
$$-|\Delta|_1 \leqq \bar{\varphi} \leqq |\Delta|_2,$$
(3)
$$-|\mathrm{H}|_1 \leqq \bar{\psi} \leqq |\mathrm{H}|_2,$$

and

(4)
$$\bar{\varphi} + \bar{\psi} = e_0.$$

Choose any two solutions $\bar{\mathsf{V}}_J$, $\bar{\mathsf{W}}_K$ of Δ, H for $E(\bar{\varphi})$, $E(\bar{\psi})$.

Then $\bar{\mathsf{U}}_I$ is the composition of $\bar{\mathsf{V}}_J$ and $\bar{\mathsf{W}}_K$ in the sense of 44.7.4.

If

(46:I:c) (1)
$$e_0 > |\Gamma|_2,$$

then Γ does not have the same solutions $\bar{\mathsf{U}}_I$ for $E(e_0)$ and U_I for $F(e_0)$. These $\bar{\mathsf{U}}_I$ and U_I are precisely those sets which obtain in the following manner: Form the two numbers $\bar{\varphi}$, $\bar{\psi}$ with

(2)
$$\bar{\varphi} > |\Delta|_2,$$
(3)
$$\bar{\psi} > |\mathrm{H}|_2,$$

which are defined by

(4)
$$e_0 - |\Gamma|_2 = \bar{\varphi} - |\Delta|_2 = \bar{\psi} - |\mathrm{H}|_2.$$

Choose any two solutions $\bar{\mathsf{V}}_J$, $\bar{\mathsf{W}}_K$ of Δ, H for $E(\bar{\varphi})$, $E(\bar{\psi})$.

Then $\bar{\mathsf{U}}_I$ is the sum of the following sets: The composition of $\bar{\mathsf{V}}_J$ and of the set of all detached $\overrightarrow{\beta}'$ (in K) with $e(\overrightarrow{\beta}') = |\mathrm{H}|_2$; the composition of the set of all detached $\overrightarrow{\alpha}'$ (in J) with $e(\overrightarrow{\alpha}') = |\Delta|_2$

and of $\bar{\mathsf{W}}_K$; the composition of the set of all detached $\overrightarrow{\alpha}'$ (in J) with $e(\overrightarrow{\alpha}') = \varphi$ and of the set of all detached $\overrightarrow{\beta}'$ (in K) with $e(\overrightarrow{\beta}') = \psi$, taking all pairs φ, ψ with

$$(5) \qquad |\Delta|_2 < \varphi < \bar{\varphi}, \qquad |\mathrm{H}|_2 < \psi < \bar{\psi},$$

and

$$(6) \qquad \varphi + \psi = e_0.$$

U_I obtains by the same process, only replacing the condition (6) by

$$(7) \qquad \varphi + \psi \leqq e_0.$$

Proof: Ad (46:I:a): This coincides with (45:O:a).

Ad (46:I:b): This is a restatement of case (a) in (46:H) except for the following modifications:

First: The identification of the E and F solutions for Γ, Δ, H. This is justified by applying (45:O:b) to Γ, Δ, H which is legitimate by (1), (2), (3) of (46:I:b).

Second: The way in which we formed $\bar{\mathsf{U}}_I = \mathsf{U}_I$ from $\bar{\mathsf{V}}_J = \mathsf{V}_J$, $\bar{\mathsf{W}}_K = \mathsf{W}_K$ which differed from the one described in (46:H) insofar as we omitted the condition (46:17). This is justified by observing that (46:17) is automatically fulfilled: $\mathsf{V}_J = \bar{\mathsf{V}}_J \subseteq E(\bar{\varphi})$, $\mathsf{W}_K = \bar{\mathsf{W}}_K \subseteq E(\bar{\psi})$, hence for $\overrightarrow{\alpha}'$ in V_J and $\overrightarrow{\beta}'$ in W_K always $e(\overrightarrow{\alpha}') = \bar{\varphi}$, $e(\overrightarrow{\beta}') = \bar{\psi}$ and so by (4) $e(\overrightarrow{\alpha}') + e(\overrightarrow{\beta}') = e_0$.

Ad (46:I:c): This is a restatement of case (b) in (46:H), except for this modification:

We consider both E and F solutions for Γ (not only F solutions as in (46:H)), and use only E solutions for Δ, H (not F solutions as in (46:H)). The way in which the former $\bar{\mathsf{U}}_I$, U_I of Γ are formed from the latter ($\bar{\mathsf{V}}_J$ of Δ, $\bar{\mathsf{W}}_K$ of H) is accordingly different from the one described in (46:H).

In order to remove these differences, one has to proceed as follows: Apply (45:I) and (45:O:c) to Γ, Δ, H which is legitimate by (1), (2), (3) of (46:I:c). Then substitute the defining for the defined in (46:H). If these manipulations are carried out on (46:H) (in the present case (46:I:c)), then precisely our above formulation results.[1]

46.7. Graphical Representation of a Part of the Result

46.7. The results of (46:I) may seem complicated, but they are actually only the precise expression of several simple qualitative principles. The reason for going through the intricacies of the preceding mathematical derivation was, of course, that these principles are not at all obvious, and that this is the way to discover and to prove them. On the other hand our result can be illustrated by a simple graphical representation.

[1] If the reader carries this out, he will see that this transformation, although somewhat cumbersome, presents absolutely no difficulty.

We begin with a more formalistic remark.

A look at the three cases (46:I:a)-(46:I:c) discloses this: While nothing more can be said about (46:I:a), the two other cases (46:I:b), (46:I:c) have some common features. Indeed, in both instances the desired solutions $\bar{U}_I$, U_I of Γ are obtained with the help of two numbers $\bar{\varphi}$, $\bar{\psi}$ and certain corresponding solutions $\bar{V}_J$, $\bar{W}_K$ of Δ, H. The quantitative elements of the representation of $\bar{U}_I$, U_I are the numbers $\bar{\varphi}$, $\bar{\psi}$. As was pointed out in footnote 2 on p. 386, they represent something like a distribution of the given excess e_0 in I between J and K.

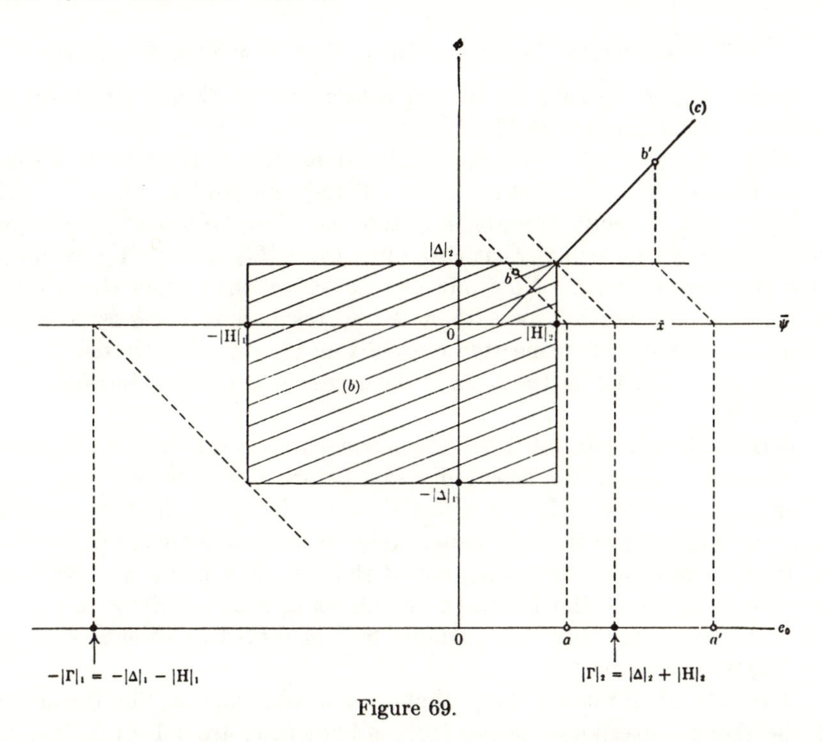

Figure 69.

$\bar{\varphi}$, $\bar{\psi}$ are characterized in the cases (46:I:b) and (46:I:c) by their respective conditions (2)-(4). Let us compare these conditions for (46:I:b) and for (46:I:c).

They have this common feature: They force the excesses $\bar{\varphi}$, $\bar{\psi}$ to belong to the same case of Δ, H as the one to which the excess e_0 belongs for Γ.

They differ, however, very essentially in this respect: In (46:I:b) they impose only one equation upon $\bar{\varphi}$, $\bar{\psi}$ while in (46:I:c) they impose two equations.[1] Of course, the inequalities too, may degenerate occasionally to equations (cf. (46:J) in 46.8.3.), but the general situation is as indicated.

The connections between e_0 and $\bar{\varphi}$, $\bar{\psi}$ are represented graphically by Fig. 69.

[1] (2), (3) are inequalities in both cases. (4) stands for one equation in (46:I:b) and for two equations in (46:I:c).

This figure shows the $\bar{\varphi}$, $\bar{\psi}$-plane and under it the e_0-line. On the latter the points $-|\Gamma|_1$, $|\Gamma|_2$ mark the division into the three zones corresponding to cases (46:I:a)-(46:I:c). The $\bar{\varphi}$, $\bar{\psi}$-domain which belongs to case (46:I:b) covers the shaded rectangle marked (b) in the $\bar{\varphi}$, $\bar{\psi}$-plane; the $\bar{\varphi}$, $\bar{\psi}$-domain, which belongs to case (46:I:c) covers the line marked (c) in the $\bar{\varphi}$, $\bar{\psi}$-plane.

Given any $\bar{\varphi}$, $\bar{\psi}$-point, following the line -------------- leads to its e_0 value—thus b, b' yield a, a', respectively. Given any e_0-value the reverse process discloses all its $\bar{\varphi}$, $\bar{\psi}$-points, thus a produces an entire interval at b, while a' yields the unique point b'.[1]

46.8. Interpretation: The Normal Zone. Heredity of Various Properties

46.8.1. Figure 69 calls for further comments, which are conducive to a fuller understanding of (46:I).

First: There have been repeated indications (for the last time in the comment following (45:O)), that the cases (46:I:a) and (46:I:c), i.e. $e_0 < -|\Gamma|_1$, and $e_0 > |\Gamma|_2$, respectively—are the "too small" or "too large" values of e_0 in the sense of 44.6.1.; i.e., that case (46:I:b), $-|\Gamma|_1 \leqq e_0 \leqq |\Gamma|_2$, is in some way the normal zone. Now our picture shows that when the excess e_0 of Γ lies in the *normal* zone, then the corresponding excesses $\bar{\varphi}$, $\bar{\psi}$ of Δ, H lie also within their respective normal zones.[2] In other words:

The normal behavior (position of the excess in (46:I:b)) is *hereditary* from Γ to Δ, H.

Second: In the case (46:I:b)—the normal zone—$\bar{\varphi}$, $\bar{\psi}$ are not completely determined by e_0, as we repeatedly saw before. In case (46:I:c), on the other hand, they are. This is pictured by the fact that the former domain is the rectangle (b) in the $\bar{\varphi}$, $\bar{\psi}$-plane, while the latter domain is only a line (c).

It is worth noting, however, that at the two ends of the case (46:I:b)—for $e_0 = -|\Gamma|_1, |\Gamma|_2$—the interval available for $\bar{\varphi}$, $\bar{\psi}$ is constricted to a point.[3] Thus the transition from the variable $\bar{\varphi}$, $\bar{\psi}$ of (46:I:b) to the fixed ones of (46:I:c) is continuous.

Third: Our first remark stated that normal behavior (i.e., that the position of the excess corresponds to (46:I:b)) is hereditary from Γ to Δ, H. It is remarkable that, in general, no such heredity holds for the vanishing of the excess, i.e. that $e_0 = 0$ [4] does not in general imply $\bar{\varphi} = 0$, $\bar{\psi} = 0$. It is precisely the vanishing of the excess which specializes our present theory (of 44.7.) to the older form (of 42.4.1. which, as we know, is equivalent to the original one of 30.1.1.). We will examine the variability of $\bar{\varphi}$, $\bar{\psi}$ when $e_0 = 0$ more closely in the last (sixth) remark. Before we do that, however, we give our attention to the connection between our present theory and the older form.

[1] We leave to the reader the simple verification that the geometrical arrangements of Fig. 69. express, indeed, the condition of (46:I:b), (46:I:c).

[2] I.e. $-|\Gamma|_1 \leqq e_0 \leqq |\Gamma|_2$ implies $-|\Delta|_1 \leqq \bar{\varphi} \leqq |\Delta|_2$, $-|H|_1 \leqq \bar{\psi} \leqq |H|_2$, cf. (46:I:b).

[3] This is one case of degeneration, alluded to at the end of 46.7.

[4] Of course, $e_0 = 0$ lies in the normal case (46:I:b): $-|\Gamma|_1 \leqq 0 \leqq |\Gamma|_2$.

Fourth: It is now evident that the present, wider form of the theory must of necessity receive consideration, even if our primary interest is with the original form alone. Indeed: in order to find the solutions of a decomposable game Γ in the original sense (for $e_0 = 0$), we need solutions for the constituent games Δ, H in the wider, new sense (for $\bar{\varphi}$, $\bar{\psi}$ which may not be zero).

This gives the remarks of 44.6.2. a more precise meaning: It is now specifically apparent how the passage from the old theory to the new one becomes necessary when the game (Δ or H) is looked upon as non-isolated. The exact formulation of this idea will come in 46.10.

46.8.2. Fifth: We can now justify the final statements about (44:D) in 44.3.2. and (44:D:a), (44:D:b) in 44.3.3. (46:I:b) shows that (44:D) is true in the case (44:I:b), if we relinquish the old theory for the new one; (44:I:c) shows that (44:D) is not true in the case (44:I:c) even at that price. Thus the desire to secure the validity of the plausible scheme of (44:D) motivates the passage to the new theory as well as the restriction to case (44:I:b)—the normal case.

If we insist interpreting (44:D), (44:D:a), (44:D:b) by the old theory, then (44:D), (44:D:a) fail,[1] while the conditional statement of (44:D:b) remains true.[2]

46.8.3. Sixth: We saw that $e_0 = 0$ does not in general imply $\bar{\varphi} = 0$, $\bar{\psi} = 0$. What does this "in general" mean?

$\bar{\varphi}$, $\bar{\psi}$ are subject to the conditions (2)-(4) of (46:I:b). As $e_0 = 0$, so (4) means that $\bar{\psi} = -\bar{\varphi}$ and permits us to express the remaining (2), (3) in terms of $\bar{\varphi}$ alone. They become this:

(46:35)
$$\begin{Bmatrix} -|\Delta|_1 \\ -|\mathrm{H}|_2 \end{Bmatrix} \leqq \bar{\varphi} \leqq \begin{Bmatrix} |\Delta|_2 \\ |\mathrm{H}|_1 \end{Bmatrix}$$

Now apply (45:E) to Δ, H. Then we see:

If Δ, H are both essential, then the lower limits of (46:35) are < 0 and the upper limits are > 0, so $\bar{\varphi}$ can really be $\neq 0$. If either Δ or H is inessential, then (46:35) implies $\bar{\varphi} = 0$ and hence $\bar{\psi} = 0$.

We state this explicitly:

(46:J) $e_0 = 0$ implies $\bar{\varphi} = 0$, $\bar{\psi} = 0$, i.e. (44:D) of 44.3.2. holds even in the sense of old theory if and only if either Δ or H is inessential.

46.9. Dummies

46.9.1. We can now dispose of the narrower type of decomposition, described in footnote 1 on p. 340—the addition of "dummies" to the game.

Consider the game Δ of the players $1', \cdots, k'$.[3] "Inflate" it by adding to it a series of "dummies" K; i.e. compose Δ with an inessential game H of the players $1'', \cdots, l''$. Then the composite game is Γ.

[1] As we may have $e_0 = 0$, $\bar{\varphi} \neq 0$, $\bar{\psi} \neq 0$. Then the decomposability requirement (44:B:a) of 44.3.1. is violated, as stated in 44.3.3.

[2] Representing the special case $e_0 = 0$, $\bar{\varphi} = 0$, $\bar{\psi} = 0$.

[3] It is now convenient to reintroduce the notations of 41.3.1. for the players.

We will use the old theory for all these games. By (31:I) in 31.2.1. there exists precisely one imputation for the inessential game H—say $\overrightarrow{\gamma}^0_K = \{\gamma^0_{1''}, \cdots, \gamma^0_{l''}\}$.[1] By (31:O) or (31:P) in 31.2.3. H possesses a unique solution: The one element set $(\overrightarrow{\gamma}^0_K)$.

Now by (46:J) and (46:I:b) the general solution of Γ obtains by composing the general solution of Δ with the general solution of H—and the latter one is unique!

In other words:

"Inflate" every imputation $\overrightarrow{\beta}_J = \{\beta_{1'}, \cdots, \beta_{k'}\}$ of J (i.e. Δ) to an imputation $\overrightarrow{\alpha}_I$ of I (i.e. Γ) by composing it with $\overrightarrow{\gamma}^0_K$, i.e. by adding to it the components $\gamma^0_{1''}, \cdots, \gamma^0_{l''}$: $\overrightarrow{\alpha}_I = \{\beta_{1'}, \cdots, \beta_{k'}, \gamma^0_{1''}, \cdots, \gamma^0_{l''}\}$. Then this process of "inflation"—i.e. of composition—produces the general solution of Γ from the general solution of Δ.

This result can be summed up by saying that the "inflation" of a game by the addition of "dummies" does not affect its solution essentially—it is only necessary to add to every imputation components representing the "dummies," and the values of these components are the plausible ones: What each "dummy" would obtain in the inessential game H, which describes their relationship to each other.

46.9.2. We conclude by adding that (46:J) states that if and only if the composition is not of the special type discussed above, the old theory ceases to have the simple properties of the new one, and its hereditary character fails, as indicated in the third remark of 46.8.1.

46.10. Imbedding of a Game

46.10.1. In the fourth remark of 46.8.1. we reaffirmed the indications of 44.6.2., according to which the passage from the old theory to the new one becomes necessary when the game is looked upon as non-isolated. We will now give this idea its final and exact expression.

It is more convenient this time to denote the game under consideration by Δ and the set of its players by J. It ought to be understood that this Δ is perfectly general—no decomposability of Δ is assumed.

We begin by introducing the concepts which are needed to treat a given game Δ as a non-isolated occurrence: This amounts to imbedding it without modifying it, into a wider setup, which it is convenient to describe as another game Γ. We define accordingly: Δ is *imbedded* into Γ, or Γ is an *imbedding of* Δ, if Γ is the composition of Δ with another game H.[2] In other words, Δ is imbedded in all those games of which it is a constituent.[3]

[1] Recall the notations of 44.2.

[2] The game H and the set of its players K are perfectly arbitrary, except that K and J must be disjunct.

[3] Since a constituent of a constituent is itself a constituent (recall the appropriate definitions, in particular (43:D) in 43.3.1.), an imbedding of an imbedding is again an imbedding. In other words: Imbedding is a transitive relation. This relieves us from considering any indirect relationships based upon it.

46.10.2. Let us now investigate the solutions of Δ viewing Δ as a non-isolated occurrence. In the light of the above, this amounts to enumerating all solutions of all imbeddings Γ of Δ, and interpreting them, as far as Δ is concerned. The last operation must be the taking of the J-constituent in the sense of 44.7.4. We know from the fifth remark in 46.8.2. that this is only feasible, if we consider no solutions from outside the normal zone (b).

One might hesitate whether the solutions of Γ should be taken in the sense of the old or the new theory. The former may appear to be more justified on the standpoint of 44.6.2.: The outside influences upon the game having been accounted for by the passage from Δ to H, there is no longer any excuse for going outside the old theory.[1] It happens, however, that we need not settle this point at all, because the result for Δ will be the same, irrespectively of which theory we use for Γ. But if we use the new theory for Γ, we must restrict ourselves to the case (46:I:b), as discussed above.

Thus the question presents itself ultimately in this form:

(46:K) Consider all imbeddings Γ of Δ, and all solutions of these Γ:
(a) in the sense of the old theory, i.e. for $E(0)$,
(b) in the sense of the new theory in the normal zone, i.e. for any $E(e_0)$ of (46:I:b).
Which are the J-constituents of the solutions?

46.10.3. The answer is very simple:

(46:L) The J-constituents (of the Γ solutions) referred to in (46:K) are precisely the following sets: All solutions for Δ in the normal zone, i.e. for any $E(\bar{\varphi})$ of (46:I:b). This is true for both (a) and (b) of (46:K).

Proof: $e_0 = 0$ belongs to case (46:I:b) (cf. footnote 4 on p. 396), hence (a) is narrower than (b). Therefore, we need only show that all the sets obtained from (b) are among the ones described above, and that all these sets can even be obtained with the help of (a).

The first assertion is only that of the hereditary character of the normal zone (b).

The second assertion follows from (46:I:b), if we can do this: Given a $\bar{\varphi}$ with $-|\Delta|_1 \leqq \bar{\varphi} \leqq |\Delta|_2$, find a game H and $\bar{\psi}$ with $-|\mathrm{H}|_1 \leqq \bar{\psi} \leqq |\mathrm{H}|_2$, such that $\bar{\varphi} + \bar{\psi} = 0$ and that H possesses solutions for $E(\bar{\psi})$. Now such an H exists, and it can even be chosen as a three-person game.

Indeed: Let H be the essential three-person game with general $\gamma > 0$. Then by (45:2) in 45.1. $|\mathrm{H}|_1 = 3\gamma$ and by (45:9) in 45.3.3. $|\mathrm{H}|_2 = \frac{1}{2}|\mathrm{H}|_1 = \frac{3}{2}\gamma$. We have required $\bar{\psi} = -\bar{\varphi}$ and what we know now amounts to

$$-3\gamma \leqq \bar{\psi} \leqq \tfrac{3}{2}\gamma.$$

[1] Besides, the transitivity pointed out in footnote 3 on p. 398, shows that any further imbedding of Γ can be regarded directly as an imbedding of Δ.

This can clearly be met by choosing γ sufficiently great. Then we also need a solution of H for $E(\bar{\psi})$. The existence of such a solution (for $-3\gamma \leqq \bar{\psi} \leqq \frac{3}{2}\gamma$) will be shown in 47.

46.10.4. To this result two more remarks should be added:

First: If we wanted to handle the process of imbedding in such a manner that the old theory remains hereditary, we would have to see to this: The composition of Γ from Δ and H has to be such that $e_0 = 0$ implies $\bar{\varphi} = 0$ (and hence $\bar{\psi} = 0$). By (46:T) this means that either Δ or H are inessential. The latter means (cf. eod), that only "dummies" are added to Δ.

Summing up:

(46:M) The old theory remains hereditary if and only if either the original game Δ is inessential, or the imbedding is restricted to the addition of "dummies" to Δ.

Second: It was suggested already in 44.6.2. to treat the outside source, which creates the excesses and paves the way for the transition of our old theory to the new one, as another player.

Our above result (44:L) justifies a slightly modified view: The outside source of 44.6.2. is the game H which is added to Δ—or rather the set K of its players.

Now we have seen that the game H must be essential, in order to achieve the desired result. Furthermore we know that an essential game must have $n \geqq 3$ participants, and the proof of (44:L) showed that a suitable H with $n = 3$ participants does indeed exist.

So we see:

(46:N) The outside source of 44.6.2. can be regarded as a group of new players—but not as one player. Indeed, the minimum effective number of members of this group is 3.

46.10.5. The foregoing considerations have justified our passage from the old theory to the new one (within the normal zone (b)) and clarified the nature of this transition. We see now that the "common sense" surmise of 44.3. fails to hold in the old theory, but that it is true in precisely that new domain to which we changed. This rounds out the theory in a satisfactory manner.

The leading principle of the discussions of 44.4.3.-46.10.4. was this: The game under consideration was originally viewed as an isolated occurrence, but then removed from this isolation and imbedded, without modification, in all possible ways into a greater game. This order of ideas is not alien to the natural sciences, particularly to mechanics. The first standpoint corresponds to the analysis of the so-called closed systems, the second to their imbedding, without interaction, into all possible greater closed systems.

The methodical importance of this procedure has been variously emphasized in the modern literature on theoretical physics, particularly in the analysis of the structure of Quantum Mechanics. It is remarkable that it could be made use of so essentially in our present investigation.

46.11. Significance of the Normal Zone

46.11.1. The result (46:I:b) defines for every solution of the composite game Γ in the normal zone—i.e. *a fortiori* for every solution in the sense of the old theory—numbers $\bar{\varphi}$, $\bar{\psi}$. This and the immediate properties of $\bar{\varphi}$, $\bar{\psi}$ in connection with the solution, appear to be so fundamental, as to deserve a fuller non-mathematical exposition.

We are considering two games Δ, H played by two disjunct sets of players J and K. The rules of these games stipulate absolutely no physical connection between them. We view them nevertheless as one game Γ—but this game, of course, is composite, with the two isolated constituents Δ, H.

Let us now find all solutions of the entire arrangement, i.e. of the composite game Γ. Since it is not desired to consider anything outside of Γ, we adhere to the original theory of 30.1.1. and 42.4.1.[1] Then we have shown that any such solution U_I determines a number $\bar{\varphi}$[2] with the following property: For every imputation $\overrightarrow{\alpha}$ of U_I the players of Δ (i.e. in J) obtain together the amount $\bar{\varphi}$, and the players of H (i.e. in K) obtain together the amount $-\bar{\varphi}$. Thus the principle of organization embodied in U_I must stipulate (among other things) that the players of H transfer under all conditions the amount $\bar{\varphi}$ to the players of Δ.

The remainder of the characterization of U_I—i.e. of the principle of organization or standard of behavior embodied in it—is this:

First: The players of Δ, in their relationship with each other, must be regulated by a standard of behavior which is stable, provided that the transfer of $\bar{\varphi}$ from the other group is placed beyond dispute.[3]

Second: The players of H, in their relationship with each other, must be regulated by a standard of behavior which is stable, provided that the transfer of $\bar{\varphi}$ to the other group is placed beyond dispute.[4]

Third: The octroyed transfer $\bar{\varphi}$ must lie between the limits (46:35) of 46.8.3.

$$(46{:}35)\qquad \begin{Bmatrix} -|\Delta|_1 \\ -|\mathrm{H}|_2 \end{Bmatrix} \leqq \bar{\varphi} \leqq \begin{Bmatrix} |\Delta|_2 \\ |\mathrm{H}|_1 \end{Bmatrix}.$$

46.11.2. The meaning of these rules is clearly that any solution, i.e. any stable social order of Γ is based upon payment of a definite tribute by one of the two groups to the other. The amount of this tribute is an integral part of the solution. The possible amounts, i.e. those which can

[1] I.e. $e_0 = 0$.

[2] Since $\bar{\varphi} + \bar{\psi} = e_0 = 0$, we do not introduce $\bar{\psi} = -\bar{\varphi}$.

[3] I.e. that the J-constituent V_J of U_I is a solution of Δ for $E(\bar{\varphi})$.

[4] I.e. that the K-constituent W_k of U_I is a solution of H for $E(-\bar{\varphi})$.

occur in solutions, are strictly determined by (46:35) above. This condition shows in particular:

First: The tribute zero, i.e. the absence of a tribute is always among the possibilities.

Second: The tribute zero is the only possible one if and only if one of the two games Δ, H is inessential (cf. the sixth remark in 46.8.3.).

Third: In all other cases both positive and negative tributes are possible —i.e. both the players of Δ and the players of H may be the tribute paying group.

The limits of (46:35) are set by both games Δ, H, i.e., by the objective physical possibilities of both groups.[1] These limits express that each group has a minimum below which no form of social organization can depress it: $-|\Delta|_1$, $-|\mathrm{H}|_1$; and, each group has a maximum, above which it cannot raise itself under any form of social organization: $|\Delta|_2$, $|\mathrm{H}|_2$.

Thus, for a particular physical background, i.e. a game, say Δ, the two numbers $|\Delta|_1$, $|\Delta|_2$ can be interpreted this way: $-|\Delta|_1$ is the worst that will be endured under any conditions and $|\Delta|_2$ is the maximum claim which may find outside acceptance under any conditions.[2]

The results (45:E) and (45:F) of 45.3.1.-2. now acquire a new significance: According to these the two numbers can only vanish together (when Δ is inessential) and their ratio always lies between definite limits.

46.12. First Occurrence of the Phenomenon of Transfer: $n = 6$

46.12. We have seen repeatedly (thus in (46:J) in 46.8.3. and in the second and third remarks in 46.11.2.) that the characteristic new element of the theory of a composite game Γ manifests itself only when both constituents Δ, H are essential. This is the occurrence of $e_0 = 0$, but

$$\bar{\varphi} = -\bar{\psi} \neq 0,$$

i.e. a non-zero tribute in the sense of 46.11.

Now we know that in order to be essential a game must have $\geqq 3$ players. If this is to be true for both Δ, H, then the composite game Γ must have $\geqq 6$ players.

Six players are actually enough as the following consideration shows: Let Δ, H both be the essential three-person games with $\gamma = 1$. Then $|\Delta|_1 = |\mathrm{H}|_1 = 3$, $|\Delta|_2 = |\mathrm{H}|_2 = \frac{3}{2}$. (Cf. in 46.10.3.). Hence for $-\frac{3}{2} \leqq \bar{\varphi} \leqq \frac{3}{2}$, both $\bar{\varphi}$ and $\bar{\psi} = -\bar{\varphi}$ lie between -3 and $\frac{3}{2}$. This implies, as will be shown in 47., the existence of solutions V_J, W_K of Δ, H for $E(\bar{\varphi})$, $E(\bar{\psi})$. Their com-

[1] But where the actual amount $\bar{\varphi}$ lies between those limits, is not determined by those objective data, but by the solution, i.e. the standard of behavior which happens to be generally accepted.

[2] It must be recalled that all this takes the value of the coalition of all players of Δ, $v(J)$, as zero; i.e., we are discussing the losses which are purely attributable to lack of co-operation among the group, and unfavorable general social organizations—and gains, which are purely attributable to lack of co-operation in outside groups and favorable general social organizations.

position U_I is then a solution of the composite game Γ with the given $\bar{\varphi}$. Since $\bar{\varphi}$ was only restricted by $-\frac{3}{2} \leqq \bar{\varphi} \leqq \frac{3}{2}$, we can choose it non-zero.

Thus we have demonstrated:

(46:O) $n = 6$ is the smallest number of players for which the characteristic new element of our theory of composite games (the possibility of $e_0 = 0$ with $\bar{\varphi} = -\bar{\psi} \neq 0$, cf. above) can be observed in a suitable game.

We have repeatedly expressed the belief that an increase in the number of players need not only cause a more involved operation of the concepts which occurred for smaller numbers, but that it also may originate qualitatively new phenomena. Specifically such occurrences were observed as the number of players successively increased to 2,3,4. It is, therefore, of interest that the same happens now as the number of players reaches six.[1]

47. The Essential Three-person Game in the New Theory

47.1. Need for This Discussion

47.1. It remains for us to discuss the solutions of the essential three-person game, according to the new theory.

This is necessary, since we have already made use of the existence of these solutions in 46.10. and 46.12., but the discussion possesses also an interest of its own. In view of the interpretation which we were induced to put on these solutions in 46.12. and also of their central role in the theory of decomposition,[2] it seems desirable to acquire a detailed knowledge of their structure. Furthermore, a familiarity with these details will lead to other interpretations of some significance. (Cf. 47.8. and 47.9.) Finally, we shall find that the principles used in determining the solutions in question are of wider applicability. (Cf. 60.3.2., 60.3.3.)

47.2. Preparatory Considerations

47.2.1. We consider the essential three-person game, to be denoted by Γ, in the normalization $\gamma = 1$. Thus $|\Gamma|_1 = 3$, $|\Gamma|_2 = \frac{3}{2}$ (cf. 46.12.). We wish to determine the solutions of this Γ for $E(e_0)$.[3] In the applications, referred to above we needed only the normal zone $-3 \leqq e_0 \leqq \frac{3}{2}$ but we prefer to discuss now all e_0.

This discussion will be carried out with the graphical method, which we used in treating the old theory in 32. We will, therefore, follow the scheme of 32. in several respects.

[1] For some other qualitatively new phenomena which emerge only when there are six players, cf. 53.2.

[2] This is the only problem of absolutely general character, of which we have a complete solution at present!

[3] We are writing Γ, e_0 although the applications employed the notations Δ, $\bar{\varphi}$ and H, $\bar{\psi}(= -\bar{\varphi})$.

Of course, the present Γ has nothing to do with the decomposable Γ considered before.

The characteristic function is the same as in 32.1.1.:

$$(47{:}1) \qquad \mathrm{v}(S) = \begin{cases} 0 \\ -1 \\ 1 \\ 0 \end{cases} \quad \text{when } S \text{ has} \quad \begin{cases} 0 \\ 1 \\ 2 \\ 3 \end{cases} \text{elements.}$$

An (extended) imputation is a vector

$$\overrightarrow{\alpha} = \{\alpha_1, \alpha_2, \alpha_3\},$$

whose three components must fulfill (44:13) in 44.7.2., which becomes now:

$$(47{:}2) \qquad \alpha_1 \geqq -1, \qquad \alpha_2 \geqq -1, \qquad \alpha_3 \geqq -1.$$

Besides, in $E(e_0)$ the excess must be e_0, according to (44:11*) in 44.7.2. and this is now

$$(47{:}3) \qquad \alpha_1 + \alpha_2 + \alpha_3 = e_0.$$ [1]

47.2.2. We wish to represent these $\overrightarrow{\alpha}$ by the graphical device of 32.1.2. But that procedure pictures only number triplets of sum zero. Therefore we define

$$(47{:}4) \qquad \alpha^1 = \alpha_1 - \frac{e_0}{3}, \qquad \alpha^2 = \alpha_2 - \frac{e_0}{3}, \qquad \alpha^3 = \alpha_3 - \frac{e_0}{3}.$$

Then (47:2), (47:3) become

$$(47{:}2^*) \quad \alpha^1 \geqq -\left(1 + \frac{e_0}{3}\right), \qquad \alpha^2 \geqq -\left(1 + \frac{e_0}{3}\right), \qquad \alpha^3 \geqq -\left(1 + \frac{e_0}{3}\right),$$

$$(47{:}3^*) \qquad \alpha^1 + \alpha^2 + \alpha^3 = 0.$$ [2]

Now the representation of 32.1.2. becomes applicable, we need only replace $\alpha_1, \alpha_2, \alpha_3$ by $\alpha^1, \alpha^2, \alpha^3$. With this reservation, Figure 52 can be used.

For these reasons, we form for every vector $\overrightarrow{\alpha} = \{\alpha_1, \alpha_2, \alpha_3\}$ of $E(e_0)$ not only its *components* in the ordinary sense but also its *quasi-components* in the sense of (47:4): $\alpha^1, \alpha^2, \alpha^3$; and with the help of the quasi-components, we utilize the graphical representation of Figure 52.

So this plane representation expresses precisely the condition (47:3*). The remaining condition (47:2*) is therefore equivalent to a restriction imposed upon the point $\overrightarrow{\alpha}$, within the plane of Figure 52. This restriction

[1] The reader should compare (47:1)-(47:3) with (32:1)-(32:3) of 32.1.1.—the sole difference lies in (47:3).

[2] Comparing these (47:2*), (47:3*) with (32:2), (32:3) of 32.1.1., it appears that (47:3*) and (32:3) coincide, and that (47:2*) and (32:2) differ only by the factor of proportionality $1 + \frac{e_0}{3}$.

obtains in the same way as the similar one in 32.1.2.: $\vec{\alpha}$ must lie within the triangle formed by the three lines $\alpha^1 = -\left(1 + \frac{e_0}{3}\right)$, $\alpha^2 = -\left(1 + \frac{e_0}{3}\right)$, $\alpha^3 = -\left(1 + \frac{e_0}{3}\right)$. This is precisely the situation of Fig. 53., except for the proportionality factor $1 + \frac{e_0}{3}$,[1] and it is represented on Figure 70. The shaded area to be called the *fundamental triangle*, represents the $\vec{\alpha}$ which fulfill (47:2*), (47:3*), i.e. those of $E(e_0)$.

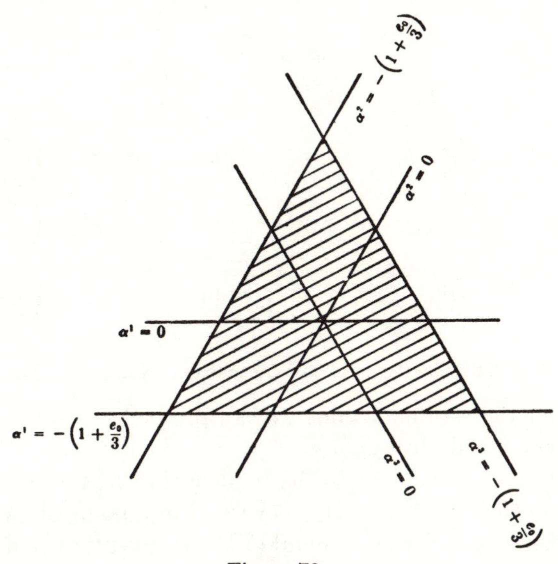

Figure 70.

47.2.3. We express the relationship of domination in this graphical representation. As we are using the new theory, the considerations of 31.1. concerning the set S of 30.1.1. for a domination $\vec{\alpha} \succ \vec{\beta}$—i.e. concerning its certainly necessary or certainly unnecessary character—no longer apply. So we discuss S *de novo*.

It is still true, that S cannot be a one- or a three-element set. In the first case $S = (i)$, so by 30.1.1. $\alpha_i \leqq v((i)) = -1$, $\alpha_i > \beta_i$, hence $\beta_i < -1$, contradicting $\beta_i \geqq -1$ by (47:2). In the second case $S = (1,2,3)$, so by 30.1.1. $\alpha_1 > \beta_1$, $\alpha_2 > \beta_2$, $\alpha_3 > \beta_3$, hence $\alpha_1 + \alpha_2 + \alpha_3 > \beta_1 + \beta_2 + \beta_3$, contradicting $\alpha_1 + \alpha_2 + \alpha_3 = \beta_1 + \beta_2 + \beta_3 = e_0$ by (47:3).

[1] Cf. footnote 2 on p. 404. Here we assume, of course, that $1 + \frac{e_0}{3} \geqq 0$, i.e.

$$e_0 \geqq -3 = -|\Gamma|_1.$$

If $1 + \frac{e_0}{3} < 0$, i.e. $e_0 < -3 = -|\Gamma|_1$, then the conditions of (47:2*), (47:3) conflict, and indeed we know from (45:A) that $E(e_0)$ is empty in this case.

Thus S must be a two-element set, $S = (i, j)$.[1] Then domination means that $\alpha_i + \alpha_j \leqq v((i, j)) = 1$ and $\alpha_i > \beta_i$, $\alpha_j > \beta_j$, i.e. that

$$\alpha^i + \alpha^j \leqq 1 - \frac{2e_0}{3},$$

and $\alpha^i > \beta^i$, $\alpha^j > \beta^j$. By (47:3*) the first condition may be written

$$\alpha^k \geqq -\left(1 - \frac{2e_0}{3}\right).$$

We restate this: Domination

$$\overrightarrow{\alpha} \succ \overrightarrow{\beta}$$

means that

$$(47{:}5)\quad \begin{cases} \text{either } \alpha^1 > \beta^1, & \alpha^2 > \beta^2 \quad \text{and} \quad \alpha^3 \geqq -\left(1 - \frac{2e_0}{3}\right), \\ \text{or} \quad \alpha^1 > \beta^1, & \alpha^3 > \beta^3 \quad \text{and} \quad \alpha^2 \geqq -\left(1 - \frac{2e_0}{3}\right), \\ \text{or} \quad \alpha^2 > \beta^2, & \alpha^3 > \beta^3 \quad \text{and} \quad \alpha^1 \geqq -\left(1 - \frac{2e_0}{3}\right). \end{cases}$$ [2]

47.3. The Six Cases of the Discussion. Cases (I)-(III)

47.3.1. After these preparations we can proceed to discuss the solutions V of Γ for $E(e_0)$, for all values of e_0.

It will be found convenient to distinguish six cases. Of these Case (I) corresponds to (45:0:a), Cases (II)-(IV) and one point of (V) to (45:0:b) (the normal zone), and Cases (V) and (VI) (without that point) to (45:0:c) (all in 45.6.1.).

47.3.2. Case (I): $e_0 < -3$. In this case $1 + \frac{e_0}{3} < 0$, so (47:2*), (47:3*) conflict and $E(e_0)$ is empty (cf. footnote 1 on p. 405) so V must be empty too.

Case (II): $e_0 = -3$. In this case $1 + \frac{e_0}{3} = 0$, so (47:2*), (47:3*) imply $\alpha^1 = \alpha^2 = \alpha^3 = 0$, i.e. $\alpha_1 = \alpha_2 = \alpha_3 = \frac{e_0}{3} = -1$,

$$\overrightarrow{\alpha} = \{-1, -1, -1\}.$$

So $E(e_0)$ is a one-element set, and V must be $= E(e_0)$ by the same argument as in the proof of (31:0) in 31.2.3. Thus the conditions are very similar to those encountered in an inessential game, cf. loc. cit.

[1] i, j, k a permutation of 1,2,3.
[2] This differs from the corresponding (32:4) in 32.1.3. only by the extra condition at the end of each line.

Case (III): $-3 < e_0 \leqq 0$. In this case $1 + \frac{e_0}{3} > 0$, so we can use Figure 70. Also $1 + \frac{e_0}{3} \leqq 1 - \frac{2e_0}{3}$, so the extra conditions of (47:5) in 47.2.3. are automatically fulfilled throughout the fundamental triangle. So (47:5) coincides with (32:4) in 32.1.3. (cf. footnote 2 on p. 406). Consequently the entire discussion of 32.1.3.-32.2.3. applies again, if the proportionality factor $1 + \frac{e_0}{3}$ is inserted.

Thus we obtain the solutions of $E(e_0)$ in this case simply by taking those described in 32.2.3., multiplying each component by $1 + \frac{e_0}{3}$, and adding $\frac{e_0}{3}$ (to pass from α^i to α_i).

47.4. Case (IV): First Part

47.4.1. Case (IV): $0 < e_0 < \frac{3}{2}$. In this case $0 < 1 - \frac{2e_0}{3} < 1 + \frac{e_0}{3}$. Consequently the lines

$$\alpha^1 = -\left(1 - \frac{2e_0}{3}\right), \ \alpha^2 = -\left(1 - \frac{2e_0}{3}\right), \ \alpha^3 = -\left(1 - \frac{2e_0}{3}\right)$$

(which bound the extra conditions of (47:5)) are situated with respect to the fundamental triangle of Figure 70 as indicated on Figure 71. They subdivide the fundamental triangle into seven areas, each of which can be characterized by stating which two-element sets S are effective in it in the sense of (47:5). The list is given below Figure 71. Now we can draw the analogue of Figure 54, indicating for each point of the fundamental triangle the shaded areas[1] which it dominates. This is done in Figure 72 according to (47:5). It is necessary to treat each one of the seven areas of Figure 71 separately, and every shaded area of Figure 72 must be continued across the entire fundamental triangle.

It is clear from Figure 72, that no point of the area ① can be dominated by a point outside that area.[2] Hence the condition (44:E:c) of 44.7.3., which characterizes the solution V for $E(e_0)$, i.e. for the entire fundamental triangle, must also hold for the part of V in ① when taken for ① (in place of the entire fundamental triangle, i.e. $E(e_0)$). But ① is a triangle like the fundamental triangle of Figure 53, except for the proportionality factor $1 - \frac{2e_0}{3}$.[3] Comparison of Figure 54 with ① in Figure 72 shows that the conditions of domination are the same.

[1] Excluding their boundaries.
[2] Including its boundary.
[3] Note that $1 - \frac{2e_0}{3} > 0$.

47.4.2. Consequently the entire discussion of 32.1.3.-32.2.3. applies to the part of V in ①, if the proportionality factor $1 - \frac{2e_0}{3}$ is inserted.

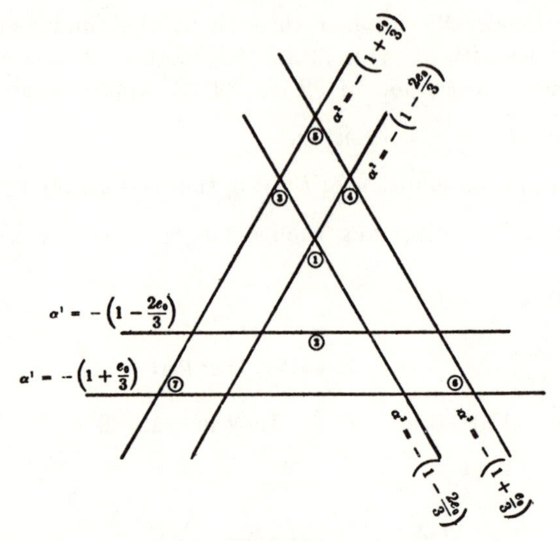

Figure 71.

Area:	Effective two-element sets S:		
①	(1, 2),	(1, 3),	(2, 3)
②	(1, 2),	(1, 3)	
③	(1, 2),		(2, 3)
④		(1, 3),	(2, 3)
⑤			(2, 3)
⑥		(1, 3)	
⑦	(1, 2)		

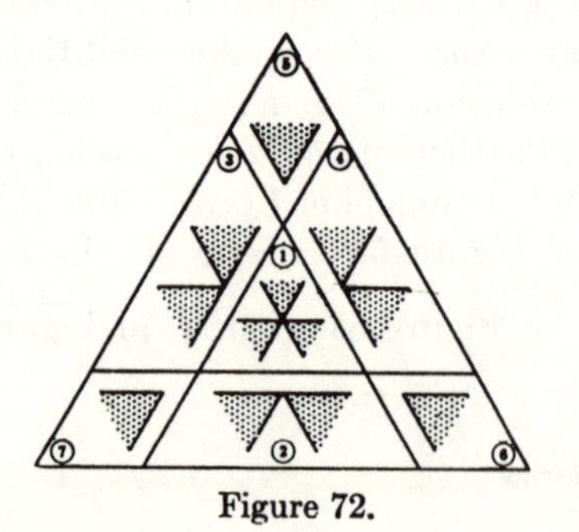

Figure 72.

Hence the part of V in ① must be either the set °₀° or the set —·—·— indicated in Figure 73. (The line —·—·—·— can be in any position below the points ° °.) However —·—·—·— must be subjected to all permutations of 1,2,3,—i.e. to rotations of the triangle by 0°, 60°, 120°,—to produce all solutions. (Cf. 32.2.3., °$_{\circ}$° is (32:B), —·—·— is (32:A) there.)

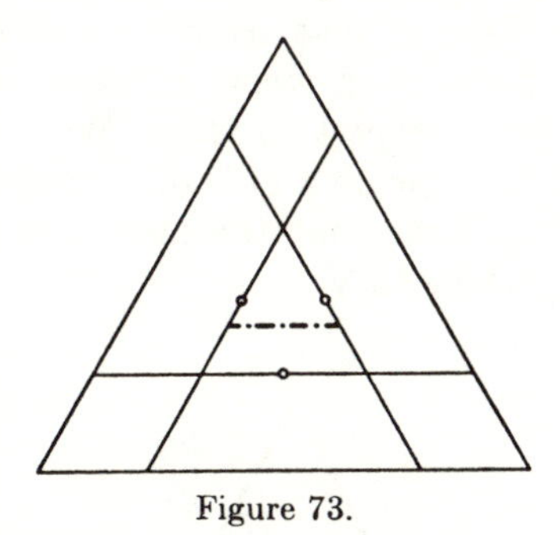

Figure 73.

Having found the part of V in ①, we proceed to determine the remainder of V. Since V is a solution, this remainder must lie in the area which is undominated by the part of V in ①. Comparison of Figure 73 with Figure 72 shows that this undominated area is the following one:

For the set °$_{\circ}$° it consists of the three ▲ triangles of Fig. 74, for the set —·—·—·— it consists of the three ▲ triangles of Figure 75.[1]

It is clear from Figure 72, that no point in any one of these triangles can be dominated by a point in another one.[2] Hence the condition (44:E:c) of 44.7.3., which characterizes the solution V for $E(e_0)$, i.e. for the entire fundamental triangle—and which holds for the part of V in ① taken for

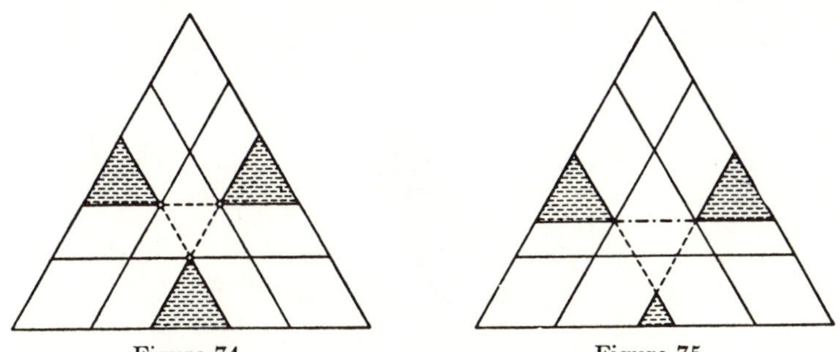

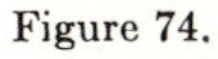

Figure 74. Figure 75.

① (in place of the entire fundamental triangle, i.e. $E(e_0)$) too—states precisely this: (44:E:c) holds for the part of V in each triangle ▲, taken for that triangle.

47.5. Case (IV): Second Part

47.5.1. Let us therefore take one of those triangles, denoting it by T. Its position in the fundamental triangle,[3] and the shaded areas dominated

[1] The position of all these triangles are clearly indicated by the drawings, except for the lower triangle in Figure 75. This triangle lies certainly outside the inner triangle (area ①)—this is equivalent to the restriction (32:8) in 32.2.2., cf. also Figure 60 there. Its position with respect to the outer (fundamental) triangle is less definite: It may shrink to a point or even disappear altogether.

It is not difficult to see that the latter phenomenon is excluded, unless the (linear) size of the inner triangle is $\geqq \frac{1}{4}$ of the outer one—this means $1 - \frac{2e_0}{3} \geqq \frac{1}{4}\left(1 + \frac{e_0}{3}\right)$, i.e. $e_0 \leqq 1$. We do not propose to discuss this subject further.

[2] All this refers to Figure 74, or all to Figure 75—but, of course, never to both in the same argument!

[3] Up to a rotation by 0°, 60°, or 120°. For the lower triangle of Figure 75 the apex does not lie on the inner triangle, but below it, (cf. footnote 1 above) but this does not alter our discussion.

by a given point in it (taken over from Fig. 72) are shown on Figure 76. We may now restrict ourselves to this triangle T, and to the concept of domination which is valid in it—and determine the solution of (44:E:c) with respect to this. We redraw T and the setup in it separately, also introducing a system of coordinates x, y in it. (Figure 77.)

Note that the apex o is undominated by points of T hence it must belong to V.[1,2]

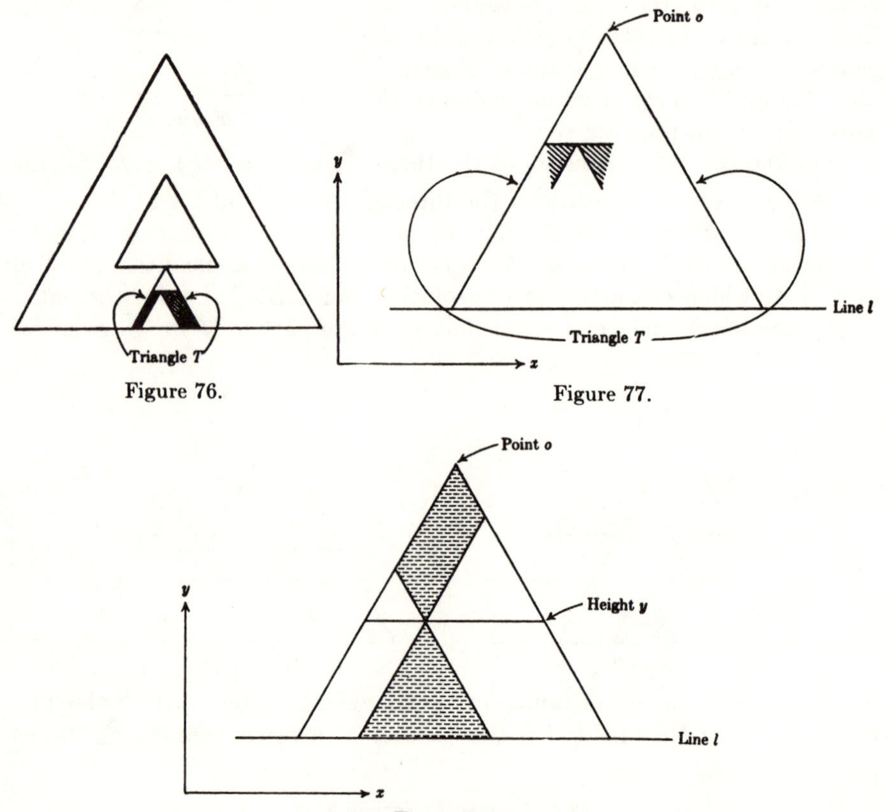

Figure 76.

Figure 77.

Figure 78.

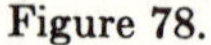

47.5.2. Now consider two points of V in T at different heights y. In order that the upper one should not dominate the lower one, the latter must not lie in the two shaded sextants, belonging to the former, i.e. the lower point must be in the middle sextant below the upper one, and *vice versa*. Thus, if a point of V in T is given, then all points of V in T at different heights y must lie in one of the two sextants ⁞⁞⁞⁞ indicated in Figure 78.

[1] For other triangles ▲ (i.e. T) than the lower one of Figure 75, this follows from another consideration, too: As Figures 74, 75 show, the apex of such a triangle lies on the border of the inner triangle (area ①) and belongs to what we know to be the part of V in ①.

[2] When the lower triangle ▲ (i.e. T) of Figure 75 degenerates to one point (cf. footnote 1 on p. 409) which is, of course, o—then this determines the part of V in T.

47.5.3. Now assume that a y_1 is the height of more than one point of V. Let then p and q be two different points of V with this height y_1 (Figure 79). Now choose a point r in the interior of the triangle Comparison of Figure 79 with Figure 77 shows that this r dominates both p and q. As p, q belong to V, r cannot belong to it. Hence there must exist a point s in V which dominates r. Now a second comparison of Figure 79 with Figure 77 shows that a point which dominates r must also dominate either p or q. Since s, p, q all belong to V this is a contradiction.

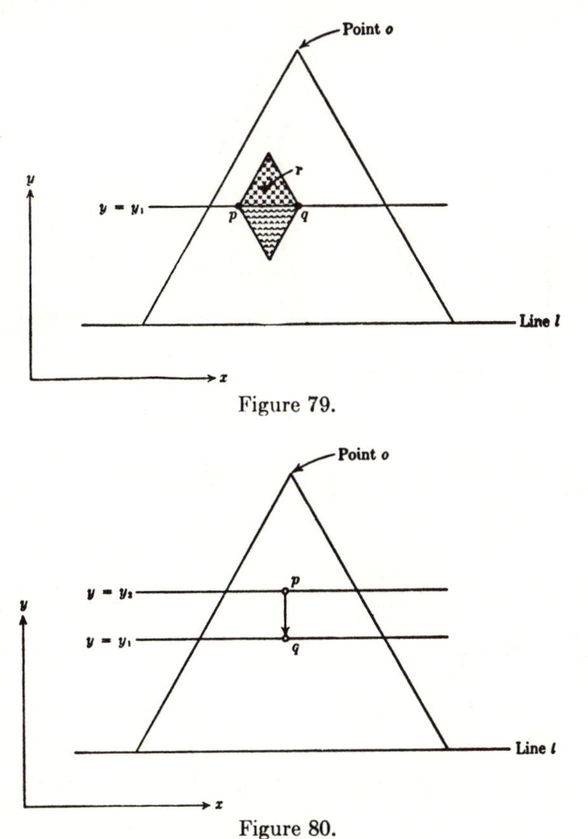

Figure 79.

Figure 80.

47.5.4. Next assume that a y_1 (in triangle T, i.e. between the base l and the apex o) is the height of no point of V. There exist certainly points of V with heights $y \geqq y_1$, e.g. the apex o is such a point. Choose a point p of V with a height $y \geqq y_1$ as low as possible, i.e. with its y minimum.[1] (Figure 80.) Denote this minimum value with $y = y_2$. Clearly $y_1 < y_2$. By the definition of y_2 no point of V has a height y with $y_1 \leqq y < y_2$ and by the above p is the only point of V with a height $y = y_2$.

Now project p perpendicularly on $y = y_1$, obtaining q. q cannot be in V hence it is dominated by an s in V. Hence this s cannot lie below q, i.e.

[1] This is possible since V is a closed set. Cf. (*) of footnote 1 on p. 384.

its height $y \geqq y_1$. Consequently $y \geqq y_2$. Comparison of Figure 80 with Figure 77 shows that p does not dominate q. Hence $s \neq p$, necessitating $y \neq y_2$. Thus $y > y_2$, i.e. s lies (definitely) above p. Now a second comparison of Figure 80 with Figure 77 shows that if a point s above p dominates q, then it must also dominate p. Since s, p both belong to V, this is a contradiction.

47.5.5. Summing up: Every y (between l and o) is the height of precisely one point of V. If y varies, then this point changes within the restrictions

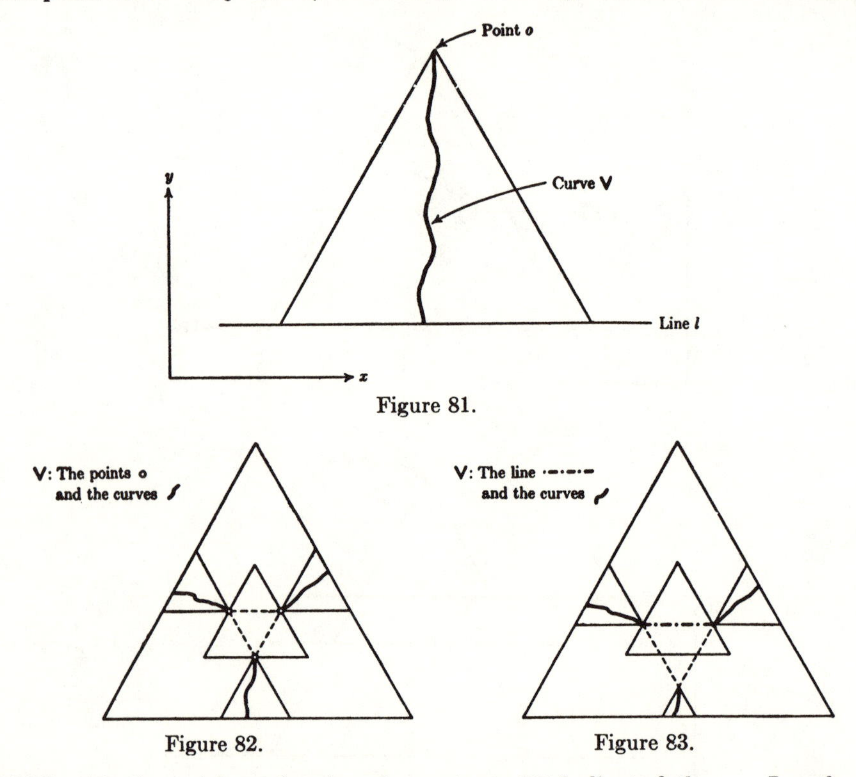

Figure 81.

Figure 82.

Figure 83.

of Fig. 78., i.e. without leaving the sextants ⁞⁞⁞ indicated there. In other words:

(47:6) V (in T) is a curve from o to l, the direction of which never deviates from the vertical by more than 30°[1] (cf. Figure 81).

Conversely, if any curve according to (47:6) is given then comparison of Figure 81 and Figure 77 makes it clear that the areas dominated by the points of V sweep out precisely the complement of V in T. So (47:6) is the exact determination of the part of V in T.[2]

We can now obtain the general solution V for $E(e_0)$ (i.e. for the fundamental triangle) by inserting curves according to Figure 81 into each triangle

[1] Hence it is continuous.

[2] It is equally true when T degenerates to a point, cf. footnote 1 on page 409.

$\blacktriangle$ of Figures 74 and 75. The results are shown on Figures 82 and 83, respectively.[1]

It will be observed that these figures show still marked similarity with those pertaining to the solutions of the essential three-person game in the old theory (cf. 32.2.3., shown in the inner triangle of Figure 73). The new element consists of the curves in the small triangles, all of which are situated in the fringe between the two major triangles of Figures 82 and 83. The width of this fringe, as shown in Figure 71, et sequ., is measured by e_0.[2] So when e_0 tends to zero, our new solutions tend to the old ones.

It is also worth pointing out that the variety of the solutions is much greater now than ever before: Entire curves can be chosen freely (within the limitations of (47:6) above). We will see later, that these curves motivate an interpretation which is of further significance. (Cf. 47.8.)

47.6. Case (V)

47.6.1. Case (V): $\frac{3}{2} \leqq e_0 < 3$. In this case $1 - \frac{2e_0}{3} \leqq 0 < 1 + \frac{e_0}{3}$ and $-2\left(1 - \frac{2e_0}{3}\right) < 1 + \frac{e_0}{3}$.[3] These inequalities express, as is easily verified, that the orientation of the inner triangle of Figure 71 is inverted, but that it is still situated entirely within the outer (fundamental) triangle, as indicated on Figure 84. The latter is again subdivided into seven areas, each of which can be characterized by stating which two element sets are effective in it in the sense of (47:5) in 47.2.3. The only difference between the present situation and that one in Case (IV) (i.e. Figure 71) is the behavior of area ①. The list is given below Figure 84.

Now we can draw the analogue of Figures 54 and 72, indicating for each point of the fundamental triangle the shaded areas[4] which it dominates. This is done in Figure 85, according to (47:5).

It is clear from Figure 85 that no point of the area ①[5] is dominated by any point.[6] Hence V must contain all of ①.

[1] The lower triangle of Figure 83 may degenerate to a point or even disappear altogether, cf. footnote 1 on p. 409.

[2] The sides of the outer (fundamental) triangle are given by $\alpha^i = -\left(1 + \frac{e_0}{3}\right)$, those of the inner triangle by $\alpha^i = -\left(1 - \frac{2e_0}{3}\right)$ (Cf. Fig. 71). The difference of $-\left(1 + \frac{e_0}{3}\right)$ and $-\left(1 - \frac{2e_0}{3}\right)$ is e_0.

[3] This latter inequality is equivalent to $e_0 < 3$.

[4] Excluding their boundaries.

[5] Including its boundary.

[6] I.e. by any $\overrightarrow{\alpha}$ in $E(e_0)$. It is easy to show that they are dominated by no $\overrightarrow{\alpha}$ at all—they are the detached imputations, by (45:D) in 45.2.4.

The points of the interior of the area ① dominate no other points either. I.e. they dominate no $\overrightarrow{\alpha}$ in $E(e_0)$. Again, it is easy to show that they dominate no $\overrightarrow{\alpha}$ at all—they are the fully detached imputations, cf. (45:C) in 45.2.4.

These statements can also be verified directly, by using the definitions of 45.2.

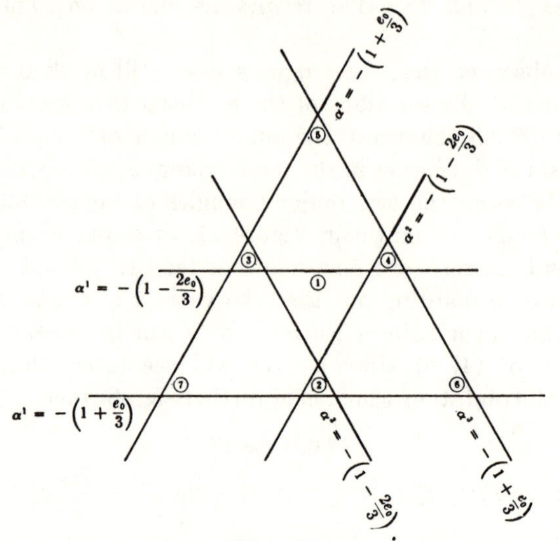

Figure 84.

Area:	Effective two-element sets S:		
①			
②	(1, 2),	(1, 3)	
③	(1, 2),		(2, 3)
④		(1, 3)	(2, 3)
⑤			(2, 3)
⑥		(1, 3)	
⑦	(1, 2)		

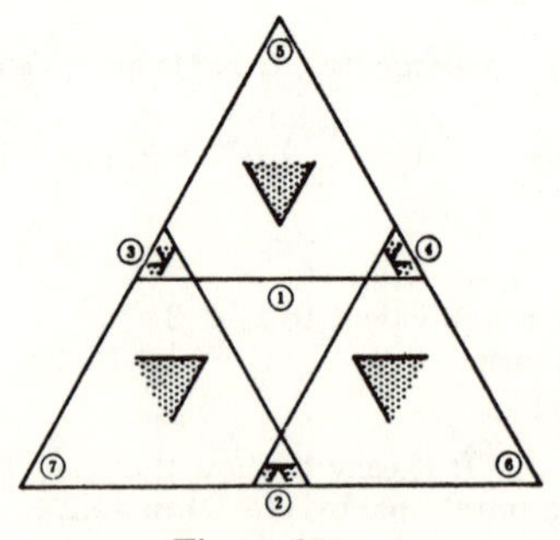

Figure 85.

47.6.2. Having found the part of V in ①, we proceed to determine the remainder of V. Since V is a solution, this remainder must lie in the area which is undominated by the already known part of V, i.e. by ①. Consideration of Figure 85 shows that this undominated area consists precisely of the three triangles ②, ③, ④.[1]

It is clear from Figure 85 that no point in any of the three triangles can be dominated by a point in another one. Hence the argument of 47.4.2. shows, that our requirement of V must be precisely this: (44:E:c) of 44.7.3. must hold for the part of V in each one of these triangles, taken for that triangle (in place of the entire fundamental triangle, i.e. $E(e_0)$).

The conditions in the triangles ②, ③, ④ are the same as those described in Figures 76, 77 for the triangle T. Hence the entire deduction of 47.5.1.-47.5.4. may be repeated literally, and the parts of V in ②, ③, ④ are curves as shown on Figure 81, characterized by (47:6) in 47.5.5.

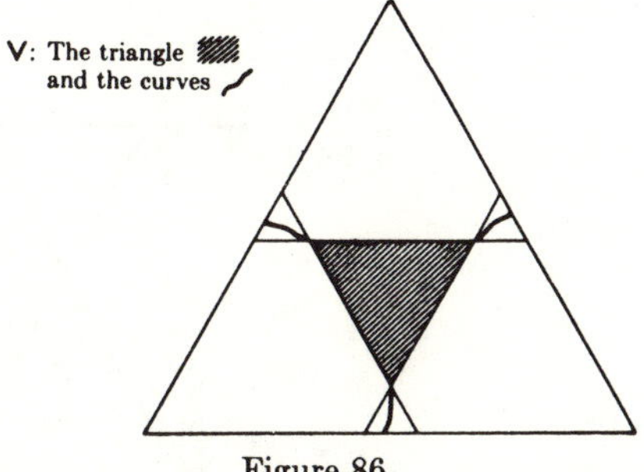

Figure 86.

We can now obtain the general solution V for $E(e_0)$ (i.e. for the fundamental triangle) by inserting such curves into ②, ③, ④ in Figure 85. The result is shown in Figure 86. For further remarks concerning these solutions cf. 47.8., 47.9.

47.7. Case (VI)

47.7. $e_0 \geqq 3$. In this case $1 - \frac{2e_0}{3} < 0 < 1 + \frac{e_0}{3}$ and $-2\left(1 - \frac{2e_0}{3}\right) \geqq 1 + \frac{e_0}{3}$.[2] These inequalities express, as is easily verified, that the inner triangle of Figure 84 has still the same orientation, but that it reaches the boundaries of the outer (fundamental) triangle, and possibly beyond,[3] as indicated on Figure 87. The only difference between the present situation and that one in Case (V) (i.e. Figure 84) is the disappearance of the areas ②, ③, ④. The list is given below Figure 87.

The analogue of Figures 54, 72 and 85 indicating the domination relations, is contained in Figure 88.

The argument of 47.6.1. can be repeated literally, proving that V contains all of ①. Consideration of Figure 88 shows that ① leaves no part of the fundamental triangle undominated.[4] Hence V is precisely ①. For further remarks concerning this solution, cf. 47.9.

[1] The remainder of the fundamental triangle is dominated by the boundary of ① which belongs to ①.

[2] This last inequality is equivalent to $e_0 \geqq 3$.

[3] When $e_0 > 3$.

[4] The remainder of the fundamental triangle is dominated by the boundary of ① which belongs to ①.

47.8. Interpretation of the Result: The Curves (One Dimensional Parts) in the Solution

47.8.1. The solutions obtained in the discussions of 47.2.-47.7. deserve a brief interpretative analysis. It is quite conspicuous that the repeated

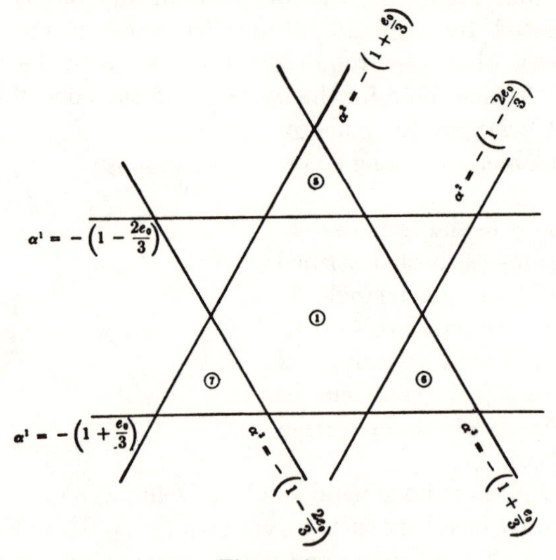

Figure 87.

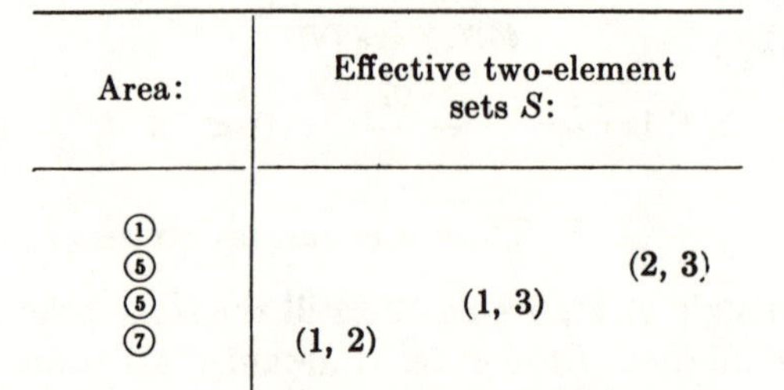

Area:	Effective two-element sets S:
①	
⑤	(2, 3)
⑤	(1, 3)
⑦	(1, 2)

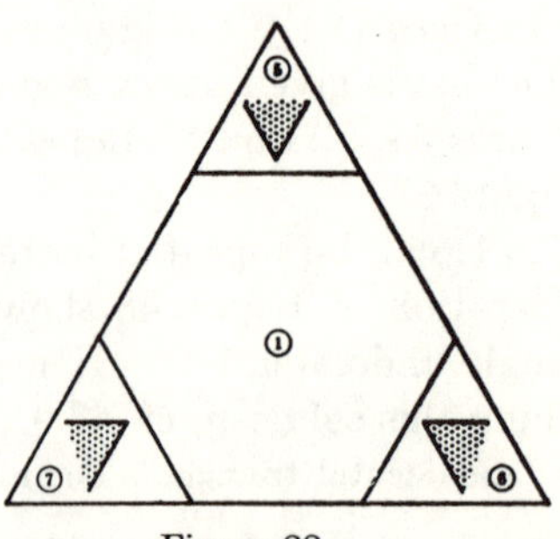

Figure 88.

appearance of a small number of qualitative features goes far in characterizing their structures—insofar as they deviate from the types familiar in the solutions of the essential three-person game of the old theory. These features are: The curves—arbitrary within the restriction (47:6) of 47.5.5.—which occur as soon as $e_0 > 0$ (and as long as $e_0 < 3$); and the two-dimensional areas, which appear when $e_0 > \frac{3}{2}$. We will now undertake their interpretation.

Consider first Case (IV): $0 < e_0 < \frac{3}{2}$ (in the "normal" zone). Let us consider those solutions of the present case which extend the non-discriminatory solution of the old theory (cf. 33.1.3. and (32:B) in 32.2.3.). Such a solution is pictured on Figure 82.

This figure shows the three points ° which form the analogue of a solution in the old theory. Taking, e.g. the lower point °, one verifies easily that there

$$\alpha^1 = -\left(1 - \frac{2e_0}{3}\right) = -1 + \frac{2e_0}{3}, \qquad \alpha^2 = \alpha^3 = \frac{1}{2}\left(1 - \frac{2e_0}{3}\right) = \frac{1}{2} - \frac{e_0}{3},$$

i.e.

$$\alpha_1 = -1 + e_0, \qquad \alpha_2 = \alpha_3 = \tfrac{1}{2}.$$

Thus these three points express an arrangement where two players have formed a coalition, obtained its total proceeds (amounting to 1), and divided them evenly—but the defeated player has not been reduced to his minimum value -1, because he retained beyond that the total available excess e_0.

Now the curves, starting from these points ° (in the fringe between the two triangles), express the situation where the total excess e_0 is not left in the indisputed possession of the defeated player. By claiming any part of the excess, the victorious coalition exacts more than the amount 1 which it can actually get in the game—i.e. it ceases to be effective. (Cf. the areas ②, ③, ④ in the Figures 71 and 72.) Therefore the conduct of affairs of this coalition—the distribution of the spoils within it—is no longer determined by the realities of the game—i.e. by the threats between the partners—but by the standard of behavior. This is expressed by the curve, which is part of the solution. The possible threats between the partners still restrict this curve to a certain extent (cf. (47:6) in 47.5.5.), but beyond that it is highly arbitrary. It must be re-emphasized that this arbitrariness is just an expression of the multiplicity of stable standards of behavior—but a definite standard of behavior, i.e. solution, means a definite curve, i.e. rule of conduct in this situation.

47.8.2. These considerations suggest the following tentative interpretation:

(47:A) In the presence of a positive excess it may happen that a coalition can obtain beyond its effective maximum also some fraction of the excess. This possibility is then due entirely to the standard of behavior and not to the physical possibilities of the game. The fraction of the excess thus obtained may vary from 0% to 100% and be left undetermined by the standard of behavior. The latter will prescribe, however, uniquely, how

> the fraction obtained is to be distributed between the members of the coalition. This rule of division will depend on which o. the many possible stable standards of behavior is chosen, and if the latter is varied, this rule will vary widely, although not quite unrestrictedly.

We have seen already, that undetermined curves according to (47:6) occur in many solutions, and they will occur again in the future. The above interpretation seems to fit them in every case.

The indefiniteness of the distribution of the excess between the victorious coalition and the defeated player (in a given solution) is an instance how certain social adjustments may be left open even within a specified social order. Our curves express the further nuance that while such an indefinite distribution is decided upon, some players can be tied to each other by definite conventions. (We will see further instances of this in the third remark of 67.2.3, 67.3.3. and in 62.6.2.)

47.9. Continuation: The Areas (Two-dimensional Parts) in the Solution

47.9.1. The interpretation (47:A) in 47.8.2. could be tested by applying it to the extension of the discriminatory solution of the old theory (cf. 33.1.3. and (32:A) in 32.2.3.) as pictured on Figure 83. This would bring up some instructive view-points, particularly with respect to the curve in the lower triangle of Fig. 83. However, we refrain from elaborating this case any further.

We turn, instead, to the Cases (V) and (VI), specifically when $e_0 > \frac{3}{2}$ (these are the "too large" excesses in the sense of 44.6.1., 45.2). These cases are characterized by the circumstance, that their solutions contain two-dimensional areas. Actually, two different situations may arise:

(a) Case (V), i.e. $\frac{3}{2} < e_0 < 3$. A solution V contains the two-dimensional area ①, but besides also curves as discussed in 47.8. (cf. Fig. 86).

(b) Case (VI), i.e. $e_0 \geqq 3$. The unique solution V is the two-dimensional area ①, and nothing else (cf. Fig. 88).

The emergence of two-dimensional areas within the solution indicates that the standard of behavior fails to contain rules of distribution at least within certain limits. In the Cases (a), (b) these limits are specified. In the case (a) the curves of 47.8. appear outside of those limits, i.e. the standard of behavior still sanctions certain coalitions—in the Case (b) this is no longer the case.

47.9.2. So we see that the "disorganizing" effect of a "too large" excess—i.e. gift from an outside source (cf. 44.6.1.)—manifests itself in two successive stages: In the Case (a) it is present in a certain central area, but does not exclude certain conventional coalitions. In the Case (b) the standard of behavior no longer allows coalitions but it sets certain limiting principles for the distribution.

We have seen that these successive stages of disorganization are reached at $e_0 = \frac{3}{2}$, 3 respectively.[1]

These considerations seem to be quite instructive in a qualitative way for the possibilities of standards of behavior and organizations. It appears likely that they will provide useful guidance in the further development of the theory. But the reader must be cautioned against drawing far reaching conclusions from the quantitative results: They apply to the three-person game with an excess,[2] which is thus shown to be the simplest model for their operation. But it must have become amply clear by now that an increase in the number of participants will affect conditions fundamentally.

[1] Note that $|\Gamma|_2 = \frac{3}{2}$.

[2] Hence also to a decomposition six-person game in the old theory, cf. 46.12.

CHAPTER X

SIMPLE GAMES

48. Winning and Losing Coalitions and Games Where They Occur

48.1. The Second Type of 41.1. Decision by Coalitions

48.1.1. The program formulated in 34.1. provided for far-reaching generalizations of the games corresponding to the 8 corners of the cube Q, introduced in 34.2.2. The corner VIII (also representative of II, III, IV) was taken up in 35.2.1. and provided the source for a generalization, leading to the theory of composition and decomposition to which all of Chapter IX was devoted. We now pass to the corner I (also representative of V, VI, VII), which we will treat in a similar fashion.

By generalizing the principle, of which a special instance can be discerned in this game, we will arrive at an extensive class of games, to be called *simple.* It will be seen that a study of this class yields a body of information which is of value for a deeper understanding of the general theory in the sense of 34.1.

48.1.2. Consider the corner I of Q, discussed in 35.1. As was brought out in 35.1.1., this game has the following conspicuous feature: The aim of the players is to form certain coalitions consisting either of player 4 and one ally, or of all three other players together. Any one of these coalitions is winning in the full sense of the word. Any coalition which falls short of these is completely defeated. I.e. the quantitative element, the payments expressed by the characteristic function, can be treated as something secondary—the primary aim in this game is to succeed in forming certain decisive coalitions.

This description suggests strongly that the number four of players and the particular scheme of decisive coalitions are special and accidental and that a more general principle can be extracted from this particular arrangement.

48.1.3. In carrying out this generalization, the following observation is useful. In our above example, the decisive coalitions—the attainment of which is the sole aim of the players—were these:

$$(1,4),\ (2,4),\ (3,4),\ (1,2,3). \tag{48:1}$$

Now it is convenient to view not only these as winning coalitions, but also all their (proper) supersets:

$$(1,2,4),\ (1,3,4),\ (2,3,4),\ (1,2,3,4). \tag{48:2}$$

The point is that although the coalitions (48:2) contain participants whose presence is not necessary in order to win, the coalition is nevertheless a winning one—i.e. the opponents are defeated.[1] These opponents form those coalitions which are the complements of the sets in (48:1), (48:2), i.e. the sets

(48:3)
$$\begin{matrix} (2,3), & (1,3), & (1,2), & (4). \\ (3), & (2), & (1), & \ominus. \end{matrix}$$

Thus (48:1), (48:2), contain the winning coalitions, and (48:3) contains the defeated ones.

It is easily verified that every subset of $I = (1,2,3,4)$ belongs to precisely one of these two classes: (48:1), (48:2), or (48:3).[2]

48.2. Winning and Losing Coalitions

48.2.1. Let us now consider a set of n players: $I = (1, \cdots, n)$. The scheme of 48.1.3. generalizes to subdividing the system of all subsets of I into two classes W and L, such that the subsets of W will represent the winning coalitions and the subsets of L will represent the losing ones. The analogues of the properties formed in 48.1.3. can be formulated as follows:

Denote the system of all subsets of I by $\bar{I}$.[3] The mapping of every subset S of I on its complement (in I):

(48:4)
$$S \to -S$$

is clearly a one-to-one mapping of $\bar{I}$ on itself. Now we have:

(48:A:a) Every coalition is either winning or defeated and not both—i.e. W and L are complementary sets in $\bar{I}$.

(48:A:b) Complementation (in I) carries a winning coalition into a losing one and *vice versa*—i.e. the mapping (48:4) maps W and L on each other.

(48:A:c) A coalition is winning, if part of it is winning—i.e. W contains all supersets of its elements.

(48:A:d) A coalition is losing, if it is part of one which is losing—i.e. L contains all subsets of its elements.

48.2.2. Before we discuss the concepts of winning and losing in their relationship to the game, we may analyze the structure of conditions (48:A:a)-(48:A:d) somewhat further.

The first conspicuous fact is that, although we need both classes W and L to interpret the game, these classes determine each other. Indeed they do this in two ways: Given one of W or L (48:A:a) as well as (48:A:b) can be used to construct the other. I.e. starting from one of these sets, the other one is obtained in this way:

According to (48:A:a): Take the given set as a whole and form its complement (in $\bar{I}$).

[1] I.e. the complements are flat in the sense of 31.1.4. Cf. the discussion in 35.1.1.

[2] (1,2,3,4) has $2^4 = 16$ subsets, of these 8 are in (48:1), (48:2), and the remaining 8 in (48:3).

[3] As I has n elements, $\bar{I}$ has 2^n elements.

According to (48:A:b): Take each element of the given set separately, and replace it by its complement (in I).[1]

It should be noted also that if the given set, W or L, possesses the property (48:A:c) or (48:A:d) respectively, then the other set—obtained from the former by (48:A:a) or by (48:A:b) will possess the other property (48:A:c) or (48:A:d).[2]

It follows from the above, that we can base the entire structure now under consideration on either one of the two sets W and L. We must only require that both transformation (48:A:a) and (48:A:b) lead from it to the same set (which is then the other one of W and L) and that it must satisfy the pertinent one of the two conditions (48:A:c) and (48:A:d) (the other condition of (48:A:c) and (48:A:d) is then automatically taken care of, according to what we have seen).

Thus we have only two conditions for W or L: First the equivalence of (48:A:a) and (48:A:b) and second (48:A:c) or (48:A:d).

The former condition means this: The non-elements of the set coincide with the complements (in I) of the elements of the set. In other words: Of two complements (in I) S, $-S$, one and only one belongs to the set.

Summing up:

The sets $W(\subseteq \bar{I})$ are characterized by these properties:

(48:W)

(48:W:a) Of two complements (in I) S, $-S$, one and only one belongs to W.

(48:W:b) W contains all supersets of its elements.

The sets $L(\subseteq \bar{I})$ are characterized by these properties:

(48:L)

(48:L:a) Of two complements (in I) S, $-S$, one and only one belongs to L.

(48:L:b) L contains all subsets of its elements.

[1] The reader will note the remarkable structure of this condition: The given set must produce the same result, irrespectively of whether complementation is applied to it as a unit, or to its elements separately.

[2] This is actually true for (48:A:a) as well as for (48:A:b), and independent of the question whether (48:A:a) and (48:A:b) produce the same set. Precisely:

(48:B) Let a set M possess the property (48:A:c) [(48:A:d)], then both sets which are obtained from it by (48:A:a) and (48:A:b)—we do not assume that they are identical—possess the other property (48:A:d) [(48:A:c)].

Proof: We must show that both transformations (48:A:a) and (48:A:b) carry (48:A:c) into (48:A:d) and *vice versa*.

Clearly (48:A:c) is equivalent to this:

(48:A:c*) If S is in M, and T is not, then $S \subseteq T$ is excluded.

Again (48:A:d) is equivalent to this:

(48:A:d*) If S is not in M, and T is, then $S \subseteq T$ is excluded.

Now the transformation (48:A:a) interchanges "being in M" and "not being in M" with each other. Hence it interchanges (48:A:c*) and (48:A:d*). The transformation (48:A:b) interchanges $\subseteq$ and $\supseteq$ (this is brought about by individual complementation for the elements S, T; besides the symbols S, T must be interchanged). Hence it, too, interchanges (48:A:c*) and (48:A:d*).

We restate:

If W [L] fulfills (48:W) [(48:L)], then (48:A:a) and (48:A:b) yield the same set L [W]. W and L fulfill (48:A:a)-(48:A:d) and L [W] fulfills (48:L) [(48:W)]. Conversely, if W, L fulfill (48:A:a)-(48:A:d), then they fulfill separately (48:W), (48:L).

49. Characterization of the Simple Games

49.1. General Concepts of Winning and Losing Coalitions

49.1.1. We now pass to the consideration of the connection between winning and losing coalitions in the game itself.

Assume, therefore, that an n-person game Γ is given. In all the considerations which follow, it is advantageous to restrict ourselves to the old theory in the sense of 30.1.1. or 42.4.1. Consequently, as pointed out in 42.5.3., we may assume Γ to be zero or constant-sum as we desire. For the present we prefer to choose Γ as a zero-sum game.

Beyond this Γ is not restricted and in particular no normalization is assumed.

49.1.2. Let us first analyze the concept of a losing coalition. Repeating essentially what was said in 35.1.1., we may argue as follows:[1] The player i, when left to himself, obtains the amount $\mathrm{v}((i))$. This is manifestly the worst thing that can ever happen to him, since he can protect himself against further losses without anyone else's help. Thus we may consider the player i when he gets this amount $\mathrm{v}((i))$ to be completely defeated. A coalition S may be considered as defeated, if it gets the amount $\sum_{i \text{ in } S} \mathrm{v}((i))$, since then each player i in it must necessarily get $\mathrm{v}((i))$.[2] Thus the criterion of defeat is

$$\mathrm{v}(S) = \sum_{i \text{ in } S} \mathrm{v}((i)).$$

In the terminology of 31.1.4., this means that the coalition S is flat. (Cf. also footnote 3 on page 296.)

We have obtained a satisfactory definition of the system L_Γ[3] of all losing (defeated) coalitions:

(49:L) $\quad L_\Gamma$ is the set of all flat sets $S(\subseteq I)$.

It is now easy to say what a winning coalition is. It is plausibly one, the opponents of which are losing, i.e. the system W_Γ[3] of all winning coalitions is this:

[1] The difference is that our present Γ is more general.

[2] Since no player i need ever accept less than $\mathrm{v}((i))$, and those in the coalition S have together $\sum_{i \text{ in } S} \mathrm{v}((i))$, this is the only way in which they can split.

[3] In order to avoid confusion, we will use the symbols W_Γ, L_Γ instead of the W, L of 48.2.2. The difference between this and the former is that 48.2.2. is a postulational discussion of the properties which appeared desirable for the concepts of "winning" and "losing" (described by W, L)—while we are now analyzing definite sets obtained from a specific game Γ.

The two viewpoints will be merged in (49:E) of 49.3.3.

(49:W) W_Γ is the set of all sets $S(\subseteq I)$ for which $-S$ is flat.

It should be conceptually clear, and is also immediately verified with the help of 27.1.1.-2., that the sets W_Γ, L_Γ are invariant under strategic equivalence.

49.1.3. We cannot expect the above W_Γ, L_Γ to fulfill the conditions (48:A:a)-(48:A:d) (for W, L) of 48.2.1. The game in its present generality need not be of the simple type referred to, where the only aim of all players is to form certain decisive coalitions and there are no other motives which require a quantitative description.[1] It will therefore be necessary to restrict in order to express the property we have in mind. Indeed, the precise formulation of this restriction is our immediate objective.

Nevertheless, we begin by determining how much of (48:A:a)-(48:A:d) holds true for the Γ in its present generality. We give the answer in several stages.

(49:A) W_Γ, L_Γ always fulfill (48:A:b)-(48:A:d)

Proof: Ad (48:A:b): Immediate by comparing (49:L) and (49:W) in 49.1.2.[2]

Ad (48:A:c), (48:A:d): Since we have (48:A:b), we can apply (48:B) in 48.2.2.[3] and therefore (48:A:c) and (48:A:d) imply each other.

But (48:A:d) coincides with (31:D:c) in 31.1.4., considering (49:L).

Thus the main difference between our present W_Γ, L_Γ and the setup of 48.2. lies in (48:A:a)—i.e. in the question whether or not W_Γ and L_Γ are complements. We can decompose this assertion into two parts:

(49:1)

(49:1:a) $$W_\Gamma \cap L_\Gamma = \ominus,^4$$

(49:1:b) $$W_\Gamma \cup L_\Gamma = \bar{I}.^5$$

(49:1:a) leads back to familiar concepts:

(49:B)

(49:B:a) (49:1:a) holds if and only if Γ is essential.

(49:B:b) If Γ is inessential, then $W_\Gamma = L_\Gamma = \bar{I}$.[6]

[1] Our discussion of the four-person game has provided many illustrations of such motives, for which the end of 36.1.2. provides a good instance. This situation is, indeed, the usual (general) one—the class of games at which we are aiming now, is in a certain sense an extreme case, cf. the concluding observation of 49.3.3.

[2] Actually the concept of "winning" was based on the concept of "losing" by just this operation of complementation.

[3] It appears now why we separated (48:A:a) from (48:A:b) in 48.2.1: We have now (48:A:b), but not (48:A:a).

[4] It may seem odd that this—no coalition can at the same time be both winning and losing—must be stated separately. The meaning of this condition will appear in (49:B) and footnote 6.

[5] This states that every coalition—i.e. every subset of I—is definitely winning or losing. This is, of course, the idea on the basis of which we wish to specialize Γ.

[6] Thus a coalition can at the same time be both winning and losing, when the game is inessential—manifestly because in this case both states are irrelevant.

Proof: Ad (49:B:a): The negation of (49:1:a) is the existence of an S such that both S and $-S$ are flat. This amounts to inessentiality by (31:E:b) in 31.1.4.

Ad (49:B:b): $W_\Gamma = L_\Gamma = \bar{I}$ means that every S in $\bar{I}$ is flat. This amounts to inessentiality by (31:E:c) in 31.1.4.

Before passing to (49:1:b) we note that W_Γ, L_Γ possess one property which did not occur in (48:A:a)-(48:A:d).

(49:C) L_Γ contains the empty set and all one-element sets.[1]

Proof: This coincides with (31:D:a), (31:D:b) in 31.1.4.

(49:C) is really a new condition, i.e. it is not a consequence of (48:A:a)-(48:A:d); we will verify this in 49.2. below. Thus our plausible discussion of 48.2. overlooked a necessary feature of the W_Γ, L_Γ. We must, therefore, make sure that the present conditions contain everything. I.e. that the conditions (48:A:b)-(48:A:d) and (49:C), together with the results of (49:B) on inessentiality, characterize the W_Γ, L_Γ completely. This will be shown in 49.3. below.

49.2. The Special Role of One-element Sets

49.2.1. We begin with the example announced above: Two systems W, L which fulfill (48:A:a)-(48:A:d),[2] but not (49:C). Actually, we can determine all such pairs.

(49:D) W, L fulfill (48:A:a)-(48:A:d), but not (49:C), if and only if they have the following form: W is the set of all S containing i_0, L is the set of all S not containing i_0, where i_0 is an arbitrary but definite player.

Proof: Sufficiency: It is immediately verified that the W, L formed as indicated fulfill (48:A:a)-(48:A:d). (49:C) is violated, since the one element set (i_0) belongs to W, and not to L.

Necessity: Assume that W, L fulfill (48:A:a)-(48:A:d), but not (49:C). Let (i_0) be a one-element set, which does not belong to L.[3] Then (i_0) belongs to W.

Every S containing i_0 has $S \supseteq (i_0)$, hence it belongs to W by (48:A:c). If S does not contain i_0, then $-S$ contains it; hence $-S$ belongs to W by the above and S belongs to L by (48:A:b).

Finally W, L are disjunct by (48:A:a), hence W is precisely the set of the S containing i_0 and L is precisely the set of the S not containing i_0.

49.2.2. It may be worth while to comment briefly upon this result.

[1] It is clearly in the spirit of our entire analysis of games that a coalition of one player is to be considered as defeated—as this player has not succeeded in finding partners for a coalition.

[2] We mentioned originally (48:A:b)—(48:A:d) only, but the above strengthening requires no extra effort.

[3] If the empty set does not belong to L, then no set can belong to L owing to (48:A:d); hence any (i_0) will do.

The W, L formed in (49:D) cannot be the W_Γ, L_Γ of any game since they violate (49:C). This may seem odd, since (49:D) appears to convey a very clear idea of the kind of "winning" and "losing" described by its W, L. Indeed, they describe the situation where a coalition wins if the player i_0 belongs to it, and loses if he does not. Why can no game be constructed to fit this specification?

The reason is that under the conditions described, "winning" would not be a matter of forming coalitions at all:[1] The player i_0 is "victorious" without anybody else's help. Still worse, in our terminology this position of i_0 is no victory—it is not the result of any strategic operation,[2] but a fixed state given him by the rules of the game.[3] A game in which coalitions involve no advantage is inessential,[4]—even if one player i_0 should have a considerable fixed advantage in it.

The reader will understand, of course, that all this is just an additional comment on results which were already rigorously established above (in (49:C), (49:D)).

49.3. Characterization of the Systems W, L of Actual Games

49.3.1. We now turn to the second subject mentioned at the end of 49.1.3. Let two systems W, $L(\subseteq \bar{I})$ be given, which fulfill the conditions (48:A:b)-(48:A:d) and (49:C), and also (49:1:a).[5] We wish to construct an essential game Γ with $W_\Gamma = W$, $L_\Gamma = L$. In doing this, we normalize Γ with $\gamma = 1$.

The sets S in L_Γ are characterized by their flatness, i.e. by $\mathrm{v}(S) = -p$, where p is the number of elements of S.[6] The sets S in W_Γ are characterized by the fact that $-S$ belongs to L_Γ, i.e. by $\mathrm{v}(-S) = -(n - p)$, owing to the above. Now $\mathrm{v}(-S) = -\mathrm{v}(S)$, hence we may write for this $\mathrm{v}(S) = n - p$.

Hence we have shown:

The desired relations $W_\Gamma = W$, $L_\Gamma = L$ are equivalent to this:

(49:2) For a q-element set S, $(q = 0, 1, \cdots, n-1, n)$

(49:2:a) $$\mathrm{v}(S) = n - q$$

if and only if S belongs to W, and

(49:2:b) $$\mathrm{v}(S) = -q$$

if and only if S belongs to L.

[1] The equivalent consideration was carried out in a special case in 35.1.4.

[2] We always consider this to be the same thing as forming appropriate coalitions.

[3] Cf. our treatment of the basic values a', b', c' in the three-person game, in 22.3.4. The entire discussion of strategic equivalence, cf. 27.1.1., was made in the same spirit: advantages like this one can be removed by strategically equivalent transformations, while those which are really due to forming coalitions, cannot.

[4] Hence its W_Γ, L_Γ are not the desired ones, described in (49:D), but those of (49:B:b).

[5] We require (49:1:a) because we aim primarily at essential games (cf. (49:B)). Subsequently we will make our discussion exhaustive—as will be seen in (49:E).

[6] Recall that all $\mathrm{v}((i)) = -\gamma = -1$.

Thus our task is to construct a game Γ (normalized and $\gamma = 1$) with a characteristic function $\mathrm{v}(S)$ which fulfills (49:2).

49.3.2. (49:2) determines $\mathrm{v}(S)$ for the S of W and L, so we need only define it for those S which belong to neither set. We try there the value 0. Accordingly we define:

$$\mathrm{v}(S) = \begin{cases} \left.\begin{matrix} n - q \text{ for } S \text{ in } W \\ -q \text{ for } S \text{ in } L \end{matrix}\right\} S \text{ a } q\text{-element set with } q = 0, 1, \cdots, n-1, n. \\ 0 \text{ otherwise}^1 \end{cases}$$

We first prove that $\mathrm{v}(S)$ is a characteristic function, i.e. that it fulfills (25:3:a)-(25:3:c) in 25.3.1. We prove these conditions in their equivalent form of (25:A) in 25.4.2.:

Case $p = 1$ with $=$: This is $\mathrm{v}(I) = 0$, immediate since I is in W, because $\ominus = -I$ is in L by (48:A:b), (49:C).

Case $p = 2$ with $=$: This is $\mathrm{v}(S_1) + \mathrm{v}(S_2) = 0$ when S_1, S_2 are complements. If both S_1, S_2 are not in W, L, then $\mathrm{v}(S_1) = \mathrm{v}(S_2) = 0$. If one of S_1, S_2 is in W or L, then the other is in L or W, respectively, by (48:A:b). Assume, by symmetry, S_1 in L, S_2 in W. Let S_1 have q elements, hence S_2 has $n - q$. Then $\mathrm{v}(S_1) = -q$, $\mathrm{v}(S_2) = q$.

So at any rate $\mathrm{v}(S_1) + \mathrm{v}(S_2) = 0$.

Case $p = 3$ with $\leqq$: This is $\mathrm{v}(S_1) + \mathrm{v}(S_2) + \mathrm{v}(S_3) \leqq 0$ when S_1, S_2, S_3 are pairwise disjunct with the sum I. If none of S_1, S_2, S_3 is in W, then $\mathrm{v}(S_1)$, $\mathrm{v}(S_2)$, $\mathrm{v}(S_3) \leqq 0$.[2] If one of S_1, S_2, S_3 is in W, we may assume by symmetry, that it is S_3. Hence $-S_3 = S_1 \cup S_2$ is in L by (48:A:b), and so S_1, S_2 are in L by (48:A:d). Let S_1 have q_1 elements, S_2 have q_2 elements, hence S_3 has $n - q_1 - q_2$. Then $\mathrm{v}(S_1) = -q_1$, $\mathrm{v}(S_2) = -q_2$, $\mathrm{v}(S_3) = q_1 + q_2$.

So at any rate $\mathrm{v}(S_1) + \mathrm{v}(S_2) + \mathrm{v}(S_3) \leqq 0$.

49.3.3. Thus $\mathrm{v}(S)$ belongs to a game Γ. We now establish the remaining assertions.

$\mathrm{v}(S)$ (i.e. Γ) is normalized and $\gamma = 1$: Indeed, by (49:C) all $\mathrm{v}((i)) = -1$.

$\mathrm{v}(S)$ fulfills (49:2): Owing to (48:A:b) and $\mathrm{v}(-S) = -\mathrm{v}(S)$, the two parts of (49:2) go over into each other if we interchange S and $-S$. We consider therefore only the second half.

If S is in L, then clearly $\mathrm{v}(S) = -q$. If S is not in L, then $\mathrm{v}(S) = -q$ would necessitate $0 = -q$,[3] or $q = 0$. But this means that S is empty, contradicting (49:C).

So the game Γ possesses all desired properties.

We are now able to prove the following exhaustive statement:

(49:E) In order that two given systems W, $L(\subseteq \bar{I})$ be the W_Γ, L_Γ of a suitable game Γ, these requirements are necessary and sufficient:

Γ inessential: $W = L = \bar{I}$.

Γ essential: (48:A:b)-(48:A:d), (49:C), (49:1:a).

[1] That the two first specifications do not conflict, is due to (49:1:a).

[2] Clearly $\mathrm{v}(S) \leqq 0$ if S is not in W.

[3] As $n - q \neq -q$, S could not be in W; hence $\mathrm{v}(S) = 0$.

Proof: Γ inessential: Immediate by (49:B:b).

Γ essential: The necessity was established in (49:A), (49:B:a), (49:C). The sufficiency is the content of the construction which we have carried out.

In concluding we mention another interpretation of (49:2). Recalling the inequalities (27:7) of 27.2. (also shown on Figure 50), which specify limitations for $v(S)$, it appears that W_Γ is the set of those S for which $v(S)$ reaches the upper limiting value, and L_Γ the set of those S for which $v(S)$ reaches the lower limiting value.

49.4. Exact Definition of Simplicity

49.4. (49:E) permits us to give a rigorous definition of that class of games to which we alluded in 48.1.2. and 48.2.1., and which was circumscribed in more detail at the beginning of 49.1.3.: Where the only aim of all players is to form certain decisive coalitions and where there are no other involved motives which require a quantitative description.

By combining the part of (49:E) which refers to essential games with (49:1), it appears that the formal expression of this idea is

(49:1:b) $$W_\Gamma \cup L_\Gamma = \bar{I}.$$

Indeed, this condition expresses that any given coalition S belongs either to the winning or to the losing category—without any further qualification.

We define accordingly: An essential game which fulfills (49:1:b) is called *simple.*

The concept of simplicity is invariant under strategic equivalence, since the sets W_Γ, L_Γ are.

49.5. Some Elementary Properties of Simplicity

49.5.1. Before we take up the detailed mathematical discussion of this concept, let us consider once more the closing remark of 49.3. In the sense of that remark an essential game is simple, if $v(S)$ lies for every S on the boundary[1] of the area assigned to it by the inequalities (27:7) in 27.2.

The variety of all essential n-person games (normalized, $\gamma = 1$) can be viewed as a geometrical configuration of a certain number of dimensions, given in Figure 65. More precisely the inequalities referred to define a convex polyhedric domain Q_n in the linear space of the dimensionality in question, and the points of this domain represent all these games.[2]

49.5.2. E.g.: For $n = 3$ the dimensionality is zero, and the domain Q_3 a single point. For $n = 4$ the dimensionality is 3, and the domain Q_4 the cube Q of 34.2.2.

Now the simple games are those for which we are on the boundary of each defining inequality. With respect to the convex polyhedric domain Q_n this means: The simple games are the vertices of Q_n, $n = 3, 4$.

[1] The boundary consists of two points: the upper limiting value $n - p$ and the lower one $-p$, $(\gamma = 1)$. $v(S)$ must be one of these two, no matter which.

[2] The reader who is familiar with n-dimensional linear geometry, will note: Since Q_n is defined by linear inequalities, it is a polyhedron. The discussion of 27.6. allows to conclude that it is convex.

E.g.: For $n = 3$ Q_3 is a single point, i.e. nothing but a vertex, so the essential 3-person game is simple.[1] For $n = 4$ Q_4 is the cube Q, so the simple games are the vertices, i.e. corners *I–VIII*.[2]

49.6. Simple Games and Their W, L. The Minimal Winning Coalitions: W^m

49.6.1. Combining (49:E) with the definition of simplicity, we obtain:

(49:F) In order that two given systems W, $L(\subseteq \bar{I})$ be the W_Γ, L_Γ of a suitable simple game Γ, these requirements are necessary and sufficient: (48:A:a)-(48:A:d), (49:C).

That the S referred to in (49:2) exhaust all subsets of I, is definitory for simplicity. Consequently it is for simple games and for these alone, that knowledge of W_Γ, L_Γ determines $v(S)$, provided that the game is normalized and $\gamma = 1$. I.e., without the last proviso, that it determines the game up to a strategic equivalence.

We restate this:

(49:G) In case of simplicity, and only then, the game Γ is determined by its W_Γ, L_Γ up to a strategic equivalence.

Consequently, according to (49:F) and (49:G) a theory of simple games is coexistensional with the theory of those pairs of systems W, L which fulfill (48:A:a)-(48:A:d), (49:C).

49.6.2. In studying the pairs W, L described above, 48.2.2. should be recalled, and particularly (48:W), (48:L) there and (49:2). According to these, it is sufficient to name either W or L in order to determine the pair W, L.

The conditions (48:A:a)-(48:A:d) are then to be replaced as follows: If W is used, by (48:W); if L is used, by (48:L).

As to (49:C), it refers to L directly. We can equally well refer it to W, by applying (48:A:b)—then the sets mentioned in it must be replaced by their complements.

For the sake of completeness, we restate (48:W) and (48:L), together with the corresponding forms of (49:C).

The sets $W(\subseteq \bar{I})$ are characterized by these properties:

(49:W*)

(49:W*:a) Of two complements (in I) S, $-S$, one and only one belongs to W.

(49:W*:b) W contains the supersets of its elements.

(49:W*:c) W contains I and all $(n - 1)$-element sets.

[1] Cf. also (50:A) in 50.1.1.

[2] As far as the corners *I*, *V*, *VI*, *VII* are concerned, this is no surprise: Our discussion started with these in 48.1 and our concept of simplicity obtained from them by generalization.

The reappearance of the corners *II*, *III*, *IV*, *VIII* is more puzzling: We treated them in 35.2. as the prototypes of decomposability. However, they are simple too, as follows easily from (50:A) and the beginning of 51.6.

The sets L ($\subseteq \bar{I}$) are characterized by these properties:

(49:L*)

(49:L*:a) Of two complements (in I) S, $-S$, one and only one belongs to L.

(49:L*:b) L contains the subsets of its elements.

(49:L*:c) L contains the empty set and all one-element sets.

As pointed out above, we could base the theory on either W with (49:W*) or on L with (49:L*).

49.6.3. Since it is more in keeping with the usual way of thinking about these matters to specify the winning rather than the losing coalitions, we shall use the first mentioned procedure.

In this connection we observe that a certain subset of W shares the importance of W. This is the set of those elements S of W of which no proper subset belongs to W. We call these S the *minimal* elements of W (i.e. W_Γ) and their set W^m (i.e. W^m_Γ).

The intuitive meaning of this concept is clear: These minimal winning coalitions are the really decisive ones, those winning coalitions in which no participant can be spared. (It will be remembered that our discussion of 48.1.3. began with the enumeration of these coalitions for the game we were then considering.)

49.7. The Solutions of Simple Games

49.7.1. The heuristic considerations which led us to the concept of simple games make it plausible, that the discussion of games belonging to this category, may turn out easier than that of (zero-sum) n-person games in general. For a corroboration of this we must examine how solutions are determined in a simple game. Since we are now considering the old form of the theory, 30.1.1. must be consulted.[1] We begin with the observation that a considerable simplification must be expected from the fact that in a simple game every set is either certainly necessary or certainly unnecessary (cf. 31.1.2.).

49.7.2. In order to establish this assertion we prove first:

(49:H) In any essential game Γ all sets S of W_Γ are certainly necessary, and all sets S of L_Γ are certainly unnecessary.

Proof: If S is in L_Γ then it is flat, hence certainly unnecessary by (31:F) in 31.1.5. If S is in W_Γ, then $-S$ is flat (because it is in L_Γ) and $S \neq \ominus$ (because $\ominus$ is in L_Γ, hence not in W_Γ). So S is certainly necessary by (31:G) in 31.1.5.

We can now fulfill our above promise concerning simple games—indeed, this can be done in two different ways.

[1] In the terminology of the new form of the theory—as introduced in 44.7.2. et sequ.—this means: We are looking for solutions for $E(0)$, i.e. the excess is being restricted to the value 0.

The significance of this restriction will become clearer in the third remark of 51.6.

(49:I) In any simple game Γ all sets S of W_Γ are certainly necessary, and all others are certainly unnecessary.

Proof: Combine (49:H) with the fact, that for a simple game L_Γ is precisely the complement of W_Γ.

(49:J) In any simple game Γ all sets S of W^m_Γ are certainly necessary, and all others are certainly unnecessary.[1]

Proof: We can replace the W_Γ of (49:I) by its subset W^m_Γ, i.e. we can transfer all S of $W_\Gamma - W^m_\Gamma$ from the certainly necessary class into the certainly unnecessary, owing to (31:C) in 31.1.3. Indeed, every S in W_Γ possesses a subset T in W^m_Γ.

Of these two criteria (49:I) and (49:J), the latter is more useful. Their importance will be established by actual determination of solutions in simple games.[2] Indeed, this analysis of simple games permits the deepest penetration yet effected into the theory of games with many participants.[3]

50. The Majority Games and the Main Solution

50.1. Examples of Simple Games: The Majority Games

50.1.1. Before going any further, it is appropriate to give some examples of simple games, i.e. of the pairs W, L of (49:F) in 49.6.1. We know from 49.6.2. that it suffices to discuss the W as characterized by (49:W*) there.

Let us therefore consider some possible ways of introducing such W—i.e. possible definitions of a concept of winning.

The principle of majority suggests itself as a particularly suitable definition of winning. Hence it is plausible to define W as the system of all those S which contain a majority of all players. It will be noticed, however, that we must exclude ties—indeed (49:W*:a) states for this W, that for every S either S or $-S$ must contain the majority of all players, thus excluding that both may contain exactly half. In other words: The total number of participants must be odd.

So if n is odd, we may define W as the set of all S with $> \frac{n}{2}$ elements.[4] The simple game which obtains in this way,[5] will be called the *direct majority game*.

[1] Comparison of (49:I) and (49:J) shows that the S of $W_\Gamma - W^m_\Gamma$ are simultaneously certainly necessary and certainly unnecessary. This is another illustration for the remark at the end of footnote 1 on p. 274.

[2] Cf. 50.5.2. and 55.2.

[3] Cf. 55.2.-55.11., and in particular the general remarks of 54.

[4] Since the smallest integer $> \frac{n}{2}$ is $\frac{n+1}{2}$ (n odd!), we may also say: S must have $\geqq \frac{n+1}{2}$ elements.

[5] Precisely: The class of strategically equivalent ones (of n participants).

The smallest n for which this can be done,[1] is 3. We know that there exists only one essential three-person game, and that for this W it consists precisely of the 2 and 3-element sets—i.e. of the sets with $> \frac{3}{2}$ elements. So we see:

(50:A) The (unique) essential three-person game is simple; it is the direct majority game of three participants.

For the subsequent n which are eligible, $n = 5, 7, \cdots$, the direct majority game is merely one possibility among many.

50.1.2. The direct majority game is only available, when n is odd, and yet simple games exist for even n as well—indeed our prototype of simple games (cf. 48.1.2., 48.1.3.) had $n = 4$.

However, the concept of majority is easily extended to cover the case of even n as well. To this end we introduce *weighted majorities* in the following manner: Let each one of the players $1, \cdots, n$ be given a *numerical weight*, say $w_1, \cdots, w_n$ respectively. Define W as the set of all those S which contain a majority of total weight. This means:

(50:1) $$\sum_{i \text{ in } S} w_i > \frac{1}{2} \sum_{i=1}^{n} w_i,$$

or equivalently,

(50:2) $$\sum_{i \text{ in } S} w_i > \sum_{i \text{ in } -S} w_i.$$

We must again take care to exclude ties. However, owing to the greater generality of our present setup, it is better to proceed immediately to a complete discussion of (49:W*).

50.1.3. Let us see, therefore, what restriction (49:W*) imposes upon the $w_1, \cdots, w_n$.

Ad (49:W*:a): Since we can express that S belongs to W by (50:2), so $-S$ belongs to W when

(50:3) $$\sum_{i \text{ in } S} w_i < \sum_{i \text{ in } -S} w_i.$$

So (49:W*:a) means that always (50:2) or (50:3) holds, but never both. This means clearly, that never

(50:4) $$\sum_{i \text{ in } S} w_i = \sum_{i \text{ in } -S} w_i,$$

or equivalently, that never

(50:5) $$\sum_{i \text{ in } S} w_i = \frac{1}{2} \sum_{i=1}^{n} w_i.$$

[1] I.e. which is odd and for which a game can be essential.

Ad (49:W*:b): Using the definition of W in the form (50:1), this requirement is clearly satisfied if all $w_i \geqq 0$.[1]

Ad (49:W*:c): Using again (50:1), it is clear that $I = (1, \cdots, n)$ belongs to W. For the general $(n-1)$-element set $S = I - (i_0)$, the condition (50:1) states that

$$w_{i_0} < \tfrac{1}{2} \sum_{i=1}^{n} w_i.$$

Summing up:

(50:B) The weights $w_1, \cdots, w_n$ can be used to define by (50:1) or (50:2) a W which satisfies (49:W*) if and only if they fulfill the following conditions:

(50:B:a) For all $i_0 = 1, \cdots, n$

$$0 \leqq w_{i_0} < \tfrac{1}{2} \sum_{i=1}^{n} w_i.$$

(50:B:b) For all $S \subseteq I$

$$\sum_{i \text{ in } S} w_i \neq \tfrac{1}{2} \sum_{i=1}^{n} w_i.$$

Verbally: A player has always non-negative weight, but never half of the total weight or more; no combination of players has precisely half of the total weight.[2]

The simple game which obtains from this W[3] will be called the *weighted majority game* (of n participants with the weights $w_1, \cdots, w_n$). We will also designate this game by the symbol $[w_1, \cdots, w_n]$.

Thus the direct majority game has the symbol $[1, \cdots, 1]$.

It will be noted that the four-person game represented by the corner I of Q, discussed in 48.1.2., 48.1.3. can be described as a weighted majority game. Indeed, the principle of winning found in 48.1.3. can be expressed by saying that players 1,2,3 have a common weight, while player 4 has the double weight. I.e. this game has the symbol [1,1,1,2].

50.2. Homogeneity

50.2.1. The introduction of majority games and their explanatory symbols $[w_1, \cdots, w_n]$ is a step in the direction of a quantitative (numerical) classification and characterization of simple games. There are good reasons to think that it would be most desirable to carry out such a program fully: Simplicity was defined in combinatorial, set-theoretical terms and it is to be expected that a numerical characterization would make them

[1] This is, of course, a perfectly plausible condition; indeed, the surprising thing is that we are not forced to require $w_i > 0$—i.e. that we can permit a weight to vanish.

[2] The first requirement obviates the difficulty of 49.2., the second excludes ties.

[3] Precisely: The class of strategically equivalent ones.

easier to handle. Such a characterization usually facilitates a more exhaustive, quantitative understanding of the notion considered. Besides, in our present problem we are ultimately searching for solutions that are defined numerically, and therefore it seems likely that a numerical characterization will correspond to them more directly than a combinatorial one.

However, this first step is far from carrying out the transition.

On the one hand, a simple game may possess more than one symbol $[w_1, \cdots, w_n]$—indeed, every simple game that has one at all has infinitely many.[1] On the other hand, we do not know whether all simple games possess such a symbol at all.[2]

We begin by considering the first deficiency. Since the same simple game may possess several symbols $[w_1, \cdots, w_n]$, the natural procedure is to single out from among them a particular one by some convenient principle of selection. It is desirable to specify in this principle such requirements which increase the significance and usefulness of the $w_1, \cdots, w_n$.

First some preliminary observations. The conditions (50:1), (50:2) suggest consideration of the difference

(50:6) $$a_S = 2 \sum_{i \text{ in } S} w_i - \sum_{i=1}^{n} w_i = \sum_{i \text{ in } S} w_i - \sum_{i \text{ in } -S} w_i.$$

This a_S expresses how much the coalition S outweighs its opponents—how much of a weighted majority it possesses. These are its immediate properties:

(50:C) $$a_S = -a_{-S}$$

Proof: Use the last form of (50:6) for a_S.

(50:D:a) $a_S > 0$ if and only if S belongs to W.
(50:D:b) $a_S < 0$ if and only if S belongs to L.
(50:D:c) $a_S = 0$ is impossible.

Proof: Ad (50:D:a): Definitory.

Ad (50:D:b): Immediate by (50:D:a) and (50:C).

Ad (50:D:c): Immediate by (50:D:a), (50:D:b) since W, L exhaust all S. It also coincides with (50:B:b).

50.2.2. Now it is natural to try to arrange the weights $w_1, \cdots, w_n$ so that the amount of a_S which secures victory be the same for each winning coalition. It would be unreasonable, however, to require this actually for all S of W: If S belongs to W, then its proper supersets T do too, and they may have $a_T > a_S$.[3] Since such a T contains participants who are not necessary for winning, it seems natural to disregard it. I.e. we require the constancy of a_S only for those S of W which are not proper supersets of other elements

[1] Obviously, sufficiently small changes of the w_i will not disturb the validity of (50:1), particularly since (50:5) is excluded by (50:B:b).

[2] We will see in 53.2. that certain simple games have none.

[3] Thus for $T = I \supseteq S$, $a_I > a_S$ unless $w_i = 0$ for all i not in S.

of W. In the terminology introduced in 49.6.3.: a_S is required to be constant for the minimal elements of W—i.e. the elements of W^m.

We define accordingly:

(50:E) The weights $w_1, \cdots, w_n$ are homogeneous, if the a_S of (50:6) have a common value, to be denoted by a, for all S of W^m.

Whenever (50:E) is valid we shall indicate this by writing $[w_1, \cdots, w_n]_h$ instead of $[w_1, \cdots, w_n]$.

Clearly $a > 0$. A common positive factor affects none of the significant properties of $w_1, \cdots, w_n$, therefore we can use this in the case of homogeneity for a final normalization: Making $a = 1$.

We conclude by observing that the games mentioned at the end of 50.1.3. are homogeneous and normalized by $a = 1$. These are the direct majority games of an odd number of participants $[1, \cdots, 1]$, and the corner I of Q $[1,1,1,2]$—which can accordingly be written $[1, \cdots, 1]_h$ and $[1,1,1,2]_h$. Indeed, the reader will verify with ease that $a_S = 1$ for all S of W^m in both instances.

50.3. A More Direct Use of the Concept of Imputation in Forming Solutions

50.3.1. The homogeneous case introduced above is closely connected with the ordinary economic concept of imputation. We propose to show this now.

More precisely: We defined in 30.1.1. a general concept of imputations and based on it a concept of solutions. In forming these we were led by the same principles of judgment which are used in economics, and therefore some relationship with the ordinary economic concept of imputation must be expected. However, our considerations took us rather far from that concept. This applies especially to the constructions which were necessary when we found that sets of imputations—i.e. solutions—and not single imputations must be the subject of our theory. It will now appear that for certain simple games the connection with the ordinary economic concept of imputation can be established somewhat more directly. One might say that for the special games in question the connection between this primitive concept and our solutions can be directly established. Actually it will provide a simple method to find a particular solution for each one of those games.

50.3.2. The two concepts of solution, i.e. the two procedures, support each other quite effectively. The ordinary economic concept provides a useful surmise as to the form of a certain solution. And then our mathematical theory may be used to determine the solutions in question and to make the requirements of the ordinary approach complete. (Cf. 50.4. on the one hand, and 50.5. et sequ. on the other.)

These considerations also serve another end: They bring out the limitations of the ordinary approach with great clarity. The ordinary approach functions in this form only for the simple games, and even there not always and not entirely unaided by our mathematical theory. Besides, it does not disclose all the solutions for the games to which it applies. (Further remarks on this subject occur throughout the discussion, and particularly in 50.8.2.)

In this connection we emphasize again that any game is a model of a possible social or economic organization and any solution is a possible stable standard of behavior in it. And the games and solutions not covered by the method referred to—i.e. by the unimproved economic concept of imputation—will prove to be quite vital ones for social or economic theory. It will be seen that the simple games which can be treated by this special method are closely connected with the homogeneous weighted majority games of which they are a generalization.

50.4. Discussion of This Direct Approach

50.4.1. Consider a simple game Γ which we assume in the reduced form with $\gamma = 1$, but which we do not yet restrict any further. Let us try to discuss it in the sense of the ordinary economic ideas without making use of our systematic theory.

Clearly, in this game the sole aim of players is to form a winning coalition, and once a minimal coalition of this kind is formed, there is no motive for its participants to admit additional members. Consequently one can assume that the minimal winning coalitions—the S of W^m—are the structures that will form. It is therefore plausible to assume that a player's fate presents only two significant alternatives: He either succeeds in joining one of the desirable coalitions or he does not. In the latter case he is defeated, hence he obtains the amount -1. In the former case he is successful and according to ordinary ideas one ought to ascribe to this success a value. This value may vary from one player to another; for player i we denote it by $-1 + x_i$ so that x_i is the margin between defeat and success for player i.[1]

50.4.2. Let us now formulate the requirements which must be imposed on these $x_1, \cdots, x_n$ in the course of a conventional economic discussion.

First: By the very meaning of the x_i necessarily

$$x_i \geqq 0. \tag{50:7}$$

Second: If it happens that no minimal winning coalition contains a certain player i, then there exists for him no alternative to the value -1, and so we need not define any x_i for him.[2]

[1] We assume here that there is only one way of winning, i.e. that the margin x_i is the same whichever (minimal winning) coalition the player succeeds in joining. This is plausible since there is only one kind of success in a simple game: the complete one—every coalition being either fully defeated or fully winning.

It will appear in 50.7.2. and 50.8.2. how far this standpoint carries. As far as it does, it can be advantageously combined with our systematical theory.

[2] For the really important simple games such i do not exist—i.e. every player belongs to some minimal winning coalition. Cf. the first observation in 51.7.1. and (51:O) in 51.7.3.

Third: If a minimal winning coalition S becomes effective, then the division between the players will be this: Each player i not in S obtains -1, each player i in S obtains $-1 + x_i$. The sum of these amounts must be zero. This means

$$0 = \sum_{i \text{ not in } S} (-1) + \sum_{i \text{ in } S} (-1 + x_i) = -n + \sum_{i \text{ in } S} x_i,$$

i.e.

$$\sum_{i \text{ in } S} x_i = n. \tag{50:8}$$

In our system of notations this distribution is described by the vector $\vec{\alpha} = \{\alpha_1, \cdots, \alpha_n\}$ with the components

$$\alpha_i = \begin{cases} -1 & \text{for } i \text{ not in } S, \\ -1 + x_i & \text{for } i \text{ in } S. \end{cases}$$

We denote this vector by $\vec{\alpha}^S$. Our first condition and the present one actually state just that $\vec{\alpha}^S$ is an imputation in the sense of 30.1.1.

50.4.3. Continuing the usual line of argument, we shall now want to determine the $x_1, \cdots, x_n$ by means of the equations and inequalities of the three above remarks. In doing this, one more point must be considered: We have stated in the third remark, that its S must be minimal winning, i.e. belong to W^m. However it may be asked whether all S of W^m can be used.

Indeed, the present procedure is nothing but the usual one to determine the imputation of values to complementary goods by means of their alternative uses.[1] Now these alternative uses may be more numerous than the different goods under consideration—i.e. W^m may have more elements than n.[2] In such a situation one might expect that some of the uses are unprofitable and need not be included in the third remark. Indeed, we already made use of this principle by taking the S of W^m only, and not all elements of W, because the S of $W - W^m$ (the non-minimal winning coalitions) are clearly wasteful. Are we now sure that all S of W^m must be considered as equivalents of profitable uses? They are clearly not wasteful in the crude sense of the S mentioned above; no participant of an S in W^m can be spared without causing defeat. But unprofitability can arise in less direct ways than this, as numerous economic examples show. Thus the question remains unanswered as to which S of W^m are to be used in the third remark.

It is clear, however, that if an S of W^m is not included there, i.e. if

$$\sum_{i \text{ in } S} x_i = n \tag{50:8}$$

[1] In this case it would be more suitable to say, services. The object considered is the total service of player i in cooperating within a coalition which he joins.

[2] Cf. The fourth remark in 53.1.

fails to hold for it, then it must be definitely unprofitable. I.e. we must have $>$ in place of $=$ in (50:8):

$$\sum_{i \text{ in } S} x_i > n. \tag{50:9}$$

Thus the question arises: By what criteria are we to determine which S of W^m fall under the third remark—i.e. for which must (50:8) hold. Denote their set by $U(\subseteq W^m)$. Then (50:9) must hold for the S of $W^m - U$. So the problem is to determine U.[1]

50.5. Connection with the General Theory. Exact Formulation

50.5.1. Instead of attempting a verbal description, let us settle this point by going back to our systematic theory. From the statement made in 50.4. we carry over this much: Consider a system of minimal winning coalitions, i.e. a set $U \subseteq W^m$ and the x_i. Form the imputations

$$\overrightarrow{\alpha}^S = \{\alpha_1^S, \cdots, \alpha_n^S\}$$

as in 50.4.

$$\alpha_i^S = \begin{cases} -1 \text{ for } i \text{ not in } S \\ -1 + x_i \text{ for } i \text{ in } S \end{cases} \quad \text{when } S \text{ is in } U.$$

That these $\overrightarrow{\alpha}^S$, S in U, are indeed imputations, is expressed, as we know, by the conditions of 50.4.

$$x_i \geqq 0, \tag{50:7}$$

$$\sum_{i \text{ in } S} x_i = n \quad \text{when } S \text{ is in } U. \tag{50:8}$$

Form the set V of the $\overrightarrow{\alpha}^S$, S in U. We will decide whether U and the x_i are satisfactory, by determining whether this V is a solution in the sense of 30.1.1.

It will be seen that the result which is obtained in this manner can be stated verbally and is perfectly reasonable from the ordinary economic point of view. But it may be questioned whether it could have been unequivocally established by the usual procedures. This may serve as an illustration of how our mathematical theory can serve as a guide even for the purely verbal discussions of the ordinary economic approach, (cf. 50.7.1.).

50.5.2. We proceed to investigate whether V is a solution.

Let us determine first, when a given imputation $\overrightarrow{\beta} = \{\beta_1, \cdots, \beta_n\}$ is dominated by a given $\overrightarrow{\alpha}^T$, T in U. Since the game is simple, the set S

[1] It would be utterly mistaken to try to define $W^m - U$ (and so U) by means of (50:9). This would not restrict the $x_1, \cdots, x_n$ sufficiently—and their determination is the real objective!

of 30.1.1. for this domination can be assumed to belong to W (or even to W^m, use (49:I) or (49:J) in 49.7.2.). For every i in S, $\alpha_i^T > \beta_i \geqq -1$; for every i not in T, $\alpha_i = -1$: hence $S \subseteq T$. Now T is in $U \subseteq W^m$, S is in W, therefore $S \subseteq T$ yields $S = T$. So we see: The set S of 30.1.1. for this domination must be our T. And T can be used there, since it is certainly necessary, as it belongs to $U \subseteq W^m \subseteq W$, cf. above. Hence the domination $\overrightarrow{\alpha}^T \succ \overrightarrow{\beta}$ amounts to this: $\alpha_i^T > \beta_i$ for i in T, i.e.

(50:10) $$\beta_i < -1 + x_i \qquad \text{for } i \text{ in } T.$$

Denote for any imputation $\overrightarrow{\beta} = \{\beta_1, \cdots, \beta_n\}$ the set of all i with

(50:11) $$\beta_i \geqq -1 + x_i$$

by $R(\overrightarrow{\beta})$. Then (50:10) states that $R(\overrightarrow{\beta})$ and T are disjunct. An alternative way of writing this is:

(50:12) $$-R(\overrightarrow{\beta}) \supseteq T.$$

We repeat:

(50:F) $\overrightarrow{\alpha}^T \succ \overrightarrow{\beta}$ is equivalent to (50:12).

From this we can infer:

(50:G) Let U^* be the set of all $R(\subseteq I)$ which possess some subset belonging to U.

Let U^+ be the set of all $R(\subseteq I)$ for which $-R$ does not belong to U^*.

Then $\overrightarrow{\beta}$ is undominated by any element of V if and only if $R(\overrightarrow{\beta})$ belongs to U^+.

Proof: That $\overrightarrow{\beta}$ is dominated by some element of V—i.e. by some $\overrightarrow{\alpha}^T$, T in U means that (50:12) holds for some T in U. This is equivalent to saying that $-R(\overrightarrow{\beta})$ is in U^*, i.e. that $R(\overrightarrow{\beta})$ is not in U^+.

Hence $R(\overrightarrow{\beta})$ belongs to U^+ if and only if $\overrightarrow{\beta}$ is dominated by no element of V.

50.5.3. Before going any further we observe four simple properties of the set U^+ of (50:G)

(50:H:a) $$U^* = U^+ = W \qquad \text{if} \qquad U = W^m$$

Proof: Assume $U = W^m$. Then U^* consists of those sets which possess a subset belonging to W^m—i.e. a minimal winning subset. Hence $U^* = W$. The operation which leads in (50:G) from U^* to U^+ is the combination of the transformation (48:A:a) and (48:A:b) in 48.2.1. Now we noted already then, that these two transformations compensate each other, when applied to W. Hence $U^* = W$ gives $U^+ = W$.

(50:H:b) U^* is a monotonic and U^+ is an antimonotonic operation. I.e. $U_1 \subseteq U_2$ implies $U_1^* \subseteq U_2^*$ and $U_1^+ \supseteq U_2^+$.

Proof: It suffices to recall the definitions in (50:G), to see that $U_1 \subseteq U_2$ implies $U_1^* \subseteq U_2^*$ and this in turn $U_1^+ \supseteq U_2^+$.

(50:H:c) All our $U \subseteq W^m$ have $U^* \subseteq W \subseteq U^+$.

Proof: Combine (50:H:a) and (50:H:b) (with U, W^m in place of U_1, U_2).

(50:H:d) Both U^* and U^+ contain all supersets of their elements.

Proof: This is obvious for U^*. The property under consideration is the same one which was formulated in (48:A:c) in 48.2.1. (W taking the place of our U^*, U^+.) Now the operation which leads in (50:G) from U^* to U^+, is the combination of the transformations (48:A:a) and (48:A:b) in 48.2.1. (Cf. the proof of (50:H:a)). Application of (48:B) in 48.2.2. to these two transformations shows that the property in question is conserved when passing from U^* to U^+.

50.5.4. Note that U^*, U^+ allow a simple verbal interpretation. If we knew only of the coalitions belonging to U that they are winning, of which coalitions could we then assert that they are certainly winning, and of which that they are not certainly defeated?

The former is the case for the coalitions with subsets in U, i.e. for those of U^*. The certainly defeated ones are the complements of these, i.e. those not in U^+. Hence U^* is the set of the first mentioned coalitions, and U^+ the set of the last mentioned ones.

Now the meaning of (50:H:a)-(50:H:c) becomes clear: For $U = W^m$, everything is unambiguous: The certainly winning coalitions are precisely those which are not certainly defeated, and they form the set W. As U decreases from W^m, the gap widens. The first set decreases through subsets of W, the second one increases through supersets of W.

The assertion of (50:H:d) is equally plausible.

50.6. Reformulation of the Result

50.6.1. (50:G) of 50.5.2. allows us to state:

(50:I) V is a solution if and only if $R(\overrightarrow{\beta})$ belongs to U^+ precisely when $\overrightarrow{\beta}$ belongs to V.

So we must only decide when (50:I) holds. For this purpose we consider an R in U^+ and determine the $\overrightarrow{\beta}$ for which $R(\overrightarrow{\beta}) = R$.

Consider the three possibilities:

$$\sum_{i \text{ not in } R} (-1) + \sum_{i \text{ in } R} (-1 + x_i) \begin{matrix} > \\ = \\ < \end{matrix} 0, \tag{50:13}$$

i.e.

(50:14)
$$\sum_{i \text{ in } R} x_i \overset{>}{\underset{<}{=}} n.$$

If a $\overrightarrow{\beta}$ with $R(\overrightarrow{\beta}) = R$ exists, then we have

(50:15)
$$0 = \sum_{i=1}^{n} \beta_i \geqq \sum_{i \text{ not in } R} (-1) + \sum_{i \text{ in } R} (-1 + x_i),$$

i.e. $\leqq$ in (50:13), (50:14). So $>$ in (50:13), (50:14) excludes the existence of any $\overrightarrow{\beta}$ with $R(\overrightarrow{\beta}) = R$. I.e. the sets R in U^+ with $>$ in (50:13), (50:14) need not be considered further. Consider on the other hand, an R in U^+ with $<$ in (50:13), (50:14). Then there are infinitely many ways of choosing $\overrightarrow{\beta}$ with $\sum_{i=1} \beta_i = 0$ and $\beta_i \geqq \begin{Bmatrix} -1 \text{ for } i \text{ not in } R \\ -1 + x_i \text{ for } i \text{ in } R \end{Bmatrix}$. For all these $R(\overrightarrow{\beta})$ necessarily $\supseteq R$. Hence it belongs to V by (50:H:d). Since V is finite, these $\overrightarrow{\beta}$ cannot all belong to V. This is a contradiction. I.e. sets R in U^+ with $<$ in (50:13), (50:14) must not exist.

50.6.2. It remains for us to consider the sets which are in U^+ with $=$ in (50:13), (50:14). According to the above, these must furnish precisely the $\overrightarrow{\beta}$ of V.

If $\overrightarrow{\beta}$ belongs to V, i.e. $\overrightarrow{\beta} = \overrightarrow{\alpha}^T$, T in U, then we have this situation: $R(\overrightarrow{\beta})$ is T plus the set of those i for which $x_i = 0$. T belongs to $U \subseteq U^* \subseteq U^+$ (for the second relation use (50:H:c)), hence $R(\overrightarrow{\beta})$ belongs to U^+. Also

$$\sum_{i \text{ in } R(\overrightarrow{\beta})} x_i = \sum_{i \text{ in } T} x_i = n.$$

So we have $=$ in (50:13), (50:14). Hence the $\overrightarrow{\beta}$ of V are all taken care of.

Conversely: Consider an R in U^+ with $=$ in (50:13), (50:14). Addition of all i for which $x_i = 0$ to R affects neither the fact that R belongs to U^+ (by (50:H:d)), nor the equation (50:14). So we may assume that R contains all these i.

If now an imputation $\overrightarrow{\beta}$ has $R(\overrightarrow{\beta}) = R$, then $\beta_i \geqq -1 + x_i$ for i in R. Always $\beta_i \geqq -1$. As $\sum_{i=1}^{n} \beta_i = 0$, this implies:

(50:16)
$$\beta_i = \begin{cases} -1 & \text{for } i \text{ not in } R, \\ -1 + x_i & \text{for } i \text{ in } R. \end{cases}$$

Conversely: (50:16) implies that $\overrightarrow{\beta}$ is an imputation with $R(\overrightarrow{\beta}) = R$. Hence our requirement in this case must be that the $\overrightarrow{\beta}$ of (50:16) be an $\overrightarrow{\alpha}^{T}$, T in U. This means, that T and R differ only in elements i for which $x_i = 0$. And this property is insensible to our original modification of R, the inclusion of all such i into R.

Summing up:

(50:J) V is a solution if and only if this is the case: Call an i indifferent when $x_i = 0$.[1]

Then we have

(50:8*)
$$\sum_{i \text{ in } T} x_i = n$$

for the T of U and, of course, also for those which differ from these only by indifferent elements.

And we must require

(50:9*)
$$\sum_{i \text{ in } T} x_i > n$$

for all other T of U^+.

In making use of this result, one may chose the set $U \subseteq W^m$ first, then attempt to determine the x_i from (50:8*) and finally verify whether these x_i fulfill the inequalities

(50:7)
$$x_i \geqq 0$$

and (50:9*).

50.7. Interpretation of the Result

50.7.1. The result (50:J) permits the verbal statement promised in 50.5.1. This is it:

A solution V is found by choosing arbitrarily the set U of those minimal winning coalitions (i.e. $U \subseteq W^m$), which are to be considered profitable. The x_i must then satisfy the corresponding equations (50:8*). But after this, it must be verified that certain other coalitions are definitely inprofitable in the sense of (50:9*). This must be required not only for those coalitions which are known to be winning, (i.e. W), but for all those which cannot

[1] These i constitute a slight complication which is further aggravated by the fact that we have no example of a game in which they actually occur. It may be that they never exist; an indifferent i characterizes a player who belongs to some minimal winning coalitions, but never receives a share.

The excluded player in a discriminatory solution of the three-person game is in this situation (cf. (32:A) in 32.2.3. with $c = 1$). But that solution is an infinite set, whereas our V must be finite.

It would be of interest to decide this existential question. At any rate we must at present provide for the indifferent i to avoid loss of generality or rigour.

be established as definitely defeated by the coalitions of U alone (i.e. U^+)—excepting, of course, the coalitions of U itself.[1]

The reader may now judge whether the concluding remark of 50.5.1. is justified by this formulation.

50.7.2. The question of finding the proper U for (50:I) is a rather delicate one. The antimonotony of U^+ (cf. (50:H:b) in 50.5.3.) makes itself felt now: Decreasing U, i.e. the number of equations, increases U^+, the number of inequalities, and *vice versa*.

In particular, if we choose U as large as possible, i.e. $U = W^m$, then the inequalities associated with U^+ create no difficulties at all. Indeed: $U = W^m$ implies $U^+ = W$ by (50:H:a) in 50.5.3. A T of W certainly possesses a subset S which is minimal in W, i.e. belongs to $U = W^m$. Now if T differs from this S by more than indifferent elements, then we have $x_i > 0$ for some i in $T - S$, hence $\sum_{i \text{ in } T} x_i > \sum_{i \text{ in } S} x_i = n$, i.e. (50:9*) as desired.

Thus $U = W^m$ always yields a solution V, if its equations (50:8*) can be solved at all (with (50:7)).

But as we pointed out in 50.4.3., we have no right to expect *a priori* that this will always be the case—especially since there may be more equations (50:8*) (i.e. elements in W^m), than variables x_i.

The last objection is not an absolute one; indeed it is easy to find a simple game for which the number of these equations exceeds the number of variables and the solution nevertheless exists.[2] On the other hand there exist simple games for which those equations have no solutions. An example of this is somewhat more hidden,[3] but the phenomenon is probably fairly general. When this occurs, one must investigate whether a solution V cannot be found by appropriate choices $U \subset W^m$. The difficulty and delicacy of this question has been commented upon already at the beginning of this section.[4]

50.8. Connection with the Homogeneous Majority Games

50.8.1. We now restrict ourselves to the case $U = W^m$. I.e. we assume that the full system of equations

(50:17) $$\sum_{i \text{ in } S} x_i = n \qquad \text{for all } S \text{ in } W^m,$$

can be solved with

(50:7) $$x_i \geqq 0.$$

[1] And those which differ from them only by indifferent elements.

[2] This happens for the first time for $n = 5$, cf. the fifth remark in 53.1.

[3] This happens for the first time for $n = 6$, cf. the fifth remark in 53.2.5.

[4] No instance of a simple game with a solution V derived from $U \subset W^m$ is known, nor is it established that none exists. The further-going question whether every simple game possesses solutions V of suitable $U \subseteq W^m$ is equally open.

The problem seems to be of some importance. It may be difficult to solve it. It appears to have some similarity with the solved questions mentioned in footnote 1 on p. 154, but it has not been possible, so far, to exploit this connection.

We saw that in this case the set V of all $\overrightarrow{\alpha}^S$, S in W^m, is a solution. In this situation and only then, we call V a *main simple solution* of the game.

There is a certain similarity between these requirements and those which characterize a homogeneous weighted majority game. Indeed, the latter are defined by

$$\text{(50:18)} \qquad \sum_{i \text{ in } S} w_i = b \qquad \text{for all } S \text{ in } W^m$$

where

$$b = \tfrac{1}{2}\left(\sum_{i=1}^{n} w_i + a\right), \qquad a > 0 \qquad \text{(combine (50:D) (50:E) of 50.2.)}$$

and

$$\text{(50:19)} \qquad w_i \geqq 0.$$

Actually, there is more than similarity. Thus, if a system of w_i, fulfilling (50:18), (50:19) is given, a system x_i fulfilling (50:17), (50:7) obtains as follows: The quantity b of (50:18) is positive.[1] Multiplication of all w_i by a common positive factor leaves everything unaffected, and by choosing this factor as n/b we can replace b in (50:18) by n. Now we can simply put $x_i \equiv w_i$ and (50:18), (50:19) become (50:17), (50:7).

If conversely a system of x_i fulfilling (50:17), (50:7) is given, there is an extra difficulty. We may put $w_i \equiv x_i$.[2] Then (50:7) becomes (50:19) and (50:17) yields (50:18) with $b = n$, i.e. $a = 2n - \sum_{i=1}^{n} w_i$. But now the question arises whether the last requirement $a > 0$ is fulfilled—i.e. whether

$$\text{(50:20)} \qquad \sum_{i=1}^{n} x_i < 2n.$$

Summing up:

(50:K) Every homogeneous, weighted majority game possesses a main simple solution.

Conversely, if a (simple) game possesses a main simple solution, homogeneous weights for the game can be derived from it if and only if (50:20) is fulfilled.

50.8.2. This connection between homogeneous weights and main simple solutions is significant. But it must be stressed that a homogeneous, weighted majority game will in general have other solutions besides the

[1] Otherwise all i occurring in the S of W^m would have $w_i = 0$ by (50:18) and (50:19). Then (50:6) of 50.2.1. and (50:19) gives $a_s \leqq 0$ for the S of W^m, hence $a \leqq 0$, which is not the case.

[2] The i which belong to no minimal winning set cause a slight disturbance, since they have no x_i (cf. the second remark in 50.4.), while we require their w_i. However, the contingency is unimportant (cf. loc. cit.) and we can put these $w_i = 0$ as is easily concluded from the references of footnote 2 on p. 436.

main simple one.[1] And a game with a main simple solution may not fulfill (50:20), i.e. there need not be $<$ in

(50:21)
$$\sum_{i=1}^{n} x_i \overset{<}{\underset{>}{=}} 2n.$$[2]

Beyond all this, finally, we must not lose sight of the main limitation of these considerations: Whether we take the concept of "ordinary" imputation in its narrower form of 50.8.1. (i.e. $U = W^m$) or in its wider original form of 50.6., 50.7.1. (i.e. $U \subseteq W^m$, cf. (50:I) in 50.6.2.), it is certainly restricted to simple games. That it is necessary to go beyond these, and beyond the special solutions described here, and that this forces us to fall back completely on the systematical theory of 30.1.1., was pointed out at the end of 50.3.

51. Methods for the Enumeration of All Simple Games

51.1. Preliminary Remarks

51.1.1. Beginning with 50.1.1. we introduced specific simple games which permitted characterization by numerical criteria instead of the original set theoretical ones (cf. the beginning of 50.2.1.). We saw, however, that these numerical procedures could be carried out in several ways and that there was no certainty that all simple games could be accounted for with their help. It is therefore desirable to devise combinatorial (set theoretical) methods that produce systematic enumeration of all simple games.

This is, indeed, indispensable in order to gain an insight into the possibilities of simple games and particularly to see how far the above mentioned numerical procedures carry us. It will appear that the decisive examples of the non-obvious possibilities obtain only for relatively high numbers of players,[3] so that a mere verbal analysis cannot be very effective.

51.1.2. We pointed out at the end of 49.6.3. that the enumeration of all simple games is equivalent to the enumeration of their sets W, i.e. of all sets W which fulfill (49:W*) in 49.6.2. We also noted there that it may be advantageous to replace the use of W (all winning coalitions), by W^m (all minimal winning coalitions).

Either procedure provides an enumeration of all simple games. The use of W is preferable from the conceptual standpoint since W has the

[1] The main simple solution of the essential three-person game ($[1,1,1]_h$, cf. the end of 50.2.) is the original solution of 29.1.2., i.e. (32:B) of 32.2.3. We know from 32.2.3. and 33.1. that other solutions exist.

The main simple solution of corner I of Q ($[1,1,1,2]_h$, cf. the end of 50.2.) is the original solution of 35.1.3. We will discuss this game, together with the more general one $[1, \cdots, 1, n-2]_h$ (n participants) in 55. and obtain all solutions.

All these references make it clear that the solutions other than the main simple one are quite significant, cf. 33.1 and 54.1.

[2] $=$ occurs for the first time for $n = 6$, cf. the fourth remark of 53.2.4. $>$ occurs for the first time for $n = 6$ or 7, cf. the sixth remark of 53.2.6.

Both these examples are quite interesting in their own right.

[3] $n = 6, 7$ cf. 53.2.

simpler definition and W^m was introduced indirectly with the help of W. For a practical enumeration of all simple games—which is our present aim—the use of W^m is preferable since W^m is a smaller set than W [1] and therefore more readily described.

We will give both procedures successively. It will appear that these discussions provide a natural application of the concepts of satisfactoriness and saturation introduced in 30.3.

51.2. The Saturation Method: Enumeration by Means of W

51.2.1. The sets W are characterized by (49:W*) in 49.6.2. i.e. by the conditions (49:W*:a)-(49:W*:c) which constitute (49:W*).

Let us for a moment disregard (49:W*:c), and consider (49:W*:a), (49:W*:b). These two conditions imply that no two elements of W can be disjunct.[2] In other words: Denote the negation of disjunctness—i.e. of $S \cap T \neq \ominus$—by $S\mathfrak{R}_1T$. Then (49:W*:a), (49:W*:b) imply $\mathfrak{R}_1$-satisfactoriness.[3] A more exhaustive statement along these lines is this:

(51:A) (49:W*:a), (49:W*:b) are equivalent to $\mathfrak{R}_1$-saturation.[3]

Proof: $\mathfrak{R}_1$-saturation of W means this:

(51:1) S belongs to W, if and only if $S \cap T \neq \ominus$ for all T of W.

(49:W*:a), (49:W*:b) imply (51:1): Let W fulfill (49:W*:a), (49:W*:b). If S belongs to W, then we know that $S \cap T \neq \ominus$ for all T of W. If S does not belong to W, then $T = -S$ belongs to W by (49:W*:a), and $S \cap T = \ominus$.

(51:1) implies (49:W*:a), (49:W*:b): Let W fulfill (51:1). We prove (49:W*:a), (49:W*:b) in the reverse order.

Ad (49:W*:b): If S meets the criterion of (51:1), then every superset of S does too. Hence W contains the supersets of its elements.

Ad (49:W*:a): Owing to the above, $-S$ is not in W if and only if no subset of $-S$ is in W. I.e. when every T of W is not $\subseteq -S$, or again, when for every T of W, $S \cap T \neq \ominus$. By (51:1) this means precisely that S is in W.

Thus, at any rate, precisely one of S, $-S$ belongs to W.

Now $S\mathfrak{R}_1T$ is clearly symmetric, hence we can apply (30:G) in 30.3.5.[4]

[1] W, L are disjunct sets. They have the same number of elements owing to (48:A:b) in 48.2.1. Together they exhaust $\bar{I}$ which has 2^n elements. Hence W as well as L has exactly 2^{n-1} elements.

The number of elements in W^m varies, but it is always considerably smaller. (Cf. the fourth remark in 53.1.)

[2] *Proof:* Let S, T belong to W, $S \cap T = \ominus$. Then $-S \supseteq T$, hence $-S$ belongs to W by (49:W*:b), thus violating (49:W*:a).

[3] Cf. the definitions of 30.3.2.

[4] It will be remembered that we also assumed in 30.3.5. the general validity of $x\mathfrak{R}x$—i.e. in this case of $S\mathfrak{R}_1S$. This means $S \neq \ominus$—so it fails for $S = \ominus$.

However, (49:W*:a), (49:W*:c) exclude $\ominus$ form W, hence we may use as domain D in the sense of 30.3.2. instead of $\bar{I}$ (the system of all subsets of I) equally well $\bar{I} - (\ominus)$ (the system of all non-empty subsets of I). This rids us of $S = \ominus$.

51.2.2. In order to discuss (49:W*) on this basis, we must take (49:W*:c) also into account. This can be done in two ways. The first way will be useful for a subsequent comparison.

(51:B) W fulfills (49:W*) if and only if it is $\mathfrak{R}_1$-saturated and contains neither $\ominus$ nor any one-element set.

Proof: (49:W*) is the conjunction of (49:W*:a), (49:W*:b) and (49:W*:c). The first two amount by (51:A) to $\mathfrak{R}_1$-saturation. Taking (49:W*:a) for granted, (49:W*:c) may be stated thus: If S is I or an $(n - 1)$-element set, then $-S$ is not in W. I.e.: Neither $\ominus$ nor any one-element set is in W.

The second way is more directly useful.

Let V_0 be the system of all sets of (49:W*:c)—i.e. of I and all $(n - 1)$-element subsets of I. Then we have:

(51:C) V is a subset of a W fulfilling (49:W*) if and only if $V \cup V_0$ is $\mathfrak{R}_1$-satisfactory.

Proof: $W \supseteq V$ and W fulfilling (49:W*) amount to this: $W \supseteq V$, W fulfills (49:W*:a), (49:W*:b)—i.e. W is $\mathfrak{R}_1$-saturated by (51:A)—W fulfills (49:W*:c)—i.e. $W \supseteq V_0$. In other words: We are looking for an $\mathfrak{R}_1$-saturated $W \supseteq V \cup V_0$—i.e. we are asking whether $V \cup V_0$ can be extended to an $\mathfrak{R}_1$-saturated set.

Now we know that (30:G) of 30.3.5. applies, and hence the considerations of the last part of 30.3.5. apply too.[1] This extensability is equivalent to the $\mathfrak{R}_1$-satisfactoriness of $V \cup V_0$.

51.2.3. We rephrase (51:C) more explicitly:

(51:D) V is a subset of a W fulfilling (49:W*) if and only if it possesses these properties:

(51:D:a) No two S, T of V are disjunct.

(51:D:b) V contains neither $\ominus$ [2] nor any one-element set.

Proof: We must express, according to (51:C), the $\mathfrak{R}_1$-satisfactoriness of $V \cup V_0$. I.e. that no two S, T of V or V_0 are disjunct.

S, T are both in V: This coincides with (51:D:a).

S, T are both in V_0: Both have $\geq n - 1$ elements, hence they cannot be disjunct.[3]

Of S, T one is in V and the other in V_0: We may assume by symmetry, S as the former and T as the latter. So an S of V must not be disjunct with I or any $(n - 1)$-element set. This is precisely (51:D:b).

[1] Note that the domain $D = \bar{I} - (\ominus)$ (cf. footnote 4 on p. 446) is finite.

[2] For this cf. also footnote 4 on p. 446.

[3] We are using that $2(n - 1) > n$ i.e. $n > 2$ i.e. $n \geq 3$. This should have been stated explicitly at the beginning—but it is a natural assumption, since simple games (i.e. sets with (49:W*)) exist only for $n \geq 3$. (Cf. 49:4, 49:5.)

(51:D) solves the question of enumerating all W: Starting with any V which fulfills (51:D:a), (51:D:b)[1] we may increase it gradually as long as this can be done without violating (51:D:a), (51:D:b). When this process cannot be continued any further, than we have a V which is maximal among the subsets of the W (with (49:W*))—i.e. we then have such a W.

In performing this gradual building up process in all possible ways, we obtain all W in question.

The reader may try this for $n = 3$ or $n = 4$. It will appear that the procedure is quite cumbersome even for small n, although it is rigorous and exhaustive for all n.

51.3. Reasons for Passing from W to W^m. Difficulties of Using W^m

51.3.1. Let us consider the sets W^m of 49.6.

We wish to characterize these W^m directly and to find some simple process to construct them all. In what follows, we will derive two different ways of characterization, both being of the saturation type. The first will be by means of an asymmetric relation, while the second will be by a symmetric one. Thus it is the second one which is suited for construction purposes, in analogy with the construction of W in 51.2.

We give nevertheless both characterizations because the equivalence is quite instructive: The first one is in some (technical) respects similar to the definition of a solution (cf. 30.3.3. and 30.3.7.), and therefore the transition to the equivalent second form is of interest since it points a way to solve problems of this type. We have mentioned before (in 30.3.7.) how desirable the corresponding transition for our concept of solution would be.

51.3.2. Let W be a system which contains all supersets of its elements: e.g. fulfilling (49:W*:b). Then the system of its minimal elements W^m determines W: Indeed, it is clear that W is the system of the supersets of all elements of W^m.

Hence if a system V is given, and we are looking for a W with (49:W*) such that $V = W^m$, then this W must necessarily be the system $\tilde{V}$ of the supersets of all elements of V.

Consequently $V = W^m$ for a W with (49:W*) if and only if these two requirements are met by $W = \tilde{V}$.[2] We are now going to transform this characterization of the $V = W^m$ into one of the saturation type.

Denote the assertion that neither $S \cap T = \ominus$ nor $S \supset T$, by $S \mathfrak{R}_2 T$. Then we have:

(51:E) $V = W^m$ for a W with (49:W*) if and only if V is $\mathfrak{R}_2$-saturated and contains neither $\ominus$ nor any one-element set.

Proof: According to the above, we must only investigate whether $W = \tilde{V}$ has the desired properties:

[1] In principle we may start with the empty set. The reader will note that the exclusion of $\ominus$ from V (cf. above) does not affect the possibility of $V = \ominus$.

[2] I.e. $W = \tilde{V}$ is the only system which can possibly meet these requirements, but even it may fail.

$V = W^m$: Let S be a minimal element of this W. Then $S \supseteq T$ for some T of V. Hence T is in W, and so the minimality of S excludes $S \supset T$. So $S = T$ i.e. S belongs to V.

Thus only the converse property must be discussed: Whether every S of V is really minimal in W. Any S of V clearly belongs to W. So the minimality means that $S \supset T'$, T' of W is impossible; i.e. that $S \supset T' \supseteq T$, T of V is impossible. This implies the impossibility of $S \supset T$, T of V and is implied by it (put $T' = T$). So we have this condition:

(51:2) Never $S \supset T$ for S, T in V.

W fulfills (49:W*): We must consider (49:W*:a), (49:W*:b), (49:W*:c) separately. We do this in a different order.

Ad (49:W*:b): Clearly $W = \tilde{V}$ contains all supersets of its elements so this is automatically fulfilled.

Ad (49:W*:c): Take (49:W*:a) for granted. (Cf. below.) Then (49:W*:c) may be stated thus: If S is I or a $(n - 1)$-element set, then $-S$ is not in W. I.e. neither $\ominus$ nor any one-element set is in W; that is, no subset of these is in V. So we have this condition:

(51:3) Neither $\ominus$ nor any one-element set is in V.

Ad (49:W*:a): We consider this in two parts:

S', $-S'$ cannot both belong to W: I.e. if S, T belong to V, then we cannot have $S \subseteq S'$, $T \subseteq -S'$. Now the existence of such an S' implies $S \cap T = \ominus$ and is implied by it (put $S' = S$). So we have this condition:

(51:4) Never $S \cap T = \ominus$ for S, T in V.

One of S, $-S$ must belong to W: Assume that neither of S, $-S$ belongs to W. This means that no T of V has $T \subseteq S$ or $T \subseteq -S$, the latter meaning $S \cap T = \ominus$. I.e. no T of V has $T = S$ or $S \supset T$ or $S \cap T = \ominus$. Or again: S is not in V and no T of V fulfills the negation of $S \mathfrak{R}_2 T$.[1]

I.e. S is not in V, but $S \mathfrak{R}_2 T$ for all T of V.

Now we have to express that this is impossible: i.e.:

(51:5) If $S \mathfrak{R}_2 T$ for all T of V, then S belongs to V.

Thus (51:2)-(51:5) are the criteria we want.

Now (51:2) and (51:4) can be stated together like this: $S \mathfrak{R}_2 T$ for all S, T of V. I.e.:

(51:6) $S \mathfrak{R}_2 T$ for all T of V, if S belongs to V.

(51:5) and (51:6) together express precisely the $\mathfrak{R}_2$-saturation of V. Hence this and (51:3) form the criterion—and this is precisely what we wanted to prove.

(51:E) is of some interest because it is a perfect analogue of (51:B). Thus these characterizations of W and W^m differ only in the replacement of

[1] This is indeed $S \supset T$ or $S \cap T = \ominus$.

$$S\mathfrak{R}_1T\text{: not } S \cap T = \ominus$$

by

$$S\mathfrak{R}_2T\text{: neither } S \cap T = \ominus \text{ nor } S \supset T.$$

But as this replaces the symmetric $\mathfrak{R}_1$ by the asymmetric $\mathfrak{R}_2$, (51:E) cannot be used in the way in which we used (51:B)—or rather the underlying (51:A).

51.4. Changed Approach: Enumeration by Means of W^m

51.4.1. We now turn to the second procedure. This consists in analyzing the following question: Given a system V, what does it mean for a W fulfilling (49:W*) that $V \subseteq W^m$?

The meaning of the $V \subseteq W^m$ is this: Every S of V is a minimal element of W. I.e. such an S must belong to W but its proper subsets must not belong to W. As W fulfills (49:W*:b), i.e. contains the supersets of all its elements, it suffices to state this for the maximal proper subsets of S only; i.e. for the $S - (i)$, i in S. As W fulfills (49:W*:a) we may say instead of $S - (i)$ not belonging to W, that $-(S - (i)) = (-S) \cup (i)$ is in W. So we see:

(51:F) $V \subseteq W^m(W$ with (49:W*)) means precisely this: For every S of V, S belongs to W; and for every i of this S, $(-S) \cup (i)$ belongs to W.

We now prove:

(51:G) V is a subset of the W^m of a W fulfilling (49:W*) if and only if it possesses these properties:

(51:G:a) No two S, T of V are disjunct.

(51:G:b) No two S, T of V have $S \supset T$.

(51:G:c) For S, T of V, $S \cup T = I$ implies that $S \cap T$ is a one-element set.

(51:G:d) Neither $\ominus$ nor any one-element set, nor I must belong to V.

Proof: Let V_1 be the set of all $(-S) \cup (i)$, S in V, i in S. Then $V \subseteq W^m$ means by (51:F), that $V \cup V_1 \subseteq W$. This is possible for some W with (49:W*) according to (51:D), if $V \cup V_1$ fulfills (51:D:a), (51:D:b).

Let us therefore formulate (51:D:a), (51:D:b) for $V \cup V_1$.

Ad (51:D:a): S, T are both in V: This coincides with (51:G:a).

S, T are both in V_1: I.e. $S = (-S') \cup (i)$, $T = (-T') \cup (j)$, S', T' in V, i in S', j in T'.

The disjunctness of S, T means these: $-S'$, $-T'$ disjunct, i.e. $S' \cup T' = I$; (i), (j) disjunct, i.e. $i \neq j$; $-S'$, (j) disjunct, i.e. j in S'; $-T'$, (i) disjunct, i.e. i in T'.

Summing up: $S' \cup T' = I$; i, j two different elements of both S' and T'—i.e. of $S' \cap T'$.

Now we must state that this is impossible. I.e. if $S' \cup T' = I$, then $S' \cap T'$ cannot possess two different elements. As $S' \cap T'$ cannot be empty by (51:G:a), this means that it must be a one-element set.

Thus precisely (51:G:c) obtains (S', T' in place of its S, T).

Of S, T one is in V and the other in V_1: We may assume by symmetry, that S is the former and T the latter. So $T = (-T') \cup (j)$, T' in V, j in T'. The disjunctness S, $(-T') \cup (j)$ means: S, $-T'$ are disjunct, i.e. $S \subseteq T'$; S, (j) are disjunct, i.e. j not in S.

Summing up: $S \subseteq T'$, j an element of T' not in S.

Now we must state that this is impossible. I.e. not $S \subset T'$. Thus precisely (51:G:b) obtains. (T', S in place of its S, T.)

Ad (51:D:b): Neither $\ominus$ nor any one-element set must belong to V nor to V_1. The latter means that neither must be a $(-S) \cup (i)$, S in V, i in S. Only a one-element set could be such a $(-S) \cup (i)$ and this would mean: $-S = \ominus$, i.e. $S = I$.

Summing up: Neither $\ominus$ nor any one-element set, nor I must belong to V. This coincides with (51:G:d).

Thus we have obtained precisely the conditions (51:G:a)-(51:G:d) as desired.

(51:G) solves the problem of enumerating all W^m in perfect analogy to the solution by (51:D) of the corresponding problem for the W: Starting with any V which fulfills (51:G:a)-(51:G:d)[1] we increase it gradually as long as this can be done without violating (51:G:a)-(51:G:d). When this process cannot be continued any further, we have a V which is maximal among the subsets of the W^m of a W with (49:W*)—i.e. we have such a W^m.

In performing such a gradual process of building up in all possible ways, we obtain all W^m in question.

51.4.2. Our last remarks show that the practical enumeration of all simple games can be based on (51:G)—and we will, indeed, undertake it in 52. But some other considerations are better carried out at first.

We now propose to analyze the assertion that (51:G) is a condition of the saturation type a little more closely.

Observe first, that as (51:G:b) refers to two arbitrary S, T of V, we can interchange these in it. I.e. we can replace (51:G:b) by this:

(51:G:b*) No two S, T of V have $S \supset T$ or $S \subset T$.

Denote the assertion that S, T fulfill (51:G:a), (51:G:b*), (51:G:c)—i.e. that neither $S \cap T = \ominus$ nor $S \supset T$, nor $S \subset T$, nor $S \cup T = I$ without $S \cap T$ being a one-element set—by $S\mathcal{R}_3T$.

Then (51:G) simply states that V is $\mathcal{R}_3$-satisfactory, together with (51:G:d). Now let the domain D be the system $\bar{\bar{I}}$ of those subsets of I which fulfill (51:G:d)—i.e. neither $\ominus$, nor a one-element set, nor I. Then the last remarks of 51.4.1. show that the W are the maximal $\mathcal{R}_3$-satisfactory subsets of $\bar{\bar{I}}$.

[1] In principle we may start with the empty set.

$S\mathfrak{R}_3T$ is clearly symmetric.[1] Hence we may apply (30:G) of 30.3.5. This gives:

(51:H) $V = W^m$ for a W with (49:W*) if and only if V is $\mathfrak{R}_3$-saturated (in $\bar{\bar{I}}$).

Comparing (51:E) with (51:H) shows that we have succeeded in passing from the asymmetric $\mathfrak{R}_2$ to the symmetric $\mathfrak{R}_3$—fulfilling the promise made in footnote 1 p. 271.

51.4.3. It is quite instructive to compare $\mathfrak{R}_2$ (in 51.3.2.) with our $\mathfrak{R}_3$:

$S\mathfrak{R}_2T$: neither $S \cap T = \ominus$, nor $S \supset T$.
$S\mathfrak{R}_3T$: neither $S \cap T = \ominus$, nor $S \supset T$, nor $S \subset T$, nor $S \cup T = I$ without $S \cap T$ being a one-element set.

Mere symmetrization of $\mathfrak{R}_2$ (cf. 30.3.2.) would give the three first parts of $\mathfrak{R}_3$, but not the last one. This last part is the essential achievement of (51:G) and (51:H) and not connected in any obvious way with the three others.

One can infer from this how recondite the operations must be by which the program of 30.3.7. might be carried out—if this proves to be feasible at all.

51.5. Simplicity and Decomposition

51.5.1. Let us consider the connections between the concept of a simple game and that of decomposition.

Assume, therefore, that Γ is a decomposable game with the constituents Δ, H (J, K complements in I). Then we must answer this question: What does it mean for Δ, H that Γ is simple?

We begin by determining the sets W, L. Since we must consider them for all three games Γ, Δ, H, it is necessary to indicate this dependence. We write therefore W_Γ, L_Γ; W_Δ, L_Δ; W_H, L_H.

It should be added that we assume neither essentiality nor any normalization for the games Γ, Δ, H. It is convenient, however, to assume them all in a zero-sum form.[2]

(51:I) $S = R \cup T$ ($R \subseteq J$, $T \subseteq K$) belongs to W_Γ [L_Γ] if and only if R belongs to W_Δ [L_Δ] and T to W_H [L_H].

Proof: Replacement of S by its complement (in I), $I - S$,[3] replaces R, T by their respective complements (in J, K). This transformation inter-

[1] And $S\mathfrak{R}_3S$ holds in $\bar{\bar{I}}$: $S \cap S = \ominus$ occurs only for $S = \ominus$, $S \supset S$ never, $S \cup S = I$ only for $S = I$—hence neither of these happen for an S of $\bar{\bar{I}}$.

[2] The reader who recalls the discussions of 46.10. may want to know at this point how the question of the excesses (in Γ, Δ, H the e_0, $\bar{\varphi}$, $\bar{\psi}$ loc. cit.) is to be handled. This question will be clarified in the discussion of 51.6.

[3] It is preferable to write the complement in this way, instead of the usual $-S$, $-R$, $-T$, since we are complementing in different sets.

changes W_Γ, W_Δ, W_H with L_Γ, L_Δ, L_H. Hence our statement concerning the W implies that one concerning the L and *vicc versa.* We are going to prove the latter.

That S belongs to L_Γ is expressed by

(51:7) $$v(S) = \sum_{i \text{ in } S} v((i))$$

since Δ, H are the constituents of Γ, we have $v(S) = v(R) + v(T)$. Hence we can write (51:7) thus:

(51:8) $$v(R) + v(T) = \sum_{i \text{ in } R} v((i)) + \sum_{i \text{ in } T} v((i)).$$

That R belongs to L_Δ and T to L_H is expressed by

(51:9) $$v(R) = \sum_{i \text{ in } R} v((i)),$$

(51:10) $$v(T) = \sum_{i \text{ in } T} v((i)).$$

The assertion which we must prove, then, is the equivalence of (51:7) and (51:9), (51:10).

Clearly (51:9), (51:10) imply (51:7); the reverse implication can be drawn since always

$$v(R) \geqq \sum_{i \text{ in } R} v((i)),$$

$$v(T) \geqq \sum_{i \text{ in } T} v((i))$$

(cf. (31:2) in 31.1.4.).

51.5.2. We are now able to prove:

(51:J) Γ is simple if and only if this is true: Of the two constituents Δ, H one is simple, and the other is inessential.

Proof: The condition is necessary: Simplicity of Γ means this:

(51:11) For any $S \subseteq I$ one and only one of these two statements is true:

(51:11:a) S is in W_Γ.

(51:11:b) S is in L_Γ.

Put $S = R \cup T$ $(R \subseteq J,\ T \subseteq K)$ and apply (51:I) to (51:11). Then this results:

(51:12) For any two $R \subseteq J$, $T \subseteq K$ one and only one of these two statements is true:

(51:12:a) R is in W_Δ and T is in W_H.

(51:12:b) R is in L_Δ and T is in L_H.

Now put $R = \ominus$, $T = K$. Then R belongs to L_Δ and T belongs to W_H. Hence for (51:12:a) W_Δ and L_Δ have a common element: R, and for (51:12:b) W_H and L_H have a common element: T. By (49:E) in 49.3.3. (applied to Δ, H instead of its Γ) the former implies that Δ is inessential, and the latter, that H is inessential.

So we see:

(51:13) If Γ is simple, then either Δ or H is inessential.

The condition is sufficient: We assume, by symmetry, that H is inessential. Then (49:E) in 49.3.3. (applied to H in place of its Γ) shows that every $T \subseteq K$ belongs to both W_H and L_H. Hence we can now reformulate the characterization (51:12) of the simplicity of Γ.

(51:14) For any $R \subseteq J$ one and only one of these two statements is true:

(51:14:a) R is in W_Δ.

(51:14:b) R is in L_Δ.

This is precisely the statement of the simplicity of Δ. So we see:

(51:15) If H [Δ] is inessential, then the simplicity of Γ is equivalent to that one of Δ [H].

(51:13), (51:15) together complete the proof.

51.6. Inessentiality, Simplicity and Composition. Treatment of the Excess

51.6. It is worth while to compare (51:J) with (46:A:c) of 46.1.1. There we found that a decomposable game is inessential if and only if its two constituents are—i.e. inessentiality is hereditary under composition. This is not true for simplicity, which as we know, is the simplest form of essentiality: By (51:J) a decomposable game is not simple if its two constituents are. (51:J) shows that a simple game Δ remains simple under composition if and only if it is combined with an inessential game H—i.e. with a set of "dummies" (cf. footnote 1 on p. 340).

In this connection four further remarks are appropriate:

First: If the simple game Γ obtains as described above from the constituent (simple) game Δ by an addition of "dummies" (i.e. of the inessential game H), then the solutions of Γ are directly obtainable from those of Δ. Indeed, this is described in detail in 46.9.[1]

Second: We stated at the beginning of 49.7. that we use the old form of the theory for simple games. It is therefore worth noting that the type of composition to which we were led (cf. the above remark) is precisely the one for which the old form of the theory is hereditary. (Cf. the end of 46.9. or (46:M) in the first remark in 46.10.4.)

[1] This is, of course, what common sense leads one to expect anyhow. The surprising turns of the theory of decomposition—cf. in particular the resumé in 46.11.—show, however, that it is unsafe to lose sight of the exact results. In this case 46.9. provides the firm ground.

Third: In this connection it becomes clearer also why we had to refrain from considering other excesses than zero—i.e. the new form of the theory in the sense of 44.7.—for the theory of simple games.

Indeed: If we had been able to carry this out successfully, then the results of 46.6. and 46.8. would enable us to deal with all compositions of simple games. Now we have seen that a composition of simple games is not a simple game. In other words: A theory of simple games with general excess would indirectly embrace non-simple games as well. It is therefore not surprising that we could not proceed in generality.[1]

Fourth: In the light of the analysis of 46.10. the above remarks concerning the excess assume the following significance: They show that the concept of simplicity does not stand the general operation of imbedding.[2] This shows that the methodical principle considered in 46.10.5. cannot be applied under all conditions.

51.7. A Criterion of Decomposability in Terms of W^m

51.7.1. In 51.5. we discussed when a decomposable game Γ is simple. We now tackle the converse problem: To decide when a simple game is decomposable.

Let a simple game Γ be given. It will appear that the following concept is of importance: An i of I is *significant* if and only if it belongs to some S of W^m.[3] Denote the set of all significant elements of I—i.e. the sum of all S in W^m—by I_0.

We now proceed in several successive steps:

(51:K) If Γ is simple and decomposable, and if the simple constituent is Δ (cf. (51:J) and the use of notations of 51.5.) then Γ and Δ have the same W^m.

Proof: According to (51:1) the $S = R \cup T$ $(R \subseteq J,\ T \subseteq K)$ of W_Γ obtain by taking any R of W_Δ and any T of W_H. H is inessential (by (51:J)), hence the T of W_H are simply all $T \subseteq K$ (cf. the proof of (51:J)). Consequently this $S = R \cup T$ is minimal—i.e. it belongs to W^m_Γ—if its R, T are minimal. This means that R belongs to W^m_Δ and that $T = \ominus$, i.e. $S = R$.

Thus W^m_Γ and W^m_Δ coincide, i.e. Γ and Δ have the same W^m.

(51:L) With the same assumptions as in (51:K), necessarily $J \supseteq I_0$.

Proof: Γ and Δ have the same W^m (by (51:K)), hence the same significant elements—therefore those of Γ, which form the set I_0, are all among the participants of Δ, which form the set J.

[1] In a certain sense this may be viewed as an application of the methodical principle referred to in footnote 3 on p. 270.

[2] Unless it is merely an addition of "dummies" as discussed above.

[3] Thus a player i is significant if there exists a minimal winning coalition to which he belongs; i.e. if there exists a conceivable essential service he may render.

It will be seen that the opposite of this is a "dummy" (cf. the end of 51.7.3.).

All this refers, of course, to simple games.

(51:M) Assume only that Γ is simple. Then I_0 is a splitting set,[1] the I_0-constituent Δ being simple, and the $(I - I_0)$-constituent H inessential (cf. (51:J)).

Proof: Consider an $S = R \cup T$, $R \subseteq I_0$, $T \subseteq I - I_0$. Then:

(51:16) S is in W if and only if R is in W.

Indeed: If R is in W, then $S \supseteq R$ is too. Conversely: Let S be in W. Then a minimal T in W with $T \subseteq S$ exists. So T is in W^m, every i of T is in I_0. Hence $T \subseteq I_0$. Thus $T \subseteq S \cap I_0 = R$, and therefore R is in W along with T.

(51:17) T is in L.

Indeed: Replace S by $T(\subseteq I - I_0)$; this replaces our R, T by $\ominus$, T. As $\ominus$ is in L, (51:16) permits to infer the same for T.

We now prove:

(51:18) $$\mathrm{v}(S) = \mathrm{v}(R) + \mathrm{v}(T).$$

Consider the S of L and of W separately:

S is in L: R, $T \subseteq S$ are also in L. Hence

$$\mathrm{v}(S) = \sum_{i \text{ in } S} \mathrm{v}((i)) = \sum_{i \text{ in } R} \mathrm{v}((i)) + \sum_{i \text{ in } T} \mathrm{v}((i)) = \mathrm{v}(R) + \mathrm{v}(T),$$

i.e. (51:18).

S in W: By (51:16), (51:17) R is in W and T in L. Hence

$$\mathrm{v}(S) = -\sum_{i \text{ not in } S} \mathrm{v}((i)),$$

$$\mathrm{v}(R) = -\sum_{i \text{ not in } R} \mathrm{v}((i)) = -\sum_{i \text{ not in } S} \mathrm{v}((i)) - \sum_{i \text{ in } T} \mathrm{v}((i)),$$

$$\mathrm{v}(T) = \sum_{i \text{ in } T} \mathrm{v}((i)),$$

and so

$$\mathrm{v}(S) = \mathrm{v}(R) + \mathrm{v}(T).$$

i.e. (51:18)

(51:18) is precisely the statement that I_0 is a splitting set. For all $T \subseteq I - I_0$ (51:17) gives

$$\mathrm{v}(T) = \sum_{i \text{ in } T} \mathrm{v}((i)),$$

hence the $I - I_0$ constituent H is inessential. Consequently the I_0-constituent Δ must be simple by (51:J).

Thus the proof is completed.

[1] In the sense of 43.1.

51.7.2. We are now able to describe the decomposibility of a simple game Γ completely—i.e. we can name its decomposition partition Π_Γ in the sense of 43.3.

(51:N) With the same assumptions as in (51:M): The decomposition partition Π_Γ consists of the set I_0 and of the one-element sets (i) for all i in $I - I_0$.

Proof: All (i), i in $I - I_0$, belong to Π_Γ: By (51:M) $I - I_0$ is a splitting set of Γ with an inessential constituent H. Hence every (i), i in $I - I_0$, is a splitting set of H (use, e.g. (43:J) in 43.4.1.) and so of Γ (use (43:D) in 43.3.1.). Being a one-element set, (i) is necessarily minimal. Hence it belongs to Π_Γ.

I_0 belongs to Π_Γ: I_0 is a splitting set by (51:M). If J is a splitting set $\neq \ominus$, then (51:L) applies to J or to $I - J$, hence either $J \supseteq I_0$ or $I - J \supseteq I_0$, $I_0 \cap J = \ominus$. Both exclude $J \subset I_0$. Thus I_0 is minimal. Hence it belongs to Π_Γ.

No further J belongs to Π_Γ: Any other J of Π_Γ must be disjunct with I_0 and with all (i), i in $I - I_0$, (use (43:F) in 43.3.2.). As the sum of these sets is I, this would necessitate $J = \ominus$—but $\ominus$ is not an element of Π_Γ (cf. the beginning of 43.3.2).

Thus the proof is completed.

51.7.3. Combination of (43:K) in 43.4.1. with (51:N) gives:[1]

(51:O) A simple game Γ is indecomposable if and only if $I_0 = I$, i.e. if and only if all its participants are significant.

We conclude by proving:

(51:P) A simple game Γ possesses precisely one J-constituent which is simple and indecomposable: That with $J = I_0$.

Proof: The I_0-constituent can be formed and is simple by (51:M).

Now consider a simple J-constituent. Then it has, by (51:K), the same W^m and the same significant elements as Γ itself,—hence the latter form the set I_0. So the indecomposability of the J-constituent is by (51:O) equivalent to $J = I_0$.

We call the I_0-constituent Δ_0 of Γ its *kernel.* All other participants—i.e. those of $I - I_0$—are "dummies." (Cf. (51:M) or (51:N), and the last part of 43.4.2.). Hence all that matters in the game Γ takes place in its kernel Δ_0; to see this, it suffices to apply the first remark in 51.6.

52. The Simple Games for Small n

52.1. Program: $n = 1, 2$ Play No Role. Disposal of $n = 3$

52.1. Our next objective is the enumeration of all simple games for the smaller values of n. We propose to push this casuistic analysis so far as is necessary to produce the examples referred to in 50.2. (cf. footnote 2 on

[1] Or more directly of (43:K) in 43.4.1. with (51:L), (51:M).

p. 434), 50.7.2., (cf. footnotes 2, 3, 4 on p. 443), 50.8.2. (cf. footnotes 1, 2, on p. 445).

Since every simple game is essential, we need only consider games with $n \geqq 3$.

For $n = 3$ the situation is this: The (unique) essential three-person game is simple and it has the symbol $[1,1,1]_h$.[1]

So we can assume from now on that $n \geqq 4$.

52.2. Procedure for $n \geqq 4$: The Two-element Sets and Their Role in Classifying the W^m

52.2.1. Let an $n \geqq 4$ be given. We wish to enumerate all simple games with this n. In order to do this it is convenient to introduce a principle of further classification of these games which is very effective for the smaller values of n.

The enumeration in question amounts to the enumeration of the sets W^m for which we have various characterizations available—e.g. that one of (51:G) in 51.4.1.

Consider the smallest sets which may belong to W^m. Since (51:G:d) loc. cit. excludes the empty set and the one-element sets from W^m, this means considering the two-element sets in W^m. These sets possess the following property:

(52:A) A two-element set belongs to W^m if and only if it belongs to W.[2]

Proof: The forward implication is obvious. Now assume conversely, that the two-element set S belongs to W. The proper subsets of S are empty or one-element sets, hence not in W. Therefore S belongs to W^m.

We propose to classify according to two-element sets in W^m.

52.2.2. Conceivably W^m may contain no two-element sets at all. We denote this possibility by the symbol C_0.

The next alternative is that W^m contains precisely one two-element set. By a permutation of the players $1, \cdots, n$ we can make this set to be (1,2). We denote this possibility by the symbol C_1.

Further, W^m may contain two or more two-element sets. Consider two of these. By (51:G:a) loc. cit. they must have a common element. By a permutation of the players $1, \cdots, n$ we can make the common element to be 1, and the two other elements of these two sets 2 and 3.

Se W^m contains (1,2) and (1,3).

We denote the possibility that W^m contains no further two-element sets by the symbol C_2.

52.2.3. Now assume that W^m does contain further two-element sets. Assume furthermore that not all of them contain 1.

Consider therefore a two-element set of W^m not containing 1. By (51:G:a) loc. cit. it must have common elements with (1,2) and (1,3)—1 being excluded, these must be 2 and 3—so the set must be (2,3).

[1] Cf. (50:A) in 50.1.1. and the last remark of 50.2.2.

[2] I.e. a non-minimal set in W must have at least three elements.

Thus (1,2), (1,3), (2,3) belong to W^m. (To this extent we have perfect symmetry in 1,2,3.)

Now consider any other two-element set which may belong to W^m. It cannot contain all three of 1,2,3; by a permutation of these players we can arrange it so that the set in question fails to contain 1. Now it must have common elements with (1,2) and (1,3)—1 being excluded, these must be 2 and 3—so the set must be (2,3), but we assumed it to be different from (2,3) (among others).

Thus W^m contains the two-element sets (1,2), (1,3), (2,3), and no others. We denote this possibility by the symbol C^*.

52.2.4. The remaining alternative is that W^m contains other two-element sets besides (1,2), (1,3), but that they all contain 1.

By a permutation of the players $4, \cdots, n$ we can make these players to be $4, \cdots, k+1$ with a $k = 3, \cdots, n-1$.

Thus W^m contains the two-element sets (1,2), (1,3), (1,4), $\cdots$ $(1,k+1)$, and no others. We denote this possibility by the symbol C_k.

52.2.5. It is convenient to bracket the cases C_0, C_1, C_2 of 52.2.2. and the C_k, $k = 3, \cdots, n-1$ of 52.2.4. together: We then have the cases

$$C_k, \; k = 0, 1, \cdots, n-1.$$

In the case C_k now W^m contains the two-element sets $(1,2), \cdots, (1, k+1)$, and no others. By an additional permutation of the players $1, \cdots, n$[1] we can replace these sets by $(1,n), \cdots, (k,n)$.

It is in this form that we are going to use the case C_k, $k = 0, 1 \cdots, n-1$. Now C_k contains the two-element sets $(1,n), \cdots, (k,n)$, and no others.

Besides these C_k the only alternative is C^* of 52.2.3. which we will not transform.

52.3. Decomposability of the Cases C^*, C_{n-2}, C_{n-1}

52.3.1. Of all these alternatives three can be disposed of immediately: C^*, C_{n-2}, C_{n-1}. We discuss these in a different order.

Ad C^*: Consider an $S \subseteq I$. If S contains two or more of 1,2,3 say (at least) 1,2, then $S \supseteq (1,2)$. (1,2) belongs to W, hence S does too. If S contains one or fewer of 1,2,3, say (at most) 1, then $S \subseteq -(2,3)$. (2,3) belongs to W, $-(2,3)$ to L, hence S to L too. So we see: W consists precisely of those S which contain two or more of 1,2,3. Hence W^m consists precisely of the sets (1,2), (1,3), (2,3).[2] So (1,2,3) is the I_0 of 51.7. for this game.

In other words: The kernel of the game under consideration is a three-person game with the participants 1,2,3, its W^m consisting again of (1,2), (1,3), (2,3). As mentioned before—for the last time in 52.1.—this game has the symbol $[1,1,1]_h$. The remaining $n-3$ players, $4, \cdots, n$ are "dummies."

[1] Namely $\begin{pmatrix} 1, 2, 3, \cdots, n \\ n, 1, 2, \cdots, n-1 \end{pmatrix}$, cf. 28.1.1.

[2] These were the two-element sets of W^m by definition—but we have now shown that they exhaust W^m completely.

So we see:

Case C^* is represented by precisely one game: The three-person game $[1,1,1]_h$, with the necessary number $(n-3)$ of "dummies."

52.3.2. Case C_{n-1}: Consider an $S \subseteq I$. Assume first, that n belongs to S. If S has no further elements, then it is the one-element set (n), and so in L. If S has further elements, say $i = 1, \cdots, n-1$, then $S \supseteq (i, n)$ Now this (i, n) belongs to W, hence S does too. In other words: if n is in S, then S belongs to W, except when $S = (n)$. Applying this to $-S$ gives: If n is not in S, then S belongs to W, when $-S$ does not, i.e. if and only if $-S = (n)$, i.e. $S = (1, \cdots, n-1)$.

Hence W consists precisely of these S: All sets containing n, except the smallest one (n); no set not containing n, except the largest one, $(1, \cdots, n-1)$. One verifies easily that this W indeed fulfills the requirements (49:W*). Also that this game can be described as a weighted majority game, all players $1, \cdots, n-1$ having a common weight, while player n has the $n-2$ fold weight. I.e. this game has the symbol $[1, \cdots, 1, n-2]$.

W^m obtains immediately from W. It consists precisely of these S: $(1, n), \cdots, (n-1, n)$ and $(1, \cdots, n-1)$.[1] It is now easy to verify that this game is homogeneous and normalized by $a = 1$. I.e. that $a_S = 1$ (cf. 50.2.) for all the S of this W^m. Hence we can write $[1, \cdots, 1, n-2]_h$.

So we see:

Case C_{n-1} is represented by precisely one game: The n-person game $[1, \cdots, 1, n-2]_h$.

52.3.3. Ad C_{n-2}: Consider an $S \subseteq I$. Assume first, that n belongs to S. If S has no further elements other than possibly $n-1$, then $S \subseteq (n-1, n)$. Now $(n-1, n)$ is not in W^m, hence not in W (by (52:A) in 52.2.1.). So S is in L along with $(n-1, n)$. If S has further elements, other than $n-1$, say $i = 1, \cdots, n-2$, then $S \supseteq (i, n)$. Now this (i, n) belongs to W hence S does too. So we see: If n is in S, then S belongs to W, except when $S = (n)$ or $(n-1, n)$. Applying this to $-S$ gives: If n is not in S then S belongs to W when $-S$ does not, i.e. if and only if $-S = (n)$ or $(n-1, n)$, i.e. $S = (1, \cdots, n-1)$ or $(1, \cdots, n-2)$.

Hence W consists precisely of these sets S: All sets containing n, except (n), and $(n-1, n)$; no set not containing n, except $(1, \cdots, n-1)$ and $(1, \cdots, n-2)$. One verifies easily that this indeed fulfills the requirement (49:W*).

W^m obtains immediately from W. It consists precisely of these S: $(1, n), \cdots, (n-2, n)$, and $(1, \cdots, n-2)$.[2] So $(1, \cdots, n-2, n)$ is the I_0 of 51.7. for this game.

[1] Thus the two-element sets in W^m are $(1, n), \cdots, (n-1, n)$, as it should be by definition. The new fact is that the only further element of W^m is $(1, \cdots, n-1)$.
Note that this last set is not a two-element set only because of $n \geqq 4$.

[2] Thus the two-element sets in W^m are $(1, n), \cdots, (n-2, n)$, as it should be by definition. The new fact is that the only further element of W^m is $(1, \cdots, n-2)$.
For $n = 4$ this last set is also a two-element set, thereby falsifying the class of the game. (It becomes C^* instead of C_{n-2}, i.e. C_2.)
Hence this class (C_{n-2}) is void, unless $n \geqq 5$.

In other words: The kernel of the game under consideration is an $(n-1)$-person game with the participants $1, \cdots, n-2, n$, its W^m consisting again of $(1, n), \cdots, (n-2, n), (1, \cdots, n-2)$. Thus this is the case C_{n-2} for $n-1$ players—the analogue of the case C_{n-1} for n players (replacing n by $n-1$!) discussed above. Hence it has the symbol $[1, \cdots, 1, n-3]_h$. The remaining player $n-1$ is a "dummy."

So we see:

Case C_{n-2} is represented by precisely one game:[1] The $(n-1)$-person game $[1, \cdots, 1, n-3]_h$ with one dummy.

52.4. The Simple Games Other than $[1, \cdots, 1, l-2]_h$ (with Dummies): The Cases C_k, $k = 0, 1, \cdots, n-3$

52.4. The results of 52.3. deserve to be considered somewhat further and to be reformulated. We saw that for every $l \geqq 4$ the homogeneous weighted majority game of l players $[1, \cdots, 1, l-2]_h$ can be formed.[2] We can even form it for $l = 3$: Then it is the direct majority game of three participants $[1,1,1]_h$. So we will use it for all $l \geqq 3$.

If $n \geqq 4$ then we can obtain a simple n-person game by forming this $[1, \cdots, 1, l-2]_h$ for any $l = 3, \cdots, n$ and adding to it the necessary number of "dummies."

The result of 52.3. was this: This game with $l = 3, n$, and (for $n \geqq 5$) $n-1$ exhausts the cases C^*, C_{n-1}, C_{n-2}.

The odd thing about this result is that these values of l do not exhaust the full set of its possibilities $l = 3, \cdots, n$ (cf. above). That is to say, they do this for $n = 4, 5$, but not for $n \geqq 6$. There remain the $l = 4, \cdots, n-2$ for $n \geqq 6$. What is their significance?

The answer is this: Consider the game $[1, \cdots, 1, l-2]_h$ (l players) with $n-l$ "dummies." Assume only $l = 3, \cdots, n$ and $n \geqq 4$. The W^m consists of $(1, l), \cdots, (l-1, l)$ and $(1, \cdots, l-1)$.[3] Hence we have case C^* when $l = 3$ and case C_{l-1} when $l = 4, \cdots, n$.[4]

Thus we have in these games specimens from the cases $C^*, C_3, \cdots, C_{n-1}$. The result of 52.3. can now be formulated like this: The cases C^*, C_{n-2}, C_{n-1} are exhausted by the pertinent ones among these games.[5]

We restate this conclusion:

(52:B) We wish to enumerate all simple n-person games $n \geqq 4$. The game $[1, \cdots, 1, l-2]_h$ (l players) with the necessary number $(n-l)$ of "dummies" is a simple n-person game for all $l = 3, 4, \cdots, n$. Its case is $C^*, C_3, \cdots, C_{n-1}$, respec-

[1] For $n \geqq 5$—it is void for $n = 4$. Cf. footnote 2 on p. 460.

[2] Cf. Case C_{n-1} above, with l in place of n.

[3] We take players $1, \cdots, l$ as the participants of the kernel $[1, \cdots, 1, l-2]_h$ and players $l+1, \cdots, n$ as "dummies." This differs from the arrangement in case C_{n-1} of 52.3.—where $l = n-1$ and player $n-1$ was "dummy"—by an interchange of players $n-1$ and n.

[4] For $l = 3$, C^* replaces C_2 since $(1, \cdots, l-1)$ is in this case a two-element set.

[5] Hence C_2 is void for $n \leqq 4$, since it occurs on the second list, but not on the first one. Cf. 52.3.

tively. All other simple n-person games (if any) are in the cases $C_0, C_1, \cdots, C_{n-3}$.[1]

52.5. Disposal of $n = 4, 5$

52.5.1. We will discuss the values $n = 4, 5$ fully and some characteristic instances in $n = 6, 7$.

$n = 4$ is easily settled. By (52:B) above, we need only investigate C_0, C_1 for this n. In these cases W^m contains $\leqq 1$ two-element sets. However this is impossible: Since the complement of a two-element set is a two-element set, there must be the same number of two-element sets in W and in L. I.e. half of the total number, which is 6. So W contains 3 two-element sets and the same is true for W^m.[2]

Thus the only simple games for $n = 4$ are those of (52:B). We state this as follows:

(52:C) Disregarding games which obtain by adding dummies to simple games of $<$ four persons,[3] there exists precisely one simple four-person game: $[1,1,1,2]_h$.

52.5.2. Consider next $n = 5$. By (52:B) above we must investigate $l = 0,1,2$. In contrast to the $n = 4$ case, all of these represent concrete possibilities.

C_0: No two-element set is in W^m i.e. in W. So they are all in L and their complements, the three-element sets, are all in W. Thus W consists of all sets of $\geqq$ three elements, and W^m of all sets of three elements. Hence this is the direct majority game $[1,1,1,1,1]_h$.

C_1: (1,2) is the only two-element set in W^m i.e. in W. Passing to the complements: (3,4,5) is the only three-element set in L—i.e. the others are in W. Thus W consists precisely of these sets: (1,2), all three-element sets but (3,4,5), all four- and five-element sets. It is easy to verify that this fulfills (49:W*) and also that its W^m consists of the following sets:

(1,2), (a,b,c), where $a = 1,2$, and $b, c =$ any two of 3,4,5.

Now one shows without difficulty, that this game has the symbol $[2,2,1,1,1]_h$.

C_2: (1,2), (1,3) are the only two-element sets in W^m, i.e. in W Passing to the complements: (3,4,5), (2,4,5) are the only three-element sets in L—i.e. the others are in W. Thus W consists precisely of these sets: (1,2), (1,3), all three-element sets but (2,4,5), (3,4,5), all four- and five-element sets. It is easy to verify that this W fulfills (49:W*) and also that its W^m consists of the following sets:

$$(1,2),\ (1,3),\ (2,3,4),\ (2,3,5),\ (1,4,5).$$

Now one shows without difficulty that this game has the symbol $[3,2,2,1,1]_h$.

[1] All those cases which we succeeded in exhausting so far were void or contained precisely one game. This is, however, not generally true. Cf. the first remark in 53.2.1.

[2] Owing to (52:A) in 52.2.1. This will be used in what follows continuously without further reference.

[3] I.e. to the unique simple three-person game $[1,1,1]_h$.

Hence the simple games for $n = 5$ are these three, and those of (52:B). We state as follows:

(52:D) Disregarding games which obtain by adding dummies to simple games of < five persons,[1] there exist precisely four simple five-person games: $[1,1,1,1,1]_h$, $[1,1,1,2,2]_h$,[2] $[1,1,2,2,3]_h$,[2] $[1,1,1,1,3]_h$.

53. The New Possibilities of Simple Games for $n \geqq 6$

53.1. The Regularities Observed for $n < 6$

53.1. Before we go further, let us draw some conclusions from the above lists.

First: All simple games which we have obtained so far, possessed a symbol, $[w_1, \cdots, w_n]_h$, i.e. they were homogeneous weighted majority games. Having verified this for $n = 4, 5$, the question arises whether it is always true. As stated in footnote 3 on p. 443, this is not so; the first counter-example comes for $n = 6$.

Second: So far every class C_k which contained any game at all, contained only one. This too fails from $n = 6$ on. (Cf. the first remark in 53.2.1.)

Third: One might think *a priori*, that there is great freedom in choosing the weights for a homogeneous weighted majority game. Our lists show, however, that the possibilities are very limited: One each for $n = 3, 4$, and four for $n = 5$.[3] We emphasize that since our lists are exhaustive, this is a rigorously established objective fact—and not a more or less arbitrary peculiarity of our procedure.

Fourth: We can verify the statement of footnote 1 on p. 446 that while the number of elements in W is determined by n (it is 2^{n-1}), that one of W^m may vary for simple games of the same n. This phenomenon begins for $n = 5$.

For $n = 3$: W has 4 elements, W^m in the unique instance has 3. For $n = 4$: W has 8 elements, W^m in the unique instance 4. For $n = 5$: W has 16 elements, W^m in the four instances 10,7,5,5, respectively.

Fifth: We can verify the statement of footnote 2 on p. 443, that the equations (50:8) of 50.4.3., 50.6.2. (with $U = W^m$) may be more numerous than their variables, and nevertheless possess a solution—i.e. a system of imputations in the ordinary sense. The former means that W^m has $> n$ elements, the latter is certainly the case for homogeneous weighted majority games ((50:K) in 50.8.1.).

We saw above that for $n = 3, 4$ W^m necessarily has n elements, but for $n = 5$, it may have 10 or 7 elements as well. And all these games are homogeneous weighted majority games.[4]

[1] I.e. to $[1,1,1]_h$ and to $[1,1,1,2]_h$.

[2] We permute the players of these games (belonging to C_1 and C_2) in order to have an increasing arrangement of weights.

[3] Disregarding permutations of the players!

[4] Thus we have the first counter-examples for $n = 5$: $[1,1,1,1,1]_h$ (the direct majority game) and $[1,1,1,2,2]_h$.

For a simple game, where these solutions do not exist, cf. the fifth remark in 53.2.5.

53.2. The Six Main Counter-examples (for $n = 6, 7$)

53.2.1. We now pass to $n = 6, 7$. A complete exhaustion of these cases, even of $n = 6$, would be rather voluminous. We forego it for this reason. We will only give some characteristic instances of simple games in $n = 6, 7$ which illustrate certain phenomena which begin—as mentioned before—at these n.

First: We mentioned in the second remark of 53.1. that for $n = 6$, a case C_k may contain several games. Indeed, it is not difficult to verify that the two homogeneous weighted majority games

$$[1,1,1,2,2,4]_h, \ [1,1,1,3,3,4]_h,$$

(cf. footnote 2 on p. 463) are different from each other and belong both to C_2.

53.2.2. Second: We mentioned in the first remark of 53.1. that for $n = 6$ a simple game exists which is not a homogeneous weighted majority game, i.e. one which does not possess any symbol $[w_1, \cdots, w_n]_h$. By (50:K) in 50.8.1. this is necessarily the case when there exists no main simple solution; i.e. no system of imputations in the ordinary sense. (Cf. the fifth remark in 53.1.)

Such a game exists indeed, and it is even possible to differentiate further: It is possible to find one which is nevertheless a weighted majority game (without homogeneity!), i.e. which possesses a symbol $[w_1, \cdots, w_n]$, and it is also possible to find one which does not even have that property.

We begin with the first mentioned alternative.

Put $n = 6$: Define W as the system of all those sets $S \subseteq I = (1, \cdots, 6)$ which either contain a majority of all players (i.e. have $\geqq 4$ elements), or which contain exactly half of all players (i.e. have 3 elements), but a majority of all the players 1,2,3 (i.e. $\geqq 2$ of these). In other words: The players 1,2,3 form a privileged group as against the players 4,5,6—but their privilege is rather limited: Normally the overall majority wins; only in case of a tie does the majority of the privileged group decide.

It is easy to verify that this W satisfies (49:W*). The game is clearly a weighted majority one: It suffices to give the members of the privileged group (1,2,3) some excess weight over those of the others (4,5,6), which must be insufficient to override an overall majority. Any symbol

$$[w, w, w, 1, 1, 1]$$

with $1 < w < 3$ will do.[1]

[1] $w > 1$ is necessary, e.g. for $S = (1,2,4)$ to defeat $-S = (3,5,6)$ (i.e. $2w + 1 > w + 2$). $w < 3$ is necessary, e.g. for $S = (3,4,5,6)$ to defeat $-S = (1,2)$ (i.e. $w + 3 > 2w$).

W^m is quickly determined; it consists of these sets:

$$
(S_1) \begin{cases} (S_1'): & (1,2,3) & \\ (S_1''): & (a,b,h) & \text{where } a,b = \text{any two of } 1,2,3, \\ & & \quad h = 4 \text{ or } 5 \text{ or } 6 \\ (S_1'''): & (a,4,5,6) & \text{where } a = 1 \text{ or } 2 \text{ or } 3.^1 \end{cases}
$$

The equations (50:8) of 50.4.3., 50.6.2. (with $U = W^m$)—which determine a main simple solution in the sense of 50.8.1.—are:

$$
(E_1) \begin{cases} (E_1'): & x_1 + x_2 + x_3 = 6, & \\ (E_1''): & x_a + x_b + x_h = 6, & \text{where } a,b = \text{any two of } 1,2,3, \\ & & \quad h = 4 \text{ or } 5 \text{ or } 6 \\ (E_1'''): & x_a + x_4 + x_5 + x_6 = 6, & \text{where } a = 1 \text{ or } 2 \text{ or } 3 \end{cases}
$$

These equations (E_1) cannot be solved.[2] Indeed, (E_1'') with $a = 1, b = 2$ and $h = 4,5,6$ shows that $x_4 = x_5 = x_6$; (E_1''') with $a = 1,2,3$ shows that $x_1 = x_2 = x_3$; now (E_1') gives $3x_1 = 6$, $x_1 = 2$; hence (E_1'') gives $4 + x_4 = 6$, $x_4 = 2$; and then (E_1''') gives $2 + 6 = 6$—a contradiction.

It should be noted that the ordinary economic aspect of this occurrence would be this: (S_1'') (i.e. (E_1'')) shows that the services of players 4,5,6 can be substituted for each other—hence they are of the same value. (S_1''') (i.e. (E_1''')) shows the same for 1,2,3. Now comparison of (S_1') and (S_1'') shows that one player of the group 1,2,3 can be substituted for one player of the group 4,5,6—and comparison of (S_1'') and (S_1''') shows that one player of the former group can be substituted for two players of the latter. Hence no substitution rate between these two groups can be defined at all. The natural way out would be to declare some of the sets of W^m enumerated in (S_1) to be "no profitable uses" of the players' services. In the sense of 50.4.3. this amounts to choosing $U \subset W^m$ (Cf. also 50.7.1. and footnote 4 on p. 443). Whether in this game a $U \subset W^m$ can have the required properties (cf. 50.7.1.) could be decided by a simple but somewhat lengthy combinatorial discussion, which has not yet been carried out. The existence of such a V is highly improbable, because it can be shown that it would have mathematically unlikely characteristics if it existed.

This game is also very peculiar in another respect: It is possible to prove that there exists no solution V which contains only a finite number of imputations and which possesses the full symmetry of the game itself; i.e. invariance under all permutations of the players 1,2,3 and under all permutations of the players 4.5,6. We do not discuss this rather lengthy proof at this place.[3] Thus the type of solution which one would term the natural one does not exist.

[1] Thus W^m has $1 + 9 + 3 = 13$ elements.

[2] They are 13 equations in 6 variables, but this in itself is not necessarily an obstacle, as the fifth remark in 53.1. shows.

[3] Whether any finite solution V exists at all, is not known. We suspect that even this question will be anwered in the negative.

This is an indication of how extremely careful one must be in terming extraordinary solutions "unnatural," or in trying to exclude them.

53.2.3. Third: Let us now consider the second alternative referred to in the second remark above: A simple game for $n = 6$, which is no majority game at all—i.e. which has no symbol $[w_1, \cdots, w_n]$. This alternative itself can be subdivided further: It is possible to find a game such that it possesses a main simple solution (cf. above)—and it is also possible to find one that has no main simple solution.

Consider the first case:

Put $n=6$. Define W as the system of all those sets $S(\subseteq I = (1, \cdots, 6))$ which either contain a majority of all players (i.e. have $\geqq 4$ elements), or which contain exactly half (i.e. have 3 elements), but an even number of the players 1,2,3 (i.e. have 0 or 2 of these). Comparing this with the example in the second remark above, this observation must be made: The players 1,2,3 still form a group of special significance, but it would be misleading to call their significance a privilege—since their absence from the tying (i.e. three-element) set S is just as advantageous as their strong representation (presence of precisely two of them), and the presence of all of them just as disastrous as their weak representation (presence of precisely one of them). They bring about a decision not by their presence in S but by an arithmetical relation:[1]

It is easy to verify that this W fulfills (49:W*) in 49.6.2.[2]

Let us now determine W^m. Since W contains all $\geqq$ four-element sets, no $\geqq$ five-element set can be in W^m. Consider now a four-element set in W.

If the number of players 1,2,3 in it is even, remove from it a player 4 or 5 or 6.[3] If the number of players 1,2,3 in it is odd, remove from it a player 1 or 2 or 3.[4] At any rate a three-element subset with an even number of players 1,2,3 obtains—i.e. one in W. So no four-element set can be in W^m. Hence W^m consists of the three-element sets in W. These are:

$$(S_2)\begin{cases} (S_2'): & (4,5,6) \\ (S_2''): & (a,b,h) \qquad \text{where } a,b = \text{any two of } 1,2,3; \\ & \qquad\qquad h = 4 \text{ or } 5 \text{ or } 6.^5 \end{cases}$$

[1] Note also that the group 4,5,6 has a similar significance: Since S must have three elements (in order that these criteria become operative), the statement that an even number of 1,2,3 is in S is equivalent to the statement that an odd number of 4,5,6 is in S.

This lends further emphasis—if any be needed—to our frequently made observation concerning the great complexity of the possible forms of social organization, and the extreme wealth of attendant phenomena.

[2] Note in particular that always one of S and $-S$ belongs to W: This is evident if one of the two has $\geqq 4$ elements (and so the other $\leqq 2$). Otherwise both S and $-S$ have 3 elements. Hence one of them contains an even number of players 1,2,3 and the other an odd one.

[3] This is possible, as 1,2,3 are only 3 players.

[4] This is possible, as 4,5,6 are only 3 players.

[5] Thus W^m has $1 + 9 = 10$ elements.

If this game had a symbol $[w_1, \cdots, w_n]$, then there would be

$$\sum_{i \text{ in } S} w_i > \sum_{i \text{ not in } S} w_i \qquad \text{for all } S \text{ in } W.$$

Apply this to the sets of W^m enumerated in (S_2). This gives in particular:

$$\begin{aligned} w_4 + w_5 + w_6 &> w_1 + w_2 + w_3, \\ w_1 + w_2 + w_6 &> w_3 + w_4 + w_5, \\ w_1 + w_3 + w_5 &> w_2 + w_4 + w_6, \\ w_2 + w_3 + w_4 &> w_1 + w_5 + w_6. \end{aligned}$$

Adding these inequalities gives:

$$2(w_1 + w_2 + w_3 + w_4 + w_5 + w_6) > 2(w_1 + w_2 + w_3 + w_4 + w_5 + w_6),$$

a contradiction.

The equations (50:8) of 50.4.3., 50.6.2. (with $U = W^m$)—which determine a main simple solution—on the other hand are:

$$(E_2) \begin{cases} (E_2'): & x_4 + x_5 + x_6 = 6, \\ (E_2''): & x_a + x_b + x_h = 6, \qquad \text{where } a,b = \text{any two of } 1,2,3; \\ & \qquad\qquad\qquad\qquad h = 4 \text{ or } 5 \text{ or } 6. \end{cases}$$

They are obviously solved by $x_1 = \cdots = x_6 = 2$.[1]

In the ordinary economic terminology one would have to say that the structural difference between the groups of players 1,2,3 and 4,5,6 cannot be expressed by weights and majorities, and that as far as values are concerned, there is no difference.

53.2.4. Fourth: Note that the above example is also suited to establish the difference between the homogeneous weighted majority principle and the existence of a main simple solution, as discussed in 50.8.2. Indeed, it is an instance of = in (50:21) loc. cit.: Since $x_1 = \cdots = x_6 = 2$ (cf. above), so

$$\sum_{i=1}^{n} x_i = 12 = 2n.$$

53.2.5. Fifth: Now consider the second case described in the third remark above: A simple game for $n = 6$, for which neither a symbol

$$[w_1, \cdots, w_n]$$

nor a main simple solution exists.

Compared with the two previous examples—given in the second and third remark above—this one is based on less transparent principles. This is not surprising since now all our simplifying criteria are to be unfulfilled.

This is the example:

Put $n = 6$. Define W as the system of all those sets $S (\subseteq I = (1, \cdots, 6))$ which contain either a majority of all players (i.e. have $\geqq 4$ elements), or

[1] It is easily seen that this is their only solution.

which contain exactly half (i.e. have 3) elements, and fulfill the following further condition: Either S contains player 1, but it is not (1,3,4) or (1,5,6)[1]—or S is (2,3,4) or (2,5,6).[2,3]

It is easy to verify that this W satisfies (49:W*) in 49.6.2.

W^m can be determined without serious difficulties. It turns out to consist of these sets:

$$(S_3)\begin{cases} (S_3'): & (1,2,b) & \text{where } b = 3 \text{ or } 4 \text{ or } 5 \text{ or } 6 \\ (S_3''): & (1,a,b) & \text{where } a = 3 \text{ or } 4,\ b = 5 \text{ or } 6^{4} \\ (S_3'''): & (2,p,q) & \text{where } p = 3,\ q = 4, \text{ or } p = 5,\ q = 6.^{4} \\ (S_3^{IV}): & (3,4,5,6)^{5} & \end{cases}$$

If this game had a symbol $[w_1, \cdots, w_n]$, then there would be

$$\sum_{i \text{ in } S} w_i > \sum_{i \text{ in } -S} w_i \qquad \text{for all } S \text{ in } W.$$

Apply this to the sets of W^m, enumerated in (S_3). This gives in particular:

$$\begin{aligned} w_1 + w_3 + w_5 &> w_2 + w_4 + w_6, \\ w_1 + w_4 + w_6 &> w_2 + w_3 + w_5, \\ w_2 + w_3 + w_4 &> w_1 + w_5 + w_6, \\ w_2 + w_5 + w_6 &> w_1 + w_3 + w_4. \end{aligned}$$

Adding these four inequalities gives:

$$2(w_1 + w_2 + w_3 + w_4 + w_5 + w_6) > 2(w_1 + w_2 + w_3 + w_4 + w_5 + w_6),$$

a contradiction.

The equations (50:8) of 50.4.3., 50.6.2. (with $U = W^m$)—which determine a main simple solution—on the other hand are:

$$(E_3)\begin{cases} (E_3'): & x_1 + x_2 + x_b = 6, & \text{where } b = 3 \text{ or } 4 \text{ or } 5 \text{ or } 6, \\ (E_3''): & x_1 + x_a + x_b = 6, & \text{where } a = 3 \text{ or } 4,\ b = 5 \text{ or } 6, \\ (E_3'''): & x_2 + x_p + x_q = 6, & \text{where } p = 3,\ q = 4, \text{ or } p = 5,\ q = 6, \\ (E_3^{IV}): & x_3 + x_4 + x_5 + x_6 = 6. & \end{cases}$$

These equations (E_3) cannot be solved.[6] Indeed (E_3'') shows that $x_3 = x_4$ and $x_5 = x_6$, hence (E_3''') gives $x_2 + 2x_3 = 6$, $x_2 + 2x_5 = 6$, therefore

[1] I.e. it is $(1,a,b)$ with $a = 2$, $b = 3$ or 4 or 5 or 6; or with $a = 3$ or 4, $b = 5$ or 6.

[2] The complements of the previously excluded sets (1,5,6) and (1,3,4).

[3] If this last exception—concerning (1,3,4), (1,5,6) and (2,3,4), (2,5,6)—were omitted, then W would be defined by this principle: The player 1 is privileged—normally the overall majority wins, but ties are decided by player 1.

It is easy to verify that this is simply the game $[2,1,1,1,1,1]_h$. I.e. this case is even simpler, than our—in some ways, analogous—example in the second remark above—since the privilege existing here has a numerical value in the conventional sense.

Thus the complicating exception—concerning (1,3,4), (1,5,6) and (2,3,4), (2,5,6)—is decisive in bringing forth the real character of our example.

[4] Note that a,b vary independently of each other, while p,q do not!

[5] Thus W^m has $4 + 4 + 2 + 1 = 11$ elements.

[6] They are 10 equations in 6 variables, cf. footnote 2 on p. 465.

$x_3 = x_5$, and so $x_3 = x_4 = x_5 = x_6$. Now (E_3^{IV}) gives $4x_3 = 6$, $x_3 = \frac{3}{2}$ whence (E_3''), (E_3''') yield $x_1 + 3 = 6$, $x_2 + 3 = 6$, i.e. $x_1 = x_2 = 3$. Finally (E_3') becomes $3 + 3 + \frac{3}{2} = 6$,—a contradiction.

As to the interpretation of this insolubility, essentially the same comments are in order as at the corresponding point of the second remark above.

53.2.6. Sixth: We have already referred to the difference between the homogeneous weighted majority principle, and the existence of a main simple solution, as discussed in 50.8.2. This was done in the fourth remark above, where an example for = in (50:21) loc. cit. was given. We will now give an example for > in (50:21) loc. cit.

Since we found that for $n \leqq 5$ all simple games were homogeneous weighted majority games, we must now assume $n \geqq 6$. We do not know whether an example of the desired kind exists for $n = 6$—the one which will be given has $n = 7$.

Put $n = 7$. Define W as the system of all those sets $S(\subseteq I = (1, \cdots, 7))$ which contain any one of the 7 following three-element sets:[1]

(S_4): (1,2,4), (2,3,5), (3,4,6), (4,5,7), (5,6,1), (6,7,2), (7,1,3)

The principle embodied in this definition can be illustrated in various ways.

This is one: The 7 sets of (S_4) obtain from the first one $-(1,2,4)$—by cyclic permutation. I.e. by increasing all its elements by any one of the numbers 0,1,2,3,4,5,6—but all three by the same one—provided that the numbers 8,9,10,11,12,13 are identified with 1,2,3,4,5,6 respectively.[2]

In other words: They obtain from the set marked × × × on Figure 89, by any one of the 7 rotations which this figure allows.

Another illustration: Figure 90 shows the players $1, \cdots, 7$ in an arrangement in which it is feasible to mark 7 sets of (S_4) directly. They are indicated by the 6 straight lines and the circle ○.[3]

The verification, that this W fulfills (49:W*) is not difficult, but we prefer to leave it to the reader if he is interested in this type of combinatorics. W^m consists obviously of the 7 sets of (S_4).

It is easy to show—along the lines given in the third and fifth remarks above—that this is not a weighted majority game. We omit this discussion.

The equations (50:8) of 50.4.3., 50.6.2. (with $U = W^m$)—which determine a main simple solution—on the other hand are:

(E_4): $x_a + x_b + x_c = 7$, where (a,b,c) runs over the 7 sets of (S_4).

They are obviously solved by $x_1 = \cdots = x_7 = \frac{7}{3}$.[4]

[1] Thus W^m has 7 elements.

[2] In the terminology of number theory: Reduced modulo 7.

[3] The reader who is familiar with projective geometry will note that Figure 90 is the picture of the so-called 7 point plane geometry. The seven sets in question are its straight lines, each one containing 3 points, and the circle ○ also rating as such.

One should add that other projective geometries do not seem to be suited for our present purpose.

[4] It is easily seen that this is their only solution.

Now we can establish that > holds in (50:21) in 50.8.2. Indeed:

$$\sum_{i=1}^{n} x_i = \tfrac{49}{3} > 14 = \dot{2n}.$$

As the games discussed in the second, third and fifth remarks, this one too corresponds to an organizational principle that deserves closer study. In this game the sets of W^m, i.e. the decisive winning coalitions are always minorities (three-element sets). Nevertheless, no player has any advantage over any other: Figure 89 and its discussion show that any cyclic permutation of the players 1, · · · , 7—i.e. any rotation of the circle of Figure 89—leaves the structure of the game unaffected. Any player can be carried in this manner into any other player's place.[1] Thus the structure of the game

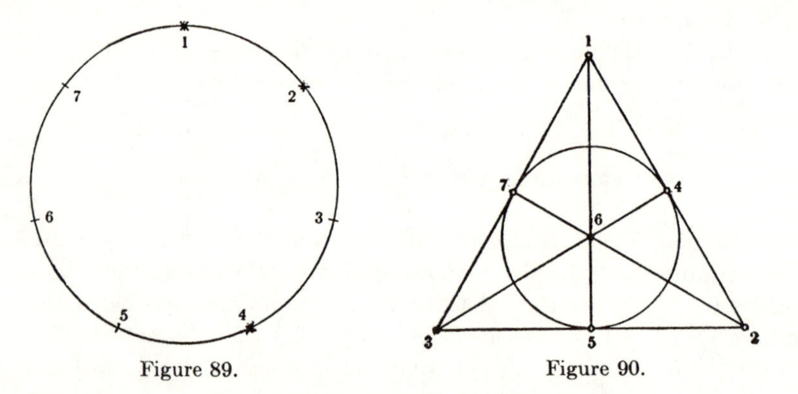

Figure 89. Figure 90.

is determined not by the individual properties of the players[2]—all are, as we saw, in exactly the same position—but by the relation among the players. It is, indeed, the understanding reached among 3 players who are correlated by (S_4)[3] which decides about victory or defeat.

54. Determination of All Solutions in Suitable Games

54.1. Reasons to Consider Other Solutions than the Main Solution in Simple Games

54.1.1. Our discussion of simple games thus far placed most emphasis upon the special kind of solutions discussed in 50.5.1.-50.7.2. and particularly on the main simple solution of 50.8.1. On the basis of what we have learned in the previous sections—especially from the examples of 53.2.—this approach does not appear to do justice to all aspects of our problem.

[1] The game is nevertheless not fair in the sense of 28.2.1., since e.g. the two three-element sets (1,2,4) and (1,3,4) act differently: The former belongs to W, the latter to L. (So in the reduced form of the game, with $\gamma = 1$, the v(S) of the former is 4, and that of the latter is -3.)

[2] Which the rules of the game might give them.

[3] There exist in this game no significant relations between any two players: It is possible to carry any two given players into any two given ones by a suitable permutation (of all players 1, · · · , 7) which leaves the game invariant.

To begin with, we have seen that we cannot expect all simple games to have solutions of the type mentioned. Already for $n = 6$ a wealth of new possibilities emerged. This is significant, since 6 is a sizeable number from the point of view of combinatorics, but a small one when viewed in the context of social organization.

But further, even when these solutions exist, indeed even for the homogeneous weighted majority games, they do not tell the whole story. For the most primitive specimen of that class, the essential three-person game which as we know has the symbol $[1,1,1]_h$, there exist many solutions. And our discussion in 33. showed that they are all essential for our understanding of the characteristics and the implications of our theory—actually some fundamental interpretations were first obtained at that point.

54.1.2. Consequently it is important to determine all solutions of a simple game and, as long as we are not able to do this for all simple games, to do it for as many simple games as possible. In particular this should be done for at least one simple game at each value of n. Such results would provide some information about the structural possibilities and principles of classification of solutions for n-participants.

It is true that this information would be equally welcome if it could be obtained for other than simple games. However the simple games possess a manifest advantage over all others when solutions are to be determined systematically: For simple games the so-called preliminary conditions of 30.1.1. cause no difficulties (cf. 31.1.2.), since there every set S is certainly necessary or certainly unnecessary (cf. 49.7.).

It is equally true that the determinations which we envisage would only provide information concerning a few isolated cases. But they would nevertheless cover all n—i.e. enable us to vary n at will. This is bound to lead to essential insights.

54.2. Enumeration of Those Games for Which All Solutions Are Known

54.2.1. Let us take inventory of the cases for which we already know all solutions of a game. There are three:

(a) All inessential games (cf. (31:P) in 31.2.3., complemented by (31:I) in 31.2.1.).

(b) The essential three-person game both in the old theory (excess zero) and in the new one (general excess). (Cf. 32.2.3. for the former and the analysis of 47.2.1.-47.7. for the latter.)

(c) All decomposable games—provided that all solutions of the constituents are known. (Cf. (46:I) in 46.6.)

Clearly we can use the device (c) to combine the games provided by (a) and (b)—thus obtaining games for which all solutions are known.[1] In this process of building up (a) furnishes only "dummies" (cf. the end of 43.4.2.), hence we may well dispense with it, since we want structural

[1] This can also be expressed in the following way:

A given game Γ is the composite of its indecomposable constituents, according to the definition of the decomposition partition at the end of 43.3. and (43:E) eod. We

information. Thus we are left with those games which are obtained by iterated application of (c) to (b). In this way we can obtain games which are the composite of essential three-person games.[1]

54.2.2. This gives $n = 3k$-person games for which we know all solutions. Since k is arbitrary, we can make n arbitrarily great. To this extent things are satisfactory. However the fact remains that such an n-person game is just a polymer of the essential three-person game—the players form in reality sets of 3 which the rules of the game fail to link to each other. It is true that our results concerning the solutions of decomposable games show that a linkage of these sets of players is nevertheless provided for in the typical solution—i.e. by the typical standard of behavior. But naturally we want to see how the ordinary type of linkage, explicitly set by the rules of the game, affects the organization of the players—i.e. the solutions or standards. And we want this for great numbers of players.

Consequently we must look for further n-person games for which it is possible to determine all solutions.

54.3. Reasons to Consider the Simple Game $[1, \cdots, 1, n-2]_h$

54.3.1. As pointed out above, we are going to look for these specimens among the simple games.[2] Now it turns out that there is a certain simple game for every $n \geqq 3$, for which this determination can be carried out. This is the only n-person game, of a general n, for which we succeeded thus far in the general determination. This obviously gives it a position of special interest. We will also see that it permits interesting interpretations in several respects.

The game in question has already occurred in 52.3. and in (52:B) of 52.4. It is the homogeneous weighted majority game $[1, \cdots, 1, n-2]_h$, (n players).

---

know from (43:L) in 43.4.2. that the sets of participants into which this partition subdivides them, are sets of 1 or $\geqq$ 3 elements.

The simplest possibility is therefore that they are all one-element sets. According to (43:J) in 43.4.1. this means that the game is inessential—i.e. it takes us back to the case (a) above.

The next simplest possibility is that they are all one- or three-element sets. These are exactly the games which we can form according to (c) from (a) and (b). I.e. it is for these that we know all solutions.

This is satisfactory since it shows that a classification based on the sizes of the indecomposable constituents (i.e. of the elements of the decomposition partition, cf. (43:L) in 43.4.2.) is a natural one: Our progress in obtaining all solutions follows precisely the lines drawn by it.

It also stresses how limited these results are: It is indeed a very special occurrence when a game is decomposable at all. (Remember the defining equations of (41:6) or (41:7) in 41.3.2., according to the criterion at the end of 42.5.2.!) The typical n-person game is indecomposable and cannot be reached by means of (c).

[1] By application of strategic equivalence we can assume them all to be in the reduced form. But denoting their γ by $\gamma_1, \cdots, \gamma_k$ respectively, we cannot expect to make them all equal to 1 by a change of unit (unless $k = 1$). Indeed, their ratios $\gamma_1 : \cdots : \gamma_k$ are unaffected by changes of unit.

[2] For this reason we use the old theory, i.e. excess zero. Cf. the third and fourth remarks in 51.6.

54.3.2. As discussed in 52.3. in this game the minimal winning coalitions are these S: $(1, n), \cdots, (n-1, n)$ and $(1, \cdots, n-1)$. I.e. player n wins as soon as he finds any ally at all, but if he remains completely isolated then he loses.[1] This result invites some remarks:

First: The statement of this rule suggests strongly that player n is in a privileged position: He needs only one ally to win, while the others need each other without exception. Actually the situation is this: Player n needs a coalition of two, the others together need one of $n-1$, hence a privilege exists only if $n-1 > 2$, i.e. $n \geqq 4$.

For $n = 3$ there is, indeed, no difference between the three players: We have then the game $[1,1,1]_h$, the unique essential three-person game which is obviously symmetric.

Second: The privilege of player n is as extensive as a privilege can be: We required that n must find at least one ally in order to win and it would not have been possible to require less.[2] It is impossible to specify that n can win without an ally, i.e. to declare that the one-element set (n) to be winning—this is incompatible with the essentiality of the game. (This was discussed extensively in 49.2.)

55. The Simple Game $[1, \cdots, 1, n-2]_h$

55.1. Preliminary Remarks

55.1. The determination of all solutions of the game which we discussed above will show that they fall into a complex array of classes, exhibiting widely varying characteristics. These create an opportunity for the interpretations we have alluded to previously. We will discuss some of them, while further discussions along the same line will probably follow in subsequent investigations.

The exact derivation of this complete list of solutions will be given in the sections which follow (55.2.-55.11.). This derivation is of not inconsiderable complexity. We are giving it in full for the same reasons as the analogous one concerning the solutions of decomposable games in Chapter IX.: The proof itself is a convenient and natural vehicle for certain interpretations. It presents at several stages an opportunity to bring out verbally the emerging structural features of the organizations under consideration. In fact this circumstance will be even more pronounced in the proofs of this chapter than in those of Chapter IX.

55.2. Domination. The Chief Player. Cases (I) and (II)

55.2.1. After these preliminaries we proceed to the systematic investigation of the game $[1, \cdots, 1, n-2]_h$ (n players). Assume that it is in the reduced form, normalized by $\gamma = 1$.

[1] As every one-element set must.

[2] We stated above that player n is not at all privileged in this game when $n = 3$—and now we state that he is as privileged as he possibly can be! Yet $n = 3$ is no exception from this statement: Since there exists only one essential three-person game, the

We begin with an immediate observation on domination:

(55:A) For $\overrightarrow{\alpha} = \{\alpha_1, \cdots, \alpha_n\}$, $\overrightarrow{\beta} = \{\beta_1, \cdots, \beta_n\}$,

$$\overrightarrow{\alpha} \succ \overrightarrow{\beta}$$

if and only if either

(55:1) $\alpha_n > \beta_n$ and $\alpha_i > \beta_i$ for some $i = 1, \cdots, n-1$,

or

(55:2) $\alpha_i > \beta_i$ for all $i = 1, \cdots, n-1$.

Proof: This coincides with (49:J) in 49.7.2., since W^m consists of the sets $(1, n), \cdots, (n-1, n)$ and $(1, \cdots, n-1)$.

Note that $\sum_{i=1}^{n} \alpha_i = \sum_{i=1}^{n} \beta_i = 0$ permit us to infer from (55:2) the validity of

(55:3) $\alpha_n < \beta_n$.

Hence:

(55:B) $\overrightarrow{\alpha} \succ\prec \overrightarrow{\beta}$ necessitate $\alpha_n \neq \beta_n$.

Proof: By symmetry we need only consider $\overrightarrow{\alpha} \succ \overrightarrow{\beta}$. We saw that this implies (55:1) or (55:3), hence at any rate $\alpha_n \neq \beta_n$.

These two results, simple as they are, deserve some interpretative comment.

We discussed in 54.3. that the player n has a privileged role in this game.[1] He is in a situation which is comparable to that of a *monopolist*, with the inescapable limitation (cf. the second remark loc. cit.) that he must find at least one ally. I.e. a general coalition of all others against him—but nothing less than that—can defeat him. We will call him the *chief player* in this game.[2]

55.2.2. These circumstances are brought out clearly in (55:1) and (55:2). One may say that (55:1) is the direct form of domination by the chief player and an arbitrary ally (any player $i = 1, \cdots, n-1$) while (55:2) may be termed a state of general cooperation against him. (55:1), (55:3) or (55:B) show that in a domination the chief player is certainly affected: Advantageously in (55:1) (the direct form of domination with the chief player), adversely in case (55:2) (the general cooperation against the chief player). Any other player can be unaffected, left aside, in a domination.[3]

---

position in which a player finds himself there may as well be called the best possible one—since it is the only one there is.

[1] Except the case $n = 3$, about which more will be said later.

[2] As to the case $n = 3$, the end of the first remark in 54.3. should be kept in mind.

[3] I.e. It may happen for an $i = 1, \cdots, n-1$ that $\overrightarrow{\alpha} \succ \overrightarrow{\beta}$ and $\alpha_i = \beta_i$. This is actually only possible when $n \geq 4$, cf. again the observations concerning $n = 3$.

55.2.3. Now consider a solution V of this game.[1] Form

$$\operatorname{Max}_{\vec{\alpha} \text{ in } V} \alpha_n = \bar{\omega},$$
$$\operatorname{Min}_{\vec{\alpha} \text{ in } V} \alpha_n = \underline{\omega}.^{2}$$

Clearly

$$-1 \leqq \underline{\omega} \leqq \bar{\omega}.$$

The meaning of $\underline{\omega}$, $\bar{\omega}$ is plain: They represent the worst and the best possible outcome for the chief player, within the solution V.

We distinguish two possibilities:

(I) $\underline{\omega} = \bar{\omega}$,

(II) $\underline{\omega} < \bar{\omega}$.

55.3. Disposal of Case (I)

55.3.1. Consider the case (I). This means that for all $\vec{\alpha}$ in V

(55:4) $\alpha_n = \bar{\omega}$,

i.e. that the chief player obtains the same amount under all conditions within the solution. In other words: (I) expresses that the chief player is segregated in the game in the sense of 33.1. Considering the central role of the chief player it is not unreasonable that the first alternative distinction in our discussion should proceed along this line.[3]

55.3.2. Let us now discuss V in case (I).

(55:C) V is precisely the set of all $\vec{\alpha}$ fulfilling (55:4).

[1] In the sense of the old theory, cf. footnote 2 on p. 472.

[2] That these quantities can be formed, i.e. that the maximum and minimum exist and are assumed can be ascertained in the same way as in footnote 1 on p. 384. Cf. in particular (*) loc. cit.

[3] The reference to 33.1. re-emphasizes that this procedure is analogous to that one of the essential three-person game.

This will appear even more natural if it is recalled that the essential three-person game is a special case of the one we consider now—pertaining to $n = 3$. (Cf. e.g. the end of the first remark in 54.3.)

Closer consideration of the case $n = 3$ shows, however, that this analogy suffers from a rather unsatisfactory limitation: In this case the game is really symmetric, and so any one of the three players could have been called the chief player. (Cf. also footnote 2 on p. 474.) In 33.1. the segregation in question was indeed applicable to any one of the three players, and now we have arbitrarily restricted it to player n!

Yet there is no way so far to apply this to the other players too if we want our discussion to cover all $n \geqq 3$ (and not only $n = 3$): For $n \geqq 4$ the chief player and his role are unique.

The only sense in which this situation can be accepted—temporarily—is that of keeping in mind that case (II) must in fine turn out to be a composite one.

Thus for $n = 3$ comparison with the classification of 32.2.3.—which is analyzed in 33.1.—shows this: Our case (I) is one of the possibilities of (32:A) there: discrimination against player 3. Our case (II), on the other hand, covers the other two possibilities of (32:A): discrimination against players 1,2—together with (32:B), the non-discriminatory solution. So (II) is really an aggregate of 3 possibilities when $n = 3$.

This scheme will, indeed, generalize for all n. Cf. (*e*) in the fourth remark of 55.12.5.

Proof: We know already that all $\overrightarrow{\alpha}$ of V fulfill (55:4). If, conversely, a $\overrightarrow{\beta}$ fulfills (55:4), then every $\overrightarrow{\alpha}$ of V has $\alpha_n = \beta_n$, hence (55:B) excludes $\overrightarrow{\alpha} \succ \overrightarrow{\beta}$. Hence $\overrightarrow{\beta}$ belongs to V.

Thus V is determined easily enough, but we must now answer the converse question: Given an $\bar{\omega} \geqq -1$, is the V defined by (55:4) (i.e. by (55:C)) a solution? I.e. does it fulfill (30:5:a), (30:5:b) of 30.1:1.?

Now (55:B) and (55:4) exclude $\overrightarrow{\alpha} \succ \overrightarrow{\beta}$ for $\overrightarrow{\alpha}, \overrightarrow{\beta}$ in V, hence (30:5:a) is automatically satisfied. Therefore we need only investigate (30:5:b) of 30.1.1. I.e. we must secure this property:

(55:5) If $\beta_n \neq \bar{\omega}$, then $\overrightarrow{\alpha} \succ \overrightarrow{\beta}$ for some $\overrightarrow{\alpha}$ with $\alpha_n = \bar{\omega}$.

More explicitly: We must determine what limitations (55:5) imposes upon $\bar{\omega}$.

The $\beta_n \neq \bar{\omega}$ of (55:5) can be classified:

(55:6) $\beta_n > \bar{\omega}$,

(55:7) $\beta_n < \bar{\omega}$.

We show first:

(55:D) In the case (55:6) condition (55:5) is automatically fulfilled.

Proof: Assume $\beta_n > \bar{\omega}$, i.e. $\beta_n = \bar{\omega} + \epsilon$, $\epsilon > 0$. Define

$$\overrightarrow{\alpha} = \{\alpha_1, \cdots, \alpha_n\}$$

by $\alpha_i = \beta_i + \frac{\epsilon}{n-1}$ for $i = 1, \cdots, n-1$, and $\alpha_n = \beta_n - \epsilon = \bar{\omega}$. $\overrightarrow{\alpha}$ is an imputation of the desired kind with $\overrightarrow{\alpha} \succ \overrightarrow{\beta}$ by (55:2).

Thus only the case (55:7) remains. Concerning this case we have:

(55:E) For $\bar{\omega} = -1$, (55:7) is impossible.

Proof: $\beta_n \geqq -1$, hence not $\beta_n < \bar{\omega} = -1$.

The possibility $\bar{\omega} > -1$ is somewhat deeper.[1]

(55:F) Assume $\bar{\omega} > -1$ and case (55:7). Then condition (55:5) is equivalent to $\bar{\omega} < n - 2 - \frac{1}{n-1}$.

[1] $\bar{\omega} = -1$ means that the chief player is not only segregated but also discriminated against (by V) in the worst possible way. (Cf. 33.1.)

Thus $\bar{\omega} = -1$ gives a solution outright, while $\bar{\omega} > -1$ necessitates the more detailed analysis of (55:F). This is not surprising: An extreme form of discrimination is a more elementary proposition and requires less delicate adjustments than an intermediate one.

Proof: Assume $\beta_n < \bar{\omega}$. For an $\overrightarrow{\alpha}$ with $\alpha_n = \bar{\omega}$, (55:3) of 55.2.1. is excluded, i.e. domination $\overrightarrow{\alpha} \succ \overrightarrow{\beta}$ must operate through (55:1) (and not (55:2)!) in (55:A). Since $\alpha_n > \beta_n$, this condition amounts merely to

(55:8) $$\alpha_i > \beta_i \quad \text{for some } i = 1, \cdots, n-1.$$

Thus (55:5) requires the existence of an imputation $\overrightarrow{\alpha}$ with $\alpha_n = \bar{\omega}$ and (55:8).

Consider first (55:8) for a fixed $i = 1, \cdots, n-1$. Then this condition and $\alpha_n = \bar{\omega}$ can be met by an imputation $\overrightarrow{\alpha}$ if and only if β_i and $\bar{\omega}$ add up with $n-2$ addends -1 to < 0. I.e. $\beta_i + \bar{\omega} - (n-2) < 0$, $\beta_i < n - 2 - \bar{\omega}$. Consequently (55:8) is unfulfillable for all $i = 1, \cdots, n-1$, if and only if

(55:9) $$\beta_i \geqq n - 2 - \bar{\omega} \quad \text{for all} \quad i = 1, \cdots, n-1.$$

(55:5) expresses that this should happen for no $\overrightarrow{\beta}$ with $\beta_n < \bar{\omega}$. I.e. no imputation $\overrightarrow{\beta}$ could have (55:9) together with $-1 \leqq \beta_n < \bar{\omega}$.[1] This means that $n-1$ addends $n - 2 - \bar{\omega}$ and one addend -1 must add up to > 0. I.e. $(n-1)(n-2-\bar{\omega}) - 1 > 0$, $n - 2 - \bar{\omega} > \frac{1}{n-1}$, and so $\bar{\omega} < n - 2 - \frac{1}{n-1}$, as desired.

Combining (55:E), (55:F) and recalling (55:D) and the statements made concerning (55:5) and (55:6), (55:7), we can summarize as follows:

(55:G) Let $\bar{\omega}$ be any number with

$$-1 \leqq \bar{\omega} < n - 2 - \frac{1}{n-1}.$$

Form the set V of all $\overrightarrow{\alpha}$ with

$$\alpha_n = \bar{\omega}.$$[2]

These are precisely all solutions V in the case (I).

[1] We are assuming that (55:9) implies $\beta_i \geqq -1$ for $i = 1, \cdots, n-1$. This means $n - 2 - \bar{\omega} \geqq -1$, $\bar{\omega} \leqq n - 1$. Indeed $\bar{\omega} > n - 1$ must be excluded since it makes (55:4) unfulfillable by imputations: $\bar{\omega}$ and $n-1$ addends -1 would then add up to > 0.

Therefore the hypotheses of (55:F) imply this: $\bar{\omega} \leqq n - 1$.

[2] In pursuance with the parallelism with the discussion of the special case $n = 3$ in 33.1.—referred to in footnote 3 on p. 475—we note that this corresponds to the c loc. cit. For $n = 3$ our $n - 2 - \frac{1}{n-1}$ becomes the $\frac{1}{2}$ occurring there.

The first values of the quantity $n - 2 - \frac{1}{n-1}$ are:

n	3	4	5	6
$n - 2 - \frac{1}{n-1}$	$\frac{1}{2} = 0.5$	$\frac{5}{3} = 1.67$	$\frac{11}{4} = 2.75$	$\frac{19}{5} = 3.8$

Figure 91.

55.3.3. The interpretation of this result is not difficult:

This standard of behavior (solution) is based on the exclusion of the chief player from the game. This makes the distribution between the other players quite indefinite—i.e. any imputation which gives the chief player the "assigned" amount $\bar{\omega}$ belongs to the solution. The upper limit of the "assigned" amount $\bar{\omega}$, $n - 2 - \frac{1}{n-1}$ could also be motivated following the lines of 33.1.2., but we will not consider this question.

55.4. Case (II): Determination of $\underline{V}$

55.4.1. We now pass to the considerably more difficult case (II). (Cf. the last part of footnote 3 on p. 475.) We then have

$$-1 \leqq \underline{\omega} < \bar{\omega}.$$

This suggests the following decomposition of V into three pairwise disjunct sets:

$$\underline{V}, \text{ set of all } \overrightarrow{\alpha} \text{ in } V \text{ with } \alpha_n = \underline{\omega},$$

$$\bar{V}, \text{ set of all } \overrightarrow{\alpha} \text{ in } V \text{ with } \alpha_n = \bar{\omega},$$

$$V^*, \text{ set of all } \overrightarrow{\alpha} \text{ in } V \text{ with } \underline{\omega} < \alpha_n < \bar{\omega}.$$

By the very nature of $\underline{\omega}$, $\bar{\omega}$ (cf. the beginning of 55.2.3) $\underline{V}$, $\bar{V}$ cannot be empty—while we cannot make such an assertion concerning V^*.[1]

55.4.2. We begin by investigating $\underline{V}$.

(55:H) If $\overrightarrow{\alpha}$ belongs to $\underline{V}$ and $\overrightarrow{\beta}$ to $\bar{V} \cup V^*$, then $\alpha_i \geqq \beta_i$ for all $i = 1, \cdots, n-1$.

Proof: Otherwise $\beta_i > \alpha_i$ for a suitable $i = 1, \cdots, n-1$. Now $\alpha_n = \underline{\omega}$, $\beta_n > \underline{\omega}$, so $\beta_n > \alpha_n$; hence $\overrightarrow{\beta} \succ \overrightarrow{\alpha}$ by (55:1), which is impossible, since $\overrightarrow{\alpha}$, $\overrightarrow{\beta}$ belong to V.

[1] V^* is actually empty in the case considered preceding (55:V).

Form

$$\underline{\alpha}_i = \text{Min}_{\overrightarrow{\alpha} \text{ in } \underline{V}} \alpha_i \qquad \text{for} \qquad i = 1, \cdots, n-1.$$[1]

Now (55:H) gives immediately:

(55:I) If $\overrightarrow{\beta}$ belongs to $\bar{V} \cup V^*$, then
$\underline{\alpha}_i \geqq \beta_i$ for all $i = 1, \cdots n-1$.[2]

We prove further

(55:J) $$\sum_{i=1}^{n-1} \underline{\alpha}_i + \underline{\omega} \geqq 0.$$[3]

Proof: Assume that $\sum_{i=1}^{n-1} \underline{\alpha}_i + \underline{\omega} < 0$. Then we can choose $\gamma_i > \underline{\alpha}_i$ for $i = 1, \cdots, n-1$, $\gamma_n = \underline{\omega}$ with $\sum_{i=1}^{n} \gamma_i = 0$, forming the imputation $\overrightarrow{\gamma} = \{\gamma_1, \cdots, \gamma_n\}$.[4]

$\bar{V}$ is not empty, choose a $\overrightarrow{\beta}$ from $\bar{V}$. Then by (55:I) $\beta_i \leqq \underline{\alpha}_i < \gamma_i$ for all $i = 1, \cdots, n-1$, hence by (55:2) $\overrightarrow{\gamma} \succ \overrightarrow{\beta}$. As $\overrightarrow{\beta}$ belongs to V, this excludes $\overrightarrow{\gamma}$ from V.

Hence there exists an $\overrightarrow{\alpha}$ in V with $\overrightarrow{\alpha} \succ \overrightarrow{\gamma}$. If $\overrightarrow{\alpha}$ belongs to $\underline{V}$, then $\alpha_n = \underline{\omega} = \gamma_n$, hence $\overrightarrow{\alpha} \succ \overrightarrow{\gamma}$ contradicts (55:B). So $\overrightarrow{\alpha}$ must belong to $\bar{V} \cup V^*$. Now by (55:I) $\alpha_i \leqq \underline{\alpha}_i < \gamma_i$ for all $i = 1, \cdots, n-1$. But both (55:1) and (55:2) in (55:A) require—since $\overrightarrow{\alpha} \succ \overrightarrow{\gamma}$—that $\alpha_i > \gamma_i$ for at least one $i = 1, \cdots, n-1$. Thus we have a contradiction.

Now the determination of $\underline{V}$ can be completed:

(55:K) $\underline{V}$ has precisely one element:

$$\overrightarrow{\underline{\alpha}}^0 = \{\underline{\alpha}_1, \cdots, \underline{\alpha}_{n-1}, \underline{\omega}\}.$$

[1] That these quantities can be formed, i.e. that these minima exist and are assumed can be ascertained in the same way as in footnote 1 on p. 384. Cf. in particular (*) loc. cit. What is stated there concerning V, is equally true for $\underline{V}$, the intersection of V with the closed set of the $\overrightarrow{\alpha}$ with $\alpha_n = \underline{\omega}$.

[2] Note that this cannot be asserted for the $\overrightarrow{\beta}$ of $\underline{V}$ since β_i may exceed the minimum value $\underline{\alpha}_i$. Cf., however, (55:L).

[3] Cf. however, (55:12) below.

[4] Note that by their definitions all $\underline{\alpha}_i \geqq -1$, $(i = 1, \cdots, n-1)$ and $\underline{\omega} \geqq -1$, hence all our $\gamma_i \geqq -1$, $(i = 1, \cdots, n-1, n)$.

Proof: Let $\vec{\alpha} = \{\alpha_1, \cdots, \alpha_{n-1}, \alpha_n\}$ be an element of $\underline{V}$. Then

$$(55:10) \qquad \begin{cases} \alpha_i \geqq \underline{\alpha}_i & \text{for} \quad i = 1, \cdots, n-1, \\ \alpha_n = \underline{\omega}, \end{cases}$$

by the definition of these quantities. Now $\sum_{i=1}^{n} \alpha_i = 0$ and by (55:J), $\sum_{i=1}^{n-1} \underline{\alpha}_i + \underline{\omega} \geqq 0$, hence $>$ is excluded from all inequalities of (55:10).

Thus

$$(55:11) \qquad \begin{cases} \alpha_i = \underline{\alpha}_i & \text{for} \quad i = 1, \cdots, n-1, \\ \alpha_n = \underline{\omega}, \end{cases}$$

i.e.

$$\{\alpha_1, \cdots, \alpha_{n-1}, \alpha_n\} = \{\underline{\alpha}_1, \cdots, \underline{\alpha}_{n-1}, \underline{\omega}\}.$$

So $\underline{V}$ can have no element other than $\{\underline{\alpha}_1, \cdots, \underline{\alpha}_{n-1}, \underline{\omega}\}$. Since $\underline{V}$ is not empty, this is its unique element.

55.4.3. Note that as $\vec{\alpha}^{\,0} = \{\underline{\alpha}_1, \cdots, \underline{\alpha}_{n-1}, \underline{\omega}\}$ belongs to $\underline{V}$, it is necessarily an imputation. So we can strengthen (55:J) to

$$(55:12) \qquad \sum_{i=1}^{n-1} \underline{\alpha}_i + \underline{\omega} = 0.$$

We can also strengthen (55:I):

(55:L) If $\vec{\beta}$ belongs to V, then
$$\underline{\alpha}_i \geqq \beta_i \quad \text{for all} \quad i = 1, \cdots, n-1.$$

Proof: For $\vec{\beta}$ in $\bar{V} \cup V^*$ this has been stated in (55:I), for $\vec{\beta}$ in $\underline{V}$ (55:K) yields even $\beta_i = \underline{\alpha}_i$.

We conclude this part of the analysis by proving:

$$(55:M) \qquad \underline{\omega} = -1.$$

Proof: Assume $\underline{\omega} > -1$, i.e. $\underline{\omega} = -1 + \epsilon$, $\epsilon > 0$. Define

$$\vec{\beta} = \{\beta_1, \cdots, \beta_{n-1}, \beta_n\}$$

by $\beta_i = \underline{\alpha}_i + \dfrac{\epsilon}{n-1}$ for $i = 1, \cdots, n-1$, and $\beta_n = \underline{\omega} - \epsilon = -1$. $\vec{\beta}$ is an imputation (cf. (55:12) above). $\beta_n < \underline{\omega}$, or equally (55:L), excludes $\vec{\beta}$ from V.

Hence there exists an $\overrightarrow{\alpha}$ in V with $\overrightarrow{\alpha} \succ \overrightarrow{\beta}$. By (55:L) $\alpha_i \leqq \underline{\alpha}_i < \beta_i$ for all $i = 1, \cdots, n-1$. But both (55:1) and (55:2) in (55:A) require—since $\overrightarrow{\alpha} \succ \overrightarrow{\beta}$—that $\alpha_i > \beta_i$ for at least one $i = 1, \cdots, n-1$. Thus we have a contradiction.

Note that now (55:12) becomes

(55:N) $$\sum_{i=1}^{n-1} \underline{\alpha}_i = 1.$$

The essential results of this analysis are (55:K), (55:L), (55:M). They can be summarized as follows:[1]

The worst possible outcome for the chief player is complete defeat (value -1). There is one and only one arrangement—i.e. imputation—(in V) which does this, and for all other players this is the best possible one (in V).

This arrangement (in V) is the state of complete cooperation against the chief player.[2]

The reader will note that while this verbal formulation is not at all complicated, it could only be established by a mathematical, not by a verbal, procedure.

55.5. Case (II): Determination of $\bar{V}$

55.5.1. We are now able to investigate $\bar{V}$.

(55:O) Consider an imputation $\overrightarrow{\beta} = \{\beta_1, \cdots, \beta_n\}$ with $\beta_i \geqq \underline{\alpha}_i$ for some $i = 1, \cdots, n-1$ and $\beta_n \geqq \bar{\omega}$. Then $\overrightarrow{\beta}$ belongs to V.

Proof: Assume that $\overrightarrow{\beta}$ does not belong to V. Then there exists an $\overrightarrow{\alpha}$ in V with $\overrightarrow{\alpha} \succ \overrightarrow{\beta}$. Hence (55:1) or (55:2) of (55:A) must hold. As $\overrightarrow{\alpha}$ is in V, $\alpha_n \leqq \bar{\omega} \leqq \beta_n$, and this excludes (55:1). By (55:L) $\alpha_i \leqq \underline{\alpha}_i$ for all $i = 1, \cdots, n-1$, hence $\alpha_i \leqq \underline{\alpha}_i \leqq \beta_i$ for at least one $i = 1, \cdots, n-1$, and this excludes (55:2). So we have a contradiction in both cases.

(55:P) $\underline{\alpha}_i \geqq n - 2 - \bar{\omega}$ for $i = 1, \cdots, n-1$.

Proof: Assume that $\underline{\alpha}_i < n - 2 - \bar{\omega}$ for a suitable $i = 1, \cdots, n-1$, i.e. that

$$-(n-2) + \underline{\alpha}_i + \bar{\omega} < 0.$$

[1] All this applies, of course, to the case (II) only.

[2] This expression was also used—in a related, but somewhat different sense—in the last part of 55.2.

Then we can choose $\beta_j \geqq -1$ ($j = 1, \cdots, n-1$, $j \neq i$, i.e. $n - 2$ values of j) $\beta_i \geqq \underline{\alpha}_i$, $\beta_n > \bar{\omega}$ with $\sum_{j=1}^{n} \beta_j = 0$, forming the imputation

$$\overrightarrow{\beta} = \{\beta_1, \cdots, \beta_n\}.$$

This $\overrightarrow{\beta}$ meets the requirements of (55:O), hence it belongs to V. But this necessitates $\beta_n \leqq \bar{\omega}$ by the definition of that quantity—contradicting $\beta_n > \bar{\omega}$.

Now put

(55:13) $$\alpha_* = \mathrm{Min}_{i=1,\ldots,n-1}\ \underline{\alpha}_i.^{1}$$

Then (55:P) states this:

(55:14) $$\alpha_* \geqq n - 2 - \bar{\omega}.^{2}$$

Denote the set of all $i (= 1, \cdots, n-1)$ with

(55:15) $$\underline{\alpha}_i = \alpha_*$$

by S_*. By its nature this set must have these two properties:

(55:Q) $S_* \subseteq (1, \cdots, n-1)$, S_* is not empty.

55.5.2. We continue:

(55:R) $$\alpha_* = n - 2 - \bar{\omega}.$$

(55:S) $\bar{\mathsf{V}}$ consists of these elements: $\overrightarrow{\alpha}^{\,i}$ where i runs over all S_*, and where $\overrightarrow{\alpha}^{\,i} = \{\alpha_1^i, \cdots, \alpha_{n-1}^i, \alpha_n^i\}$ with

$$\alpha_j^i = \begin{cases} \underline{\alpha}_i = \alpha_* & \text{for } j = i, \\ \bar{\omega} & \text{for } j = n, \\ -1 & \text{otherwise.} \end{cases}$$

Proof of (55:R) and (55:S): We begin by considering an element $\overrightarrow{\beta}$ of $\bar{\mathsf{V}}$.

If $\beta_i < \underline{\alpha}_i$ for all $i = 1, \cdots, n-1$, then (55:2) gives $\overrightarrow{\alpha}^{\,0} \succ \overrightarrow{\beta}$ since $\overrightarrow{\alpha}^{\,0} = \{\underline{\alpha}_1, \cdots \underline{\alpha}_{n-1}, \underline{\omega}\}$. As $\overrightarrow{\alpha}^{\,0}$ belongs to V by (55:K), so $\overrightarrow{\alpha}^{\,0}$, $\overrightarrow{\beta}$ are both in V—hence this is impossible. So

(55:16) $\beta_i \geqq \underline{\alpha}_i \geqq \alpha_*$ for some $i = 1, \cdots, n-1$.

Necessarily

(55:17) $\beta_j \geqq -1$ for all $j = 1, \cdots, n-1$, $j \neq i$,

and since $\overrightarrow{\beta}$ is in $\bar{\mathsf{V}}$, so

(55:18) $$\beta_n = \bar{\omega}.$$

[1] This time the minimum is formed with respect to a finite domain!
[2] Cf. however, (55:R) below.

Now $\sum_{j=1}^{n} \beta_j = 0$ and by (55:14) $-(n-2) + \alpha_* + \bar{\omega} \geqq 0$, hence $>$ is excluded from all inequalities of (55:16), (55:17). Hence $\underline{\alpha}_i = \alpha_*$, i.e. i belongs to S_*. And

$$\beta_i = \begin{cases} \underline{\alpha}_i = \alpha_* & \text{for } j = i, \\ \bar{\omega} & \text{for } j = n, \\ -1 & \text{otherwise,} \end{cases}$$

i.e. $\overrightarrow{\beta} = \overrightarrow{\alpha}^i$ as defined above.

So we see:

(55:19) Every $\overrightarrow{\beta}$ of $\bar{\mathsf{V}}$ is necessarily an $\overrightarrow{\alpha}^i$ with i in S_*.

Now $\bar{\mathsf{V}}$ is not empty, hence an $\overrightarrow{\alpha}^i$ in V (i in S_*) exists. Consequently this $\overrightarrow{\alpha}^i$ is an imputation, hence $\sum_{j=1}^{n} \alpha^i_j = 0$, i.e. $-(n-2) + \alpha_* + \bar{\omega} = 0$. This is equivalent to (55:R).

Consider finally any i of S_*. Since (55:R) is true, we have

$$-(n-2) + \alpha_* + \bar{\omega} = 0.$$

Hence $\sum_{j=1}^{n} \alpha^i_j = 0$, i.e. $\overrightarrow{\alpha}^i$ is an imputation. But $\alpha^i_i = \underline{\alpha}_i = \alpha_*$, $\alpha^i_n = \bar{\omega}$, hence (55:O) guarantees that $\overrightarrow{\alpha}^i$ belongs to V. And since $\alpha^i_n = \bar{\omega}$, $\overrightarrow{\alpha}^i$ is even in $\bar{\mathsf{V}}$. I.e.

(55:20) Every $\overrightarrow{\alpha}^i$ with i in S_* is an imputation and belongs to $\bar{\mathsf{V}}$.

(55:19) and (55:20) together establish (55:S). (55:R) was demonstrated above. Thus the proof is completed.

55.5.3. The essential results of this analysis are (55:R), (55:S) together with the introduction of the set S_*. It is again possible to give a verbal summary.[1]

The best possible outcome for the chief player assigns to him a certain value $\bar{\omega}$. In order to achieve this he needs precisely one ally who can be selected at will from a certain set S_* of players. This set consists of those among the players $1, \cdots, n-1$ who are least favored in the state of complete cooperation against the chief player, referred to at the end of 55.4.

Thus the arrangements which the players $1, \cdots, n-1$ make between themselves, when they combine to defeat the chief player completely, determine his conduct in those cases where he achieves complete success. This "interaction" between fundamentally different situations is worth

[1] All this applies, of course, to the case (II) only.

noting.[1] It is also of interest that the natural allies of the chief player, when he aims at complete success, are the least favored members of a potential absolute opposition against him.[2]

The concluding remark of 55.4. concerning the contrast of formulation and proof applies again.

55.6. Case (II): $\mathfrak{A}$ and S_*

55.6. We determined in 55.4., 55.5. the two parts $\underline{V}$, $\bar{V}$ of V.[3] It is therefore time to turn to the last remaining part of V: V^*.

Let $\mathfrak{A}$ be the set of all $\overrightarrow{\alpha}$ with $\alpha_i = \underline{\alpha}_i = \alpha_*$ for all i in S_*. Then we have:

(55:T) $$\underline{V} \cup V^* \subseteq \mathfrak{A}.$$

Proof: Consider an $\overrightarrow{\alpha}$ in $\underline{V} \cup V^*$. We must prove $\alpha_i = \underline{\alpha}_i$ for all i in S_*.

Now $\alpha_i \leqq \underline{\alpha}_i$ for all $i = 1, \cdots, n-1$ by (55:L). Hence we need only exclude $\alpha_i < \underline{\alpha}_i$ when i is in S_*.

For i in S_* form the $\overrightarrow{\alpha}^i$ of (55:S). It belongs to $\bar{V}$, so $\alpha^i_n = \bar{\omega}$; $\overrightarrow{\alpha}$ belongs to $\underline{V} \cup V^*$, so $\alpha_n < \bar{\omega}$. Hence $\alpha^i_n > \alpha_n$. Now $\alpha_i < \underline{\alpha}_i$ means $\alpha^i_i = \underline{\alpha}_i > \alpha_i$, hence $\overrightarrow{\alpha}^i \succ \overrightarrow{\alpha}$ by (55:1)—and this is impossible, since $\overrightarrow{\alpha}^i$, $\overrightarrow{\alpha}$ both belong to V.

(55:U) $\bar{V} \subseteq \mathfrak{A}$ if and only if S_* is a one-element set or $\alpha_* = -1$; otherwise $\bar{V}$ and $\mathfrak{A}$ are disjunct.

Proof: Consider an $\overrightarrow{\alpha}$ in $\bar{V}$. Then $\overrightarrow{\alpha} = \overrightarrow{\alpha}^i$ (from (55:S)), i in S_*. Comparing the definitions of $\overrightarrow{\alpha}^i$ and $\mathfrak{A}$ makes it clear that this belongs to $\mathfrak{A}$ if and only if S_* has a unique element i or $\alpha_* = -1$.

[1] In 4.3.3. we insisted on the influence exercised by the "virtual" existence of an imputation—i.e. of its belonging to a certain standard of behavior (solution)—on all other imputations of the same standard. Almost all solutions of $n \geqq 3$—person games which we found can be used to illustrate this principle. A specific reference to it was made at an early stage of the discussion, in 25.2.2. The present instance, however, is particularly striking.

[2] Political situations to illustrate this principle are well known and in connection with them its general validity is frequently asserted. It is difficult to deny, however, that the case which can be made purely verbally for this principle is no better than that which could be made for a number of other conflicting ones.

The point is that for the particular game—i.e. social structure—we consider at present, this and no other principle is valid. To establish it a mathematical proof of some complexity was needed. All purely verbal plausibility arguments would have been inconclusive and ambiguous.

[3] The set S_* is still unknown, although restricted by (55:Q). The numbers $\underline{\alpha}_1, \cdots, \underline{\alpha}_{n-1}$ are also unknown, but restricted by (55:N). They determine α_* (their minimum). ω, $\bar{\omega}$ are given by (55:M), (55:R). The determination of these unknowns will be attended to later. Cf. (55:O′) (i.e. (55:L′), (55:N′) and (55:P′)).

Nevertheless, the form of $\underline{V}$ and of $\bar{V}$ has been found, and the remaining uncertainties are of a less fundamental character.

The verbal meaning of (55:T), (55:U) is this: Each player of the least favored group (S_*, cf. the end of 55.5.) reaches his optimum[1] in every apportionment in which the chief player is not fully successful (i.e. in $\underline{V} \cup V^*$). When the chief player is fully defeated (i.e. in $\underline{V}$), this is even true for all players $1, \cdots, n-1$ (cf. the end of 55.4.). When the chief player is fully successful then this is true for one and only one player, who may be any member of the least favored group (S_*, cf. the end of 55.5.).

55.7. Cases (II′) and (II″). Disposal of Case (II′)

55.7.1. Consider the case $S_* = (1, \cdots, n-1)$, to be called case (II′). In this case $\underline{\alpha}_i = \alpha_*$ for all $i = 1, \cdots, n-1$, so (55:N) gives $(n-1)\alpha_* = 1$, i.e., $\alpha_* = \frac{1}{n-1}$, and (55:R) gives $\bar{\omega} = n - 2 - \frac{1}{n-1}$. If $\overrightarrow{\alpha}$ belongs to $\mathfrak{A}$ then $\alpha_i = \underline{\alpha}_i = \alpha_* = \frac{1}{n-1}$ for $i = 1, \cdots, n-1$. Hence $\alpha_n = -1$, i.e. $\overrightarrow{\alpha} = \left\{\frac{1}{n-1}, \cdots, \frac{1}{n-1}, -1\right\}$. By (55:T) this is equally true for all $\overrightarrow{\alpha}$ in $\underline{V} \cup V^*$.

This $\overrightarrow{\alpha}$ is clearly the unique element $\overrightarrow{\alpha}^0$ of $\underline{V}$ by (55:K), hence V^* is empty. Hence $V = \underline{V} \cup \bar{V}$, and now (55:K), (55:S) give

(55:V) V consists of these elements:

(a) $\overrightarrow{\alpha}^0 = \left\{\frac{1}{n-1}, \cdots, \frac{1}{n-1}, -1\right\}$,

(b) $\overrightarrow{\alpha}^i$,

where $i = 1, \cdots, n-1$ and where

$$\overrightarrow{\alpha}^i = \{\alpha_1^i, \cdots, \alpha_{n-1}^i, \alpha_n^i\}$$

with

$$\alpha_j^i = \begin{cases} \frac{1}{n-1} & \text{for } j = i, \\ n - 2 - \frac{1}{n-1} & \text{for } j = n, \\ -1 & \text{otherwise.} \end{cases}$$

(55:V) determines the only possible solution V in the case (II′). This does not necessarily imply, however, that this V is either a solution or in case (II′). Indeed, if it failed to meet any one of these two requirements, then we would only have shown—although in a rather indirect way—that

[1] His individual optimum within the given standard—i.e. solution—V. For the player $i (= 1, \cdots, n-1)$ this optimum (maximum) is $\underline{\alpha}_i$ owing to (55:L)—although $\underline{\alpha}_i$ was originally defined as his pessimum (minimum) in the part $\underline{V}$ of V.

no solution in the case (II') exists. We will prove, therefore, that both requirements are met.[1]

55.7.2.

(55:W) The V of (55:V) is the unique solution in the case (II').

Proof: We need only show that this V is a solution in the case (II')—the uniqueness then follows from the above, i.e. from (55:V).

Case (II') is easily established: Clearly, for this V

$$\underline{\omega} = -1, \qquad \bar{\omega} = n - 2 - \frac{1}{n-1},$$

$$\underline{\alpha}_1 = \cdots = \underline{\alpha}_{n-1} = \alpha_* = \frac{1}{n-1}, \qquad S_* = (1, \cdots, n-1).$$

It remains for us to prove that V is a solution, i.e. to verify (30:5:c) in 30.1.1. To this end we must determine the imputations $\overrightarrow{\beta}$ which are undominated by elements of V.

For $\overrightarrow{\alpha}^0 \succ \overrightarrow{\beta}$ (55:1) is excluded since $\alpha^0_n = -1$. So this domination can operate through (55:2) alone, hence it amounts to $\alpha^0_i > \beta_i$, i.e. $\beta_i < \frac{1}{n-1}$ for $i = 1, \cdots, n-1$.

For $\overrightarrow{\alpha}^k \succ \overrightarrow{\beta}$, $k = 1, \cdots, n-1$, (55:1) is excluded when $i \neq k$ and (55:2) is excluded since $\alpha^k_i = -1$ for $i \neq k$. So this domination can operate through (55:1) with $i = k$ alone, hence it amounts to $\alpha^k_j > \beta_j$ for $j = k, n$, i.e. $\beta_k < \frac{1}{n-1}$, $\beta_n < n - 2 - \frac{1}{n-1}$.

Hence $\overrightarrow{\beta}$ is undominated by elements of V if and only if this is true: $\beta_i \geqq \frac{1}{n-1}$ holds for some $i = 1, \cdots, n-1$ and it holds even for all these i in case that $\beta_n < n - 2 - \frac{1}{n-1}$.

Thus $\beta_n < n - 2 - \frac{1}{n-1}$ necessitates $\beta_1, \cdots, \beta_{n-1} \geqq \frac{1}{n-1}$. Also $\beta_n \geqq -1$. Hence $\sum_{i=1}^{n} \beta_i = 0$ yields $=$ for all these $\geqq$ relations, i.e. $\overrightarrow{\beta} = \overrightarrow{\alpha}^0$. On the other hand $\beta_n \geqq n - 2 - \frac{1}{n-1}$ necessitates $\beta_i \geqq \frac{1}{n-1}$ for one $i(= 1, \cdots, n-1)$ and $\beta_j \geqq -1$ for the other $n-2$ values of j. Hence $\sum_{j=1}^{n} \beta_j = 0$ again yields $=$ for all these $\geqq$ relations, i.e. $\overrightarrow{\beta} = \overrightarrow{\alpha}^i$.

[1] Cf. this situation with (55:G), where the case (I) was settled. No such secondary considerations were needed there, because (55:G) was *ab ovo* necessary and sufficient.

So the $\overrightarrow{\beta}$ undominated by V are $\overrightarrow{\alpha}^0$ and $\overrightarrow{\alpha}', \cdots, \overrightarrow{\alpha}^{n-1}$; i.e. precisely the elements of V, as desired.

55.7.3. This solution is of importance because it is a finite set—as we shall see it is the only solution with that property. If the general coalition against the chief player is formed, the $n-1$ participants share in it equally—as described by $\overrightarrow{\alpha}^0$. If the chief player finds an ally, he gives him the same amount as $\overrightarrow{\alpha}^0$ and retains the remainder—as described by $\overrightarrow{\alpha}', \cdots, \overrightarrow{\alpha}^{n-1}$.
All this is perfectly reasonable and non-discriminatory.[1] Nevertheless this is not the only possible solution—we found another one in 55.3. (cf. (55:G)) and more will emerge in the sections which follow.

55.8. Case (II''): $\mathfrak{A}$ and V'. Domination

55.8.1. Consider next the case $S_* \neq (1, \cdots, n-1)$, to be called case (II'').

Using (55:Q) we may also formulate this as follows:

(55:X) $\qquad S_* \subset (1, \cdots, n-1), \qquad S_*$ not empty.

We can also say: Cases (II') and (II'') are characterized respectively by the absence and by the presence of discrimination within the possible general coalition against the chief player.

As we enter upon the discussion of case (II'') the following remark is indicated:

The argumentation of 55.4.-55.7. was mathematical, but the (intermediate) results which were obtained there allowed simple verbal formulations. I.e. it was possible to work into the mathematical deduction relatively frequent interruptions, giving verbal illustrations of the stages attained successively.

This situation changes now, insofar as a longer mathematical deduction is needed to carry us to the next point (in 55.12.) where a verbal interpretation is again appropriate.

55.8.2. We now proceed to give this deduction.

Write $\mathsf{V}' = \mathfrak{A} \cap \mathsf{V}$ (the part of V in $\mathfrak{A}$). By (55:T), (55:U) $\mathsf{V}' = \underline{\mathsf{V}} \cup \mathsf{V}^*$ or $\mathsf{V}' = \underline{\mathsf{V}} \cup \mathsf{V}^* \cup \bar{\mathsf{V}} = \mathsf{V}$, according to whether the condition of $(5\bar{5}$:U) is not or is satisfied.

(55:Y) $\qquad$ The condition (30:5:c) holds for V' in $\mathfrak{A}$.

Proof: Replace (30:5:c) by the equivalent (30:5:a), (30:5:b) in 30.1.1.

Ad (30:5:a): Since $\mathsf{V}' \subseteq \mathsf{V}$, elements of V' cannot dominate each other because the same is true for V.

[1] The special cases $n = 3, 4$ of this solution are familiar: For $n = 3$ it is the non-discriminatory solution of the essential three-person game; for $n = 4$ it was discussed in 35.1.

Ad (30:5:b): Let $\vec{\beta}$ in $\mathcal{A}$ not be in V'. Then we must find an $\vec{\alpha}$ in V' with $\vec{\alpha} \succ \vec{\beta}$.

To begin with, $\vec{\beta}$ is even not in V. Hence an $\vec{\alpha}$ in V with $\vec{\alpha} \succ \vec{\beta}$ exists. This $\vec{\alpha}$ must be in V' if it is not in $\bar{\mathsf{V}}$ (cf. the remarks preceding (55:Y)), and this would establish our statement. So we need only to exclude $\vec{\alpha}$ from $\bar{\mathsf{V}}$.

Assume that $\vec{\alpha}$ is in $\bar{\mathsf{V}}$, i.e. (by 55:S) $\vec{\alpha} = \vec{\alpha}^k$, k in S_*. We have $\vec{\alpha}^k \succ \vec{\beta}$. (55:1) is excluded when $i \neq k$, and (55:2) is excluded since $\alpha_i^k = -1$ for $i \neq k$ $(i = 1, \cdots, n-1)$. So this domination can only operate through (55:1) with $i = k$, hence it implies $\alpha_k^k > \beta_k$, i.e. $\beta_k < \underline{\alpha}_k = \alpha_*$. However, this is impossible, since $\vec{\beta}$ belongs to $\mathcal{A}$.

55.8.3. Thus our task is now to find all solutions (i.e. all sets fulfilling (30:5:c) in 30.1.1.) for $\mathcal{A}$. This necessitates determining the nature of domination in $\mathcal{A}$.

(55:Z) For $\vec{\alpha}$, $\vec{\beta}$ in $\mathcal{A}$, $\vec{\alpha} \succ \vec{\beta}$ is equivalent to this: $\alpha_n > \beta_n$ and $\alpha_i > \beta_i$ for some i in $(1, \cdots, n-1) - S_*$.

Proof: For $\vec{\alpha} \succ \vec{\beta}$ (55:1) is excluded when i is in S_*, and (55:2) is excluded since $\alpha_k = \beta_k (= \underline{\alpha}_k = \alpha_*)$ for all k of S_*.

So this domination can only operate through (55:1) with i in $(1, \cdots, n-1) - S_*$. And this means $\alpha_n > \beta_n$ and $\alpha_i > \beta_i$, as asserted.

We have replaced the set of all imputations by $\mathcal{A}$ and the concept of domination described in (55:A) by that one described in (55:Z). Otherwise the problem of finding all solutions has remained the same. The progress is that the concept of domination in (55:Z) can be worked more easily than that one in (55:A) as will be seen in what follows.

55.9. Case (II''): Determination of V'

55.9.1. Let p be the number of elements in S_*.

Then we have:

(55:A')
$$1 \leqq p \leqq n - 2.$$

Proof: Immediate by (55:X).

(55:B')
$$-1 \leqq \alpha_* < \frac{1}{n-1}.$$

Proof: $-1 \leqq \alpha_*$ is evident. Next $\alpha_i = \alpha_*$ for i in S_*, $\alpha_i > \alpha_*$ for i in $(1, \cdots, n-1) - S_*$, and by (55:A') neither set is empty. So

$\sum_{i=1}^{n-1} \alpha_i > (n-1)\alpha_*$, and hence (55:N) gives $1 > (n-1)\alpha_*$, $\alpha_* < \frac{1}{n-1}$, as desired.

An $\vec{\alpha}$ in $\mathfrak{A}$ has p fixed components: the $\alpha_i(=\alpha_i=\alpha_*)$, i in S_*; and $n-p$ variable ones: the α_i, i in $(1, \cdots, n) - S_*$. These are subject to the conditions

$$(55:21) \qquad \alpha_i \geqq -1 \qquad \text{for } i \text{ in } (1, \cdots, n) - S_*,$$

and $\sum_{i=1}^{n} \alpha_i = 0$, i.e.

$$(55:22) \qquad \sum_{i \text{ in } (1,\ldots,n)-S_*} \alpha_i = -p\alpha_*.$$

The lower limits in (55:21) add up to less than the sum prescribed in (55:22), i.e.—$(n-p) < -p\alpha_*$. Indeed, this means $\alpha_* < \frac{n-p}{p} = \frac{n}{p} - 1$. And by (55:A′) $p < n-1$, so $\frac{n}{p} - 1 > \frac{n}{n-1} - 1 = \frac{1}{n-1}$, and (55:B′) guarantees $\alpha_* < \frac{1}{n-1}$.

So we see:

(55:C′) The domain $\mathfrak{A}$ is $(n-p-1)$-dimensional.

55.9.2. We now proceed to a closer analysis of V' and of $\mathfrak{A}$.[1] Put

$$(55:23) \qquad \omega^* = n - p - 1 - p\alpha_*.$$

By (55:R) we can write

$$(55:24) \qquad \omega^* = \bar{\omega} - (p-1)(\alpha_* + 1).$$

(55:D′) $\omega^* = \bar{\omega}$ if and only if S_* is a one-element set (i.e. $p = 1$) or $\alpha_* = -1$ i.e. if and only if the condition of (55:U) is unfulfilled; otherwise $\omega^* < \bar{\omega}$.

Proof: Since $p \geqq 1$, $\alpha_* \geqq -1$ by (55:A′), (55:B′), this is immediate from (55:24).

$$(55:E') \qquad \operatorname{Max}_{\vec{\alpha} \text{ in } \mathfrak{A}} \alpha_n = \omega^*.$$

[1] The lemmas (55:D′)-(55:P′) which follow are the analytical equivalent of the graphical deduction of 47.5.2.-47.5.4. The technical background differs, but the analogies between the two proofs are nevertheless very marked—the interested reader may follow them up step by step.

(55:C′) shows that a graphical discussion would have to take place in a $(n-p-1)$ dimensional space (by (55:A′) this is $\geqq 1$, $\leqq n-2$). This is the reason why we use an analytical one. (The graphical proof referred to above took place in a plane, i.e. it required 2 dimensions.)

(55:F') This maximum is assumed for precisely one $\overrightarrow{\alpha}$ in $\mathfrak{A}$:

$$\overrightarrow{\alpha}^* = \{\alpha_1^*, \cdots, \alpha_{n-1}^*, \alpha_n^*\}$$

with

$$\alpha^* = \begin{cases} \alpha_i = \alpha_* & \text{for } i \text{ in } S_*, \\ \omega^* & \text{for } i = n, \\ -1 & \text{otherwise.}^1 \end{cases}$$

Proof of (55:E') *and* (55:F'): It is clear from the definition of $\mathfrak{A}$ that for the $\overrightarrow{\alpha}$ of $\mathfrak{A}$ the variable component α_n assumes its maximum when the other variable components—α_i, i in $(1, \cdots, n-1) - S_*$—assume their minima. These minima are -1. So for this maximum

$$\alpha_i = \begin{cases} \alpha_i = \alpha_* & \text{for } i \text{ in } S_*, \\ -1 & \text{for } i \text{ in } (1, \cdots, n-1) - S_*. \end{cases}$$

Now $\alpha_n = -\sum_{i=1}^{n-1} \alpha_i = -p\alpha_* + (n - 1 - p) = n - p - 1 - p\alpha_*$. By (55:23) this means $\alpha_n = \omega^*$.

This proves all our assertions.

(55:G') $\overrightarrow{\alpha}^*$ belongs to V'.

Proof: $\overrightarrow{\alpha}^*$ belongs to $\mathfrak{A}$, for any $\overrightarrow{\alpha}$ of $\mathfrak{A}$ (55:E'), (55:F') give

$$\alpha_n \leqq \alpha_n^* (= \omega^*).$$

So (55:Z) excludes $\overrightarrow{\alpha} \succ \overrightarrow{\alpha}^*$, and therefore (55:Y) necessitates that $\overrightarrow{\alpha}^*$ belong to V'.

55.9.3. After these preparations the decisive part of the deduction follows:

(55:H') If $\overrightarrow{\alpha}$, $\overrightarrow{\beta}$ belong to V', then $\alpha_n = \beta_n$ implies $\overrightarrow{\alpha} = \overrightarrow{\beta}$.

Proof: Consider two $\overrightarrow{\alpha}$, $\overrightarrow{\beta}$ in V' with $\alpha_n = \beta_n$.

Put $\gamma_i = \text{Min}\,(\alpha_i, \beta_i)(i = 1, \cdots, n-1, n)$ and assume first that $\sum_{i=1}^{n} \gamma_i < 0$, say $\sum_{i=1}^{n} \gamma_i = -\epsilon$, $\epsilon > 0$.

Put $\overrightarrow{\delta} = \{\delta_1, \cdots, \delta_{n-1}, \delta_n\}$ where

$$\delta_i = \begin{cases} \gamma_i & \text{for } i \text{ in } S_*, \\ \gamma_i + \dfrac{\epsilon}{n-p} & \text{for } i \text{ in } (1, \cdots, n-1, n) - S^*. \end{cases}$$

[1] Comparison of this definition with (55:D') shows that this $\overrightarrow{\alpha}^*$ is an $\overrightarrow{\alpha}^i$, i in S—i.e. that it belongs to $\bar{\mathsf{V}}$—if and only if the condition of (55:U) is fulfilled.

Since $\overrightarrow{\alpha}^*$ belongs to $\mathfrak{A}$ this is in agreement with the result of (55:U).

This $\overrightarrow{\delta}$ is clearly an imputation, and as i in S_* gives $\delta_i = \gamma_i = \alpha_i = \beta_i = \underline{\alpha}_i = \alpha_*$, so $\overrightarrow{\delta}$ belongs to $\mathfrak{A}$. We have $\delta_n > \gamma_n = \alpha_n = \beta_n$, and for i in $(1, \cdots, n-1) - S_*$, $\delta_i > \gamma_i = \alpha_i$ or β_i, hence $\overrightarrow{\delta} \succ \overrightarrow{\alpha}$ or $\overrightarrow{\delta} \succ \overrightarrow{\beta}$. Since $\overrightarrow{\alpha}$, $\overrightarrow{\beta}$ belong to V', this excludes $\overrightarrow{\delta}$ from V'. Hence there exists an $\overrightarrow{\eta}$ in V' with $\overrightarrow{\eta} \succ \overrightarrow{\delta}$.

Now by (55:Z) $\eta_n > \delta_n$ and $\eta_i > \delta_i$ for an i in $(1, \cdots, n-1) - S_*$. *A fortiori* $\eta_n > \delta_n > \gamma_n = \alpha_n = \beta_n$, $\eta_i > \delta_i > \gamma_i = \alpha_i$ or β_i. Thus $\overrightarrow{\eta} \succ \overrightarrow{\alpha}$ or $\overrightarrow{\eta} \succ \overrightarrow{\beta}$. As $\overrightarrow{\alpha}$, $\overrightarrow{\beta}$, $\overrightarrow{\eta}$ all belong to V', this is a contradiction.

Consequently $\sum_{i=1}^{n} \gamma_i < 0$ is impossible, so

(55:25) $$\sum_{i=1}^{n} \gamma_i \geqq 0.$$

Now $\gamma_i \leqq \alpha_i$, $\gamma_i \leqq \beta_i$ and $\sum_{i=1}^{n} \alpha_i = \sum_{i=1}^{n} \beta_i = 0$. Hence (55:25) yields $=$ for all these $\leqq$ relations, i.e. $\gamma_i = \alpha_i = \beta_i$. This proves $\overrightarrow{\alpha} = \overrightarrow{\beta}$ as desired.

(55:I') The values of the α_n for all $\overrightarrow{\alpha}$ in V' make up precisely the interval

$$-1 \leqq \alpha_n \leqq \omega^*.$$

Proof: For an $\overrightarrow{\alpha}$ in V', $\alpha_n \geqq -1$ is evident, and $\alpha_n \leqq \omega^*$ follows from (55:E'). Hence we need only exclude the existence of a y_1 in

$$-1 \leqq y_1 \leqq \omega^*,$$

such that $\alpha_n \neq y_1$ for all $\overrightarrow{\alpha}$ in V'.

There exist certainly elements $\overrightarrow{\alpha}$ of V' with $\alpha_n \geqq y_1$: Indeed $\overrightarrow{\alpha}^*$ belongs to V' by (55:G'), and $\alpha_n^* = \omega^* \geqq y_1$. Form

$$\operatorname{Min}_{\overrightarrow{\alpha} \text{ in } V' \text{ with } \alpha_n \geqq y_1} \alpha_n = y_2,[1]$$

[1] In this case it is not necessary to form the exact minimum, but the procedure which achieves this is somewhat longer than the one used below. That this minimum can be formed, i.e. that it exists and is assumed, can be ascertained in the same way as in footnote 1 on p. 384. Cf. in particular (*) loc. cit. What is stated there for V is equally true for the analogous set V' in $\mathfrak{A}$ and for the intersection of V' with the closed set of the $\overrightarrow{\alpha}$ with $\alpha_n \geqq y_1$.

Because of this need for closure we must use the condition $\alpha_n \geqq y_1$ and not $\alpha_n > y_1$ although we are really aiming at the latter. But the two will be seen to be equivalent in the case under consideration. (Cf. (55:26) below.)

and choose an $\overrightarrow{\alpha}^+$ in V' with $\alpha_n^+ \geqq y_1$ for which this minimum is assumed: $\alpha_n^+ = y_2$. By (55:H') this $\overrightarrow{\alpha}^+$ is unique.

So $y_2 \geqq y_1$, and since necessarily $\alpha_n^+ \neq y_1$, so $y_2 \neq y_1$, i.e.

(55:26) $$y_1 < y_2.$$

It follows from the definition of y_2 that

(55:27) $$y_1 \leqq \alpha_n < y_2 \qquad \text{for no } \overrightarrow{\alpha} \text{ in } V'.$$

Now put $y_1 = y_2 - \epsilon$, $\epsilon > 0$ and form the imputation

$$\overrightarrow{\beta} = \{\beta_1, \cdots, \beta_{n-1}, \beta_n\}$$

with $\beta_n = \alpha_n^+ - \epsilon = y_2 - \epsilon = y_1$, $\beta_i = \alpha_i^+ = \alpha_i = \alpha_*$ for i in S_*, $\beta_i = \alpha_i^+ + \frac{\epsilon}{n - 1 - p}$ for i in $(1, \cdots, n-1) - S_*$. Clearly $\overrightarrow{\beta}$ belongs to $\mathcal{A}$ and $\beta_n = y_1$ excludes $\overrightarrow{\beta}$ from V'. Hence there exists a $\overrightarrow{\gamma}$ in V' with $\overrightarrow{\gamma} \succ \overrightarrow{\beta}$.

By (55:Z) this means $\gamma_n > \beta_n$ and $\gamma_i > \beta_i$ for an i in $(1, \cdots, n-1) - S_*$.

Now $\gamma_n > \beta_n = y_1$ necessitates by (55:27) $\gamma_n \geqq y_2$. $\gamma_n = y_2$ would imply $\overrightarrow{\gamma} = \overrightarrow{\alpha}^+$ (by (55:H'), cf. above). Hence $\gamma_i = \alpha_i^+ < \beta_i$ for the above i in $(1, \cdots, n-1) - S_*$, and not $\gamma_i > \beta_i$ as required. Hence $\gamma_n > y_2$.

Thus $\gamma_n > y_2 = \alpha_n^+$ and $\gamma_i > \beta_i > \alpha_i^+$ for the above i in $(1, \cdots, n-1) - S_*$. So $\overrightarrow{\gamma} \succ \overrightarrow{\alpha}^+$, and as $\overrightarrow{\gamma}$, $\overrightarrow{\alpha}^+$ both belong to V' this is a contradiction.

55.9.4. By (55:I'), (55:H') we see: For every y in

$$-1 \leqq y \leqq \omega^*$$

there exists a unique $\overrightarrow{\alpha}$ in V' with $\alpha_n = y$. Denote this $\overrightarrow{\alpha}$ by

$$\overrightarrow{\alpha}(y) = \{\alpha_1(y), \cdots, \alpha_{n-1}(y), \alpha_n(y)\}.$$

Clearly $\alpha_n(y) = y$ and $\alpha_i(y) = \alpha_i = \alpha_*$ for i in S_*. So the functions which matter are the $\alpha_i(y)$ for i in $(1, \cdots, n-1) - S_*$.

Combining this with (55:I') gives:

(55:J') V' consists of these elements:

$$\overrightarrow{\alpha}(y)$$

where y runs over the interval

$$-1 \leqq y \leqq \omega^*,$$

and where $\overrightarrow{\alpha}(y) = \{\alpha_1(y), \cdots, \alpha_{n-1}(y), \alpha_n(y)\}$ with

$$\alpha_i(y) = \begin{cases} \alpha_i = \alpha_* & \text{for } i \text{ in } S_*, \\ y & \text{for } i = n, \\ \text{a suitable function of } y \text{ (and } i\text{) for } i \text{ in} \\ (1, \cdots, n-1) - S_*. \end{cases}$$

55.9.5. And to conclude:

(55:K') The functions $\alpha_i(y)$, i in $(1, \cdots, n-1) - S_*$, of (55:J') fulfill the following conditions:

(55:K':a) The domain of $\alpha_i(y)$ is the interval

$$-1 \leqq y \leqq \omega^*.$$

(55:K':b) $y_1 \leqq y_2$ implies $\alpha_i(y_1) \geqq \alpha_i(y_2)$.[1]

(55:K':c) $\alpha_i(-1) = \underline{\alpha}_i$.

(55:K':d) $\alpha_i(\omega^*) = -1$.

(55:K':e) $\sum_{i \text{ in } (1,\cdots,n-1)-S_*} \alpha_i(y) = -p\alpha_* - y$.[2,3]

[1] I.e. $\alpha_i(y)$ is an antimonotonic function of y.

[2] From these relations the continuity of all functions $\alpha_i(y)$, i in $(1, \cdots, n-1) - S_*$ follows. Indeed, we can even prove more, the so-called *Lipschitz condition:*

(55:28) $$|\alpha_i(y_2) - \alpha_i(y_1)| \leqq |y_2 - y_1|.$$

Proof: This relation is symmetric in y_1, y_2 hence we may assume $y_1 \leqq y_2$. Now application of (55:K':e) to $y = y_1$ and $y = y_2$ and subtraction give

$$\sum_{i \text{ in } (1,\cdots,n-1)-S_*} \{\alpha_i(y_1) - \alpha_i(y_2)\} = y_2 - y_1.$$

By (55:K':b) all these addends $\alpha_i(y_1) - \alpha_i(y_2)$ are $\geqq 0$, hence they are also $\leqq$ than their sum $y_2 - y_1$. Thus

$$0 \leqq \alpha_i(y_1) - \alpha_i(y_2) \leqq y_2 - y_1.$$

These inequalities make it also clear that the middle term is $|\alpha_i(y_2) - \alpha_i(y_1)|$ and that the last term is $|y_2 - y_1|$. Hence we have

$$|\alpha_i(y_2) - \alpha_i(y_1)| \leqq |y_2 - y_1|.$$

as desired.

The reader will note that we never assumed any continuity—we proved it! This is quite interesting from the technical mathematical point of view.

[3] Note that (55:K':c), (55:K':d) do not conflict with (55:K':e). Indeed:

For $y = -1$ (55:K':e) gives $\sum_{i \text{ in } (1,\cdots,n-1)-S_*} \alpha_i(-1) = -p\alpha_* + 1$, hence (55:K':c) requires $\sum_{i \text{ in } (1,\cdots,n-1)-S_*} \underline{\alpha}_i = -p\alpha_* + 1$, $\sum_{i=1}^{n-1} \underline{\alpha}_i = 1$, agreeing with (55:N).

For $y = \omega^*$ (55:K':e) gives $\sum_{i \text{ in } (1,\cdots,n-1)-S_*} \alpha_i(\omega^*) = -p\alpha_* - \omega^*$, hence (55:K':d) requires $-(n-p-1) = -p\alpha^* - \omega^*$, $\omega^* = n - p - 1 - p\alpha^*$, agreeing with (55:23).

Proof: Ad (55:K′:a): Contained in (55:J′).

Ad (55:K′:b): Assume the opposite: $y_1 \leqq y_2$ and $\alpha_i(y_1) < \alpha_i(y_2)$ (for a suitable i in $(1, \cdots, n-1) - S_*$). This excludes $y_1 = y_2$, so $y_1 < y_2$. Then $\overrightarrow{\alpha}(y_2) \succ \overrightarrow{\alpha}(y_1)$, which is impossible since $\overrightarrow{\alpha}(y_1)$, $\overrightarrow{\alpha}(y_2)$ both belong to V′.

Ad (55:K′:c): This is a restatement of the fact that $\overrightarrow{\alpha}^0$ belongs to V′, indeed it belongs to $\underline{\text{V}}$. (Cf. (55:K), (55:M).)

Ad (55:K′:d): This is a restatement of the fact that $\overrightarrow{\alpha}^*$ belongs to V′ (cf. (55:G′)).

Ad (55:K′:e): $\overrightarrow{\alpha}(y)$ is an imputation, hence $\sum_{i=1}^{n} \alpha_i(y) = 0$.

By (55:J′) this means that $\sum_{i \text{ in } (1,\cdots,n-1)-S_*} \alpha_i(y) + p\alpha_* + y = 0$, i.e. that $\sum_{i \text{ in } (1,\cdots,n-1)-S_*} \alpha_i(y) = -p\alpha_* - y$, as desired.

55.10. Disposal of Case (II″)

55.10.1. The results obtained in 55.8.-55.9. contain a complete description of the solution V. Indeed: As we saw at the beginning of 55.8.2. $\text{V} = \text{V}' \cup \bar{\text{V}}$, although the addend $\bar{\text{V}}$ may be omitted (because it is $\subseteq \text{V}'$) if and only if the condition of (55:U) is satisfied. $\bar{\text{V}}$ is described in (55:S), V′ in (55:J′). These characterizations make use of the parameters

$$\underline{\alpha}_i\ (i = 1, \cdots, n-1),\ \alpha_*,\ S_*,\ \bar{\omega},\ \omega^*,$$
$$\alpha_i(y)\,(i \text{ in } (1, \cdots, n-1) - S_*,\ -1 \leqq y \leqq \omega^*),$$

which are subject to the restrictions stated in (55:N); (55:13), (55:15) in 55.5.1.; (55:R); (55:23), (55:24) in 55.9.2.; (55:K′).

Since this material is dispersed over seven sections, it is convenient to restate the complete result in one place:

(55:L′)

(55:L′:a) S_* is a set $\subset (1, \cdots, n-1)$, not empty. Let p be the number of elements of S_*, so that $1 \leqq p \leqq n-2$.

(55:L′:b) $\underline{\alpha}_1, \cdots, \underline{\alpha}_{n-1}$ are numbers $\geqq -1$, with $\sum_{i=1}^{n-1} \underline{\alpha}_i = 1$.

(55:L′:c) For all i in S_* $\underline{\alpha}_i = \alpha_*$, for all i in $(1, \cdots, n-1) - S_*$ $\underline{\alpha}_i > \alpha_*$.

(55:L′:d) Put $\bar{\omega} = n - 2 - \alpha_*$, $\omega^* = n - p - 1 - p\alpha_*$, so that $\bar{\omega} - \omega^* = (p-1)(\alpha_* + 1)$.

(55:L′:e) $\alpha_i(y)$ is defined for i in $(1, \cdots, n-1) - S_*$,

$$-1 \leqq y \leqq \omega^*.$$

These functions satisfy the conditions (55:K′:a)-(55:K′:e). V consists of these elements:

(a) $\overrightarrow{\alpha}(y)$ where y runs over the interval $-1 \leqq y \leqq \omega^*$, and where

$$\overrightarrow{\alpha}(y) = \{\alpha_1(y), \cdots, \alpha_n(y)\},$$

with

$$\alpha_i(y) = \begin{cases} \alpha_i = \alpha_* & \text{for } i \text{ in } S_*, \\ y & \text{for } i = n, \\ \text{the } \alpha_i(y) \text{ of (55:L′:e) for } i \text{ in} \\ \quad (1, \cdots, n-1) - S_*. \end{cases}$$

(b) $\overrightarrow{\alpha}^i$ where i runs over all S_* and where

$$\overrightarrow{\alpha}^i = \{\alpha_1^i, \cdots, \alpha_{n-1}^i, \alpha_n^i\}$$

with

$$\alpha_j^i = \begin{cases} \underset{\sim}{\alpha}_i = \alpha_* & \text{for } j = i, \\ \bar{\omega} & \text{for } j = n, \\ -1 & \text{otherwise.} \end{cases}$$

Remark: If $p = 1$ (S_* a one-element set) or $\alpha_* = -1$, then $\bar{\omega} = \omega^*$ and the $\overrightarrow{\alpha}^i$ of (b) coincide with $\overrightarrow{\alpha}(y)$ of (a) for $y = \omega^*$. If this is not the case—i.e. $p \geqq 2$, $\alpha_* > -1$, then $\bar{\omega} > \omega^*$ and the $\overrightarrow{\alpha}^i$ of (b) are disjunct from the $\overrightarrow{\alpha}(y)$ of (a).

The reader will verify with little difficulty that all these statements are nothing but reformulations of the results referred to above.

55.10.2. (55:L′) must be followed by similar considerations as (55:V). We must investigate whether all V obtained from (55:L′) are solutions and in the case (II″). Those of them which meet both these requirements form the complete system of all solutions in the case (II″). We will prove that all V of (55:L′) meet these requirements.

(55:M′) The V of (55:L′) are precisely all solutions in the case (II″).

Proof: We need only show that every V of (55:L′) is a solution in the case (II″)—that these V are precisely all such solutions then follows from (55:L′).

Case (II″) is easily established: Clearly for this V, $\underset{\sim}{\omega} = -1$, and

$$\bar{\omega}, \underset{\sim}{\alpha}_1, \cdots, \underset{\sim}{\alpha}_{n-1}, S_*$$

(in the sense of their definitions given in 55.2.-55.5.) are precisely the quantities designated in (55:L′) by these symbols,[1] hence

$$S_* \subset (1, \cdots, n-1)$$

by (55:L′:a).

It remains for us to prove that V is a solution. In the present case we will do this by proving that V fulfills (30:5:a), (30:5:b) in 30.1.1.

Ad (30:5:a): Assume $\vec{\alpha} \succ \vec{\beta}$ for $\vec{\alpha}, \vec{\beta}$ in V. We must distinguish to which cases (a), (b) of (55:L′) $\vec{\alpha}, \vec{\beta}$ belong. There are four possible combinations:

$\vec{\alpha}, \vec{\beta}$ in (a): I.e. $\vec{\alpha} = \vec{\alpha}(y_1)$, $\vec{\beta} = \vec{\alpha}(y_2)$ and so $\vec{\alpha}(y_1) \succ \vec{\alpha}(y_2)$. Now (55:1) is excluded when i is in S_* and (55:2) is excluded since $\alpha_i(y_1) = \alpha_i(y_2) = \underline{\alpha}_i = \alpha_*$ for i in S_*. So this domination can operate through (55:1) with i in $(1, \cdots, n-1) - S_*$ only. By (55:L′:e) this means $\alpha_n(y_1) > \alpha_n(y_2)$, $y_1 > y_2$, and $\alpha_i(y_1) > \alpha_i(y_2)$ for a suitable i in $(1, \cdots, n-1) - S_*$, contradicting (55:K′:b).

$\vec{\alpha}$ in (a), $\vec{\beta}$ in (b): I.e. $\vec{\alpha} = \vec{\alpha}(y)$, $\vec{\beta} = \vec{\alpha}^i$ (i in S_*), and so $\vec{\alpha}(y) \succ \vec{\alpha}^i$. Now (55:1) is excluded, since $\alpha_n(y) = y \leqq \omega^* \leqq \bar{\omega} = \alpha^i_n$, and (55:2) is excluded, since $\alpha_i(y) = \alpha^i_i = \underline{\alpha}_i = \alpha_*$. So we have a contradiction.

$\vec{\alpha}$ in (b), $\vec{\beta}$ in (a): I.e. $\vec{\alpha} = \vec{\alpha}^i$ (i in S_*), $\vec{\beta} = \vec{\alpha}(y)$, and so $\vec{\alpha}^i \succ \vec{\alpha}(y)$. Now $\alpha^i_i = \alpha_i(y) = \underline{\alpha}_i = \alpha_*$ and for $j \neq i, n$, $\alpha^i_j = -1 \leqq \alpha_i(y)$, i.e. $\alpha^i_j \leqq \alpha_j(y)$ for all $j = 1, \cdots, n-1$. This excludes both (55:1), (55:2), and gives a contradiction.

$\vec{\alpha}, \vec{\beta}$ in (b): I.e. $\vec{\alpha} = \vec{\alpha}^i$, $\vec{\beta} = \vec{\alpha}^k$ (i, k in S_*), and so $\vec{\alpha}^i \succ \vec{\alpha}^k$. Now $\alpha^i_n = \alpha^k_n = \bar{\omega}$, thus contradicting (55:B)

Ad (30:5:b): Assume that $\vec{\beta}$ is undominated by the elements of V. We wish to prove that this implies that $\vec{\beta}$ belongs to V—which establishes (30:5:b).

Assume first that $\beta_n \geqq \bar{\omega}$. If $\beta_i < \underline{\alpha}_i = \alpha_*$ for all $i = 1, \cdots, n-1$, then $\vec{\alpha}(-1) \succ \vec{\beta}$, contradicting our assumption. Hence $\beta_i \geqq \underline{\alpha}_i$ for some $i = 1, \cdots, n-1$. Now the argument used in the proof of (55:R) shows that necessarily i in S_* and $\vec{\beta} = \vec{\alpha}^i$. Therefore $\vec{\beta}$ belongs to V in this case.

[1] $\bar{\omega}$ obtains from (b), $\underline{\alpha}_1, \cdots, \underline{\alpha}_{n-1}$ from (a) with $y = -1$, and then α_*, S_* from (55:L′:c).

Assume next that $\beta_n < \bar{\omega}$. If $\beta_i < \underline{\alpha}_i = \alpha_*$ for some i in S_*, then clearly $\overrightarrow{\alpha}^i \succ \overrightarrow{\beta}$ contradicting our assumption. So $\beta_i \geqq \underline{\alpha}_i = \alpha_*$ for all i in S_*.

Now $\sum_{i=1}^{n} \beta_i = 0$ gives $\beta_n = -\sum_{i=1}^{n-1} \beta_i \leqq n - p - 1 - p\alpha_* = \omega^*$, i.e. $-1 \leqq \beta_n \leqq \omega^*$. Put $y = \beta_n$.

Assume that $\beta_i \geqq \alpha_i(y)$ for all i in $(1, \cdots, n-1) - S_*$. Then we have clearly $\beta_i \geqq \alpha_i(y)$ for all $i = 1, \cdots, n$. (For i in S_* and $i = n$ we have even $=$, cf. above.) Hence $\sum_{i=1}^{n} \beta_i = \sum_{i=1}^{n} \alpha_i(y) = 0$ necessitates that we have $=$ in all these $\geqq$ relations. So $\overrightarrow{\beta} = \overrightarrow{\alpha}(y)$. Therefore $\overrightarrow{\beta}$ belongs to V in this subcase too.

There remains the possibility that $\beta_i < \alpha_i(y)$ for a suitable i in $(1, \cdots, n-1) - S_*$. A sufficiently small increase of y (from $y = \beta_n$ to some $y > \beta_n$) will not affect this relation $\beta_i < \alpha_i(y)$.[1] For this new y we have $y > \beta_n$, $\alpha_i(y) > \beta_i$, and therefore $\overrightarrow{\alpha}(y) \succ \overrightarrow{\beta}$ contradicting our assumption.

Thus all possibilities are accounted for.

55.11. Reformulation of the Complete Result

55.11.1. These three cases (I), (II′), (II″)—into which we subdivided our problem—have been completely settled by (55:G), (55:W), (55:M′) respectively. Let us now see to what extent these three classes of solutions are related to each other.

Among the undetermined parameters occurring in (55:L′)—i.e. in (55:M′), describing case (II″)—is the set S_*. According to (55:L′:a) this is any set $\subseteq (1, \cdots, n-1)$ with the exception of $(1, \cdots, n-1)$ and $\ominus$. This raises the question whether it is not possible to find some interpretation for these excluded cases $S_* = (1, \cdots, n-1)$ and $S_* = \ominus$ also.

For $S_* = (1, \cdots, n-1)$ the answer is easy. If we use this S_* (disregarding (55:L′:a) to this extent), then we obtain (using all other parts of (55:L′)): $p = n - 1$ by (55:L′:a), $\underline{\alpha}_1 = \cdots = \underline{\alpha}_{n-1} = \alpha_* = \frac{1}{n-1}$ by (55:L′:b), (55:L′:c), $\bar{\omega} = n - 2 - \frac{1}{n-1}$, $\omega^* = -1$, by (55:L′:d). There is no occasion to introduce the functions $\alpha_i(y)$ of (55:L′:e), since $(1, \cdots, n-1) - S_*$ is empty. Inasmuch as the interval $-1 \leqq y \leqq \omega^*$ plays a role (in (a) of (55:L′:e)), it must be noted that it shrinks to the point $y = -1$ (since $\omega^* = -1$). Now comparison with (55:V) discloses that under these conditions (55:L′) coincides with (55:V).

[1] $\alpha_i(y)$ is continuous! Cf. footnote 2 on p. 493.

So we have:

(55:N′) If we include in (55:L′:a) $S_* = (1, \cdots, n-1)$ (hence $p = n - 1$) also, then (55:L′) enumerates all solutions in the cases (II′) and (II″): Case (II′) corresponds to $S_* = (1, \cdots, n-1)$ and case (II″) to $S_* \neq (1, \cdots, n-1)$.

55.11.2. After this result one might feel inclined to correlate the remaining exception $S_* = \ominus$ with the remaining case (I). However, inspection of (55:L′) with $S_* = \ominus$ and comparison with (55:G) show that this is not possible—at least not in this direct way.

Indeed: Use of (55:L) with $S_* = \ominus$ (hence $p = 0$) gives an empty (b), so a V coinciding with (a)—i.e. V is the set of all

$$\overrightarrow{\alpha}(y) = \{\alpha_1(y), \cdots, \alpha_{n-1}(y), y\},$$

$-1 \leqq y \leqq \omega^*$, with suitable functions $\alpha_1(y), \cdots, \alpha_{n-1}(y)$. Disregarding other maladjustments[1] we note: In this arrangement the α_n of an $\overrightarrow{\alpha}$ in V determines its $\alpha_1, \cdots, \alpha_{n-1}$; while in (55:G) α_n was constant and

$$\alpha_1, \cdots, \alpha_{n-1}$$

arbitrary![2]

Summing up:

(55:O′) All solutions V are enumerated by (55:G)—Case (I)—and (55:N′)—Cases (II′) and (II″). (55:N′) coincides with (55:L′), when (55:L′:a) is widened to include all $S_* \subseteq (1, \cdots, n-1)$ with $S_* \neq \ominus$. The exclusion of $S_* = \ominus$ is necessary; this choice would produce a V which is not the solution of (55:G), and indeed is no solution at all.

55.11.3. We conclude with the following observations:

(55:P′)

(55:P′:a) In case (II′), i.e. $S_* = (1, \cdots, n-1)$, $p = n - 1$, we have: $\omega^* = -1$, i.e. the interval $-1 \leqq y \leqq \omega^*$ of (55:L′:e) shrinks to a point. Also $\alpha_* = \dfrac{1}{n-1}$.

(55:P′:b) In Case (II″), i.e. $S_* \subset (1, \cdots, n-1)$, $p < n - 1$, we have $\omega^* > -1$, i.e. the interval $-1 \leqq y \leqq \omega^*$ of (55:L′:e) does not shrink to a point. Also $\alpha_* < \dfrac{1}{n-1}$.

[1] Owing to $p = 0$ (55:23) now gives $\bar{\omega} - \omega^* = -(\alpha_* + 1)$, hence we may have $\omega^* > \bar{\omega}$, and so $\text{Max}_{\overrightarrow{\alpha} \text{ in V}} \alpha_n = \text{Max}_{-1 \leqq y \leqq \omega^*} y = \omega^*$, although it should be $\bar{\omega}$!

For $S_* \neq \ominus$, (55:L′:b), (55:L′:c) gave $\text{Min}_{i=1,\ldots,n-1} \alpha_i = \alpha_*$; for $S = \ominus$ they give $\text{Min}_{i=1,\ldots,n-1} \alpha_i > \alpha_*$, although the former was the definition of α_*!

[2] The V of (55:L′) with $S = \ominus$ is thus not a set from our list of solutions, hence it is no solution at all. It would have been easy to verify this directly.

Proof: Ad (55:P′:a): We proved these statements immediately preceding (55:N′).

Ad (55:P′:b): We saw in the proof of (55:B′) that $\alpha_* < \frac{n-p}{p}$, hence $\omega^* + 1 = n - p - p\alpha_* > 0, \omega^* > -1$. $\alpha_* < \frac{1}{n-1}$ was stated in (55:B′).

55.12. Interpretation of the Result

55.12.1. We can now begin to interpret this result. It is hardly possible to do this in an exhaustive way for two reasons. First the final result—contained in (55:O′), i.e. in (55:G), (55:K′), (55:L′)—is rather involved, hence a precise statement must necessarily be mathematical and not verbal. Any verbal formulation would fail to do justice to some of the numerous nuances expressed by the mathematical result. Second we still lack the necessary experience and perspective for a really thoroughgoing interpretation of a situation like the present one. The game which we consider here is a characteristic n-person game in some significant ways, as we set forth in 54.1.2. and 54.3. But our success in determining all of its solutions is still an isolated occurrence (the case of 54.2.1. notwithstanding). It will take many more discussions like this one before one can attempt really exhaustive interpretations of characteristic n-person games.

It is nevertheless useful to do a certain amount of interpreting—without any claim of completeness. We have seen in several previous instances that such interpretations give valuable guidance for the further progress of the theory. Besides, this procedure does throw some light on the significance of our rather complicated mathematical result.

Since we do not try to be complete, the interpretation is best made in the form of several remarks.

55.12.2. First: The solution of Case (I) described in (55:G) is an infinite set of imputations. The same is true for the solutions of (II″), described in (55:L′) (cf. (55:N′)) since the y mentioned there varies over an entire interval which does not shrink to a point. (Cf. (55:P′:b).) On the other hand the solution of Case (II′) is a finite set of imputations as was already observed at the end of 55.7.[1] This solution also has the attractive property of sharing the full symmetry of the game—i.e. invariance under all permutations of the players $1, \cdots, n-1$.

Thus it is in several ways the simplest solution of our game. Heuristic discussions of its special cases $n = 3, 4$ (in 22., 35.1. respectively) led to this solution and it is easy to extend them to the general n.[2] It takes the full machinery of our formal theory to find the other solutions.

It will be sufficiently clear to the reader by now that these other solutions can in no way be disregarded. Besides, the existence and the uniqueness

[1] The reader may compare to the same effect (55:P′:a), (55:P′:b).

[2] The (heuristic) argument would run as follows: The chief player needs an ally to win, with any such ally he obtains $n - 2$. Thus if he wishes to retain the amount ω (this corresponds to the $\bar{\omega}$ of our exact deductions) he can concede each ally $n - 2 - \omega$. If

of a finite solution is a favorable contingency in the present game, but by no means general.[1]

55.12.3. Second: The above solution corresponded to the largest possible $S_*:(1, \cdots, n-1)$. The other extreme is the solution which we associated with $S_* = \ominus$ (cf. preceding (55:O')). This is the solution in Case (I) described in (55:G). Like that one in the preceding remark it possesses the full symmetry of the game. Indeed these two—the Cases (I) and (II')—are the only ones with this symmetry.[2]

On the other hand this solution is infinite. As we saw in 55.3. it expresses the organizational principle that the chief player is segregated in the game in the sense of 33.1. Inspection of (55:G) discloses that this standard of behavior—i.e. solution offers absolutely no principle of division among the other players—i.e. all imputations where the chief player receives the prescribed amount belong to it. This is perfectly reasonable by common sense: The chief player being excluded, the other players can only combine with each other unanimously. All quantitative checks in their relationships (i.e. the possibility of siding with the chief player) being forbidden, there is no telling what the outcome of their bargaining with each other will be.

55.12.4. Third: The remaining solutions are those in Case (II''), described in (55:L') (cf. (55:N')), i.e. those with $S_* \neq \ominus, (1, \cdots, n-1)$. They form a more complicated group than the two solutions dealt with above. Indeed, they took up a considerable part—and the most involved one—of our mathematical deductions. Their interpretation, too, is more difficult and complicated. We will indicate the main points only.

We described in (55:L') in detail how in all imputations of a standard behavior—i.e. a solution—of this category the players of $(1, \cdots, n-1) - S_*$

---

his $n-1$ potential allies together can make more than that, i.e. if

$$(n-1)(n-2-\omega) < 1,$$

then his chances of finding an ally are destroyed—and this is the only limit to his exactions.

Thus ω is only limited by $(n-1)(n-2-\omega) \geqq 1$, i.e. $\omega \leqq n-2-\frac{1}{n-1}$. So $\omega = n-2-\frac{1}{n-1}$.

So the chief player obtains $n-2-\frac{1}{n-1}$, if he succeeds in forming a coalition, and of course -1 if he does not. For the other players the corresponding amounts are $\frac{1}{n-1}$ and -1.

The reader can now verify that this is just the solution arrived at in (55:V), i.e. Case (II').

[1] As to the uncertainty concerning the existence cf. the end of the second remark in 53.2.2. An instance where the uniqueness fails is analyzed in 38.3.1.

[2] Any other solution belongs to Case (II'') and so has an $S_* \neq \ominus, (1, \cdots, n-1)$. Hence an appropriate permutation of the players $1, \cdots, n-1$ will carry an element of S_* into one outside, thus changing S_* and with it the solution under consideration.

are causally linked to the chief player. I.e. how the respective amounts which they get are uniquely determined by the amount assigned to the chief player. This connection was expressed by definite functions.[1] These functions could be chosen in different ways, thus yielding different standards of behavior—i.e. solutions—but a definite standard meant a definite choice of these functions. Thus the uncorrelatedness of the players $1, \cdots, n-1$, so prominent in the second remark, is now gone. There is obviously some kind of indefinite bargaining going on between the chief player and those of $(1, \cdots, n-1) - S_*$,[2] but the relationship of the latter players to each other is completely determined by the standard.

It is worth while to emphasize once more this difference between the situation described in the second remark and in the present one—i.e. between the Cases (I) and (II''). In the former case there was bargaining between all players except the chief player with absolutely no rules or correlations to cover it,[3] so that the standard of behavior had to make no provision in this respect. Now we have bargaining between the chief player and some of the others, but this time the standard must provide definite correlations and rules for the opponents of the chief player. Accordingly there is a multiplicity of possible standards.

The qualitative types of indefiniteness arising in the Cases (I) and (II''), as discussed above, are a more general form of that one which we investigated in 47.8., 47.9. The remarks made there about the 2-dimensional (area) and one-dimensional (curve) parts of those solutions are indeed applicable to our present Cases (I) and (II''), respectively.

While it is possible to motivate this difference by verbal arguments of some plausibility, they are all far from convincing. The mathematical deduction alone, such as we gave it, gives the real reason—and its relative complication shows how difficult it must be to translate it into ordinary language. This is another instance of a result which can be expressed, but scarcely demonstrated, verbally.

55.12.5. Fourth: The situation of the remaining players—those in S_*—has also its interesting aspects.

Inspection of (55:L') shows that in every imputation of our solution either all these players get the amount α_*, or one of them gets α_* and the others the amounts -1. From this one infers immediately:

(a) If S_* is a one-element set then the player in S_* gets always the same amount: α_*

(b) If $\alpha_* = -1$ then each player of S_* always gets the same amount: -1.

(c) If neither of (a) or (b) is the case—i.e. if the condition of (55:U) (also referred to in (55:D')) is fulfilled—then each player in S_*

[1] The $\alpha_i(y)$, i in $(1, \cdots, n-1) - S_*$.

[2] This corresponds to the variability of y in (55:L':e). Cf. also (55:P':b).

[3] Except for the assignment to the chief player who is segregated.

always gets one of the two different amounts α_* and -1, and neither can be omitted.[1]

From these we can draw the following interpretative conclusions:

(d) In the two cases (a) and (b), but not in (c), the players of S_* are segregated in the sense of 33.1.

(e) The case (a) where S_* is a one-element set: $S_* = (i)$, $i = 1, \cdots, n-1$, expresses the segregation of the player i alone. The value α_* which is then assigned to him, is limited by (55:B′):

$$(55:29) \qquad -1 \leqq \alpha_* < \frac{1}{n-1}.$$

This is a satisfactory complement to the segregation of the chief player, Case (I), described in the second remark.[2] The value $\bar{\omega}$ which was then assigned to the chief player was limited by (55:G):

$$(55:30) \qquad -1 \leqq \bar{\omega} < n - 2 - \frac{1}{n-1}.$$

(f) If S_* is not a one-element set, then there is within the cases (a), (b) only the possibility (b): $\alpha_* = -1$.
In other words:
If more than one player is to be segregated, then their set must not contain the chief player, nor all other players, and the segregated players must all be assigned the value:

$$(55:31) \qquad \alpha_* = -1.$$

(g) We conclude from (e), (f) that those sets of players which can be segregated are precisely the sets of L[3]—the defeated sets.

(h) If only one player is segregated, then (e) shows that he need not be discriminated against in an absolutely disadvantageous way. I.e. he may be assigned more than -1. (55:29), (55:30) also state the upper limit of what this assignment can be: It is clearly the same amount which this player would get in the finite solution of Case (I), discussed in the first remark.[4] It is very satisfactory that this extends the result of 33.1.2. from $n = 3$ to all n.

(i) If, on the other hand, more than one player is segregated[5] then (55:31) shows that there can be no concessions: They must all be given the absolute minimum -1.

[1] I.e. both occur in appropriate imputations of the solution.
[2] This resolves the difficulty pointed out in footnote 3 on p. 475.
[3] This is best verified by recalling the enumeration of the elements of W and so of L—in the case C_{n-1} in 52.3.
[4] $n - 2 - \frac{1}{n-1}$ for the chief player, $\frac{1}{n-1}$ for the others. The assignment must be less than these amounts.
[5] I.e. the number p of elements in S_* is $\geqq 2$. Since $p \leqq n - 2$ (cf. (55:L:a)) this can happen only when $n - 2 \geqq 2$ i.e. $n \geqq 4$. This is the reason why the phenomena of (i) and (j) were not observed in the discussion of $n = 3$.

(j) This assertion must be qualified to the following extent: If S_* has more than one element, the α_* of (55:29) are still all possible—indeed (55:L′) with (55:B′) allows for them explicitly. But the situation of the players in S_* is then described by (c), and can no longer be termed segregation: They may join coalitions and thereby improve their status.

It is clear that these remarks, particularly (g), (h), (i), invite further comment. However we will now restrict ourselves to these indications and return to the subject at another occasion.

55.12.6. Fifth: We found a great number of solutions, characterized by numerous parameters, some of which were even functions which could be chosen with considerable freedom. The main classification, however, was rather simple: It was affected by the set $S_* \subseteq (1, \cdots, n-1)$.[1] The pairs S_*, $-S_*$ exhaust obviously all partitions of $I = (1, \cdots, n)$ into two sets. Possibly this is the first indication of a general principle. In a simple game a partition into two complements seems to decide everything, since one of them is necessarily winning and the other necessarily defeated. In general games partitions into more sets may be equally important. At any rate the role of S_* in the present special case gives the first idea of what may be a general classifying principle in all games.

We are not in a position, as yet, to give this surmise a more precise form.

[1] We use as in the second remark $S_* = \ominus$ to symbolize the case (I), the discussion preceding (55:O′) notwithstanding.

CHAPTER XI

GENERAL NON-ZERO-SUM GAMES

56. Extension of the Theory

56.1. Formulation of the Problem

56.1.1. Our considerations have reached the stage at which it is possible to drop the zero-sum restriction for games. We have already relaxed this condition once to the extent of considering constant-sum games—with a sum different from zero. But this was not a really significant extension of the zero-sum case since these games were related to it by the isomorphism of strategic equivalence (cf. 42.1. and 42.2.). We now propose to go the whole way and abandon all restrictions concerning the sum.

We pointed out before that the zero-sum restriction weakens the connection between games and economic problems quite considerably.[1] Specifically, it emphasizes the problem of apportionment to the detriment of problems of "productivity" proper (cf. 4.2.1., particularly footnote 2 on p. 34; also 5.2.1.). This is especially clear in the case of the one-person game: behavior in this situation is manifestly a matter of production alone, with no conceivable imputation (apportionment) between players. And indeed the one-person game offers no problem at all in the zero-sum case, and a perfectly good maximum problem in the non-zero-sum case (cf. 12.2.1.).

Accordingly our present program of extending the theory to all non-zero-sum games must be expected to bring us into closer contact with questions of the familiar economic type. In the discussions which follow, the reader will soon observe a change in the trend of the illustrative examples and of the interpretations: we shall begin to deal with questions of bilateral monopoly, oligopoly, markets, etc.

56.1.2. Complete abandonment of the zero-sum restriction for our games means, as was pointed out in 42.1., that the functions $\mathcal{H}_k(\tau_1, \cdots, \tau_n)$ which characterized it in the sense of 11.2.3. are now entirely unrestricted. I.e., that the requirement

$$\sum_{k=1}^{n} \mathcal{H}_k(\tau_1, \cdots, \tau_n) \equiv 0 \tag{56:1}$$

[1] It should be noted that zero-sum games not only cover the type of games played for entertainment (cf. 5.2.1.), but also that many of them describe quite adequately relationships of a definitely social nature. The reader who has progressed up to this point and recalls the interpretations which we have made in numerous cases will be fully aware of the validity of this statement.

Thus the distinction between zero-sum and non-zero-sum games reflects to a certain extent the distinction between purely social and social-economic questions. (The next sentence in the text expresses the same idea.)

of 11.4. and 25.1.3. is dropped with nothing to take its place. Accordingly we proceed on this basis from now on.

This change necessitates a complete reconsideration of our theory with all the attendant concepts on which it is based. Characteristic functions, domination, solutions,—all these concepts are no longer defined when (56:1) is dropped. We emphasize the fact that the problem which arises here is a conceptual one, and not merely technical as were all those treated in Chapters VI–X, on the basis of our theory of the zero-sum games.[1]

56.1.3. The prospect of having to start all over again would be very discouraging: we have already spent considerable effort on these concepts and the theory based on them. Furthermore we face a conceptual problem and the qualitative principles on which our theory was based do not seem to carry beyond the zero-sum case. Thus this final generalization—the passage from the zero-sum to the non-zero-sum case—would seem to nullify all our past efforts. We must find therefore a way to avoid this difficulty.

At this point one might recall the comparable situation which arose in 42.2. There our transition from the zero-sum case to the constant-sum case threatened—on a narrower scale—with similar consequences. They were avoided by an appropriate use of the isomorphisms of strategic equivalence, as effected in 42.3. and 42.4.

The usefulness of this particular device was, however, exhausted by the application referred to: strategic equivalences extend the family of all zero-games precisely to the family of all constant-sum games and no further. (This should be clear from the considerations of 42.2.2., 42.2.3. or 42.3.1.)

So we must find some other procedure to link the theory of the non-zero-sum games to the established theory of the zero-sum games.

56.2. The Fictitious Player. The Zero-sum Extension $\bar{\Gamma}$

56.2.1. Before going further we have to clarify a point of terminology. The games which we shall now consider are those where—as stated in 56.1.2. —condition (56:1) is dropped without anything else taking its place. We talked of these as non-zero-sum games, but it is important to realize that this expression is meant in the neutral sense,—i.e. that we do not wish to exclude those games for which (56:1) happens to be true. It is therefore preferable to use a less negative name for these games. Accordingly we shall call the games with entirely unrestricted $\mathcal{H}_k(\tau_1, \cdots, \tau_n)$ *general games*.[2]

We have formulated the program of linking the theory of the general games in some way to the theory of the zero-sum games. It will actually be possible to do more: any given general game can be re-interpreted as a zero-sum game.

[1] Among these technical problems was one which we preferred to treat by a method involving a certain conceptual generalization: the case of the constant-sum games, which will be referred to further below in the text.

[2] This is in agreement with 12.1.2.

This may seem paradoxical since the general games form a much more extensive family than the zero-sum games. However, our procedure will be to interpret an n-person general game as an $n + 1$-person zero-sum game. Thus the restriction caused by the passage from general games to zero-sum games will be compensated for—indeed made possible—by the extension due to the increase in the number of participants.[1]

56.2.2. The procedure by which a given general n-person game is re-interpreted as an $n + 1$-person zero-sum game is a very simple and natural one.

It consists of introducing a—fictitious—$n + 1$-st player who is assumed to lose the amount which the totality of the other n—real—players wins and *vice versa*. He must, of course, have no direct influence on the course of the game.

Let us express this mathematically: Consider the general n-person game Γ of the players $1, \cdots, n$, with the functions $\mathcal{H}_k(\tau_1, \cdots, \tau_n)$ $(k = 1, \cdots, n)$ in the sense of 11.2.3. We introduce the *fictitious player* $n + 1$ by defining

$$(56{:}2) \qquad \mathcal{H}_{n+1}(\tau_1, \cdots, \tau_n) \equiv - \sum_{k=1}^{n} \mathcal{H}_k(\tau_1, \cdots, \tau_n).$$

The variables $\tau_1, \cdots, \tau_n$ are controlled by the—real—players $1, \cdots, n$, respectively. They represent their influence on the course of the game. Since it is intended that the fictitious player have no influence on the course of the game, a variable τ_{n+1} which he controls, was not introduced.[2]

In this way we obtain a zero-sum $n + 1$-person game, the *zero-sum extension* of Γ, to be denoted by $\bar{\Gamma}$.

56.3. Questions Concerning the Character of $\bar{\Gamma}$

56.3.1. In stating that we have re-interpreted the general n-person game Γ as the zero-sum $n + 1$-person game $\bar{\Gamma}$, we imply *prima facie* that the entire theory of $\bar{\Gamma}$ has validity for Γ. This assertion requires, of course, closest scrutiny.

We shall now undertake this investigation. It must be understood that this cannot be a purely mathematical analysis, like the analyses in the preceding chapters which were based on a definite theory. We are analyzing once more the foundations of a proposed theory. Consequently the analysis must in the main be in the nature of plausibility arguments,—even if intermixed with subsidiary mathematical considerations. The situation is exactly the same as in those earlier instances where we made our decisions

[1] This may serve as a further illustration of the principle stated repeatedly that any increase in the number of participants necessarily entails a generalization and complication of the structural possibilities of the game.

[2] The formalism of 11.2.3. provided a variable τ_k for every player k. (In order to replace it for the present case we must replace its n by our $n + 1$.) Hence one might insist on the appearance of the variable τ_{n+1} of the fictitious player $n + 1$.

This requirement is easy to meet. It suffices to introduce a variable τ_{n+1} with only one possible value (i.e. to put $\beta_{n+1} = 1$, loc. cit.). Actually one could even use any domain of τ_{n+1} (i.e. any β_{n+1}) as long as all $\mathcal{H}_k(\tau_1, \cdots, \tau_n, \tau_{n+1})$ are independent of τ_{n+1},—so that they are really functions $\mathcal{H}_k(\tau_1, \cdots, \tau_n)$ as used in the text.

concerning the theories of the zero-sum two-, three-, n-person games. (Cf. 14.1.-14.5., 17.1.-17.9. for the zero-sum two-person game; Chap. V for the zero-sum three-person game; 29., 30.1., 30.2. for the zero-sum n-person game. For the general n-person game—i.e. the relationship between the theories of $\bar{\Gamma}$ and Γ—the equivalent sections begin with 56.2., and go on to 56.12.)

The result of our analysis will be that it is not the entire theory of $\bar{\Gamma}$—as a zero-sum $n + 1$-person game, in the sense of 30.1.1.—which applies to Γ, but only a part of it which we shall determine. In other words, we shall find that not the system of all solutions for $\bar{\Gamma}$, but only a certain subsystem produces what will be interpreted as the solutions of Γ.

56.3.2. The fictitious player was introduced as a mathematical device to make the sum of the amounts obtained by the players equal to zero. It is therefore absolutely essential that he should have no influence whatever on the course of the game. This principle was duly observed in the definition of $\bar{\Gamma}$ as given in 56.2.2. We must nevertheless put to ourselves the question whether the fictitious player is absolutely excluded from all transactions connected with the game.

This caveat is not at all superfluous. As soon as $\bar{\Gamma}$ involves three or more persons[1] the game is ruled by coalitions, as we observed at an early stage of our analysis. A participation of the fictitious player in any coalition—which is likely to involve the payment of compensations between the participants—would be completely contrary to the spirit in which he is introduced. Specifically: the fictitious player is no player at all, but only a formal device for a formal purpose. As long as he takes no part in the game in any direct or indirect form, this is permissible. But as soon as he begins to interfere, his introduction into the game—i.e. the passage from Γ to $\bar{\Gamma}$—ceases to be legitimate. That is, $\bar{\Gamma}$ cannot then be regarded as an equivalent, or a re-interpretation of Γ, since the real players of $\bar{\Gamma}$, $1, \cdots, n$, may have to provide against dangers or may profit by possibilities which certainly do not exist in Γ.

56.3.3. One might think that this objection is void due to the way in which the fictitious player was introduced. Indeed, the amounts

$$\mathcal{H}_1, \cdots, \mathcal{H}_n,$$

which the real players $1, \cdots, n$ obtain at the end of the play, do not depend on any variable which he controls[2]—i.e. he has no moves in the play. How can he then be a desirable partner in a coalition?

It may appear at first that this argument has some merit. The conditions described make it seem that any coalition of real players is just as well

[1] I.e. when $n + 1 \geq 3$ which means $n \geq 2$. Thus only the general one-person game is free from the objections which follow. This is in harmony with the fact which we have emphasized repeatedly, that the general n-person game is a pure maximum problem only when $n = 1$.

[2] Nor does the amount $\mathcal{H}_{n+1} = -\sum_{k=1}^{n} \mathcal{H}_k$ which he obtains.

off without the fictitious player as with him. Is he anything but a dummy? If this were so, the theory of Γ could be applied without any further qualifications to $\bar{\Gamma}$. However this is not the case.

It is true that the fictitious player, having no moves to influence the course of the game, is not a desirable partner for any coalition. I.e. no player or group of players will pay a (positive) compensation for his cooperation. However he himself may have an interest in finding allies. The amount which he gets at the end of the play—$\mathcal{H}_{n+1}(\tau_1, \cdots, \tau_n)$—depends on the moves of the other players—on $\tau_1, \cdots, \tau_n$—and it may be worth his while to pay one or more among the players a (positive) compensation for ceasing to cooperate with the others. It is important not to misunderstand this: As long as Γ is played, i.e. as long as the fictitious player is really a formalistic fiction, no such thing will happen; but if the game really played is $\bar{\Gamma}$, i.e. if the fictitious player behaves as a real player would in his position, then his offer of compensations to the others must be expected.

56.3.4. As soon as the fictitious player begins to offer compensations to other players for cooperation with him—which, as we saw above, amounts to their non-cooperation with others—he is an influence to be reckoned with. He offers to join coalitions and to pay a price for this privilege and his willingness to pay is fully as good as a direct influence on the game exercised by ability to make significant moves.

Thus the fictitious player gets into the game in spite of his inability to influence its course directly by moves of his own. Indeed it is just this impotence which determines his policy of offering compensations to others, and thus sets the above mechanism into motion.

For a better understanding of the situation it may be helpful to give a specific example.

56.4. Limitations of the Use of $\bar{\Gamma}$

56.4.1. Consider a general two-person game in which each of the players 1,2 if left to himself can secure for himself only the amount -1, while the two together can secure the amount 1. It is easy to specify definite rules for a game to bring this about.[1] A particularly simple combinatorial arrangement which does it, is as follows:[2]

Each player will, by a personal move, choose one of the numbers 1,2. Each one makes his choice uninformed about the choice of the other player.

After this the payments will be made as follows: if both players have chosen the number 1, each gets the amount $\frac{1}{2}$, otherwise each gets the amount -1.[3]

[1] Thus it will be seen in 60.2., 61.2., 61.3. that the *bilateral monopoly* corresponds to just this.

[2] This construction should be compared with the one used in defining the simple majority game of three persons in 21.1., with which it has a certain similarity.

[3] With the notations of all 11.2.3.: $\beta_1 = \beta_2 = 2$ and

$$\mathcal{H}_1(\tau_1, \tau_2) = \mathcal{H}_2(\tau_1, \tau_2) = \begin{cases} \frac{1}{2} & \text{for } \tau_1 = \tau_2 = 1, \\ -1 & \text{otherwise.} \end{cases}$$

It is easy to verify that this game possesses the desired properties.

Let us now consider the fictitious player 3 and form the game as defined in 56.2.2., with its characteristic function $v(S)$, $S \subseteq (1,2,3)$. According to what we said above

$$v((1)) = v((2)) = -1,$$
$$v((1,2)) = 1.$$

Obviously

$$v(\ominus) = 0,$$

and by the general properties of the characteristic function (of a zero-sum game)

$$\begin{aligned} v((3)) &= -v((1,2)) = -1, \\ v((1,3)) &= -v((2)) = 1, \\ v((2,3)) &= -v((1)) = 1, \\ v((1,2,3)) &= -v(\ominus) = 0. \end{aligned}$$

Summing up:

$$(56{:}3) \qquad v(S) = \begin{cases} 0 \\ -1 \\ \\ 1 \\ 0 \end{cases} \quad \text{when } S \text{ has} \quad \begin{cases} 0 \\ 1 \\ \\ 2 \\ 3 \end{cases} \text{elements.}$$

This formula (56:3) is precisely the (29:1) of 29.1.2.; i.e. $\bar{\Gamma}$ is the essential zero-sum three-person game in its reduced form, with $\gamma = 1$. Thus it coincides equally with the simple majority game of three persons, which was discussed in 21.[1]

Now we learned previously from the heuristic discussions of 21.-23. that this game is nothing but a competition of all players for coalitions. Indeed, this is immediately obvious, considering the nature of the simple majority game of three persons (cf. 21.2.1.). Hence a fictitious player will certainly show a strong tendency to enter into coalitions. In fact the game $\bar{\Gamma}$ is, as far as the characteristic function is concerned, completely symmetric with respect to its three players. Consequently the two real players 1,2 play exactly the same role as the fictitious player 3, and so there is no reason why their ability to enter coalitions should be at all different from his.[2]

56.4.2. We can also revert to the argument used in the last part of 56.3.3. and apply it to this game: If the fictitious player 3 in $\bar{\Gamma}$ behaves as a real one would, he has every reason to try to prevent the formation of a couple of the players 1,2, since he loses the amount -1 if this couple is formed, and wins

[1] Of course all these games coincide only as far as their characteristic functions are concerned, but the entire theory of 30.1.1. is based on the characteristic functions alone.

[2] To avoid misunderstandings we re-emphasize this: The rules of the game $\bar{\Gamma}$, fully expressed by the $\mathcal{H}_k$, are not at all symmetric with respect to the players 1,2,3; $\mathcal{H}_k$ depends on τ_1, τ_2 but not on τ_3. It is only the characteristic function $v(S)$, $S \subseteq (1,2,3)$, which is symmetric in 1,2,3. But we know that $v(S)$ alone matters. (Cf. footnote 1, above.)

the amount 2 if it is not formed.[1] Hence he will offer player 1 or player 2 a compensation for disrupting this couple, i.e. for choosing τ_1 or τ_2, respectively, equal to 2 instead of 1. This compensation can be determined by the considerations of 22., 23., and turns out to be $\frac{3}{2}$.[2] The reader may verify this for himself, together with the fact that this procedure leads to the known results concerning the simple majority game of three persons.

56.4.3. The example of 56.4.1. gives substance to the objection formulated in 56.3.3. and 56.3.4. Thus the fictitious player $n + 1$ can influence the game $\bar{\Gamma}$ not directly through personal moves but indirectly by offering compensations and thereby modifying the conditions and the outcome of the competition for coalitions. As pointed out at the end of 56.3.3., this does not mean that any such thing happens in Γ, i.e. as long as the fictitious player is a mere formalistic fiction. It does happen in $\bar{\Gamma}$ if the theory of 30.1.1. is applied to it literally,—i.e. if the fictitious player is permitted to behave (in offering compensations) as if he were a real one. In other words, the considerations of the last paragraphs do not mean that we want to attribute to the fictitious player abilities conflicting with the spirit in which he was introduced. They served only to show that an uncompromising application of our original theory to $\bar{\Gamma}$ brings us into such a conflict. Hence we must conclude that the zero-sum game $\bar{\Gamma}$ cannot be considered an unqualified equivalent of the general game Γ.

What are we then to do? In order to answer this question, it is best to return to the analysis of the specific example of 56.4.1. where the difficulty was expressed fully.

56.5. The Two Possible Procedures

56.5.1. One might try to escape from our present difficulty by observing that it was brought about in 56.4.1. by the exclusive use of the characteristic function. Indeed the game $\bar{\Gamma}$ there coincided with the simple majority game of three persons—where the mechanism of coalition formation is beyond doubt—only to the extent that they had the same characteristic functions, but not the same $\mathcal{H}_k$ (cf. in particular footnotes 1 and 2 on p. 509) Thus a possible expedient might be to abandon the claim that the characteristic function alone matters, and to base the theory on the $\mathcal{H}_k$ themselves.

At closer inspection, however, this suggestion appears to be entirely without merit—at least for the problem under consideration.

First: Abandonment of the characteristic function $v(S)$ in favor of the underlying $\mathcal{H}_k$ would deprive us of all means to handle the problem. For

[1] By footnote 3 on p. 508 and by (56:2):

$$\mathcal{H}_3(\tau_1, \tau_2) = -\mathcal{H}_1(\tau_1, \tau_2) - \mathcal{H}_2(\tau_1, \tau_2) = \begin{cases} -1 & \text{for } \tau_1 = \tau_2 = 1, \\ 2 & \text{otherwise.} \end{cases}$$

[2] This is the compensation which brings up the player 1 or 2 (who joined the fictitious player 3) from the loss -1, to a gain $\frac{1}{2}$ which is what he would obtain in a couple, i.e. in a coalition of the players 1 and 2. It also brings the fictitious player's gain from 2 down to $\frac{1}{2}$, which is indeed what it should be.

zero-sum games we possess no general theory other than that of 30.1.1., based on $v(S)$ exclusively. Thus the adoption of this program would make our passage from the general game Γ to the zero-sum game $\bar{\Gamma}$ entirely useless, since it would render us just as incapable of handling zero-sum games as, originally, general games. Hence this sacrifice of our entire existing theory would be reasonable only if it were quite certain that despite its adequacy in all other respects there was no other avenue of escape. However, neither of these two conditions is fulfilled.

Second: The retrogression from the characteristic function to the underlying $\mathcal{H}_k$ does not meet the objections of the previous paragraphs. Indeed, at the end of 56.3.2. as well as in 56.4.2. we did operate in a manner which took the $\mathcal{H}_k$ into account. We established the necessity for the fictitious player in $\bar{\Gamma}$ to offer compensations in a direct manner,—a necessity which was in no way dependent upon a replacement of $\bar{\Gamma}$ by a different game with the same characteristic function.[1]

Third: It will appear from the discussion which follows that it is not necessary to sacrifice the theory based on the characteristic function, but rather that the objections can be met by a simple restriction of its scope.

56.5.2. Reconsideration of 56.3.2.-56.4.2. shows that we were not justified in placing the blame for our present difficulties, with regard to the behavior of the fictitious player, entirely upon the theory of 30.1.1.

The considerations of 56.3.2.-56.3.4. and 56.4.2. were entirely heuristic. This is particularly important in the case 56.4.2. where the undesirable result was obtained in a definite way for a specific instance. Indeed, the treatment of 56.4.2. referred to the "preliminary" heuristic discussion of the essential zero-sum three-person game in 21.-23., and not to its exact theory in 32.

What happened there—in 56.4.2. as well as in 56.4.1.—can be described in the terminology of the exact theory as follows: the general two-person game Γ of 56.4.1. led to a zero-sum three-person game $\bar{\Gamma}$ which coincides with the simple majority game of three persons. The exact theory of 30.1.1. provided various solutions for this game, which were classified and analyzed in 33.1. Now the considerations of 56.4.1. and 56.4.2. amounted to selecting a particular one from among these solutions: the non-discriminatory solution of 33.1.3.

Consequently we must ask ourselves: Was it reasonable to select just this—the non-discriminatory—solution? Is it not possible that another one among the solutions—i.e. a discriminatory one in the sense of 33.1.3.—is free from the objections which hold us up?

56.6. The Discriminatory Solutions

56.6.1. If we had approached the essential zero-sum three-person game—i.e. the simple majority game of three persons—from any other angle, and if it had been necessary to select a particular one from among its solutions,

[1] We made repeated use of such replacements in 56.4.1., but not in the subsequent argument of 56.4.2.!

there would have been a strong presumption in favor of the non-discriminatory one. This solution—i.e. the standard of behavior which it represents—gives all three players equal possibilities to compete for coalitions, and in the absence of any definite motive for discrimination one is tempted to treat it as the most "natural" solution of this game.[1]

However, in our present situation there is every reason to discriminate: In the game $\bar{\Gamma}$, players 1,2 are real players, the original participants of Γ, while player 3 is, as repeatedly emphasized, just a formalistic fiction. Throughout the discussion of the preceding paragraphs we have stressed that this player should not compete for a coalition, and that he should not be treated like the others. In other words, if we expect to be able at all to apply the theory of 30.1.1. to this situation, then there is an absolute necessity of discriminating against the fictitious player 3,—i.e. to choose one of those solutions which were termed discriminatory in 33.1., the excluded player being the fictitious player 3.

We saw, loc. cit., that these discriminatory solutions are characterized by the fact that the excluded player—whom the solution, i.e. the standard of behavior, disqualifies from competing for a coalition—is assigned a fixed amount c in all imputations of the solution. It appeared in 33.1.2. that this amount need not be the minimum at which the excluded player can maintain himself alone,—i.e. not necessarily $c = -1$. Actually c could be chosen from a certain interval: $-1 \leqq c < \frac{1}{2}$.

56.6.2. At this point it may be useful to interrupt the discussion for a moment in order to comment briefly upon the discriminatory solution which excludes the fictitious player in the worst possible situation,—i.e. with $c = -1$. According to 33.1.1. this solution consists of precisely those imputations in which the fictitious player 3 gets -1, and each one of the two real players gets $\geqq -1$.

As pointed out loc. cit. this means that the solution—i.e. the standard of behavior—restricts in no way the division of the proceeds between the two real players. The reason given there is now valid in a much more fundamental way: the bargaining of the players 1,2 has become entirely unrestricted, not only because the accepted standard of behavior excludes the interference of player 3—which was the only normative influence in the relationship of players 1,2—but also for the still better reason that player 3 does not exist. It is easy to see that this removes the threat that player 1 or 2 will forsake cooperation with the other if his "fair share" is not conceded by his partner and that instead he will cooperate with player 3 and obtain a compensation from that source.

56.7. Alternative Possibilities

56.7.1. Let us now continue the discussion where it left off at the end of 56.6.1.

[1] Of course the other solutions are just as good in the rigorous sense of 30.1.1., but the above statement is nevertheless reasonable *prima facie*.

It may seem questionable whether we should insist upon $c = -1$, or allow the entire variability $-1 \leqq c < \frac{1}{2}$. The first alternative is the more plausible *prima facie*. Indeed, $c > -1$ means that the real players do not exploit the fictitious player to the utmost of their possibilities, i.e. that they do not endeavor to gain as much (as a totality) as feasible. One might view such a self-denial as a compensation paid to the fictitious player by virtue of the accepted stable standard of behavior. And since we have decided to exclude any participation of the fictitious player in the interplay of coalitions and compensations, there is some justification in forbidding this.

It must be conceded, however, that this argument is not altogether cogent. A (positive) compensation paid by the fictitious player is a qualitatively different thing from one paid to him. The former is a patent absurdity, since the fictitious player does not exist and will therefore not pay compensations. The latter, on the other hand, is not absurd at all. It merely expresses a self-denial in exploiting a possible collective advantage, and we have had several instances showing that a stable standard of behavior can require such conduct.[1] It is not *a priori* evident that such a self-denial is out of the question in the present situation.[2] To exclude it would mean that a stable standard of behavior—in the presence of complete information—necessarily entails attainment of the maximum collective benefit. The reader who is familiar with the existing sociological literature will know that the discussion of this point is far from concluded.

We shall nevertheless succeed in settling this question within the framework of our theory by showing that c must be restricted to its minimum value.[3]

56.7.2. For the moment, however, we must develop both alternatives concurrently.

To this end we return to the general n-person game Γ, and the corresponding zero-sum $(n + 1)$-person game $\bar{\Gamma}$. We are now able to formulate the relevant concepts rigorously.

(56:A:a) Denote the set of all solutions $\bar{V}$ of $\bar{\Gamma}$ by Ω.

(56:A:b) Given a number c, denote the system of those solutions $\bar{V}$ of $\bar{\Gamma}$ for which every imputation $\overrightarrow{\alpha} = \{\alpha_1, \cdots, \alpha_n, \alpha_{n+1}\}$ of $\bar{V}$ has $\alpha_{n+1} = c$,[4] by Ω_c.

[1] This is, of course, just another way to express the possibility of 33.1.2. Another instance, in a zero-sum four-person game, is given in (38:F) of 38.3.2. Still another obtains for all decomposable games in 46.11. (In this last instance the self-denial is exercised by the players of Δ when $\bar{\varphi} < 0$, and by those of H when $\bar{\varphi} > 0$, cf. loc. cit.)

We emphasize that such self-denial is exercised under the pressure of the accepted standard of behavior, although the players are assumed—as always in our theory—to be informed fully about the possibilities of the game.

[2] However if it occurred, it would be regarded normally as an inefficient—though stable—form of social organization.

[3] I.e. the self-denial in question does not occur and the maximum social benefit is always obtained. This result is not as sweeping as it may seem, since we are assuming a numerical and unrestrictedly transferable utility, as well as complete information.

[4] I.e. where the fictitious player gets the same amount c in all imputations of the solution.

(56:A:c) Denote the sum of all sets Ω_c by Ω'.

(56:A:d) Denote the Ω_c of $c = \mathrm{v}((n+1)) = -\mathrm{v}((1, \cdots, n))$ by Ω''.[1]

In connection with (56:A:c) we note this:

For some c the set Ω_c is empty. These c may obviously be omitted when Ω' is formed. Thus $\alpha_{n+1} \geqq \mathrm{v}((n+1)) = -\mathrm{v}((1, \cdots, n))$ necessitates $c \geqq -\mathrm{v}((1, \cdots, n))$, otherwise Ω_c is empty. Again

$$\alpha_{n+1} = -\sum_{k=1}^{n} \alpha_k \leqq -\sum_{k=1}^{n} \mathrm{v}((k))$$

necessitates $c \leqq -\sum_{k=1}^{n} \mathrm{v}((k))$, otherwise Ω_c is empty. So c is subject to the restriction

$$(56:4) \qquad -\mathrm{v}((1, \cdots, n)) \leqq c \leqq -\sum_{k=1}^{n} \mathrm{v}((k)).$$

Actually it is usually even more restricted.[2]

The Ω'' of (56:A:d) belongs to the minimum c of (56:4).

56.8. The New Setup

56.8.1. Our discussion of 56.3.2.-56.4.3. showed that the solutions of Ω are certainly not all significant for Γ. The analysis of 56.6.1. restricted those solutions further, but it left the question unanswered whether the system of all significant solutions is Ω' or Ω''.

Thus the systems Ω' and Ω'' correspond to the two alternatives referred to.

We now proceed to differentiate between Ω' and Ω''.

Consider the imputations

$$(56:5) \qquad \overrightarrow{\alpha} = \{\alpha_1, \cdots, \alpha_n, \alpha_{n+1}\}$$

of the game Γ. Among the components $\alpha_1, \cdots, \alpha_n, \alpha_{n+1}$ the n first ones, $\alpha_1, \cdots, \alpha_n$ express realities: the amounts which the real players, $1, \cdots, n$ respectively, are to obtain from this imputation. The last component, α_{n+1} on the other hand, expresses a fictitious operation: the amount attributed to the fictitious player $n+1$. Further, this component

[1] I.e. where the fictitious player gets, in all imputations of the solution, only that amount which he could obtain for himself even in opposition to all others. This means—as we know—that the real players obtain together the maximum collective benefit.

[2] Thus in the essential zero-sum three-person game, (56:4) gives

$$-1 \leqq c \leqq 2,$$

while we know from 32.2.2. that the exact domain of c (with non-empty Ω_c) is

$$-1 \leqq c < \tfrac{1}{2}.$$

α_{n+1} is not only fictitious in the interpretation of $\bar{\Gamma}$, but it is also mathematically unnecessary,—i.e. it is determined if the $\alpha_1, \cdots, \alpha_n$ are known. Indeed (since the sum of all components of the imputation $\overrightarrow{\alpha}$ must be zero)

(56:6) $$\alpha_{n+1} = - \sum_{k=1}^{n} \alpha_k.$$

Consequently it may be preferable to express $\overrightarrow{\alpha}$ by specifying its components $\alpha_1, \cdots, \alpha_n$ only, always remembering that α_{n+1} can be obtained—if desired—from (56:6). Thus we shall write

(56:7) $$\overrightarrow{\alpha} = \{\{\alpha_1, \cdots, \alpha_n\}\}.$$

We observe that this notation is not intended to supersede the original one,—i.e. we wish to be free to use both (56:5) and (56:7), whichever may be more suitable. Indeed, it is in order to avoid misunderstandings which might ensue from this double notation, that we are using the double brackets $\{\{\quad\}\}$ in (56:7) instead of the simple ones $\{\quad\}$ in (56:5).[1]

56.8.2. The imputation $\overrightarrow{\alpha}$ in its form (56:5) was subject to the zero-sum restriction, and also to the restriction

(56:8) $$\alpha_i \geqq \mathrm{v}((i)) \qquad \text{for} \qquad i = 1, \cdots, n, n+1.$$

We must express (56:8) for (56:7) (with (56:6)).

Now for $i = 1, \cdots, n$ (56:8) is unaffected by the transition from (56:5) to (56:7), but for $i = n + 1$ we must make use of (56:6). So it becomes

$$\sum_{i=1}^{n} \alpha_i \leqq -\mathrm{v}((n+1)) = \mathrm{v}((1, \cdots, n)).$$

[1] Of course we could have done this all along, i.e. for the original zero-sum n-person games. Here an imputation

$$\overrightarrow{\alpha} = \{\alpha_1, \cdots, \alpha_n\}$$

is determined if only its components α_i, $i \neq i_0$ are given (for any fixed i_0), since

$$\alpha_{i_0} = - \sum_{i \neq i_0} \alpha_i.$$

In conformity with this we have observed already in (31:I) in 31.2.1. that the imputations of the (essential) zero-sum n-person game form an $(n-1)$-dimensional, and not an n-dimensional, manifold.

However, there was no particular advantage to be gained by getting rid of an α_{i_0}, and there was no way to decide which α_{i_0} should be eliminated, if any. In the graphical discussion of the essential zero-sum three-person game we actually made an effort to keep all α_i in the picture. (Cf. 32.1.2.)

The situation now is altogether different, considering the special role of α_{n+1}. The elimination of α_{n+1} will be essential for our subsequent deductions.

Thus (56:8) goes over into this:

(56:9) $\alpha_i \geqq \mathrm{v}((i))$ for $i = 1, \cdots, n;$

(56:10) $\sum_{i=1}^{n} \alpha_i \leqq \mathrm{v}((1, \cdots, n)).$

56.9. Reconsideration of the Case Where Γ is a Zero-sum Game

56.9.1. Let us stop for a moment to interpret these restrictions.

(56:9) is not new. It expresses again what we had already for the zero-sum games, namely that nobody would accept in any case less than he can get for himself in opposition to all others. (56:10), however, appears for the first time. Its meaning becomes transparent if we consider the quantity $\mathrm{v}((1, \cdots, n))$ more closely.

$\mathrm{v}((1, \cdots, n))$ is the value of the game for the composite player comprising all real players $1, \cdots, n$, and playing against the fictitious player $n + 1$. The amount which this composite player gets at the end of a play is, of course

$$\sum_{k=1}^{n} \mathcal{H}_k(\tau_1, \cdots, \tau_n).$$

He controls the variables $\tau_1, \cdots, \tau_n$, i.e. all the variables which occur in this expression. Thus in the zero-sum two-person game the real players control all moves the fictitious player having no influence on the course of the game.

In comparing this with the zero-sum two-person scheme described in 14.1.1. our $\sum_{k=1}^{n} \mathcal{H}_k$ corresponds to $\mathcal{H}$ there, all our variables $\tau_1, \cdots, \tau_n$ to the one variable τ_1 there, while no domain of variability in our present set-up corresponds to the variable τ_2 there.

It is intuitively clear that the value of such a game (for the first player) obtains by maximizing with respect to all variables (since they are all controlled by him). This is

(56:11) $\mathrm{Max}_{\tau_1, \ldots, \tau_n} \sum_{k=1}^{n} \mathcal{H}_k(\tau_1, \cdots, \tau_n)$

in our set up, the corresponding expression in the scheme of 14.1.1. being

(56:12) $\mathrm{Max}_{\tau_1} \mathcal{H}(\tau_1, \tau_2)$ (τ_2 is really absent).

Of course the systematic theory of 14., 17. gives the same result: $\mathrm{v}_1, \mathrm{v}_2$ in 14.4.1. are equal to each other and to (56:12), since the operation Min_{τ_2} is void. So the game is strictly determined and has the value (56:12) in the sense of 14.4.2. and 14.5. Consequently the general theory of 17. yields necessarily the same value.

So we see:

$$(56:13) \qquad \mathrm{v}((1, \cdots, n)) = \mathrm{Max}_{\tau_1, \ldots, \tau_n} \sum_{k=1}^{n} \mathcal{H}_k(\tau_1, \cdots, \tau_n).$$

Consequently (56:10) expresses this: No imputation should offer all (real) players together more than the totality can expect in the most favorable case, i.e. assuming complete co-operation and the best possible strategy.[1,2]

Summing up:

(56:B) The imputations of (56:7) are subject to the following restrictions:

(56:B:a) No real player must be offered less than he can obtain for himself even in opposition to all other players (cf. (56:9)).

(56:B:b) All real players together must not be offered more than the totality can expect in the most favorable case, i.e. assuming complete co-operation and the best possible strategy (cf. (56:10) and (56:13)).

This formulation makes the common-sense meaning of our restrictions (56:9), (56:10) (i.e. (56:B:a), (56:B:b)) quite clear: A violation of (56:9) (i.e. of (56:B:a)) means that one of the (real) players receives an offer which is more unfavorable than what can be enforced against him. A violation of (56:10) (i.e. of (56:B:b)) means that the totality of all (real) players receives an offer which is more favorable than it could ever expect to achieve. It seems reasonable to consider these as precisely the conditions under which players who act rationally will refuse to consider a distribution scheme (an imputation) because it is manifestly absurd.

56.9.2. Before proceeding any further we must once more retrace our steps and compare our present set up with the previous one, in the cases where both apply.

Specifically: Assume that we are applying the procedure of the past sections to an n-person game Γ which is already zero-sum. Accordingly we form for this game the zero-sum $n + 1$-person game $\bar{\Gamma}$ as described in 56.2.2., and then proceed as in 56.8.2.

[1] Note that the concept of a best strategy for the totality of all real players is clearly defined: if there is complete co-operation, then the totality faces a pure maximum problem.

[2] If the game in its original form—i.e. before the normalization of 12.1.1. and 11.2.3. is performed—contains chance moves, then the "most favorable case" referred to above must not be taken to include these too. I.e. only co-operation and optimal choice of strategies is to be assumed, while the chance moves must be accounted for by forming expectation values. Indeed, it is in this way that we passed in 11.2.3. from the

$$\mathcal{G}_k(\tau_0, \tau_1, \cdots, \tau_n)$$

(τ_0 representing the influence of all chance moves) to the $\mathcal{H}_k(\tau_1, \cdots, \tau_n)$ which we are using now.

It is important not to misunderstand the meaning of this operation. Obviously the operations of 56.2.2. and 56.8.2. are entirely unnecessary if Γ itself is a zero-sum game; we possess a theory which disposes of this case. But if a more general theory, valid for all games, is to be constructed on this basis, then we must demand that it agree with the (more special) old theory as far as the latter goes. I.e. in the domain of the old theory, where the new theory is superfluous, the new must agree with the old.[1]

56.9.3. That Γ is a zero-sum n-person game means

$$\sum_{k=1}^{n} \mathcal{H}_k(\tau_1, \cdots, \tau_n) \equiv 0,$$

i.e. $\mathcal{H}_{n+1}(\tau_1, \cdots, \tau_n) \equiv 0$. Thus $v(S)$ is not affected if the fictitious player $n + 1$ is added to (or removed from) the set S. I.e.:

(56:14) $\qquad v(S) = v(S \cup (n + 1)) \qquad$ for $\qquad S \subseteq (1, \cdots, n)$.

The special cases $S = \ominus, (1, \cdots, n)$ give

(56:15) $$v((n + 1)) = 0,$$
(56:16) $$v((1, \cdots, n + 1)) = 0.$$

(56:14), (56:15) together show that the game $\bar{\Gamma}$ is decomposable with the splitting sets $(1, \cdots, n)$ and $(n + 1)$. Its $(1, \cdots, n)$ constituent is the original game Γ, while the fictitious player $n + 1$ is a dummy.[2] (For the decomposition cf. the end of 42.5.2. as well as 43.1. For the dummies cf. footnote 1 on p. 340, and the end of 43.4.2.)

Now we can observe:

56.9.4. First: Since $\bar{\Gamma}$ obtains from Γ by the addition of a dummy, the solutions of Γ and $\bar{\Gamma}$ (in the old theory) correspond to each other, the only difference being that the latter takes care of the dummy (the fictitious player $n + 1$) also, assigning him the amount $v((n + 1))$, i.e. zero. (Cf. 46.9.1. or (46:M) in 46.10.4.)

Our proposed new theory would obtain the solutions for Γ from the (old theory) solutions of $\bar{\Gamma}$. Hence the above consideration proves that all the new solutions to be obtained for Γ will be among the old ones. Furthermore we see that in this case we can—indeed must—take the entire system Ω of (56:A:a) in 56.7.2. It must be noted, however, that in this case all solutions of Ω automatically assign the fictitious player $n + 1$ the amount $v((n + 1))$. I.e. here $\Omega = \Omega_c$ with $c = v((n + 1))$, i.e. $\Omega = \Omega''$. (Cf.

[1] This is a well known methodological principle of mathematical generalization.

[2] The reader should recall that the fictitious player is not, in general, a dummy in the game $\bar{\Gamma}$. This may sound paradoxical, but it was established in 56.3. for the very special case of the general two-person games Γ. Indeed, it is just because the rules of the game $\bar{\Gamma}$ do not in general assign him the role of a dummy that we must restrict the solution $\bar{V}$ of $\bar{\Gamma}$ to those which do restrict him to such a role. This is the meaning of the discussions of 56.3.2.-56.6.2.

We shall determine in 57.5.3. which properties of Γ are necessary and sufficient in order that the fictitious player be a dummy.

(56:A:b) and (56:A:d), loc. cit.) Consequently any sets we might define between Ω and Ω''—in particular, both Ω' and Ω'' of (56:A:c) and (56:A:d), loc. cit.—coincide with Ω and are equally acceptable for our purpose.

In other words: the choice between Ω' and Ω'' which is still ahead of us is of no significance in this case. Both alternatives here are in agreement with the old theory; indeed, there is no need here to abandon the old theory at all.[1]

56.9.5. Second: The imputations for a zero-sum n-person game were defined in the old theory in this way:

(56:C:a) $$\overrightarrow{\alpha} = \{\alpha_1, \cdots, \alpha_n\};$$

(56:C:b) $$\alpha_i \geqq \mathrm{v}((i)) \quad \text{for} \quad i = 1, \cdots, n;$$

(56:C:c) $$\sum_{i=1}^{n} \alpha_i = 0.$$

Our new arrangement of (56:7) in 56.8.1. differs from this. Here we have

(56:C:a*) $$\overrightarrow{\alpha} = \{\{\alpha_1, \cdots, \alpha_n\}\};$$

and by (56:9), (56:10) and (56:16),

(56:C:b*) $$\alpha_i \geqq \mathrm{v}((i)) \quad \text{for} \quad i = 1, \cdots, n;$$

(56:C:c*) $$\sum_{i=1}^{n} \alpha_i \leqq 0.$$

We already know from the preceding remark that there can be no real difference in the present case between the old theory and the new one.[2] It is nevertheless useful to see directly that (from the point of view of the old theory) the two procedures (56:C:a)-(56:C:c) and (56:C:a*)-(56:C:c*) contain really no discrepancy.

The only difference between these two arrangements lies in (56:C:c) and (56:C:c*). Recalling the definitions of 44.7.2., we see that the difference between (56:C:a)-(56:C:c) and (56:C:a*)-(56:C:c*) can also be stated in this way: The first amounts to considering solutions for $E(0)$; the second, to considering solutions for $F(0)$. Now we have noted in 46.8.1. that 0 lies in the "normal" zone of the game Γ, and by (45:O:b) in 45.6.1., $E(0)$ and $F(0)$ have the same solutions. Thus we have a perfect agreement.

These two remarks made systematic use of the theory of composition and decomposition of Chapter IX, in order to analyze the influence of our contemplated new procedure on the zero-sum games Γ. This procedure consisted mainly in the passage from Γ to $\bar{\Gamma}$, which as we saw amounted to the addition of a dummy to Γ. This is considerably more special than the general compositions dealt with loc. cit. The specific results used could

[1] The necessity of restricting Ω was deduced in 56.5.-56.6. by considering a non-zero sum game Γ.

[2] Or rather any new one built along the lines contemplated—we have not yet made the decision between Ω' and Ω''.

accordingly have been obtained with less effort than by the use of the far more general theorems referred to. We shall not enter into this subject further since the general results of Chapter IX are available in any case, and because the above treatment projects our present considerations more clearly upon their proper background.

56.10. Analysis of the Concept of Domination

56.10.1. We now return to the general n-person game Γ, its zero-sum extension $\bar{\Gamma}$, and the new treatment of imputations as introduced in 56.8.

Certainly all solutions of $\bar{\Gamma}$ in general cannot be used to define a satisfactory concept of solutions for Γ. This was established by the consideration of a special case—i.e. by casuistic procedure—in 56.5.-56.6. Let us now approach this problem systematically; i.e. apply to the game the formal definition of a solution as given in 30.1.1. and try to determine in full generality which of its features are unsatisfactory and require modification.

In doing this we shall use the concept of imputation (of $\bar{\Gamma}$) in the new arrangement (56:7) in 56.8.1. The important point about this arrangement is that it stresses *ab initio* the primary importance of the real players in $\bar{\Gamma}$,—i.e. directs our attention more to Γ than to $\bar{\Gamma}$. This does not impair, of course, the fact that we apply the formal theory of 30.1.1. to the zero-sum $n + 1$-person game $\bar{\Gamma}$, and not to the general n-person game Γ (which would not be possible).

The concepts of 30.1.1. are all based on that of domination. We therefore begin by expressing the meaning of domination as defined loc. cit. for imputations (of $\bar{\Gamma}$) with the new arrangement of (56:7) in 56.8.1.

Consider two imputations

$$\overrightarrow{\alpha} = \{\{\alpha_1, \cdots, \alpha_n\}\}, \qquad \overrightarrow{\beta} = \{\{\beta_1, \cdots, \beta_n\}\}.$$

Domination

$$\overrightarrow{\alpha} \succ \overrightarrow{\beta}$$

means that there exists a non-empty set $S \subseteq (1, \cdots n, n + 1)$ which is effective for $\overrightarrow{\alpha}$, i.e.

$$\sum_{i \text{ in } S} \alpha_i \leqq \mathrm{v}(S), \tag{56:17}$$

such that

$$\alpha_i > \beta_i \qquad \text{for all } i \text{ in } S. \tag{56:18}$$

We wish to express this in terms of the α_i, β_i with $i = 1, \cdots, n$ alone. It is therefore necessary to distinguish between two possibilities:

56.10.2. First: S does not contain $n + 1$. Then

$$S \subseteq (1, \cdots, n), \qquad S \text{ not empty}. \tag{56:19}$$

The conditions (56:17), (56:18) above need not be reformulated since they involve only the α_i, β_i with $i = 1, \cdots, n$. Besides $S \subseteq (1, \cdots, n)$ in the $v(S)$ of (56:17).

Second: S does contain $n + 1$. Put $T = S - (n + 1)$. Then

(56:20) $$T \subseteq (1, \cdots, n), \qquad T \text{ may be empty.}$$

The conditions (56:17), (56:18) above must be reformulated since they involve α_{n+1}, β_{n+1}.

It is natural to form $-S$ in $(1, \cdots, n, n + 1)$, i.e. as $(1, \cdots, n, n + 1) - S$; and $-T$ in $(1, \cdots, n)$; i.e. as $(1, \cdots, n) - T$. These two sets are clearly equal, but it is nevertheless useful to have symbols for both. We denote the first by $\perp S$ and the second by $-T$.

Since $\sum_{i=1}^{n+1} \alpha_i = 0$, so

$$\sum_{i \text{ in } S} \alpha_i = - \sum_{i \text{ in } \perp S} \alpha_i = - \sum_{i \text{ in } -T} \alpha_i,$$
$$v(S) = -v(\perp S) = -v(-T).$$

Hence (56:17) becomes

(56:21) $$\sum_{i \text{ in } -T} \alpha_i \geqq v(-T).$$

This involves only the α_i with $i = 1, \cdots, n$. Besides $-T \subseteq (1, \cdots, n)$ in the $v(-T)$ of (56:21). Next (56:18) becomes

(56:22) $$\alpha_i > \beta_i \qquad \text{for all } i \text{ in } T,$$
and
$$\alpha_{n+1} > \beta_{n+1}.$$

This last inequality means that

(56:23) $$\sum_{i=1}^{n} \alpha_i < \sum_{i=1}^{n} \beta_i.$$

(56:22), (56:23) also involve only the α_i, β_i with $i = 1, \cdots, n$.

Summing up:

(56:D) $\overrightarrow{\alpha} \succ \overrightarrow{\beta}$ means that there exists either
(56:D:a) an S with (56:19) and (56:17), (56:18);
or
(56:D:b) a T with (56:20) and (56:21), (56:22), (56:23).

Note that these criteria involve only sets S, T, $-T \subseteq (1, \cdots, n)$ and the α_i, β_i with $i = 1, \cdots, n$. I.e. they refer only to the original game Γ and to the real players $1, \cdots, n$.

56.10.3. The criterion (56:D) of domination was obtained by a literal application of the original definition of 30.1.1., the application being made directly to $\bar{\Gamma}$ and then translated in terms of Γ. This rigorous operation having been carried out, let us now examine the result from the point of view of interpretation; i.e. let us see whether the conditions of (56:D) produce a reasonable definition of domination for the present case.

According to (56:D) domination holds in two cases (56:D:a) and (56:D:b).

(56:D:a) is merely a restatement of the original definition of 30.1.1.[1] It expresses that there exists a group of (real) players (the set S of (56:19)), each of whom prefers his individual situation in $\overrightarrow{\alpha}$ to that in $\overrightarrow{\beta}$ (this is (56:18)), and who know that they are able as a group, i.e. as an alliance, to enforce this preference of theirs (this is (56:17)).

(56:D:b) on the other hand is, when viewed in terms of Γ and of the real players alone, something entirely new. It requires again that there exist a group of (real) players (the set T of (56:20)) each of whom prefers his own individual situation in $\overrightarrow{\alpha}$ to that in $\overrightarrow{\beta}$ (this is (56:22)). The ability of this group to enforce the preference in question (i.e. (56:17)) is not required. Instead we have the condition that the real players left out of this group must not be able to block the preferred imputation in question, that is insofar as it affects them (this is (56:21)).[2]

Finally there is the peculiar condition that the totality of all (real) players—i.e. society as a whole—must be worse off under the (preferred) regime $\overrightarrow{\alpha}$ than under the (rejected) regime $\overrightarrow{\beta}$ (this is (56:23)).

56.10.4. This strange alternative (56:D:b) was, of course, obtained by treating the fictitious player $n + 1$ as a real entity. If we refrain from

[1] Applied, however, to the general game Γ, for which that theory was not intended!

[2] The (real) players left out, i.e. those of $-T$, could block the preferred imputation $\overrightarrow{\alpha}$ if they could get for themselves separately more than $\overrightarrow{\alpha}$ assigns to them together; i.e. if

$$\sum_{i \text{ in } -T} \alpha_i < \text{v}(-T).$$

(Note that we had to exclude equality here, since that would not block $\overrightarrow{\alpha}$.) The negation of this is indeed:

$$\text{(56:21)} \qquad \sum_{i \text{ in } -T} \alpha_i \geqq \text{v}(-T).$$

This may be compared with the expression of the ability of the original group T to enforce its preference, i.e.

$$\text{(56:17)} \qquad \sum_{i \text{ in } T} \alpha_i \leqq \text{v}(T).$$

It should be noted that neither of (56:17), (56:21) implies the other: It is perfectly possible that the group T can enforce $\overrightarrow{\alpha}$ as far as it affects the members of T, and that at

doing that and try to appraise matters in terms of realities—i.e. of the real players—then it becomes very difficult to interpret (56:D:b). The best one can say of it is that it seems to assume the effective operation of an influence which is definitely set to injure society as a whole (i.e. the totality of all real players). Specifically in this case domination is asserted when all players of a certain group (of real players) prefer their individual situation in $\overrightarrow{\alpha}$ to that in $\overrightarrow{\beta}$, if the remaining (real) players cannot block this arrangement, and if it is definitely injurious to society as a whole.

In comparing this domination (56:D:b) to the ordinary one (56:D:a) the following differences are particularly conspicuous: First, that in (56:D:a) the ability to enforce one's preference is essential, while in (56:D:b) the essential point is the ability of the others to block it. Second, that in (56:D:a) the active group had to be a non-empty set, whereas in (56:D:b) it could also be an empty set (cf. (56:19) and (56:20)). Third, the antisocial viewpoint figures in (56:D:b), but not at all in (56:D:a).

The reader will have noticed by now that (56:D:b) is of a rather irrational character, but nevertheless not altogether unfamiliar. It would be easy to enlarge upon the images and allegories of which (56:D:b) is an exact formalization. There is no need to dwell further upon this subject here. What matters is that we have every reason to see in the alternative (56:D:b) the general cause for those difficulties for which a special case was analyzed in 56.5.-56.6. Clearly (56:D:b) is not an immediately plausible approach to the concept of domination in the sense in which (56:D:a) is.

We shall therefore try to resolve our difficulties by the simple expedient of rejecting (56:D:b) altogether.

56.11. Rigorous Discussion

56.11.1. We have decided to redefine domination by rejecting (56:D:b) and retaining (56:D:a) in (56:D) of 56.10.2. This new concept of *domination* can be stated in two ways which both seem to deserve consideration.

First: As pointed out at the beginning of 56.10.3., (56:D:a) amounts to a repetition of the corresponding definition of 30.1.1. The only difference

---

the same time the group $-T$ can block $\overrightarrow{\alpha}$ as far as it affects the members of $-T$. On the other hand it is also possible that neither group can enforce or prevent anything.

However, if Γ is a zero-sum game, and if we require (as in the old theory) $\sum_{i=1}^{n} \alpha_i = 0$, then (56:17) and (56:21) are equivalent. Indeed, in this case

$$v(T) + v(-T) = v((1, \cdots, n)) = 0,$$

so

$$\sum_{i \text{ in } -T} \alpha_i = - \sum_{i \text{ in } T} \alpha_i, \qquad v(-T) = -v(T),$$

from which the equivalence follows as asserted.

is that then Γ was a zero-sum n-person game, while now it is a general n-person game.

Thus our present procedure means that we extend to the present case the definition of domination in 30.1.1. unchanged, irrespective of the fact that the game is no longer required to be of zero sum.[1]

Second: Let us now view the restriction to (56:D:a) from the standpoint of $\bar{\Gamma}$ rather than of Γ. Our original discussion in 56.10. yielded the two cases (56:D:a) and (56:D:b) depending upon the following disjunction. In the sense of 30.1.1. the domination in $\bar{\Gamma}$ had to be based on a set S. Now (56:D:a) obtains when $n + 1$ does not belong to S, while (56:D:b) obtains when it does. Hence the restriction to (56:D:a) amounts to requiring that the set S must not contain $n + 1$.

We repeat: Our new concept of domination means, in terms of $\bar{\Gamma}$, that in the definition of domination in 30.1.1. we add to the conditions (30:4:a)-(30:4:c) imposed upon the set S, the further condition that S must not contain a specified element, namely $n + 1$.

This can also be construed as a restriction on the concept of effectivity loc. cit.: We regard a set S as effective only if it does not contain $n + 1$. (Of course, the original condition (30:3) loc. cit. is also required.)

56.11.2. We now proceed to study the new concept of solution for $\bar{\Gamma}$, i.e. for Γ, based upon the new concept of domination, introduced in 56.11.1. In analyzing it we shall rely upon the game $\bar{\Gamma}$ and the form (56:5) of imputations (rather than upon the game Γ and the form (56:7) of imputations) and the definition of domination as formulated in the second remark of 56.11.1.

We obtain our result by proving four successive lemmas:

(56:E) If $\bar{V}$ is a solution for $\bar{\Gamma}$ in the new sense, then every

$$\overrightarrow{\alpha} = \{\alpha_1, \cdots, \alpha_n, \alpha_{n+1}\}$$

of $\bar{V}$ has $\alpha_{n+1} = v((n + 1))$.

Proof: Assume the opposite. Necessarily $\alpha_{n+1} \geqq v((n + 1))$, hence there would exist an $\overrightarrow{\alpha} = \{\alpha_1, \cdots, \alpha_n, \alpha_{n+1}\}$ in $\bar{V}$ with $\alpha_{n+1} > v((n + 1))$. Put $\alpha_{n+1} = v((n + 1)) + \epsilon$, $\epsilon > 0$. Define $\overrightarrow{\beta} = \{\beta_1, \cdots, \beta_n, \beta_{n+1}\}$ with

$$\beta_i = \alpha_i + \frac{\epsilon}{n} \quad \text{for} \quad i = 1, \cdots, n;$$
$$\beta_{n+1} = \alpha_{n+1} - \epsilon = v((n + 1)).$$

[1] It may seem peculiar that it took us so long to reach this simple principle,—in fact we need the further considerations of 56.11.2. before we accept it finally. However, the act of taking over the definition of 30.1.1. without any alternatives, in spite of the extremely wide generalization which is now performed, requires most careful attention. The detailed inductive approach given in these paragraphs seemed to be best suited for this purpose.

Since $\sum_{i=1}^{n} \beta_i = -\beta_{n+1} = -\mathrm{v}((n+1)) = \mathrm{v}((1, \cdots, n))$ and $\beta_i > \alpha_i$ for $i = 1, \cdots, n$, so the use of $S = (1, \cdots, n)$ establishes $\overrightarrow{\beta} \succ \overrightarrow{\alpha}$.[1] As $\overrightarrow{\alpha}$ belongs to $\bar{V}$, $\overrightarrow{\beta}$ cannot belong to it. Hence there exists a $\overrightarrow{\gamma}$ in $\bar{V}$ with $\overrightarrow{\gamma} \succ \overrightarrow{\beta}$. Now consider the set S which enforces this domination. Since S does not contain $n + 1$, we have $S \subseteq (1, \cdots, n)$. As $\beta_i > \alpha_i$ for $i = 1, \cdots, n$, $\overrightarrow{\gamma} \succ \overrightarrow{\beta}$ implies $\overrightarrow{\gamma} \succ \overrightarrow{\alpha}$. But $\overrightarrow{\gamma}, \overrightarrow{\alpha}$ are both in $\bar{V}$; hence we have a contradiction.

(56:F) If $\bar{V}$ is a solution for $\bar{\Gamma}$ in the new sense, then it is so also in the old sense.

Proof: We must show that (30:5:a), (30:5:b) of 30.1.1., with domination in the new sense here imply the same with domination in the old sense. Now domination in the new sense implies domination in the old sense; hence our assertion concerning (30:5:b) is immediate. So only (30:5:a) requires closer inspection.

Assume therefore that (30:5:a) is invalid in the old sense, i.e. that for two $\overrightarrow{\alpha}, \overrightarrow{\beta}$ in $\bar{V}$, $\overrightarrow{\alpha} \succ \overrightarrow{\beta}$ in the old sense. Let S be the set which enforces this domination. By (56:E) $\alpha_{n+1} = \beta_{n+1}$ $(= \mathrm{v}((n+1)))$, hence $n + 1$ cannot belong to S. Consequently $\overrightarrow{\alpha} \succ \overrightarrow{\beta}$ in the new sense, i.e. (30:5:a) fails in the new sense too. This completes the proof.

(56:G) If $\bar{V}$ is a solution for $\bar{\Gamma}$ in the old sense, and if every

$$\overrightarrow{\alpha} = \{\alpha_1, \cdots, \alpha_n, \alpha_{n+1}\}$$

of $\bar{V}$ has $\alpha_{n+1} = \mathrm{v}((n+1))$, then $\bar{V}$ is also a solution in the new sense.

Proof: We must show that (30:5:a), (30:5:b) of 30.1.1. with domination in the old sense here imply the same with domination in the new sense. Now domination in the old sense is implied by domination in the new sense; hence this time our assertion concerning (30:5:a) is immediate. So only (30:5:b) requires closer inspection.

Consider therefore an $\overrightarrow{\alpha} = \{\alpha_1, \cdots, \alpha_n, \alpha_{n+1}\}$ not in $\bar{V}$. As (30:5:b) holds in the old sense, there exists a $\overrightarrow{\beta} = \{\beta_1, \cdots, \beta_n, \beta_{n+1}\}$ in $\bar{V}$ with $\overrightarrow{\beta} \succ \overrightarrow{\alpha}$ in the old sense. Let S be the set which enforces this domination. Necessarily $\alpha_{n+1} \geqq \mathrm{v}((n+1))$, and since $\overrightarrow{\beta}$ belongs to $\bar{V}$ by assumption, $\beta_{n+1} = \mathrm{v}((n+1))$. Hence $\beta_{n+1} \leqq \alpha_{n+1}$ and so $n + 1$ cannot belong to S. Consequently $\overrightarrow{\beta} \succ \overrightarrow{\alpha}$ in the new sense, i.e. (30:5:b) holds in the new sense too. This completes the proof.

[1] This domination as well as all others in this proof are in the new sense.

(56:H) $\bar{V}$ is a solution of $\bar{\Gamma}$ in the new sense, if and only if it belongs to the system Ω'' of (56:A:d) in 56.7.2.

Proof: The forward implication results from (56:E) and (56:F), the inverse implication from (56:G).

56.11.3. In interpreting the result of (56:H) we must remember that the discussion originated from the necessity of restricting the system Ω of all solutions of $\bar{\Gamma}$ for the purposes of the theory of Γ. We saw in 56.7. that the plausible result of this restriction should be the set Ω' or Ω'' (or possibly some set between the two). Thereafter our effort was directed to making a decision between these two possibilities. Furthermore we concluded in 56.10.-56.11.1. that a modification of the concept of domination in $\bar{\Gamma}$ might answer our problem. Now the statement of (56:H) is that this modification of the concept of domination leads precisely to the set Ω''. By these concurrent results the decision is clearly indicated. We accept Ω'' as the system of all solutions for Γ.

56.12. The New Definition of a Solution

56.12. We reformulate this together with references to the main results on which the decision was based:

(56:I)

(56:I:a) For a general n-person game Γ, a *solution* is any solution (in the original sense of 30.1.1.) of its zero-sum extension, the zero-sum $n + 1$-person game $\bar{\Gamma}$, for which all

$$\overrightarrow{\alpha} = \{\alpha_1, \cdots, \alpha_n, \alpha_{n+1}\}$$

in $\bar{V}$ have

(56:24) $$\alpha_{n+1} = \mathrm{v}((n + 1)).$$

These solutions form precisely the set Ω'' of (56:A:d) in 56.7.2.

(56:I:b) Using the form (56:7), $\overrightarrow{\alpha} = \{\{\alpha_1, \cdots, \alpha_n\}\}$ for these imputations (i.e. emphasizing Γ and its players rather than $\bar{\Gamma}$), transforms (56:24) above into

(56:25) $$\sum_{i=1}^{n} \alpha_i = \mathrm{v}((1, \cdots, n)).$$

This is clearly a strengthened form of (56:10) in 56.8.2.

(56:I:c) In the special case in which Γ is itself a zero-sum game, our new concept of solutions (for Γ) coincides with the old one,—i.e. the unmodified application of 30.1.1. (Cf. the first remark in 56.9.4.) Thus it is no longer necessary to distinguish between the old theory and the new one. (Cf. also footnote 1 on p. 518.)

(56:I:d) For a general n-person game Γ, the solutions can also be obtained by applying the definitions of 30.1.1. (which were intended for zero-sum games only) to Γ directly and without any modification. The concept of imputations for Γ must then be used in the form (56:7). (Cf. the first remark in 56.11.1.)

(56:I:e) The validity of (56:I:d) means that nothing must be added to the characterization of the imputations in the form (56:7) as given in 56.8.2. However, by (56:I:b) the equation (56:25) will then automatically hold in each solution $\bar{V}$. Hence we may, if we wish, add (56:25),—i.e. strengthen (56:10) of 56.8.2. to (56:25).[1]

(56:I:f) The restriction imposed in (56:I:a) upon the solutions of $\bar{\Gamma}$ can also be expressed by modifying the concept of domination for $\bar{\Gamma}$ but then allowing all solutions in the modified sense. This modification consists of imposing upon effective sets (in the sense of 30.1.1.) the further requirement that they must not contain $n + 1$. (Cf. the second remark of 56.11.1.)

57. The Characteristic Function and Related Topics

57.1. The Characteristic Function: The Extended and the Restricted Forms

57.1. We are now in the possession of a theory which applies to all games and—like the theory of 30.1.1. for zero-sum games, of which it is an extension—is based exclusively upon the characteristic function. I.e. the functions $\mathcal{H}_k(\tau_1, \cdots, \tau_n)$, $k = 1, \cdots, n$ of 11.2.3., which actually define the game, do not affect the theory directly, but only through the characteristic function $v(S)$.[2]

There is, however, a difference between the use of the characteristic function $v(S)$ for a zero-sum game, and for a general game. For a zero-sum n-person game Γ the characteristic function $v(S)$ is defined for all sets $S \subseteq (1, \cdots, n)$ and for only these. (Cf. 25.1.) For a general n-person game Γ we had to form its zero-sum extension, the zero-sum $(n + 1)$-person game $\bar{\Gamma}$, and the characteristic function $v(S)$ was actually formed like the characteristic function (in the old sense) of $\bar{\Gamma}$. (This is the $v(S)$ which

[1] This permissibility of restricting

$$(56:10) \qquad \sum_{i=1}^{n} \alpha_i \leqq v((1, \cdots, n))$$

to

$$(56.25) \qquad \sum_{i=1}^{n} \alpha_i = v((1, \cdots, n))$$

is analogous to (but more general than) the equivalence of $E(0)$ and $F(0)$ referred to in the second remark of 56.9.5.

[2] Of course $v(S)$ is defined with the help of the $\mathcal{H}_k(\tau_1, \cdots, \tau_n)$. Cf. 25.1.3. and 58.1.

figured in all of our recent discussions, particularly throughout 56.4.1., 56.5.1., 56.7.2., 56.8.2., 56.9.1., 56.9.3.-56.10.3., 56.11.2.-56.12.) Accordingly $v(S)$ is now defined for all sets $S \subseteq (1, \cdots, n, n+1)$ and for only these. We may, however, if we wish, consider $v(S)$ for the sets $S \subseteq (1, \cdots, n)$ only. When this is done, we shall speak of the *restricted characteristic function;* while $v(S)$ in its original domain, embracing all $S \subseteq (1, \cdots, n, n+1)$ is the *extended characteristic function.*

From this we conclude in the special case of a zero-sum game: Here the characteristic function of the old theory is the restricted characteristic function of the new one.[1]

Returning to the general games, we see that the characteristic function is the basis of our entire present theory. Among the equivalent formulations of that theory, (56:I:a) in 56.12. uses the extended characteristic function, while (56:I:d) uses the restricted one.

Consequently our next objective is necessarily the determination of the nature of these characteristic functions, and of their relationship to each other.

57.2. Fundamental Properties

57.2.1. Consider a general n-person game Γ, and its two characteristic functions, as defined above: The restricted one, $v(S)$ defined for all subsets S of $I = (1, \cdots, n)$ and the extended one, $v(S)$ defined for all subsets S of $\bar{I} = (1, \cdots, n, n+1)$.[2]

In what follows we must distinguish between two possibilities in our notations for $-S$, as in the second remark in 56.10.2. For $S \subseteq \bar{I} = (1, \cdots, n, n+1)$ we can form $-S$ in $\bar{I}$, i.e. as $\bar{I} - S$, while for $S \subseteq I = (1, \cdots, n)$ we can also form $-S$ in I, i.e. as $I - S$.[3] We denote again the first set by $\perp S$ and the second one by $-S$.

We propose to determine the essential properties of both characteristic functions of the general n-person game—just as we did in 25.3. and 26., for the characteristic function of the zero-sum n-person game.

Consider the extended characteristic function first. Since it is the characteristic function in the old sense for the zero-sum $(n+1)$-person game $\bar{\Gamma}$, it must have the properties (25:3:a)-(25:3:c) formulated in 25.3.1.—only with $\bar{I} = (1, \cdots, n+1)$ in place of the $I = (1, \cdots, n)$ there. In this way we obtain:

(57:1:a) $$v(\ominus) = 0,$$

[1] All these distinctions and definitions cannot and do not affect the rigorously established fact that for all zero-sum games the two theories are equivalent to each other. (Cf. (56:I:c) in 56.12.)

[2] We denote them by the same letter, v, since they have the same value wherever both are defined.

[3] Formed for the same S (of course $S \subseteq I$), these two sets are clearly different. Loc. cit. we claimed that they are equal, but there we formed them for two different sets S and T.

(57:1:b) $$v(\perp S) = -v(S),$$

(57:1:c) $$v(S \cup T) \geqq v(S) + v(T) \quad \text{if} \quad S \cap T = \ominus. \quad (S, T \subseteq \bar{I}).$$

Consider next the restricted characteristic function. We obtain conditions for it from (57:1:a)-(57:1:c), by restricting ourselves to subsets of I. This is immediately feasible for (57:1:a), (57:1:c), but it is impossible for (57:1:b).[1] In this way we obtain:

(57:2:a) $$v(\ominus) = 0,$$

(57:2:c) $$v(S \cup T) \geqq v(S) + v(T) \quad \text{if} \quad S \cap T = \ominus. \quad (S, T \subseteq I)$$

Note that we cannot replace (57:1:b) by something equivalent for $-S$. Indeed, all we can do with $-S$, is to put $T = -S$ in (57:1:c). This gives

(57:2:b) $$v(-S) \leqq v(I) - v(S).$$

Even if $v(I) = 0$, which need not be the case, (57:2:b) becomes merely

(57:2:b*) $$v(-S) \leqq -v(S),$$

but not the equivalent of (25:3:b) in 25.3.1.

$$v(-S) = -v(S).$$

(57:1:a)-(57:1:c) as well as (57:2:a), (57:2:c) are, by virtue of their derivation, only necessary properties of the (extended or restricted) characteristic functions. We must now see whether they are sufficient as well.

57.2.2. If $\bar{\Gamma}$ were an arbitrary zero-sum $(n + 1)$-person game, then we could conclude from the result of 26.2. that any $v(S)$ which fulfills (57:1:a)-(57:1:c) is the (old sense) characteristic function of a suitable $\bar{\Gamma}$,—i.e. the extended characteristic function of a suitable general n-person game Γ. In other words: This would prove that the conditions (57:1:a)-(57:1:c) are necessary and sufficient—that they contain a complete mathematical characterization of the characteristic functions of all possible general n-person games Γ.

However, $\bar{\Gamma}$ is not at all arbitrary. As we saw in 56.2.2., the (fictitious) player $n + 1$ has no influence on the course of the game—i.e. he has no personal moves; the $\mathcal{H}_k(\tau_1, \cdots, \tau_n, \tau_{n+1})$ do not really depend on his variable τ_{n+1}. Furthermore, it is clear from 56.2.2. that this is the only restriction to which $\bar{\Gamma}$ must be subjected: If in a zero-sum $n + 1$-person game Γ the player $n + 1$ has no influence on the course of the game, then we can view $\bar{\Gamma}$ as the zero-sum extension of a general n-person game Γ played by the remaining players $1, \cdots, n$.[2]

[1] S, $\perp S$ cannot be both $\subseteq I = (1, \cdots, n)$ since one of them must necessarily contain $n + 1$.

[2] I.e. we can treat $n + 1$ as if he were a fictitious player—as far as the rules of the game are concerned. We know, of course, that there are solutions $\bar{V}$ for $\bar{\Gamma}$ which bring out the fact that he is a real player (those in Ω but not in Ω'', cf. (56:A:a)-(56:A:d) in 56.7.2. and (56:I:a) in 56.12.; recall also 56.3.2., 56.3.4.).

Consequently the following question arises: (57:1:a)-(57:1:c) are necessary and sufficient conditions for the characteristic functions in the old sense of all zero-sum $n + 1$-person games. How must they be strengthened so as to do the same thing for the (old sense) characteristic functions of all those zero-sum $n + 1$-person games in which the player $n + 1$ has no influence on the course of the game?

Answering this question would amount to giving a complete mathematical characterization for the extended characteristic functions of all general $n + 1$-person games. But then the problem of doing the same for the restricted characteristic functions would still remain.

It will be seen that by attacking the last problem first, a somewhat more advantageous arrangement obtains: the first problem can be solved in a few lines with the help of the latter one. However, our approach will be dominated by the above considerations.

57.3. Determination of All Characteristic Functions

57.3.1. We proceed to prove that the necessary conditions (57:2:a), (57:2:c) are also sufficient: For any numerical set function $v(S)$ which fulfills (57:2:a), (57:2:c) there exists a general n-person game Γ of which this $v(S)$ is the restricted characteristic function.[1]

In order to avoid confusion it is better to denote the given numerical set-function which fulfills (57:2:a); (57:2:c) by $v_0(S)$. With its help we shall define a certain general n-person game Γ, and denote the restricted characteristic function of this Γ by $v(S)$. It will then be necessary to prove that $v(S) = v_0(S)$.

Let therefore a numerical set-function $v_0(S)$ which fulfills (57:2:a), (57:2:c) be given. We define the general n-person game Γ as follows:[2]

Each player $k = 1, \cdots, n$ will, by a personal move, choose a subset S_k of I which contains k. Each one makes his choice independently of the choice of the other players.

After this the payments are made as follows:

Any set S of players for which

(57:3) $$S_k = S \quad \text{for every } k \text{ belonging to } S$$

is called a *ring*. Any two rings with a common element are identical. In other words: The totality of all rings (which actually have formed in a play) is a system of pairwise disjunct subsets of I.

Each player who is contained in none of the rings thus defined forms by himself a (one-element) set which is called a *solo set*. Thus the totality of all rings and solo sets (which actually have formed in a play) is a decomposition of I, i.e. a system of pairwise disjunct subsets of I with the sum I. Denote these sets by $C_1, \cdots, C_p$ and the respective numbers of their elements by $n_1, \cdots, n_p$.

[1] The construction which follows has much in common with that of 26.1

[2] The reader should now compare the details with those in 26.1.2.

Now consider a player k. He belongs to precisely one of these sets $C_1, \cdots, C_p$, say to C_q. Then the player k gets the amount

(57:4) $$\frac{1}{n_q} v_0(C_q).$$

This completes the description of the game Γ. Γ is clearly a general n-person game, and it is clear what its zero-sum extension $\bar{\Gamma}$ is. We emphasize in particular that in $\bar{\Gamma}$ the fictitious player $n + 1$ gets the amount

(57:5) $$-\sum_{q=1}^{p} v_0(C_q).$$[1]

We are now going to show that Γ has the desired restricted characteristic function $v_0(S)$.

57.3.2. Denote the restricted characteristic function of Γ by $v(S)$. Remember that (57:2:a), (57:2:c) hold for $v(S)$ because it is a restricted characteristic function, and for $v_0(S)$ by hypothesis.

If S is empty, then $v(S) = v_0(S)$ by (57:2:a). So we may assume that S is not empty. In this case a coalition of all players belonging to S can govern the choices of its S_k so as with certainty to make S a ring. It suffices for every k in S to choose his $S_k = S$. Whatever the other players (in $-S$) do, S will thus be one of the sets (rings or solo sets) $C_1, \cdots, C_p$ say C_q. Each k in $C_q = S$ gets the amount (57:4), hence the entire coalition gets the amount $v_0(S)$. Consequently

(57:6) $$v(S) \geqq v_0(S).$$

Now consider the complement $-S$. A coalition of all players k belonging to $-S$ can govern the choices of its k so as with certainty to make S a sum of rings and solo sets. If $-S$ is empty, then this is automatically true, since then $S = I$. If $-S$ is not empty, then it suffices for every k in $-S$ to choose his $S_k = -S$. Hence $-S$ is a ring, and therefore S is a sum of rings and solo sets.

Thus S is the sum of some among the sets $C_1, \cdots, C_p$ say of

$$C_{1'}, \cdots, C_{r'}$$

($1', \cdots, r'$ are some among the numbers $1, \cdots, p$). Each k in C_q ($q = s' = 1', \cdots, r'$) gets the amount (57:4), hence the n_q players in C_q together get the amount $v_0(C_q)$, and so all players of S together get the amount $\sum_{s=1}^{r} v_0(C_{s'})$. Since the $C_{1'}, \cdots, C_{r'}$ are pairwise disjunct sets

[1] The n_q players in C_q get together the amount $v_0(C_q)$ by (57:4); hence all players $1, \cdots, n$, i.e. all players in $C_1, \cdots, C_p$, get together the amount $\sum_{q=1}^{p} v_0(C_q)$. Now (57:5) ensues.

with the sum S, repeated application of (57:2:c) gives $\sum_{s=1}^{r} v_0(C_{s'}) \leqq v_0(S)$. I.e.: Whatever the players of S do, together they get an amount $\leqq v_0(S)$. Consequently

$$v(S) \leqq v_0(S). \tag{57:7}$$

Now (57:6), (57:7) together give

$$v(S) = v_0(S), \tag{57:8}$$

as desired.

57.3.3. Let us now consider the extended characteristic functions. Here we know that the conditions (57:1:a)-(57:1:c) are necessary. We shall prove that they are also sufficient: That for *any* numerical set function $v(S)$ which fulfills (57:1:a)-(57:1:c) there exists a general n-person game Γ of which this $v(S)$ is the extended characteristic function.

In order to avoid confusion, it is again better to denote the given numerical set function which fulfills (57:1:a)-(57:1:c) by $v_0(S)$. The extended characteristic function of the general n-person game Γ which we shall use will be denoted by $v(S)$.

Let therefore a numerical set function $v_0(S)$ which fulfills (57:1:a)-(57:1:c) be given. Consider it for a moment for the sets $S \subseteq I = (1, \cdots, n)$ only, then it fulfills (57:2:a), (57:2:c). Hence our construction of 57.3.1., 57.3.2. can be applied to this $v_0(S)$. So a general n-person game Γ obtains, such that its restricted characteristic function has always $v(S) = v_0(S)$[1] and so its extended characteristic function has $v(S) = v_0(S)$ for $S \subseteq I$. I.e., if we revert to the natural domain of these S,[2] then we have:

$$v(S) = v_0(S) \qquad \text{if } n+1 \text{ is not in } S. \tag{57:9}$$

Now let $n + 1$ be in S. Then it is not in $-S$. Hence (57:9) gives $v(-S) = v_0(-S)$. (57:1:a)-(57:1:c) hold for $v(S)$ because it is an extended characteristic function, and for $v_0(S)$ by hypothesis. Therefore (57:1:b) gives $v(-S) = -v(S)$, $v_0(-S) = -v_0(S)$. All these equations combine to

$$v(S) = v_0(S) \qquad \text{if } n+1 \text{ is in } S. \tag{57:10}$$

Now (57:9), (57:10) together give

$$v(S) = v_0(S) \tag{57:11}$$

unrestrictedly, as desired.

57.3.4. To sum up: We have obtained complete mathematical characterizations of both the restricted and the extended characteristic functions $v(S)$ of all possible general n-person games Γ. The former are described by (57:2:a), (57:2:c), and the latter by (57:1:a)-(57:1:c).

[1] The "always" in this case refers, of course, only to the $S \subseteq I$.
[2] Which in this case consists of all $S \subseteq \bar{I}$.

We follow therefore the comparable procedure of 26.2., and call the functions which satisfy these conditions *restricted characteristic functions* or *extended characteristic functions*, respectively—even when they are viewed in themselves, without reference to any game.

57.4. Removable Sets of Players

57.4.1. The result which we obtained for extended characteristic functions can also be stated as follows: Every characteristic function (in the old sense) of any zero-sum $(n + 1)$-person game is also the extended characteristic function of a suitable general n-person game.[1] Remembering the discussion of 57.2.2., this means: Every characteristic function of any zero-sum $n + 1$-person game is also the characteristic function of a suitable zero-sum $n + 1$-person game in which the player $n + 1$ has no influence on the course of the play.

Let us replace in this statement $n + 1$ by n, obtaining the equivalent for zero-sum n-person games and the role of the player n. In order to formulate this result, it is convenient to define:

(57:A) Let a zero-sum n-person game Γ and a set $S \subseteq I = (1, \cdots, n)$ be given. Then we call S *removable* for Γ, if it is possible to find another zero-sum n-person game Γ', which has the same characteristic function as Γ but in which no player belonging to S has an influence upon the course of the game.

Using this definition, our assertion becomes that the set $S = (n)$ is removable. Given any player $k = 1, \cdots, n$, we can interchange the roles of the players k and n, hence the set $S = (k)$ is also removable. So we see:

(57:B) Every one-element set S is removable in every game Γ.

Now it should be noted that according to our theory the entire strategy of coalitions and compensations in a game depends only on its characteristic function. Consequently the two games Γ and Γ' of (57:A) are exactly alike from that point of view.

Hence (57:B) can be interpreted as follows: The role of any one player in any zero-sum n-person game—insofar as the strategic possibilities of coalitions and compensations are concerned—can be duplicated exactly in an arrangement which deprives him of all direct influence upon the course of the game. Here we mean his "role" in the most extended sense: including his relationship to all other players, and his influence on their relationships to each other.

In other words: We described in 56.3.2.-56.3.4. a mechanism by which a player who has no direct influence upon the course of the game can nevertheless influence the negotiations for coalitions and compensations. We have now shown in (57:B), that this mechanism is perfectly adequate

[1] Indeed the conditions (57:1:a)-(57:1:c) coincide with (25:3:a)-(25:3:c) of 25.3.1. with $\bar{I} = (1, \cdots, n, n + 1)$ in place of its $I = (1, \cdots, n)$.

to describe the influence which any player in any game could have in this respect. This statement must be taken absolutely literally: Our result guarantees that all conceivable details and nuances will be reproduced.

57.4.2. By (57:B) every player $k = 1, \cdots, n$ is removable individually —i.e. the one-element set $S = (k)$ is—but this does not mean that all these players are removable simultaneously—i.e. that the set

$$S = I = (1, \cdots, n)$$

is. Indeed we have:

(57:C) The set $S = I$ is removable if and only if the game Γ is inessential.

Proof: That no player $k = 1, \cdots, n$ has an influence on the course of the game Γ', means that all functions $\mathcal{H}'_k(\tau_1, \cdots, \tau_n)$ are independent of all their variables $\tau_1, \cdots, \tau_n$—i.e. that they are constants

$$\mathcal{H}'_k(\tau_1, \cdots, \tau_n) = \alpha_k. \tag{57:12}$$

From this

$$\text{v}(S) = \sum_{k \text{ in } S} \alpha_k \qquad \text{for all} \qquad S \subseteq I. \tag{57:13}$$

Conversely, if (57:13) is required, it can be secured by (57:12).

Hence (57:13) is the characteristic function of a game Γ for which such a Γ' exists—and (57:13) is precisely the definition of inessentiality.

For $n = 1,2$ every game Γ is inessential, hence there the set $S = I$— and with it every set—is removable.[1] For $n \geqq 3$ there exist essential games, and therefore $S = I$ is in general not removable.

Therefore this question arises:

(57:D) Which are the removable sets for an essential game Γ?

(57:B), (57:C) contain a partial answer: The one-element sets are removable, the n-element set ($S = I$) is not. Where is the dividing line?

57.4.3. The upper extreme is reached when all $(n - 1)$-element sets— and with them all sets except I—are removable. We call such a game *extreme*. It is worth while to visualize what this property entails: The strategic situation in such a game is equivalent to that where only one player has an influence on the course of the game, and the role of all others consists merely in trying to influence his decisions. The means of influencing him is, of course, offering him compensations; the motive is to induce

[1] The main result concerning zero-sum two-person games, according to which each game of this type has a definite value for each player (say v, $-$v cf. the discussion of 17.8., 17.9.), means just this: It states that the game is equivalent to the fixed payments v, $-$v to the two players—and this is an arrangement where neither of them can influence anything.

In every essential game, on the other hand, there exists the interplay of negotiations for coalitions and compensations—and this excludes the simultaneous removability of all players.

him to make decisions which are favorable to the player or players who make the offer.

Now we can prove:

(57:E) For $n = 3$: The essential zero-sum three-person game is extreme.

(57:F) For $n = 4$: There exist extreme as well as non-extreme essential zero-sum four-person games.

More in detail:

(57:E*) For the essential zero-sum three-person game, all two-element sets are removable.

(57:F*) For an essential zero-sum four-person game all three-element sets are removable, or all but one.[1,2]

The proofs of these statements present no serious difficulties, but we do not propose to give them here.

The results (57:B), (57:C), (57:E), (57:F) show that a general theory of removable sets and extreme games is not likely to be very simple. It will be considered systematically in a subsequent publication.

57.5. Strategic Equivalence. Zero-sum and Constant-sum Games

57.5.1. We have exhausted the usefulness of the zero-sum extension $\bar{\Gamma}$ of the general n-person game Γ, and therefore from now on we shall discuss the theory of general n-person games without referring to that concept. Consequently, hereafter we shall use only the game Γ itself and its restricted characteristic function,—unless explicitly stated to the contrary. For this reason the qualification "restricted" will be dropped, and we shall speak simply of the *characteristic function* of Γ. This is also in harmony with our preceding terminology for zero-sum n-person games, since now the old and the new use of the concept of a characteristic function are concordant. (Cf. the remarks next to the end of 57.1.)

Considering these arrangements, the definition of the concept of a solution must be that described in (56:I:d) of 56.12. Imputations are best defined as described in (56:I:b) and in the last part of (56:I:e) id. It seems worth while to restate this latter definition explicitly:

An imputation is a vector

(57:14) $$\overrightarrow{\alpha} = \{\{\alpha_1, \cdots, \alpha_n\}\}$$

the components $\alpha_1, \cdots, \alpha_n$ being subject to the conditions

(57:15) $$\alpha_i \geqq \mathrm{v}((i)) \qquad \text{for} \qquad i = 1, \cdots, n;$$

[1] Every two-element set is a subset of two three-element sets (remember that $n = 4$), and by the above at least one of these is removable. Hence every two-element set is removable in any event.

[2] The parts of the cube Q of 34.2.2. which correspond to these various alternatives can be explicitly determined.

(57:16) $$\sum_{i=1}^{n} \alpha_i = v(I).$$ [1]

We can now extend the concept of strategic equivalence to the present setup. This will be done exactly as in 42.2. and 42.3.1., i.e. in analogy to 27.1.1.:

Given a general n-person game Γ with the functions $\mathcal{H}_k(\tau_1, \cdots, \tau_n)$ and a set of constants $\alpha_1^0, \cdots, \alpha_n^0$ we define a new game Γ' with the functions $\mathcal{H}'_k(\tau_1, \cdots, \tau_n)$ by

(57:17) $$\mathcal{H}'_k(\tau_1, \cdots, \tau_n) \equiv \mathcal{H}_k(\tau_1, \cdots, \tau_n) + \alpha_k^0.$$

From this we conclude, exactly as before, that the characteristic functions $v(S)$ and $v'(S)$ of these two games are connected by

(57:18) $$v'(S) = v(S) + \sum_{k \text{ in } S} \alpha_k^0.$$

We call two such games, as well as their characteristic functions, *strategically equivalent.*

Since we are free of all zero-sum restrictions, the constants $\alpha_1^0, \cdots, \alpha_n^0$ are unrestricted, just as in (42:B) in 42.2.2.

We note that this strategic equivalence induces an isomorphism of the imputations of Γ and Γ' exactly as in the two previous instances referred to above. Specifically the considerations and conclusions of 31.3.3. and of 42.4.2. carry over to the present case unchanged, so that it seems unnecessary to reformulate them explicitly.

57.5.2. The domain of all characteristic functions (of all general n-person games) was characterized by the conditions (57:2:a), (57:2:c), which we restate:

(57:2:a) $$v(\ominus) = 0,$$

(57:2:c) $$v(S \cup T) \geqq v(S) + v(T) \quad \text{for} \quad S \cap T = \ominus.$$

Among these the characteristic functions of zero-sum games and of constant-sum games form two special classes. The former are characterized by (25:3:a)-(25:3:c) of 25.3.1. (Cf. 26.2.) I.e. we must add to our (57:2:a), (57:2:c) (which coincide with the (25:3:a), (25:3:c) mentioned) the further condition

(57:19) $$v(-S) = -v(S).$$

The latter are characterized by (42:6:a)-(42:6:c) in 42.3.2. (cf. id.). I.e. we must add to our (57:2:a), (57:2:c) (which coincide with the (42:6:a), (42:6:c)

[1] As was pointed out loc. cit. we could have used equivalently

$$\sum_{i=1}^{n} \alpha_i \leqq v(I).$$

Indeed this is the original form of this condition. However, we prefer (57:16).

mentioned) the further condition

(57:20) $$v(S) + v(-S) = v(I).$$

Since the zero-sum games are a special case of the constant-sum games, (57:20) must be a consequence of (57:19), always assuming (57:2:a), (57:2:c). This is indeed so; we can actually prove somewhat more, namely:

(57:G) (57:19) is equivalent to the conjunction of (57:20) with $v(I) = 0$.

Proof:[1] Assuming $v(I) = 0$, (57:19) and (57:20) are clearly the same assertion. Hence it suffices to show that (57:19) implies $v(I) = 0$. Indeed, (57:2:a), (57:19) give $v(I) = v(-\ominus) = -v(\ominus) = 0$.

Note that (57:20) is the assertion that equality holds in (57:2:c) when $S \cup T = I$.[2] Thus the $v(S)$ of constant-sum games are characterized by the property that the merger of two distinct coalitions S and T produces no further profit if together they contain all players.

For the $v(S)$ of zero-sum games the further requirement $v(I) = 0$ must be added.

To conclude, we emphasize that the extra conditions (57:19) or (57:20) do not mean that any game with such a characteristic function is necessarily a zero-sum or a constant-sum game. They imply only that such a characteristic function must belong—among others—to at least one zero-sum or constant-sum game. It can happen that a game without being zero-sum (or constant-sum) itself has such a characteristic function, i.e. the characteristic function of a zero-sum (or constant-sum) game. In this case it will behave from the point of view of the strategy of coalitions, and the compensations like a zero-sum (or constant-sum) game without actually being one.

57.5.3. We are now in the position to settle a question which was in the foreground several times in our discussions. The analysis of 56.3.2.-56.4.3. was concerned already with the fact that the fictitious player—in spite of his unreality—is not *ipso facto* a dummy. I.e. not one in the sense of the extended characteristic function and the decomposition theory of the zero-sum extension $\bar{\Gamma}$.[3] This subject came up again at the beginning of 56.9.3., where we noted that he is a dummy for zero-sum games Γ.

The question which we will answer now is accordingly this: For which general games Γ is the fictitious player a dummy?[4] We prove:

(57:H) The fictitious player is a dummy if and only if Γ has the same characteristic function as a constant-sum game—i.e. if (57:20) is fulfilled.

[1] Essentially this argument was made in 42.3.2.

[2] Indeed $S \cup T = I$ and the usual hypothesis of (57:2:c) $S \cap T = \ominus$ mean that $T = -S$.

[3] We had to exclude him from the game by explicitly restricting the solutions from Ω to Ω''.

[4] The argument of the first remark in 56.9.4., shows then that for these games Ω and Ω'' coincide—i.e. the restriction of the solutions of $\bar{\Gamma}$ is unnecessary.

Proof: As observed at the end of 43.4.2. a player is a dummy, if and only if he forms (as a one-element set) a constituent of the game. We must apply this to the fictitious player $n + 1$ in the zero-sum game $\bar{\Gamma}$. That $(n + 1)$ is a constituent, means obviously that

$$(57{:}21) \quad \mathrm{v}(S) + \mathrm{v}((n+1)) = \mathrm{v}(S \cup (n+1)) \qquad \text{for all} \qquad S \subseteq (1, \cdots, n).$$

Now we have

$$\mathrm{v}((n+1)) = -\mathrm{v}(I),$$
$$\mathrm{v}(S \cup (n+1)) = -\mathrm{v}(\perp S \cup (n+1)) = -\mathrm{v}(-S).$$

Hence (57:21) becomes

$$\mathrm{v}(S) - \mathrm{v}(I) = -\mathrm{v}(-S),$$

i.e.

$$(57{:}22) \qquad \mathrm{v}(S) + \mathrm{v}(-S) = \mathrm{v}(I).$$

And this is precisely the condition (57:20).

58. Interpretation of the Characteristic Function

58.1. Analysis of the Definition

58.1. We have arrived at a formulation of the theory of the general n-person game, and found that the concept of the characteristic function is just as fundamental in it as it was in the preceding theory of the zero-sum n-person game. It is therefore appropriate to survey the meaning of this concept once more, putting its mathematical definition into an explicit form and adding some interpretative remarks.

Consider accordingly a general n-person game Γ, described by the functions $\mathcal{H}_k(\tau_1, \cdots, \tau_n)$ $(k = 1, \cdots, n)$ in the sense of 11.2.3. The value $\mathrm{v}(S)$ of the characteristic function for a set $S \subseteq I = (1, \cdots, n)$ obtains by forming this quantity for the zero-sum $n + 1$-person game $\bar{\Gamma}$—the zero-sum extension of Γ.[1] Hence we can express it by means of the definitory formulae of 25.1.3.:

$$(58{:}1) \qquad \mathrm{v}(S) = \mathrm{Max}_{\vec{\xi}}\, \mathrm{Min}_{\vec{\eta}}\, \mathrm{K}(\vec{\xi}, \vec{\eta}) = \mathrm{Min}_{\vec{\eta}}\, \mathrm{Max}_{\vec{\xi}}\, \mathrm{K}(\vec{\xi}, \vec{\eta}),$$

where we have:

$\vec{\xi}$ is a vector with the components ξ_{τ^S},

$$\xi_{\tau^S} \geqq 0, \qquad \sum_{\tau^S} \xi_{\tau^S} = 1;$$

[1] We restrict ourselves to the $S \subseteq I = (1, \cdots, n)$ i.e. to the restricted characteristic function. The use of all $S \subseteq \bar{I} = (1, \cdots, n + 1)$, i.e. of the extended characteristic function, is contrary to our present standpoint. (Cf. the beginning of 57.5.1.)

$\overrightarrow{\eta}$ is a vector with the components $\eta_{\tau^{-S}}$

$$\eta_{\tau^{-S}} \geqq 0, \qquad \sum_{\tau^{-S}} \eta_{\tau^{-S}} = 1;$$

τ^S is the aggregate of the variables τ_k, k in S; τ^{-S} is the aggregate of the variables τ_k, k in $-S$;[1] and finally

$$(58{:}2) \qquad \mathrm{K}(\overrightarrow{\xi}, \overrightarrow{\eta}) = \sum_{\tau^S, \tau^{-S}} \bar{\mathcal{H}}(\tau^S, \tau^{-S}) \xi_{\tau^S} \eta_{\tau^{-S}},$$

where

$$(58{:}3) \qquad \bar{\mathcal{H}}(\tau^S, \tau^{-S}) = \sum_{k \text{ in } S} \mathcal{H}_k(\tau_1, \cdots, \tau_n).^2$$

58.2. The Desire to Make a Gain vs. That to Inflict a Loss

58.2.1. $\mathrm{K}(\overrightarrow{\xi}, \overrightarrow{\eta})$ is obviously the expectation value of a play of the game Γ for the coalition S, if the coalition S uses the mixed strategy $\overrightarrow{\xi}$ and the opposing coalition $-S$ [3] uses the mixed strategy $\overrightarrow{\eta}$. Hence (58:1) defines $\mathrm{v}(S)$, the value of a play for the coalition S under the assumption that the coalition S wants to maximize the expectation value $\mathrm{K}(\overrightarrow{\xi}, \overrightarrow{\eta})$, while the opposing coalition $-S$ wants to minimize it, — and they choose their respective (mixed) strategies $\overrightarrow{\xi}, \overrightarrow{\eta}$ accordingly.

Now this principle is certainly correct in the zero-sum $n + 1$-person game $\bar{\Gamma}$,[4] but we are really dealing with the general n-person game Γ—

[1] $-S$ denotes $I - S$. Since we are dealing with $\bar{\Gamma}$, we should have formed $\perp S$ which is $\bar{I} - S$. (Cf. the beginning of 57.2.1.) However, this is immaterial, because no variable τ_{n+1} exists. (Cf. the end of 56.2.2.)

[2] We use only the original $\mathcal{H}_k$, $k = 1, \cdots, n$, i.e. the $\mathcal{H}_{n+1}$ of (56:2) in 56.2.2.

$$(58{:}4) \qquad \mathcal{H}_{n+1}(\tau_1, \cdots, \tau_n) \equiv - \sum_{k=1}^{n} \mathcal{H}_k(\tau_1, \cdots, \tau_n)$$

does not occur here. This is, of course, due to the fact that $S \subseteq I = (1, \cdots, n)$.

It must be remembered that formula (58:3) above is the first formula of (25:2) in 25.1.3. The second formula of (25:2) loc. cit. gives

$$(58{:}5) \qquad \bar{\mathcal{H}}(\tau^S, \tau^{-S}) \equiv - \sum_{k \text{ in } \perp S} \mathcal{H}_k(\tau_1, \cdots, \tau_n).$$

(Note that we must now definitely use $\perp S = \bar{I} - S$ for the $-S$ loc. cit., since we are dealing with $\bar{\Gamma}$. Cf. also footnote 1 above.) Since $n + 1$ is not in S, it is in $\perp S$; hence the sum $\sum_{k \text{ in } \perp S}$ of (58:5) does contain the $\mathcal{H}_{n+1}$ of (58:4). However, (58:4) guarantees, as it must, the identity of the right-hand sides of (58:3) and (58:5).

[3] The observations of footnote 1 above apply again.

[4] I.e. if we view $-S = I - S$ as really representing $\perp S = \bar{I} - S$.

Γ is merely a "working hypothesis"! And in Γ the desire of the coalition $-S$ to harm its opponent, the coalition S, is by no means obvious. Indeed, the natural wish of the coalition $-S$ should be not so much to decrease the expectation value $K(\vec{\xi}, \vec{\eta})$ of the coalition S as to increase its own expectation value $K'(\vec{\xi}, \vec{\eta})$. These two principles would be identical if every decrease of $K(\vec{\xi}, \vec{\eta})$ were equivalent to an increase of $K'(\vec{\xi}, \vec{\eta})$. This is of course the case when Γ is a zero-sum game,[1] but it need not at all be so for a general game Γ.

I.e. in a general game Γ the advantage of one group of players need not be synonymous with the disadvantage of the others. In such a game moves—or rather changes in strategy—may exist which are advantageous to both groups. In other words, there may exist an opportunity for genuine increases of productivity, simultaneously in all sectors of society.

58.2.2. Indeed, this is more than a mere possibility—the situations to which it refers constitute one of the major subjects with which economic and social theory must deal. Hence the question arises: Does our approach not disregard this aspect altogether? Did we not lose this cooperative side of social relationships because of the great emphasis which we placed on their opposite, antagonistic, side?

We think that this is not so. It is difficult to present a complete case, since the validity of a theory is *ultima analysi* only established by success in the applications—and we have made no applications in our discussion thus far. We will suggest therefore only the main points which seem to

[1] This is so because when Γ is zero-sum then

$$K(\vec{\xi}, \vec{\eta}) + K'(\vec{\xi}, \vec{\eta}) \equiv 0. \tag{58:6}$$

This is clear by common sense; a formal proof obtains in this way: Clearly

$$K'(\vec{\xi}, \vec{\eta}) \equiv \sum_{\tau^S, \tau^{-S}} \bar{\mathcal{H}}'(\tau^S, \tau^{-S}) \xi_{\tau^S} \eta_{\tau^{-S}}, \tag{58:7}$$

where

$$\bar{\mathcal{H}}'(\tau^S, \tau^{-S}) \equiv \sum_{k \text{ in } -S} \mathcal{H}_k(\tau_1, \cdots, \tau_n). \tag{58:8}$$

(Note that this is not the $\sum_{k \text{ in } \perp S} \mathcal{H}_k(\tau_1, \cdots, \tau_n)$ which occurs in (58:5)). Now comparison of (58:2) with (58:7) shows that (58:6) is equivalent to

$$\bar{\mathcal{H}}(\tau^S, \tau^{-S}) + \bar{\mathcal{H}}'(\tau^S, \tau^{-S}) \equiv 0, \tag{58:9}$$

and (58:3), (58:5) imply that (58:9) amounts to

$$\sum_{k=1}^{n} \mathcal{H}_k(\tau_1, \cdots, \tau_n) \equiv 0,$$

i.e. the zero-sum condition for Γ.

support our procedure, and then refer to the applications which provide a definite corroboration.

58.3. Discussion

58.3.1. The following considerations deserve particular attention in this connection:

First: Inflicting losses on the adversary may not be directly profitable in a general (i.e. not necessarily zero-sum) game, but it is the way to exert pressure on him. He may be induced by such threats to pay a compensation, to adjust his strategy in a desired way, etc. Hence it is not *a limine* unreasonable that this category of strategic possibilities should be taken into account; and our procedure in forming the characteristic function, as analyzed above, might be the proper one to do just that. It must be admitted, however, that this is not a justification of our procedure—it merely prepares the ground for the real justification which consists of success in applications.

Second: A further consideration pointing in the same direction is this. We have seen that in our theory all solutions correspond to attainment of the maximum collective profit by the totality of all players.[1] When this maximum is reached, any further gain of one group of players must be compensated by an at least equal loss of the others. True, there could be overcompensation: i.e. one group might obtain a gain by inflicting a greater loss on the others. However, we have assumed complete information for all players, and a perfect interplay of threats, counterthreats and compensations among them.[2] Hence one may assume that such possibilities will be effective only as threats, and that the corresponding actions will be obviated always by negotiations and compensations. By this we do not mean that these threats are "bluffs" which are never "called." Since there exists complete information for all players, there can never be any doubt. But when an action is threatened by which one party gains less than the other one loses, then there exists *ipso facto* the possibility of avoiding it by compensations in a way which is advantageous to both sides.[3] And when this happens it is again true that one side gains exactly what the other loses.

If this argument is accepted as generally valid, then our difficulties disappear.

58.3.2. Third: It may be said that the argumentation of the two preceding remarks is too sketchy and that it does not justify our theory in the exact form in which we propose to use it. This is true, but our very detailed

[1] Cf. the end of 56.7.1., particularly footnote 3 on p. 513.

[2] Our entire attitude towards coalitions and compensations was based on this, already in the theory of the zero-sum games.

[3] We do not propose to determine here the amount of the compensation—i.e. the nature of the compromise. This is the task of the exact theory which we possess already. It will be the main subject in each application. (Cf. the various interpretations in 61.-63.) At this point we want only to show that actions which would lead to a loss for the totality of all players, can be avoided by the mechanism described above.

motivation of that theory, as given in 56.2.2.-57.1. meets the latter requirement. If the reader reconsiders those sections in the light of the two preceding remarks, then he will see that the detailed justification in the desired sense was their subject. Indeed, the possibility of the objection now under consideration was our reason for making the discussion of our theory so detailed, and avoiding plausible shortcuts.[1]

Fourth: In spite of all this, the reader may feel that we have overemphasized the role of threats, compensations, etc., and that this may be a one-sidedness of our approach which is likely to vitiate the results in applications. The best answer to this is, as repeatedly pointed out before, the examination of those applications.

We shall therefore consider definite applications which correspond to familiar economic problems. Their study will disclose that our theory leads to results which are, up to a certain point, in satisfactory agreement with the usual common-sense views on these matters. This is the case as long as the two following conditions are fulfilled: First that the setup is simple enough to allow a purely verbal analysis, not making use of any mathematical apparatus. Second that those factors which are inseparable from our theory, but often excluded in the ordinary, verbal approach—coalitions and compensations—have not come essentially into play. This situation will be found to exist in the application of 61.2.2.-61.4. Indeed, that example provides the decisive corroboration of our procedure.

Beyond this point, where the first condition is still satisfied, but not the second, we shall find discrepancies just in the direction and to the extent to which the difference in standpoint justifies it. This will be particularly clear in the applications of 61.5.2., 61.6.3. and 62.6.

Finally, as even the first condition fails, because the problem is no longer elementary, we gradually reach ground where the theoretical procedure necessarily takes over the leading role from the ordinary, purely verbal one.[2]

59. General Considerations

59.1. Discussion of the Program

59.1.1. We can now proceed to the applications of our theory of the general n-person game. The best way of starting such applications is a systematic discussion of all general n-person games for small values of n. It will appear that we can carry this out in absolute completeness for the same n as for the zero-sum games: For the $n \leqq 3$. The discussion for the greater values, i.e. for $n \geqq 4$, is necessarily at least as difficult as it was for

[1] A possible one would have been to define the characteristic function as in 58.1. and to come out then with a flat generalization of the theory for zero-sum games, i.e. with (56:I:d) in 56.12.

[2] This gradual transfer of the emphasis from corroboration of the theory by the reliable common-sense results in the simple case, to overriding any untheoretical approach by the theory in the complicated ones, is, of course, quite characteristic in the formation of scientific theories.

the zero-sum games where we could only dispose of special cases of various kinds.

We propose to do considerably less in the way of analyzing games with $n \geqq 4$ this time. We can afford to be considerably briefer now than we were in discussing the zero-sum games: The detailed discussion there was necessary in order to reassure ourselves of the propriety of our procedure, and of the general ideas and methodical principles underlying it. At the stage which we have reached now, the general setup of the theory appears to be justified, and we want only to gain assurance concerning the one generalizing step carried out in this chapter. For this purpose a less extensive analysis of applications should suffice.

Further, it will be possible already to connect the general games with $n \leqq 3$ with some typical economic problems (bilateral monopoly, duopoly versus monopoly, etc.) which allow judgment of the appropriateness of our theory in the sense indicated before.

More detailed investigations of general games with $n \geqq 4$ will be undertaken in subsequent publications.

59.1.2. The systematic application of our new theory is best introduced by a general discussion, similar to that of 31. It will not be necessary, however, to carry out the equivalent considerations in detail; we must only analyze to what extent the results obtained there carry over to the present situation, or what modifications are required.

We need not discuss again the role of strategic equivalence, as expounded in 31.3., since this subject has already been dealt with satisfactorily in 57.5.1. On the other hand, we shall take up certain matters originating elsewhere than in 31.: reduced forms, inequalities which hold for the characteristic function, inessentiality and essentiality (cf. 27.1.-27.5.); further, the absolute values $|\Gamma|_1$, $|\Gamma|_2$ (cf. 45.3.), and finally some remarks concerning the theory of decomposition of Chapter IX.

59.2. The Reduced Forms. The Inequalities

59.2.1. The concept of strategic equivalence, as introduced in 57.5.1. can be used to define reduced forms for all characteristic functions, along the lines of 27.1.

Given a characteristic function $\mathrm{v}(S)$ its general strategically equivalent transformation is given by (57:18) in 57.5.1., i.e. by

$$\text{(59:1)} \qquad \mathrm{v}'(S) = \mathrm{v}(S) + \sum_{k \text{ in } S} \alpha_k^0.$$

This is precisely (27:2) in 27.1.1., but the $\alpha_1^0, \cdots, \alpha_n^0$ are now completely unrestricted, while they were subject loc. cit. to the condition (27:1): $\sum_{k=1}^{n} \alpha_k^0 = 0$. Hence the $\alpha_1^0, \cdots, \alpha_n^0$ are now n independent parameters, while they represented only $n - 1$ independent parameters formerly (cf. 27.1.3.).[1]

It would be erroneous to assume, however, that this leads to more restrictive possibilities of normalization than we found in 27.1.4. Indeed, we desired loc. cit. to obtain a particular $v'(S)$—to be denoted by $\bar{v}(S)$—which fulfills the $n - 1$ conditions (27:3):

$$\bar{v}((1)) = \bar{v}((2)) = \cdots = \bar{v}((n)). \tag{59:2}$$

Yet, the characteristic functions considered at that time belonged to zero-sum games; hence we had automatically

$$\bar{v}((1, \cdots, n)) = 0. \tag{59:3}$$

In imposing this as a normalizing requirement, we now have n conditions: (59:2) and (59:3). So we obtain

$$v(I) + \sum_{k=1}^{n} \alpha_k^0 = 0, \tag{59:4}$$

$$v((1)) + \alpha_1^0 = v((2)) + \alpha_2^0 = \cdots = v((n)) + \alpha_n^0. \tag{59:5}$$

(59:4) expresses (59:3); (59:5) expresses (59:2). These equations correspond to (27:1*), (27:2*) loc. cit., and it is easy to verify that they are solved by precisely one system of $\alpha_1^0, \cdots, \alpha_n^0$:

$$\alpha_k^0 = -v((k)) + \frac{1}{n}\left\{\sum_{k=1}^{n} v((k)) - v(I)\right\}.[2] \tag{59:6}$$

So we can say:

(59:A) We call a characteristic function $\bar{v}(S)$ *reduced* if and only if it satisfies (59:2), (59:3).[3] Then every characteristic function $v(S)$ is in strategic equivalence with precisely one reduced $\bar{v}(S)$. This $\bar{v}(S)$ is given by the formulae (59:1) and (59:6), and we call it the *reduced form* of $v(S)$.

59.2.2. Another possible requirement for the n parameters $\alpha_1^0, \cdots, \alpha_n^0$ consists in requiring for $v'(S)$—to be denoted by $\tilde{v}(S)$—the n conditions

$$\tilde{v}((1)) = \tilde{v}((2)) = \cdots = \tilde{v}((n)) = 0. \tag{59:7}$$

[1] Our present standpoint in this respect is similar to that which we took for the constant-sum games in 42.2.2.

[2] *Proof:* Denote the joint value of n terms in (59:5) by β. Then (59:5) amounts to $\alpha_k^0 = -v((k)) + \beta$ and so (59:4) becomes $v(I) - \sum_{k=1}^{n} v((k)) + n\beta = 0$, i.e. $\beta = \frac{1}{n}\left\{\sum_{k=1}^{n} v((k)) - v(I)\right\}$.

[3] This is precisely the definition of 27.1.4.

This means

(59:8) $$\mathrm{v}((1)) + \alpha_1^0 = \mathrm{v}((2)) + \alpha_2^0 = \cdots = \mathrm{v}((n)) + \alpha_n^0 = 0,$$

i.e.

(59:9) $$\alpha_k^0 = -\mathrm{v}((k)).$$

So we can say:

(59:B) We call a characteristic function $\mathrm{v}(S)$ *zero reduced* if and only if it satisfies (59:7). Then every characteristic function $\mathrm{v}(S)$ is in strategic equivalence with precisely one zero-reduced $\tilde{\mathrm{v}}(S)$. This $\tilde{\mathrm{v}}(S)$ is given by the formulae (59:1) and (59:9), and we call it the *zero-reduced form* of $\mathrm{v}(S)$.

59.2.3. Let us consider the reduced characteristic function $\bar{\mathrm{v}}(S)$. We denote the joint value of the n terms in (59:2) by $-\gamma$, i.e.

(59:10) $$-\gamma = \bar{\mathrm{v}}((1)) = \bar{\mathrm{v}}((2)) = \cdots = \bar{\mathrm{v}}((n)).$$

Hence $-\gamma = \mathrm{v}((k)) + \alpha_k^0$ and so (59:6) gives

(59:11) $$\gamma = \frac{1}{n}\left\{\mathrm{v}(I) - \sum_{k=1}^{n} \mathrm{v}((k))\right\}.$$

If we use the zero-reduced form $\tilde{\mathrm{v}}(S)$ of the same $\mathrm{v}(S)$ then we have $\tilde{\mathrm{v}}(I) = \mathrm{v}(I) + \sum_{k=1}^{n} \alpha_k$, hence by (59:9) $\tilde{\mathrm{v}}(I) = \mathrm{v}(I) - \sum_{k=1}^{n} \mathrm{v}((k))$, i.e. using (59:11)

(59:12) $$n\gamma = \tilde{\mathrm{v}}(I).$$

Returning to the reduced form $\bar{\mathrm{v}}(S)$, we see that some equalities and all inequalities of 27.2. are still valid.

To begin with, (59:10) can be stated as follows:

(59:13) $$\bar{\mathrm{v}}(S) = -\gamma \quad \text{for every one-element set } S.$$

This coincides with (27:5*) loc. cit., while (27:5**) id. fails, since we saw in 57.2.1. that the equivalent of (25:3:b) in 25.3.1. is now missing, and this was required to derive (27:5**) from (27:5*) there.

Repeated application of (57:2:c) in 57.2.1. to the sets $(1), \cdots, (n)$ gives by (59:13) $-n\gamma \leqq 0$, i.e.:

(59:14) $$\gamma \geqq 0.$$

This coincides with (27:6) in 27.2.

Consider next an arbitrary subset S of I. Let p be the number of its elements: $S = (k_1, \cdots, k_p)$. Repeated application of (57:2:c) in 57.2.1.

to the sets $(k_1), \cdots, (k_p)$ gives by (59:13)

$$\bar{v}(S) \geqq -p\gamma.$$

Apply this to $-S$ which has $n - p$ elements. Owing to (57:2:b) in 57.2.1. and (59:3), we have

$$\bar{v}(-S) \leqq -\bar{v}(S)^1$$

hence the preceding inequality now becomes

$$\bar{v}(S) \leqq (n - p)\gamma.$$

Combining these two inequalities gives:

(59:15) $\quad -p\gamma \leqq \bar{v}(S) \leqq (n - p)\gamma \qquad$ for every p-element set S.

This coincides with (27:7) in 27.2.

(59:13) and $\bar{v}(\ominus) = 0$ (i.e. (57:2:a) in 57.2.1.) can also be formulated as follows:

(59:16) $\qquad$ For $p = 0, 1$ we have $=$ in the first relation of (59:15).

This coincides with (27:7*) in 27.2. $\bar{v}(I) = 0$ (i.e. (59:3)) can also be formulated as follows:

(59:17) $\qquad$ For $p = n$ we have $=$ in the second relation of (59:15).

This coincides with (27:7**) loc. cit., except that $p = n - 1$ is missing, for the same reason for which the equivalent of (27:5**) id. is missing (cf. the remark following our (59:13)).

59.3. Various Topics

59.3.1. These inequalities can now be treated in the same way as in 27.3.1.

There are two alternatives, based on (59:14):

First case: $\gamma = 0$. Then (59:15) gives $\bar{v}(S) = 0$ for all S. This is precisely the *inessential* case discussed in 27.3.1., with all the attributes enumerated there. Considering (59:A), the inessential games are precisely those which are equivalent to the game with $\bar{v}(S) \equiv 0$,—the game which is perfectly "vacuous."

Second case: $\gamma > 0$. By a change of unit we could make $\gamma = 1$, with the consequences pointed out in 27.3.2. And just as there, we refrain from doing this immediately. For the same reasons as pointed out there, the strategy of coalitions is decisive in such a game. We call a game in this case *essential*.

The criteria (27:B), (27:C), (27:D) of 27.4. for inessentiality and essentiality are again valid: In (27:B) $\sum_{k=1}^{n} v((k))$ must be replaced by

[1] Note that in our present application this inequality replaces this missing equality (25:3:b) in 25.3.1., which was used in 27.2.

$$\sum_{k=1}^{n} v((k)) - v(I),$$

while (27:C), (27:D) are completely unaffected. Indeed, it is easy to verify that the proofs given there carry over to the present case, their bases being provided in 59.2.1.

We leave it to the reader to apply the considerations of 27.5.—for the essential case, with the normalization $\gamma = 1$—to the present situation.

59.3.2. We can now pass to the considerations which correspond to those of 31.

The remarks of 31.1.1.-31.1.3. concerning the structure of the concept of domination and *certainly necessary* and *certainly unnecessary* sets can be repeated without any change. The concepts of *convexity* and of *flatness* can be introduced as in 31.1.4. The conclusions of 31.1.4.-31.1.5. are also unaffected, except for (31:E:b) in 31.1.4. and (31:G) in 31.1.5., as well as (31:H) id. for $p = n - 1$. These are the only ones where (25:3:b) of 25.3.1. (cf. 57.2.1.) is used.

Finally, the remark at the end of 31.1.5. must be modified. Owing to what we said above, the value $p = n - 1$ is just as dubious as those included in (31:8) loc. cit. I.e. the p for which the necessity of S is in doubt, are restricted to $p \neq 0, 1, n$, i.e. to the interval

(59:18) $$2 \leqq p \leqq n - 1.$$

Thus this interval begins to play a role when $n \geqq 3$,—not only, as loc. cit., when $n \geqq 4$.[1]

Consider next the results of 31.2. The reader who consults that section will have no difficulty in verifying the following: (31:I), (31:J), (31:K) are unaffected. In (31:L) the construction of $\vec{\beta}$ with the help of $\vec{\alpha}$ can be carried out without any change; the first assertion, $\vec{\beta} \succ \vec{\alpha}$, cannot be maintained, since it uses that part of (31:H) in 31.1.5. which is no longer valid; the second assertion, not $\vec{\alpha} \succ \vec{\beta}$, is unaffected. This weakening of (31:L) removes (31:M). (31:N) remains true because it uses the intact part of (31:L) only. (31:O), (31:P) are unaffected.

59.3.3. To conclude, let us consider some of the concepts of Chapter IX. We defined there the two numbers $|\Gamma|_1$, $|\Gamma|_2$ the former in 45.1., the latter in 45.2.3., and we discussed their properties in 45.3.

Both definitions—i.e. the pertinent considerations of 45.1., 45.2.—carry over literally. There are, however, essential changes in 45.3.: In (45:F) only the second part of the proof is valid, but not the first part, since that—and that alone—makes use of (25:3:b) in 25.3.1. (cf. 57.2.1.).

[1] This is in agreement with the connection of general n-person games and zero-sum $n + 1$-person games, which was prominent throughout 56.2-56.12.

Specifically: We still have

(59:19) $$|\Gamma|_2 \leqq \frac{n-2}{2}|\Gamma|_1,$$

and so we can evaluate $|\Gamma|_2$ in terms of $|\Gamma|_1$; but we do not have

(59:20) $$|\Gamma|_1 \leqq (n-1)|\Gamma|_2,$$ [1]

nor can we evaluate $|\Gamma|_1$ in terms of $|\Gamma|_2$ at all. Indeed, we shall see in 60.2.1. that

(59:21) $$|\Gamma|_1 > 0, \qquad |\Gamma|_2 = 0$$

occurs for certain games.

In consequence of this, the remarks of 45.3.3.-45.3.4. become pointless. The same holds for 45.3.1., i.e. its result (45:E) fails in so far as it concerns $|\Gamma|_2$. It is true for $|\Gamma|_1$ but this is merely a restatement of the definitions. Considering this, and (59:19), (59:21) above, we see that (45:E) must be weakened as follows:

(59:C) If Γ is inessential, then $|\Gamma|_1 = 0$, $|\Gamma|_2 = 0$.
If Γ is essential, then $|\Gamma|_1 > 0$, $|\Gamma|_2 \geqq 0$.

The theory of *composition* and *decomposition*, which is the main object of Chapter IX, can be extended in its essential parts to our present set-up. The difference between the behavior of $|\Gamma|_1$ and $|\Gamma|_2$ discussed above, necessitates some minor changes, but these are easily applied. Of course the theory of excesses and of solutions in the sets $E(e_0)$ and $F(e_0)$ (cf. there) must be extended to the present case—but this, too, entails no real difficulties.

A detailed analysis of this subject would lengthen our exposition beyond the limits that we set ourselves in 59.1.1. Furthermore, the interpretative value of the results would not differ materially from what was already obtained in Chapter IX, when considering zero-sum games.

60. The Solutions of All General Games with $n \leqq 3$

60.1. The Case $n = 1$

60.1. We proceed to the systematic discussion of all general n-person games with $n \leqq 3$, as announced in 59.1.1.

Consider first $n = 1$. This case has already been considered (and, for practical purposes, settled) in 12.2. In particular, we pointed out in 12.2.1. that in this (and only in this) case we deal with a pure maximum problem. It is nevertheless desirable to verify that our general theory produces in this (trivial) special case the common-sense result.[2] We apply therefore the general theory in complete mathematical rigor.

[1] (59:20) and (59:19) express the two parts of (45:F), respectively.
[2] This brings us back to the fourth remark in 58.3.2.

A general game Γ with $n = 1$ is necessarily inessential: This is clear by considering the characteristic function $\bar{v}(S)$ of its reduced form, since then (59:16) and (59:17) in 59.2.3. give (for $p = 1 = n$) $-\gamma = 0$, i.e. $\gamma = 0$. We may also use—without reducing—any one criterion (27:B), (27:C), (27:D) of 27.4. (cf. 59.3.1.). E.g. (27:C) loc. cit. is clearly satisfied, with $\alpha_1 = v((1))$. Note that this is $v(I)$, i.e. by (56:13) in 56.9.1. (with the notation of 12.2.1.) $\text{Max}_\tau \mathcal{H}(\tau)$. We restate this:

(60:1) $$\alpha_1 = v((1)) = v(I) = \text{Max}_\tau \mathcal{H}(\tau).$$

Since Γ is inessential, we can apply (31:O) or (31:P) in 31.2.3. (cf. 59.3.2.). This gives:

(60:A) Γ possesses precisely one solution, the one-element set $(\overrightarrow{\alpha})$ where

$$\overrightarrow{\alpha} = \{\{\alpha_1\}\}$$

with the α_1 of (60:1).

This is obviously the "common-sense" result of 12.2.1.—as it should be.

60.2. The Case $n = 2$

60.2.1. Consider next $n = 2$. The main fact is that a general game with $n = 2$ need not be inessential—thus differing from the zero-sum games with $n = 2$. (The latter are inessential by the first remark in 27.5.2.)

Indeed: The characteristic function $\bar{v}(S)$ of its reduced form is completely determined by (59:16) and (59:17) in 59.2.3. It is

(60:2) $$\bar{v}(S) = \begin{cases} 0 \\ -\gamma \\ 0 \end{cases} \quad \text{when } S \text{ has} \quad \begin{cases} 0 \\ 1 \\ 2 \end{cases} \text{elements.}$$

Now one verifies immediately that a $\bar{v}(S)$ of (60:2) fulfills the conditions (57:2:a), (57:2:c) of 57.2.1., i.e. that it is the characteristic function of a suitable Γ (cf. 57.3.4.), if and only if $\gamma \geqq 0$. This is precisely the condition (59:14) in 59.2.3. So we see: the $\gamma \geqq 0$ of (59:14) in 59.2.3. are precisely the possibilities in (60:2).

Thus $\gamma > 0$, i.e. essentiality, is among the possibilities, as asserted. In the case of essentiality we may further normalize $\gamma = 1$, thereby completely determining (60:2). Thus there exists only one type of essential general two-person games.

Note that while $|\Gamma|_1 = 2\gamma$ may thus be > 0, there is always (for $n = 2$) $|\Gamma|_2 = 0$. It suffices to prove this for the reduced form, i.e. for (60:2).

Indeed: Recalling the definitions of 45.2.1. and 45.2.3. we see that $\overrightarrow{\alpha} = \{\{\alpha_1, \alpha_2\}\}$ is detached when $\alpha_1, \alpha_2 \geqq -\gamma$, $\alpha_1 + \alpha_2 \geqq 0$, and that the minimum of the corresponding $e = \alpha_1 + \alpha_2$ is 0.[1] Hence $|\Gamma|_2 = 0$ as desired.

[1] It is assumed e.g. for $\alpha_1 = \alpha_2 = 0$.

Summing up: For $n = 2$ a zero-sum game must be inessential, a general game need not be. Accordingly the former must have $|\Gamma|_1 = 0$; the latter may have $|\Gamma|_1 > 0$ too. But both have always $|\Gamma|_2 = 0$.

We leave it to the reader to interpret this result in the light of previous discussions, and particularly of 45.3.4.

60.2.2. The solutions for a general game Γ with $n = 2$ are easily determined.

By the valid part of (31:H) in 31.1.5. (cf. the pertinent observations in 59.3.2.) all sets $S \subseteq I$ with 0, 1 or n elements are certainly unnecessary—but since $n = 2$, these exhaust all subsets. Hence we may determine the solutions of Γ as if domination never held. Consequently a solution is simply defined by the property that no imputation can be outside of it. I.e. there exists precisely one solution: the set of all imputations.

The general imputation is given in this case as $\overrightarrow{\alpha} = \{\{\alpha_1, \alpha_2\}\}$, subject to the conditions (57:15), (57:16) in 57.5.1., which now become:

$$(60:3) \qquad \alpha_1 \geqq \mathrm{v}((1)), \qquad \alpha_2 \geqq \mathrm{v}((2)),$$

$$(60:4) \qquad \alpha_1 + \alpha_2 = \mathrm{v}((1,2)) = \mathrm{v}(I).$$

We restate the result:

(60:B) Γ possesses precisely one solution, the set of all imputations. These are the

$$\overrightarrow{\alpha} = \{\{\alpha_1, \alpha_2\}\}$$

with the α_1, α_2 of (60:3), (60:4).

Note that (60:3), (60:4) determine a unique pair α_1, α_2 (i.e. $\overrightarrow{\alpha}$) if and only if

$$(60:5) \qquad \mathrm{v}((1)) + \mathrm{v}((2)) = \mathrm{v}((1,2)).$$

By the criteria of 27.4. this expresses precisely the inessentiality of Γ. This result is, as it should be, in harmony with (31:P) in 31.2.3. (cf. 59.3.2.).

Otherwise

$$(60:6) \qquad \mathrm{v}((1)) + \mathrm{v}((2)) < \mathrm{v}((1,2)),$$

and there exist infinitely many α_1, α_2,—i.e. $\overrightarrow{\alpha}$. This is the case of essentiality for Γ.

The interpretation of these results will be given in 61.2.-61.4.

60.3. The Case $n = 3$

60.3.1. Consider finally $n = 3$. These games include the essential zero-sum three-person game for which $|\Gamma|_1 > 0$ and $|\Gamma|_2 > 0$ (cf. 45.3.3.). So we see:

For $n = 3$ a zero-sum game as well as a general game may be essential, and both $|\Gamma|_1 > 0$ and $|\Gamma|_2 > 0$ are possibilities.

The case where Γ is inessential is taken care of by (31:O) or (31:P) in 31.2.3. (cf. 59.3.2.). We assume therefore that Γ is essential.

Use the reduced form of Γ in the normalization $\gamma = 1$. Then we can describe its characteristic function $\bar{v}(S)$ with the help of (59:16) and (59:17) in 59.2.3. as follows:

(60:7) $$\bar{v}(S) = \begin{cases} 0 \\ -1 \\ 0 \end{cases} \quad \text{when } S \text{ has} \quad \begin{cases} 0 \\ 1 \text{ elements,} \\ 3 \end{cases}$$

and

(60:8) $$\bar{v}((2,3)) = a_1, \qquad \bar{v}((1,3)) = a_2, \qquad \bar{v}((1,2)) = a_3 \qquad \text{when } S \text{ has 2 elements.}$$

And it is verified immediately that a $\bar{v}(S)$ of (60:7), (60:8) fulfills the conditions (57:2:a), (57:2:c) of 57.2.1., i.e. that it is the characteristic function of a suitable Γ (cf. 57.3.4.), if and only if

(60:9) $$-2 \leqq a_1, a_2, a_3 \leqq 1.$$

Note that this Γ can be chosen zero-sum, i.e. that (25:3:b) of 25.3.1. holds if and only if

(60:10) $$a_1 = a_2 = a_3 = 1.$$

In other words: The domain (60:9) represents all general games, while its upper boundary point (60:10) represents the (unique) zero-sum game of our case.

60.3.2. Let us now determine the solutions of this (essential) general three-person game.

The general imputation is given in this case as $\vec{\alpha} = \{\{\alpha_1, \alpha_2, \alpha_3\}\}$, subject to the conditions (57:15), (57:16) in 57.5.1., which now become:

(60:11) $$\alpha_1 \geqq -1, \qquad \alpha_2 \geqq -1, \qquad \alpha_3 \geqq -1,$$
(60:12) $$\alpha_1 + \alpha_2 + \alpha_3 = 0.$$

These conditions are precisely those of 32.1.1. for $\alpha_1, \alpha_2, \alpha_3$ (cf. (32:2), (32:3) there), i.e. those used in the theory of the essential zero-sum three-person game. They agree also, apart from the factor $1 + \frac{e_0}{3}$, with the conditions of 47.2.2. for $\alpha^1, \alpha^2, \alpha^3$ (cf. (47:2*), (47:3*) there), i.e. with those used in the theory of the essential zero-sum three-person game with excess. Consequently we can use the graphical representation described in 32.1.2., in particular in Figure 52. We obtain the domain of the $\vec{\alpha}$ as the fundamental triangle in 32.1.2. in Figure 53. It is also similar to that in 47.2.2. in Figure 70.

We express the relationship of domination in this graphical representation. Concerning the set S of 30.1.1. for a domination $\vec{\alpha} \succ \vec{\beta}$, the following can be said. By the valid part of (31:H) in 31.1.5. (cf. the pertinent observations in 59.3.2.) all sets $S \subseteq I$ with 0, 1 or n elements are certainly unnecessary—but since $n = 3$, this restricts our analysis to two-element sets S.

Put therefore $S = (i, j)$.[1] Then domination means that

$$\alpha_i + \alpha_j \leqq \bar{v}((i, j)) = a_k \qquad \text{and} \qquad \alpha_i > \beta_i, \qquad \alpha_j > \beta_j.$$

By (60:12) the first condition may be written $\alpha_k \geqq -a_k$.

We restate this: Domination

$$\overrightarrow{\alpha} \succ \overrightarrow{\beta}$$

means that

$$\text{(60:13)} \qquad \begin{cases} \text{either } \alpha_1 > \beta_1, & \alpha_2 > \beta_2 \quad \text{and} \quad \alpha_3 \geqq -a_3; \\ \text{or} \quad \alpha_1 > \beta_1, & \alpha_3 > \beta_3 \quad \text{and} \quad \alpha_2 \geqq -a_2; \\ \text{or} \quad \alpha_2 > \beta_2, & \alpha_3 > \beta_3 \quad \text{and} \quad \alpha_1 \geqq -a_1.^2 \end{cases}$$

The circumstances described in (60:13) can now be added to the picture of the fundamental triangle. The similarity is now more with 47. than with 32. The operation corresponds to the transition from Figure 70 to Figures 71, 72, or to Figures 84, 85, or to Figures 87, 88. Indeed, the difference as against Figures 71, 84, 87 (which all describe the same operation, in the successive Cases (IV), (V), (VI)) is only this:

The six lines

$$\text{(60:14)} \qquad \begin{cases} \alpha^1 = -\left(1 + \frac{e_0}{3}\right), & \alpha^2 = -\left(1 + \frac{e_0}{3}\right), & \alpha^3 = -\left(1 + \frac{e_0}{3}\right), \\ \alpha^1 = -\left(1 - \frac{2e_0}{3}\right), & \alpha^2 = -\left(1 - \frac{2e_0}{3}\right), & \alpha^3 = -\left(1 - \frac{2e_0}{3}\right), \end{cases}$$

which form the configuration there, are now replaced by the six lines

$$\text{(60:15)} \qquad \begin{cases} \alpha_1 = -1, & \alpha_2 = -1, & \alpha_3 = -1, \\ \alpha_1 = -a_1, & \alpha_2 = -a_2, & \alpha_3 = -a_3, \end{cases}$$

respectively. Hence the second triangle (formed by the three last lines) which appears in the fundamental triangle (formed by the three first lines) need not be placed symmetrically with respect to the latter, as it is in the three figures mentioned.

[1] i, j, k a permutation of 1,2,3.

[2] This is quite similar to (47:5) in 47.2.3., except that we have there $1 - \frac{2e_0}{3}$ in place of all three a_1, a_2, a_3. There is also the change of scale by the factor $1 + \frac{e_0}{3}$ referred to after (60:11), (60:12).

The relation to (32:4) in 32.1.3. is the same as for (47:5) in 47.2.3., cf. footnote 2 on p. 406.

60.3.3. It is convenient to distinguish two cases, according to whether the

$$(60:16) \qquad \alpha_1 \geqq -a_1, \qquad \alpha_2 \geqq -a_2, \qquad \alpha_3 \geqq -a_3$$

sides of the three last lines of (60:15) (where the three domination relations of (60:13) are valid) intersect in a common area, or not. Owing to (60:12) the former means that

$$(60:17:a) \qquad a_1 + a_2 + a_3 > 0,$$

while the latter means that

$$(60:17:b) \qquad a_1 + a_2 + a_3 \leqq 0.$$

We call these cases (a) and (b), respectively.

Case (a): We have the conditions of Figures 71, 72, except that the inner triangle need not be placed symmetrically with respect to the fundamental triangle, as it is there. If this is borne in mind, then the discussion of Case (IV), as given in 47.4.-47.5. can be repeated literally. The solutions are therefore, with the same qualification, those depicted in Figures 82, 83.

We note that if an $a_i = 1$, then the corresponding sides of the inner and the fundamental triangle coincide (cf. (60:15)), and the corresponding curve disappears.[1]

Case (b): We have essentially the conditions of Figures 84, 85—of which those of Figures 87, 88 are but a variant—with the same proviso for asymmetry as in Case (a) above.

We redraw the arrangement of Figure 84, the fundamental triangle being marked by / and the inner triangle by \\: Figure 92. The arrangement has several variants, because the inner triangle can stick out from the fundamental triangle in various ways.[2] Figures 92–95 depict these variants.[3,4]

If these circumstances are borne in mind, then the discussion of Case (V) as given in 47.6. can be repeated literally.[5] The solutions are therefore,

[1] Thus in the zero-sum case, where $a_1 = a_2 = a_3 = 1$, none of these curves occur—in accord with the result of 32.

[2] By (60:9) $-2 \leqq a_i \leqq 1$. This means, as the reader may easily verify for himself, that each side of the "inner" triangle must pass between the corresponding side of the fundamental triangle and its opposite vertex. Our Figures 92–95 exhaust all possibilities within this restriction.

[3] The only ones which can occur in a zero-sum game, i.e. for $a_1 = a_2 = a_3 = 1$, are those which can be symmetric: Figures 92, 95. Of these, Figure 92 corresponds to Figure 84, and Figure 95 corresponds to Figure 87.

[4] The Figures 92–95 differ from each other by the successive disappearance of the areas ②, ③, ④. Besides one or more of the areas ① and ⑤, ⑥, ⑦ may degenerate to a linear interval or even to a point. It is sometimes not quite easy to distinguish between the "disappearance" mentioned above, and this "degeneration." A rule which allows differentiation between the four cases corresponding to the Figures 92–95 and in which this difficulty does not present itself, is this: Figures 92–95 correspond respectively to the cases where the "inner" triangle meets 0, 1, 2, 3 sides of the fundamental triangle. (Meeting a vertex counts as meeting both sides to which it belongs.)

[5] The discussion of Case (VI) in 47.7. may also be considered as such a repetition—under much simpler conditions.

with the necessary qualifications of asymmetry and the possible disappearance or degeneration of some areas ①–⑦ (cf. Figures 92–95 and footnote 4 on p. 553), those depicted in Figure 86.

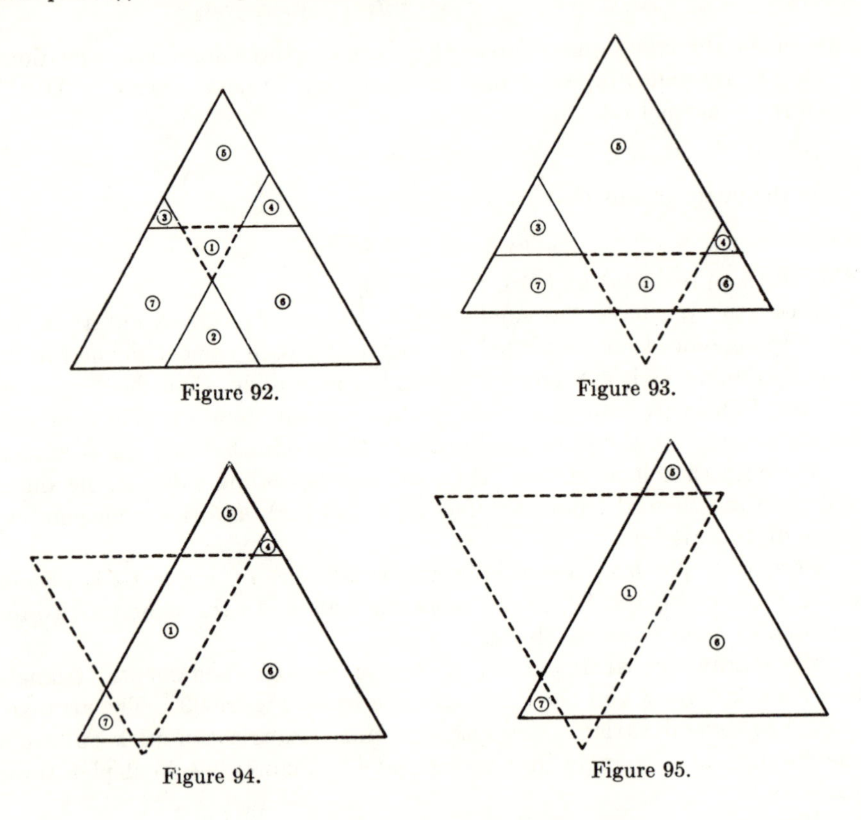

Figure 92.

Figure 93.

Figure 94.

Figure 95.

60.4. Comparison with the Zero-sum Games

60.4.1. We have determined all solutions of the general n-person games with $n = 3$ in a rigorous way, but we have not yet made an attempt to analyze the meaning of our results. We therefore pass now to this analysis.

Let us begin with some remarks of a rather formal nature. We have seen that the smallest n for which a general game can be essential is $n = 2$, while for the zero-sum games the corresponding number was $n = 3$. We have also seen that there exists (assuming reduction and normalization $\gamma = 1$) precisely one essential general game for $n = 2$, whereas for the zero-sum games the same thing was true for $n = 3$. Again the essential general games for $n = 3$ (under the same assumptions as above) form a three parameter manifold, while for the zero-sum games this was true for $n = 4$. All this indicates an analogy between general n-person games and zero-sum $n + 1$-person games. Of course, we know the reason: The zero-sum extensions of the general n-person games are zero-sum $n + 1$-person games and we saw that every zero-sum $n + 1$-person game can be obtained in this way.[1]

60.4.2. It must be remembered, however, that while the zero-sum $n + 1$-person games are exhausted by this procedure, their solutions are not—the solutions of a general n-person game form only a subset of those of its zero-sum extension (cf. e.g. (56:I:a) in 56.12.).

Thus our determination of all solutions of all general three-person games means only that we know some, but not all solutions of all zero-sum four-person games. Indeed, the voluminous and yet incomplete discussion of Chapter VII shows that determining all solutions of all zero-sum four-person games is a task of considerably greater size. Our results concerning the general three-person games imply, however, this much: There exist solutions for every zero-sum four-person game. (The casuistic discussion of Chapter VII did not reveal this.)

61. Economic Interpretation of the Results for $n = 1,2$

61.1. The Case $n = 1$

61.1. We now come to the main objective of our present analysis: The interpretation of our results for $n = 1,2,3$.

Consider first $n = 1$: What matters in this case was already stated or referred to in 60.1. Our result was, as it had to be, a repetition of the simple maximum principle which characterizes this case—and this case only, which therefore describes the "Robinson Crusoe" or completely planned communistic economy.

61.2. The Case $n = 2$. The Two-person Market

61.2.1. Consider next $n = 2$: Our result for this case, obtained in 60.2.2. can be stated verbally as follows:

There exists precisely one solution. It consists of all those imputations where each player gets individually at least that amount which he can secure for himself, while the two get together precisely the maximum amount which they can secure together.

Here the "amount which a player can get for himself" must be understood to be the amount which he can get for himself, irrespective of what his opponent does, even assuming that his opponent is guided by the desire to inflict a loss rather than to achieve a gain.[2]

In examining the solution we find the opportunity to fulfill the promise contained in the fourth remark in 58.3.2.: We must see whether our above definition of the "amount which a player can get for himself"—based on a hypothetical desire of the opponent to inflict a loss rather than to achieve

[1] Precisely: It is strategically equivalent to one which is so obtainable. (Cf. the beginning of 57.4.1.)

[2] Cf. the detailed discussion at the end of 58.2.1. and in 58.3. The amount which the player k can secure for himself is, of course, $v((k))$.

a gain—leads to common-sense results.[1] In order to compare the result of our theory in this way with "common-sense," it is desirable to present the general two-person game in a form which is easily accessible to ordinary intuition. Such a form is readily found by considering some fundamental economic relationships which can exist between two persons.

61.2.2. Accordingly we consider the situation of two persons in a market, a seller and a buyer. We wish to analyze one transaction only and it will appear that this is equivalent to the general two-person game. It is obviously also equivalent to the simplest form of the classical economic problem of *bilateral monopoly*.

The two participants are 1,2: the seller 1 and the buyer 2. The transaction which we consider is the sale of one unit A of a certain commodity by 1 to 2. Denote the value of the possession of A to 1 by u and for 2 by v. I.e., u represents the best alternative use of A for the seller, while v is the value to the buyer, after the sale.

In order that such a transaction have any sense, the value of A for the buyer must exceed that one for the seller. I.e. we must have

$$u < v. \tag{61:1}$$

It is convenient to use the state of the buyer when no sale occurs—i.e. his original financial position—as the zero of his utility.[2]

Let us now describe this as a game. In doing this it is best to omit A from the picture altogether and to deal instead with the value connected with its transfer or its alternative uses. We may then formulate the rules of the game as follows.

1 offers 2 a "price" p, which 2 may "accept" or "decline." In the first case 1, 2, get the amounts p, $v - p$. In the second case they get the amounts u, 0.[3]

The common-sense result is that the price p will have some value between the limits set by the alternative valuations of the two participants,

[1] The reader will understand that we do not ascribe this desire to the opponent. It is only that our theory can be formulated as if he had this desire. What matters is not this possible formulation, but the results of the theory.

Indeed, this "malevolent" behavior of the opponent determines only some, but not all features of the solution: It gives the lower limit of what each player must obtain individually, but what both get together can only be described by the opposite hypothesis of perfect cooperation. (Cf. above.)

This is just a special case of the general fact, that only the entire, rigorous theory is a reliable guide under all conditions, while the verbal illustrations of its parts are of limited applicability and may conflict with each other.

All this can be brought out even better by the detailed discussion of 58.3.

[2] We are purposely disregarding the possibility of describing a sale as an exchange of goods for goods. Our theory forces us, for reasons which we have stated repeatedly, to use an unrestrictedly transferable numerical utility, which we may as well describe in terms of money.

We shall deviate from this standpoint only in Chapter XII.

[3] We leave it to the reader to formulate this in terms of our original combinatorial definition of games.

i.e. that

$$u \leqq p \leqq v. \tag{61:2}$$

Where p will actually be between the limits of (61:2) depends on factors not taken into account in this description. Indeed, this rule of the game provides for one bid only, which must be accepted or declined—this is clearly the final bid of the transaction. It may have been preceded by negotiating, bargaining, higgling, contracting and recontracting, about which we said nothing. Consequently a satisfactory theory of this highly simplified model should leave the entire interval (61:2) available for p.

61.3. Discussion of the Two-person Market and Its Characteristic Function

61.3.1. Before going any further we add two remarks concerning this description of the game, which is our model for the economic set-up under consideration.

First: It would be possible to use more elaborate models allowing for greater (but limited) numbers of alternative bids, etc.

There is a *prima facie* evidence for considering such variants, since all existing markets are governed by more or less elaborate rules for successive bids by all participants, which appear to be essential for the understanding of their character. Besides, we did investigate in detail the game of Poker in 19. This game is based on the interplay of the bids of all participants and we saw loc. cit. that the sequence and arrangement of these bids was of decisive importance for its structure and theory. (Cf. in particular the descriptive part 19.1.-19.3., the variants discussed in 19.11.-19.14. and the concluding summary of 19.16.)

A closer inspection shows, however, that in our present setup these details do not become decisive. The situation is altogether different from Poker which is a zero-sum game and where any loss of one player is a gain for the other one.[1] Specifically, the reader may discuss any more complicated market (but with only two participants!) in the same way as we shall do it for our simple version in 61.3.3. He will find the same characteristic function as we obtain (61:5), (61:6) of 61.3.3. Indeed, the deductions given there apply *mutatis mutandis* in any market (of two participants!): The reader who carries out this comparison will observe that all that matters in those proofs[2] is that the seller (or the buyer) may, if he wishes, insist absolutely on the particular price mentioned there, irrespective of the counter-offers he may get and the number of successive bids required.[3]

These elaborations lead essentially to the same results as our simple model. We refrain therefore from considering them.

[1] This applies directly to Poker as a two-person game, as considered in 19. If more than two persons participate, then our treatment by means of coalitions brings about the same situation.

[2] The significant one is the proof of (61:5) in 61.3.3.

[3] Returning to our previous remarks concerning Poker: The reader may verify for himself how a corresponding simple overall policy would not work there—due to the

61.3.2. Second: On the other hand our model could also be simplified further. Indeed, the mechanism of compensations between (co-operating) players, which we assumed in all parts of our theories is perfectly adequate to replace bids of prices. I.e. it is not necessary to introduce offering, accepting or declining of prices as part of the rules of the game. The mechanism of compensations is fully able to take care of this, including the preliminary negotiating, bargaining, higgling, contracting and recontracting.

Such a simplified game could be described as follows: Both players 1,2 may choose to exchange or not. If either one chooses not to exchange, then 1,2 get the amounts u, 0. If both chose to exchange then they get the amounts u', u''—where u', u'' are two arbitrary but fixed quantities with the sum v.[1]

In other words: The rules of the game may provide for an arbitrary "price" $p = u'$ (then $v - p = u''$), which the players cannot influence—they will nevertheless bring about any other price they desire by appropriate compensations.

Thus it appears that the arrangement chosen in 61.2.2. is neither the simplest nor the most complete one. We are using it because it seems to be best suited to bring out the essential traits of the situation without unnecessary details.

61.3.3. The "common-sense" result of 61.2.2. amounts in the terminology of imputations to this: There exists precisely one solution and this is the set of all imputations

$$\overrightarrow{\alpha} = \{\{\alpha_1, \alpha_2\}\},$$

with

$$\alpha_1 \geqq u, \qquad \alpha_2 \geqq 0, \tag{61:3}$$

$$\alpha_1 + \alpha_2 = v. \tag{61:4}$$

Comparing this with the application of our theory in 60.2.2., we see that agreement obtains when (61:3), (61:4) coincide with (60:3), (60:4) there. This means that we must have

$$\mathrm{v}((1)) = u, \qquad \mathrm{v}((2)) = 0, \tag{61:5}$$

$$\mathrm{v}((1,2)) = v. \tag{61:6}$$

It is easily verified that (61:5), (61:6) are indeed true. For the sake of completeness we do this for both arrangements of 61.2.2. and 61.3.1.,

---

penalties which the rules of that game inflict upon any prohibitive, excessive, or in any other simple way uniform scheme of bidding.

One could, of course, incorporate similar provisions into the rules governing a market. Indeed, there are certain traditional forms of transactions which are possibly of this type, such as options. But it does not seem advisable to include them in this first, elementary survey of the problem.

[1] The characteristic functions of both arrangements (that of 61.2.2. and the one above) will be determined in 61.3.3. and they will be found to be identical.

61.3.2., the first being dealt with in the text, and the variants required for the second alternative in brackets [ ].

Ad (61:5): Player 1 can make sure to obtain u by offering the price $p = u$ [by choosing not to exchange]. Player 2 can make sure that player 1 obtains u by declining every price [by choosing not to exchange]. Hence $\mathrm{v}((1)) = u$.

Replacement of $p = u$ by $p = v$ [the same conduct of both players] yields in the same way that $\mathrm{v}((2)) = 0$.

Ad (61:6): The two players together get either u or v—the latter arising from $p + (v - p)$ [from $u' + u''$]. By (61:1) v is preferable; hence

$$\mathrm{v}((1,2)) = v.$$

61.4. Justification of the Standpoint of 58.

61.4. The coincidence of the values of the characteristic function $\mathrm{v}(S)$ with the u, 0, v as observed in 61.3.3. may appear fairly trivial. There is, however, one significant point about it: It was obtained with our definition of the characteristic function to which the criticisms of 58.3. and 61.2. apply. I.e. it is dependent upon each player ascribing to his opponent—in a certain part of the theory but not in all of it—the desire to inflict a loss rather than to achieve a gain.

It is important to realize that this dependence is really significant, i.e. that modification of this assumption would alter the result, and therefore falsify it, since the result was seen to be correct. This is best done with the arrangement of 61.2.2.

Indeed, assume that player 2 would under certain conditions prefer to make a profit for himself rather than to inflict a loss upon player 1. Assume that these conditions exist, e.g. when player 1 offers a certain price $p_0 > u$ but $< v$. In this case player 2 obtains $v - p_0$ if he accepts, and 0 if he declines. Hence he gains by accepting. On the other hand player 1 obtains p_0 if player 2 accepts and u if he declines. Hence player 2 inflicts a loss (upon player 1) by declining. Consequently our present assumption concerning the intentions of player 2 means that he will accept.

Thus under these conditions player 1 can count upon obtaining the amount p_0. This conflicts with our previous result according to which the entire price interval (61:2) should be permissible, and we saw in 61.2.2. that it is the latter result that must be considered the natural one.

Summing up: The discussion of the general two-person game which we carried out in 61.2.-61.4. has shown that the general two-person game is crucial with regard to the decision whether the characteristic function should be formed as used in our theory. The setup was simple enough as to allow a "common-sense" prediction of the result—and any change in the procedure of forming the characteristic function would have altered the theoretical result significantly. In this way we have obtained by the application of the theory, a corroboration in the sense of the fourth remark in 58.3.

61.5. Divisible Goods. The "Marginal Pairs"

61.5.1. The discussion of 61.2.-61.4. referred to a very elementary case but it nevertheless sufficed for the task of "corroboration" which we had set for ourselves. Besides, by interpreting one essential general two-person game, all were interpreted, since all of them are strategically equivalent to one reduced form (which could be normalized to $\gamma = 1$).

So far everything is satisfactory. But it is still desirable to verify that our theory can do equal justice to somewhat less trivial economic setups. For this purpose we will first extend the description of the two-person market somewhat. It will be seen that this yields nothing really new. Then we shall turn to the general three-person games. There we will find genuinely new corroborations and opportunities for more fundamental interpretations.

61.5.2. Let us return to the situation described in 61.2.2: the seller 1 and the buyer 2 in a market. We allow now for transactions involving any or all of s (indivisible and mutually substitutable) units $A_1, \cdots, A_s$ of a commodity.[1] Denote the value of the possession of $t(= 0, 1, \cdots, s)$ of these units for 1 by u_t and for 2 by v_t. Thus the quantities

$$(61:7) \qquad u_0 = 0, \qquad u_1, \cdots, u_s,$$

$$(61:8) \qquad v_0 = 0, \qquad v_1, \cdots, v_s,$$

describe the variable utilities of these units to each participant. As in 61.2.2. we use for the buyer his original position as the zero of his utility.

There is no need to repeat the considerations of 61.2.2., 61.3.1., 61.3.2. concerning the rules of the game which models this setup.

It is easy to see, what its characteristic function must be. Since each player can block all sales,[2] it follows as in 61.3.3. that

$$(61:9) \qquad \mathrm{v}((1)) = u_s, \qquad \mathrm{v}((2)) = 0.$$

Since the two players together can determine the number of units to be transferred and since with a transfer of t units they obtain together $u_{s-t} + v_t$, therefore

$$(61:10) \qquad \mathrm{v}((1,2)) = \mathrm{Max}_{t=0,1,\ldots,s}\,(u_{s-t} + v_t).$$

This $\mathrm{v}(S)$ is a characteristic function, hence it must fulfill the inequalities (57:2:a), (57:2:c) of 57.2.1. Considering (61:9), (61:10), the only one which is not immediately obvious is

$$(61:11) \qquad \mathrm{v}((1,2)) \geqq \mathrm{v}((1)).$$

This obtains, by observing that the left-hand side is $\geqq u_s + v_0 = u_s$ by (61:10) (use $t = 0$), and the right-hand side is $= u_s$.

[1] We could also allow for continuous divisibility, but this would make no material difference.

[2] 1 by offering an inacceptibly high price, 2 by declining every price.

61.5.3. Consider now the t for which the maximum in (61:10) is assumed, say $t = t_0$. It is characterized by $u_{s-t_0} + v_{t_0} \geqq u_{s-t} + v_t$ for all t. This need only be stated for the $t \neq t_0$ and we can state it for the $t \gtrless t_0$ separately. We may write these inequalities as follows:

$$(61:12) \qquad u_{s-t_0} - u_{s-t} \geqq v_t - v_{t_0} \quad \text{for} \quad t > t_0,$$
$$(61:13) \qquad u_{s-t} - u_{s-t_0} \leqq v_{t_0} - v_t \quad \text{for} \quad t < t_0.$$

Specialize (61:12) to $t = t_0 + 1$ (except when $t_0 = s$ in which case (61:12) is vacuous):

$$(61:14) \qquad u_{s-t_0} - u_{s-t_0-1} \geqq v_{t_0+1} - v_{t_0},$$

and (61:13) to $t = t_0 - 1$ (except when $t_0 = 0$ in which case (61:13) is vacuous):

$$(61:15) \qquad u_{s-t_0+1} - u_{s-t_0} \leqq v_{t_0} - v_{t_0-1}.$$

Note that (6:12), (6:13) (without the specialization $t = t_0 \pm 1$ that led to (6:14), (6:15)) can be written as follows

$$(61:16) \qquad \sum_{i=t_0+1}^{t} (u_{s-i+1} - u_{s-i}) \geqq \sum_{j=t_0+1}^{t} (v_j - v_{j-1}) \quad \text{for} \quad t > t_0,$$

$$(61:17) \qquad \sum_{i=t+1}^{t_0} (u_{s-i+1} - u_{s-i}) \leqq \sum_{j=t+1}^{t_0} (v_j - v_{j-1}) \quad \text{for} \quad t < t_0.$$

In general we can say that (61:14), (61:15) is necessary only, while (61:16), (61:17) is necessary and sufficient. However, we may now profitably introduce the assumption of *decreasing utility*—that is, that the utility of each additional unit decreases, as the total holding increases, for both participants 1,2. As a formula

$$(61:18) \qquad u_1 - u_0 > u_2 - u_1 > \cdots > u_s - u_{s-1},$$
$$(61:19) \qquad v_1 - v_0 > v_2 - v_1 > \cdots > v_s - v_{s-1}.$$

This implies

$$(61:20) \qquad \begin{cases} \left.\begin{aligned} &\sum_{i=t_0+1}^{t} (u_{s-i+1} - u_{s-i}) \geqq (t - t_0)(u_{s-t_0} - u_{s-t_0-1}) \\ &\sum_{j=t_0+1}^{t} (v_j - v_{j-1}) \leqq (t - t_0)(v_{t_0+1} - v_{t_0}) \end{aligned}\right\} & \text{for } t > t_0, \\ \left.\begin{aligned} &\sum_{i=t+1}^{t_0} (u_{s-i+1} - u_{s-i}) \leqq (t_0 - t)(u_{s-t_0+1} - u_{s-t_0}) \\ &\sum_{j=t+1}^{t_0} (v_j - v_{j-1}) \geqq (t_0 - t)(v_{t_0} - v_{t_0-1}) \end{aligned}\right\} & \text{for } t < t_0. \end{cases}$$

ence now (61:14), (61:15) imply (61:16), (61:17). Consequently (61:14),

(61:15) too are necessary and sufficient. Combining (61:14), (61:15) with part of (61:18), (61:19) we may also write:

(61:21) $\begin{cases} \text{Each one of} \\ u_{s-t_0} - u_{s-t_0-1},\ v_{t_0} - v_{t_0-1} \\ \text{is greater than each one of} \\ u_{s-t_0+1} - u_{s-t_0},\ v_{t_0+1} - v_{t_0}.\text{[1]} \end{cases}$

According to the usual ideas, the maximizing $t = t_0$ is the number of units actually transferred. We have shown that it is characterized by (61:21), and the reader will verify that (61:21) is precisely Böhm-Bawerk's definition of the "marginal pairs."[2]

So we see:

(61:A) The size of the transaction, i.e. the number t_0 of units transferred, is determined in accord with Böhm-Bawerk's criterion of the "marginal pairs."

To this extent we may say that the ordinary common-sense result has been reproduced by our theory.

It may be noted, to conclude, that the case when this game is inessential has a simple meaning. Inessentiality means here

$$\mathrm{v}((1,2)) = \mathrm{v}((1)) + \mathrm{v}((2)),$$

i.e. by (61:9) equality in (61:11). Considering (61:9), (61:10) this means that the maximum in the latter is assumed at $t = 0$, i.e. that $t_0 = 0$. So we see:

(61:B) Our game is inessential if and only if no transfers take place in it—i.e. when $t_0 = 0$.[3]

61.6. The Price. Discussion

61.6.1. Let us now pass to the determination of the price in this set-up. In order to provide an interpretation in this respect we must consider more closely the (unique) solution of our game, as provided by the considerations of 60.2.2.

Mathematically the present set-up is no more general than the earlier one analyzed in 61.2.-61.4.: both represent essential general two-person games, and we know that there exists only one such game. Nevertheless, that set-up was only a special case of our present one: Corresponding to $s = 1$. This difference will be felt as we now pass to the interpretation.

[1] Comparing the first term of the first line with the second term of the second line is (61:14); comparing similarly second and first is (61:15). Comparing first and first is an inequality from (61:18); second and second, one from (61:19).

[2] *E. von Böhm-Bawerk:* Positive Theorie des Kapitals, 4th Edit. Jena 1921, p. 266ff.

[3] Note that in our earlier arrangement of 61.2.2. we forced the occurrence of a transfer by requiring (61:1). Our present set-up leaves both possibilities open.

Comparison of (61:5), (61:6) in (61.3.3.) with (61:9), (61:10) in 61.5.2. shows that the mathematical identity of these two setups rests upon substituting the u, v of the former according to

$$(61:22) \qquad u = u_s, \qquad v = \mathrm{Max}_{t=0,1,\ldots,s}\,(u_{s-t} + v_t).$$

The (unique) solution consists, therefore, of all imputations

$$\vec{\alpha} = \{\{\alpha_1, \alpha_2\}\}$$

fulfilling (61:3), (61:4) in 61.3.3. In terms of α_2 this means

$$(61:23) \qquad 0 \leqq \alpha_2 \leqq v - u.^1$$

Let us now formulate this in terms of the ordinary concept of prices—instead of the imputations which are the means of expression of our theory.[2] Since, as we concluded in 61.5.3., t_0 units will have been transferred to the buyer 2, there must be

$$(61:24) \qquad v_{t_0} - t_0 p = \alpha_2,$$

if the price paid was p per unit. Consequently (61:23) means, in terms of p, that

$$(61:25) \qquad \frac{1}{t_0}(u_s - u_{s-t_0}) \leqq p \leqq \frac{1}{t_0} v_{t_0}.^3$$

This can also be written as

$$(61:26) \qquad \frac{1}{t_0}\sum_{i=1}^{t_0}(u_{s-i+1} - u_{s-i}) \leqq p \leqq \frac{1}{t_0}\sum_{j=1}^{t_0}(v_j - v_{j-1}).$$

61.6.2. Now the limits in (61:26) are not at all those which the Böhm-Bawerk theory provides. According to that theory, the price must lie between the utilities of the two marginal pairs named in (61:21) of 61.5.3., i.e. in the interval

$$(61:27) \qquad \left\{\begin{matrix} u_{s-t_0+1} - u_{s-t_0} \\ v_{t_0+1} - v_{t_0} \end{matrix}\right\} \leqq p \leqq \left\{\begin{matrix} u_{s-t_0} - u_{s-t_0-1} \\ v_{t_0} - v_{t_0-1} \end{matrix}\right\}.$$

This can also be written as

$$(61:28) \qquad \mathrm{Max}\,(u_{s-t_0+1} - u_{s-t_0},\ v_{t_0+1} - v_{t_0}) \leqq p \leqq \mathrm{Min}\,(u_{s-t_0} - u_{s-t_0-1},\ v_{t_0} - v_{t_0-1})$$

In order to compare this interval with (61:26), it is convenient to form a further interval

$$(61:29) \qquad u_{s-t_0+1} - u_{s-t_0} \leqq p \leqq v_{t_0} - v_{t_0-1}.$$

[1] We could base our discussion equally well on α_1, but the present procedure is better suited to be repeated in the case of the three-person market.

[2] It may be worth re-emphasizing: This is interpretation, and not the theory itself!

[3] Note that by (61:22) $u = u_s$, $v = u_{s-t_0} + v_{t_0}$.

The two last inequalities of (61:20) in 61.5.3. (with $t = 0$) yield that the lower limit of (61:29) is $\geqq$ that of (61:26), and that the upper limit of (61:29) is $\leqq$ that of (61:26). Hence the interval (61:29) is contained in the interval (61:26). Again, (61:29) obviously contains the interval (61:27), i.e. (61:28). Summing up: The intervals (61:26), (61:29), (61:28) contain each other, in this order.

So we see:

(61:C) The price p per unit is limited to the interval (61:26) only, while Böhm-Bawerk's theory restricts it to the narrower interval (61:28).

61.6.3. The two results (61:A) and (61:C) give a precise picture of the relation of our theory, in the present application, to the ordinary common sense standpoint.[1] They show that there is complete agreement concerning what will happen in fact—i.e. the number of units transferred—but a divergence as to the conditions under which it will take place—i.e. the price per unit. Specifically, our theory provided a wider interval for that price than the ordinary viewpoint.

That the divergence should come at this point and in this direction is readily understandable. Our theory is essentially dependent upon assuming (among other things) a complete mechanism of compensations among the players. This amounts to possible payments of varying premiums or rebates in connection with the various units transferred. Now the narrow price interval of the ordinary standpoint (defined by Böhm-Bawerk's "marginal pairs") is notoriously dependent upon the existence of a unique price—equally valid for all transfers which occur. Since we are actually allowing premiums and rebates, as indicated above, the unique price is obliterated. Our price p per unit is merely an average price—indeed it was defined as such by (61:24) in 61.6.1.—and it is therefore quite natural that we obtained a wider interval than the one defined by "marginal pairs."

To conclude, we observe that such abnormalities in the formation of the price structure are also quite in agreement with the fact that the market under consideration is a bilaterally monopolistic one.

62. Economic Interpretation of the Results for $n = 3$: Special Case

62.1. The Case $n = 3$, Special Case. The Three-person Market

62.1.1. Consider finally $n = 3$. We propose to obtain an interpretation in the same sense as was outlined in 61.2.1. This will be done by extending the model of 61.2.2. dealing with two persons in a market to one dealing with three persons.

[1] We took Böhm-Bawerk's treatment as representative for that standpoint. Indeed, the views of most other writers on this subject since Carl Menger are essentially the same as his.

As we have pointed out before, the first mentioned discussion could not fail to be exhaustive, since there exists only one essential general two-person game. On the other hand we know that the essential general three-person games form a 3 parameter family and their detailed discussion in 60.3.2. forced us to distinguish numerous alternatives.[1] Accordingly several models would be required to account for all possibilities of the essential general three-person game. We shall restrict ourselves to the discussion of one typical class. An exhaustive discussion would be somewhat lengthy and would not contribute proportionally to our understanding of the theory —but it would not present any additional difficulties.

62.1.2. We consider accordingly the situation of three-persons in a market, one seller and two buyers. The discussion of two sellers and one buyer would lead to the same mathematical setup and to corresponding conclusions. For the sake of definiteness we discuss the first form of the problem and leave it to the reader to carry out the parallel discussion of the second form.

The three participants are 1,2,3—the seller 1, the (prospective) buyers 2,3. We shall consider successively the special arrangement of 61.2.2. and the more general one of 61.5.2. In contrast to what we found there, the latter will now provide a real generalization of the former.

Let us begin with the setup of 61.2.2.: The transaction which we consider is the sale of one (indivisible) unit A of a certain commodity by 1 to either 2 or 3. Denote the value of the possession of A for 1 by u, for 2 by v, and for 3 by w.

In order that these transactions should make sense for all participants, the value of A for each buyer must exceed that for the seller. Also, unless the two buyers 2,3 happen to be in exactly equal positions, one of them must be stronger than the other—i.e. able to derive a greater utility from the possession of A. We may assume that in this case the stronger buyer is 3. These assumptions mean that we have

$$u < v \leqq w. \tag{62:1}$$

As in 61.2.2. and 61.5.2. we use for each buyer his original position as the zero of his utility.

As in 61.5., there is no need to repeat the considerations of 61.2.2., 61.3. concerning the rules of the game which models this setup.

It is easy to see what its characteristic function must be: Since each buyer can block sales to him, and the seller as well as both buyers together can block all sales (cf. 61.5.2.), it follows as in 61.3.3. that

$$\text{v}((1)) = u, \qquad \text{v}((2)) = \text{v}((3)) = 0, \tag{62:2}$$

$$\text{v}((1,2)) = v, \qquad \text{v}((1,3)) = w, \qquad \text{v}((2,3)) = 0, \tag{62:3}$$

$$\text{v}((1,2,3)) = w.^{2} \tag{62:4}$$

[1] The two main cases (a) and (b), the latter being subdivided into the four subcases represented by the Figures 92–95.

[2] Of course this makes use of $u < v \leqq w$.

This $v(S)$ is a characteristic function, hence it must fulfill the inequalities of (57:2:a), (57:2:c) in 57.2.1. The verification can be carried out with little trouble, and is left to the reader.

By the nature of things the game to which $v(S)$ belongs is not constant sum,[1] hence it is *a fortiori* essential.

62.2. Preliminary Discussion

62.2. We can now apply the results obtained in 60.3. concerning the essential general three-person game, to obtain all solutions for our present problem. We shall again compare the mathematical result with what the application of ordinary common sense methods gives.

The agreement will turn out to be better than in 61.5.2.-61.6.3. up to a certain point—specifically the limits to be derived for the price will be the same with both methods. This is probably ascribable to the fact that we are dealing now with one unit only, just as in 61.2.2. When we pass to s units, in 63.1.-63.6., the complications of 61.5.2.-61.6.3. will reappear.

Beyond the point referred to, however, there will be a qualitative discrepancy between our theory and the ordinary view point. It will be seen that this is due to the possibility of forming coalitions. This possibility becomes a reality for the first time for three participants, and it must be expected that our theory will do it full justice—while the ordinary approach usually neglects it. Thus the divergence of the two procedures will also turn out to be a legitimate one from the point of view of our theory.

62.3. The Solutions: First Subcase

62.3.1. We proceed to the application of 60.3.1., 60.3.2. to the $v(S)$ of (62:2)-(62:4) above.

The imputations in this setup are the

$$\vec{\alpha} = \{\{\alpha_1, \alpha_2, \alpha_3\}\}$$

with

$$\alpha_1 \geqq u, \qquad \alpha_2 \geqq 0, \qquad \alpha_3 \geqq 0, \tag{62:5}$$

$$\alpha_1 + \alpha_2 + \alpha_3 = w. \tag{62:6}$$

In order to apply 60.3.1., 60.3.2., it is necessary to bring this to its reduced form, and then to normalize $\gamma = 1$.

The first operation corresponds to the replacement of our $\alpha_1, \alpha_2, \alpha_3$ by the $\alpha'_1, \alpha'_2, \alpha'_3$ of

$$\alpha'_k = \alpha_k + \alpha^0_k \tag{62:7}$$

as mentioned in 57.5.1. and discussed in 31.3.2. and in 42.4.2. The $\alpha^0_1, \alpha^0_2, \alpha^0_3$ obtain as described in the discussion which leads to (59:A) in 59.2.1.

[1] *Proof:* (57:20) in 57.5.2. is violated, e.g. by

$$v((1)) + v((2,3)) = u < w = v((1,2,3)).$$

Specifically,

$$(62:8) \quad \alpha'_1 = \alpha_1 - \frac{w + 2u}{3}, \qquad \alpha'_2 = \alpha_2 - \frac{w - u}{3}, \qquad \alpha'_3 = \alpha_3 - \frac{w - u}{3}.$$

The corresponding changes on $v(S)$ are given by (59:1) in 59.2.1.; they carry (62:2)-(62:4) into

$$(62:9) \qquad v'((1)) = v'((2)) = v'((3)) = -\frac{w - u}{3},$$

$$(62:10) \quad v'((1,2)) = \frac{3v - 2w - u}{3}, \qquad v'((1,3)) = \frac{w - u}{3},$$

$$v'((2,3)) = -\frac{2(w - u)}{3},$$

$$(62:11) \qquad v'((1,2,3)) = 0.$$

Thus $\gamma = \frac{w - u}{3}$, and so the second operation consists of dividing everything by this quantity. Instead of doing this, we prefer to apply 60.3.1., 60.3.2. directly, inserting everywhere (where $\gamma = 1$ was assumed) the proportionality factor $\frac{w - u}{3}$.[1]

Comparison with (60:8) in 60.3.1. shows that

$$a_1 = -\frac{2(w - u)}{3}, \qquad a_2 = \frac{w - u}{3}, \qquad a_3 = \frac{3v - 2w - u}{3}.$$

The six lines of (60:15) in 60.3.2., which describe the triangle from which we derived our solutions, become now:

$$(62:12) \quad \begin{cases} \alpha'_1 = -\frac{w - u}{3}, & \alpha'_2 = -\frac{w - u}{3}, & \alpha'_3 = -\frac{w - u}{3},^{2} \\ \alpha'_1 = \frac{2(w - u)}{3}, & \alpha'_2 = -\frac{w - u}{3}, & \alpha'_3 = -\frac{3v - 2w - u}{3}.^{3} \end{cases}$$

62.3.2. We can now discuss this configuration in the sense of 60.3.3. Clearly

$$a_1 + a_2 + a_3 = v - w \leqq 0,$$

hence we have (60:17:b) loc. cit.—i.e. we have the Case (b) id., and it remains to decide which one of its four subcases, represented by Fig-

[1] The procedure is analogous to that used in the discussion of the essential zero-sum three-person game with excess in 47., in particular in 47.2.2. and 47.3.2. (Case (III)), 47.4.2. (a certain phase of Case (IV)).

[2] The -1 in (60:15) loc. cit. stands for $-\gamma$, so we must multiply it by the proportionality factor $\frac{w - u}{3}$ mentioned above.

[3] The $-a_1$, $-a_2$, $-a_3$ in (60:15) loc. cit., which reappear here, include already the factor $\frac{w - u}{3}$.

ures 92–95, is present. Therefore we proceed from here on by graphical representation.

For this representation we use, as before, the plane of Figure 52. Representing the six lines of (62:12) as those of (60:15) in 60.3.2. were represented

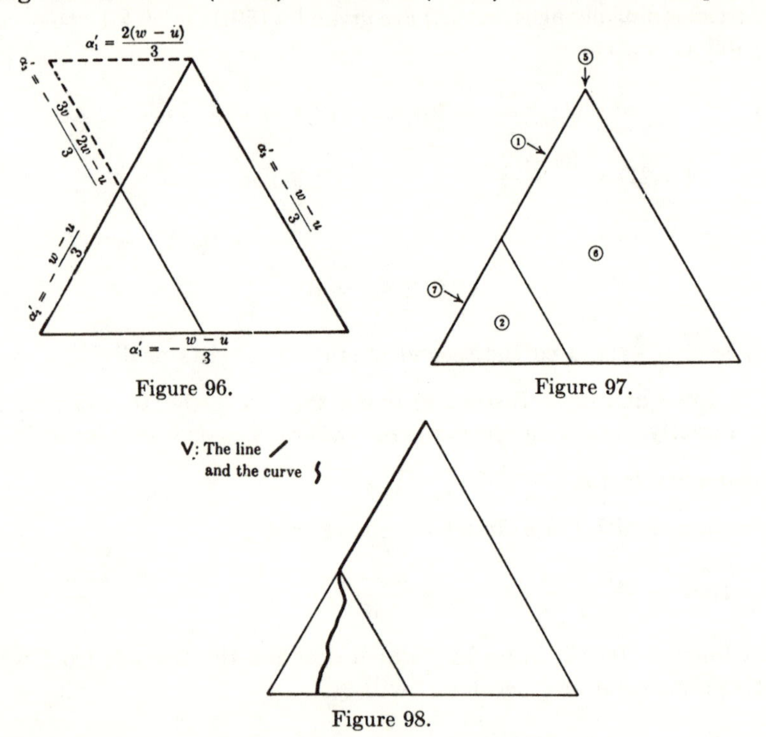

Figure 96.

Figure 97.

Figure 98.

by Figures 92–95, we obtain Figure 96. The qualitative features of this figure follow from the following considerations:

(62:A:a) The second α'_1-line goes through the intersection of the first α'_2- and α'_3-lines. Indeed:

$$\frac{2(w-u)}{3} - \frac{w-u}{3} - \frac{w-u}{3} = 0.$$

(62:A:b) The two α'_2-lines are identical.

(62:A:c) The second α'_3-line is to the left of the first one. Indeed: It has a greater α'_3-value, since

$$-\frac{3v-2w-u}{3} + \frac{w-u}{3} = w - v \geqq 0.$$

Comparison of this figure with Figures 92–95 shows that it is a (rotated and) degenerate form of Figure 94:[1] The area ⑤ is degenerated to a point (the upper vertex of the fundamental triangle △), the areas ①, ⑦ also

[1] For this and the remarks which follow, cf. footnote 4 on p. 553.

degenerated, but to two linear intervals (the upper and the lower part of the left side of the fundamental triangle $\triangle$), while the areas ⑥, ② are still undegenerated (the trapezon and the smaller triangle, into which the fundamental triangle $\triangle$ is divided on our figure). This disposition of the five areas of Figure 94 is shown in Figure 97. The general solution V now obtains, as stated at the end of 60.3.3., by fitting the picture of Figure 86 into the situation described by Figure 97. Figure 98 shows the result.[1]

62.4. The Solutions: General Form

62.4. Before we go any further we note that Figure 97 is of general validity, assuming

(62:13) $$u < v \leqq w,$$

but the picture it gives refers qualitatively to

(62:14) $$v < w.$$

When

(62:15) $$v = w,$$

then the area ① in Figure 97—i.e. the upper interval on the left side of the fundamental triangle—degenerates to a point. (Cf. (62:A:c) in 62.3.2.) Hence in this case Figure 98 assumes the appearance of Figure 99.

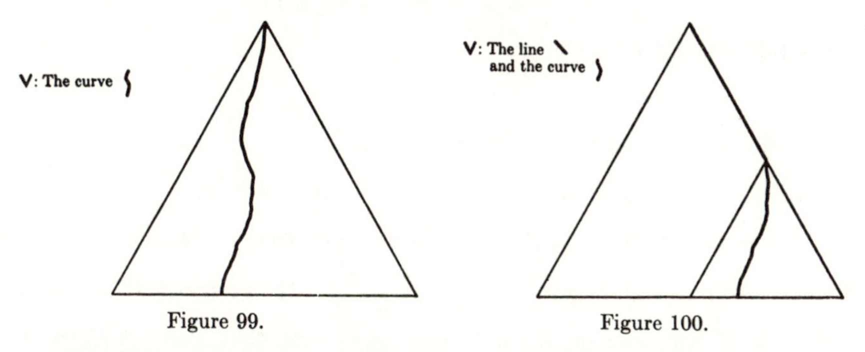

Figure 99. Figure 100.

This discussion can be rendered quite symmetric with respect to the players 2,3—the two buyers—by:

Assuming (62:14) or (62:15), we may replace (62:13) by the weaker condition

(62:16) $$u < v, w.$$

Let us therefore assume (62:16) only—and not (62:13) with (62:14), (62:15). This means that each buyer derives a higher utility from the possession of A than the seller, but it does not place the buyers with respect to each other. (Cf. the discussion in the first part of 62.1.2.)

[1] The curve in Figure 98 is like those in Figure 86, subject to the restriction stated there: (47:6) in 47.5.5.

Now (62:16) leaves three possibilities open: (62:14), (62:15), and

(62:17) $$v > w.$$

The solutions of (62:14), (62:15) were given by Figures 98, 99. (62:17) obtains from (62:14) by interchanging the two players 2,3—the two buyers —and v, w. This means that Figure 98 must be reflected on its vertical middle line (after interchanging v, w). This is shown in Figure 100.

Summing up:

(62:B) Assuming (62:16), the general solution V is given by Figures 98, 99, 100 for $v <, =, > w$, respectively.

62.5. Algebraical Form of the Result

62.5.1. The result expressed by Figure 98 can be stated algebraically as follows:[1]

The solution V consists of the upper part of the left side of the fundamental triangle, and the curve $\sim$.

The first part of V is characterized by

$$\alpha_2' = -\frac{w-u}{3}, \qquad -\frac{3v-2w-u}{3} \geqq \alpha_3' \geqq -\frac{w-u}{3}.$$

Owing to (62:8) in 62.3.1., this means that

$$\alpha_2 = 0, \qquad w - v \geqq \alpha_3 \geqq 0.$$

Now (62:6) in 62.3.1. gives

$$\alpha_1 = w - \alpha_3,$$

hence the above condition can be written as

(62:18) $$v \leqq \alpha_1 \leqq w, \qquad \alpha_2 = 0, \qquad \alpha_3 = w - \alpha_1.$$

The second part of V (the curve) extends from the smallest α_1' above to the absolute minimum of α_1' $\left(-\frac{w-u}{3}\right)$. Its geometrical shape (cf. (47:6) in 47.5.5.) can be characterized by stating that along it α_2', α_3' are both monotonic decreasing functions of α_1'. We may again pass from α_1', α_2', α_3' to α_1, α_2, α_3 by (62:8) in 62.3.1. Then α_1 varies from its minimum in (62:18) above (v) to its absolute minimum (u), and α_2, α_3 are again both monotone decreasing functions of α_1. So we have:

(62:19) $u \leqq \alpha_1 \leqq v$, α_2, α_3 are monotonic decreasing functions of α_1.[2,3]

Thus the general solution V is the sum of the two sets given by (62:18) and

[1] Note that it holds whenever $v \leqq w$, (62:B) notwithstanding.

[2] They must, of course, fulfill (62:5), (62:6) in 62.3.1.

[3] As Figure 98 shows, the lowest point on the line / coincides with the highest point on the curve. I.e. the point $\alpha_1 = v$ of (62:18) and of (62:19) is the same.

Hence we could exclude $\alpha_1 = v$ from either (but not from both!) of (62:18), (62:19).

(62:19). It will be noted that the functions mentioned in (62:19) are arbitrary (within certain limits), but that a definite solution (i.e. a definite standard of behavior) corresponds to a definite choice of these functions. This situation is entirely similar to those analyzed in (47:A) of 47.8.2. and in 55.12.4.

62.5.2. (62:18), (62:19) can be used whenever $v \leqq w$ (cf. footnote 1 on p. 570). For $v = w$ (62:18) simplifies to

(62:20) $$\alpha_1 = v, \qquad \alpha_2 = \alpha_3 = 0.$$

We shall therefore use (62:18), (62:19) only when $v < w$, and (62:20), (62:19) when $v = w$.[1]

If $v > w$, then we can utilize (62:18), (62:19) by interchanging the players 2,3—the two buyers—and v, w. Then (62:18), (62:19) become

(62:21) $$w \leqq \alpha_1 \leqq v, \qquad \alpha_2 = v - \alpha_1, \qquad \alpha_3 = 0.$$[2]

(62:23) $u \leqq \alpha_1 \leqq w$, α_2, α_3 are monotonic decreasing functions of α_1.[3]

Summing up:

(62:C) Assuming (62:16), the general solution is given by (62:18), (62:19); (62:20), (62:19); (62:21), (62:23) for $v <, =, > w$ respectively.

62.6. Discussion

62.6.1. Let us now apply the ordinary, common-sense analysis to the market of one seller and two buyers and one indivisible unit of a good, in order to compare its result with the mathematical one stated in (62:C).

The lines of this common-sense procedure are clearly laid down: we are actually dealing here with one of the simplest special cases of the theory of "marginal pairs." The argument runs as follows:

The seller offers only one indivisible unit of the good under consideration and there are two buyers. Hence one will be included in the transaction, and one will be excluded. Clearly the stronger buyer will be in the first position—except when the two buyers happen to be equally strong, in which case either is eligible. Accordingly the price at which the transaction takes place will lie between the limits of the included and the excluded buyer—and if they happen to be equally strong, the price must be precisely their common limit. The limit of the seller, which must be assumed to be

[1] The observation of footnote 3 on p. 570 concerning (62:18), (62:19) applies also to (62:20), (62:19). Hence we could omit (62:20) altogether, but it is more convenient to keep it, for the sake of the interpretation in 62.6.

[2] Note that, owing to the above interchange, (62:4) in 62.1.2. becomes

(62:22) $$\mathrm{v}((1,2,3)) = v,$$

and so (62:6) in 62.3.1. becomes

(62:6*) $$\alpha_1 + \alpha_2 + \alpha_3 = v.$$

[3] The observation of footnote 3 on p. 570 concerning (62:18), (62:19) applies to (62:21), (62:23) also. Of course we must replace its v by w.

lower than that of either buyer in order to have a genuine three-person market, comes in no case into play.

In our mathematical formulation the limits of the seller and of the two buyers were u, v, w. The above remark means

$$(62:16) \qquad u < v, w.$$

The statements concerning the price amount to

$$(62:24) \qquad v \leqq p \leqq w \quad \text{for} \quad v < w,$$
$$(62:25) \qquad p = v \quad \text{for} \quad v = w,$$
$$(62:26) \qquad w \leqq p \leqq v \quad \text{for} \quad v > w.$$

A buyer who is excluded, finishes where he started—i.e. in our normalization of utility at zero.

Consequently our present statements correspond exactly to (62:18), (62:20), (62:21), as provided for by (62:C).

So far the mathematical and the common-sense results agree. But the limit of this agreement is also in evidence: (62:C) provided for the further imputations of (62:19), (62:23), and there is no trace of these in the ordinary treatment, as presented above.

What then is the meaning of (62:19), (62:23)? Do they express a conflict between our theory and the common-sense standpoint?

It is easy to answer these questions, and to see that there exists no real conflict, but that (62:19), (62:23) represent a perfectly proper extension of the common-sense standpoint.

62.6.2. The amount obtained by the seller in a given imputation, α_1, is clearly the price p envisaged when that imputation is offered. In (62:19), (62:23), α_1 varies from u to v or w (according to which is smaller)—i.e. the price varies from the seller's limit to the weaker buyer's limit. There is also a definite (monotonic) functional connection between the (variable) amounts obtained by the two buyers.[1]

These two facts strongly suggest giving (62:19), (62:23) the following verbal interpretation: The two buyers have formed a coalition, based on a definite rule of division for any profit obtained, and are bargaining with the seller. The rule of division is embodied in the monotonic functions that occur in (62:19), (62:23). No bargaining can depress the seller under his own limit.[2] On the other hand a price above the limit of the weaker buyer would exclude him from any possibility of exerting influence.

The specific rules contained in (62:19), (62:23), and the roles of all participants in these situations may be given more extended verbal treatment. We shall not do this here, since the above should suffice to establish our main point: On the one hand (62:18), (62:20), (62:21) (i.e. the upper parts of V in Figures 98–100) correspond to the competition of the two buyers for the transaction—in which the stronger player, if one exists, is

[1] All these "amounts" refer to the utility in which we reckon—of the goods under consideration there exists only one, indivisible, unit.

[2] The seller's limit is his best alternative use (instead of a sale) for A.

sure to win. On the other hand, (62:19), (62:23) (i.e. the lower part of V in Figures 98–100, the curves) correspond to a coalition of the two buyers against the seller.

Thus it appears that the classical argument—at least in the form used in 62.6.1.—gives the first possibility only, disregarding coalitions. Our theory, to which the coalitions contributed decisively from the beginning, is necessarily different in this respect: It embraces both possibilities, indeed it gives them welded together, as a unit, in the solutions which it produces. The separation, according to schemes with and without coalitions, appears only as a verbal comment on the relatively simple three-person game—there is no reason to believe that it can be carried out for all games, while the mathematical theory applies rigorously in all situations.

63. Economic Interpretation of the Results for $n = 3$: General Case

63.1. Divisible Goods

63.1.1. It remains for us to extend the three-person setup of 62.1.2. in the same way that the two-person setup of 61.2.2. was extended in 61.5.2., 61.5.3.

Let us accordingly return to the situation described in 62.1.2.: the seller 1 and the (prospective) buyers 2, 3 in a market. We allow now for transactions involving any or all of s (indivisible and mutually substitutable) units $A_1, \cdots, A_s$ of a particular good. (Cf. also footnote 1 on p. 560.) Denote the value of t $(= 0, 1, \cdots, s)$ of these units by u_t for 1, by v_t for 2, and by w_t for 3. Thus the quantities

$$(63:1) \qquad u_0 = 0, \qquad u_1, \cdots, u_s,$$
$$(63:2) \qquad v_0 = 0, \qquad v_1, \cdots, v_s,$$
$$(63:3) \qquad w_0 = 0, \qquad w_1, \cdots, w_s,$$

describe the variable utilities of these units for each participant.

As before, we use for each buyer his original position as the zero of his utility.

As in 61.5.2., 61.5.3. and 62.1.2., we need not repeat the considerations of 61.2.2., 61.3.1., 61.3.2. concerning the rules of the game which models this setup.

It is easy to see what its characteristic function must be: Since each buyer can block sales to him, and the seller as well as both buyers together can block all sales (cf. 61.5.2. and 62.1.2.), it follows as in 61.3.3. that

$$(63:4) \qquad \mathrm{v}((1)) = u_s, \qquad \mathrm{v}((2)) = \mathrm{v}((3)) = 0,$$
$$(63:5) \qquad \mathrm{v}((2,3)) = 0.$$

Denoting the number of units transferred from the seller 1 to the buyers 2, 3 by t, r, respectively, it is easy to express what the remaining coalitions (1,2), (1,3), (1,2,3)—i.e. the seller with either one or both buyers—can achieve. The familiar arguments give

(63:6)
$$v((1,2)) = \mathrm{Max}_{t=0,1,\ldots,s}(u_{s-t}+v_t),$$
$$v((1,3)) = \mathrm{Max}_{r=0,1,\ldots,s}(u_{s-r}+w_r),$$

(63:7)
$$v((1,2,3)) = \mathrm{Max}_{\substack{t,r=0,1,\ldots,s \\ t+r\leqq s}}(u_{s-t-r}+v_t+w_r).$$[1]

This $v(S)$ is a characteristic function. We leave it to the reader to verify the inequalities which are implied thereby.

The discussion as to when this game is essential can be carried out as in 61.5.2., 61.5.3. and is left to the reader.[2] It is also possible to determine when one of the two buyers 2,3 becomes a dummy in the sense of our theory of decomposition. We shall not consider this either; the result is not difficult to obtain, and though not surprising it is not uninteresting.

63.1.2. Restricting r in the maximum of (63:7) to the value 0 converts this into the first maximum of (63:6). Restricting t there to the value 0 converts it into u_s. By each one of these operations the value becomes $\leqq$, i.e. we have

(63:8)
$$v((1)) \leqq v((1,2)) \leqq v((1,2,3)).$$

If we do the same to r, t in the reverse order, we obtain similarly

(63:9)
$$v((1)) \leqq v((1,3)) \leqq v((1,2,3)).$$

Consider the first inequality in (63:8). To have equality there means that the first maximum in (63:6) is assumed for $t = 0$. According to the usual ideas on the subject this means that the seller and buyer 2, in the absence of buyer 3, would effect no transfers. I.e., that buyer 2, in the absence of buyer 3, is unable to make the market function.

Consider the second inequality in (63:8). To have equality there means that the maximum in (63:7) is assumed for $r = 0$. According to the usual ideas on the subject, this means that the seller and buyer 3, in the presence of buyer 2, would effect no transfers. I.e., that buyer 3, in the presence of buyer 2, is unable to participate in the market.

Summing these up, together with the corresponding statements for (63:9) which obtain by interchanging buyers 2, 3, we have:

(63:A) Equality in any one of the four inequalities of (63:8), (63:9) is a sign of some weakness of one of the buyers.

In the first inequality of (63:8), [(63:9)] it means that buyer 2 [3], in the absence of buyer 3 [2], is unable to make the market function. In the second inequality of (63:8), [(63:9)] it means, that buyer 3 [2], in the presence of buyer 2 [3], is unable to influence the market.

[1] The extra condition $t + r \leqq s$ under this Max expresses that the number $t + r$ of units sold cannot exceed the number of units originally possessed by the seller.

[2] The discussion of the relationship with (62:1) in 62.1.2. or with (62:16) in 62.4., when $s = 1$, can also be carried out easily. The discussion of (61:B) at the end of 61.5.3. and footnote 3 on p. 562 should be remembered.

The really interesting case arises, obviously, when all these weaknesses are excluded. It is therefore reasonable to make the hypotheses which express this:

(63:B:a) We have $<$ in the first inequalities of both (63:8) and (63:9).

(63:B:b) We have $<$ in the second inequalities of both (63:8) and (63:9).

63.2. Analysis of the Inequalities

63.2.1. Assume for a moment (63:B:a), but the negation of (63:B:b). This means that one of the two players is absolutely stronger than the other. More precisely: That he is at least as strong as the other player, even when he tries to exclude the other player completely from the market.

Hence we may expect in this case a result which is similar to that obtained in 62.1.2.-62.5.2., when only one (indivisible) unit A was available. I.e. the divisibility of the supply into units $A_1, \cdots, A_s$ which we have here, should now become effective.

This is indeed the case. To prove it, introduce the quantities u, v, w by

(63:10) $$\mathrm{v}((1)) = u, \qquad \mathrm{v}((1,2)) = v, \qquad \mathrm{v}((1,3)) = w.$$

Then the second inequality of (63:8) and of (63:9) and the negation of (63:B:b) give

(63:11) $$\mathrm{v}((1,2,3)) = \mathrm{Max}\,(v, w)$$

while the first inequality of (63:8) and of (63:9) and (63:B:a) give

(63:12) $$u < v, w.$$

Now we have precisely the conditions of 62.1.2.-62.5.2.: (63:12) coincides with (62:16) in 62.4., while (63:4), (63:10) give (62:2), (62:3) in 62.1.2., and (63:11) gives (62:4) in 62.1.2. (when $v \geqq w$) or (62:22) in 62.5.2. (when $v \leqq w$).

Consequently the results of 62.4. and 62.5.2. are valid, with the u, v, w of (63:10). The general solution obtains as described, e.g. in (62:B) in 62.4., according to Figures 98–100.

63.2.2. From now on we assume that (63:B:a), (63:B:b) are both valid. We introduce the quantities u, v, w, z by

(63:13) $$\mathrm{v}((1)) = u, \qquad \mathrm{v}((1,2)) = v, \qquad \mathrm{v}((1,3)) = w,$$

(63:14) $$\mathrm{v}((1,2,3)) = z.$$

Then (63:8), (63:9) and (63:B:a), (63:B:b) state that

(63:15) $$u < \begin{Bmatrix} v \\ w \end{Bmatrix} < z.$$

This arrangement differs from that of 62.1.2., but it is nevertheless worth while to compare them in detail: (63:15) corresponds to (62:1) loc. cit., and (63:4), (63:13), (63:14) correspond to (62:2)-(62:4) id.

It is now convenient to introduce again the assumption of *decreasing utility*, already utilized in 61.5.2., 61.5.3. In fact we need it now at a somewhat earlier stage than we did then: It is now (at least in part) useful in the mathematical part of the theory,[1] while we needed it there only in the interpretative part.

We state the decrease of utility for all three participants 1,2,3,:

(63:16) $$u_1 - u_0 > u_2 - u_1 > \cdots > u_s - u_{s-1},$$

(63:17) $$v_1 - v_0 > v_2 - v_1 > \cdots > v_s - v_{s-1},$$

(63:18) $$w_1 - w_0 > w_2 - w_1 > \cdots > w_s - w_{s-1}.$$

In the immediate application only (63:16) will be required. This is it:

(63:19) $$v + w > z + u.$$[2]

Proof: Owing to (63:6), (63:7) and (63:13), (63:14), the assertion (63:19) can be written as follows:

$$\text{Max}_{t=0,1,\ldots,s}(u_{s-t} + v_t) + \text{Max}_{r=0,1,\ldots,s}(u_{s-r} + w_r) > \text{Max}_{\substack{t,r=0,1,\ldots,s\\ t+r \leqq s}}(u_{s-t-r} + v_t + w_r) + u_s.$$

Consider the t, r for which the maximum on the right-hand side is assumed. Since we have (63:B:b), i.e. < in the second inequalities of (63:8), (63:9), we can conclude from the argumentation of 63.1.2. that these t, r are $\neq 0$. We denote them by t_0, r_0. Hence our assertion is this

$$\text{Max}_{t=0,1,\ldots,s}(u_{s-t} + v_t) + \text{Max}_{r=0,1,\ldots,s}(u_{s-r} + w_r) > u_{s-t_0-r_0} + v_{t_0} + w_{r_0} + u_s.$$

I.e., we claim: There exist two t, r with

$$u_{s-t} + v_t + u_{s-r} + w_r > u_{s-t_0-r_0} + v_{t_0} + w_{r_0} + u_s.$$

Now this is actually the case for $t = t_0$, $r = r_0$. The above inequality may then be written as

(63:20) $$u_{s-r_0} - u_{s-t_0-r_0} > u_s - u_{s-t_0}.$$

It should be conceptually clear that this follows from our assumption of decreasing utilities. Formally it obtains from (63:16) in this way: (63:20) states that

[1] But not necessary; the absence of this property would only complicate the discussion somewhat.

[2] In 62.1.2. this was trivially true. Indeed using (63:13), (63:14) we obtain in that case

$$u < v \leqq w = z$$

and this gives (63:19) immediately.

(63:21) $$\sum_{i=1}^{t_0} (u_{s-r_0-i+1} - u_{s-r_0-i}) > \sum_{i=1}^{t_0} (u_{s-i+1} - u_{s-i}).$$

(63:16) implies

$$u_{s'} - u_{s'-1} > u_{s''} - u_{s''-1}$$

whenever $s' < s''$, hence in particular

$$u_{s-r_0-i+1} - u_{s-r_0-i} > u_{s-i+1} - u_{s-i},$$

and from this (63:21) follows.

63.3. Preliminary Discussion

63.3. We now apply 60.3.1., 60.3.2. to the present setup. This will prove to be quite similar to the application carried out in 62.3. for the setup of 62.1.2. The exposition which follows will therefore be more concise, and is best read parallel with the corresponding parts of 62.3.

As to the comparison of the mathematical result with that of the ordinary, common sense approach, the remarks of 62.2. apply again. We indicated already there what complications the present setup produces. We shall consider the situation only briefly, although it is a rather important one. The general viewpoints were sufficiently illustrated by our earlier, simpler examples, and the specific, detailed interpretative analysis of this setup—and other, even more general ones—will be taken up *suo jure* in a subsequent publication.

63.4. The Solutions

63.4.1. The imputations in the present setup are the

$$\overrightarrow{\alpha} = \{\alpha_1, \alpha_2, \alpha_3\}$$

with

(63:22) $$\alpha_1 \geqq u, \qquad \alpha_2 \geqq 0, \qquad \alpha_3 \geqq 0,$$
(63:23) $$\alpha_1 + \alpha_2 + \alpha_3 = z.$$

It is again necessary to introduce the reduced form. This amounts to a transformation

(63:24) $$\alpha'_k = \alpha_k + \alpha^0_k$$

as described in 62.3. The processes discussed there determine the α^0_1, α^0_2, α^0_3, so that (63:24) now becomes

(63:25) $$\alpha'_1 = \alpha_1 - \frac{z + 2u}{3}, \qquad \alpha'_2 = \alpha_2 - \frac{z - u}{3}, \qquad \alpha'_3 = \alpha_3 - \frac{z - u}{3}.$$

The corresponding changes on $v(S)$ are again given by (59:1) in 59.2.1.; they carry (63:4), (63:13), (63:14) into

$$(63:26) \qquad v'((1)) = v'((2)) = v'((3)) = -\frac{z-u}{3},$$

$$(63:27) \qquad v'((1,2)) = \frac{3v-2z-u}{3}, \qquad v'((1,3)) = \frac{3w-2z-u}{3},$$
$$v'((2,3)) = -\frac{2(z-u)}{3},$$

$$(63:28) \qquad v'((1,2,3)) = 0.$$

Thus $\gamma = \frac{z-u}{3}$, and we again refrain from passing to the normalization $\gamma = 1$.

Hence we must again insert a proportionality factor when applying 60.3.1., 60.3.2. as described in 62.3. This proportionality factor is now $\frac{z-u}{3}$.

Comparison with (60:8) in 60.3.1. shows that now

$$a_1 = -\frac{2(z-u)}{3}, \qquad a_2 = \frac{3w-2z-u}{3}, \qquad a_3 = \frac{3v-2z-u}{3}.$$

The six lines of (60:15) in 60.3.2. which describe the triangles from which we derived our solutions, become now:

$$(63:29) \qquad \begin{cases} \alpha'_1 = -\frac{z-u}{3}, & \alpha'_2 = -\frac{z-u}{3}, & \alpha'_3 = -\frac{z-u}{3}, \\ \alpha'_1 = \frac{2(z-u)}{3}, & \alpha'_2 = -\frac{3w-2z-u}{3}, & \\ & & \alpha'_3 = -\frac{3v-2z-u}{3} \end{cases}$$

63.4.2. Applying the criterium of 60.3.3., we find that

$$a_1 + a_2 + a_3 = v + w - 2z \leqq 0.$$

Hence we have again (60:17:b) loc. cit.—i.e. the Case (b) id., and it remains to be decided which one of its four subcases, represented by Figures 92–95, is present.

Following the same procedure of graphical representation as in 62.3., we obtain Figure 101. The qualitative features of this figure follow from the following considerations:

(63:C:a) The second α'_1-line goes through the intersection of the first α'_2- and α'_3-lines. Indeed:

$$\frac{2(z-u)}{3} - \frac{z-u}{3} - \frac{z-u}{3} = 0.$$

(63:C:b) The second α'_2- [α'_3-] line is to the right [left] of the first one.
(63:C:c) Indeed: It has a greater α'_2- [α'_3-] value, since

$$-\frac{3w - 2z - u}{3} + \frac{z - u}{3} = z - w > 0,$$

$$-\frac{3v - 2z - u}{3} + \frac{z - u}{3} = z - v > 0.$$

(63:C:d) The first α_1'-line lies below the intersection of the second α_2' and α_3'-lines. Indeed:

$$-\frac{z - u}{3} - \frac{3w - 2z - u}{3} - \frac{3v - 2z - u}{3} = z + u - v - w < 0,$$

by (63:19) in 63.2.2.

Comparison of this figure with Figures 92–95 shows that it is again a (rotated and) degenerate form of Figure 94 (cf. footnote 1 on p. 568), although less degenerate than the corresponding Figure 96 in 62.3.: The

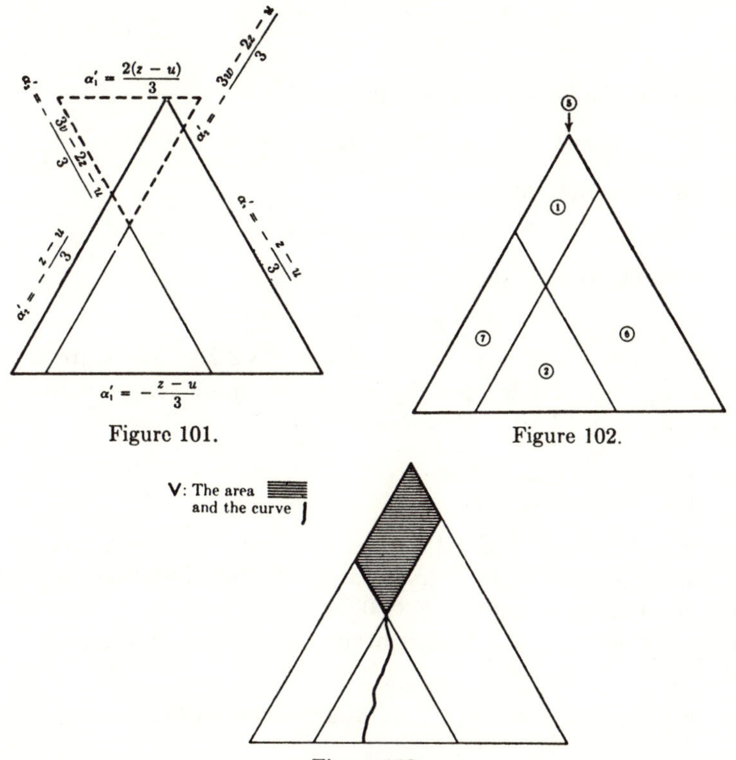

Figure 101.

Figure 102.

Figure 103.

area ⑤ is again degenerated to a point (the upper vertex of the fundamental triangle $\triangle$), but the areas ①, ②, ⑥, ⑦ are still undegenerated (the four areas into which the fundamental triangle is divided in our figure). This disposition of the five areas of Figure 94 is shown in Figure 102. The general solution now obtains as stated at the end of 60.3.3., by fitting the picture of Figure 86 into the situation described by Figure 102. Figure 103 shows the result (cf. footnote 1 on p. 569).

Summing up:

(63:D) Assuming (63:B:a), (63:B:b) and (63:16), the general solution V is given by Figure 103.

Comparison of this figure with those of 62.3.–4. shows that Figure 103 is a form intermediate between those of Figures 98–100, and those figures are in turn degenerate forms of Figure 103.

63.5. Algebraic Form of the Result

63.5. The result expressed by Figure 103 can be stated algebraically, in the same way as was done for Figure 98 in 62.5.1.

In Figure 103 the solution V consists of the area ≡ and the curve ∼.

The first part of V is characterized by

$$-\frac{3w-2z-u}{3} \geqq \alpha_2' \geqq -\frac{z-u}{3}, \qquad -\frac{3v-2z-u}{3} \geqq \alpha_3' \geqq -\frac{z-u}{3}.$$

Owing to (63:25) in 63.4.1., this means that

$$z - w \geqq \alpha_2 \geqq 0, \qquad z - v \geqq \alpha_3 \geqq 0.$$

Now (63:23) in 63.4.1. gives

$$\alpha_1 = z - \alpha_2 - \alpha_3,$$

and so the exact range of α_1 is

$$v + w - z \leqq \alpha_1 \leqq z.$$

(Recall that $v + w - z > u$ by (63:19) in 63.2.2.) We state all these conditions together, the result being somewhat more complicated than its analogue (62:18) in 62.5.1. It is this:

$$\text{(63:30)} \quad \begin{cases} v + w - z \leqq \alpha_1 \leqq z, \qquad 0 \leqq \alpha_2 \leqq z - w, \qquad 0 \leqq \alpha_3 \leqq z - v, \\ \alpha_1 + \alpha_2 + \alpha_3 = z. \end{cases}$$

The ranges in the first line of (63:30) are the precise ones for $\alpha_1, \alpha_2, \alpha_3$.

The second part of V (the curve) can be discussed literally as in 62.5.1.: α_1 varies from its minimum in (63:30) above $(v + w - z)$ to its absolute minimum (u), and α_2, α_3 are monotonic decreasing functions of α_1. So we have:

(63:31) $u \leqq \alpha_1 \leqq v + w - z,$ α_2, α_3 are monotonic decreasing functions of α_1.[1,2]

Thus the general solution V is the sum of the two sets given by (63:30) and (63:31). It will be noted that the role of the functions in (63:31) is the same as that discussed at the end of 62.5.1.

[1] They must, of course, fulfill (63:22), (63:23) in 63.4.1.

[2] As Figure 103 shows, the lowest point in the area ≡ coincides with the highest point on the curve. I.e. the point $\alpha_1 = v + w - z$ of (63:30) and of (63:31) is the same.

Hence we could exclude $\alpha_1 = v + w - z$ from either one (but not from both!) of (63:30), (63:31).

Summing up:

(63:E) Assuming (63:B:a), (63:B:b) and (63:16), the general solution V is given by (63:30), (63:31).

63.6. Discussion

63.6.1. Let us now perform the equivalent of 62.6. and apply the ordinary common-sense analysis to the market of one seller and two buyers and s indivisible units of a particular good, in order to compare its result with the mathematical one stated in (63:E).

Actually the interpretation which ought to be carried out now must combine the ideas of 61.5.2.-61.6.3. with those of 62.6.: the former apply because we have divisibility into s units; the latter because the market is one of three persons. As indicated in 63.3., we do not propose to go fully into all details on this occasion.

The two parts (63:30), (63:31), of which our present solution consists are closely similar to the two parts (62:18), (62:19) (or (62:20), (62:19), or (62:21), (62:23)) obtained in 62.5. (Cf. also (63:E) in 63.5. with (62:C) in 62.5.2.) It appears most reasonable, therefore, to interpret them in the same way as we did in the corresponding situation in 62.6.2.: (63:30) describes the situation where the two buyers compete for the s units in the seller's possession, while (63:31) describes the situation where they have formed a coalition and face the seller united. The reader will have no difficulty in amplifying the details, in parallel to 62.6.2.

These being accepted, there is nothing new to be said about (63:31), the situation in which the buyers have combined and do not compete. (63:30) however, which describes their competition, still deserves some attention.

Let us consider the imputations belonging to (63:30), and let us formulate their contents in terms of the ordinary concept of prices. This is the equivalent of what we did at the corresponding point in 61.6.1., 61.6.2.

We introduce again the t, r for which the maximum in

$$(63:7) \qquad v((1,2,3)) = \mathrm{Max}_{\substack{t,r=0,1,\ldots,s \\ t+r\leqq s}} (u_{s-t-r} + v_t + w_r)$$

is assumed: t_0, r_0. Since our imputations

$$\overrightarrow{\alpha} = \{\{\alpha_1, \alpha_2, \alpha_3\}\} \qquad \text{with} \qquad \alpha_1 + \alpha_2 + \alpha_3 = v((1,2,3))$$

actually distribute the amount $v((1,2,3))$, these t_0, r_0 must represent the numbers of units actually transferred by the seller to the buyers 2,3, respectively.

The analysis of 61.5.2., 61.5.3. leading up to (61:A), could now be repeated mutatis mutandis. It would show that numbers of units transferred, t_0, r_0, can be described in accord with Böhm-Bawerk's criterion of "marginal pairs"—just as was done loc. cit. for the corresponding number of transfers t_0. Since this discussion would bring up nothing new, we shall not dwell upon the point any further.

63.6.2. We now turn to the question of prices. The buyers 2,3 received, as we saw, t_0, r_0 units, respectively. The imputation $\overrightarrow{\alpha}$ on the other hand, ascribes them the amounts α_2, α_3. These two descriptions can be harmonized only by establishing the equations

$$(63{:}32) \qquad v_{t_0} - t_0 p = \alpha_2,$$

$$(63{:}33) \qquad w_{r_0} - r_0 q = \alpha_3,$$

and interpreting p, q as the prices paid per unit by the buyers 2,3 respectively. (63:32), (63:33) are the equivalents of (61:24) in 61.6.1., but it must be emphasized that we obtained two different prices for the two buyers!

Now (63:30) can be stated in terms of p, q [1] as follows:

$$(63{:}34) \qquad \frac{1}{t_0}(v_{t_0} - z + w) \leqq p \leqq \frac{1}{t_0} v_{t_0},$$

$$(63{:}35) \qquad \frac{1}{r_0}(w_{r_0} - z + v) \leqq q \leqq \frac{1}{t_0} w_{r_0}.$$

These inequalities are the analogues of (61:25) in 61.6.1. We could treat them in a way similar to that there, and compare them with those limits which result from the application of Böhm-Bawerk's theory. We shall not carry this out in detail for the reasons stated in 63.3. A few remarks may nevertheless be appropriate.

The intervals (63:34) and (63:35) are again wider than those of Böhm-Bawerk's theory—just as in 61.6. (cf. (61:C) id.). Some numerical examples show, however, that the difference tends to be smaller. It is therefore possible—although nothing has been proven in this respect—that a further increase in the number of buyers may tend to obliterate this discrepancy in that part of the solution which corresponds to no coalition between the buyers. This surmise must, however, be considered with the greatest caution, since we know too well how rapidly the complication of solutions increases with the number of participants and how difficult the interpretation of different parts of the solution may then become.

It will be observed also that we had to introduce two (possibly) different prices for the two buyers, and this in spite of our, still valid, assumption of complete information. This is perfectly in harmony with the interpretations of 61.6.3.: We saw there that what we called prices were really only average prices of several different transactions, that the seller and the buyers must have been operating with premiums and rebates—and all this is necessarily conducive to a differentiation between the two buyers.

Finally, we may state the equivalent of the last remark of 61.6.3. All these abnormalities in the formation of the price structure are also quite in

[1] I.e. its statements concerning α_2, α_3 can be translated by means of (63:32), (63:33) into statements on p, q.

The statement of (63:30) concerning α_1 is merely a consequence of those on α_2, α_3 using $\alpha_1 + \alpha_2 + \alpha_3 = z$. Hence it need not be considered.

agreement with the fact that the market under consideration is one of a monopoly versus duopoly.

64. The General Market

64.1. Formulation of the Problem

64.1.1. The markets which we have considered so far were very restricted: They consisted of two or of three participants. We shall now go a step further and consider a more general market, which consists of $l + m$ participants: l sellers and m buyers. This is, of course, still not the most general arrangement: That would have to provide—among other things—for the possibility that each participant can choose whether he will buy or sell; or again, that he may be seller for one class of goods, and buyer for another. In this study, however, we shall content ourselves with the above case.

Further, we propose to consider one kind of goods only, of which s units $A_1, \cdots, A_s$ are available.

It is convenient to denote the sellers by $1, \cdots, l$, their set by

$$L = (1, \cdots, l);$$

the buyers by $1^*, \cdots, m^*$, their set by

$$M = (1^*, \cdots, m^*);$$

and the set of all participants by

$$I = L \cup M = (1, \cdots, l, 1^*, \cdots, m^*).^1$$

Denote the number of units of the goods under consideration, originally in the possession of the seller i by s_i. Then clearly

$$\sum_{i=1}^{l} s_i = s. \tag{64:1}$$

Denote the utility of $t(= 0, 1, \cdots, s_i)$ units of the goods to the seller i by u_t^i and the utility of $t(= 0, 1, \cdots, s)$ units of the goods to the buyer j^* by v_t^{j*}. Thus the quantities

$$u_0^i = 0, u_1^i, \cdots, u_{s_i}^i \qquad (i = 1, \cdots, l), \tag{64:2}$$

$$u_0^{j*} = 0, v_1^{j*}, \cdots, v_s^{j*} \qquad (j = 1^*, \cdots, m^*), \tag{64:3}$$

describe the variable utilities of these units to each participant.

As before, we use for each buyer his original position as the zero of his utility.

As in 61.5.2., 61.5.3., 62.1.2. and 63.2.1., we need not repeat the considerations of 61.2.2., 61.3.1., 61.3.2. concerning the rules of the game which models this setup.

[1] We use this notation instead of the conventional $1, \cdots, l, l + 1, \cdots, l + m$.

64.1.2. The determination of the characteristic function $v(S)$ of this game is easy:

Clearly $S \subseteq I = L \cup M$. We now consider successively three alternative possibilities.

First: $S \subseteq L$. In this case S consists of sellers only, who can carry out no transactions among themselves. One sees immediately that $v(S)$ merely states their original position:

$$(64{:}4) \qquad v(S) = \sum_{i \text{ in } S} u^i_{s_i}.$$

Second: $S \subseteq M$. In this case S consists of buyers only, who are equally unable to carry out any transactions among themselves. One sees again that $v(S)$ merely states their original position:

$$(64{:}5) \qquad v(S) = 0.$$

Third: Neither $S \subseteq L$ nor $S \subseteq M$—i.e. S has elements in common both with L and with M. In this case S contains sellers as well as buyers, hence transactions between these are definitely possible. On the basis of these, the following formula obtains:

$$(64{:}6) \qquad v(S) = \mathrm{Max}_{\substack{t_i = 0,\, 1,\, \cdots,\, s_i (i \text{ in } S \cap L) \\ r_{j^*} = 0,\, 1,\, \cdots,\, s (j^* \text{ in } S \cap M) \\ \sum_{i \text{ in } S \cap L} t_i + \sum_{j^* \text{ in } S \cap M} r_{j^*} = \sum_{i \text{ in } S \cap L} s_i}} \left(\sum_{i \text{ in } S \cap L} u^i_{t_i} + \sum_{j^* \text{ in } S \cap M} v^j_{r_{j^*}} \right)$$

In this expression $S \cap L$ is the set of all sellers in S, $S \cap M$ the set of all buyers in S, t_i the number of units transferred from the seller i (in $S \cap L$), r_{j^*} the number of units transferred to the buyer j^* (in $S \cap M$).[1] The reader will now have no difficulty in verifying the formula (64:6).

64.2. Some Special Properties. Monopoly and Monopsony

64.2.1. We are far from being able to discuss the theory of this game—the market of l sellers and m buyers—exhaustively. We have at present only some fragmentary information on special cases and beyond this only a few surmises concerning wider areas. The problems which arise in this connection seem to be of definite mathematical interest, aside from their economic importance. It would seem premature, however, to discuss this subject before the investigation has penetrated deeper.

Instead we shall draw some immediate conclusions from the two simpler of our equations: (64:4), (64:5). They are as follows:

(64:A) All sets $S \subseteq L$ and all sets $S \subseteq M$ are flat.

[1] There is no need to state here which seller is transferring each particular unit to which buyer: The resulting utilities—which alone enter into $v(S)$—are not affected by this.

All negotiations between individuals, coalitions, compensation, etc., must be automatically taken care of by the application of our theory.

Proof: This means that

$$v(S) = \sum_{k \text{ in } S} v((k)) \qquad \text{for} \qquad S \subseteq L \qquad \text{and for} \qquad S \subseteq M.$$

which follows immediately from (64:4), (64:5).

(64:B) The game is constant sum if and only if it is inessential.

Proof: Sufficiency: Inessentiality clearly implies constant-sum.
Necessity: Assume that the game is constant-sum.
As L, M are complementary sets

$$v(I) = v(L) + v(M). \tag{64:7}$$

Now by (64:A) (with $S = L, M$)

$$v(L) = \sum_{k \text{ in } L} v((k)), \qquad v(M) = \sum_{k \text{ in } M} v((k)). \tag{64:8}$$

Combining (64:7) and (64:8) we obtain

$$v(I) = \sum_{k \text{ in } I} v((k)). \tag{64:9}$$

Now the modification of (27:B) in 27.4. which applies according to 59.3.1. in our case, gives just (64:9) as a criterion of inessentiality.

It may be worth noting that the criterion of inessentiality (64:9) becomes, when stated explicitly by means of (64:4)-(64:6), this: The maximum in (64:6) is equal to $\sum_{i \text{ in } L} u^i_{s_i}$. Now this is the value of the expression maximized in (64:6) when $t_i \equiv s_i$, $r_{j^*} \equiv 0$. So the statement becomes, that the maximum in (64:6) is assumed when $t_i \equiv s_i$, $r_{j^*} \equiv 0$, i.e. when no transactions take place.

Hence (64:B) can also be formulated as follows:

(64:B*) The fact that the individual utilities of the sellers and buyers are such that no transactions take place at all—i.e. that the maximum in (64:6) is assumed when $t_i \equiv s_i$, $r_{j^*} \equiv 0$—is equivalent to these: that the game is constant-sum; or equivalently (in this case!) that it is inessential.

The salient point of this result is that our game, representing a market, can be constant-sum only at the price of the market being absolutely ineffective. Hence this problem belongs quite intrinsically to games of nonconstant-sum.

64.2.2. We now continue in a somewhat different direction.

(64:C) Consider two imputations,

$$\overrightarrow{\alpha} = \{\{\alpha_1, \cdots, \alpha_l, \alpha_{1^*}, \cdots, \alpha_{m^*}\}\},$$
$$\overrightarrow{\beta} = \{\{\beta_1, \cdots, \beta_l, \beta_{1^*}, \cdots, \beta_{m^*}\}\}.$$

Assume that

$$\vec{\alpha} \succ \vec{\beta},$$

S being the set of 30.1.1. for this domination. Then neither $S \cap L$ nor $S \cap M$ can be empty.[1]

Proof: Otherwise we should have $S \subseteq M$ or $S \subseteq L$. Hence S is flat by (64:A) and therefore certainly unnecessary (cf. 59.3.2.).

We conclude from (64:C) that in this case

(64:10) $\alpha_i > \beta_i$ for at least one i in L,

(64:11) $\alpha_{j^*} > \beta_{j^*}$ for at least one j^* in M.

These formulae (64:10) and (64:11) have a role of some interest, when either L or M is a one-element set: $l = 1$ or $m = 1$. This means that there exists precisely one seller or precisely one buyer,—i.e. that we have *monopoly* or *monopsony*.

In these cases the i of (64:10) or the j^* of (64:11) is uniquely determined: $i = 1$ or $j^* = 1^*$. Se we have:

(64:D) $\vec{\alpha} \succ \vec{\beta}$ implies

(64:12) $\alpha_1 > \beta_1$ when $l = 1$,

(64:13) $\alpha_{1^*} > \beta_{1^*}$ when $m = 1$.

The remarkable thing is that both (64:12) and (64:13) are transitive relations, while domination $\vec{\alpha} \succ \vec{\beta}$ is not. There is, of course, no contradiction in this,—(64:12) or (64:13) is merely a necessary condition for $\vec{\alpha} \succ \vec{\beta}$. But it is nevertheless the first time that the domination concept in an actual game is so closely linked to a transitive relation.

This connection seems to be a quite essential feature of the monopolistic (or monopsonistic) situations.[2] It will play a role of some importance in 65.9.1.

[1] I.e. S must contain both sellers and buyers.

[2] The verbal interpretation of (64:12), (64:13) is simple and plausible: No effective domination is possible without the monopolist (or monopsonist).

CHAPTER XII

EXTENSIONS OF THE CONCEPTS OF DOMINATION AND SOLUTION

65. The Extension. Special Cases

65.1. Formulation of the Problem

65.1.1. Our mathematical considerations of the n-person game beginning with the definitions of 30.1.1. made use of the concepts of imputation, domination and solution, which were then unambiguously established. Nevertheless in the subsequent development of the theory there occurred repeatedly instances where these concepts underwent variations. These instances were of three kinds:

First: It happened in the course of our mathematical deductions, based strictly on the original definitions, that concepts rose to importance which were obviously analogous to the original ones (of imputation, domination, solution) but not exactly identical with them. In this case it was convenient to designate them by those names, necessarily remembering the differences. Examples of this are to be found in the investigation of the essential three-person game with excess in 47.3.-47.7. where the discussion of the fundamental triangle is reduced to that of one of the various smaller triangles in it. Another example is offered by the investigation of a special simple n-person game in 55.2.-55.11., where the discussion of the original domain is reduced to that one of V' in $\mathfrak{A}$ (cf. the analysis of 55.8.2., 55.8.3.).

Second: In the course of our considerations on decomposability in Chapter IX, we explicitly re-defined (generalized) the concepts of imputation, domination and solution in 44.4.2.-44.7.4. This corresponded to an extension of the theory from zero-sum to constant-sum games. Throughout what followed we emphasized that we were investigating a new theory, analogous to, but not identical with, the original one of 30.1.1.

Actually these two types of variations of our concepts are not fundamentally different: The second type can be subsumed under the first one. Indeed, the new theory was introduced in order to handle the problem of decomposition of the original one more effectively. This motive was stressed throughout the heuristic considerations which led to this generalization. In the analysis of imbedding in 46.10., particularly in (46:K) and (46:L) there, we established rigorously that the new theory can be subordinated to the original one precisely in this sense.

Third: The concepts of imputation, domination and solutions were again re-defined (generalized) in Chapter XI, specifically in 56.8., 56.11., 56.12.

This corresponded to the final extension of the theory to general games. We again emphasized that from there on we were investigating a new theory analogous to, but not identical with, the preceding ones.

This generalization was, however, fundamentally different from the two preceding ones: It represented a real conceptual widening of the theory and not a mere technical convenience.

65.1.2. Throughout the changes referred to above it was in evidence that while the concepts of imputation, domination and solution varied (particularly regarding extension), some connection among them remained invariant. In order to acquire a general insight into these changes—and other analogous ones which may follow—it is necessary to find a precise formulation of this invariant connection. When this is done we can permit complete generality in all respects and reformulate the theory on that basis.

By recalling the instances enumerated in 65.1.1., it will appear that this invariant connection is the process by which the concept of a solution is derived from those of imputation and domination. This is the condition (30:5:c) (or the equivalent ones (30:5:a) and (30:5:b)) in 30.1.1. Hence we reach perfect generality if we release the notions of imputation and domination from all restrictions, but define the solutions in the way indicated.

In accordance with this program we proceed as follows:

Instead of imputations we consider the elements of an arbitrary but fixed domain (set) D.

Instead of domination we consider an arbitrary but fixed relation $\mathcal{S}$ between the elements x, y of D.[1]

Now a *solution* (in D for $\mathcal{S}$) is a set $\mathsf{V} \subseteq D$ which fulfills the following condition:

(65:1) The elements of V are precisely those elements y of D for which $x\mathcal{S}y$ holds for no element x of V.[2]

65.2. General Remarks

65.2. These definitions provide the basis for a more general theory in the sense indicated.

It should be noted that our present concept of solution bears the same relation to that one of saturation analyzed in 30.3. and particularly in 30.3.5., as the original concept of 30.1.1. In particular our (65:1) should be compared with the fourth example in 30.3.3., our present $\mathcal{S}$ corresponding to the negation of the $\mathcal{R}$ there. It is especially significant that in the search for solutions all difficulties connected with the lack of symmetry of the relation considered, arise again. I.e., the remarks made in 30.3.6. and 30.3.7. to this effect apply once more.

[1] $x\mathcal{S}y$ expresses that this relation holds between the specific elements x and y. The reader should recall the discussions at the beginning of 30.3.2.

[2] This is the equivalent of (30:5:c) in 30.1.1., as promised.

We shall see subsequently how these difficulties can be resolved at least in some specific cases.[1]

In order to acquire a better understanding of the entire situation, we must consider some specializations of the relation $x\mathcal{S}y$. Indeed, in our present exposition $\mathcal{S}$ is entirely unrestricted and we cannot expect to find any particularly deep result while $\mathcal{S}$ remains in this generality. On the other hand, the original concept of a solution, as defined in 30.1.1., remains the most important application of $\mathcal{S}$ and it seems very difficult to discover any simple distinguishing properties of this particular relation. Therefore there is no apparent way to introduce specialization, however desirable this would be.

We will nevertheless discuss three frequently used schemes of specialization for relations $x\mathcal{S}y$ and finally find a fourth one which possesses a certain limited applicability to our problem proper. In order to carry this out, we need a few mathematical preparations which follow.

65.3. Orderings, Transitivity, Acyclicity

65.3.1. We first consider such relations $x\mathcal{S}y$ (with the domain D) which share the essential features of the concepts "greater" and "smaller." This order of ideas has received detailed and careful considerations in the mathematical literature and there exists today rather general agreement to the effect that a complete list of these properties runs as follows:

(65:A:a) For any two x, y of D one and only one of the three following relations holds:

$$x = y,\ x\mathcal{S}y,\ y\mathcal{S}x.$$

(65:A:b) $x\mathcal{S}y$, $y\mathcal{S}z$ together imply $x\mathcal{S}z$.[2]

We call a relation $\mathcal{S}$ with these properties a *complete ordering* of D.

Examples of complete orderings are easy to give and conform to ordinary intuition: The usual concept of "greater" for the set of all real numbers or for any part of it.[3] The concept of "smaller" under the same conditions. Even the points of the plane possess complete orderings, e.g. this one: $x\mathcal{S}y$ means that x must have a greater ordinate than y or the same one, but then x must have a greater abscissa than y.[4]

65.3.2. The concept of complete ordering can be weakened considerably so that a significant concept still remains. This, too, has received attention in mathematical literature[5] and is of importance in the theory of utilities. It obtains by weakening (65:A:a) above, but retaining (65:A:b) unchanged.

[1] Cf. the results of 65.4., 65.5., and the less superficial ones of 65.6.-65.7.

[2] The reader who substitutes the ordinary "greater" relation $x > y$ for $x\mathcal{S}y$ in (65:A:a), (65:A:b), will verify that these are indeed the basic properties of "greater."

[3] E.g. the integers, or any interval, etc.

[4] Without this last proviso, our $\mathcal{S}$ would fall under the next section.

[5] Cf. *G. Birkhoff:* Lattice Theory, loc. cit. Chapt. I. In this book orderings, partial orderings and similar topics are discussed in the spirit of modern mathematics. Extensive references to literature are given there.

I.e.:

(65:B:a) For any two x, y of D at most one of the three following relations holds:

$$x = y, \; x\mathbb{S}y, \; y\mathbb{S}x.$$

(65:B:b) $x\mathbb{S}y$, $y\mathbb{S}z$ together imply $x\mathbb{S}z$.

We call a relation $\mathbb{S}$ with these properties a *partial ordering* of D.[1] Two x, y of D for which none of the three relations enumerated in (65:B:a) holds (since the ordering is partial, this is a possibility) are called *incomparable* (with respect to $\mathbb{S}$).

Examples of partial orderings are easy to give: The points of the plane, $x\mathbb{S}y$ meaning that the ordinate of x is greater than that of y (cf. footnote 4 on p. 589). We may also define that $x\mathbb{S}y$ means that the ordinate and the abscissa of x are both greater than their counterparts for y.[2] Another good example obtains in the domain of positive integers, $x\mathbb{S}y$ meaning that x is divisible by y excluding equality.

65.3.3. The two preceding concepts of ordering maintained (65:A:b) in the same form, while (65:A:a) was modified (weakened) to (65:B:a). This emphasized the importance of (65:A:b), the property of *transitivity*.[3] We will now undertake to weaken the combination of (65:B:a) and (65:A:b) further, so that (65:A:b) is essentially affected, too.

Note first that (65:B:a) is equivalent to these two conditions:

(65:C:a) Never $x\mathbb{S}x$.

(65:C:b) Never $x\mathbb{S}y$, $y\mathbb{S}x$ together.

Indeed (65:B:a) excludes these three combinations: $x = y$, $x\mathbb{S}y$; $x = y$, $y\mathbb{S}x$; $x\mathbb{S}y$, $y\mathbb{S}x$. Now the first and the second are merely two ways of writing (65:C:a), while the third is precisely (65:C:b).

We now prove:

(65:D) Consider the assertion:

(A_m) Never $x_1\mathbb{S}x_0$, $x_2\mathbb{S}x_1$, $\cdots$, $x_m\mathbb{S}x_{m-1}$, where $x_0 = x_m$ and $x_0, x_1, \cdots, x_{m-1}$ belong to D.

Then we have:

(65:D:a) (65:B:a) is equivalent to (A_1), (A_2) together.

(65:D:b) (65:B:a), (65:A:b) together imply all (A_1), (A_2), (A_3), $\cdots$

Proof: Ad (65:D:a): Clearly (A_1) is (65:C:a) and (A_2) is (65:C:b).

Writing the relations of (A_m) in the reverse order, and applying (65:A:b) $m - 1$ times gives $x_m\mathbb{S}x_0$. As $x_m = x_0$, this means $x_0\mathbb{S}x_0$, contradicting (65:B:a).

[1] Note that the word partial is used in the neutral sense, i.e., a complete ordering is a special case of the partial ones, since (65:A:a) implies (65:B:a).

[2] Note that this is close to a plausible type of partially ordered utilities in the sense of the last remark of 3.7.2. Each imagined event may be affected with two numerical characteristics, both of which must be increased in order to produce a clear and reproducible preference.

[3] Some other important relations, not at all in the nature of an ordering, also possess this property: E.g. equality, $x = y$.

This result suggests considering the total aggregate of all conditions (A_1), (A_2), (A_3), $\cdots$. They are implied by (65:B:a), (65:A:b), i.e. by partial ordering, and represent, as will appear, a further weakening of this property.

We define accordingly:

(65:D:c) A relation $\mathcal{S}$ is *acyclic* if it fulfills all conditions (A_1), (A_2), (A_3), $\cdots$

The reader will understand why we call this acyclicity: If any (A_m) should fail, there would be a chain of relations

$$x_1 \mathcal{S} x_0,\ x_2 \mathcal{S} x_1,\ \cdots\ x_m \mathcal{S} x_{m-1},$$

which is a cycle, since its last element, x_m, coincides with its first one, x_0.

We have already remarked that acyclicity is implied by partial ordering (this is, of course, the content of (65:D:b)), and hence *a fortiori* by complete ordering. It remains to show that it is actually a broader concept than partial ordering, i.e., that a relation can be acyclical without being an ordering (partial or complete).

These are examples of the latter phenomenon: Let D be the set of all positive integers, and $x \mathcal{S} y$ the relation of immediate succession, i.e., $x = y + 1$. Or, let D be the set of all real numbers, and $x \mathcal{S} y$ the relation of being greater than, but not by too much—say by no more than 1—i.e., the relation $y + 1 \geqq x > y$.

We conclude this section by observing that our examples of complete and of partial orderings and of acyclical relations could easily be multiplied. Space forbids us to go into this here, but it may be suggested to the reader as a useful exercise. The references to the literature in footnote 1 on page 62 and footnote 5 on page 589 can also be consulted to advantage.

65.4. The Solutions: For a Symmetric Relation. For a Complete Ordering

65.4.1. Let us now discuss the schemes of specialization referred to at the end of 65.2.

First: $\mathcal{S}$ is symmetric in the sense of 30.3.2. In this case it is expedient to go back to the connection with saturation, pointed out at the beginning of 65.2. Owing to the symmetry of $\mathcal{S}$ it will provide all information about solutions which we desire.

Second: $\mathcal{S}$ is a complete ordering. In this case we define as usual: x is a *maximum* of D if no y with $y \mathcal{S} x$ exists. It is sometimes convenient to indicate the connection with a complete ordering by calling it an *absolute maximum* of D. (Cf. this with the corresponding place in the next remark.) Clearly D has either no maximum or precisely one.[1]

Now we have:

(65:E) V is a solution if and only if it is a one-element set, consisting of the maximum of D.

[1] *Proof:* If x, y are both maxima of D, then $y \mathcal{S} x$ and $x \mathcal{S} y$ being excluded, (65:A:a) necessitates $x = y$.

Proof: Necessity: Let V be a solution. Since D is not empty, V is not empty either.

Consider a y in V. If $x \mathcal{S} y$, then x cannot be in V, hence a u in V with $u \mathcal{S} x$ exists. The transitivity gives $u \mathcal{S} y$ which is impossible, since u, y are both in V. So no x (in D!) exists with $x \mathcal{S} y$,[1] and y must be a maximum of D.

So D has a maximum which must be unique (cf. above). Hence V is a one-element set, consisting of it.

Sufficiency: Let x_0 be the maximum of D, $\mathsf{V} = (x_0)$. Given a y (of D!), the validity of $x \mathcal{S} y$ for no x of V amounts simply to the negation of $x_0 \mathcal{S} y$. Since $y \mathcal{S} x_0$ is excluded, this negation is equivalent to $y = x_0$. So these y form the set V. Hence V is a solution.

65.4.2. Thus there exists no solution V if D has no maximum, while a solution exists and is unique if D has a maximum.

If D is finite, then the latter is certainly the case. This is intuitively quite plausible and also easy to prove. For the sake of completeness and also to make the parallelism with the corresponding parts of the next remark more evident, we nevertheless give the proof in full:

(65:F) If D is finite, then it has a maximum.

Proof: Assume the opposite, i.e., that D has no maximum. Choose any x_1 in D, then an x_2 with $x_2 \mathcal{S} x_1$, then an x_3 with $x_3 \mathcal{S} x_2$ etc., etc. By (65:A:b) $x_m \mathcal{S} x_n$ for $m > n$, hence by (65:A:a) $x_m \neq x_n$. I.e., the $x_1, x_2, x_3, \cdots$ are all distinct from each other, and so D is infinite.

These results show that both the existence and uniqueness of V parallel those of the maximum of D.

65.5. The Solutions: For a Partial Ordering

65.5.1. Third: $\mathcal{S}$ is a partial ordering. In this case we take over literally the definition of a *maximum* of D from the preceding remark. It is sometimes convenient to indicate the connection with a partial ordering by calling it a *relative maximum* of D. (Cf. this with the corresponding place in the preceding remark. This contrast is quite useful, footnote 2 below notwithstanding.) D may have no maximum, it may have one and it may have several.[2] Thus relative maxima are not necessarily unique, while the absolute ones are.[3]

[1] A similar situation was already discussed in 4.6.2.

[2] The argument of footnote 1 on p. 591 fails, since it depends on (65:A:a) which is now weakened to (65:B:a).

E.g., take for D the unit square in the plane and define in it a partial ordering by either one of the two processes in the two first examples at the end of 65.3.2. Then the maxima of D form its entire upper edge, or the upper and the right edges together, respectively.

[3] The reader is warned against mixing up our notion of a relative maximum with that one which occurs in the theory of functions: There a local maximum is frequently called a relative one. Since the quantities involved there are numerical, hence completely ordered, this has nothing to do with our present considerations.

The question of existence also plays a different role for relative maxima than for absolute ones. It will appear that the decisive property now is this:

(65:G) If y in D is not a maximum, then a maximum x with $x \mathcal{S} y$ exists.

For absolute maxima—i.e., if $\mathcal{S}$ is a complete ordering—(65:G) expresses precisely the existence of one.[1] For relative maxima this need not be the case, i.e. for a partial ordering the mere existence of some (relative) maxima need not imply (65:G). Examples of this are easy to give, but we will not pursue this matter further. Suffice it to say, that (65:G) will prove to be the proper extension of the existence of an absolute maximum (cf. the preceding remark) to the case of relative maxima (cf. below).

Now we have:

(65:H) V is a solution if and only if (65:G) is fulfilled (by D and $\mathcal{S}$!) and V is the set of all (relative) maxima.

Proof: Necessity: Let V be a solution.

If y is not in V, then an x in V with $x \mathcal{S} y$ exists, hence y is not a maximum. So all maxima belong to V.

If y is in V, then the argument given in the proof of (65:E) in the preceding remark can be repeated literally, showing that y is a maximum.

So V is precisely the set of all maxima.

If y is not a maximum, i.e., not in V, then an x in V, i.e., a maximum, with $x \mathcal{S} y$ exists, so (65:G) is fulfilled (by D and $\mathcal{S}$).

Sufficiency: Assume that (65:G) is fulfilled, and let V be the set of all maxima.

For x, y in V, $x \mathcal{S} y$ is impossible, since y is a maximum. If y is not in V, i.e., not a maximum, then by (65:G) an x which is a maximum, i.e., in V, with $x \mathcal{S} y$ exists. So V is a solution by (65:1).

The reader should verify how this result (65:H) specializes to (65:E) of the preceding remark when the ordering is complete.

Our result (65:H) shows that there exists no solution V if D and $\mathcal{S}$ do not fulfill the condition (65:G), while a solution exists and is unique if this condition is fulfilled.

65.5.2. If D is finite then the latter is certainly the case. We give the proof in full:

(65:I) If D is finite, then it fulfills the condition (65:G).

Proof: Assume the opposite, i.e., that D does not fulfill (65:G). Call a y *exceptional*, if it is not a maximum and $x \mathcal{S} y$ holds for no maximum x. The failure of (65:G) means that exceptional y exist.

Consider an exceptional y. Since it is not a maximum, an x with $x \mathcal{S} y$ exists. Since y is exceptional, this x is not a maximum. If a maximum u

[1] *Proof:* Since D is not empty (65:G) implies the existence of a maximum.

Conversely: Let x_0 be the maximum of D. Then for every y not a maximum, i.e., $y \neq x_0$, the exclusion of $y \mathcal{S} x_0$ and the validity of (65:A:a) (complete ordering!) give $x_0 \mathcal{S} y$.

with $u\mathcal{S}x$ existed, this would give by (65:B:b) $u\mathcal{S}y$ contradicting the exceptional character of y. Hence no such u exists, i.e., x too is exceptional. I.e.:

(65:J) If y is exceptional then there exists an exceptional x with $x\mathcal{S}y$.

Now choose an exceptional x_1, and exceptional x_2 with $x_2\mathcal{S}x_1$, an exceptional x_3 with $x_3\mathcal{S}x_2$ etc., etc. By (65:B:b) $x_m\mathcal{S}x_n$ for $m > n$, hence by (65:B:a) $x_m \neq x_n$. I.e., $x_1, x_2, x_3, \cdots$ are all distinct from each other and so D is infinite.

(Cf. the last part of this argument with the proof of (65:F) in the preceding remark. Observe, that we could replace its (65:A:a) by the weaker (65:B:a).)

These results show that the existence of a solution now does not correspond to the existence of a maximum, but to the condition (65:G). This is quite remarkable considering the concluding part of the preceding remark in 65.4.2. It corroborates our earlier observation that in the present case of partial ordering (65:G) is the proper substitute for the existence of a maximum.

The uniqueness of the solution is even more remarkable. In the light of the last part of our preceding remark, it would have seemed natural for this uniqueness to be connected with that one of the maximum. But we see now that the solution is unique, while the (relative) maximum need not be, as was already mentioned.[1]

65.6. Acyclicity and Strict Acyclicity

65.6.1. Fourth: $\mathcal{S}$ is acyclic. We know that this case comprises the two preceding ones, i.e., that it is more general than both.

In those two cases we determined the necessary and sufficient conditions for the existence of a solution and we also found that when they are satisfied the solution is unique. (Cf. (65:E) and (65:H).) Furthermore, it was seen that when D is finite these conditions are certainly satisfied. (Cf. (65:F) and (65:I).)

In the acyclic case we will find conditions which are similar to these in many ways and in some respects we will gain deeper insights than before. It will be necessary, however, to vary our standpoint somewhat in the course of our discussion and our results will be subject to certain limitations. The case of a finite D will again be settled in an exhaustive and satisfactory way.

It is again convenient to introduce the concept of maxima,[2] and not only for D itself but also for its subsets. So we define: x is a *maximum* of $E(\subseteq D)$ if x belongs to E and if no y in E with $y\mathcal{S}x$ exists. We denote the set of all maxima of E by $E^m(\subseteq E)$.

[1] (65:H) shows that the solution V is not connected with any particular (non-unique) maximum, but with the (unique) set of all maxima.

[2] Since we used the qualification "absolute" in the second, and "relative" in the third remark, we should now employ another still weaker one. It seems unnecessary, however, to bring in such a terminological innovation at this occasion.

Our discussions will show that it is of decisive importance whether D and $\mathcal{S}$ possess this property:

(65:K) $E \neq \ominus$ (for $E \subseteq D$) implies $E^m \neq \ominus$.

I.e.: Every non-empty subset of D possesses maxima.[1] *Prima facie* (65:K) does not appear to be related in any way to acyclicity, but there exists actually a very close connection. Before we attack our proper objective, the role of solutions in the present case, we investigate this connection.

65.6.2. For this purpose we drop all restrictions concerning D and $\mathcal{S}$, even that of acyclicity.

It is convenient to introduce a property which is a variation of the (A_m) of (65:D) in 65.3.3., and which will turn out intrinsically connected with them:

(A_∞) Never $x_1 \mathcal{S} x_0$, $x_2 \mathcal{S} x_1$, $x_3 \mathcal{S} x_2$, $\cdots$, where $x_0, x_1, x_2, \cdots$ belong to D.[2]

We define, for reasons which shall appear soon:

A relation $\mathcal{S}$ is *strictly acyclic* if it fulfills the condition (A_∞).

We now clarify the relationship of strict acyclicity—i.e. of (A_∞)—both to (65:K) and to acyclicity, by proving the five lemmas which follow. The essential results are (65:O) and (65:P); (65:L)-(65:N) are preparatory for (65:O).

(65:L) Strict acyclicity implies acyclicity.

Proof: Assume that $\mathcal{S}$ is not acyclic. Then there exist $x_0, x_1, \cdots, x_{m-1}$ and $x_m = x_0$ in D, such that $x_1 \mathcal{S} x_0$, $x_2 \mathcal{S} x_1$, $\cdots$, $x_m \mathcal{S} x_{m-1}$. Now extend this sequence $x_0, x_1, \cdots, x_{m-1}$ to an infinite one $x_0, x_1, x_2, \cdots$ by putting

$$\begin{aligned} x_0 &= x_m = x_{2m} = \cdots, \\ x_1 &= x_{m+1} = x_{2m+1} = \cdots, \\ &\cdots \\ x_{m-1} &= x_{2m-1} = x_{3m-1} = \cdots. \end{aligned}$$

Then clearly $x_1 \mathcal{S} x_0$, $x_2 \mathcal{S} x_1$, $x_3 \mathcal{S} x_2$, $\cdots$ etc., etc., and so strict acyclicity fails.

(65:M) Acyclicity without strict acyclicity implies this:

(B^*_∞) There exists a sequence $x_0, x_1, x_2, x_3, \cdots$[3] in D with this property:

[1] Even if $\mathcal{S}$ is a complete ordering, this property (65:K) is of great importance in set theory. Those readers who are familiar with that theory will observe, that (65:K) is precisely the fundamental concept of *well ordering*. (In this case $\mathcal{S}$ must be interpreted as "before" instead of "greater.") For literature cf. *A. Fraenkel*, loc. cit. p. 195ff, and 299ff, and *F. Hausdorff*, loc. cit. p. 55ff, both in footnote 1 on p. 61; also *E. Zermelo* loc. cit. in footnote 2 on p. 269. It is remarkable that the same property plays a role in connection with our concept of solution for arbitrary relations. The major part of the considerations which make up the remainder of this chapter deals with this property and its consequences.

Actually this subject and its ramifications appear to deserve considerable further study from the mathematical point of view.

[2] The sequence $x_0, x_1, x_2, \cdots$ should be infinite in the sense that the indices must go on *ad infinitum*, but the x_i themselves need not all be different from each other.

[3] Cf. this with footnote 2 above, and the last part of this lemma.

For $x_p \mathcal{S} x_q$, $p = q + 1$ is sufficient and $p > q$ is necessary.[1]
(B^*_∞) implies that the $x_0, x_1, x_2, \cdots$ are pairwise different from each other and therefore D must be infinite in this case.

Proof: Since $\mathcal{S}$ is not strictly acyclic, there exist $x_0, x_1, x_2, \cdots$ in D, such that $x_1 \mathcal{S} x_0$, $x_2 \mathcal{S} x_1$, $x_3 \mathcal{S} x_2$, $\cdots$. Hence $p = q + 1$ is sufficient for $x_p \mathcal{S} x_q$.

Now assume that $x_p \mathcal{S} x_q$. We wish to prove the necessity of $p > q$. Assume the opposite: $p \leqq q$. Now $x_{p+1} \mathcal{S} x_p$, $x_{p+2} \mathcal{S} x_{p+1}$, $\cdots$, $x_q \mathcal{S} x_{q-1}$,[2] $x_p \mathcal{S} x_q$ and these relations contradict (A_m) with $m = q - p + 1$: It suffices to replace its $x_0, x_1, \cdots, x_{m-1}$ and $x_m = x_0$ by our $x_p, x_{p+1}, \cdots, x_q$ and x_p. This conflicts with the acyclicity of $\mathcal{S}$.

Thus all parts of (B^*_∞) are established.

Now the consequences of (B^*_∞): If the $x_0, x_1, x_2, \cdots$ were not pairwise distinct, then $x_p = x_q$ would occur for some $p > q$. By (B^*_∞) $x_{q+1} \mathcal{S} x_q$, hence $x_{q+1} \mathcal{S} x_p$; by (B^*_∞) this implies $q + 1 > p$, i.e. $q \geqq p$, but $q < p$. So the $x_0, x_1, x_2, \cdots$ are pairwise distinct and therefore D must be infinite.

(65:N) Non-acyclicity implies this:
For some $m(= 1, 2, \cdots)$ we have:
(B^*_m) There exist $x_0, x_1, \cdots, x_{m-1}$ and $x_m = x_0$ in D with this property:
For $x_p \mathcal{S} x_q$, $p = q + 1$ is necessary and sufficient.[3]

Proof: Since $\mathcal{S}$ is not acyclic, there exist $x_0, x_1, \cdots, x_{m-1}$ and $x_m = x_0$ in D such that $x_1 \mathcal{S} x_0$, $x_2 \mathcal{S} x_1$, $\cdots$, $x_m \mathcal{S} x_{m-1}$. Choose such a system with its $m(= 1, 2, \cdots)$ as small as possible.

Clearly $p = q + 1$ is sufficient for $x_p \mathcal{S} x_q$. We wish to prove that it is necessary too. Assume therefore $x_p \mathcal{S} x_q$ but $p \neq q + 1$.

Now a cyclical rearrangement of the $x_0, x_1, \cdots, x_{m-1}, x_m = x_0$ does not affect their properties and we can apply this so as to make x_p the last element—i.e., to carry p into m. I.e., there is no loss of generality in assuming $p = m$. Now $p \neq q + 1$, i.e., $q \neq m - 1$. We can also assume that $q \neq m$, since $q = m$ could be replaced by $q = 0$. So $q \leqq m - 2$. After these preparations we can replace $x_0, x_1, \cdots, x_{m-1}, x_m = x_0$ by $x_0, x_1, \cdots, x_q, x_m = x_0$ [4] without affecting their properties. This replaces m by $q + 1$, which is $< m$, and this contradicts the assumed minimum property of m.

Thus all parts of (B^*_m) are established.

65.6.3. Summing up:

(65:O)
(65:O:a) Acyclicity is equivalent to the negation of all (B^*_1), (B^*_2), $\cdots$.

[1] In connection with this result cf. also 65.8.3.
[2] These are precisely $q - p$ relations, hence they do not appear if $p = q$.
[3] Observe that the characterization of the interrelatedness of the $x_0, x_1, x_2, \cdots$ is complete in (B^*_m), but not in (B^*_∞). This will be of importance below.
[4] I.e., omit $x_{q+1}, \cdots, x_{m-1}$.

(65:O:b) Strict acyclicity is equivalent to the negation of all (B_1^*), (B_2^*), $\cdots$ and of (B_∞^*).

(65:O:c) Strict acyclicity implies acyclicity for all D but it is equivalent to it for the finite D.

Proof: Ad (65:O:a): The condition is necessary since (B_m^*) contradicts (A_m), hence acyclicity. The condition is sufficient by (65:N).

Ad (65:O:b): The condition is necessary since non-acyclicity contradicts strict acyclicity by (65:L), and (B_∞^*) contradicts (A_∞), hence strict acyclity. The condition is sufficient since the negation of strict acyclicity permits the application of (65:M) in case of acyclicity, and the application of (65:O:a) above in case of non-acyclicity.

Ad (65:O:c): The forward implication was stated in (65:L). If D is finite the reverse implication—and hence the equivalence—results from the last remark in (65:M).

Finally we establish the connection with (65:K):

(65:P) (65:K) is equivalent to strict acyclicity.

Proof: Necessity: Assume that $\mathcal{S}$ is not strictly acyclic. Choose $x_0, x_1, x_2, \cdots$ in D with $x_1 \mathcal{S} x_0$, $x_2 \mathcal{S} x_1$, $x_3 \mathcal{S} x_2$, $\cdots$. Then $E = (x_0, x_1, x_2, \cdots)$ is $\subseteq D$ and $\neq \ominus$, and it possesses clearly no maxima. So (65:K) fails.

Sufficiency: Assume that (65:K) fails. Choose a nonempty $E \subseteq D$ without maxima.[1] Choose an x_0 in E. x_0 is not a maximum in E, so choose an x_1 in E with $x_1 \mathcal{S} x_0$. x_1 is not a maximum in E, so choose an x_2 in E with $x_2 \mathcal{S} x_1$, etc., etc. In this way a sequence $x_0, x_1, x_2, \cdots$ in E, hence in D, obtains and $x_1 \mathcal{S} x_0$, $x_2 \mathcal{S} x_1$, $x_3 \mathcal{S} x_2$, $\cdots$. This contradicts strict acyclicity.

So we see: Strict acyclicity is the exact equivalent of the property (65:K), which we expect to be fundamental. Acyclicity and strict acyclicity are closely related to each other. The particular role of the finite D begins already to make itself felt: For finite D the two above concepts are equivalent.

65.7. The Solutions: For an Acyclic Relation

65.7.1. We now turn to our main objective: The investigation of the solutions in D for $\mathcal{S}$. It is at this point that it will appear, why we attribute to the property (65:K) such a fundamental importance: (65:K) will turn out to be quite intimately connected with the existence of precisely one solution.

We begin by showing that there exists precisely one solution (in D for $\mathcal{S}$) if (65:K) is fulfilled. In proving this we will restrict ourselves to finite sets D, in which case the solution can even be obtained by an explicit construction. This construction is effected by *finite induction*. The finiteness of D is not really necessary, but for an infinite set D the construction in question would be more complicated.[2]

[1] The reader should compare this proof with that one of (65:F) in 65.4.2.

[2] It would be necessary to make use of more advanced set-theoretical concepts (cf.

Since we must assume (65:K), this means by (65:P) that D must be strictly acyclic. Since D is finite, this is by (65:O:c) indistinguishable from ordinary acyclicity. So it does not matter for the moment, whether we state that we require acyclicity or strict acyclicity of D. It is nevertheless appropriate to remember that we are using (65:K), i.e., strict acyclicity, and that the assumption of finiteness, which obliterates the distinction in question, could be removed.

We repeat: For the remainder of this paragraph finiteness of D is assumed and the property (65:K)—i.e., acyclicity, i.e., strict acyclicity.

Let us now carry out the inductive construction referred to. This will be done first and the announced properties will be established afterwards.

We define for every $i = 1, 2, 3, \cdots$ three sets A_i, B_i, C_i (all $\subseteq D$) as follows: $A_1 = D$. If for an i ($= 1, 2, 3, \cdots$) A_i is already known, then B_i, C_i and A_{i+1} obtain in this way: $B_i = A_i^m$ i.e., B_i is the set of those y in A_i for which $x\mathcal{S}y$ for no x in A_i. C_i is the set of those y in A_i for which $x\mathcal{S}y$ for some x in B_i. Finally $A_{i+1} = A_i - B_i - C_i$.

Now we prove:

(65:Q) B_i, C_i are disjunct.

Proof: Immediate by their definitions.

(65:R) $A_i \neq \ominus$ implies $A_{i+1} \subset A_i$.[1]

Proof: $A_i \neq \ominus$ implies $B_i = A_i^m \neq \ominus$ by (65:K),[2] hence

$$A_{i+1} = A_i - B_i - C_i \subset A_i.$$

(65:S) There exists an i with $A_i = \ominus$.

Proof: Otherwise by (65:R) $D = A_1 \supset A_2 \supset A_3 \supset \cdots$, contradicting the finiteness of D.

(65:T) Let i_0 be the smallest i of (65:S), then

$$D = A_1 \supset A_2 \supset A_3 \supset \cdots \supset A_{i_0-1} \supset A_{i_0} = \ominus.$$

Proof: Restatement of (65:R) and (65:S).

(65:U) $B_1, \cdots, B_{i_0-1}, C_1, \cdots, C_{i_0-1}$, are disjunct sets, with the sum D.

Proof: By the definition of A_{i+1} we have $B_i \cup C_i = A_i - A_{i+1}$. Hence $B_1 \cup C_1, \cdots, B_{i_0-1} \cup C_{i_0-1}$, are pairwise disjunct and their sum is

$$A_1 - A_{i_0} = D - \ominus = D.$$

Combining this with (65:Q) shows that $B_1, C_1, \cdots, B_{i_0-1}, C_{i_0-1}$, i.e.,

the references of footnote 2 on p. 269 and footnote 1 on p. 595), in particular of *transfinite induction* or some equivalent technique.

These matters will be considered elsewhere.

[1] The point is that we have $\subset$ and not merely $\subseteq$!

[2] This is the only—but decisive!—use we make of (65:K).

$B_1, \cdots, B_{i_0-1}, C_1, \cdots, C_{i_0-1}$, are pairwise disjunct, and that their sum is also D.

65.7.2. We now put

(65:2) $$\mathsf{V}_0 = B_1 \cup \cdots \cup B_{i_0-1}.$$

Then (65:U) gives

(65:3) $$D - \mathsf{V}_0 = C_1 \cup \cdots \cup C_{i_0-1}.$$

Now we prove:

(65:V) If V is a solution (in D for $\mathcal{S}$), then $\mathsf{V} = \mathsf{V}_0$.

Proof: We begin by showing that $B_i \subseteq \mathsf{V}$ for all $i = 1, \cdots, i_{0-1}$.

Assume the opposite and consider the smallest i for which $B_i \subseteq \mathsf{V}$ fails to be true. Let z be an element of this B_i not in V. Then $y\mathcal{S}z$ for some y in V. z is a maximum in A_i hence y is not in A_i. Consider the smallest k for which y is not in A_k. Then $k \leqq i$ and as y is in $D = A_1$, so $k \neq 1$. Put $j = k - 1$, then $1 \leqq j < i$. y is in A_j but not in $A_{j+1} = A_k$, hence it is in $B_j \cup C_j = A_j - A_{j+1}$.

z is in $B_i \subseteq A_i \subset A_j$. So if y were in B_j, $y\mathcal{S}z$ would imply that z is in C_j. This is not so, since z is in B_i. Hence y is in C_j.

Now necessarily an x in B_j with $x\mathcal{S}y$ exists. Since y is in V, this excludes x from V. Thus $B_j \subseteq \mathsf{V}$ cannot hold. As $j < i$, this contradicts the assumed minimum property of i.

So we see:

(65:4) $$B_i \subseteq \mathsf{V} \quad \text{for all} \quad i = 1, \cdots, i_0 - 1.$$

If y is in C_i, then an x in B_i with $x\mathcal{S}y$ exists. Since this x is in V by (65:4), y cannot be in V.

So we see:

(65:5) $$C_i \subseteq -\mathsf{V} \quad \text{for all} \quad i = 1, \cdots, i_0 - 1.$$

Comparing (65:4), (65:5) with (65:2), (65:3) above shows that V must coincide with V_0, as asserted.

(65:W) V_0 is a solution (in D for $\mathcal{S}$).

Proof: We prove this in two steps:

If x, y belong to V_0 then $x\mathcal{S}y$ is excluded: Assume the opposite: x, y in V_0, $x\mathcal{S}y$.

x, y belong to V_0, say x to B_i and y to B_j. If $i \leqq j$, then y is in $B_j \subseteq A_j \subseteq A_i$. x is in B_i, so $x\mathcal{S}y$ implies that y is in C_i. This is not so, since y is in B_j. If $i > j$, then x is in $B_i \subseteq A_i \subset A_j$. y is a maximum in A_j, hence $x\mathcal{S}y$ is impossible.

Thus we have a contradiction in any event.

If y is not in V_0, then $x \mathcal{S} y$ for some x in V_0: y is in $-V_0$, hence in some C_i. Hence $x \mathcal{S} y$ for an x in B_i, and this x is in consequence in V_0.

This completes the proof.

Combining (65:V) and (65:W) we can state:

(65:X) There exists one and only one solution (in D for $\mathcal{S}$), the V_0 of (65:2) above.

65.8. Uniqueness of the Solutions, Acyclicity and Strict Acyclicity

65.8.1. Let us reconsider the last three remarks, still retaining for a moment the assumption of finiteness, in order to avoid further complications. It is conspicuous that they all yielded the same result, although under varying assumptions. In each case we proved the existence of a unique solution, but the hypothesis was first complete ordering, then partial ordering, and finally (ordinary or strict) acyclicity—i.e., it was weakened at every step.

This being so, it is natural to ask whether we have reached with the last remark the limit of this weakening—or whether acyclicity could be replaced by even less without impairing the existence of a unique solution.

It must be admitted, that this line of investigation takes us away from the theory of games. Indeed, in that theory the existence of solutions was of primary importance, but we have learned that there could be no question of uniqueness.

Nevertheless, since we now have some results on existence with uniqueness, we will continue to study this case. We will see later, that it has even indirectly a certain bearing on the theory of games. (Cf. 67.)

In the sense outlined we should ask therefore this: Which properties of the relation $\mathcal{S}$ are necessary and sufficient in order that there exist a unique solution? It is easy to see, however, that this question is not likely to have a simple and satisfactory answer. Indeed, the solution (in D for $\mathcal{S}$) discloses only little about the structure of D (together with $\mathcal{S}$). The acyclical case is less suited to judge this, since it is somewhat complicated, but the cases of complete or partial ordering make the point quite clear. There the solution is only related to the maxima of D and it does not express at all what the properties of the other elements of D are.

It is not difficult to eliminate this objection. Consider a set $E \subseteq D$ instead of D. The relation $\mathcal{S}$ in D is also a relation in E and if it was a complete ordering or a partial ordering or (ordinarily or strictly) acyclic in D, then it will be the same in E.[1] Hence our result (65:X) implies that in every $E \subseteq D$ there exists a unique solution (for $\mathcal{S}$). Now these solutions, when formed for all $E \subseteq D$, tell much more about the structure of D. It is best to restrict ourselves again to the cases of (complete or partial) ordering. Clearly the knowledge of the maxima of E for all sets $E \subseteq D$ gives a very detailed information about the structure of D (together with $\mathcal{S}$).

[1] I.e., at least the same—it can happen that a partial ordering in D is complete in E—or that an acyclic relation in D is an ordering in E.

65.8.2. Thus we arrive at the following question: Which properties of the relation $\mathcal{S}$ are necessary and sufficient in order that there exist for each $E \subseteq D$ a unique solution (in E for $\mathcal{S}$)? We can show that here acyclicity and strict acyclicity are the significant concepts, although the subject is not completely exhausted. The two lemmas which follow contain what we can assert on this matter.

(65:Y) In order that there exist for each $E \subseteq D$ a unique solution (in E for $\mathcal{S}$), strict acyclicity is sufficient.

For finite D this follows from (65:X), and strict acyclicity may be replaced by acyclicity, owing to (65:O:c).

For infinite D this is dependent upon the extension of (65:X) to infinite sets (cf. the beginning of 65.7.1.)

Proof: If D is (ordinarily or strictly) acyclic, then the same is true of all $E \subseteq D$ (cf. above). Now all assertions of our lemma become obvious.

(65:Z) In order that there exist for each $E \subseteq D$ a unique solution (in E for $\mathcal{S}$) acyclicity is necessary.

Proof: If D is not acyclic, then (65:O:a) yields the validity of a (B_m^*), $m = 1, 2, \cdots$, in (65:N). Form its $x_0, x_1, \cdots, x_{m-1}$ and $x_m = x_0$ and put $E = (x_0, x_1, x_2, \cdots, x_{m-1})$. Then $E \subseteq D$ and (B_m^*) describes $\mathcal{S}$ in E completely. Let us consider the solution V in E (for $\mathcal{S}$).

Consider such a solution V. If x_i is in V, then x_{i+1} is not, since $x_{i+1}\mathcal{S}x_i$. If x_i is not in V then there exists a y in V with $y\mathcal{S}x_i$, i.e., $y = x_j$ with $x_j\mathcal{S}x_i$. This means $j = i + 1$,[1] so $y = x_{i+1}$, and hence x_{i+1} is in V. So we see:

(65:6) x_i is in V if and only if x_{i+1} is not.

Iteration of (65:6) gives:

(65:7) If k is even, then x_0 is in V if and only if x_k is.
If k is odd, then x_0 is in V if and only if x_k is not.

As $x_0 = x_m$, (65:7) involves a contradiction if m is odd. Hence there exists no solution in E (for $\mathcal{S}$) if m is odd. If m is even, then (65:7) implies that V is either the set of all x_k with an even k or the set of all x_k with an odd k. And it is easy to verify that both these sets are indeed solutions in E (for $\mathcal{S}$).

So we have:

(65:8) The number of solutions in $E = (x_0, x_1, \cdots, x_{m-1})$ for $\mathcal{S}$ (with the $x_0, x_1, \cdots, x_{m-1}$ from (B_m^*)) is 2 or 0 according to whether m is even or odd.

Consequently there is in no case a unique solution in this $E(\subseteq D)$.

Combining (65:Y) and (65:Z) we see: The existence of a unique solution (in E for $\mathcal{S}$) for all $E \subseteq D$ is completely characterised for finite sets: For these

[1] If $i = m$, then replace it by $i = 0$.

it is equivalent to acyclicity, i.e., to strict acyclicity, which in this case is the same thing. For infinite sets D we can only say that acyclicity is necessary and strict acyclicity is sufficient.

65.8.3. The gap which exists in this case can only be bridged by a study of the acyclic, but not strictly acyclic (infinite) sets D and their subsets E. By comparing (65:O:a), (65:O:b) we see that such a D satisfies (B^*_∞). Form its $x_0, x_1, x_2, \cdots$ and put $D^* = (x_0, x_1, x_2, \cdots)$. This is also acyclic but not strictly acyclic, hence we may study it in place of D.

Thus the question has become this:

(65:9) Assume that $D^* = (x_0, x_1, x_2, \cdots)$ fulfills (B^*_∞). Will then every $E \subseteq D^*$ possess a unique solution (in E for $\mathcal{S}$)?

The answer to (65:9) cannot be given immediately, because (B^*_∞) describes the relation $x \mathcal{S} y$ in D^*—i.e. $x_p \mathcal{S} x_q$—only incompletely. The corresponding question for (B^*_m) $(m = 1, 2, \cdots)$ was answered in the proof of (65:Z) in the negative, but (B^*_m) described the relation $x \mathcal{S} y$ in its set—i.e. $x_p \mathcal{S} x_q$—completely. Thus the answer to (65:9) requires an exhaustive analysis of all possible forms of the relation $x_p \mathcal{S} x_q$ which fulfill (B^*_∞). The problem appears to be one of considerable difficulty.[1]

65.9. Application to Games: Discreteness and Continuity

65.9.1. Our above results on acyclicity and on strict acyclicity have, as pointed out before, no direct bearing on the theory of games.

As regards strict acyclicity, it suffices to emphasize its equivalence to (65:K) (by (65:P)), and to remember that in the theory of games even D itself (the set of all imputations) possesses no maxima (i.e., undominated elements).[2]

Ordinary acyclicity too is violated, e.g., already in the essential three person game.[3]

Nevertheless there were situations that arose during the mathematical discussion of certain games, where the concept of acyclicity could have been applied. These situations are to be regarded in the spirit of the first remark of 65.1.1. and they are specifically among the examples referred to there.

Thus in the triangles T discussed in 47.5.1. we have an acyclical concept of domination, as the inspection of figures 76, 77 shows.[4] Further in the set $\mathcal{A}$ described in 55.8.2. there is an acyclic concept of domination as the criterion (55:Z) makes apparent.[5]

Finally, in the market discussed in 64. there is an acyclic concept of domination in the case of monopoly or monoposony, as the discussion at the

[1] It lies on the boundary line of combinatorics and set theory, and seems to deserve further attention.

[2] This holds for all essential games. Cf. (31:M) in 31.2.3.

[3] The reader is invited to verify this, e.g., on the diagram of Figure 54. It is easy to ascertain that (B^*_m) holds (and (A_m) fails) for all $m \geqq 3$.

[4] Here domination implies having a greater ordinate.

[5] Here domination implies having a greater n-component, and from this acyclicity obviously follows.

end of 64.2.2. and in particular (64:12), (64:13) there shows.[1] We may accentuate the concluding remark made there by observing that an intrinsic connection is to be surmised between the monopolistic situations in the economic sphere and the mathematical concept of the acyclicity of domination.

It is very remarkable, therefore, that in all these cases particularly extensive families of solutions were found to exist. Indeed, not only numerical parameters, but even highly undetermined curves or functions entered into those solutions. For this cf. 47.5.5. and Fig. 81. in the first instance, and the fifth remark in 55.12. in the second one. In the third instance we can only refer to the mathematical discussion of a special case: The three-person market—monopoly versus duopoly—which was analyzed in 62.3., 62.4. and 63.4.

65.9.2. The great number of solutions in the acyclic situations referred to above may seem natural, if the infiniteness of these D (the set of the imputations under consideration) is emphasized. After all, it was only for finite sets D that acyclicity implied uniqueness of the solution, for the infinite ones strict acyclicity became the crucial concept. (Cf. the last part of 65.8., in particular 65.8.2.) And all these examples are, of course, not strictly acyclic, as can be verified with ease.

The situation is nevertheless paradoxical, for the following reason: Modifications of the concept of utility, which will be considered in 67.1.2. can be applied in such a manner as to make the sets in question finite. Then the acyclical games mentioned will have unique solutions. Now these finite modifications can be made to resemble arbitrarily closely to the original, unmodified games. Hence the original acyclic games with many solutions (infinite D!) can be approximated arbitrarily closely by the modified acyclical games with unique solutions (finite D!). How can the unique solutions be "arbitrarily close" approximations of the non-unique ones?

This paradoxical situation will be described in detail in 67. The analysis which we are going to give there will clarify this lack of continuity and present an opportunity for some interpretations of a certain interest.

66. Generalization of the Concept of Utility

66.1. The Generalization. The Two Phases of the Theoretical Treatment

66.1.1. In the past sections we have generalized the concept of a solution—based on a relation $\mathcal{S}$, which takes the role of domination—in a most extensive way. These generalizations should be used in our theory as follows: Our concepts of imputation, domination and solutions rest upon the more fundamental one of utility. Now if we desire to vary the formalism

[1] Here the domination implies having a greater 1- (or 1*-) component, and from this acyclicity obviously follows.

If neither monopoly nor monopsony exist, i.e., if with the notations loc. cit. $l, m > 1$, then (64:10), (64:11) apply instead of (64:12), (64:13) eod. It is easy to verify that in this case acyclicity does not prevail.

used to describe the latter, we can try to render these variations adequately by appropriate generalizations of the former concepts.

Of course, we do not wish to carry out generalizations for their own sake, but there are certain modifications which would make our theory more realistic. Specifically: We have treated the concept of utility in a rather narrow and dogmatic way. We have not only assumed that it is numerical —for which a tolerably good case can be made (cf. 3.3. and 3.5.)—but also that it is substitutable and unrestrictedly transferable between the various players (cf. 2.1.1.). We proceeded in this way for technical reasons: The numerical utilities were needed for the theory of the zero-sum two-person game—particularly because of the role that expectation values had to play in it. The substitutability and transferability were necessary for the theory of the zero-sum n-person game in order to produce imputations that are vectors with numerical components and characteristic functions with numerical values. All these necessities present themselves implicitly in every subsequent construction built upon the preceding ones—and so *in fine* in our theory of the general n-person game.

Thus a modification of our concept of utility—in the nature of a generalization—appears desirable, but at the same time it is clear that definite difficulties must be overcome in order to carry out this program.

66.1.2. Our theory of games divides clearly into two distinct phases: The first one comprising the treatment of the zero-sum two-person game and leading to the definition of its value, the second one dealing with the zero-sum n-person game, based on the characteristic function, as defined with the help of the values of the two-person games. We pointed out above, how each of these two phases makes use of specific properties of the utility concept. Therefore, if any of these properties are to be generalized, modified or abandoned, we must study the effect of such a change in each phase. It is therefore indicated to analyze these two phases separately.

66.2. Discussion of the First Phase

66.2.1. The difficulties of generalizing the first phase are very serious. The theory of the zero-sum two-person game as expounded in Chapter III makes full use of the numerical character of utility.

Specifically: It is difficult to see how a definite value can be assigned to a game, unless it is possible for each player to decide in all cases which of the various situations that may arise is preferable from his point of view. This means that individual preference must define a complete ordering of the utilities.

Next the operation of combining utilities with numerical probabilities cannot be dispensed with either. We have seen that the rules of the game may explicitly require such operations, if they provide for chance moves. But even when this is not the case, the theory of Chapter III leads in general to the use of mixed strategies with the same effect. (Cf. 17.)

Now it is well known that the completely ordered character of utilities does not imply the numerical one. But we have seen in 3.5. that complete ordering in conjunction with the possibility of combining utilities with numerical probabilities implies the numerical character of utility.

Thus we have at present no way to adscribe a zero-sum two-person game a value unless numerical utilities are available.

In the n-person game the characteristic function is defined with the help of the value in various (auxiliary) zero-sum two-person games. Our reduction of the general n-person games to the zero-sum ones, in addition, made use of the transferability of utilities from one player to another. Indeed, constructions like $\mathcal{H}_{n+1} \equiv -\sum_{k=1}^{n} \mathcal{H}_k$ in 56.2.2. can hardly be given any other meaning. Thus the definition of the characteristic function in an n-person game is technically tied up with the numerical nature of utility in a way from which we cannot at present escape.

The values $v(S)$ of the characteristic function of such a game are the values of the corresponding sets—coalitions—of players S. Hence our conclusion can also be stated in this way: Our general method to adscribe a value to every possible coalition of players is essentially dependent upon the numerical nature of utility and we are at present not able to remedy this.

We have pointed out before that the hypothesis of the numerical nature of utility is not as special as it is generally believed to be. (Cf. the discussion of 3.) Besides, we can avoid all conceptual difficulties by referring our considerations to a strictly monetary economy. Nevertheless it would be more satisfactory, if we could free our theory of these limitations—and it must be conceded that the possibility of doing this has not been established thus far.

66.2.2. In spite of this inadequacy in general, there are many games where the difficulty of defining the characteristic function is never serious. Thus the examples of 26.1. and of 57.3. were such that the characteristic function could be determined directly, without real need for the elaborate considerations of the theory of the zero-sum two-person games. It is true that these were examples synthesized in order to obtain a known, preassigned characteristic function—hence the ease with which they can be handled in this respect is scarcely surprising. However, there exist other instances of the same phenomenon which are of a certain significance: Thus the characteristic function causes no difficulty whatever throughout the theory of simple games of Chapter X.[1] Again the various markets considered in 61.2.-64.2. all had characteristic functions which were easily and directly obtained.

In these cases it would be easy to replace the numerical utilities by more general concepts. We propose to take them up on another occasion.

[1] These games were defined by stating which are the winning coalitions and this implied an implicit determination of the characteristic function.

66.3. Discussion of the Second Phase

66.3.1. If the characteristic function is taken for granted, we can pass to the second phase.

Here the necessity for a numerical utility can be entirely circumvented. We do not propose to describe this in complete detail, since the entire subject does not seem to be mature yet for a final mathematical formalization. Indeed, the first phase is obstructed by unsolved difficulties as described above. Besides, there appears to be some justification for believing that a more unified form of the theory, of which we can at present see only the outlines, might lead us to the desired goal.

We shall therefore give only some general indications relative to the treatment of the second phase.

To begin with, when we renounce the transferability of utilities, as well as when we renounce their numerical character, concepts like zero-sum or constant-sum games are not immediately defined. Hence it is best to deal directly with general games.

Let us therefore consider a general n-person game. Since we possess the theory of Chapter XI, we may forget its origin in the theory of zero-sum games and try to extend it directly to the case of more general (non-numerical, non-transferable) utilities.

The imputations

$$\vec{\alpha} = \{\{\alpha_1, \cdots, \alpha_n\}\}$$

will still be vectors, but their components $\alpha_1, \cdots, \alpha_n$ may not be numbers. It must be noted, that if we give up the numerical character of utility, it is best to concede that each participant $i (= 1, \cdots, n)$ has a domain of individual utilities $\mathfrak{U}_i$ of his own. I.e., the $\mathfrak{U}_1, \cdots, \mathfrak{U}_n$ will in general be different. In this setup the component α_i must belong to $\mathfrak{U}_i$.

It must be noted that even if all utilities are numerical—i.e., if $\mathfrak{U}_1, \cdots, \mathfrak{U}_n$ coincide with each other and with the set of all real numbers—we may still omit the assumption of transferability. Also we may consider the case where transferability exists, but subject to certain restrictions. Indeed, an example of this will be discussed in detail in 67.

66.3.2. Now the restrictions on these components α_i must be considered. They are of two kinds: First the domain of all imputations was defined in 56.8.2. by

$$(66:1) \qquad \alpha_i \geqq \mathrm{v}((i)) \quad \text{for} \quad i = 1, \cdots, n,$$

$$(66:2) \qquad \sum_{i=1}^{n} \alpha_i \leqq \mathrm{v}((1, \cdots, n)).^1$$

Second we defined domination with the help of a concept of effectivity based on

[1] We prefer to use (56:10) here instead of the alternatively possible (56:25) of (56:I:b) in 56.12.

(66:3) $$\sum_{i \text{ in } S} \alpha_i \leqq \mathrm{v}(S),$$

which is (30:3) of 30.1.1.

All these inequalities belong to a common type: A certain set T is given ($T = (i)$ in (66:1), $T = (1, \cdots, n) = I$ in (66:2), $T = S$ in (66:3)) and the imputation $\overrightarrow{\alpha}$ is required to place the set—coalition—T into a position which is at least as good (in (66:1)) or at most as good (in (66:2) and in (66:3)) as that one stated by $\mathrm{v}(T)$.

The position of the coalition T—i.e., the composite position of all its participants—is expressed in all these inequalities by the sum of their components: $\sum_{k \text{ in } T} \alpha_k$. For non-numerical utilities the domains $\mathcal{U}_1, \cdots \mathcal{U}_n$ may be different from each other, and besides, there may exist no addition in them—thus rendering formations like $\sum_{k \text{ in } T} \alpha_k$ senseless. But even if the utilities are numerical, the use of $\sum_{k \text{ in } T} \alpha_k$ in the above context is clearly equivalent to assuming unrestricted transferability. Indeed, the position of a coalition can be described by the sum of the amounts given to its members—without any reference to the individual amounts themselves—only when those members are able to distribute that sum among themselves in any way in which they all agree, i.e., if there are no physical obstructions to transfers.

In general, therefore, we shall have to forego the use of $\sum_{k \text{ in } T} \alpha_k$. Instead, we must introduce the domain of utilities for the composite person, consisting of all members of a given coalition T. Denote this domain by $\mathcal{U}(T)$. Clearly, $\mathcal{U}((k))$ is the same thing as $\mathcal{U}_k$. $\mathcal{U}(T)$ must be obtainable by some process of synthesis from the $\mathcal{U}_k$ of all k in T. It is not at all difficult to devise the proper mathematical procedure required for this process, but we propose to discuss it on another occasion.

The aggregate of the α_k, k in T, as well as the value $\mathrm{v}(T)$ of the characteristic function must be elements of this system. The inequalities (66:1), (66:2), (66:3) refer then to preferences in that system of utilities.

66.4. Desirability of Unifying the Two Phases

66.4. In the hope that the reader will not find the analysis of 66.3. too sketchy, we now indicate in which way the desired unification of the two phases may be looked for. Our theory of the zero-sum two-person game was really based on the same general principles as the subsequent structure of imputations, domination and solutions for zero-sum n-person games and even for general n-person games. Specifically, the decisive discussion of the inter-relatedness of various strategies in a zero-sum two-person game carried out in 14.5., 17.8., 17.9.—i.e., the analysis of the concept of a good strategy—is in many ways analogous to our use of dominations of imputations.

Now it would seem that the weakness of our present theory lies in the necessity to proceed in two stages: To produce a solution of the zero-sum two-person game first and then, by using this solution, to define a characteristic function in order to be able to produce a solution of the general n-person game, based on the characteristic function. General experience in mathematics and in the physical sciences indicates that such a two stage procedure with an intermediary halt—represented in our case by the characteristic function—has two essential aspects. In the early stages of the investigation it may be advantageous, since it divides the difficulties. In the later stages, however, when full conceptual generality is desired, it can be a handicap. The requirement of producing a sharply defined quantity in the middle of our procedure—in our case the characteristic function—might be an unnecessary technicality, saddling the main problem with an extraneous difficulty.

To apply this specifically to our experience with games: We had to divide the difficulties in order to overcome them and to consider successively zero-sum two-person games with strict determinateness, zero-sum two-person games with general strict determinateness, zero-sum n-person games, general n-person games. However, all these steps but two were finally merged into the general theory: Only the zero-sum two-person game and the general n-person game remained. Our insistence on the characteristic function amounts to insisting that for the zero-sum two-person game an intermediary result be obtained which is much sharper than that one which we accepted as satisfactory for the n-person game.[1] Of course, we were able to fulfill this requirement in the case of a numerical, unrestrictedly transferable utility. However, this may be different when these assumptions concerning utility are discarded. And it seems rather plausible that our difficulty with the n-person game may be ascribed to our continued insistence on this special setup for the zero-sum two-person game. Our present technical procedure forces us to insist in this respect, but this insistence may nevertheless be misplaced.

A unified treatment for the entire theory of the n-person game—without the (as it now appears) artificial halt at the zero-sum two-person game and the characteristic function—may therefore *in fine* prove to be the remedy for these difficulties.

67. Discussion of an Example

67.1. Description of the Example

67.1.1. We shall now discuss an example in which the concepts of utility and transferability are modified. These modifications do not represent a

[1] For the zero-sum two-person game we obtained a unique value—i.e., imputation. For the general n-person game (as well as for the zero-sum one) we had only a—usually not unique—solution, and even the individual solution is a set of imputations!

particularly significant broadening of our standpoint with respect to those concepts. The interest of our example is rather that it permits of an application of our results concerning acyclicity and thereby yields conclusions which throw some new light on the subject discussed at the end of 65.9. Specifically, it is hoped that procedures of this kind will provide a more adequate mathematical approach to the phenomenon of bargaining.

67.1.2. The modification to be considered is this: We assume that utility—or its monetary equivalent—is made up of indivisible units. I.e. we do not question its numerical character but require that its value be—in appropriate units—an integer. Thus transfers too are necessarily restricted to integers, but we do not restrict them further. We propose to use the characteristic function as before, but also with integer values. The concepts of domination and solution, after this, are unaltered.

If this standpoint is applied to general one and two-person games, no significant changes occur; i.e., everything remains essentially as in our old theory. It is therefore unnecessary to enter upon the details of these cases. The three-person game, on the other hand, offers some new features, even in its old zero-sum form. It gives rise to some quite peculiar difficulties which appear to be of considerable interest, but are not yet sufficiently analyzed. We therefore prefer to postpone this discussion for a later occasion.

This excludes an exhaustive discussion of the general three-person game in the new setup. We shall, however, analyze a special case which bears directly upon the nature of bargaining. This is the three-person market, consisting of one seller and two buyers.

67.1.3. We obtained in our previous analysis of this case various solutions, depending on whether we assumed that only one (individual) transaction could take place or several, also depending on the relative strength of the two buyers. These solutions were described in (62:C) of 62.5.2. and in (63:E) of 63.5. In all these cases it appeared that the general solution was made up of two parts: (62:18) (or (62:20), (62:21), (63:30)) and (62:19) (or (62:23), (63:31)). Our discussion there showed that the parts of the type (62:18) correspond to the situation where the two buyers are competing with each other, while the parts of the type (62:19) correspond to the situation where they have formed a coalition against the seller. The type (62:18) part was uniquely determined and in essential agreement with the ordinary, common sense economic ideas on the subject. The type (62:19) part, on the other hand, was defined with the help of some highly arbitrary functional connections. These expressed, as we saw in 62.6.2., the various possibilities to set up a rule of division between the allied buyers for any profit obtained. I.e. they constituted their standard of behavior within their coalition. Our present discussion is going to provide some additional information concerning the functioning of this part of the social mechanism.

In order to do this effectively, it is reasonable to eliminate from our problem all those elements which do not contribute to this aspect. I.e. we wish to get rid of the type (62:18) part of the solution. We know from 62.5.2., 62.6.1. that this part is of the smallest size—and indeed could be omitted altogether (cf. footnote 1 on p. 571)—when $v = w$ in the notations loc. cit. This means that only one (indivisible) transaction can take place and that the two buyers are of exactly equal strength. The solution is then given by (62:20) and (62:19) of 62.5., ((62:20) being superfluous, cf. above), or equivalently by Figure 99.

So we assume $v = w$ in the scheme of 62.1.2. We can simplify the situation further, without any significant loss, by putting the "alternative use for the seller" $u = 0$. In this way the (62:2)-(62:4) of 62.1.2., defining the characteristic function, simplify to

$$\text{(67:1)} \qquad \begin{cases} \mathrm{v}((1)) = \mathrm{v}((2)) = \mathrm{v}((3)) = 0, \\ \mathrm{v}((1,2)) = \mathrm{v}((1,3)) = w, \qquad \mathrm{v}((2,3)) = 0, \\ \mathrm{v}((1,2,3)) = w. \end{cases}$$

The imputations are now defined by

$$\overrightarrow{\alpha} = \{\{\alpha_1, \alpha_2, \alpha_3\}\}$$

with

$$\text{(67:2:a)} \qquad \alpha_1 \geqq 0, \ \alpha_2 \geqq 0, \ \alpha_3 \geqq 0,$$

$$\text{(67:2:b)} \qquad \alpha_1 + \alpha_2 + \alpha_3 \leqq w.$$[1]

67.1.4. We now assume all these quantities to be integers—i.e. the given w and all permissible $\alpha_1, \alpha_2, \alpha_3$ of (67:2:a), (67:2:b).

We define domination as before, i.e. following 56.11.1.—which means that we repeat the definitions of 30.1.1. literally.

It is therefore necessary to determine the character of the sets $S \subseteq I = (1,2,3)$ with respect to their role in defining a domination. It is easy to show that the sets

$$S = (1,2), (1,3)$$

are certainly necessary, and all others certainly unnecessary.[2] Thus we

[1] Note that we are using (67:2:b) with $\leqq$ and not with $=$. This is the standpoint taken in the discussion of (66:2) in 66.3.2. In the terminology of (56:I:b) in 56.12., it amounts to using (56:10) and not (56:25). The reason for this procedure is that the former condition is the original one (cf., e.g., 56.8.2.), and the equivalence of the two, made use of in 56.12., fails in the setup to be used now.

It will be seen in the first remark of 67.2.3. that the $\leqq$ and the $=$ in (67:2:b) must produce different results, but that this divergence nevertheless fits into the general picture. Besides, the use of $=$ instead of $\leqq$ in (67:2:b) would lead to results which differ only in details of secondary importance from those which we are going to obtain.

[2] The conditions for certainly necessary and certainly unnecessary sets were derived in 31.1., and reconsidered in 59.3.2. Since our standpoint has changed again (cf. above, and particularly footnote 1), it would be necessary to reconsider these things once more. It seems simpler to take them up *de novo:*

Owing to (67:2:a) above, and the condition (30:3) in 30.1.1., every S with $\mathrm{v}(S) = 0$ is certainly unnecessary. This disposes of $S = (1), (2), (3), (2,3)$. Again (67:1), (67:2:a), (67:2:b) above give $\alpha_1 + \alpha_2 \leqq w = \mathrm{v}((1,2))$, $\alpha_1 + \alpha_3 \leqq w = \mathrm{v}((1,3))$, hence $S = (1,2), (1,3)$ are certainly necessary. And since (31:C) in 31.1.3. is clearly still valid, this renders $S = (1,2,3)$ certainly unnecessary.

can use the definition of domination with $S = (1,2), (1,3)$. I.e.:

$$\vec{\alpha} \succ \vec{\beta}$$

means that

(67:3:a) $$\alpha_1 > \beta_1$$

and

(67:3:b) $$\alpha_2 > \beta_2 \quad \text{or} \quad \alpha_3 > \beta_3.$$

Thus domination implies (67:3:a), and therefore it is clearly acyclical. (Cf. the corresponding discussion of 65.9.) Furthermore, the domain (67:2:a), (67:2:b) of the $\vec{\alpha}$ is finite, because the components $\alpha_1, \alpha_2, \alpha_3$ must be integers.[1]

Now we can apply (65:X) of 65.7.2.: There exists one and only one solution V_0 which is characterized by the formulae (65:2), (65:3), id.

67.2. The Solution and Its Interpretation

67.2.1. In order to apply the formulae (65:2), (65:3) of 65.7.2., we must determine the sets B_i, C_i defined at the beginning of 65.7.1. Let us do this for B_1, C_1.

B_1 is the set of those $\vec{\alpha}$ which cannot be dominated. To dominate $\vec{\alpha}$ we must increase α_1 and α_2 or α_3 without violating (67:2:a), (67:2:b) in 67.1.3. These increases are by 1 at least, while the other one of α_2, α_3 may be decreased as far as to 0. Hence $\vec{\alpha}$ can be dominated, if either

$$(\alpha_1 + 1) + (\alpha_2 + 1) \leqq w \quad \text{or} \quad (\alpha_1 + 1) + (\alpha_3 + 1) \leqq w.$$

So B_1 is defined by

(67:4) $$(\alpha_1 + 1) + (\alpha_2 + 1) > w, \qquad (\alpha_1 + 1) + (\alpha_3 + 1) > w.$$

By (67:2:a), (67:2:b) this implies $\alpha_3 < 2$, $\alpha_2 < 2$, i.e. $\alpha_2, \alpha_3 = 0, 1$. Now (67:4) gives, in conjunction with (67:2:a), (67:2:b), the following possibilities:

(67:A) $$\alpha_2 = \alpha_3 = 0, \qquad \alpha_1 = w, w - 1;$$

(67:B) $$\left\{\begin{matrix} \alpha_2 = 1, \alpha_3 = 0 \\ \text{or } \alpha_2 = 0, \alpha_3 = 1 \end{matrix}\right\}, \qquad \alpha_1 = w - 1;$$

(67:C) $$\alpha_2 = \alpha_3 = 1, \qquad \alpha_1 = w - 2.$$

C_1 is the set of those $\vec{\alpha}$ which are dominated by elements of B_1,—i.e. by those in (67:A)-(67:C). It is easy to verify that these are characterized

[1] This was, of course, not the case in the original continuum setup.

by

(67:D) $$\left\{\begin{matrix} \alpha_2 = 0 \\ \text{or} \\ \alpha_3 = 0 \end{matrix}\right\}, \qquad \alpha_1 \leqq w - 2.$$

67.2.2. Now it is better to deviate from the scheme of (65:2), (65:3) of 65.7.2.; that is, not to continue by determining $B_2, C_2, B_3, C_3, \cdots$, but to use an inductive process which is better suited to this particular case. This process goes as follows:

Consider the $\overrightarrow{\alpha}$ with

(67:E) $$\alpha_2 = 0 \qquad \text{or} \qquad \alpha_3 = 0.$$

They make up exactly (67:A), (67:B), (67:D). We know that among these V_0 contains precisely the (67:A), (67:B). The remaining $\overrightarrow{\alpha}$ are those with

(67:F) $$\alpha_2, \alpha_3 \geqq 1;$$

hence undominated by (67:A), (67:B). So we form V_0 by taking (67:A), (67:B) outside of (67:F), and repeating the process of finding a solution in (67:F).

Compare (67:F) with (67:2:a), (67:2:b) in 67.1.3. The only difference is that α_2, α_3 are increased by 1. Hence w must be treated as if it were $w - 2$. Thus V_0 now contains further

(67:G) $$\alpha_2 = \alpha_3 = 1, \qquad \alpha_1 = w - 2, w - 3;$$

(67:H) $$\left\{\begin{matrix} \alpha_2 = 2, \alpha_3 = 1 \\ \text{or} \\ \alpha_2 = 1, \alpha_3 = 2 \end{matrix}\right\}, \qquad \alpha_1 = w - 3;$$

and we must repeat the process of finding a solution in

(67:I) $$\alpha_2, \alpha_3 \geqq 2.$$

Repetition of this procedure assigns

(67:J) $$\alpha_2 = \alpha_3 = 2, \qquad \alpha_1 = w - 4, w - 5;$$

(67:K) $$\left\{\begin{matrix} \alpha_2 = 3, \alpha_3 = 2 \\ \text{or} \\ \alpha_2 = 2, \alpha_3 = 3 \end{matrix}\right\}, \qquad \alpha_1 = w - 5;$$

to V_0, and requires us to repeat the process of finding a solution in

(67:L) $$\alpha_2, \alpha_2 \geqq 3,$$

etc., etc.

Thus V_0 consists of (67:A), (67:B), (67:G), (67:H), (67:J), (67:K), $\cdots$ This set can be characterized as follows:

(67:M) $$\alpha_1 = 0, 1, \cdots w;$$

(67:N) $$\alpha_2 = \alpha_3 = \frac{w - \alpha_1}{2} \quad \text{if } w - \alpha_1 \text{ is even;}$$

(67:O) $$\left\{\begin{array}{c} \alpha_2 = \alpha_3 = \dfrac{w - 1 - \alpha_1}{2} \\ \text{or} \\ \alpha_2 = \dfrac{w + 1 - \alpha_1}{2},\ \alpha_3 = \dfrac{w - 1 - \alpha_1}{2} \\ \text{or} \\ \alpha_2 = \dfrac{w - 1 - \alpha_1}{2},\ \alpha_3 = \dfrac{w + 1 - \alpha_1}{2} \end{array}\right\} \quad \text{if } w - \alpha_1 \text{ is odd.}$$

67.2.3. The results (67:M)-(67:O) suggest these remarks:

First: The values of $\alpha_1 + \alpha_2 + \alpha_3$ in this solution are w and $w - 1$. Thus we cannot replace the $\leqq$ in (67:2:b) of 67.1.3. by $=$, the result stated in (56:I:b) of 56:12 is no longer true. The maximum social benefit is not necessarily obtained—and this appears as the direct consequence of the existence of an indivisible unit of utility.[1]

Second: This "discrete" utility scale converges toward our usual, continuous one, if $w \to \infty$. (Cf. the corresponding considerations concerning discrete and continuous "hands" in Poker, in 19.12.) The difference of $\alpha_1 + \alpha_2 + \alpha_3$ and w, mentioned above, is at most 1. So it becomes more and more insignificant as $w \to \infty$, i.e. this aspect of the situation tends to what it was in the continuous case.

Third: α_2, α_3 differ from each other by at most 1. So this difference too tends to insignificance as $w \to \infty$. I.e., when we approach the continuous case the solution tends to look like this:

(67:P) $$0 \leqq \alpha_1 \leqq w,$$

(67:Q) $$\alpha_2 = \alpha_3 = \frac{w - \alpha_1}{2}.$$

As pointed out in the first part of 67.1.3., this solution must be compared with (62:19) in 62.5.1., using the values $u = 0$, $v = w$. The two solutions are indeed similar, but our solution covers only one special case of (62:19): The monotonic decreasing functions of α_1 mentioned there coincide with each other and with $\frac{w - \alpha_1}{2}$.

Those functions describe, as discussed in 62.6.2., the rule of division upon which the two buyers agreed when forming their coalition (which is expressed by (62:19)). In the continuous case this rule was highly arbitrary. But now, in the discrete case, we find that it is completely determined—the two buyers must be treated exactly alike!

What is the meaning of this symmetry? Are the other distribution rules—i.e. the other choices of the functions in (62:19)—really impossible in the "discrete" case?

[1] Cf. this with footnote 3 on p. 513.

67.3. Generalization: Different Discrete Utility Scales

67.3.1. In order to answer the above questions, we shall try to destroy the symmetry (between the two buyers), but conserve the "discreteness."

This will be done by altering the setup of 67.1. in so far that we assign the indivisible unit of utility for the buyer 2 a value different from that one for the buyer 3. Specifically: Let us prescribe that the values of α_1, α_2 must be integers, while those of α_3 must be even integers. Apart from this, everything in 67.1. remains unaltered.

We now carry out the equivalent of the considerations of 67.2. Accordingly, we begin by determining the sets B_1, C_1 of 65.7.

B_1 is the set of those $\overrightarrow{\alpha}$ which cannot be dominated. To dominate $\overrightarrow{\alpha}$ we must increase α_1 and α_2 or α_3 without violating (67:2:a), (67:2:b) in 67.1.3. These increases are 1 (for α_1, α_2) or 2 (for α_3) at least, while the other one of α_2, α_3 may be decreased as far as to 0. Hence $\overrightarrow{\alpha}$ can be dominated if either $(\alpha_1 + 1) + (\alpha_2 + 1) \leqq w$ or $(\alpha_1 + 1) + (\alpha_3 + 2) \leqq w$. So B_1 is defined by

$$(67:5) \quad (\alpha_1 + 1) + (\alpha_2 + 1) > w, \qquad (\alpha_1 + 1) + (\alpha_3 + 2) > w.$$

By (67:2:a), (67:2:b) this implies $\alpha_3 < 2$, $\alpha_2 < 3$, i.e. $\alpha_2 = 0, 1, 2$, $\alpha_3 = 0$. Now (67:5) gives, in conjunction with (67:2:a), (67:2:b), the following possibilities:

$$(67:R) \qquad \alpha_2 = 0, \quad \alpha_3 = 0, \quad \alpha_1 = w, w - 1;$$
$$(67:S) \qquad \alpha_2 = 1, \quad \alpha_3 = 0, \quad \alpha_1 = w - 1, w - 2;$$
$$(67:T) \qquad \alpha_2 = 2, \quad \alpha_3 = 0, \quad \alpha_1 = w - 2.$$

C_1 is the set of those $\overrightarrow{\alpha}$, which are dominated by elements of B_1—i.e. by those in (67:R)-(67:T). It is easy to verify that these are characterized by

$$(67:U) \qquad \alpha_2 = 0, \quad \alpha_1 \leqq w - 2;$$
$$(67:V) \qquad \alpha_2 = 1, \quad \alpha_1 \leqq w - 3.$$

67.3.2. Now we repeat the variant of 67.2.2.: Instead of determining B_2, C_2, B_3, C_3, $\cdots$, we use a different inductive process.

Consider the $\overrightarrow{\alpha}$ with

$$(67:W) \qquad \alpha_2 = 0, 1.$$

They make up exactly (67:R), (67:S), (67:U), (67:V).[1] We know that among these, V_0 contains precisely the (67:R), (67:S). The remaining $\overrightarrow{\alpha}$ are those with

$$(67:X) \qquad \alpha_2 \geqq 2;$$

[1] Note that α_3 cannot be 1, since it must be even.

hence undominated by (67:R), (67:S). So we form V_0 by taking (67:R), (67:S) outside of (67:X), and repeating the process of finding a solution in (67:X).

Compare (67:X) with (67:2:a), (67:2:b) in 67.1.3. The only difference is that α_2 is increased by 2. Hence w must be treated as if it were $w - 2$.[1] Thus V_0 now contains further

(67:Y) $\qquad \alpha_2 = 2, \quad \alpha_3 = 0, \quad \alpha_1 = w - 2, w - 3;$

(67:Z) $\qquad \alpha_2 = 3, \quad \alpha_3 = 0, \quad \alpha_1 = w - 3, w - 4;$

and we must repeat the process of finding a solution in

(67:A′) $\qquad \alpha_2 \geqq 4.$

Repetition of this procedure assigns

(67:B′) $\qquad \alpha_2 = 4, \quad \alpha_3 = 0, \quad \alpha_1 = w - 4, w - 5;$

(67:C′) $\qquad \alpha_2 = 5, \quad \alpha_3 = 0, \quad \alpha_1 = w - 5, w - 6;$

to V_0 and requires us to repeat the process of finding a solution in

(67:D′) $\qquad \alpha_2 \geqq 6,$

etc., etc.

Thus V_0 consists of (67:R), (67:S), (67:Y), (67:Z), (67:B′), (67:C′), $\cdots$. This set can be characterized as follows:

(67:E′) $\qquad \alpha_1 = 0, 1, \cdots w;$

(67:F′) $\qquad \alpha_2 = w - \alpha_1, w - 1 - \alpha_1$ (excluding the second one when $\alpha_1 = w$);

(67:G′) $\qquad \alpha_3 = 0.$

67.3.3. The results (67:E′)-(67:G′) suggest these remarks:

First and second: Concerning the sum $\alpha_1 + \alpha_2 + \alpha_3$ and its relation to w we may repeat literally the corresponding parts of 67.2.3.

Third: Here things are altogether different from 67.2.3. We have identically $\alpha_3 = 0$. Approaching continuity, i.e. for $w \to \infty$ the solution tends to look like this:

(67:H′) $\qquad 0 \leqq \alpha_1 \leqq w,$

(67:I′) $\qquad \alpha_2 = w - \alpha_1,$

(67:J′) $\qquad \alpha_3 = 0.$

Repeating the comparison to (62:19) in 62.5.1., as made in the corresponding part of 67.2.3., we see that the situation is now this: The monotonic functions of (62:19), which describe the rule of division between the two allied buyers (cf. loc. cit.) are again completely determined—but this time we find (instead of the equal treatment they received in 67.2.3.) the entire advantage going to buyer 2!

[1] Note the difference between this and the corresponding step in 67.2.2. following (67:F) there.

We must now compare this result with the corresponding one in 67.2.3. and interpret the entire phenomenon.

67.4. Conclusions Concerning Bargaining

67.4. The conclusion from the results of 67.2.3., 67.3.3. is evident. In the former case the two buyers had exactly equal powers of discernment—i.e. equal units of utility—and the rule of distribution was found to treat them equally. In the latter case buyer 2 had a better power of discernment than buyer 3—i.e. 2's unit of utility was half of 3's—and in the rule of division the advantage went in its entirety to buyer 2. Clearly, if their abilities had been reversed, the result would have been also. We may also say: The advantage in the rule of division between allied buyers is equally divided if they have equally fine utility scales, and goes entirely to the one with the finer utility scale otherwise.[1]

This is true in the discrete case where each participant has a definite utility scale and the rule of division (i.e. the solution) is uniquely determined. In the continuous case the "fineness" of the utility scale is undefined and the rule of division can be chosen in many different ways, as we have seen.

So we observe for the first time how the ability of discernment of a player—specifically the fineness of his subjective utility scale—has a determining influence on his position in bargaining with an ally.[2] It is therefore to be expected that problems of this type can only be settled completely when the psychological conditions referred to are properly and systematically taken into account. The considerations of the last paragraph may be a first indication of the appropriate mathematical approach.

[1] It is possible to consider more subtle arrangements: We can assign to α_2 and to α_3 ranges of varying density. In this case we have still a unique solution for the same reasons as before. The correlation of α_2, α_3 when plotted in the α_2, α_3—plane will be a combination of the three types described above: Symmetric in α_2, α_3, i.e. parallel to the bisectrix of the two coordinate axes; parallel to the α_3-axis; parallel to the α_2-axis.

It is actually possible to bring about any desired combination of these elements by choosing the ranges of α_2 and α_3 appropriately. Any desired shape of the curve can be approximated arbitrarily well in this manner. In this way the original generality of the continuous case is recovered.

We do not propose to consider this matter, and various related ones, in detail here.

[2] This occurs, of course, only when the theory with continuous utilities allows several different rules of division between allies—which is plainly the case where bargaining plays a role.

APPENDIX. THE AXIOMATIC TREATMENT OF UTILITY

A.1. Formulation of the Problem

A.1.1. We will prove in this Appendix, that the axioms of utility enumerated in 3.6.1. make utility a number up to a linear transformation.[1] More precisely: We will prove that those axioms imply the existence of at least one mapping (actually, of course, of infinitely many) of the utilities on numbers in the sense of 3.5.1., with the properties (3:1:a), (3:1:b); and we will also prove that any two such mappings are linear transforms of each other, i.e. connected by a relation (3:6).

Before we undertake this analysis of the axioms (3:A)-(3:C) of 3.6.1., two further remarks concerning them may be useful in dispelling possible misunderstandings.

A.1.2. The first remark is this: These axioms, specifically the group (3:A), characterize the concept of *complete ordering*, based on the relations $>$, $<$. We do not axiomatize the relation $=$, but interpret it as *true identity*. The alternative procedure, to axiomatize $=$ also, would be mathematically perfectly sound, but so is our procedure too. The two procedures are trivially equivalent and represent only variants in taste. The practice in the relevant mathematical and logical literature is not uniform and we have therefore adhered to the simpler procedure.

The second remark is this: As pointed out at the beginning of 3.5.1., we are using the symbol $>$ both for the "natural" relation $u > v$ affecting utilities u, v and for the numerical relation $\rho > \sigma$ affecting numbers ρ, σ; also we are using the symbol $\alpha \cdots + (1 - \alpha) \cdots$ both for the "natural" operation $\alpha u + (1 - \alpha)v$ affecting utilities u, v and for the numerical operation $\alpha\rho + (1 - \alpha)\sigma$ affecting numbers ρ, σ (α is a number in either case). One might object that this practice can lead to misunderstandings and to confusion; however, it does not, provided that one keeps always in evidence whether the quantities involved are utilities (u, v, w) or numbers (α, β, γ, $\cdots$, ρ, σ). This identification of the designations for relations and operations in the two cases ("natural" and numerical) has a certain simplicity and facilitates keeping track of the "natural" and numerical pairs of analogs. For these reasons it is fairly generally accepted in similar situations in the mathematical literature, and we propose to make use of it.

A.1.3. The deductions which follow in A.2. are rather lengthy and may be somewhat tiring for the mathematically untrained reader. From the purely technical-mathematical viewpoint there is the further objection, that they cannot be considered deep—the ideas that underly the deductions are

[1] I.e. without fixing a zero or a unit of utility

quite simple, but unfortunately the technical execution had to be somewhat voluminous in order to be complete. Possibly a shorter exposition might be found later.

At any rate, we are now forced to use the esthetically not quite satisfactory mode of exposition which follows in A.2.

A.2. Derivation from the Axioms

A.2.1. We now proceed to carry out our deductions from the axioms (3:A)-(3:C) of 3.6.1. The deduction will be broken up into several successive steps and it will be carried out in this section and the four next ones. The final result will be stated in (A:V), (A:W).

(A:A) If $u < v$, then $\alpha < \beta$ implies

$$(1 - \alpha)u + \alpha v < (1 - \beta)u + \beta v.$$

Proof: Clearly $\alpha = \gamma\beta$ with $0 < \gamma < 1$. By (3:B:a) (applied to u, v, $1 - \beta$ in place of u, v, α) $u < (1 - \beta)u + \beta v$, and hence by (3:B:b) (applied to $(1 - \beta)u + \beta v$, u, γ in place of u, v, α)

$$(1 - \beta)u + \beta v > \gamma((1 - \beta)u + \beta v) + (1 - \gamma)u.$$

By (3:C:a) this can be written

$$(1 - \beta)u + \beta v > \gamma(\beta v + (1 - \beta)u) + (1 - \gamma)u.$$

Now by (3:C:b) (applied to v, u, γ, β, $\alpha = \gamma\beta$ in place of u, v, α, β, $\gamma = \alpha\beta$) the right hand side is $\alpha v + (1 - \alpha)u$, hence by (3:C:a) $(1 - \alpha)u + \alpha v$. Thus $(1 - \alpha)u + \alpha v < (1 - \beta)u + \beta v$, as desired.

(A:B) Given two fixed u_0, v_0 with $u_0 < v_0$, consider the mapping

$$\alpha \rightarrow w = (1 - \alpha)u_0 + \alpha v_0.$$

This is a one-to-one and monotone mapping of the interval $0 < \alpha < 1$ on part of the interval $u_0 < w < v_0$.[1]

Proof: The mapping is on part of the interval $u_0 < w < v_0$: $u_0 < w$ coincides with (3:B:a) (applied to u_0, v_0, $1 - \alpha$ in place of u, v, α), $w < v_0$ coincides with (3:B:b) (applied to v_0, u_0, α in place of u, v, α).

One-to-one character: Follows from the monotony, which we establish next.

Monotone character: Coincides with (A:A).

(A:C) The mapping of (A:B) actually maps the α of $0 < \alpha < 1$ on all the w of $u_0 < w < v_0$.

Proof: Assume that this were not so, i.e. that some w_0 with $u_0 < w_0 < v_0$ were omitted. Then for all α in $0 < \alpha < 1$ $(1 - \alpha)u_0 + \alpha v_0 \neq w_0$, i.e. $(1 - \alpha)u_0 + \alpha v_0 \lessgtr w_0$. According to whether we have $<$ or $>$, let α

[1] It will appear in (A:C), that this part is actually the whole interval $u_0 < w < v_0$.

belong to class I or II. Thus the classes I, II, which are clearly mutually exclusive, exhaust together the interval $0 < \alpha < 1$. Now we observe:

First: Class I is not empty. This is immediate by (3:B:c) (applied to $u_0, w_0, v_0, 1 - \alpha$ in place of u, w, v, α).

Second: Class II is not empty. This is immediate by (3:B:d) (applied to v_0, w_0, u_0, α in place of u, w, v, α).

Third: If α is in I and β is II, then $\alpha < \beta$. Indeed, since I and II are disjunct, necessarily $\alpha \neq \beta$. Hence the only alternative would be $\alpha > \beta$. But then the monotony of the mapping of (A:B) would imply, that since α is in I, β too must be in I—but β is in II. Hence only $\alpha < \beta$ is possible.

Considering these three properties of I, II, there must exist an α_0 with $0 < \alpha_0 < 1$ which separates them, i.e. such that all α of I have $\alpha \leqq \alpha_0$, and all α of II have $\alpha \geqq \alpha_0$.[1]

Now α_0 itself must belong to I or to II. We distinguish accordingly:

First: α_0 in I. Then $(1 - \alpha_0)u_0 + \alpha_0 v_0 < w_0$. Also $w_0 < v_0$. Applying (3:B:c) (with $(1 - \alpha_0)u_0 + \alpha_0 v_0, w_0, v_0, \gamma$ in place of u, w, v, γ) we obtain a γ with $0 < \gamma < 1$ and $\gamma((1- \alpha_0)u_0 + \alpha_0 v_0) + (1 - \gamma)v_0 < w_0$, i.e. by (3:C:b) (with $u_0, v_0, \gamma, 1 - \alpha_0, 1 - \alpha = \gamma(1 - \alpha_0)$ in place of $u, v, \alpha, \beta, \gamma = \alpha\beta$) $(1 - \alpha)u_0 + \alpha v_0 < w_0$. Hence $\alpha = 1 - \gamma(1 - \alpha_0)$ belongs to I. However $\alpha > 1 - (1 - \alpha_0) = \alpha_0$, although we should have $\alpha \leqq \alpha_0$.

Second: α_0 in II. Then $(1 - \alpha_0)u_0 + \alpha_0 v_0 > w_0$. Also $u_0 < w_0$. Applying (3:B:d) (with $(1 - \alpha_0)u_0 + \alpha_0 v_0, w_0, u_0, \gamma$ in place of u, w, v, α) we obtain a γ with $0 < \gamma < 1$ and $\gamma((1 - \alpha_0)u_0 + \alpha_0 v_0) + (1 - \gamma)u_0 > w_0$, i.e. by (3:C:a) $\gamma(\alpha_0 v_0 + (1 - \alpha_0)u_0) + (1 - \gamma)u_0 > w_0$, hence by (3:C:b) (with $v_0, u_0, \gamma, \alpha_0, \alpha = \gamma\alpha_0$ in place of $u, v, \alpha, \beta, \gamma = \alpha\beta$) $\alpha v_0 + (1 - \alpha)u_0 > w_0$, i.e. by (3:C:a) $(1 - \alpha)u_0 + \alpha v_0 > w_0$. Hence $\alpha = \gamma\alpha_0$ belongs to II. However $\alpha < \alpha_0$, although we should have $\alpha \geqq \alpha_0$.

Thus we obtain a contradiction in each case. Therefore the original assumption is impossible, and the desired property is established.

A.2.2. It is worth while to stop for a moment at this point. (A:B) and (A:C) have effected a one-to-one mapping of the utility interval $u_0 < w < v_0$ (u_0, v_0 fixed with $u_0 < v_0$, otherwise arbitrary!) on the numerical interval $0 < \alpha < 1$. This is clearly the first step towards establishing a numerical representation of utilities. However, the result is still significantly incomplete in several respects. These seem to be the major limitations:

First: The numerical representation was obtained for a utility interval $u_0 < w < v_0$ only, not for all utilities w simultaneously. Nor is it clear, how the mappings which go with different pairs u_0, v_0 fit together.

Second: The numerical representation of (A:B), (A:C) has not yet been correlated with our requirements (3:1:a), (3:1:b). Now (3:1:a) is clearly

[1] This is intuitively fairly plausible. It is, furthermore, a perfectly rigorous inference. Indeed, it coincides with one of the classical theorems effecting the introduction of irrational numbers, the theorem concerning the Dedekind cut. Details can be found in texts on real function theory or on the foundations of analysis. Cf. e.g. *C. Carathéodory* loc. cit. footnote 1 on p. 343. Cf. there p. 11, Axiom VII. Our class I should be substituted for the set {a} mentioned there. The set {A} mentioned there then contains our class II.

satisfied: It is just another way of expressing the monotony that is secured by (A:B). However the validity of (3:1:b) remains to be established.

We will fulfill all these requirements jointly. The procedure will primarily follow a course suggested by the first remark, but in the process the requirements of the second remark and the appropriate uniqueness results will also be established.

We begin by proving a group of lemmata which is more in the spirit of the second remark and of the uniqueness inquiry; however it is basic in order to make progress towards the objectives of the first remark too.

(A:D) Let u_0, v_0 be as above: u_0, v_0 fixed, $u_0 < v_0$. For all w in the interval $u_0 < w < v_0$ define the numerical function $f(w) = f_{u_0,v_0}(w)$ as follows:

(i) $f(u_0) = 0$.

(ii) $f(v_0) = 1$.

(iii) $f(w)$ for $w \neq u_0, v_0$, i.e. for $u_0 < w < v_0$, is the number α in $0 < \alpha < 1$ which corresponds to w in the sense of (A:B), (A:C).

(A:E) The mapping

$$w \rightarrow f(w)$$

has the following properties:

(i′) It is monotone.

(ii′) For $0 < \beta < 1$ and $w \neq u_0$

$$f((1-\beta)u_0 + \beta w) = \qquad \beta f(w).$$

(iii′) For $0 < \beta < 1$ and $w \neq v_0$

$$f((1-\beta)v_0 + \beta w) = 1 - \beta + \beta f(w).$$

(A:F) A mapping of all w with $u_0 \leqq w \leqq v_0$ on any set of numbers, which possesses the properties (i), (ii) and either (ii′) or (iii′), is identical with the mapping of (A:D)

Proof: (A:D) is a definition; we must prove (A:E) and (A:F).

Ad (A:E): Ad (i′): For $u_0 < w < v_0$ the mapping is monotone by (A:B). All w of this interval are mapped on numbers > 0, < 1, i.e. on numbers $>$ than the map of u_0 and $<$ than the map of v_0. Hence we have monotony throughout $u_0 \leqq w \leqq v_0$.

Ad (ii′): For $w = v_0$: The statement is $f((1-\beta)u_0 + \beta v_0) = \beta$, and this coincides with the definition in (A:B) (with β in place of α).

For $w \neq v_0$, i.e. $u_0 < w < v_0$: Put $f(w) = \alpha$, i.e. by (A:B)

$$w = (1-\alpha)u_0 + \alpha v_0.$$

Then by (3:C:b) (with v_0, u_0, β, α in place of u, v, α, β, and using (3:C:a)) $(1-\beta)u_0 + \beta w = (1-\beta)u_0 + \beta((1-\alpha)u_0 + \alpha v_0) = (1-\beta\alpha)u_0 + \beta\alpha v_0$. Hence by (A:B) $f((1-\beta)u_0 + \beta w) = \beta\alpha = \beta f(w)$, as desired.

Ad (iii′): For $w = u_0$: The statement is $f((1-\beta)v_0 + \beta u_0) = 1 - \beta$, and this coincides with the definition in (A:B) (with $1-\beta$ in place of α and using (3:C:a)).

For $w \neq u_0$, i.e. $u_0 < w < v_0$: Put $f(w) = \alpha$, i.e. by (A:B)

$$w = (1 - \alpha)u_0 + \alpha v_0.$$

Then by (3:C:b) (with $u_0, v_0, \beta, 1 - \alpha$ in place of u, v, α, β, and using (3:C:a))

$$(1 - \beta)v_0 + \beta w = (1 - \beta)v_0 + \beta((1 - \alpha)u_0 + \alpha v_0) = \beta(1 - \alpha)u_0 + (1 - \beta(1 - \alpha))v_0,$$

hence by (A:B)

$$f((1 - \beta)v_0 + \beta w) = 1 - \beta(1 - \alpha) = 1 - \beta + \beta\alpha = 1 - \beta + \beta f(w),$$

as desired.

Ad (A:F): Consider a mapping

(A:1) $$w \to f_1(w)$$

with (i), (ii) and either (ii′) or (iii′). The mapping

(A:2) $$w \to f(w)$$

is a one-to-one mapping of $u_0 \leqq w \leqq v_0$ on $0 \leqq \alpha \leqq 1$, hence it can be inverted:

(A:3) $$\alpha \to \psi(\alpha).$$

Now combine (A:1) with (A:3), i.e. with the inverse of (A:2):

(A:4) $$\alpha \to f_1(\psi(\alpha)) = \varphi(\alpha).$$

Since both (A:1) and (A:2) fulfill (i), (ii), we obtain for (A:4)

(A:5) $$\varphi(0) = 0, \qquad \varphi(1) = 1.$$

If (A:1) fulfills (ii′) or (iii′), then, as (A:2) fulfills both (ii′) and (iii′), we obtain for (A:4)

(A:6) $$\varphi(\beta\alpha) = \beta\varphi(\alpha),$$

or

(A:7) $$\varphi(1 - \beta + \beta\alpha) = 1 - \beta + \beta\varphi(\alpha).$$

Now putting $\alpha = 1$ in (A:6) and using (A:5) gives

(A:8) $$\varphi(\beta) = \beta,$$

and putting $\alpha = 0$ in (A:7) and using (A:5) gives $\varphi(1 - \beta) = 1 - \beta$. Replacing β by $1 - \beta$ gives again (A:8).

Thus (A:8) is valid at any rate. (ii′), (iii′) restrict it to the β with $0 < \beta < 1$. However (A:5) extends it to $\beta = 0, 1$ too, i.e. to all β with $0 \leqq \beta \leqq 1$. Considering the definition of $\varphi(\alpha)$ by (A:3), (A:4), the general validity of (A:8) expresses the identity of (A:1) and (A:2), which is precisely what we wanted to prove.

(A:G) Let u_0, v_0 be as above: u_0, v_0 fixed, $u_0 < v_0$. Let also two fixed α_0, β_0 with $\alpha_0 < \beta_0$ be given. For all w in the interval $u_0 \leqq w \leqq v_0$ define the numerical function $g(w) = g_{u_0,v_0}^{\alpha_0,\beta_0}(w)$ as follows:

$$g(w) = (\beta_0 - \alpha_0)f(w) + \alpha_0,$$

($f(w) = f_{u_0,v_0}(w)$ according to (A:D)).

We note:

(i) $g(u_0) = \alpha_0$,

(ii) $g(v_0) = \beta_0$.

(A:H) This mapping

$$w \to g(w)$$

has the following properties:

(i') It is monotone.

(ii') For $0 < \beta < 1$ and $w \neq u_0$

$$g((1 - \beta)u_0 + \beta w) = (1 - \beta)\alpha_0 + \beta g(w).$$

(iii') For $0 < \beta < 1$ and $w \neq v_0$

$$g((1 - \beta)v_0 + \beta w) = (1 - \beta)\beta_0 + \beta g(w).$$

(A:I) A mapping of all w with $u_0 \leqq w \leqq v_0$ on any set of numbers which possesses the properties (i), (ii) and either (ii') or (iii'), is identical with the mapping of (A:G).

Proof: Using the correspondence between functions

$$g_1(w) = (\beta_0 - \alpha_0)f_1(w) + \alpha_0,$$

or equivalently

$$f_1(w) = \frac{g_1(w) - \alpha_0}{\beta_0 - \alpha_0}$$

(for $f_1(w)$, $g_1(w)$, and also for $f(w)$, $g(w)$), the statements of (A:G)-(A:I) go over into the statements of (A:D)-(A:F). Hence (A:G)-(A:I) follow from (A:D)-(A:F).

(A:J) Assuming (i), (ii) in (A:G), the equation

$$g((1 - \beta)u + \beta v) = (1 - \beta)g(u) + \beta g(v)$$

$(u_0 \leqq u < v \leqq v_0)$ with $u = u_0$, $v \neq u_0$ is equivalent to (ii') in (A:I), and with $u \neq v_0$, $v = v_0$ it is equivalent to (iii') in (A:I).

Proof: Ad (ii'): Put u_0, w, β in place of u, v, β.

Ad (iii'): Put w, v_0, $1 - \beta$ in place of u, v, β.

A.2.3. In (A:G)-(A:J) the mapping of a utility interval $u_0 \leqq w \leqq v_0$ on a numerical interval $\alpha_0 \leqq \alpha \leqq \beta_0$ has been given its technically adequate form, with the necessary uniqueness properties. We can now begin to fit the various mappings

$$w \to g(w) = g^{\alpha_0,\beta_0}_{u_0,v_0}(w)$$

together.

(A:K) Consider $g^{\alpha_0,\beta_0}_{u_0,v_0}$ and a w_0 with $u_0 \leqq w_0 \leqq v_0$. Put

$$\gamma_0 = g^{\alpha_0,\beta_0}_{u_0,v_0}(w_0).$$

Then $g^{\alpha_0,\beta_0}_{u_0,v_0}(w)$ coincides with $g^{\alpha_0,\gamma_0}_{u_0,w_0}(w)$ in the latter's domain $u_0 \leqq w \leqq w_0$ (if $w_0 \neq u_0$, i.e. $u_0 < w_0$), and $g^{\alpha_0,\beta_0}_{u_0,v_0}(w)$ coincides with $g^{\gamma_0,\beta_0}_{w_0,v_0}(w)$ in the latter's domain $w_0 \leqq w \leqq v_0$ (if $w_0 \neq v_0$, i.e. $w_0 < v_0$).

Proof: Ad $g^{\alpha_0,\gamma_0}_{u_0,w_0}(w)$: $g^{\alpha_0,\beta_0}_{u_0,v_0}(w)$ possesses the properties (i), (ii') (of (A:G), (A:H)) for α_0, γ_0, u_0, w_0, because they coincide with those for α_0, β_0, u_0, v_0 (since they involve only the lower end α_0, u_0). It also possesses (ii) (of (A:G)) for α_0, γ_0, u_0, w_0, because $g^{\alpha_0,\beta_0}_{u_0,v_0}(w_0) = \gamma_0$. Hence it follows from (A:I) that $g^{\alpha_0,\beta_0}_{u_0,v_0}$ fulfills within $u_0 \leqq w \leqq w_0$ a unique characterization of $g^{\alpha_0,\gamma_0}_{u_0,w_0}$.

Ad $g^{\gamma_0,\beta_0}_{w_0,v_0}$: $g^{\alpha_0,\beta_0}_{u_0,v_0}$ possesses the properties (ii), (iii') (of (A:G), (A:H)) for γ_0, β_0, w_0, v_0 because they coincide with those for α_0, β_0, u_0, v_0 (since they involve only the upper end β_0, v_0). It also possesses (i) (of (A:G)) for γ_0, β_0, w_0, v_0, because $g^{\alpha_0,\beta_0}_{u_0,v_0}(w_0) = \gamma_0$. Hence it follows from (A:I) that $g^{\alpha_0,\beta_0}_{u_0,v_0}$ fulfills within $w_0 \leqq w \leqq v_0$ a unique characterization of $g^{\gamma_0,\beta_0}_{w_0,v_0}$.

(A:L) Consider a $g^{\alpha_0,\beta_0}_{u_0,v_0}$ and two u_1, v_1 with $u_0 \leqq u_1 < v_1 \leqq v_0$. Put $\alpha_1 = g^{\alpha_0,\beta_0}_{u_0,v_0}(u_1)$, $\beta_1 = g^{\alpha_0,\beta_0}_{u_0,v_0}(v_1)$. Then $g^{\alpha_0,\beta_0}_{u_0,v_0}(w)$ coincides with $g^{\alpha_1,\beta_1}_{u_1,v_1}(w)$ in the latter's domain $u_1 \leqq w \leqq v_1$.

Proof: Apply first (A:K) to $g^{\alpha_0,\beta_0}_{u_0,v_0}$ and $g^{\alpha_0,\beta_1}_{u_0,v_1}$ (i.e. with u_0, v_0, α_0, β_0, v_1, β_1 in place of u_0, v_0, α_0, β_0, w_0, γ_0; note that $\beta_1 = g^{\alpha_0,\beta_0}_{u_0,v_0}(v_1)$)—this shows that $g^{\alpha_0,\beta_0}_{u_0,v_0}(w)$ coincides with $g^{\alpha_0,\beta_1}_{u_0,v_1}(w)$ in the latter's domain $u_0 \leqq w \leqq v_1$. Apply next (A:K) to $g^{\alpha_0,\beta_1}_{u_0,v_1}$ and $g^{\alpha_1,\beta_1}_{u_1,v_1}$ (i.e. with u_0, v_1, α_0, β_1, u_1, α_1 in place of u_0, v_0, α_0, β_0, w_0, γ_0; note that $\alpha_1 = g^{\alpha_0,\beta_0}_{u_0,v_0}(u_1) = g^{\alpha_0,\beta_1}_{u_0,v_1}(u_1)$)—this shows that $g^{\alpha_0,\beta_1}_{u_0,v_1}(w)$, and hence also $g^{\alpha_0,\beta_0}_{u_0,v_0}(w)$, coincides with $g^{\alpha_1,\beta_1}_{u_1,v_1}(w)$ in the latter's domain $u_1 \leqq w \leqq v_1$.

(A:L) has to be combined with a second line of reasoning. At this point we also assume that two u^*, v^* with $u^* < v^*$ have been chosen; we will consider them as fixed from now on until we get to (A:V) and (A:W).

We now prove:

(A:M) If $u_0 \leqq u^* < v^* \leqq v_0$, then there exists one and only one $g^{\alpha_0,\beta_0}_{u_0,v_0}(w)$ such that

(i) $g^{\alpha_0,\beta_0}_{u_0,v_0}(u^*) = 0$,

(ii) $g^{\alpha_0,\beta_0}_{u_0,v_0}(v^*) = 1$.

We denote this $g^{\alpha_0,\beta_0}_{u_0,v_0}(w)$ by $h_{u_0,v_0}(w)$.

Proof: Form the $f(w) = f_{u_0,v_0}(w)$ of (A:D). As $u^* < v^*$, so $f(u^*) < f(v^*)$. For variable α_0, β_0 (A:G) gives $g^{\alpha_0,\beta_0}_{u_0,v_0}(w) = (\beta_0 - \alpha_0)f(w) + \alpha_0$. Hence the above (i), (ii) mean that $(\beta_0 - \alpha_0)f(u^*) + \alpha_0 = 0$, $(\beta_0 - \alpha_0)f(v^*) + \alpha_0 = 1$, and these two equations determine α_0, β_0 uniquely.[1] Hence the desired

[1] $\alpha_0 = -\dfrac{f(u^*)}{f(v^*) - f(u^*)}$, $\beta_0 = \dfrac{1 - f(u^*)}{f(v^*) - f(u^*)}$.

$g^{\alpha_0,\beta_0}_{u_0,v_0}(w)$ exists and is unique.

(A:N) If $u_0 \leqq u_1 \leqq u^* < v^* \leqq v_1 \leqq u_1$, then $h_{u_0,v_0}(w)$ coincides with $h_{u_1,v_1}(w)$ in the latter's domain $u_1 \leqq w \leqq v_1$.

Proof: Put $\alpha_1 = h_{u_0,v_0}(u_1)$, $\beta_1 = h_{u_0,v_0}(v_1)$. Then, by (A:L), $h_{u_0,v_0}(w)$ coincides with $g^{\alpha_1,\beta_1}_{u_1,v_1}(w)$ in the latter's domain $u_1 \leqq w \leqq v_1$. Applying this to $w = u^*$ and to $w = v^*$ gives $g^{\alpha_1,\beta_1}_{u_1,v_1}(u^*) = h_{u_0,v_0}(u^*) = 0$ and $g^{\alpha_1,\beta_1}_{u_1,v_1}(v^*) = h_{u_0,v_0}(v^*) = 1$. Hence by (A:M), $g^{\alpha_1,\beta_1}_{u_1,v_1}(w) = h_{u_1,v_1}(w)$. Consequently $h_{u_0,v_0}(w)$ coincides with $h_{u_1,v_1}(w)$ in the latter's domain $u_1 \leqq w \leqq v_1$.

We can now establish the decisive fact: The functions $h_{u_0,v_0}(w)$ all fit together to one function. Specifically:

(A:O) Given any w, it is possible to choose u_0, v_0 so that $u_0 \leqq u^* < v^* \leqq v_0$ and $u_0 \leqq w \leqq v_0$. For all such choices of u_0, v_0, $h_{u_0,v_0}(w)$ has the same value. I.e. $h_{u_0,v_0}(w)$ depends on w only. We denote it therefore by $h(w)$.

Proof: Existence of u_0, v_0: $u_0 = \text{Min}\ (u^*, w)$ and $v_0 = \text{Max}\ (v^*, w)$ obviously possess the desired properties.

$h_{u_0,v_0}(w)$ depends on w only: Choose two such pairs u_0, v_0 and u'_0, v'_0: $u_0 \leqq u^* < v^* \leqq v_0$, $u_0 \leqq w \leqq v_0$ and $u'_0 \leqq u^* < v^* \leqq v'_0$, $u'_0 \leqq w \leqq v'$. Put $u_1 = \text{Max}\ (u_0, u'_0)$, $v_1 = \text{Min}\ (v_0, v'_0)$. Then $u_0 \leqq u_1 \leqq u^* < v^* \leqq v_1 \leqq v_0$, $u_1 \leqq w \leqq v_1$, and $u'_0 \leqq u_1 \leqq u^* < v^* \leqq v_1 \leqq v'_0$, $u_1 \leqq w \leqq v_1$. Now two applications of (A:N) (first with u_0, v_0, u_1, v_1, w, then with u'_0, v'_0, u_1, v_1, w) give $h_{u_0,v_0}(w) = h_{u_1,v_1}(w)$ and $h_{u'_0,v'_0}(w) = h_{u_1,v_1}(w)$. Hence

$$h_{u_0,v_0}(w) = h_{u'_0,v'_0}(w)$$

as desired.

A.2.4. The function $h(w)$ of (A:O) is defined for all utilities and it has numerical values. We can now show with little trouble that it possesses all the properties that we need.

This is most easily done with the help of two auxiliary lemmata.

(A:P) Given any two u, v with $u < v$, there exist two u_0, v_0 with $u_0 \leqq u^* < v^* \leqq v_0$, $u_0 \leqq u < v \leqq v_0$.

Proof: Put $u_0 = \text{Min}\ (u^*, u)$, $v_0 = \text{Max}\ (v^*, v)$.

(A:Q) Given any two u, v with $u < v$, put $h(u) = \alpha$, $h(v) = \beta$. Then $\alpha < \beta$, and $h(w)$ coincides with $g^{\alpha,\beta}_{u,v}(w)$ in the latter's domain $u \leqq w \leqq v$.

Proof: Choose u_0, v_0 as indicated in (A:P). By (A:M) $h_{u_0,v_0}(w)$ is a $g^{\alpha_0,\beta_0}_{u_0,v_0}(w)$ with two suitable α_0, β_0. By (A:O) $h(w)$ coincides with $h_{u_0,v_0}(w)$, i.e. with $g^{\alpha_0,\beta_0}_{u_0,v_0}(w)$, in the latter's domain $u_0 \leqq w \leqq v_0$. Applying this to

$w = u$ and to $w = v$ gives $g^{\alpha_0,\beta_0}_{u_0,v_0}(u) = h(u) = \alpha$ and $g^{\alpha_0,\beta_0}_{u_0,v_0}(v) = h(v) = \beta$. Since $g^{\alpha_0,\beta_0}_{u_0,v_0}(w)$ is monotone, this implies $\alpha < \beta$. Next by (A:L) (with $u_0, v_0, \alpha_0, \beta_0, u, v, \alpha, \beta$ in place of $u_0, v_0, \alpha_0, \beta_0, u_1, v_1, \alpha_1, \beta_1$) $g^{\alpha_0,\beta_0}_{u_0,v_0}(w)$ coincides with $g^{\alpha,\beta}_{u,v}(w)$ in the latter's domain $u \leqq w \leqq v$. Consequently the same is true for $h(w)$.

After these preparations we establish the relevant properties of $h(w)$

(A:R) The mapping

$$w \rightarrow h(w)$$

of all w on a set of numbers has the following properties:
(i) $h(u^*) = 0$.
(ii) $h(v^*) = 1$.
(iii) $h(w)$ is monotone,
(iv) For $0 < \gamma < 1$ and $u < v$

$$h((1 - \gamma)u + \gamma v) = (1 - \gamma)h(u) + \gamma h(v)$$

(A:S) A mapping of all w on any set of numbers, which possesses the properties (i), (ii) and (iv) is identical with the mapping of (A:R).

Proof: Ad (A:R): Ad (i), (ii): Immediate by (A:O) and (A:M).

Ad (iii): Contained in (A:Q).

Ad (iv): Choose u, v according to (A:P) and then α, β and $g^{\alpha,\beta}_{u,v}(w)$ according to (A:Q). Now by (A:H), (ii′) (with u, v, v, γ in place of u_0, v_0, w, γ) $g^{\alpha,\beta}_{u,v}((1 - \gamma)u + \gamma v) = (1 - \gamma)g^{\alpha,\beta}_{u,v}(u) + \gamma g^{\alpha,\beta}_{u,v}(v)$. Hence by (A:Q)

$$h((1 - \gamma)u + \gamma v) = (1 - \gamma)h(u) + \gamma h(v)$$

as desired.

Ad (A:S): Consider a mapping

$$w \rightarrow h_1(w)$$

of all utilities w on numbers, which fulfills (i), (ii) and (iv). Choose two u_0, v_0 with $u_0 \leqq u^* < v^* \leqq v_0$, and put $\alpha_0 = h_1(u^*)$, $\beta_0 = h_1(v^*)$. Then, by (A:I), $h_1(w)$ coincides with $g^{\alpha_0,\beta_0}_{u_0,v_0}(w)$ in the latter's domain $u_0 \leqq w \leqq v_0$. Putting $w = u^*$ and $w = v^*$ we get $g^{\alpha_0,\beta_0}_{u_0,v_0}(u^*) = h_1(u^*) = 0$, $g^{\alpha_0,\beta_0}_{u_0,v_0}(v^*) = h_1(v^*) = 1$. Hence by (A:M) $g^{\alpha_0,\beta_0}_{u_0,v_0}$ is h_{u_0,v_0}. Thus $h_1(w)$ coincides with $h_{u_0,v_0}(w)$, i.e. with $h(w)$, in $u_0 \leqq w \leqq v_0$. By (A:O) this means that $h_1(w)$ and $h(w)$ are altogether identical.

A.2.5. (A:R), (A:S) give a mapping of all utilities on numbers, which possesses plausible properties and is uniquely characterized by them, and therefore we might let the matter rest there. However, we are not yet quite satisfied, for the following reasons: The characterization in (A:R) does not coincide with that one by (3:1:a), (3:1:b)—(A:R) goes less far in (iv) (this is asserted in (3:1:b) for all u, v, in (iv) only for those with $u < v$); and it introduces an arbitrary normalization in (i), (ii) (by means of the arbitrary u^*, v^*). In what follows, we will eliminate these maladjustments. This will prove fairly easy.

We first extend (iv) in (A:R).

(A:T) Always $(1 - \gamma)u + \gamma u = u$.

Proof: For $u \lessgtr (1 - \gamma)u + \gamma u$ say that γ belongs to class I (upper case) or II (lower case). If γ is in class I or II and if $0 < \beta < 1$, then

$$u \lessgtr (1 - \beta)u + \beta((1 - \gamma)u + \gamma u) \lessgtr (1 - \gamma)u + \gamma u$$

by (3:B:a) and (3:B:b). (For γ in class I or II, respectively: First, u, $(1 - \gamma)u + \gamma u$, $1 - \beta$ in place of u, v, α in (3:B:a) or (3:B:b). Second, $(1 - \gamma)u + \gamma u$, u, β in place of u, v, α in (3:B:b) or (3:B:a).) By(3:C:a) and (3:C:b) (with u, u, β, γ in place of u, v, α, β)

$$(1 - \beta)u + \beta((1 - \gamma)u + \gamma u) = (1 - \beta\gamma)u + \beta\gamma u.$$

Hence $u \lessgtr (1 - \beta\gamma)u + \beta\gamma u \lessgtr (1 - \beta)u + \beta u$. Put $\delta = \beta\gamma$. Since β is free in $0 < \beta < 1$, therefore δ is free in $0 < \delta < \gamma$. Assuming $0 < \gamma < 1$, $0 < \delta < 1$, we have therefore:

(A:9) If γ is in class I or II, then every $\delta < \gamma$ is in the same class I or II.

(A:10) Under the conditions of (A:9)

$$(1 - \delta)u + \delta u \lessgtr (1 - \gamma)u + \gamma u,$$

respectively.

The expression $(1 - \gamma)u + \gamma u$ is unchanged if we replace γ by $1 - \gamma$. As $1 - \gamma < 1 - \delta$ is equivalent to $\gamma > \delta$, we can put $1 - \gamma$, $1 - \delta$ in place of γ, δ in (A:9). Then (A:9) and (A:10) become this:

(A:11) If γ is in class I or II, then every $\delta > \gamma$ is the same class I or II.

(A:12) Under the condition of (A:11)

$$(1 - \delta)u + \delta u \lessgtr (1 - \gamma)u + \gamma u,$$

respectively.

Now (A:9) and (A:11) show, that if γ is class I or II, then every δ($< \gamma$ or $= \gamma$ or $> \delta$) is in the same class I or II. I.e. if either class I or II is not empty, then it contains all δ with $0 < \delta < 1$. Assume this to be the case (for class I or II), and consider two γ, δ with $\gamma < \delta$. Then by (A:10) $(1 - \delta)u + \delta u \lessgtr (1 - \gamma)u + \gamma u$, and by (A:12) (with δ, γ in place of γ, δ)

$(1 - \delta)u + \delta u \gtrless (1 - \gamma)u + \gamma u$. Hence at any rate both $<$ and $>$ hold in $(1 - \delta)u + \delta u \lesseqgtr (1 - \gamma)u + \gamma u$. This is a contradiction. Therefore both classes I and II must be empty.

Consequently never $u \lessgtr (1 - \gamma)u + \gamma u$, i.e. always $(1 - \gamma)u + \gamma u = u$, as desired.

(A:U) Always

$$h((1 - \gamma)u + \gamma v) = (1 - \gamma)h(u) + \gamma h(v)$$

$(0 < \gamma < 1$, any $u, v)$.

Proof: For $u < v$ this is (A:R), (iv). For $u > v$ it obtains from the former by putting $v, u, 1 - \gamma$ in place of u, v, γ. For $u = v$ it follows from (A:T).

We can now prove the existence and uniqueness theorem in the desired form, i.e. corresponding to (3:1:a) and (3:1:b). At this point we also drop the assumed fixed choice of u^*, v^*, which was introduced before (A:M).

(A:V) There exists a mapping

$$w \rightarrow \mathrm{v}(w)$$

of all w on a set of numbers possessing the following properties:
(i) Monotony.
(ii) For $0 < \gamma < 1$ and any u, v

$$\mathrm{v}((1 - \gamma)\, u + \gamma v) = (1 - \gamma)\mathrm{v}(u) + \gamma \mathrm{v}(v).$$

(A:W) For any two mappings $\mathrm{v}(w)$ and $\mathrm{v}'(w)$ possessing the properties (i), (ii), we have

$$\mathrm{v}'(w) = \omega_0 \mathrm{v}(w) + \omega_1,$$

with two suitable but fixed ω_0, ω_1 and $\omega_0 > 0$.

Proof: Let u^*, v^* be two different utilities,[1] $u^* \lessgtr v^*$.

If $u^* > v^*$, then interchange u^* and v^*. Thus at any rate $u^* < v^*$. Use these u^*, v^* for the construction of $h(w)$, i.e. for (A:L)-(A:U). We now prove:

Ad (A:V): The mapping

$$w \rightarrow h(w)$$

fulfills (i) by (A:R), (iii), and (ii) by (A:U).

Ad (A:W): Consider $\mathrm{v}(w)$ first. By (i) $\mathrm{v}(u^*) < \mathrm{v}(v^*)$. Put

$$h_1(w) = \frac{\mathrm{v}(w) - \mathrm{v}(u^*)}{\mathrm{v}(v^*) - \mathrm{v}(u^*)}.$$

[1] Strictly speaking, the axioms permit that there should be no two different utilities. This possibility is hardly interesting, but it is easily disposed of. If there are no two different utilities, then their number is zero or one. In the first case our assertions are vacuously fulfilled. Assume therefore the second case: There exists one and only one utility w_0. A function is just a constant $\mathrm{v}(w_0) = \alpha_0$. Any such function fulfills (i), (ii) in (A:V). In (A:W), with $\mathrm{v}(w) = \alpha_0$, $\mathrm{v}'(w) = \alpha_0'$, choose $\omega_0 = 1$ and $\omega_1 = \alpha_0' - \alpha_0$.

Then $h_1(w)$ fulfills (i), (ii) in (A:R) automatically, and (iii), (iv) in (A:R) by (i), (ii) above. Hence by (A:S) $h_1(w) = h(w)$, i.e.

$$\text{(A:13)} \qquad v(w) = \alpha_0 h(w) + \alpha_1,$$

where α_0, α_1 are fixed numbers: $\alpha_0 = v(v^*) - v(u^*) > 0$, $\alpha_1 = v(u^*)$. Similarly for $v'(w)$:

$$\text{(A:14)} \qquad v'(w) = \alpha_0' h(w) + \alpha_1',$$

where α_0', α_1' are fixed numbers: $\alpha_0' = v'(v^*) - v'(u^*) > 0$, $\alpha_1' = v(u^*)$. Now (A:13) and (A:14) give together

$$\text{(A:15)} \qquad v'(w) = \omega_0 v(w) + \omega_1,$$

where ω_0, ω_1 are fixed numbers: $\omega_0 = \dfrac{\alpha_0'}{\alpha_0} > 0$, $\omega_1 = \dfrac{\alpha_0\alpha_1' - \alpha_1\alpha_0'}{\alpha_0}$. This is the desired result.

A.3. Concluding Remarks

A.3.1. (A:V) and (A:W) are clearly the existence and uniqueness theorems called for in 3.5.1. Consequently the assertions of 3.5.–3.6. are established in their entirety.

At this point the reader is advised to reread the analysis of the concept of utility and of its numerical interpretation, as given in 3.3. and 3.8. There are two points, both of which have been considered or at least referred to loc. cit., but which seem worth reemphasizing now.

A.3.2. The first one deals with the relationship between our procedure and the concept of complementarity. Simply additive formulae, like (3:1:b), would seem to indicate that we are assuming absence of any form of complementarity between the things the utilities of which we are combining. It is important to realize, that we are doing this solely in a situation where there can indeed be no complementarity. As pointed out in the first part of 3.3.2., our u, v are the utilities not of definite—and possibly coexistent—goods or services, but of imagined events. The u, v of (3:1:b) in particular refer to alternatively conceived events u, v, of which only one can and will become real. I.e. (3:1:b) deals with either having u (with the probability α) or v (with the remaining probability $1 - \alpha$)—but since the two are in no case conceived as taking place together, they can never complement each other in the ordinary sense.

It should be noted that the theory of games does offer an adequate way of dealing with complementarity when this concept is legitimately applicable: In calculating the value $v(S)$ of a coalition S (in an n-person game), as described in 25., all possible forms of complementarity between goods or between services, which may intervene, must be taken into account. Furthermore, the formula (25:3:c) expresses that the coalition $S \cup T$ may be worth more than the sum of the values of its two constituent coalitions S T, and hence it expresses the possible complementarity between the services of the members of the coalition S and those of the members of the coalition T. (Cf. also 27.4.3.)

A.3.3. The second remark deals with the question, whether our approach forces one to value a loss exactly as much as a (monetarily) equal gain, whether it permits to attach a utility or a disutility to gambling (even when the expectation values balance), etc. We have already touched upon these questions in the last part of 3.7.1. (cf. also the footnotes 2 and 3 eod.). However, some additional and more specific remarks may be useful.

Consider the following example: Daniel Bernoulli proposed (cf. footnote 2 on p. 28), that the utility of a monetary gain dx should not only be proportional to the gain dx, but also (assuming the gain to be infinitesimal—that is, asymptotically for very small gains dx) inversely proportional to the amount x of the owner's total possessions, expressed in money. Hence (using a suitable unit of numerical utility), the utility of this gain is $\frac{dx}{x}$. The excess utility of owning x_1, over owning x_2, is then $\int_{x_2}^{x_1} \frac{dx}{x} = \ln \frac{x_1}{x_2}$. The excess utility of gaining the (finite) amount η over losing the same amount is $\ln \frac{x+\eta}{x} - \ln \frac{x}{x-\eta} = \ln \left(1 - \frac{\eta^2}{x^2}\right)$. This is < 0, i.e. of equal gains and losses the latter are more strongly felt than the former. A 50%–50% gamble with equal risks, is definitely disadvantageous.

Nevertheless Bernoulli's utility satisfies our axioms and obeys our results: However, the utility of possessing x units of money is proportional to $\ln x$, and not to x![1,2]

Thus a suitable definition of utility (which in such a situation is essentially uniquely determined by our axioms) eliminates in this case the specific utility or disutility of gambling, which prima facie appeared to exist.

We have stressed Bernoulli's utility, not because we think that it is particularly significant, or much nearer to reality than many other more or less similar constructions. The purpose was solely to demonstrate, that the use of numerical utilities does not necessarily involve assuming that 50%–50% gambles with equal monetary risks must be treated as indifferent, and the like.[3]

It constitutes a much deeper problem to formulate a system, in which gambling has under all conditions a definite utility or disutility, where numerical utilities fulfilling the calculus of mathematical expectations cannot be defined by any process, direct or indirect. In such a system some of our

[1] The 50%–50% gamble discussed above involved equal risks in terms of x, but not in terms of $\ln x$.

[2] That the utility of x units of money may be measurable, but not proportional to x, was pointed out in footnote 3 on p. 18.

[3] As stated in remark (1) in 3.7.3., we are disregarding transfers of utilities between several persons. The stricter standpoint used elsewhere in this book, as outlined in 2.1.1., specifically, the free transferability of utilities between persons, does force one to assume proportionality between utility and monetary measures. However, this is not relevant at the present stage of the discussion.

axioms must be necessarily invalid. It is difficult to foresee at this time, which axiom or group of axioms is most likely to undergo such a modification.

A.3.4. There are nevertheless some observations which suggest themselves in this respect.

First: The axiom (3:A)—or, more specifically, (3:A:a)—expresses the completeness of the ordering of all utilities, i.e. the completeness of the individual's system of preferences. It is very dubious, whether the idealization of reality which treats this postulate as a valid one, is appropriate or even convenient. I.e. one might want to allow for two utilities u, v the relationship of *incomparability*, denoted by $u \parallel v$, which means that neither $u = v$ nor $u > v$ nor $u < v$. It should be noted that the current method of indifference curves does not properly correspond to this possibility. Indeed, in that case the conjunction of "neither $u > v$ nor $u < v$," corresponding to the disjunction of "either $u = v$ or $u \parallel v$," and to be denoted by $u \approx v$, can be treated as a mere broadening of the concept of equality (of utilities, cf. also the remark concerning identity in A.1.2.).

Thus if $u \parallel u'$, $v \parallel v'$, then u', v' can replace u, v in any relationship, e.g. in this case $u < v$ implies $u' < v'$. Hence in particular $u \parallel u'$ and $v = v'$ have this consequence, and $u = u'$ and $v \parallel v'$ have this consequence. I.e., writing v, w, u for u, v, u' and u, v, w for u, v, v', respectively:

(A:16) $$u \parallel v \text{ and } v < w \text{ imply } u < w.$$

(A:17) $$u < v \text{ and } v \parallel w \text{ imply } u < w.$$

However, for the really interesting cases of partially ordered systems neither (A:16) nor (A:17) is true. (Cf. e.g. the second example at the end of 65.3.2., which is also dealt with in footnote 2 on page 590, where the connection with the concept of utility is pointed out. This is the ordering of a plane so that $u > v$ means that u has a greater ordinate than v as well as a greater abscissa than v.)

Second: In the group (3:B) the axioms (3:B:a) and (3:B:b) express a property of monotony which it would be hard to abandon. The axioms (3:B:c) and (3:B:d), on the other hand express what is known in geometrical axiomatics as the *Archimedean property:* No matter how much the utility v exceeds (or is exceeded by) the utility u, and no matter how little the utility w exceeds (or is exceeded by) the utility u, if v is admixed to u with a sufficiently small numerical probability, the difference that this admixture makes from u will be less than the difference of w from u. It is probably desirable to require this property under all conditions, since its abandonment would be tantamount to introducing infinite utility differences.[1]

[1] For a statement of the Archimedean property in an axiomatization of geometry, where it originated, cf. e.g. *D. Hilbert*, loc. cit. footnote 1 on page 74. Cf. there Axiom V.1. The Archimedean property has since been widely used in axiomatizations of number systems and of algebras.

There is a slight difference between our treatment of the Archimedean property and its treatment in most of the literature we are referring to. We are making free use of the concept of the real number, while this is usually avoided in the literature in question. Therefore the conventional approach is to "majorise" the "larger" quantity by successive

In this connection it is also worth while to make the following observation: Let any completely ordered system of utilities $\mathcal{U}$ be given, which does not allow the combination of events with probabilities, and where the utilities are not numerically interpreted. (E.g. a system based on the familiar ordering by indifference curves. Completeness of this ordering obtains, as indicated in the first remark above, by extending the concept of equality—i.e. by treating the concept $u \approx v$, that we introduced there, as equality. In this case $u \approx v$ means, of course, that u and v lie on the same indifference curve.) Now introduce events affected with probabilities. This means that one introduces combinations of, say, n $(= 1, 2, \cdots)$ events with respective probabilities $\alpha_1, \cdots, \alpha_n$ $(\alpha_1, \cdots, \alpha_n \geqq 0, \sum_{i=1}^{n} \alpha_i = 1)$. This requires the introduction of the corresponding (symbolic) utility combinations $\alpha_1 u_1 + \cdots + \alpha_n u_n$ $(u_1, \cdots, u_n$ in $\mathcal{U})$. It is possible to order these $\alpha_1 u_1 + \cdots + \alpha_n u_n$ (any $n = 1, 2, \cdots$ and any $\alpha_1, \cdots, \alpha_n$ and $u_1, \cdots, u_n$ subject to the above conditions) completely, and without making them numerical—if the ordering is allowed to be non-Archimedean. Indeed, comparing, say, $\alpha_1 u_1 + \cdots + \alpha_n u_n$ and $\beta_1 v_1 + \cdots + \beta_m v_m$ we may assume that $n = m$ and that the $u_1, \cdots, u_n$ and the $v_1, \cdots, v_m$ coincide (write $\alpha_1 u_1 + \cdots + \alpha_n u_n + 0v_1 + \cdots + 0v_m$ and $0u_1 + \cdots + 0u_n + \beta_1 v_1 + \cdots + \beta_m v_m$ for $\alpha_1 u_1 + \cdots + \alpha_n u_n$ and $\beta_1 v_1 + \cdots + \beta_m v_m$, and then replace $n + m$; $u_1, \cdots, u_n, v_1, \cdots, v_m$; $\alpha_1, \cdots, \alpha_n, 0, \cdots, 0$; $0, \cdots, 0, \beta_1, \cdots, \beta_m$ by n; $u_1, \cdots, u_n$; $\alpha_1, \cdots, \alpha_n$; $\beta_1, \cdots, \beta_n$). Then we compare $\alpha_1 u_1 + \cdots + \alpha_n u_n$ and $\beta_1 u_1 + \cdots + \beta_n u_n$. Next make, by an appropriate permutation of $1, \cdots, n$, $u_1 > \cdots > u_n$. After these preparations define $\alpha_1 u_1 + \cdots + \alpha_n u_n > \beta_1 u_1 + \cdots + \beta_n u_n$ as meaning that for the smallest $i (= 1, \cdots, n)$ for which $\alpha_i \neq \beta_i$, say $i = i_0$, there is $\alpha_{i_0} > \beta_{i_0}$.

It is clear that these utilities are non-numerical. Their non-Archimedean character becomes clear if one visualizes that here an arbitrary small excess probability $\alpha_{i_0} - \beta_{i_0}$ affecting u_{i_0} will outweigh any potential opposite excess probabilities $\beta_i - \alpha_i$ of the remaining u_i, $i = i_0 + 1, \cdots, n$, i.e. of utilities $< u_{i_0}$. (This then excludes the application of criteria like that one in footnote 1 on page 18.) Obviously, they violate our axioms (3:B:c) and (3:B:d).

Such a non-Archimedean ordering is clearly in conflict with our normal ideas concerning the nature of utility and of preference. If, on the other

---

additions of the "smaller" one (cf. e.g. Hilbert's procedure loc. cit.), while we "minorise" the "smaller" entity (the utility discrepancy between w and u in our case) by a suitable small numerical multiple (the α-fold in our case) of the "larger" entity (the utility discrepancy between v and u in our case).

This difference in treatment is purely technical and does not affect the conceptual situation. The reader will also note that we are talking of entities like "the excess of v over u" or the "excess of u over v" or (to combine the two former) the "discrepancy of u and v" (u, v, being utilities) merely to facilitate the verbal discussion—they are not part of our rigorous, axiomatic system.

hand, one desires to define utilities (and their ordering) for the probability-including system, satisfying our axioms (3:A)-(3:C)—and hence possessing the Archimedean property—then the utilities would have to be numerical, since our deduction of A.2. applies.

Third: It seems probable, that the really critical group of axioms is (3:C)—or, more specifically, the axiom (3:C:b). This axiom expresses the combination rule for multiple chance alternatives, and it is plausible, that a specific utility or disutility of gambling can only exist if this simple combination rule is abandoned.

Some change of the system (3:A)-(3:C), at any rate involving the abandonment or at least a radical modification of (3:C:b), may perhaps lead to a mathematically complete and satisfactory calculus of utilities, which allows for the possibility of a specific utility or disutility of gambling. It is hoped that a way will be found to achieve this, but the mathematical difficulties seem to be considerable. Of course, this makes the fulfillment of the hope of a successful approach by purely verbal means appear even more remote.

It will be clear from the above remarks, that the current method of using indifference curves offers no help in the attempt to overcome these difficulties. It merely broadens the concept of equality (c.f. the first remark above), but it gives no useful indications—and *a fortiori* no specific instructions—as to how one should treat situations that involve probabilities, which are inevitably associated with expected utilities.

Afterword

ARIEL RUBINSTEIN

During the past ten years Princeton University Press has done a remarkable job of republishing, in a beautiful and eye-catching format, many of the seminal works from the early days of game theory at Princeton. This new printing of *Theory of Games and Economic Behavior*, marking the book's sixtieth anniversary, continues the celebration of game theory. Since the original publication of the book, game theory has moved from the fringe of economics into its mainstream. The distinction between economic theorist and game theorist has virtually disappeared. The 1994 Nobel Prize awarded to John Nash, John Harsanyi, and Reinhard Selten was viewed not just as recognition of three great scholars but also as a victory for game theory as a discipline. Evidence of the immense importance of this book in the development of game theory is the fact that, notwithstanding the intense search for the ancient origins of its ideas, there is consensus that the book was the first major publication in the field. It set the tone for the subsequent half century of research, after which game theory joined the Walrasian economy as one of the major paradigms within economics.

The game theorist reading this book does not need another lecture on the importance of the book and the development of game theory. Very few other books in economics have been as highly praised and influential. Only a handful of topics have received as much attention or been surveyed as intensively in contemporary economics as game theory. The reader of the book who is not a scholar of game theory, and is interested in catching up with the development of the discipline since the book was published, can choose from a number of excellent introductory books. They are written in a variety of styles and levels of mathematical sophistication and are directed at laypersons as well as scholars of economics, law, political science, management theory, mathematics, and biology.

So, what remains to be said? I have chosen to offer some skeptical observations about game theory. Overall, it is skepticism which makes the game of academic discourse much more interesting.

Whoever came up with the name "game theory" was a genius not only in mathematics but also in public relations. Imagine if it had been called the "Theory of Rationality and Decision Making in Interactive Economic Situations." Would this book and the theory as a whole have enjoyed the same degree of popularity? The word "game" has a young and familiar sound. All of us play games—board games, computer games, political games. But game theory is not a box of magic tricks that can help us play games more successfully. There are very few insights from game theory that would improve one's game of chess or poker. These games are only used in game theory as convenient illustrations.

So is game theory useful in any way? The popular literature is full of nonsensical claims to that effect. But within the community of game theorists there is sharp disagreement over its meaning and potential usefulness. There are those who believe that the goal of game theory is ultimately to provide a good prediction of behavior in strategic situations and even if we are not "there" (yet) we will get "there" once we have enriched the models with additional parameters and come up with better ways of measuring the considerations of real players. I am not sure on what this vision is based. Most situations can be analyzed in a number of ways, which usually yield contradictory "predictions." Furthermore, we need also to deal with the fundamental difficulty of predicting behavior in the social sciences, where prediction itself is part of the game and forecasters are themselves players.

Then there are those who believe in the power of game theory to improve performance in real-life strategic interactions. I have never been persuaded that there is a solid foundation for this belief. The fact that academics have a vested interest in it makes it even less credible. There seems to be some regularity in strategic behavior which becomes apparent in game theoretical experiments. It is gratifying to sometimes find similar distributions of modes of behavior across societies. But are these regularities related to the classical predictions of game theory?

Others (including myself) think that the object of game theory is primarily to study the considerations used in decision making in interactive situations. It identifies patterns of reasoning and investigates their implications on decision making in strategic situations. According to this opinion, game theory does not have normative implications and its empirical significance is very limited. Game theory is viewed as a cousin of logic. Logic does not allow us to screen out true statements from false ones and does not help us distinguish right from wrong. Game theory does not tell us which action is preferable or predict what other people will do. If game theory is nevertheless useful or practical, it is only indirectly so. In any case, the burden of proof is on those who use game theory to make policy recommendations, not on those who doubt the practical value of game theory in the first place.

And, by the way, I sometimes wonder why people are so obsessed in looking for "usefulness" in economics generally and in game theory in particular. Should academic research be judged by its usefulness?

Game theory is responsible for some new terms in our language. For example, the wide use of the term "zero-sum game" is attributed to the influence of game theory, although it is often used by speakers simply to demonstrate their level of sophistication (or lack thereof . . .). game theory popularized the term "Prisoner's Dilemma," which is widely used in the popular press and by politicians. However, it is used to express a rather trivial idea: that there are situations in which selfish behavior can ultimately hurt all participants.

I view economics (and even more generally, all social sciences) as culture. It is a collection of terms, considerations, models, and theories used by people who think about economic interactions. Game theory changed the culture of economics. Most contemporary economists use game theory as an essential tool for transferring their assumptions about a situation into outcomes. Game theory has essentially become a toolbox from which economists select, often mechanically, the tools for transforming assumptions into predictions.

Personally, I am not sure that game theory "improves the world." Overall, economics, and game theory in particular, is not just a description of human behavior. When we teach game theory we may be affecting the way people think and behave in economic and strategic interactions. Is it impossible that the study of game theoretical considerations in economics makes people more manipulative or more selfish?

Game theory's appeal is also a result of its language. Terms such as "strategy" and "solution" were not intended to be arbitrary names for mathematical concepts. Are they used in an appropriate fashion? This is not an easy question to answer since we do not possess objective tools for judging the interpretation of a formal notion in game theory (or more generally in the social sciences). The evaluation of the connection between a formal model and its interpretation relies entirely on common sense. For example, my own view is that the key term "strategy" is used most often in a way which is difficult to reconcile with its natural interpretation as a "course of action." The use of the term "solution" in game theory may create expectations that game theory can provide solutions to real world problems. In fact, a solution in game theory is no more than a systematic principle for analyzing classes of games; game theory encompasses many different solution concepts which yield conflicting predictions. The ambiguity of the language used in game theory is problematic and potentially misleading. Have we traded the ambiguity of words in the natural language for the confusing interpretations of formal concepts?

This book was a landmark in the transition of economics into a mathematical discipline. The advantage of making economics more mathematical is that it introduces order, precision, and a sense of objectivity into what would otherwise be considered a vague social science. But of course there are also disadvantages. The heavy use of mathematics limits the number of people who are able to comprehend the material. There is sometimes a feeling that there exists a small sect of "high priests" who have mastered the material while the rest are left to wonder whether mathematical sophistication is being used to hide assumptions and pull a rabbit out of a hat. Does game theory require such a high level of mathematics? In the future will revolutionary ideas be stated in mathematical terms or will part of the revolution be the return to everyday language?

As to the state of the theory, it is my impression that the well of game theory is relatively dry. This is in spite of its success and probably because of it. Game theory has become a primary tool in the economist's toolbox. However, the last decade has seen few new ideas in game theory. Thus, the stage is set for a new unconventional work which will shake economics like this one did sixty years ago. Of course, original ideas cannot simply be ordered. Nevertheless, it is the responsibility of the profession to create an environment that will attract unconventional individuals with a broad educational base and the mental approach which can generate innovative ideas. The playing of games is dependent on abilities that game theory does not capture well, such as memory, the ability to process information and the quality of associations. The assimilation of these concepts constitutes one of the main challenges for the future. Will we see a new concept added to those of competitive equilibrium and Nash equilibrium as an additional pillar of economic thought?

Finally, I can not help noticing that the book was written during the Second World War and published in 1944, a year of loss and tragedy. This coincidence and the role later played by certain institutions (which had been involved in security matters) in the development of game theory led some people to the ridiculous conclusion that "game theory is a plot." I often find myself wondering how such intellectual progress could have been made during a period of such turbulent events. Perhaps we feel more of an urgency to understand the world when things are unstable. In any case, we should feel privileged that we can play games not only as children but also as academics—but we need to keep in mind that the challenges facing the world today are far too complex to be captured by any matrix game.

REVIEWS

The American Journal of Sociology (May 1945)
HERBERT A. SIMON
Volume 50, No. 6, pp. 558–60

The *Theory of Games* is a rigorous mathematical development of a theory of games of strategy and an application of that theory to certain simple problems in economics. Although no explicit applications are made to sociology or political science, the schema is of such generality and breadth that it can undoubtedly make contributions of the most fundamental nature to those fields.

In the Foreword the authors, quite correctly, state: "The mathematical devices used are elementary in the sense that no advanced algebra, or calculus occurs. . . . However, the reader . . . will have to familiarize himself with the mathematical way of reasoning definitely beyond its routine, primitive phases." What is required of the reader, then, is not training in mathematics so much as "mathematical maturity." The reviewer found the *Theory of Games* at all times a model of clear and careful exposition.

Social scientists have for decades carried on a largely sterile debate as to the applicability of mathematical modes of thought to their discipline. Like all arguments about methodology, this one must in the long run be settled by results. To date, with the notable exception of mathematical economics, mathematical reasoning (as distinguished from the use of quantitative data, statistics) has not much to show in the way of results in the social sciences. To be sure, the single exception is a notable one—progress in economic theory is becoming more and more dependent upon the application of the calculus, and most of the gains of the last fifty years in the development of marginal analysis must be attributed to economists trained in mathematical modes of thought, even if they sometimes translated their thought into the more acceptable literary form for presentation.

In sociology and political science, attempts at a mathematical theory can be numbered on the fingers of one hand, and the results have heretofore certainly been negligible. In spite of the use of symbolism and so-called "equations," neither Parsons' *The Structure of Social Action* nor Dodd's *Dimensions of Society* would be acknowledged by any mathematician to be "mathematical"—although the latter volume evidently aspires to be. The "mathematics" of Zipf's *National Unity and Disunity* does not go beyond simple arithmetic, nor do the conclusions Zipf draws have any relation to the arithmetic on which they are "based." The sole remaining examples with which this reviewer is familiar are the single article by Karl Menger which appeared in this *Journal* in March, 1938, and the series of papers by Rashevsky which have been appearing in *Psychometrika* and elsewhere.[1] The former, while suggestive, did not progress very far; the latter, apparently drawing their sociological inspiration from the works of Sorokin, attempt, by the introduction of heroic assumptions, to solve complicated problems which are certainly not solvable by the techniques and theory as yet available to us.

The *Theory of Games* is both more modest and infinitely more impressive in its results than any of these earlier attempts. It seeks merely to develop in systematic and rigorous manner a theory of rational human behavior. Now the simplest setting in which human rationality is exercised is in the playing of games; hence the authors select as their starting-point a theory of games which von Neumann originated as early as 1928.

While most attempts at mathematization in the social sciences have employed the tools of the calculus and differential equations, von Neumann moves in an entirely different direction and employs, instead, the mathematics of point-set theory and of topology. As a matter of fact, von Neumann insists—and his eminence among mathematicians lends great weight to his opinion on this point—that the lack of success of mathematics in the social sciences (which he certainly exaggerates, at least in relation to economics) has been due in large part to the use of tools which were developed in intimate connection with the growth of mathematical physics and which are not applicable to social theory. "It is therefore to be expected," he says, "that mathematical discoveries of a stature comparable to that of calculus will be needed in order to produce decisive success in this field. . . . it is unlikely that a mere repetition of the tricks which served us so well in physics will do for the social phenomena too" (p. 6).

The first step in the theory of games is to construct a formal, mathematical description of a game. This is very beautifully accomplished in chapter ii, which is essentially based on von Neumann's 1928 publication. The reviewer considers this to be the most important contribution of the entire theory to social science. Sociology has been forced to treat of human behavior (at least in its rational aspects) in terms of "ends" and "means"; for example, these are fundamental categories in *The Structure of Social Action.* It could easily be shown that these two terms complicate rather than simplify the analysis of human rationality, and it is to be

[1]See the references in *Psychometrika,* IX (September, 1944), 215.

hoped that they will now be discarded, both in sociology and in ethics, in favor of the schema of "alternatives," "consequences," and "values" attached to "consequences" (the terminology here is the reviewer's and not that of the *Theory of Games*) which the description of games of strategy provides. This schema quite obviously owes its origins to the utility calculus of economics, but in its generality it can be applied, at least descriptively, to all behavior, whether rational or not.

The schema of chapter ii also exhibits for the first time a system of *social* behavior in which the consequences of the behavior of each member of the system is explicitly dependent on the behaviors of the other members. Within this system it is possible to define unambiguously and to analyze the concepts of "competition" and "co-operation" which have become such important categories of sociological, political, and economic theory. Here, too, the reviewer believes, is to be found the beginnings of a sound theory of administrative organization.

The succeeding chapters of the *Theory of Games* are concerned with the description of a "good strategy"—i.e., rational conduct in playing a game—and with the analysis of the stability of behavior systems of the players of a game. The concept of "stability" as defined in the study (see particularly pp. 261 ff.) is perhaps not entirely free from objection in its details, but it certainly points in a proper direction. Furthermore, it leads to most important sociological results—notably to a demonstration that in games with more than two persons coalitions (organizations of two or more persons who co-ordinate their behavior) will in general appear.

It is here that the most immediate applications of the theory suggest themselves. The authors have themselves undertaken certain simple applications in the field of economics—specifically to the problems of bilateral monopoly and duopoly—but there are many other possibilities in the fields of sociology and political science. For example, it should be possible to identify the theory of revolutions with the theory of stability and instability of "standards of behavior" in certain games. For this purpose the theory will probably have to be developed from a static to a dynamic one, however. In the field of politics one might construct games which would illustrate the formation of two-party or multi-party systems, respectively, and this could lead to a comparison of the circumstances favoring one or another type of equilibrium.

The reviewer does not wish to be overoptimistic here about the immediate application of the *Theory of Games* to "burning questions" but simply to point out that many of the research problems which appear to social scientists to be significant lend themselves rather directly to translation into the theory of games and, hence, to rigorous treatment. Certainly previous "literary" analyses of these same problems have not been so conclusive and rigorous that we can dismiss attempts at mathematical treatment as superfluous.

The reviewer can only hope that the foregoing remarks will serve to indicate the potential importance of this volume and to encourage every social scientist who is convinced of the necessity for mathematizing social theory—as well as those unconverted souls who are still open to persuasion on this point—to undertake the task of mastering the *Theory of Games*. The student of the *Theory of Games* will learn from it the directions his own mathematical education must take if he is to make contributions to formal social theory. He will come away from the volume with a wealth of ideas for application and for development of the theory into a fundamental tool of analysis for the social sciences.

Bulletin of the American Mathematical Society (July 1945)
ARTHUR H. COPELAND
Volume 51, No. 7, pp. 498–504

Posterity may regard this book as one of the major scientific achievements of the first half of the twentieth century. This will undoubtedly be the case if the authors have succeeded in establishing a new exact science—the science of economics. The foundation which they have laid is extremely promising. Since both mathematicians and economists will be needed for the further development of the theory it is in order to comment on the background necessary for reading the book. The mathematics required beyond algebra and analytic geometry is developed in the book. On the other hand the non-mathematically trained reader will be called upon to exercise a high degree of patience if he is to comprehend the theory. The mathematically trained reader will find the reasoning stimulating and challenging. As to economics, a limited background is sufficient.

The authors observe that the give-and-take of business has many of the aspects of a game and they make an extensive study of the strategy of games with this similarity in mind (hence the title of this book). In the game of life the stakes are not necessarily monetary; they may be merely *utilities*. In discussing utilities the authors find it advisable to replace the questionable marginal utility theory by a new theory which is more suitable to their analysis. They note that in the game of life as well as in social games the players are frequently called upon to choose between alternatives to which probabilities rather than certainties are attached. The authors show that *if a player can always arrange such fortuitous alternatives in the order of his preferences, then it is possible to assign to each alternative a number or numerical utility expressing the degree of the player's preference for that alternative. The assignment is not unique but two such assignments must be related by a linear transformation.*

The concept of a game is formalized by a set of postulates. Even the status of information of each player on each move is accounted for and is characterized by a partition of a certain set. *The amount which player k receives at the conclusion of the play is a function* $\mathfrak{F}_k(\sigma_1, \sigma_2, \ldots, \sigma_\nu)$ *of the moves* $\sigma_1, \sigma_2, \ldots, \sigma_\nu$ *where some of the* σ's *may be the moves of chance* (dealing cards, throwing dice, and so on).

The concept of a game admits of a rather drastic simplification which practically relieves the players of the necessity of playing. Imagine that all possible strategies of all players have been catalogued. Then player k can tell his secretary that he wishes to play strategy τ_k. When she looks up this strategy she finds a complete prescription determining every move for every possible eventuality. Thus the secretaries could get together and determine the outcome of the game if they could only find an equitable method of accounting for the moves of chance. But chance enters into the game very much as one of the players. Thus we can imagine a cataloguing of the possible strategies of chance. Suppose for the moment that the strategy τ_0 of chance has been decided upon and that the players have chosen respectively the strategies $\tau_1, \tau_2, \ldots, \tau_n$. Then the strategies determine the moves. Hence $\mathfrak{F}_k(\sigma_1, \sigma_2, \ldots, \sigma_\nu)$ *is a function* $\mathfrak{G}_k(\tau_0, \tau_1, \ldots, \tau_n)$ *of the strategies* and the outcome of the game is determined. But how should τ_0 be selected? Instead of selecting τ_0 the secretaries could *assign to each player k the amount* $\mathfrak{H}_k(\tau_1, \tau_2, \ldots, \tau_n)$ *which he would receive on the average if strategies* $\tau_1, \tau_2, \ldots, \tau_n$ *of the players were chosen.* The amount $\mathfrak{H}_k$ is the mathematical expectation of $\mathfrak{G}_k$. It is computed in terms of the probabilities of the various strategies τ_0 and these probabilities are in turn computed in terms of the probabilities of the moves of chance.

The game has now been reduced to one in which each player makes just one move—the selection of a strategy. Each player makes his move in complete ignorance of the moves of the other players. The authors have accomplished this simplification of the game with complete rigor and with complete adherence to the rules laid down by the postulates.

A 1-*player game corresponds to the economy of a man on a desert island.* It is the Robinson Crusoe economy or a strictly regimented communism. If the player is wise, he will choose his strategy τ_1 so that $\mathfrak{H}_1(\tau_1)$ is a maximum.

This is the only case where a game is settled by simple maximum considerations.

An n-player zero-sum game is one for which the sum of the $\mathfrak{F}_k$'s is zero for all choices of $\sigma_1, \sigma_2, \ldots, \sigma_\nu$ and hence one for which the sum of the $\mathfrak{H}_k$'s is zero for all choices of $\tau_1, \tau_2, \ldots, \tau_n$. Social games are zero-sum but the game of economics is decidedly not zero-sum since society as a whole can improve its status if all members behave properly. *However an arbitrary n-player game can be reduced to a zero-sum* $(n + 1)$*-player game* by introducing a fictitious player $n + 1$ who receives the amount $\mathfrak{H}_{n+1}(\tau_1, \tau_2, \ldots, \tau_n)$ which is the negative of the total received by the remaining n players. Note that the functions $\mathfrak{H}_k$ do not contain the variable τ_{n+1}, that is, the fictitious player is not permitted to choose a strategy. It will appear later that further restrictions on the activities of this player are necessary to prevent his influencing the outcome of the game.

A zero-sum 2-player game Γ can be characterized by a single function $\mathfrak{H}(\tau_1, \tau_2) = \mathfrak{H}_1(\tau_1, \tau_2)$ since $\mathfrak{H}_2(\tau_1, \tau_2) = -\mathfrak{H}(\tau_1, \tau_2)$ by virtue of the relation $\mathfrak{H}_1 + \mathfrak{H}_2 = 0$. In this game player 1 will attempt to maximize $\mathfrak{H}$ (or $\mathfrak{H}_1$) whereas player 2 will attempt to minimize $\mathfrak{H}$ (or maximize $\mathfrak{H}_2$). Since these are diametrically opposed tendencies, it looks as though nothing could be decided. However we can gain insight into the problem by considering a modified game Γ_1 which is the same as Γ except that player 1 moves first and player 2 knows 1's move. In Γ_1 after player 1 chooses τ_1, player 2 will choose τ_2 so as to minimize $\mathfrak{H}$. It is therefore advisable for 1 to choose τ_1 so as to maximize $\min_{\tau_2} \mathfrak{H}(\tau_1, \tau_2)$ where $\min_{\tau_2} \mathfrak{H}(\tau_1, \tau_2)$ is the minimum with respect to τ_2 of $\mathfrak{H}(\tau_1, \tau_2)$. Player 1 will then receive

$$v_1 = \max_{\tau_1} \quad \min_{\tau_1} \mathfrak{H}(\tau_1, \tau_2)$$

and 2 will receive $-v_1$. Next consider a third game Γ_2 which is the same as Γ except that 2 moves first and 1 knows 2's move. If both players of Γ_2 are skillful, then 1 will receive the amount

$$v_2 = \min_{\tau_2} \quad \max_{\tau_1} \mathfrak{H}(\tau_1, \tau_2)$$

and 2 will receive $-v_2$. In the original game Γ if both players are skillful, 1 will receive at least v_1 and at most v_2 whereas 2 will receive at least $-v_2$ and at most $-v_1$. *Hence* $v_1 \leqq v_2$ *and these quantities are bounds for the outcome of the game.* If $v_1 = v_2$, the game is determined but in general this is not the case.

Note that Γ reduces to Γ_1 if 2 discovers 1's strategy whereas Γ becomes Γ_2 if 1 discovers 2's strategy. Hence it is advisable for the players to conceal their strategies. The concealment is accomplished by using probabilities.

Thus 1 chooses τ_1 with probability ξ_{τ_1} and 2 chooses τ_2 with probability η_{τ_2}. *The average outcome* $K(\xi, \eta)$ *for player 1 is the mathematical expectation of* $\mathfrak{H}(\tau_1, \tau_2)$ *with respect to the probabilities* ξ_{τ_2} *and* η_{τ_2} where ξ is the vector with components $\xi_1, \xi_2, \ldots$ and η is the vector with components $\eta_1, \eta_2, \ldots$. The introduction of these probabilities modifies Γ and consequently modifies Γ_1, Γ_2 and the bounds v_1, v_2. The new bounds become

$$v_1' = \min_{\xi} \quad \max_{\eta} K(\xi, \eta)$$

and

$$v_2' = \min_{\eta} \quad \max_{\xi} K(\xi, \eta).$$

It is easily shown that $v_1 \leqq v_1' \leqq v_2' \leqq v_2$, that is, that each player is at least as well off as before the probabilities were introduced. Moreover it can be shown that

$$v_1' = v_2' = v$$

and hence that the game is determined. The proof of the latter result depends on the fact that the numbers $x_{\tau_2} = \Sigma_{\tau_1}\mathfrak{H}(\tau_1, \tau_2)\xi_{\tau_1}$ are components of a vector χ which depends on ξ and that the tips of the vectors χ for all possible ξ's constitute a convex set of points.

Next consider an n-player game in which the players divide into two hostile groups called S and $-S$. This can be interpreted as a 2-player game between the players S and $-S$. If probabilities are employed in the manner described above, then S will receive

$$v(S) = v_1' = v_2' = v$$

and $-S$ will receive

$$v(-S) = -v(S).$$

If I is the set of all players, then $v(I) = 0$, that is, the game is zero-sum. Finally

$$v(S + T) \geqq v(S) + v(T)$$

if S and T are mutually exclusive groups. That is, the players of $S + T$ can obtain at least as much by cooperating as they can by splitting up into two groups S and T. The function $v(S)$ satisfying the above relations is called a characteristic function. Corresponding to any function satisfying these relations there exists a game having this $v(S)$ as its characteristic function. The construction of such a game involves partitions of I into subsets called rings and solo sets.

If the equality $v(S + T) = v(S) + v(T)$ always holds, that is, if $v(S)$ is additive, then the coalitions will be ineffective and the game will be determined. This is the case for $n = 2$. Moreover two characteristic functions (whether additive or not) which differ by an additive function will produce the same strategies of coalitions. If $v(S)$ is not additive, it can be modified by a suitable additive function and a suitable scale factor so that $v(S) = -1$ for all 1-element sets. Thus for $n = 3$, $v(S)$ is given by the following table

$$v(S) = \begin{cases} 0 \\ -1 \\ +1 \\ 0 \end{cases} \textit{for the} \begin{cases} \text{0-}\textit{element set } (-I \textit{ or the complement of } I) \\ \text{1-}\textit{element sets} \\ \text{2-}\textit{element sets } (\textit{complements of } \text{1-}\textit{element sets}) \\ \text{3-}\textit{element set } (I). \end{cases}$$

For $n \geqq 4$, $v(S)$ is no longer determined and the number of possibilities becomes almost bewildering. The reader will begin to realize that there is never a dull moment with these games. We have seen that for each of the cases $n = 1, 2, 3, 4$ a new situation appears. For $n = 5$ no new phenomenon has as yet been discovered but for $n \geqq 6$ we first meet the possibility of a game which splits into two or more games which are in some respects quite distinct but which nevertheless exert potent influences on one another. This phenomenon has the counterpart of nations whose economies are distinct yet interdependent.

It remains to consider what coalitions can be expected to form in a given game and how the stakes will be divided in the presence of such coalitions. A division of stakes is called an *imputation* and is represented by a vector α with components $\alpha_1, \alpha_2, \ldots, \alpha_n$ where α_k is the amount the kth player receives. One could imagine that if a group of novices were playing one of these games a certain chaos would result. Coalitions would be made and broken as each player sought to improve his own status. Finally as the players became more acquainted with the game certain imputations would come to be trusted because of the stability of the corresponding coalitions and because of the profitableness to an effective group of players. There would thus emerge *a set V of trusted imputations.* There would of course be players who were dissatisfied with any given trusted imputation but they would not be strong enough to force a change unless they could bribe some of the favored players to desert their coalitions. Nor would such bribery be effective since the potential recipient of the bribe would realize that the chaos produced by his desertion would eventually leave him in a less favorable position. Thus V corresponds to a group behavior pattern. It is an institution or a morality arising from enlightened self interest.

But how can V be described mathematically? We begin with a definition. *We say that an imputation α dominates an imputation β if there is an effective group of players each of which is better off under α than under β. The group is effective provided it can guarantee for its members the stakes prescribed by α against any opposition from without the group.* A set V of imputations is called a solution provided every imputation outside of V is dominated by some imputation of V and no imputation in V is dominated by any other imputation in V. Thus *V is a maximal set of mutually undominated imputations.* Unfortunately dominance does not produce even a partial ordering of the set of all imputations. It is not a transitive relation. This makes the discovery of solutions a difficult task. We shall however outline a method of finding solutions for the case $n = 3$.

When $n = 3$ we have

$$\alpha = (\alpha_1, \alpha_2, \alpha_3) \text{ with } \alpha_1 + \alpha_2 + \alpha_3 = 0,$$

that is, the game is zero-sum. Thus the tip of α lies in a plane which passes through the origin and is equally inclined to the coordinate axes. This plane is divided into six congruent sectors by the traces of the coordinate planes. Next $\alpha_k \geqq -1$ (for $k = 1, 2, 3$) since each player can obtain at least -1 without the benefit of any coalition (see the above table). These inequalities require the tip of α to lie within an equilateral triangle whose center is at the common intersection of the traces of the coordinate planes and whose sides are parallel to these traces. An imputation α dominates those imputations which are represented by points interior to three parallelograms each of which has two sides in common with the above equilateral triangle and one vertex at the tip of α. On the basis of these geometrical considerations it is easy to find solutions V. We first look for a V whose imputations do not all lie on a line $\alpha_k =$ a constant (that is, a line parallel to a trace). There is only one such solution, namely,

$$V\colon (1/2, 1/2, 0), \quad (1/2, 0, 1/2), \quad (0, 1/2, 1/2).$$

We next look for a V whose imputations do lie on a line, say, $\alpha_3 = c$. The corresponding solutions are

$$V_c\colon (a_1, -a - c, c)$$

where a and c are required to satisfy certain inequalities. Thus V_c contains a continuum of solutions corresponding to values of the parameter a. This exhausts the possible solutions. The first solution V seems quite reasonable whereas V_c seems unnatural and difficult to interpret but let us return to this question later.

Let us consider the following non-zero-sum 2-player game. Each player (1 or 2) chooses either the number 1 or the number 2. If both players choose 1, then each receives the stake 1/2. Otherwise each receives -1. If we reduce this game to a zero-sum 3-player game by the introduction of a fictitious player 3, then the characteristic function becomes the one given in the above table. Now if we take the first solution V, we discover that the fictitious player may play an active part in the formation of coalitions. Hence if we wish to retain the 2-player character of the game, we must choose the solution V_c and it is reasonable to assign to c the value -1.

The authors apply this theory of games to the analysis of a market consisting of one buyer and one seller and also of a market consisting of two buyers and one seller.

The book leaves much to be done but this fact only enhances its interest. It should be productive of many extensions along the lines of economic interpretation as well as of mathematical research. In fact the authors suggest a number of directions in which research might profitably be pursued.

The American Economic Review (December 1945)[1]
LEONID HURWICZ
Volume 35, No. 5, pp. 909–25

Had it merely called to our attention the existence and exact nature of certain fundamental gaps in economic theory, the *Theory of Games and Economic Behavior* by von Neumann and Morgenstern would have been a book of outstanding importance. But it does more than that. It is

[1] Cowles Commission Papers, New Series, No. 13A.

The author, on leave from Iowa State College where he is associate professor, is now on a Guggenheim Memorial Fellowship, working with the Cowles Commission for Research in Economics, of which he is a research associate.

The tables and figures used in this article were drawn by Mrs. D. Friedlander of the University of Chicago.

essentially constructive: where existing theory is considered to be inadequate, the authors put in its place a highly novel analytical apparatus designed to cope with the problem.

It would be doing the authors an injustice to say that theirs is a contribution to economics only. The scope of the book is much broader. The techniques applied by the authors in tackling economic problems are of sufficient generality to be valid in political science, sociology, or even military strategy. The applicability to games proper (chess and poker) is obvious from the title. Moreover, the book is of considerable interest from a purely mathematical point of view. This review, however, is in the main confined to the purely economic aspects of the *Theory of Games and Economic Behavior.*

To a considerable extent this review is of an expository[2] nature. This seems justified by the importance of the book, its use of new and unfamiliar concepts, and its very length which some may find a serious obstacle.

The existence of the gap which the book attempts to fill has been known to the economic theorists at least since Cournot's work on duopoly, although even now many do not seem to realize its seriousness. There is no adequate solution of the problem of defining "rational economic behavior" on the part of an individual when the very rationality of his actions depends on the probable behavior of other individuals: in the case of oligopoly, other sellers. Cournot and many after him have attempted to sidetrack the difficulty by assuming that every individual has a definite idea as to what others will do under given conditions. Depending on the nature of this expected behavior of other individuals, we have the special, well-known solutions of Bertrand and Cournot, as well as the more general Bowley concept of the "conjectural variation."[3] Thus, the individual's "rational behavior" is determinate *if* the pattern of behavior of "others" can be assumed *a priori* known. But the behavior of "others" cannot be known *a priori* if the "others," too, are to behave rationally! Thus a logical *impasse* is reached.

The way, or at least *a* way,[4] out of this difficulty had been pointed out by one of the authors[5] over a decade ago. It lies in the rejection of a narrowly interpreted maximization principle as synonymous with rational

[2] The exposition is mostly carried out by means of comparatively simple numerical examples. This involves loss of generality and rigor, but it may be hoped that it will make the presentation more accessible.

[3] More recent investigations have led to the idea of a kinked demand curve. This, however, is a special—though very interesting—case of the conjectural variation.

[4] *Cf.* reference to von Stackelberg in footnote 17 and some of the work quoted by von Stackelberg, *op. cit.*

[5] J. von Neumann, "Zur Theorie der Gesellschaftsspiele," *Math. Annalen* (1928).

behavior. Not that maximization (of utility[6] or profits) would not be desirable if it were feasible, but there can be no true maximization when only one of the several factors which decide the outcome (of, say, oligopolistic competition) is controlled by the given individual.

Consider, for instance, a duopolistic situation[7] where each one of the duopolists A and B is *trying* to maximize his profits. A's profits will depend not only on his behavior ("strategy") but on B's strategy as well. Thus, *if* A could control (directly or indirectly) the strategy to be adopted by B, he would select a strategy for himself and one for B so as to maximize his own profits. But he cannot select B's strategy. Therefore, he can in no way make sure that by a proper choice of his own strategy his profits will actually be unconditionally maximized.

It might seem that in such a situation there is no possibility of defining rational behavior on the part of the two duopolists. But it is here that the novel solution proposed by the authors comes in. An example will illustrate this.

Suppose each of the duopolists has three possible strategies at his disposal.[8] Denote the strategies open to duopolist A by A_1, A_2, and A_3, and those open to duopolist B by B_1, B_2, and B_3. The profit made by A, to be denoted by a, obviously is determined by the choices of strategy made by the two duopolists. This dependence will be indicated by subscripts attached to a, with the first subscript referring to A's strategy and the second subscript to that of B; thus, *e.g.*, a_{13} is the profit which will be made by A if he chooses strategy A_1 while B chooses the strategy B_3. Similarly, b_{13} would denote the profits by B under the same circumstances. The possible outcomes of the "duopolistic competition" may be represented in the following two tables:

Table 1a shows the profits A will make depending on his own and B's choice of strategies. The first row corresponds to the choice of A_1, etc.; columns correspond to B's strategies. Table 1b gives analogous information regarding B's profits.

[6]A side-issue of considerable interest discussed in the *Theory of Games* is that of measurability of the utility function. The authors need measurability in order to be able to set up tables of the type to be presented later in the case where utility rather than profit is being maximized. The proof of measurability is not given; however, an article giving the proof is promised for the near future and it seems advisable to postpone comment until the proof appears. But it should be emphasized that the validity of the core of the *Theory of Games* is by no means dependent on measurability or transferability of the utilities and those who feel strongly on the subject would perhaps do best to substitute "profits" for "utility" in most of the book in order to avoid judging the achievements of the *Theory of Games* from the point of view of an unessential assumption.

[7]It is assumed that the buyers' behavior may be regarded as known.

[8]Actually the number of strategies could be very high, perhaps infinite.

B's choice of strategies / A's choice of strategies	B_1	B_2	B_3
A_1	a_{11}	a_{12}	a_{13}
A_2	a_{21}	a_{22}	a_{23}
A_3	a_{31}	a_{32}	a_{33}

Table 1a. A's Profits

B's choice of strategies / A's choice of strategies	B_1	B_2	B_3
A_1	b_{11}	b_{12}	b_{13}
A_2	b_{21}	b_{22}	b_{23}
A_3	b_{31}	b_{32}	b_{33}

Table 1b. B's Profits

In order to show how A and B will make decisions concerning strategies we shall avail ourselves of a numerical example given in Tables 2a and 2b.

B's choice of strategies / A's choice of strategies	B_1	B_2	B_3
A_1	2	8	1
A_2	4	3	9
A_3	5	6	7

Table 2a. A's Profits

B's choice of strategies / A's choice of strategies	B_1	B_2	B_3
A_1	11	2	20
A_2	9	15	3
A_3	8	7	6

Table 2b. B's Profits

Now let us watch A's thinking processes as he considers his choice of strategy. First of all, he will notice that by choosing strategy A_3 he will be sure that his profits cannot go down below 5, while either of the remaining alternatives would expose him to the danger of going down to 3 or even to 1. But there is another reason for his choosing A_3. Suppose there is a danger or a "leak": B might learn what A's decision is before he makes his own. Had A chosen, say, A_1, B—if he knew about this—would obviously choose B_3 so as to maximize his own profits; this would leave A with a profit of only 1. Had A chosen A_2, B would respond by selecting B_2, which again would leave A with a profit below 5 which he could be sure of getting if he chose A_3.

One might perhaps argue whether A's choice of A_3 under such circumstances is the only way of defining rational behavior, but it certainly is *a* way of accomplishing this and, as will be seen later, a very fruitful one. The reader will verify without difficulty that similar reasoning on B's part will make him choose B_1 as the optimal strategy. Thus, the outcome of the duopolistic competition is determinate and can be described as follows: A will choose A_3, B will choose B_1, A's profit will be 5, B's 8.

An interesting property of this solution is that neither duopolist would be inclined to alter his decision, even if he were able to do so, after he found out what the other man's strategy was.

To see this, suppose B has found out that A's decision was in favor of strategy A_3. Looking at the third row of Table 2b, he will immediately see that in no case could he do better than by choosing B_1, which gives him the highest profit consistent with A's choice of A_3. The solution arrived at is of a very stable nature, independent of finding out the other man's strategy.

But the above example is artificial in several important respects. For one thing, it ignores the possibility of a "collusion" or, to use a more neutral term, coalition between A and B. In our solution, yielding the strategy combination (A_3, B_1), the joint profits of the two duopolists amount to 13; they could do better than that by acting together. By agreeing to choose the strategies A_1 and B_3 respectively, they would bring their joint profits up to 21; this sum could then be so divided that both would be better off than under the previous solution.

A major achievement of the *Theory of Games* is the analysis of the conditions and nature of coalition formation. How that is done will be shown below. But, for the moment, let us eliminate the problem of coalitions by considering a case which is somewhat special but nevertheless of great theoretical interest: the case of *constant sum* profits. An example of such a case is given in Tables 3a and 3b.

Table 3a is identical with Table 2a. But figures in Table 3b have been selected in such a manner that the joint profits of the two duopolists always amount to the same (10), no matter what strategies have been chosen. In such a case, A's gain is B's loss and *vice versa*. Hence, it is intuitively obvious (although the authors take great pains to show it rigorously) that no coalition will be formed.

The solution can again be obtained by reasoning used in the previous case and it will again turn out to be (A_3, B_1) with the respective profits 5 and 5 adding up to 10. What was said above about stability of

B's choice of strategies / A's choice of strategies	B_1	B_2	B_3
A_1	2	8	1
A_2	4	3	9
A_3	5	6	7

Table 3a. A's Profits

B's choice of strategies / A's choice of strategies	B_1	B_2	B_3
A_1	8	2	9
A_2	6	7	1
A_3	5	4	3

Table 3b. B's Profits

solution and absence of advantage in finding the opponent[9] out still applies.

There is, however, an element of artificiality in the example chosen that is responsible for the determinateness of the solution. To see this it will suffice to interchange 5 and 6 in Table 3a. The changed situation is portrayed in Table 4 which gives A's profits for different choices of strategies.[10]

B's choice of strategies / A's choice of strategies	B_1	B_2	B_3
A_1	2	8	1
A_2	4	3	9
A_3	6	5	7

Table 4. A's Profits

There is no solution now which would possess the kind of stability found in the earlier example. For suppose A again chooses A_3; then if B should find that out, he would obviously "play" B_2 which gives him the highest possible profit consistent with A_3. But then A_3 would no longer be A's optimum strategy: he could do much better by choosing A_1; but if he does so, B's optimum

[9] In this case the interests of the two duopolists are diametrically opposed and the term "opponents" is fully justified; in the previous example it would not have been.

[10] The table for B's profits is omitted because of the constant sum assumption. Clearly, in the constant sum case, B may be regarded as minimizing A's profits since this implies maximization of his own.

strategy is B_3, not B_2, etc. There is no solution which would not give at least one of the opponents an incentive to change his decision if he found the other man out! There is no stability.[11]

What is it in the construction of the table that insured determinateness in the case of Table 3 and made it impossible in Table 4? The answer is that Table 3 has a *saddle point* ("minimax") while Table 4 does not.

The saddle point has the following two properties: it is the highest of all the row minima and at the same time it is lowest of the column maxima. Thus, in Table 3a the row minima are respectively 1, 3, and 5, the last one being highest among them (*Maximum Minimorum*); on the other hand, the column maxima are respectively 5, 8, and 9 with 5 as the lowest (*Minimum Maximorum*). Hence the combination (A_3, B_1) yields both the highest row minimum and the lowest column maximum, and, therefore, constitutes a saddle point. It is easy to see that Table 4 does *not* possess a saddle point. Here 5 is still the *Maximum Minimorum,* but the *Minimum Maximorum* is given by 6; the two do not coincide, and it is the absence of the saddle point that makes for indeterminateness in Table 4.

Why is the existence of a unique saddle point necessary (as well as sufficient) to insure the determinateness of the solution? The answer is inherent in the reasoning used in connection with the earlier examples: if A chooses his strategy so as to be protected in case of any leakage of information concerning his decision, he will choose the strategy whose row in the table has the highest minimum value, *i.e.,* the row corresponding to the *Maximum Minimorum*—A_3 in case of Table 4—for then he is sure he will not get less than 5, even if B should learn of this decision. B, following the same principle, will choose the column (*i.e.,* strategy) corresponding to the *Minimum Maximorum*—B_1 in Table 4—thus making sure he will get at least 4, even if the information does leak out.

In this fashion both duopolists are sure of a certain minimum of profit—5 and 4, respectively. But this adds up to only 9. The residual—1—is still to be allocated and this allocation depends on outguessing the opponent. It is this residual that provides an explanation, as well as a measure, of the extent of indeterminacy. Its presence will not surprise economists familiar with this type of phenomenon from the theory of bilateral monopoly. But there are cases when this residual does equal zero, that is, when the *Minimum Maximorum* equals the *Maximum Minimorum,* which (by definition) implies the existence of the saddle point and complete determinacy.

[11] There is, however, a certain amount of determinateness, at least in the negative sense, since certain strategy combinations are excluded: *e.g.* (A_2, B_1); A would never choose A_2 if he knew B had chosen B_1, and *vice versa.*

At this stage the authors of the *Theory of Games* had to make a choice. They could have accepted the fact that saddle points do not always exist so that a certain amount of indeterminacy would, in general, be present. They preferred, however, to get rid of the indeterminacy by a highly ingenious modification of the process which leads to the choice of appropriate strategy.

So far our picture of the duopolist making a decision on strategy was that of a man reasoning out which of the several possible courses of action is most favorable ("pure strategy"). We now change this picture and put in his hands a set of dice which he will throw to determine the strategy to be chosen. Thus, an element of chance is introduced into decision making ("mixed strategy").[12] But not everything is left to chance. The duopolist A must in advance formulate a rule as to what results of the throw—assume that just one die is thrown—would make him choose a given strategy. In order to illustrate this we shall use a table that is somewhat simpler, even if less interesting than those used previously. In this new table (Table 5)[13] each duopolist has only two strategies at his disposal.

An example of a rule A might adopt would be:

If the result of the throw is 1 or 2, choose A_1;

if the result of the throw is 3, 4, 5, or 6, choose A_2.

If this rule were followed, the probability that A will choose A_1 is 1/3, that of his choosing A_2 is 2/3. If a different rule had been decided upon (say, one of choosing A_1 whenever the result of the throw is 1, 2, or 3), the probability of choosing A_1 would have been 1/2. Let us call the fraction giving the probability of choosing A_1 A's *chance coefficient*; in the two examples, A's chance coefficients were 1/3 and 1/2 respectively.[14]

As a special case the value of the chance coefficient might be zero (meaning, that is, definitely choosing strategy A_2) or one (meaning that A is definitely choosing strategy A_2); thus in a sense "pure strategies" may be regarded as a special case of mixed strategies. However, this last statement is subject to rather important qualifications which are of a complex nature and will not be given here.

[12] The authors' justification for introducing "mixed strategies" is that leaving one's decision to chance is an effective way of preventing "leakage" of information since the individual making the decision does not himself know which strategy he will choose.

[13] In Table 5 there is no saddle point.

[14] Since the probability of choosing A_2 is always equal to one minus that of choosing A_1, specification of the probability of choosing A_1 is sufficient to describe a given rule. However, when the number of available strategies exceeds two, there are several such chance coefficients to be specified.

B's choice of strategies / A's choice of strategies	B_1	B_2	ROW MINIMA
A_1	5	3	3 (MAXIMUM MINIMORUM)
A_2	1	5	1
COLUMN MAXIMA	5	5	

MINIMUM MAXIMORUM (5, 5)

Table 5. A's Profits

Now instead of choosing one of the available strategies the duopolist A must choose the optimal (in a sense not yet defined) chance coefficient. How is the choice of the chance coefficient made? The answer lies in constructing a table which differs in two important respects from those used earlier. Table 6 provides an example. Each row in the table now corresponds to a possible value of A's chance coefficient; similarly, columns correspond to possible values of B's chance coefficient. Since the chance coefficient may assume any value between zero and one (including the latter two values), the table is to be regarded merely as a "sample." This is indicated by spaces between rows and between columns.

The numbers entered in the table are the average values (mathematical expectations) corresponding to the choice of chance coefficients indicated by the row and column.[15] (One should mention that Table 6 is only an

[15] To see this we shall show how, *e.g.*, we have obtained the value in the second row and third column of Table 6 (*viz.*, 3). We construct an auxiliary table (valid only for this

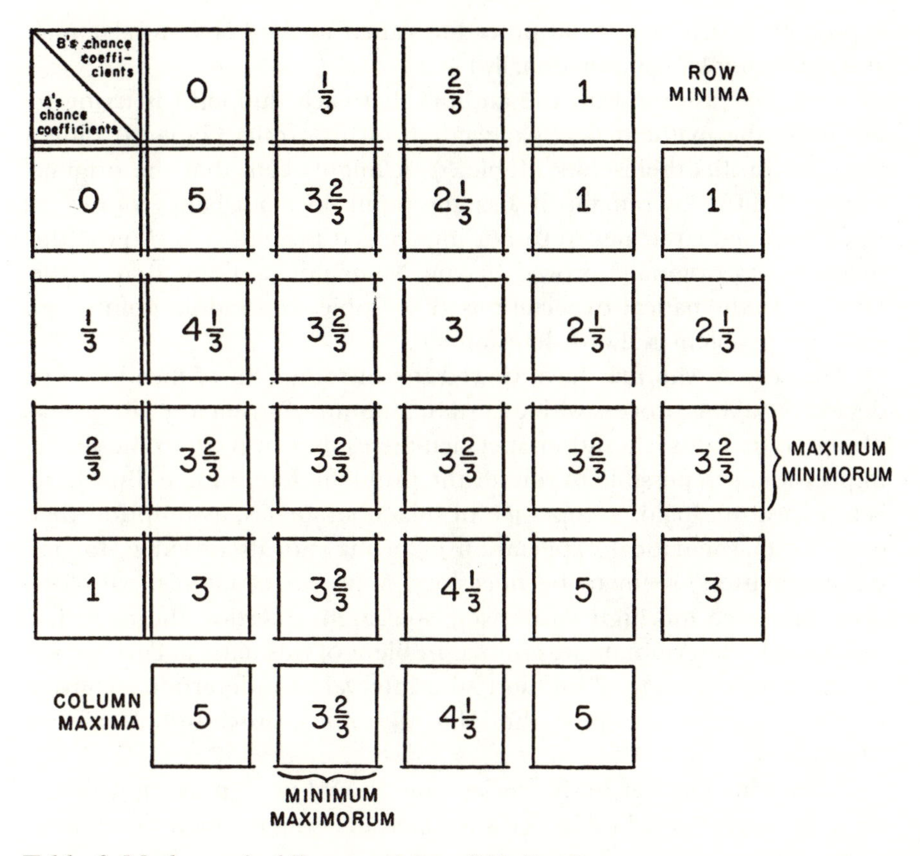

B's chance coefficients / A's chance coefficients	0	1/3	2/3	1	ROW MINIMA
0	5	3 2/3	2 1/3	1	1
1/3	4 1/3	3 2/3	3	2 1/3	2 1/3
2/3	3 2/3	3 2/3	3 2/3	3 2/3	3 2/3 MAXIMUM MINIMORUM
1	3	3 2/3	4 1/3	5	3
COLUMN MAXIMA	5	3 2/3 MINIMUM MAXIMORUM	4 1/3	5	

Table 6. Mathematical Expectations of A's Profits

---

particular combination of chance co-efficients (A's 1/3, B's 2/3).

This table differs from Table 5 only by the omission of row maxima and column minima and by the insertion of the probabilities of choosing the available strategies corresponding to the second row third column of Table 6. The computation of the mathematical expectation is indicated in Table 6.

A's choice of strategies / B's choice of strategies	B's chance coefficients / A's chance coefficients	B_1	B_2
		2/3	1/3
A_1	1/3	5	3
A_2	2/3	1	5

$$\tfrac{1}{3} \times \tfrac{2}{3} \times 5 + \tfrac{1}{3} \times \tfrac{1}{3} \times 3 + \tfrac{2}{3} \times \tfrac{2}{3} \times 1 + \tfrac{2}{3} \times \tfrac{1}{3} \times 5 = \tfrac{27}{9} = 3$$

Table 7. Computation of the Mathematical Expectation for the 2nd Row, 3rd Column in Table 6

expository device: the actual procedures used in the book are algebraic and much simpler computationally.)

If we now assume with the authors that each duopolist is trying to maximize the mathematical expectation of his profits (Table 6) rather than the profits themselves (Table 5), it might seem that the original source of difficulty remains if a saddle point does not happen to exist. But the mixed strategies were not introduced in vain! It is shown (the theorem was originally proved by von Neumann in 1928) that in the table of mathematical expectations (like Table 6) a saddle point *must* exist; the problem is always determinate.[16]

The reader who may have viewed the introduction of dice into the decision-making process with a certain amount of suspicion will probably agree that this is a rather spectacular result. Contrary to the initial impression, it *is* possible to render the problem determinate. But there is a price to be paid: acceptance of mixed strategies, assumption that only the mathematical expectation of profit (not its variance, for instance) matters, seem to be necessary. Many an economist will consider the price too high. Moreover, one might question the need for introducing determinateness into a problem of this nature. Perhaps we should consider as the "solution" the interval of indeterminacy given by the two critical points: the *Minimum Maximorum* and *Maximum Minimorum*.

As indicated earlier in this review, one should not ignore, in general, the possibility of a collusion. This is especially evident when more complex economic situations are considered.

We might, for instance, have a situation where there are two sellers facing two buyers. Here a "coalition" of buyers, as well as one of sellers, may be formed. But it is also conceivable that a buyer would bribe a seller into some sort of coöperation against the other two participants. Several other combinations of this type can easily be found.

When only *two* persons enter the picture, as in the case of duopoly (where the rôle of buyers was ignored), it was seen that a coalition would not be formed if the sum of the two persons' profits remained constant. But when the number of participants is *three* or more, subcoalitions can profitably be formed even if the sum of all participants' profits is constant; in the above four-person example it might pay the sellers to combine against the buyers even if (or, perhaps, especially if) the profits of all four always add to the same amount.

[16] In Table 6 the saddle point is in the third row second column; it is to be stressed that Table 5 has no saddle point.

Hence, the formation of coalitions may be adequately treated without abandoning the highly convenient constant-sum assumption. In fact, when the sum is known to be non-constant, it is possible to introduce (conceptually) an additional fictitious participant who, by definition, loses what all the real participants gain and *vice versa.* In this fashion a non-constant sum situation involving, say, three persons may be considered as a special case of a constant-sum four-person situation. This is an additional justification for confining most of the discussion (both in the book and in the review) to the constant-sum case despite the fact that economic problems are as a rule of the non-constant sum variety.

We shall now proceed to study the simplest constant-sum case which admits coalition formation, that involving three participants. The technique of analysis presented earlier in the two-person case is no longer adequate. The number of possibilities increases rapidly. Each of the participants may be acting independently; or else, one of the three possible two-person coalitions (A and B *vs.* C, A and C *vs.* B, B and C *vs.* A) may be formed. Were it not for the constant-sum restriction, there would be the additional possibility of the coalition comprising all three participants.

Here again we realize the novel character of the authors' approach to the problem. In most[17] of traditional economic theory the formation—or absence—of specific coalitions is *postulated.* Thus, for instance, we discuss the economics of a cartel without rigorously investigating the necessary and sufficient conditions for its formation. Moreover, we tend to exclude *a priori* such phenomena as collusion between buyers and sellers even if these phenomena are known to occur in practice. The *Theory of Games,* though seemingly more abstract than economic theory known to us, approaches reality much more closely on points of this nature. A complete solution to the problems of economic theory requires an answer to the question of coalition formation, bribery, collusion, etc. This answer is now provided, even though it is of a somewhat formal nature in the more complex cases; and even though it does not always give sufficient insight into the actual workings of the market.

[17] In his *Grundlagen einer reinen Kostentheorie* (Vienna, 1932) H. von Stackelberg does point out (p. 89) that "the competitors [duopolists] must somehow unite; they must . . . supplement the economic mechanics, which in this case is inadequate, by economic politics." But no rigorous theory is developed for such situations (although an outline of possible developments is given). This is where the *Theory of Games* has made real progress.

Let us now return to the case of three participants. Suppose two of them are sellers, one a buyer. Traditional theory would tell us the quantity sold by each seller and the price. But we know that in the process of bargaining one of the sellers might bribe the other one into staying out of the competition. Hence the seller who refrained from market operations would make a profit; on the other hand, the nominal profit made by the man who did make the sale would exceed (by the amount of bribe) the actual gain made.

It is convenient, therefore, to introduce the concept of *gain*: the bribed man's gain is the amount of the bribe, the seller's gain is the profit made on a sale minus the bribe, etc. A given distribution of gains among the participants is called an *imputation*. The imputation is not a number: it is a set of numbers. For instance, if the gains of the participants in a given situation were g_A, g_B, g_C it is the set of these three g's that is called the imputation. The imputation summarizes the outcome of the economic process. In any given situation there are a great many possible imputations. Therefore, one of the chief objectives of economic theory is that of finding those among all the possible imputations which will actually be observed under rational behavior.

In a situation such as that described (three participants, constant-sum) each man will start by asking himself how much he could get acting independently, even if the worst should happen and the other two formed a coalition against him. He can determine this by treating the situation as a two-person case (the opposing coalition regarded as one person) and finding the relevant *Maximum Minimorum*, or the saddle point, if that point does exist; the saddle point would, of course, exist if "mixed strategies" are used. Next, the participant will consider the possibility of forming a coalition with one of the other two men. Now comes the crucial question: under what conditions might such a coalition be formed?

Before discussing this in detail, let us summarize, in Table 8, all the relevant information.

TABLE 8

I.	If A acts alone, he can get	5
	If B acts alone, he can get	7
	If C acts alone, he can get	10.
II.	If A and B form a coalition, they can get	15
	If A and C form a coalition, they can get	18
	If B and C form a coalition, they can get	20.
III.	If A, B, and C act together, they can get	25.

Among the many possible imputations, let us now consider the three given in Table 9.

TABLE 9

	A	B	C
#1	6.5	8.3	10.2
#2	5.0	9.5	10.5
#3	4.0	10.0	11.0

It will be noted that under imputation #1, B and C are each better off than if they had been acting individually: they get respectively 8.3 and 10.2 instead of 7 and 10. Hence, there is an incentive for B and C to form a coalition since without such a coalition imputation #1 would not be possible. But once the coalition is formed, they can do better than under #1; *viz.*, under #2, where each gets more (9.5 and 10.5 instead of 8.3 and 10.2, respectively). In such a case we say that imputation #2 *dominates* imputation #1. It might seem that #3, in turn, dominates #2 since it promises still more to both B and C. But it promises too much: the sum of B's and C's gains under #3 is 21, which is more than their coalition could get (*cf.* Table 8)! Thus #3 is ruled out as unrealistic and cannot be said to dominate any other imputation.

Domination is an exceptionally interesting type of relation. For one thing, it is not transitive: we may have an imputation i_1 dominating the imputation i_2 and i_2 dominating i_s, without thereby implying that i_1 dominates i_s; in fact, i_1 might be dominated by i_s.[18] Moreover, it is easy to construct examples of, say, two imputations, neither of which dominates the other one.[19]

To get a geometric picture of this somewhat unusual situation one may turn to Figure 1, where points on the circle represent different possible imputations. (The reader must be cautioned that this is merely a geometrical analogy, though a helpful one.) Let us now say that point #1 dominates point #2 if #2 is less than 90° (clockwise) from #1. It is easy to see in Figure 1 that #1 dominates #2 and #2 dominates #3, but in spite of that, #1 does not dominate #3.

[18] *I.e.*, domination may be a *cyclic* relation. For instance, consider the following three imputations in the above problem: #1 and #2 as in Table 9, and #4, where

	A	B	C
#4	6.0	7.0	12.0

Here #2 (as shown before) dominates #1 (for the coalition B, C), #4 dominates #2 (for coalition A, C), but at the same time #1 dominates #4 (for the coalition A, B): the cycle is completed.

[19] For instance, #2 and #3 in Table 9.

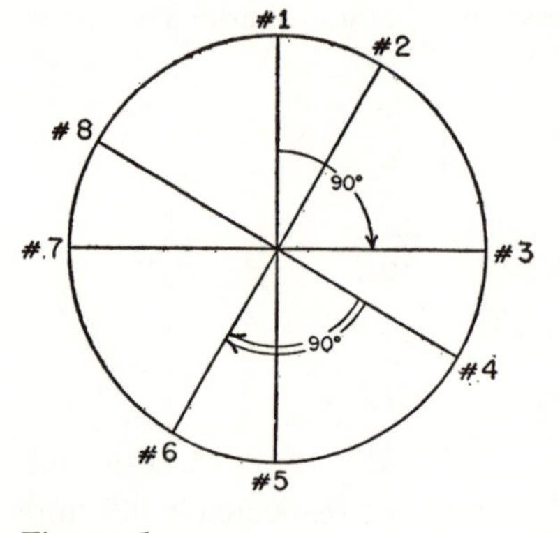

Figure 1

This geometrical picture will help define the very fundamental concept of a *solution.*

Consider the points (imputations) #1, 3, 5, and 7 in Figure 1. None of them dominates any other since any two are either *exactly* or more than 90° apart. But any other point on the circle is dominated by at least (in this case: exactly) one of them: all points between #1 and #3 are dominated by #1, etc. There is no point on the circle which is not dominated by one of the above four points. Now we *define* a solution as a set of points (imputations) with two properties: (1) no element of the set dominates any other element of the set, and (2) any point outside the set must be dominated by at least one element within the set.

We have seen that the points #1, 3, 5, 7 do have both of these properties; hence, the four points together form a solution. It is important to see that none of the individual points by itself can be regarded as a solution. In fact, if we tried to leave out any one of the four points of the set, the remaining three would no longer form a solution; for instance, if #1 were left out, the points between #1 and #3 are not dominated by any of the points #3, 5, 7. This violates the second property required of a solution and the three points by themselves are not a solution. On the other hand, if a fifth point were added to #1, 3, 5, 7, the resulting five element set would not form a solution either; suppose #2 is the fifth point chosen; we note that #2 is dominated by #1 and it also dominates #3. Thus, the first property of a solution is absent.

Contrary to what would be one's intuitive guess, an element of the solution may be dominated by points outside the solution: #1 is dominated by #8, etc.

There can easily be more than one solution. The reader should have no trouble verifying the fact that #2, 4, 6, 8 also form a solution, and it is clear that infinitely many other solutions exist.

Does there always exist at least one solution? So far this question remains unanswered. Among the cases examined by the authors none has been found without at least one solution. But it has not yet been proved that there must always be a solution. To see the theoretical possibility of a case without a solution we shall redefine slightly our concept of domination (*cf.* Figure 2): #1 dominates #2 if the angle between them (measured clockwise) does not exceed 180°.

Hence, in Figure 2 point #1 dominates #3, but not #4, etc. It can now be shown that in this case *no* solution exists. For suppose there is one; then we may, without loss of generality, choose #1 as one of its points. Clearly, #1 by itself does not constitute a solution, for there are points on the circle (*e.g.*, #4) not dominated by #1; thus the solution must have at least two points. But any other point on the circle either is dominated by #1 (*e.g.*, #2), or it dominates #1 (*e.g.*, #4), or both (#3), which contradicts the first requirement for the elements of a solution. Hence there is no solution consisting of two points either. *A fortiori*, there are no solutions containing more than two points. Hence we have been able to construct an example without a solution. But whether this type of situation could arise in economics (or in games, for that matter) is still an open question.

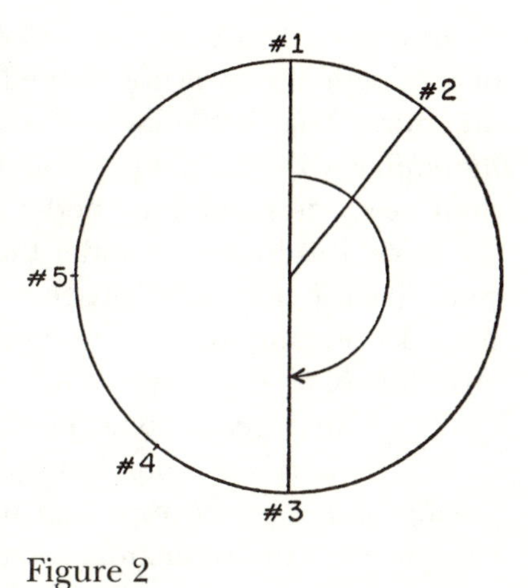

Figure 2

Now for the economic interpretation of the concept of solution. Within the solution there is no reason for switching from one imputation to another since they do not dominate each other. Moreover, there is never a good reason for going outside a given solution: any imputation outside the solution can be "discredited" by an imputation within the solution which dominates the one outside. But, as we have seen, the reverse is also usually true: imputations within the solution may be dominated by those outside. If we are to assume that the latter consideration is ignored, the given solution acquires an institutional, if not accidental, character. According to the authors, a solution may be equivalent to what one would call the "standards of behavior" which are accepted by a given community.

The multiplicity of solutions can then be considered as corresponding to alternative institutional setups; for a given institutional framework only one solution would be relevant. But even then a large number of possibilities remains since, in general, a solution contains more than one imputation. More indeterminacy yet would be present if we had refrained from introducing mixed strategies.

It would be surprising, therefore, if in their applications von Neumann and Morgenstern should get no more than the classical results without discovering imputations hitherto neglected or ignored. And there are some rather interesting "unorthodox" results pointed out, especially in the last chapter of the book.

In one case, at least, the authors' claim to generality exceeding that of economic theory is not altogether justified in view of the more recent literature. That is the case of what essentially corresponds to bilateral monopoly (p. 564, proposition 61:C). The authors obtain (by using their newly developed methods) a certain interval of indeterminacy for the price; this interval is wider than that indicated by Böhm-Bawerk, because (as the authors themselves point out) of the dropping of Böhm-Bawerk's assumption of a unique price. But this assumption has been abandoned, to give only one example, in the theories of consumer's surplus, with analogous extension of the price interval.

It will stand repeating, however, that the *Theory of Games* does offer a greater generality of approach than could be attained otherwise. The existence of "discriminatory" solutions, discovered by purely analytical methods, is an instance of this. Also, the possibility of accounting for various types of deals and collusions mentioned earlier in connection with the three-person and four-person cases go far beyond results usually obtained by customarily used methods and techniques of economic theory.

The potentialities of von Neumann's and Morgenstern's new approach seem tremendous and may, one hopes, lead to revamping, and enriching in realism, a good deal of economic theory. But to a large extent they are only potentialities: results are still largely a matter of future developments.

The difficulties encountered in handling, even by the more powerful mathematical methods, the situations involving more than three persons are quite formidable. Even the problems of monopoly and monopsony are beyond reach at the present stage of investigation. The same is true of perfect competition, though it may turn out that the latter is not a "legitimate" solution since it excludes the formation of coalitions which may dominate the competitive imputations. A good deal of light has been thrown on the problem of oligopoly, but there again the results are far from the degree of concreteness desired by the economic theorist.

The reviewer therefore regards as somewhat regrettable some of the statements made in the initial chapter of the book attacking (rather indiscriminately) the analytical techniques at present used by the economic theorists. True enough, the deficiencies of economic theory pointed out in the *Theory of Games* are very real; nothing would be more welcome than a model giving the general properties of a system with, say, m sellers and n buyers, so that monopoly, duopoly, or perfect competition could simply be treated as special cases of the general analysis. Unfortunately, however, such a model is not yet in sight. In its absence less satisfactory, but still highly useful, models have been and no doubt will continue to be used by economic theorists. One can hardly afford to ignore the social need for the results of economic theory even if the best is rather crude. The fact that the theory of economic fluctuations has been studied as much as it has is not a proof of "how much the attendant difficulties [have been] underestimated" (p. 5). Rather it shows that economics cannot afford the luxury of developing in the theoretically most "logical" manner when the need for the results is as strong as it happens to be in the case of the ups and downs of the employment level!

Nor is it quite certain, though of course conceivable, that, when a rigorous theory developed along the lines suggested by von Neumann and Morgenstern is available, the results obtained in the important problems will be sufficiently remote from those obtained with the help of the current (admittedly imperfect) tools to justify some of the harsher accusations to be found in the opening chapter of the book. It must not be forgotten, for instance, that, while theoretical derivation of coalitions to be formed is of great value, we do have empirical knowledge which can be used as a substitute (again imperfect) for theory. For example, cartel formation may be so clearly "in the cards" in a given situation that the economic theorist will simply include it as one of his assumptions while von Neumann and Morgenstern would (at least in principle) be able to *prove* the formation of the cartel without making it an additional (and logically unnecessary) assumption.

The authors criticize applications of the mathematical methods to economics in a way which might almost, in spite of protests to the contrary, mislead some readers into thinking that von Neumann and Morgenstern are not aware of the amount of recent progress in many fields of economic theory due largely to the use of mathematical tools. They also seem to ignore the fact that economics developed in literary form is, implicitly, based on the mathematical techniques which the authors criticize. (Thus it is not the methods of mathematical economics they are really questioning, but rather those elements of economic theory which literary and mathematical economics have in common.) While it is true that even mathematical treatment is not always sufficiently rigorous, it is as a rule more so than the corresponding literary form, even though the latter is not infrequently more realistic in important respects.

There is little doubt in the reviewer's mind that nothing could have been further from the authors' intentions than to give aid and comfort to the opponents of rigorous thinking in economics or to increase their complacency. Yet such may be the effect of some of the vague criticisms contained in the first chapter; they hardly seem worthy of the constructive achievements of the rest of the book.

Economists will probably be surprised to find so few references to more recent economic writings. One might almost form the impression that economics is synonymous with Böhm-Bawerk plus Pareto. Neither the nineteenth century pioneers (such as Cournot) nor the writers of the last few decades (Chamberlin, Joan Robinson, Frisch, Stackelberg) are even alluded to. But, perhaps, the authors are entitled to claim exemption from the task of relating their work to that of their predecessors by virtue of the tremendous amount of constructive effort they put into their opus. One cannot but admire the audacity of vision, the perseverance in details, and the depth of thought displayed on almost every page of the book.

The exposition is remarkably lucid and fascinating, no matter how involved the argument happens to be. The authors made an effort to avoid the assumption that the reader is familiar with any but the more elementary parts of mathematics; more refined tools are forged "on the spot" whenever needed.

One should also mention, though this transcends the scope of the review, that in the realm of strategic games proper (chess, poker) the results obtained are more specific than some of the economic applications. Those interested in the nature of determinacy of chess, in the theory of "bluffing" in poker, or in the proper strategy for Sherlock Holmes in his famous encounter with Professor Moriarty, will enjoy reading the sections of the book which have no direct bearing on economics. The reader's views on optimum military or diplomatic strategies are also likely to be affected.

Thus, the reading of the book is a treat as well as a stage in one's intellectual development. The great majority of economists should be able to go through the book even if the going is slow at times; it is well worth the effort. The appearance of a book of the caliber of the *Theory of Games* is indeed a rare event.

Economica (May 1946)
T. BARNA
New Series, Volume 13, No. 50, pp. 136–38

Professors Neumann and Morgenstern have written a book designed to become a fundamental textbook of economic theory. The essence of the book is not a refinement or summary of mathematical economics but an outright condemnation of the particular mathematical methods used in economics, and the substitution for them of an entirely different mathematical approach to the central problems of economic theory.

According to the authors, the unsuccessful use of mathematics in economics (in comparison with other sciences) was due not to inherent causes but to the fact that an incorrect mathematical technique has been used. For the solution of economic problems it is necessary to remove two preliminary obstacles: the inadequate clarity in the formulation of economic problems, and the insufficiency of the empirical background. While the removal of these obstacles is necessary, this book is on a more abstract level, being concerned with the mathematical treatment of common human behaviour of economic importance. The main possibilty of progress is seen in the quantitative treatment of factors which were hitherto labelled "psychological" and considered to be outside the scope of economics. The particular type of mathematical technique (that of infinitesimal calculus) which has been applied to economics, is well suited to the "Robinson Crusoe" type of problem; here the problem is clearly that of maximisation. But when we deal with an exchange economy, with two or more participants, the nature of the problem changes because now each individual is attempting to maximise something which is also dependent on the action of others.

An exchange economy contains various interests, sometimes parallel, sometimes conflicting, and economic equilibrium is the result of the interplay of those interests. The description of such a system requires the application of a mathematical technique different from that successful in physics and other natural sciences; in the abstract sense an exchange economy resembles games of strategy. Hence the first step in working out an economic theory is the creation of a complete theory of games. The latter requires a new mathematical technique, scarcely applied in other sciences yet, on which Professor Neumann was working for over fifteen years. It is now for the first time that the theory of games is published in its completeness, and this forms the bulk of the book.

The mathematical technique is that of combinatorics and set theory. On the face of it this technique looks more difficult than the usual method of infinitesimal calculus, but probably this is only due to its unfamiliarity, not only to economists but to most scientists. On the other hand the mastering of the new technique does not require previous knowledge of higher mathematics; in fact the book explains all the mathematical concepts introduced from the beginning. There is therefore reason to hope that it may be taken up by economists; otherwise progress by this method of analysis remains unlikely.

The theory of games starts with two-person games, and slowly develops into tackling games with a great number of participants. In most games, of course, gains balance losses, while in economic society there is a positive difference corresponding to production. This difficulty is ingeniously overcome by introducing a "dummy" player, and by demonstrating that a game with a given number of participants where gains do not balance losses raises the same problems as a game with one more participant (the "dummy") where gains balance losses. As the theory of games proceeds from two-person to *n*-person games, the theory of economics advances from bilateral monopoly to the case of perfect competition, as a special case.

It is the main advantage of this approach that it gives to mathematics a more fundamental place in economics, from which it is possible to arrive at new truths, instead of merely translating literal economics into symbols. The fact that it starts with the case of monopolistic competition and arrives at the case of perfect competition only when the theory is made general, may seem to make it a more realistic approach. If we remember the misgivings of Professor Hicks when considering the "destructive consequences for economic theory" of the assumption of monopoly in perhaps the most finished product of the approach attacked in this book,[1] the contrast is even greater. But the main advantage of the new technique is its ability to deal with "coalitions" of participants, explain the reasons for the formation of "coalitions", and describe the result. It is demonstrated clearly that the *quality* of the problem changes as the number of participants increases and not only its complexity.

What is the outcome of the new approach? We should not expect new results in relatively simple cases, such as bilateral monopoly on the one hand or perfect competition on the other. But we can expect a most rigorous treatment of the different problems and an exhaustive enumeration of the possible alternative results. It will also be feasible to scrutinise the assumptions of perfect competition and examine whether the results of the old approach can be reaffirmed. Now and then interesting conclusions emerge; for instance, the exact definition of "discrimination" and the indeterminacy of the result with discrimination present.[2]

Professors Neumann and Morgenstern of course accept some of the axioms of economic theory, notably that of the profit motive. Their theory deals with economic statics only, and we must accept their assurance of its promises when applied to dynamic problems. The problem of

[1] *Value and Capital*, p. 83.

[2] Incidentally, utility, that fundamental concept of economics, is treated as a numerically measurable quantity. The reason is that any two events in the individual's complex of preferences can be equated in the preference map by assigning to them certain probabilities of occurrence. Though the place of this argument is not paramount in the book, it is no doubt of great interest to economists.

economics, however, is considered from the point of view of the mathematician only, that is as a formal problem. Is it not the case that the backwardness of the science of economics is due not only to the lack of success of mathematical economics but also to the essential difference between the social and natural sciences? Namely, controlled experiments are not, or are very rarely, possible in social sciences, and therefore the formulation of hypotheses and their empirical testing, which proved so fruitful in other fields, cannot go hand in hand. Though the empirical background has the potentialities of great progress, the fact remains that there are no constants in economic dynamics and thus the perfection of the natural sciences must needs be unattainable. But it is too soon to pronounce judgment on a book of this nature. It is certainly an important book which will be read and studied for some time. It can be considered successful, however, only if economists are able to build on its foundations, and make use of it in the ultimate field of Applied Economics.[3]

Pyschometrika (March 1951)
WALTER A. ROSENBLITH
Volume 16, No. 1, pp. 141–46

A review of this book at this late date requires explanation. The reviewer is aware of the existence of a good many reviews, and he does not hesitate to admit that some of these reviews are good reviews. Good in that they distill much of the spirit of this 650-page opus in a mere 15–20 page paper of an expository nature [1], [2]. As is understandable, most of the reviews were published in journals of economics or mathematics. However, the

[3]It might occur to the reader, without trying to distract from the merits of the book, that having developed a complete theory of games, economics was perhaps not the most fruitful field for its application. It may be that the application to politics (both party politics and power politics) might be more interesting. Perhaps the two, three, etc., party systems correspond to bilateral, trilateral, etc., monopolies, reaching more complicated patterns in the international field. Or it may be that we have the germs of a theory capable of taking into account economic and political factors simultaneously.

impact of this book transcends these fields: The problems it raises for scientists of any quantitative denomination are analogous to those raised for staff officers in the critique of maneuvers in which a new weapon has been tested.

With this in mind let us confess without further ado that this is a difficult book. What makes it difficult is not what is expected from the reader in terms of mathematical background or familiarity with the facts and problems of economics. The book is difficult because it expresses new ideas and uses new unfamiliar techniques to buttress these ideas. The authors try to develop a general theory of economic behavior de novo. In their concern for rigor they proceed in a manner that makes it hard for them to give the reader, through examples, a feeling for the power and the beauty of the new approach.

The authors—John von Neumann, a mathematician's mathematician and Oskar Morgenstern, a well known economist of the Austrian school—start out by stating their credo: Traditional mathematical economics has been unsuccessful because of the tools it has used. They were the tools of the differential calculus forged in the birth pangs of Newtonian Physics. Now the complexity of social phenomena is at least equal to the complexity of those encountered in Physics. "It is therefore to be expected—or feared—that mathematical discoveries of a stature comparable to that of calculus will be needed in order to produce decisive success in this field." The immediate task of social science (economics being the prototype of social science) is thus twofold: (1) continuation in the direction of the descriptive approach ("our knowledge of the relevant facts of economics is incomparably smaller than that commanded in physics at the time when the mathematization of that subject was achieved") and (2) development of a mathematical precision tool for a limited field. The authors' scholarly modesty as well as their scientific and social philosophy are expressed in these sentences: "The great progress in every science came when, in the study of problems that were modest as compared with ultimate aims, methods were developed which could be extended further and further. . . . The sound procedure is to obtain first utmost precision and mastery in a limited field, and then to proceed to another, somewhat wider one, and so on. This would do away with the unhealthy practice of applying so-called theories to economic or social reform where they are in no way useful." This is an important point for the methodology and strategy of a good many incipient scientific disciplines.

The authors' advice is then briefly this: turn away from the "burning" questions, concern with them merely delays progress. Find out as much as you can about the behavior of the individual and the simplest forms of exchange. Develop gradually a theory based on a "careful analysis of the ordinary everyday interpretation of economic facts." This is a heuristic procedure: you are just groping your way from unmathematical plausibility considerations to a formal structure. Too bad! But this is the way to proceed if you want your final theory to be mathematically rigorous and conceptually general. In its first applications your results will appear trivial since they were never in doubt. But continue, work with more complicated problems until finally you will score your real successes when you can mathematically predict what will happen.

The reviewer, who is not an economist, must confess here to a certain bewilderment. He has no doubt that such a strategy has proved successful in the natural sciences where controlled experimentation led theory down a primrose path. He feels, however, less sanguine about the possibility of keeping "burning" or controversial issues out of the construction-job marked "theory of economics." The economic facts of life seem much too interwoven with the behavior—rational or otherwise—of the individual and of society. The simple recipe of chanting "an economic fact is a fact is a fact" may not constitute a powerful enough incantation to dispel the next fellow's or the next society's plausibility considerations. The authors were undoubtedly aiming at scientific neutrality. Still at least one reviewer "doubts whether their method based essentially on a capitalist form of production covers all rational economics" [3]. It is safe to say that the goals and needs of a society interact with the building of an economic theory in a complex manner.* The way in which the *Theory of Games* has caught on in the fields of economic and military strategy makes it relatively safe to state that this mathematical structure too is keeping its appointment with burning questions [4]. These somewhat critical remarks do not detract from the intrinsic value of the book and from the real enjoyment that is felt by the serious reader willing to dig through a prose heavily loaded with footnotes and references to earlier sections.† The student will soon find himself fascinated by the ease with which combinatorics, set theory, and linear algebra are developed under the very nose of unsuspecting penny pitchers and poker players.

*Some of these interaction problems were considered in recent papers read before the Boston meetings of the Institute for the Unity of Science. See in particular the papers by Dr. A. Kaplan on "Scientific Method and Social Policy" and Prof. Philipp Frank on "The Logical and the Sociological Aspects of Science." Dr. Kaplan was mainly concerned with the role of perspective, programmatic, and methodic scientism while Prof. Frank attempted to analyze the extra-scientific factors responsible for the acceptance of a particular theory.

†A sample from section 15.4.3 (page 119): "The interpretation which we are now going to give to the result of 13.5.3 is based on our considerations of 14.2–14.5—particularly those of 14.5.1, 14.5.2—and for this reason we could not propose it in 13.5.3."

The *Theory of Games* approaches economic theory from the viewpoint of the individual. It must therefore make certain assumptions concerning his motives. The authors do not hesitate to accept the traditional view according to which the consumer wants to obtain a maximum of satisfaction and the entrepreneur a maximum of profits. Once maximization of utility has been stated as the principle of rational behavior a further assumption is necessary before we can manipulate the variable "utility" numerically (for simplicity's sake we might for instance decide to use monetary units to measure utility). We must "accept the picture of an individual whose system of preferences is all-embracing and complete, i.e., who for any two imagined events (or combination of events with stated probabilities) possesses a clear intuition of preference." In their axiomatic treatment of utility von Neumann and Morgenstern combine *this* condition of a complete system of preferences with the condition of transitivity of preference relations into the single concept of complete ordering.* Our authors emphasize that they are dealing only with utilities experienced by one person with no implications concerning the comparison of utilities belonging to different individuals. Nobody should therefore expect to simply open the book in order to find weighting functions that would permit him to determine the utility function for a social group. In this connection von Neumann and Morgenstern state that the social maxim of "the greatest possible good for the greatest possible number" is self-contradictory, since "a guiding principle cannot be formulated by the requirement of maximizing two or more functions at once."

We are now almost ready to take a look at what constitutes a solution, i.e., a set of rules for a participant in an economic game. Since we cannot explore all possible types of games let us first see if we cannot categorize the types of possible economic situations an individual might encounter. A brief summary of these categories resembles a primitive system of counting: one, two, many. In the Robinson Crusoe economy the mathematics is, theoretically, simple: there are a certain number of wants and a certain number of commodities and the problem is to obtain maximum satisfaction. Obviously an ordinary (though admittedly multi-variable) maximum problem.—Now take the case of two participants in a social exchange economy. This case turns out to have particular significance in the formulation of the whole theory. It has still certain elements in common with a maximum problem but certain radically

*The axiomatic treatment leaves utility a number determined up to a linear transformation. For a discussion of the relation of transformation groups to psychological scales see the Chapter by S. S. Stevens on "Mathematics, Measurement and Psychophysics" in the forthcoming *Handbook of Experimental Psychology* (S. S. Stevens, editor).

new features have been added. Each participant attempts to maximize a function of which he does not control all the variables.

As the number of players on the economic stage increases beyond two a new concept comes to the fore, the concept of *coalition*. By means of coalitions we can attempt to reduce a more complex exchange economy to what is essentially a two-person game, but this task is neither easy nor can it always be carried out convincingly. The hope still persists that, as in physics, it will some day be easier to deal statistically with an economy of 150 million people than with the problems involving exchanges between the butcher, the baker, and the candlestick maker. But the authors strongly insist that "only after the theory for moderate number of participants has been satisfactorily developed will it be possible to decide whether extremely great numbers of participants simplify the situation." They stress that the analogy with the celestial vs. statistical mechanics situation tends to be fallacious. There the *general* theory of the mechanics of several bodies is well known. The difficulties that stem from the special, computational application of the theory to, let us say, the solar system are greater than those encountered in predicting the overall behavior of for instance 1025 freely moving particles.

We have now seen under what circumstances the concept of utility can be handled numerically and we have some ideas concerning the types of economic situations we have to investigate. The aim of the inquiry is to find the mathematically complete principles that define "rational behavior" for the participants in a social economy. While these principles ought to be perfectly general, it might be easier to start out by finding solutions for certain characteristic special cases. The next question is: how will we recognize a solution when we see one? Here is the intuitively plausible concept of a solution according to our authors: Each participant must have a set of rules telling him how to behave in *every* situation that may arise (in other words these rules make allowance for irrational behavior on the part of others).

This is a point at which the great similarity between economics and the "everyday concept of games" is driven home. Games become now formally mathematical models for social and economic problems. They constitute ideal theoretical constructs: they are amenable to precise, exhaustive, and not too complicated definitions; they further bear a resemblance to reality in the traits judged *essential* for the purposes at hand. The solution of the game which is derived from these constructs is in general an involved combinatorial catalogue. Its summary for the individual answers the question of *how much* he will get if he behaves "rationally." This is the minimum he can get; if others make mistakes, he gets more.

In particularly simple games the solution will consist of a single imputation, i.e., a single statement as to how the total proceeds are to be distributed among the participants. However, as soon as we get into more complicated games our solution undergoes parthenogenesis: the single imputation is replaced by a set of imputations: It turns out that this set of imputations is not ordered; in other words no single imputation is superior to ("dominates") all others. To our authors this lack of transitivity is a most typical phenomenon in social organizations. If the dominance relations between various states of society are of a cyclical nature (*B* is superior to *A*, *C* is superior to *B*, and finally *A* is superior to *C*; compare this with the "paper form" of horses, or baseball teams if you wish), then we have not only different possible equilibrium positions but also a possibility of passing from one of these equilibrium states to another.

This brings us to the next important point which is the static nature of this theory. The authors are aware of the fact that a dynamic theory would be more complete and therefore preferable. Yet they feel that it is futile to try to build such a dynamic theory as long as the analysis of equilibrium states is not yet thoroughly understood. This static character of the theory is of course a very serious handicap if these models are to be used for the study of adaptive or learning processes.

Much of what we have discussed up to this point comes from the introductory chapter. Chapter 2 furnishes a general formal, set-theoretical description of games of strategy. Many of the important terms are defined in these 40 pages. The unifying concept of a player's strategy emerges: it constitutes the plan of action which specifies what choices he will make in every possible situation.

This formal model is now put to work. If we leave Robinson Crusoe playing solitaire on his island, the simplest of the remaining games is the zero-sum two-person game. Zero-sum games are an important classification of all possible games. As the name implies the sum of all payments involved, by all players at the end of the game, is always zero. Two-person games are simple because of the absence of coalitions. Under these circumstances the main problems can be formulated as follow: How does a player plan his strategy? How much information does he possess and what role does the amount of information play in determining his moves? The zero-sum two-person game constitutes in other words a good dry run for the entire theory.

Let us see what happens in this two-person game. Jim's moves are determined by the rules of the game and by his desire to win as much as possible. But he proceeds cautiously; he assumes that his strategy has already been found out by his opponent Joe. This is obviously the worst that could have happened to Jim. Jim chooses therefore a strategy that will assure him a gain that is not less than a certain amount (or a loss not greater than a certain amount). If Joe actually is not as smart as Jim gives him credit for, Jim will be better off than he anticipates.

The core of the zero-sum two-person game is constituted by the Min-Max (or Minimax) problem. With its help the authors show that games in which perfect information prevails (like chess, for example) are particularly rational or strictly determined. For these games permanently optimal strategies exist. In this sense if the theory of chess were really fully known there would be nothing left to play.

Then what about non-strictly determined games like Matching Pennies? Is there any hope that the "common-sense" behavior of players will yield a clue to a solution? Since it is hard to find out the intentions of your opponent, the next best thing in this game is to concentrate on avoiding having your own intentions found out. This can be done by playing a statistical or "mixed strategy" (play "tails" or "heads" with a probability mixture of 50:50) to protect yourself against loss. Our solution can therefore be couched in terms of mixed strategies.

Before we go on to further theoretical considerations we get a chance to play a few elementary games like Stone, Paper, Scissors. We also watch Sherlock Holmes escape Professor Moriarty by helping him pick a good strategy. And as a special test we get initiated into the intricacies of a rather formalized stud poker. The emphasis is on bluffing.

With the theory of the zero-sum two-person game as our base of operations we attack the zero-sum three-person game. The analysis is here dominated by the concept of coalitions: genesis, internal arrangements and understandings, strength and stability. From here we go on to the general treatment of the zero-sum n-person game, and finally to the most general type of game by removing the restriction of the zero-sum. Here we leave the realm of games played for entertainment to enter the realm of economic reality since the sum of all payments, or the social product, is in general different from zero.

A non-zero-sum game of n persons can be shown to be reducible to a zero-sum $(n + 1)$-person game by the introduction of a fictitious player, Mr. $(n + 1)$. He turns out to be a handy mathematical gadget and a sad character at the same time. He obligingly permits us to make the sum of the amounts obtained by the players equal to zero by picking up the total check. In order to be able to do this he must have no influence whatsoever on the course of the game and remain excluded from all transactions connected with the game.

We have now run the whole economic gamut from Robinson Crusoe's simple maximum problem to Mr. $(n + 1)$'s strange market place. In the remaining chapter von Neumann and Morgenstern discuss generalizations of their concepts of solution, domination, and utility.

The reviewer does not want to imply that all readers will get to these ultimate extensions of the theory. He feels that the reader will feel well rewarded if he works through the first five chapters, i.e., up to and including the zero-sum three-person game. Most of the fundamental ideas are developed in these 240 pages, including the classic Minimax problem and the treatment of coalitions. He may by that time be able to think of some relevant applications of the theory which has unfolded before him or he may have gained enough courage to strike out in new directions.

The reviewer does not feel competent to evaluate the promise the book holds for the future, nor to prescribe it as a wonder drug to those who are dealing with difficult quantitative problems in this area. He feels, however, that the effect of the von Neumann-Morgenstern opus will not be that of strong medicine but rather that of a beneficial catalyst in the thinking of social scientists.

References

1. Hurwicz, L., *Amer. econ. Rev.*, 1945, 35, 909–925.
2. Marshak, J., *J. pol. Econ.*, 1946, 54, 97–115.
3. Gumbel, E. J., *Ann. Amer. Acad. pol. soc. Sci.*, 1945, 239, 209–210.
4. McDonald, J., A theory of strategy, *Fortune*, 1949, June, pp. 100–110.

Heads, I Win, and Tails, You Lose

PAUL SAMUELSON

This classic in the history of intellectual thought is now 20 years old and available in paperback form. Representing the collaboration of a mathematical genius and a gifted economist, the book not only has provided aesthetic delight to thousands of readers and a fertile field for subsequent mathematical research but also it has provided direct stimulus to the related fields of personal probability, decision-making in statistics and operations research, linear programming and more general optimizing. Indeed the book has accomplished everything except what it started out to do—namely, revolutionize economic theory.

Nevertheless, *Theory of Games* is a work of genius which, because it throws some light on a conundrum of the ages, will be remembered a thousand years from now, whatever be the forms of economic life in that distant future. Its more than 600 pages bristle with mathematical symbolisms, which will make it seem like Greek to the vast majority of educated men until we move into C. P. Snow's happy state when men of learning know more than one culture. One pities the tone-deaf person to whom a Mozart symphony seems mere noise. In the same way, one cannot help but feel pity for the person so innocent of mathematics as to be precluded from entering into the promised land of modern science and philosophy.

Contrary to folklore, mathematical ability is not a rare gift uncorrelated with other intellectual abilities: testing demonstrates that the child good with words and logic is most likely to have native potentiality for mathematics also. That schools, along with destroying our pleasures in Shakespeare, should turn us out ignorant of and resentful of mathematics, is a crime. And not because, in this age of Sputnik and automation, mathematical proficiency is a prerequisite of national prosperity and survival, but rather because of the sheer fun that people miss who cannot enjoy a Mozart symphony or a neat problem in elementary probability.

C. P. Snow set up understanding of the second law of thermodynamics as an acid test for 20th-century literacy. Game theory might, I suggest, be as significant a criterion. Here are a few of its basic notions!

First, von Neumann points out the basic difference between a one-person maximizer and a struggle between more than one person. A typical maximum problem Robinson Crusoe faces might be: Given a mile of fence, what four-sided figure will make it surround the largest area?

Simple algebra or calculus (or symmetry considerations) gives the answer: Make the figure a square with four equal sides of fence. An equilateral triangle will do better than any other triangle, but worse than a square. Evidently, a regular equal-sided polygon does better than any figure of the same number of sides; and the more sides to the regular polygon the greater the enclosed area. Hence, as Queen Dido of Carthage knew, a perfect circle is the best figure of all if you wish to maximize area surrounded by prescribed perimeter.

But what happens when two rational minds meet in conflict, with the destiny of each dependent upon the decision of both? A game of chess, or of tic-tac-toe, provides a simple example.

If I make my cross first, you can never beat me at tic-tac-toe if I play right. If you play right, I can't beat you. A draw is the game's solution. On the other hand, in the game where we each put a nickel on a round table until one of us finds no space to hold a nickel, I can be sure of winning if I can go first: I put my first nickel in the exact center and thereafter merely put a nickel opposite your last position; I am never at a loss for a safe move and you eventually will be.

Chess, being a game of perfect information, turns out to be a trivial game like these two. Two perfect players would either tie, have victory go to the first mover, or to the second mover. We don't know which of these holds for chess but game theory, by working backward from the end of the game, proves chess has this trivial property.

If chess is trivial, penny-matching is not. To match you, I will turn up heads always if you will turn up heads. But if you know I will turn up heads, you will turn up tails—which means I should turn up tails—which means you will reverse your strategy. So it goes in an infinite regress.

Now here is where the genius of von Neumann, the French mathematician Borel, and the English statistician Sir Ronald Fisher managed to rout the infinite regress. "If you don't want your secret found out, don't know it yourself. Let a flip of the coin determine whether you turn up heads or tails. Against this *probability* strategy, your rational opponent can never win more than half the time even if he finds you out."

Von Neumann proved the basic theorem of 2-person games where Man 1 tries to maximize his average gain and Man 2 tries to minimize 1's average gain, thereby maximizing his *own* average gain. Each man should employ a *probability strategy* that maximizes his most vulnerable points—for a chain is only as strong as its weakest link (at least against a rational adversary). Here are a few consequences of this far-reaching theorem.

1. Always to bluff in poker is to be called with losing hands. Never to bluff is to invite others to drop out where you have good hands.

There is an optimal probability of bluffing that will maximize your mean gain against a rival who is following optimal probability strategy to maximize his gain.

2. In quizzing students, I select questions from the textbook at random so that students must prepare all the work.
3. Forbidding strikes and lockouts must change the outcome of collective bargaining.

Beyond the two-person case of perfect opposition, game theory becomes more complicated and less definite. (When three sadists step into a room, two will gang up on the third but we can't predict which two.) Still the book abounds with pregnant suggestions.

Game theory, and its extension, can help a little in playing the horses, buying stocks, perhaps negotiating with the Russians in a world involving the Chinese. Can it help you decide whether to have your only child operated on when the outcome can be death or complete cure? This is an unsettled question, perhaps to be debated forever. Mathematics alone won't answer the philosophical issue, but, without a program and dictionary, you can't even begin to follow the ball game.

Big D

PAUL CRUME

As we predicted some time back, King Hill made All-America, and now this column is finally and flatly going out of the sports prediction business. You can push this psychic stuff just so far before the ghosts rebel.

For a long time, though, we have meant to do a piece on the science of predicting generally. A mathematical genius named von Neumann has already tried it. He wrote a book called the *Theory of Games,* which is a best seller among five or six people these days. As we get his theory, if something happens once in five times and something else happens five in seven, something quite different will happen if you add the two together, provided it is a square root.

As you can see, this is an obvious oversimplification.

Von Neumann's approach fails because it controverts a Great Natural Law, as follows: If you have to add numbers to get the answer to anything, you will never get it right.

. . . . .

Also, anything that has to have its square root taken is already in pretty shaky shape. You'd better discard it and go on to something else.

. . . . .

All life is governed by certain Great Natural Laws like the one—Parkinson's Law, we think—which goes: "If you keep fooling around something, sooner or later it will come off."

The root Great Natural Law for successful predictions is as follows: Things which are caused by something are really caused by something else.

Let us forecast what you will have for dinner. What are the causative factors? You have had ground beef for lunch. You do not much like ground beef, and the brown gravy on this was bad. So you determine on a wonderful dinner to make up, maybe lobster thermidor. Lobster is cheap, a dime a dozen, and you can save a lot of grocery money. You already have a bottle of fine wine to go with it. The trimmings of the meal are in the kitchen. You love lobster thermidor. These are the causative factors of what you will have for dinner.

Now, let's add the imponderables. The youngster comes home from school with a sudden need for a new spelling tablet. On the way out the door on the way to the store, the cat runs out between your feet. The neighbor's dog chases him up a light pole. The fire truck runs over the neighboring boy's bicycle. The neighbor has a good left. Your lawyer is out of town for two hours.

There was ground beef in the ice box.

What do you think happened?

.....

All this is in accord with the Theory of Logical Repulsion, which says that a force set in motion automatically creates an equal force in a direction you hadn't thought of.

.....

There are certain Lesser Natural Laws that enter into predicting. One is: Things that never happen usually do. Things that always happen never do. Another: Things equal to the same thing usually mess up.

Square roots have nothing to do with it. As a matter of fact, people have shown us square roots they said they took out of something, but they've never been able to prove it.

Mathematics of Games and Economics

E. ROWLAND

This book is based on the theory that the economic man attempts to maximize his share of the world's goods and services in the same way that a participant in a game involving many players attempts to maximize his winnings. The authors point out that the maximization of individual wealth is not an ordinary problem in variational calculus, because the individual does not control, and may even be ignorant of, some of the variables. The general theory of social games, in their view, offers a simplified conceptual model of economic behaviour, and a study of that theory can do much to throw light on certain basic concepts of economics, for example, that of utility.

The first chapter of the book introduces this basic idea, and seeks to establish the dualism between standards of economic behaviour and strategies in social games. It is stated that various authors have tried to build up a mathematical theory of economic behaviour, with a conspicuous lack of success, due to a combination of unfavourable circumstances. The most important of these is that the economic problems are not formulated clearly, and are often stated in such vague terms as to make any attempt at a mathematical solution hopeless. Also the empirical background is definitely inadequate. Another difficulty is that the problems which have been attempted have frequently been dealt with by methods so complicated that the so-called proofs are really only assertions, which are of less value than the same assertions in literary form. The great progress in every science came when, in the study of problems which were modest as compared with ultimate aims, methods were developed which could be extended further and further. The sound procedure is to obtain first utmost precision and mastery in a limited field, and then to proceed to another somewhat wider one, and so on, until finally the field of real success is reached, that is, prediction by theory. The authors state that there is no reason why economics can expect any easier fate than that which befell other sciences such as physics and chemistry, and that it will not be possible to build up a complete mathematical theory for many years to come.

Having formulated the basic economic problems, the book now turns to the mathematics of games themselves. The mathematics are self-contained, in the sense that theory is developed from primitive ideas, but an extensive use is made of set-theory, and of course mathematical notation. Forty pages are spent in formulating the general description of a game, introducing all possible complications into the theory, and finally describing the game in terms of ten axioms. It is then shown that these can be reduced to three, the two definitions being rigorously equivalent. The two definitions are designated as 'Extensive' and 'Normalised' respectively, and it is sometimes found convenient to use one, and sometimes the other in the problems which follow.

The next chapter starts with a discussion of one-person zero-sum games, for example, patience, then advances to zero-sum two-person games, the discussion being dominated by illustrations from chess, poker, bridge, etc., and not from cartels, markets, oligopolies. The theory now passes on to three-person zero-sum games. It is shown that whereas a one-person zero-sum game is merely a simple maximum problem, the passage from a one-person zero-sum game to a two-person zero-sum game obliterates the maximum problem, and the game is designated by a clear-cut opposition of interest. Similarly the passage to the three-person zero-sum game obliterates the opposition of interest. Here independent coalitions may arise, and the relationships between two players may be manifold. The game can be reduced to three two-person zero-sum games.

The theory now advances to the n-person zero-sum games. Little penetration below the surface has been made, however, and the theory comes back to a discussion of the case $n = 4$. Even here, however, the problem is soluble only in certain special cases. Some aspects of $n \geqslant 5$ are now considered, but the problem becomes so complicated that it seems rather hopeless to push this approach beyond $n = 5$. On the other hand, it is absolutely vital to get some insight into the conditions which prevail for the greater values of n. Quite apart from the fact that these are most important for the hoped-for economic and sociological applications, there is also this to consider: with every increase of n, qualitatively new phenomena appeared. This was clear for each of $n = 2, 3, 4$. It was not observed for $n = 5$, but this may have been due to lack of detailed information about the case. It will appear later that very important qualitative phenomena make their first appearance for $n = 6$.

For these reasons it is imperative that some technique should be evolved for the attack on games with higher n. In the present state of affairs nothing systematic or exhaustive can be hoped for. Consequently, the natural procedure is to find some special classes of games involving many participants which can be decisively dealt with. Two families of such special cases are then formulated and discussed, each considered as generalizations of four-person games.

The trend of the book now alters. The zero-sum restriction is dropped, and the theory comes into close contact with questions of the familiar economic type. After a short discussion of the general case which consists mainly of plausible arguments, an economic interpretation is given for $n = 1$ and 2, and for a special case of $n = 3$.

From the point of view of notation alone, it is doubtful whether anyone but a fairly competent mathematician would be able to follow the arguments put forward. This was unavoidable, but seems a pity as the book will remain closed to economists who are unfamiliar with mathematical ideas and methods.

Theory of Games

CLAUDE CHEVALLEY

Attempts to apply mathematical methods in economics have been frequent, but until now not altogether successful. The reason for this, according to J. von Neumann and O. Morgenstern, is that too often one has tried to follow the pattern indicated by the mechanical or physical theories, where the spotlight is taken by differential equations expressing the immediate future of a system in terms of its present condition. The approach of the present authors is radically different; they treat economical life as a game played by a finite number of players according to certain rules, and they investigate the possible types of behavior compatible with the rules of the game and with the principle that each player tries to get the maximum advantage.

A prerequisite for a satisfactory mathematical analysis of any situation is to have a system of axioms, which adequately describes the situation in mathematical terms. Given such a system of axioms, it is then possible to deduce results from it in a purely logical way, without taking into account the special meaning which the mathematical entities one operates on have in terms of the concrete situation to which the theory refers. It is only at the end of this process of logical deduction that one reintroduces the concrete significance of the mathematical symbols, thus transforming the theorems in the theory into statements of fact about the given situation, statements which may then be compared with the results of experimentation or observation. Following this procedure, the authors first give a detailed axiomatic description of the notion of a game.

This mathematical description of a game being posited, the main object of the book is the theory of the mathematical object which has from now on taken the place of a concrete game. The direction in which this theory is developed is naturally motivated by what one desires to learn about the given concrete game; but this is only a motivation—the reader who would know nothing about the game in question could not object logically to any of the steps which are taken, although he would often fail to understand why they are taken.

The first notion to be constructed is that of a strategy for a certain player. Roughly speaking, a strategy is a rule which tells the player what he should do in every circumstance. Now, if each player follows a certain strategy, the course of the game is determined and it is known how much each player will receive at the end.

But, of course, each player has only control over his own strategy, not on the strategies of his opponents. The first question is therefore the following: each player being ignorant of the strategies followed by his opponents, which strategy will he follow in order to get the maximum possible advantage for himself?

This question is solved in the case of games with two players in which each player receives exactly what the other is losing. It is shown that in games (such as chess) where perfect information prevails (i.e. each player knows at all times everything which has already happened in the game) there exists for each player a best possible strategy. This means that one of the following two possibilities will be realized: either one of the players has a strategy which guarantees his winning, or else each player has a strategy which insures him absolutely against anything worse than a tie. The situation is more complicated in the case of imperfect information. Nevertheless, the question is solved in this case too by the introduction of "mixed strategies." To follow a mixed strategy consists in following sometimes one strategy, sometimes another, with certain probabilities. It can be proved that, by following a suitable mixed strategy, the first player can make certain of receiving at least a certain sum v (the value of the game for the first player), while the second player can always prevent the first one from obtaining more than v. (The number v may of course be positive, negative or 0; if v is negative, what we said means that the first player can insure himself against a loss exceeding $-$v.)

The question of games with two players being thus solved, the authors attack the problem of a game with more than two players. In such a game, it may happen that the players agree to divide in two coalitions, the coalitions playing against each other. In this case, the game really becomes a game between two players, and the previous results can be applied. Thus, each coalition has a certain value, which represents the minimal amount which this coalition is certain to obtain even if all other players unite against it. The whole discussion of the game is based by the authors upon the study of the values of these coalitions. One tries of course to obtain indications as to which coalitions are likely to form and as to how the profits won by a coalition will be divided among its members. The final outcome of the game will be described as an imputation which states how much each player finally receives, either in the form of returns prescribed by the rules of the game or in the form of compensations payed him by other players desirous for instance to draw him into a coalition. Which imputation will actually be realized cannot be decided by the theory, but certain systems of imputations (the "solutions" of the game) play a preferential role. The interpretation of this is that the choice of a solution depends on factors extraneous to the game itself, such as tradition, ethical considerations, etc.

The different solutions describe the general standards of behavior which are accepted in the society of the players. Within these standards of behavior, the solution describes which imputations are likely to be realized. The criterium for such a solution is that no group of players has any definite reason (valid for each player of the group) for preferring one imputation of the solution to another, while any imputation not in the solution is definitely less advantageous for a certain group of players than some imputation in the solution. The question whether there always exist such stable standards of behavior remains unfortunately open; on the other hand, particular examples show that several such standards can exist for the same game.

The book is written in such a way as to require nothing from the reader beyond very elementary knowledge of algebra. Those more involved mathematical notions which are needed in the course of the exposition are defined and discussed in detail in the book itself. The general theories are introduced and followed by careful discussions of special examples. The authors seize every opportunity to explain verbally the meaning of the results of their mathematical analysis of the situation. All these factors contribute to make the book highly interesting even for mathematically untrained readers.

To conclude, I would like to express the hope that this book will be instrumental in helping economics to emerge from its actual condition of vagueness and confusion to the rank of a body of precise statements bearing on precisely defined situations.

Mathematical Theory of Poker Is Applied to Business Problems

WILL LISSNER

A new approach to economic analysis that seeks to solve hitherto insoluble problems of business strategy by developing and applying to them a new mathematical theory of games of strategy like poker, chess and solitaire has caused a sensation among professional economists.

The theory has been worked out in its beginnings by Dr. John von Neumann, Professor of Mathematics at the Institute for Advanced Study, Princeton, N. J., and Dr. Oskar Morgenstern, Professor of Economics at Princeton University. In its present form the theory represents fifteen years of research, apart from the years spent by Dr. von Neumann before 1928 in working out the basic theory of games.

Dr. von Neumann, a collaborator of Albert Einstein, who did mathematical work important in the development of the atomic bomb, is recognized by his colleagues as one of the great original workers of the day in mathematics. He is the author of "Mathematical Foundations of Quantum Mechanics." Dr. Morgenstern, former director of the Austrian Institute for Business Cycle Research at the University of Vienna, is considered one of the world's leading mathematical economists. He is the author of "Economic Forecasting."

They published the results of their years of research in a 625-page book, "Theory of Games and Economic Behavior," with nearly every page studded with formulae, mainly in the theory of sets and groups and in linear geometry. The book was issued by the Princeton University Press in September, 1944.

Because the book was so "formidable" intellectually, as one professional critic put it, involving, as the authors admitted, "mathematical deductions" that "are frequently intricate," with "the logical possibilities" extensively exploited, the work went unnoticed for nearly a year and a half by the technical economic journals, with the exception of the Annals of the American Academy of Political and Social Science.

The current issue of the American Economic Review, journal of the American Economic Association, devotes a full-length article to the book hailing it as "indeed a rare event," and the critics of other professional journals have pronounced the work an outstanding contribution.

Big Potentialities Seen

Prof. Leonid Hurwicz of Iowa State College, now on leave to the Cowles Commission for Research in Economics at the University of Chicago, writes in the American Economic Review that "the potentialities of von Neumann's and Morgenstern's new approach seem tremendous and may, one hopes, lead to revamping and enriching in realism a good deal of economic theory."

Professor Hurwicz says "it would be doing the authors an injustice to say that theirs is a contribution to economics only."

"The scope of the book is much broader," he adds. "The techniques applied by the authors in tackling economic problems are of sufficient generality to be valid in political science, sociology, or even military strategy."

In mathematics as well as in economics the Neumann-Morgenstern approach represents a break with the traditional development. The mathematical theory of games was based on the calculus of probabilities. Thus it was concerned with the probable distribution of cards at any stage of the game.

It was shown, for example, that the odds against any individual poker player's holding any one of the forty classes of straight flushes before the draw are 64,973 to 1; that drawing to two pairs, the chances are 11 to 1 against getting the third to either pair. In the eighteenth century classic, "Hoyle's Games," the bible of the card and the horse player, many of these probabilities are given for card games and races, although the calculus then was in its infancy.

However, the present mathematical methods, essentially those of the differential and integral calculus, were not adequate for solving some of the problems of games. One unsolved problem, cited by Prof. E. J. Gumbel of the New School for Social Research, is that of a game with several participants, each trying to maximize his gain without being able to control the activity of the other participants.

To solve these problems Professor von Neumann, in his preliminary work, "Zur Theorie der Gesellschaftsspiele," published in the Annals of Mathematics in 1928, and in the present work developed with Dr. Morgenstern a novel analytical apparatus drawn mostly from mathematical logic, set theory and functional analysis.

Strategies Analyzed

Using these new mathematical tools, they analyzed the strategies open to players in games of chance like dice-throwing and matching pennies and in games of strategy like poker, bridge and chess. They proved that there might be no best strategy for a player to employ, or there might be several equally good ones.

In matching pennies, for example, where the opponent shakes up his pennies and lets them fall into an order fixed by chance, the best strategy is to adopt none. For Professors von Neumann and Morgenstern have proved that if the opponent "stacks" his pennies according to any preconceived plan, one can win by matching chance against strategy; that is, by shaking up the pennies and letting chance determine their order.

In chess, however, he proved that a good strategy existed because this is a game of perfect information, each player knowing at every stage of the game all the moves made previously.

Similarly for games of mixed strategies, where information is imperfect, they proved that a good strategy was possible. Each game they showed, has its own characteristic function. Thus, as Louis Weisner of Hunter College has pointed out, for games of more than two players like poker, where the players may form coalitions that are decisive to the strategies, "all questions relating to coalitions such as the forces which impel the formation of coalitions, the desertion of some members of a coalition to another, mergers or fights between coalitions, the division of profits or losses among members of a coalition are answerable in terms of the characteristic function."

Practical Use in Games

The solutions are too technical for brief exposition and cannot be given here. A brief exposition of a few takes seventeen pages in The American Economic Review. They will be of practical use in games of chance and strategy, however, to experts in higher mathematics.

Example of Analysis of Business Strategy

To make the mathematical thinking of Dr. John von Neumann and Dr. Oskar Morgenstern more easily understandable, Leonid Hurwicz of the Cowles Commission for Research in Economics, with the aid of Mrs. D. Friedlander of the University of Chicago, has converted several of the mathematical formulas into numerical examples. These tables are one of them, as presented in The American Economic Review.

They illustrate a duopolistic situation, one in which the market is dominated by two sellers, A and B, where each one of the duopolists is trying to make the largest profits. The first table shows the profits seller A will make, depending on his own and B's choice of strategies. The strategies open to seller A are denoted by A_1, etc.; the columns correspond to the strategies of seller B, and are denoted by B_1, etc. Analogous information regarding B's profits is given in the other table.

B's choice of strategies / A's choice of strategies	B_1	B_2	B_3
A_1	2	8	1
A_2	4	3	9
A_3	5	6	7

A's Profit

B's choice of strategies / A's choice of strategies	B_1	B_2	B_3
A_1	11	2	20
A_2	9	15	3
A_3	8	7	6

B's Profit

A will notice that that by choosing strategy A_3 he will be sure that his profits cannot go down below 5, while either of the remaining alternatives would expose him to the danger of going down to 3 or even to 1. Another reason for choosing A_3 is that if there is a danger of a "leak," that B might learn of his decision, and if A had chosen strategy A_1, B obviously would choose strategy B_3 to maximize his own profits, and this would leave A with a profit of only one. Had A chosen strategy A_2, B would respond by selecting B_2, leaving A with less than the 5 he would get by choosing A_3.

Similar reasoning on B's part would make him choose strategy B_1 as the optimal strategy. Thus, von Neumann and Morgenstern reason, the outcome of duopolistic competition is determinate, A will choose strategy A_3, B will choose strategy B_1, and neither would be inclined to alter his decision even if he found out what the other's strategy was. The illustration, of course, loses the generality and the rigor of the formula.

How Drs. von Neumann and Morgenstern discovered that the theory of games of chance could be used to analyze economic situations is not reported in their book. In a letter to this writer while the book was in preparation, Dr. Morgenstern explained that in carrying on his work in mathematical economics, he encountered many problems that could not be handled by the traditional tools drawn from the calculus, and hence he reached the conclusion that economics was following a blind alley in depending mainly on them.

They explain in the book, however, that they are not making a mere analogy between games and economics but seek to establish "that the typical problems of economic behavior become strictly identical with the mathematical notions of suitable games of strategy." Their aim is the development of a theory that is mathematically rigorous and conceptually general, so that the science can go on, as all sciences have gone on, to genuine prediction by theory.

Five-sixths of the book is devoted to the mathematical theory, one-sixth to the economic applications. Among the latter are situations characterized by duopolistic and oligopolistic competition, where the market is dominated by two or a few sellers; the formation of coalitions, where buyers or sellers combine against other groups of buyers or sellers; and by organization of cartels. Not only "rational" but "irrational" economic behavior is comprehended by the theory.

Professor Hurwicz holds that reading the book will be a new stage in their education for the great majority of economists; the others may not be able to follow it.

Reader's Views Seen Affected

He says that because the noneconomic results are more specific, mathematical readers "interested in the nature of determinacy of chess, in the theory of 'bluffing' in poker, or in the proper strategy for Sherlock Holmes in his famous encounter with Professor Moriarty" will enjoy the sections on strategic games proper. He adds, "the reader's views on optimum military or diplomatic strategies are also likely to be affected."

One defect of the theory is that it applies to capitalist economics only. This was pointed out by Mr. Gumbel in The Annals, and by Mr. Weisner in Science and Society. This criticism applies to the economic interpretation in terms of marginal utility economics, however, and not to the mathematics, the foundations of which Mr. Weisner found "firmly laid," arousing his admiration.

In sociological circles opinion was divided. The reviewer of The American Journal of Sociology "found the 'Theory of Games' at all times a model of clear and careful exposition." The reviewer of Social Forces reported "this may be an extremely valuable new direction or it may be an interesting but useless development." While very few problems of a strictly sociological nature were attempted, Drs. von Neumann and Morgenstern believe that sociometric problems are best approached from this direction.

A Theory of Strategy

JOHN McDONALD

Mathematics has been turned from the physical world to the affairs of men—economic and military—with some surprising results.

In the Spartan surroundings of a Pentagon office a young scientist attached to the Air Force said, "We hope it will work, just as we hoped in 1942 that the atomic bomb would work." What he hoped and in some sense implied will work is a newly created theory of strategy that many scientists believe has important potentialities in military affairs, economics, and other social sciences. The theory is familiarly known to the military as "Games," though its high security classification wherever it has actual content is a sign that its intent is anything but trifling. Before too much is expected, it should be said that "Games" is spectacular, if that is the word, only in an intellectual sense, and only in that sense can it be appreciated. It was originated by one of the chief participants in the development of the bomb, the young and already great contemporary mathematician, John von Neumann, whose work in games was given a preliminary exploration in an essay on poker in *Fortune* (March, 1948).

This story began more than a year ago when the author innocently looked into the game of poker with the idea of providing *Fortune*'s readers with some diverting comments on the national indoor game of strategy. When the first part of the story was published (March, 1948), however, it seemed that we had the bear by the tail. Mathematician John von Neumann, unknown to the poker-playing fraternity, had got there first and really made something out of it. Von Neumann's poker had become the essence of strategy—so much so that on further inquiry, U.S. military security officials began to get nervous. The story, a sequel to last year's, was finally written and cleared. It is recommended to the game-minded and theory-minded, for whom it promises some remarkable illuminations.

The theory of games is essentially a theory of strategy.* It concerns the elusive nature of the constant thread of opposition and conflict that runs through various interrelationships of men, from the striking of a bargain in the market to the dread clash of war. It has none of the naiveté popularly associated with longhairs. It is more *avant-garde* than Sartre, more subtle than a Jesuit, and as honest as one can safely be. Although deeply involved in the mysteries of value, it is formal and neutral. Mathematics, qua mathematics, doesn't take sides.

In the literature of social conflict, an exact description of the nature of strategy has been wanting, Machiavelli and Clausewitz notwithstanding. No dictionary defines the term. Von Neumann's conception is more than a definition, it is a theory complete on the level of pure science. Its application in military science is one of the preoccupations of the U.S. Air Force's "Project Rand," which is conducted by The Rand Corporation, an independent, non-profit, research organization of ranking scientists in Santa Monica, California. The theory has also penetrated the Navy and is having its Army genesis in the "Eisenhower Advanced Study Group," where three specially gifted general-staff officers have been turned loose for deep thought. It is closer to application in military science than anywhere, and it has shown the mark of its importance by its impact on other social disciplines, notably economic and statistical theory.

"Games" was in fact originated by von Neumann and developed in collaboration with the eminent economist, Oskar Morgenstern, specifically for application to economic theory, and in that field its effect has been highly stimulating if not incipiently revolutionary. It is a standard topic of conversation in the economics departments of many universities today. The severest of its critics doubt its capacity to put a new foundation under the science of economics, but concede that it has raised new problems, given new insights, and challenged most existing theory, particularly with respect to the nature of markets. It is not and never can be available in its entirety to the layman, for it is written largely in the special notation of the mathematician. But its outlines are clear, and the intrepid reader may find its ideas rewarding.

Skill and Chance

Von Neumann has described the strategical situation as the interaction between two or more persons, each of whose actions is based on an expectation concerning the actions of others over whom he has no

**Theory of Games and Economic Behavior*, by John von Neumann and Oskar Morgenstern, Princeton University Press, 1944. Second revised edition, 1947, $10.

control. The outcome is dependent upon the personal moves of the participants. The policy followed in making these moves is strategy. Both the military strategist and the businessman act continuously in this state of suspended animation. And regardless of the amount of information given them—short of perfect information—they generally act in the final analysis on hunch; that is, they gamble without being able to calculate the risk. As the great British economist, John Maynard Keynes, once observed, "Businessmen play a mixed game of skill and chance, the average results of which to the players are not known by those who take a hand." Von Neumann's theory is designed to narrow this gamble to the point where it would be irrational to act otherwise than by a known, optimum policy (strategy). It tries to make the imponderable ponderable.

For example: A duel. Two men back to back, each with a pistol and one bullet. They walk away from each other a certain number of paces, turn and . . . fire or begin walking back toward each other. At the farthest distance apart the chances of a hit are less than with each pace back toward the opponent. To fire and miss would be to give the opponent a sure kill at any range he chooses. To hold fire would be to risk being shot before having an opportunity to shoot. Thus two maximums conflict: to have the first shot or to have a better shot. The dueler cannot have both. The question then is when to shoot with optimum strategical value.

Given the distances and marksmanships of the duelers, *this problem has been solved in precise mathematical terms* in one of nearly a hundred papers and memoranda written by the learned at Project Rand on the subject of the theory of games. The traditional duel is a model for a number of military situations, not the least of which is that of two airplanes entering each other's range in combat. It is also the problem of dueling tanks, where the situation may be complicated by "silent guns," which hide the actual shooting until a hit is made. "Silent guns" covers all cases in which a combatant does not know when his opponent has fired, whether the guns are actually noiseless or not. Theoretically "Games" might be presumed to bear upon the problem of the atomic bomb, the strategy for which is likely to involve feints, or bluffs.

In economics the basic example is the exchange relationship of a buyer and a seller. Each tries to "maximize" his gain and, of course, both may gain; for each gets rid of something of which he has an excess and obtains something of which he is short. But the degree of economic gain is always at stake in the strategical problem of arriving at a price. The seller cannot maximize his gain pure and simple, for he must relate his asking price to an expectation concerning the buyer. Likewise the buyer cannot maximize his gain except by getting the goods for nothing. He bids according to an expectation concerning the seller. Each must take the other fellow's thoughts into account. The opposition between bid and ask is resolved in a price that represents not a maximum but a resolution of conflicting (maximum) desires in what each hopes is his optimum. All buyers and sellers have intuitive exchange policies, or strategies, for arriving at such optimums. They are the supreme talent of the businessman.

In one sense this is self-evident. It corresponds to reality. Yet it has always confounded economists in their effort to describe economic activity. No conventional formula for describing how the price of manufactured goods is arrived at can explain, for example, such a commonplace event as G.M.'s lowering its car prices just when Chrysler raised its prices.

An intricate picture of conflicting maximums was described in the story of the House of Seagram in the September, 1948, issue of *Fortune*. Samuel Bronfman of Seagram had used up a large part of his whiskey inventory in gaining record brand sales. Facing a forthcoming shortage, he could either buy scarce bulk whiskey at a losing price and maintain brand sales with a decline in profits or maintain his rate of profit out of his remaining inventory with a decline in brand sales—or follow a mixed policy. In any event he could not maximize both brand sales and profits. His archcompetitor, Lewis Rosenstiel of Schenley, had lower brand sales and a larger inventory. He could sell the inventory to Bronfman at high bulk prices and no sales cost and thus maximize profits in the immediate circumstance; or he could put the inventory into Schenley brands with an expensive and risky sales campaign to raise brand sales at less profit. Like Bronfman, but for different reasons, he could not maximize both brand sales and profits. Nor could the policies of the antagonists fail to conflict; for Rosenstiel, who, having the inventory, had the dominant choice, could reconcile their difference only by taking profit and giving Bronfman the market (brand sales), i.e., trade short-term gains for long-term losses. Within limits, Bronfman would have been glad to buy the market (short-term losses for long-term gains)—if Rosenstiel would let him. But Rosenstiel wouldn't. Each sought an optimum move on a strategical basis.* This kind of strategical problem, too, has baffled the economist. It is recognized in modern theory in a general way as a problem of "monopolistic competition" or "imperfect competition," i.e., it has elements of both monopoly (market or price control) and competition; and furthermore, the competition takes place within small numbers of competitors on a strategical plane. Seagram and Schenley, it should be noted,

* *Fortune*'s story gave their first moves, their choices, and risks. How these turned out will be reviewed in a future issue.

were treated in the story—with appropriate acknowledgment—somewhat as if they were a "duopoly" (two sellers confronting many buyers). Actually the liquor-distilling industry, like the automobile-manufacturing industry and many others in the U.S., approximates an "oligopoly" (several sellers, each of whom must take into account the market influence of the others). The semblance of duopoly was occasioned by a contest between the top two for the leadership. Theoretically, Seagram and Schenley had another alternative, namely, to combine. There are good reasons why they would not, yet each of these corporate individuals had grown large through combination.

Competition or Monopoly?

The necessary cause of such combinations, typical of modern economic life, has likewise escaped theoretical formulation by economists. Only in the past twenty years has "modern" economic theory recognized "monopolistic competition" as central to the competitive economic system—rather than as a discrepancy in the concept of pure competition. Classical economics makes its calculations on the assumption of pure competition. Under pure competition, it is generally held, no single individual produces a large enough amount of the total supply to effect a relevant price change. This situation may be known to the farmer (and may not be), but is rarely known to the manufacturer. (For a noteworthy exception, the ladies' garment industry, see "Adam Smith on 7th Ave.," *Fortune*, January, 1949.) Free competition among many individuals is being steadily displaced by the competition and struggle of smaller numbers of combinations of individuals in such forms as corporations, trade unions, and various associations.

The theory of games approaches economic life from the standpoint of the individual (one person or one unit). It is microscopic in contrast to the three historic macroscopic systems of Smith (explicit among his followers), Marx, and Keynes, and as far from them as nuclear physics is from astronomy. Somewhere in economic thought the macroscopic and microscopic approaches may meet, as astronomy and earthbound physics did in Isaac Newton. Keynes in his major work, *General Theory*, in 1936 made the last heroic effort to cope with economic life on the basis of such aggregates as national income, consumption, savings, and investment. Keynes rested his case on an equilibrium of aggregates (e.g., income equals consumption plus investment), but to protect the equilibrium at a high level against the dynamic "stagnation" of capital (similar to Marx's "falling rate of profit"), manipulation of at least one variable was required (such as Roosevelt's credit policies). That is as close as Keynesian economics comes to the integration of economics with political economy. Among the things it does not explain are the strategical combinations that are formed at every level of the economy for the purpose, in part, of manipulating aggregates. It does not answer, for example, Walter Lippmann's recent question as to why the Eightieth Congress through tax relief gave the gun to consumption in a period when supply was still short and prices were still rising. Wise or unwise, it was not the act of a man from Mars, but a strategical act brought on, in part at least, by components of the economy. Both the cause and the effect may be guessed at, but the cause is not accounted for integrally in economic theory, and the effect is only vaguely seen.

The authors of the theory of games are not the first to examine individual economic behavior. Many economists ("marginalists") make efforts to understand the subjective basis of individual preferences but do not pursue the individual far into his objective economic acts. They find, plausibly, that each additional unit of any good brings a person less satisfaction than the preceding unit. But they do not examine strategical relationships. They have generally assumed, rather, that each person independently seeks his maximum satisfaction. The theory of games, which is objective, is meeting opposition from those marginalists who believe that utility (the capacity of goods to satisfy desires) can be ordered but not measured. The authors of the theory of games show that if your preferences are consistent and in a sense orderly, then, for you, utility can be expressed numerically. This need not be in terms of money (does \$5 mean the same thing to a salesman with \$5, another with \$5,000, and another with \$50,000?). However, if all the players in a game play for money alone, the theory applies most simply since "valuation" in dollars is easy.

From the standpoint of the individual there are three possible economic situations on earth: one man alone, two men, or three or more men. One man is Robinson Crusoe alone on a desert island. He maximizes his gain uncomplicated by anything but the forces of nature, which are predictable, at least in such terms as probable rainfall and the probability of a tornado.* If there are two men on earth, the individual enters into the relation of exchange (buyer-seller), in which the problem of the other fellow appears; he can no longer maximize his gain but must seek an optimum. But if there are three men, a novelty occurs: two of them may gain more by combining against the third (two sellers against a buyer, or vice versa). These three situations are found in games.

*Robinson Crusoe is also theoretically pure communism in which the distribution of wealth is fixed by definition and society proceeds as one against nature; as we know, Stalin's particular kind of Communism requires a considerable coercion to maintain the fixity.

Von Neumann and Morgenstern approach the problem of the economic individual with simple game models because science always begins with simple models. Games are human endeavors that isolate strategies identical in their essential character with the strategies of economic life. In games, strategies are simple, observable, and susceptible to abstraction without interference in their operation. Games themselves are sets of rules, abstractly comparable to the rules that define economic life. Around these rules, strategies are exercised. Unlike economics, games are nonproductive, but that is no obstacle to a mathematician: his games may have the device of a fictitious player who must supply the real players with their gains from productivity. Ordinary games are less complicated than the market, but games covered by mathematics are infinitely complex. The real difficulty with the market—from the standpoint of constructing economic theory—is not so much its complexity but the fact that one does not know the exact rules. If the rules were known to the last comma, it would be a "game."

If the theory of games were nothing else, its contribution to the literature of games would make it a notable work. For it has accomplished the feat of separating strategical from nonstrategical games and has made obsolete the traditional theory of probability as the only theoretical basis of games of strategy. (The theory of probability, incidentally, originated in the study of games of chance.) The one-man game, like solitaire, can be played on the basis of the theory of probability; for the cards won't talk back or get up and walk away. Craps and roulette, when played fairly, are nonstrategical and involve simple probabilities. Bridge is a strategical game, which requires the perfecting of information by signaling between partners (the rest can be probability calculation). Chess is a nonstrategical game in which perfect information prevails on both sides of the table; ". . . if the theory of Chess were really fully known there would be nothing left to play." Von Neumann's theory suggests that the seeking of information is central to the nature of strategy. Poker is a strategical game based on imperfect information and inverted signaling (i.e., bluffing to obscure information).

Solitaire corresponds to the economics of Robinson Crusoe. It is one man against the cards, or nature. The theory of probability suffices for either. The simplest two-man strategical game is matching pennies. Nothing can be done in this game to influence the long-run results except to observe the opponent's pattern of play and obscure one's own. (Establish a pattern of heads and tails and you may be found out and beaten.) The only way to obscure the pattern completely is by some method of random choice. The optimum outcome is to break even. This game introduces an important element in strategy, namely, random choice; for even in this simple game a good player takes risks that depend on chance alone.

How people choose between different risks is one of the questions that have troubled economists. And to further complicate the matter, one is not alone in taking risks. For one player in judging the value of a definite play must consider the possibility that his opponent will use a chance reply. He must be able to judge the total value to himself of various chances of winning various amounts. In assuming that utility can be measured, the mathematician assumes that the utility of a 10 per cent chance of getting a Lincoln and a 90 per cent chance of getting a Chevrolet has a definite combined value that can be compared with the certainty of getting a DeSoto or Chrysler. In military applications it is said that the question is not how much you gain on the average but rather how often you win. But the mathematician would say merely that someone wants a very large value given to the difference between winning and losing.

Poker Strategy

The theory of games originated in poker, and that game remains the ideal model of the basic strategical problem. Von Neumann strips the game to its bare structure. He restricts the players to two, the bets to two (high or low) with the alternative of passing, and the raises to one. Like many real poker players, he creates his own variation of the game. The cards are dealt face down but there is no draw. He chooses to call it "stud." The essentials of poker remain: imperfect information concerning the opponent, maintenance of the imperfection by bluffing representations of strength or weakness, the conflict of two maximum intents that can be reconciled only in a bet and a call, after which the outcome is discovered by comparing hands. Each player has the problem of coming to terms with the other player who is unpredictable, even in terms of probabilities, since he, too, can think and plan. The core of the problem then is, how can a player guarantee himself a certain minimum return regardless of the other player's action? This is the problem that has made any action in the market (buyer-seller) or in a contest (the duel) an irrational gamble, and that by the same token has confounded economic and social theory. The crux of von Neumann's theory lies at this point.

There is imagination but no magic in von Neumann's theory. It is an act of logic with an unusual twist, which can be followed to the borderline of mathematical computation. At its foundation is the remarkable requirement *that the player assume in advance that he is found out*: that the particular strategy he will follow is already known to his opponent. Such apparent pessimism is antipathetic to the gambling instinct. But it does eliminate one well-known, well-founded anxiety: that if one *is* found out, he will be surprised and defeated. However, this "rational" player prevents both his opponent and to some extent himself from knowing the specific applications of the strategy by making use of random choices. The extent to which this denial and self-denial of information is useful varies from game to game (accounting for it is a feature of the theory of games). The penny matcher, for example, distributes his choices on a random, fifty-fifty basis. The poker player likewise distributes his bluffs irregularly on a controlled-probability basis. He might, for example, choose to bluff 10 per cent of the time, but when these bluffs would occur even he could not tell you. Now it is von Neumann's contention, which he has proved with his "Persian poetry,"* that if strategy is broadly defined, no loss will be sustained through the "pessimistic" assumption that one's strategy has been found out. For the player who so reconciles himself will, "if he plays well, fare no worse than he would if he 'found out' the opponent's—assuming that the opponent too, played well." There's the twist. And it is evident from the assumptions. For the best you can do in strategical competition is to find out your opponent and act accordingly. But if your opponent has acted on the assumption that he is already found out, you gain nothing more by finding him out. Thus you can already see your lowest maximum gain against his best policy. You can also see your highest minimum gain. In von Neumann's poker, the highest minimum and the lowest maximum meet at the same point. Von Neumann proved mathematically that this is equally true without exception for all two-person games like poker (games in which one's losses equal the other's gains). In other words, when playing against a good player, the assumption that one will be found out (to the extent that one may be found out at all) does one no harm. In fact, both players should make this assumption.

This theory is called "max-min" or "minimax." (See box p. 706 for a graphic illustration.) On it von Neumann's theory stands or falls. It is a new and important term in the lexicon of science, and one of the most talked-about novelties in learned circles today. No mathematician has contested

*The central theorem of the theory of games reads, in mathematical form, as follows:

$$\mathrm{Max}_{\vec{\xi}}\,\mathrm{Min}_{\vec{\eta}}\,K(\vec{\xi},\vec{\eta}) = \mathrm{Min}_{\vec{\eta}}\,\mathrm{Max}_{\vec{\xi}}\,K(\vec{\xi},\vec{\eta})$$

Mathematicians take aesthetic pleasure in gazing at this.

its mathematical proofs under the game conditions in which it was constructed. It is the only theory that defines how to proceed rationally in what has classically been considered an irrational situation. It carries the corollary that if your opponent departs from the theory and chooses to act irrationally, you can drive him back to the rational on the pain of gaining more in some proportion to his irrationality.

How the minimax strategy protects itself from an opponent's departure from good strategy is demonstrated in poker. Poker players will understand the following verbal rendering of von Neumann's mathematical proofs. For others the next eight paragraphs of this story are an avoidable digression.

Describing the rudiments of the game, he observes that

> a player with a strong hand is likely to make high bids [bets]—and numerous overbids—since he has good reason to expect that he will win. Consequently a player who has made a high bid, or overbid, may be assumed by his opponent—*a posteriori!*—to have a strong hand. This may provide the opponent with a motive for "Passing." However, since in the case of "Passing" the hands are not compared, even a player with a weak hand may occasionally obtain a gain against a stronger opponent by creating the (false) impression of strength by a high bid, or by overbid—thus conceivably inducing his opponent to pass.
>
> This maneuver is known as "Bluffing." It is unquestionably practiced by all experienced players. Whether the above is its real motivation may be doubted; actually a second interpretation is conceivable. That is if a player is known to bid high only when his hand is strong, his opponent is likely to pass in such cases. The player will, therefore, not be able to collect on high bids, or on numerous overbids, in just those cases where his actual strength gives him the opportunity. Hence it is desirable for him to create uncertainty in his opponent's mind as to this correlation—i.e., to make it known that he does occasionally bid high on a weak hand.
>
> To sum up: Of the two possible motives for Bluffing, the first is the desire to give a (false) impression of strength in (real) weakness; the second is the desire to give a (false) impression of weakness in (real) strength. Both are instances of inverted signaling . . . —i.e., of misleading the opponent. It should be observed however that the first type of Bluffing is most successful when it succeeds, i.e., when the opponent actually "passes," since this secures the desired gain; while the second is most successful when it "fails," i.e., when the opponent "sees," since this will convey to him the desired confusing information. . . . The

> possibility of such indirectly motivated—hence apparently irrational—bids has also another consequence. Such bids are necessarily risky, and therefore it can conceivably be worth while to make them riskier by appropriate countermeasures—thus restricting their use by the opponent.

First of all, the basic good strategy must be always to bet high on a high hand and mostly low on a low hand, "but with occasional, irregularly distributed Bluffs." This strategy von Neumann demonstrates to be sound regardless of all but one significant deviation, namely, incorrect bluffing. Any other deviation, for example, betting regularly low on a low hand, will bring losses. But incorrect bluffing against the correct strategy, for example, bluffing more often than is proper on a low hand, may bring momentary gains; but then the opponent could drive you back to the correct strategy by deviating appropriately and penalizing the incorrect bluffs. This leads to a surprising conclusion regarding poker play: "The importance of 'Bluffing' lies not in the actual play, played against a good player, but in the protection which it provides against the opponent's potential deviations from the good strategy."* Anyone who has played tight poker with strong players knows the validity of this principle of the primacy of defensive bluffing.

The principle is inherent in the elementary practices of good players playing for substantial stakes, as, for example, in stud, where the first rule is that when you can't beat what you can see, drop; or, again, in draw, where the first rule is don't draw to "shorts" (pairs lower than jacks). Such conservatism of good play precludes bluffing of an aggressive character, except when deliberately intended for effect and preferably at moderate cost. It is, however, necessary to be on the lookout for the opponent's deviation from such proper practices and sometimes to pay well to keep him in line.

In short, von Neumann distinguishes "two varieties of 'Bluffing': an aggressive one practiced by the player who has the initiative and a defensive one—'seeing' irregularly, even with a medium hand, the opponent who is suspected of 'Bluffing.'" Defensive bluffing enforces the minimax strategy. If your opponent leaves it, you win.

*The poker player who wishes to speculate intuitively might consider the following rule for taking action against a player who is bluffing contrary to good strategy: "if the opponent 'Bluffs' too much for a [given] hand . . . then he can be punished by the following deviations from the good strategy: 'Bluffing' less for hands weaker than [the given hand], and 'Bluffing' more for hands stronger than [the given hand]." The reverse also holds. Thus the defensive counterattack is to imitate your opponent's mistake for stronger hands and to take contrary action for weaker hands.

Although these are real contributions to poker, von Neumann has not produced the complete poker player. The real game, like the market place, of which it is the model, has many additional complications.

One of the complications of expanding the treatment of games beyond two-man poker led to the second important illumination in the theory. In the two-man game, which brought about the principle of minimax, the two players are unable to reconcile their differences except at a "price" (the bet and the call). In this they are, as suggested, like the buyer and seller who cannot reconcile themselves except at a price (the outcome of bid and ask). The significance of the three-man game is the way it destroys this pure opposition of interest. For the three-man game, the authors invent a prototype game with no known counterpart. But any strategical three-man game with similar rules will show the same results.

In place of the single situation of pure opposition, three kinds of situation now appear. First, all three players may combine to form a one-man game (against the cards or against nature). Second (if the rules permit) two may combine against one to form a two-person game (with three possible situations of two and one). And third, a pure three-person game with each player on his own. The most important of these three possibilities is the second, the formation of coalitions. Hence the necessity to understand the two-person game as a prelude to the understanding of the inner workings and outcome of the three-person game. The theory of games shows conclusively how coalitions should be formed if there is an advantage in forming them and the rules do not forbid it. Any player, in fact, who fails to attempt a coalition in such circumstances will lose or, more exactly, will gain less. On the economic premise that man seeks gain, it follows that he must attempt to form a coalition. This is significant, for the economics of a competitive economy where the "instinct" to combine in the absence of preventive rules may be largely a law of survival; that is, not to combine is to risk a loss by a competitor's combination. (Antitrust laws have never been definitive enough to withstand this pressure against them.) In games such reasoning is wholly deductive, and it brings out the full implications of the basic assumptions. It is remarkable how closely their validity is confirmed by the inductive experience of economic affairs.

A good game model for coalitions, showing how they form and reform under changing circumstances, is the card game "Set Back," a variation of pitch, played in the mountains of northwest North Carolina. It is an easier game to follow than von Neumann's mathematical model, and shows the basic idea of coalitions. Set Back is an auction game with bids up to five points (high, low, jack, joker, and "game"). If you get the bid, you declare trumps and proceed to try to take the five points. Failing to make, you are set back the amount of your bid. Eleven points take what might be called a rubber. With three players, one generally gets ahead of the other two, threatening to go out from the vantage point of six or more points. Immediately the other two players are forced to give each other tricks to keep them away from the advanced player, and the game is momentarily two against one (von Neumann's old two-man game).

The crucial question for all coalitions then arises: how divide the coalition's gain between the two members? In von Neumann's game he has turned up some extraordinary revelations, for the price of remaining in the coalition for any one player must be no less than the cost of defection that the opposed single player could pay to break the coalition. In Set Back the coalition must give itself tricks, on pain of giving the third player the rubber; enforcement of the coalition is thus clear and definite. Very soon, however, one of the coalition members, profiting from the arrangement, will himself get six points or more, and threaten to take the rubber. The first coalition instantly breaks up and the two strong players, fearing each other, form a peculiar negative coalition based on their opposition, and both then give tricks to the weak player to keep them away from each other. This brings out another unique aspect of coalitions, that under certain conditions weakness is not a disadvantage. For that reason, it does not follow that under all circumstances the "fittest" will survive.

In economics this principle often takes the form of big organizations allowing or encouraging smaller ones to stay in business to maintain the minimum characteristics of the competitive situation—so as to avoid various dangers of monopoly. In some industries today that are dominated by a few companies, it is a custom almost approaching an ethic, to use kid gloves in competition, and to speak well of one's competitors. In the theory of games the influence of custom, prejudice, and the like on coalitions is studied as complicating "standards of behavior." To complete the Set Back game, then: when all three are within striking distance of the eleven-point goal and there is no further incentive to combine, coalitions disappear and pure competition returns. The rules of the game (and standards of good sportsmanship) prevent any two from combining by special arrangement to give one of them the game.

The importance of the three-man game for economics is its dissection of the basic structure of monopolistic competition, that area between pure competition and pure monopoly, which has baffled economic thought.

Unlike any others, von Neumann and Morgenstern build coalitions integrally into their theory. It has a further importance, for if the nature of coalitions could be completely understood it might throw light on the nature of two other great, related, modern problems—economic cycles and economic planning.

The complications that one extra player adds to game solutions are both good and bad for the theory of games (the four-person game, incidentally, becomes typically two against two, or sometimes three against one—again the inevitable return to the two-man game). The good thing is what the extra player reveals about the laws of coalitions, which are of benefit to economic, military, and political speculation. By the same token, the extra player shows the fantastic complications involved in extending actual mathematical computations to situations involving large numbers. Even in a ten-person game, the players can be split into two opposing coalitions in 511 ways.

It is by no means necessary, however, that such multitudinous combinations come into existence. The trade-union movement shows how large numbers of economic individuals can group themselves back into small numbers of strategy-minded units. Yet the problem of large numbers is the greatest challenge the authors of "Games" have to meet. Large numbers in actual economy are also accompanied by dynamic changes. The theory of games not only does not deal with large numbers, but in its present form, it is avowedly static, i.e., it does not deal with change. These two problems have been raised as objections to the theory. To the authors they are not valid objections, but limitations in the present stage of the theory's development. If the theory is ever able to cope with dynamically large numbers, it will have reached the ideal scientific goal of prediction. It is a long way from that.

Large numbers in themselves, however, do not bother mathematicians. In physical science the theory of gases, involving unbelievably large numbers of molecules, is an exact science just because of the large numbers. This has suggested that ideal free competition might be the truly calculable situation, the most rational circumstances in which free enterprise could find itself; for this conclusion parallels Adam Smith's implied assumption that large numbers of individuals, each pursuing his own interests, would on the whole bring about full employment of resources. It is an attractive if utopian thought, as revolutionary in its way as Stalin's game of solitaire. For the present type of "monopolistic competition" the theory might answer the question whether the optimum profit is identical with optimum production, and whether this is identical with optimum social and individual well-being. It would be interesting to know whether pure or monopolistic competition produces the fuller employment of resources. The theory makes no promises. But it finds man, not inanimate nature, making the decisions, and studies them.

The military application of "Games" was begun early in the past war, some time in fact before the publication of the complete theory, by ASWORG (Anti-Submarine Warfare Operations Research Group, predecessor of the Navy's present Operations Evaluation Group). Mathematicians in the group had got hold of von Neumann's first paper on poker, published in 1928. The success of their work led to the present Naval and Air Force (Project Rand) applications, which are hidden behind military security. Something, however, can be said about them.

Most military problems can be reduced to game types such as the Duel, mentioned earlier; Deployment (see Colonel Blotto's problem in box), and the Search. Problems of the latter type, involving a contest between airplane and submarine, first preoccupied ASWORG when enemy submarines were destroying our transport.

Colonel Blotto Game: A Deployment Problem

The "Colonel Blotto" game is a military deployment problem, found in Caliban's *Weekend Problems Book.* The version shown here illustrates in elementary, graphic form the basic idea of von Neumann's theory of games. With patience it is not too difficult for the layman to follow.

The problem is given to Colonel Blotto by his general as a test of competence. Blotto has four units of armed force with which to oppose an enemy of three units. Between the opposing forces is a mountain with four passes; in each pass is a fort. War is declared in the evening. The issue will be decided in the morning. Decisions are based on who outnumbers the other at each pass. One point is scored for each fort taken, and one point for each unit taken. In the original form of this game the opponent's units turn up where they do in the night as a matter of luck. Blotto thus deployed his forces against a known probability that his opponent's forces would be grouped as follows: one unit in each of three passes, two in one and one in another, or all three in one pass.

In Princeton during the past war, two mathematicians, Charles P. Winsor and John W. Tukey, working on practical military problems, spent odd moments bringing Colonel Blotto's problem closer to reality.

They allowed the units of Blotto's opponent to communicate and employ counterstrategies. The illustration below, shown here by courtesy of Professor Tukey, indicates the solution. Here is the way it goes:

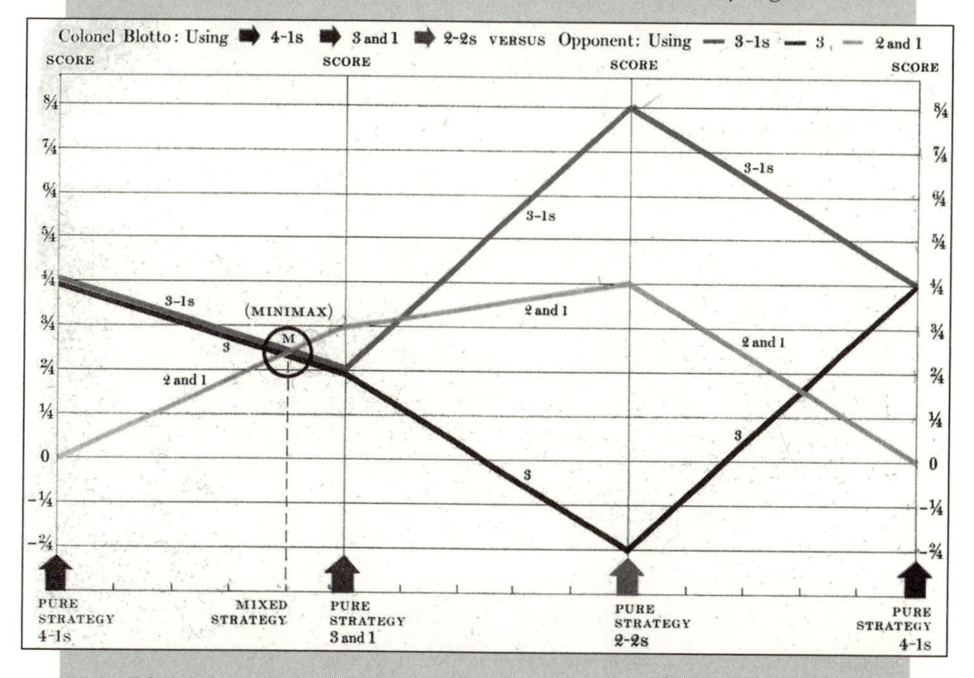

Blotto has available four forms of "pure" strategy (pure as distinguished from mixed): he may deploy his four units singly, one to a pass; or three and one; or two and two; or all four in one pass. To simplify the problem for the purposes of this illustration, a special rule is introduced forbidding Blotto to send all four units into one pass. This leaves him three possible pure strategies. His opponent likewise has three possible pure strategies.

Beginning at the left-hand side of the graph, Blotto's pure strategy "4–1's" (one in each pass) is tried against his opponent's pure strategy "2 and 1." They would meet in the mountain passes like this:

1 1 1 1
2 1

In the first pass Blotto will lose one fort and one unit for a loss of two points; the second, one against one, is a standoff; in the third Blotto will gain a fort for one point and in the fourth another fort for a second point. The result will be the same no matter which passes his opponent enters with this strategy (if there were a difference as there is when other deployments are used, all scoring positions would simply be averaged to get the score for that strategy). In this instance the score is two for Blotto and two for his opponent: or zero. No gain for either. This is indicated on the graph where pure-strategy "4–1's" meets pure-strategy "2 and 1."

The same simple arithmetic will show that when Blotto deploys his forces in the same way (singly) against his opponent's deployment of three ones (singly) or three together, the average result will be one point for Blotto (shown as four-fourths, since Blotto's vertical line is measured for convenience in fourths).

The second vertical line represents Blotto's pure strategy "3 and 1." Arithmetic will show that Blotto gains an average of one-half a point (two-fourths) in contact with his opponent's deployments of "3–1's" and "3"; and he will gain an average of three-fourths of a point in contact with his opponent's "2 and 1."

The third vertical line continues Blotto's score with pure-strategy two two's against his opponent's three deployments. The fourth vertical line is a repetition of the first. Of the three pure strategies available to Blotto, then, he will do best with "3 and 1," with a possible high score of three-fourths or a low of two-fourths or one-half. But can he do better?

The alternative, on von Neumann's principle, is to give up pure strategy for a mixed strategy. (The "mixture" could be made by rolling a five-sided die with one side marked "3 and 1" and the others marked "4–1's.") When lines are drawn between the average scoring points for each contact of strategies, all three lines intersect at the point "M." If then he mixes strategies to achieve this point, Blotto can win an average of more than two-fourths, no matter what strategy his opponent chooses. This in fact is also the best that he can do, assuming his opponent uses his best strategy. This is "minimax"—Blotto's highest minimum and lowest maximum.

An example of a similar problem in economics is the distribution (deployment) of spare parts. Thus, as Professor Tukey observes, "games of strategy . . . may have considerable practical value in diverse fields."

The search problem is described by Philip M. Morse, director of ASWORG, in a bulletin of the American Mathematical Society (July, 1948), as follows:

> An observer, equipped with some means of making contact with a target, moves or is moved over an area of volume in some more or less regular pattern of search; the problem is to find the pattern which most efficiently results in contact under specified circumstances. The problem is applicable to many cases: the means of contact may be visual, by radar or sonar; the means of transport of the observer may be by aircraft or beneath the water; the "observer" may be a proximity-fused guided missile, and the "contact" desired may be the destruction of the target; peacetime applications to geological prospecting are obvious, and so on. . . .
>
> The problem usually can be divided into a number of parts: the *contact problem,* dealing with the relationship between the physical properties of the detection equipment and the probability of contact with the target, when observer and target are in given relative positions; the *track* or *pattern problem,* dealing with the determination of the optimum pattern of search for given conditions; and the *tactical problem,* dealing with the reciprocal effects occurring when the target is also provided with detection equipment. Analyses of the tactical problem often require the techniques developed by von Neumann for his theory of games. All aspects of the search problem involve fundamental concepts and techniques of the theory of probability, expressed from a point of view enough different from that of classical probability theory as to cast new light on some of its concepts and techniques.

Airplane and submarine play a two-man "game," like poker. The submarine passes through a channel, whose length makes occasional surfacing necessary. The plane flies back and forth over the channel in an effort to detect the submarine when it surfaces. The distances are such that the plane's trips are limited. What pattern of search should the plane follow and what plan of submergence should the submarine follow? This relation of tactic to countertactic is common to most military problems. Should the plane, for example, fly the widest part of the channel with fewer trips, or the narrower parts with more trips? If the pilot chose one or the other exclusively and were found out, he would never see a submarine. The problem, in light of the theory of games, is for the pilot to find a strategy that would survive discovery and meanwhile try to find out the plans of the submarine commander in case the latter should be less discreet. This would involve a certain amount of random choice. If both are "rational," they should be able to enforce the correct strategy on each other (as the poker player did when his opponent departed from correct bluffing tactics). The correct minimax strategy for the search is very complicated—so much so that a sample solution cannot be given verbally—yet it is regarded as a relatively simple military problem. In fact, whether practical military problems can be solved mathematically by the theory of games is itself a military secret.

Speculation would indicate, however, in the case of the duel as a model for two airplanes in combat, that the planes, if they have a limited number of shots, should, on approaching each other's range, proceed a certain distance toward each other before shooting, and then shoot at certain intervals determined by random choice. It is hardly likely that a pilot would have time to cast dice in this operation, and so it appears to follow that such an operation would have to be mechanized with a chance gadget in the mechanism. And since the solution would involve vast calculations based on the functional relationship of distance and accuracy, complicated with velocities, it is hard to see how they could be worked out except by computational machines in advance of actual combat. Perhaps, in that case, the whole operation would be mechanized and there wouldn't even be a pilot. Idealized it would be a mathematician's Fourth of July show.

War, however, even push-button war, is not likely to be idealized. War is a kind of experiment, a test that generally would not be made if the opposing forces could measure accurately one another's strength. Unless there is a great disparity in the sides, the possibility of such measuring seems like daydreaming.

Even such a matter as "linear programming"—relating a specific maximum military desire for supplies to available resources—is only in the early stages of solution. Take this example on which the Air Force is working with game theory: The Air Force cannot always drop all the bombs it would like; military engagements have optimum, maximum backgrounds. At what levels, then, must various interacting activities be operated to maximize the number of tons of bombs dropped during a particular period, in view of certain constraints such as matériel and limits on the expansibility of the training and supply establishments? This can be cast as a two-person game. A typical problem of the Air Force generally requires the "maximizing" of a certain quantity, which depends on approximately 500 factors. A solution of such a problem may require several billion multiplications. That, among other things, is what the new, high-speed electronic calculating machines are for.

Strategy of the Bomb

A very serious problem with which coastal defenses are concerned is the fact that a continent cannot be protected on a maximum basis from assault by rockets containing atomic bombs. Such rockets must be intercepted far from their targets in a very small amount of time over vast spaces. It cannot be assumed that every rocket contains an important bomb—a large proportion of them are likely to be bluffs. To mobilize full resources to destroy every one would be likely to cost more than the damage inflicted. Hence, again, an optimum would have to be achieved: an optimum system of detection and an optimum amount of interception, making it too costly for the enemy to chance the onslaught. War is chance and minimax must be its modern philosophy.

The theory of games may sometime have practical applications in many social fields. And yet even now, as theory, it has performed an important service in illuminating the meaning of strategy.

The Collaboration between Oskar Morgenstern and John von Neumann on the Theory of Games

OSKAR MORGENSTERN

Time and again since the publication of the *Theory of Games and Economic Behavior* in 1944 the question has been asked how it came about that von Neumann, one of the greatest mathematicians of our age, and I met and worked together on what turned out to be a major piece in both our lives [20, (1944) 1953]. Recently I have been pressed by many to set down the history of the collaboration. And so I shall try to give a brief account of our mutual involvement. A fuller account with precise dates may follow some other time.

My first book, entitled *Wirtschaftfprognose*, was published in 1928 by Springer in Vienna, Austria [10, 1928]. In that book, written in 1926–27 while a Fellow of the Laura Spelman Rockefeller Memorial and an Honorary Research Fellow at Harvard University, the whole question of economic forecasting was examined epistemologically, and the difficulties and virtual impossibilities of prediction were studied to the best of my then-existing knowledge. In my general scientific outlook I was strongly influenced by the work of Hermann Weyl, Bertrand Russell, and others in the mathematical and physical sciences. I also struggled hard with Ludwig Wittgenstein's *Tractatus Logico-Philosophicus* of 1921 [26, (1921) 1955]. At Harvard, I frequently participated in the private seminars held by the great philosopher-mathematician Alfred North Whitehead, who had just published his *Science and the Modern World* [25, 1925] in which, however, he began to veer more towards metaphysics than was to my then-developing taste.

When I became a Rockefeller Fellow, I was a product of the Austrian School of Economics, having obtained my doctorate in 1925 with a piece on marginal productivity. But I was constantly troubled by the fact that Böhm-Bawerk's theory of bargaining and of the "marginal pairs," while dealing with fundamentals, could not be considered completed. This also led me, while still in Vienna, to Edgeworth's contract curve in his *Mathematical Psychics* [2, 1881]. On my way to the United States in 1925 I visited the aged Edgeworth in Oxford. I expressed great pleasure at the publication of his collected papers, but urged him repeatedly to republish the *Mathematical Psychics*, then totally out of print. His death intervened before he could carry out my suggestion, which he had accepted.[1]

In my book[2] I showed among other things that one is confronted in economics with two kinds of variables, which I called "dead" and "live," the former being those that do *not* reflect decisions by other economic subjects, the second, those that do. In that connection, even the word "game" occurs. I also showed that the mere increase in the size of an isolated "simple economy" (in the sense of the Austrian School, meaning a household isolated by itself) was a less complicating factor than complications encountered by a simple economy, no matter what size, when involved with others; the latter, and the latter alone, would have to deal also with "live" variables, as I called them, *i.e.*, with decisions made by others. This states exactly one of the basic tenets of game theory where one can maximize only in the first case when the variables of nature are "dead," but one is confronted with a conceptually different matter in the second case, since the "live" variables represent other "wills," other "economic acts," which may interfere with, or enhance, one's own plans, as I expressed the matter then.

One of the problems that naturally sprang to my attention was that of the influence of predictions on the predicted events, a typical case in the social sciences. I examined this area from many points of view. I analyzed several cases of predictions: the first a single prediction becoming known to, and being believed by, everyone, with their reactions influencing the predicted events; then the case of several differing predictions with different distributions of acceptance and consequently different influences upon the behavior of individuals and therefore upon the future events, *etc.* In the course of these studies I produced the example of the pursuit of Sherlock Holmes by Professor Moriarty [10, 1928, p. 98]. I showed in some detail in particular that the pursuit developing between these two could never be resolved on the basis of one of them out-thinking the other ("I think he thinks that I think! ! . . ."), but that a resolution could only be achieved by an "arbitrary decision," and that it was a problem of *strategy*.[3]

The problems touched in that book never left me, in spite of my involvement in business cycle theory and statistics. In 1935 I published

[1] Today the contract-curve and the core play a great role as one manner of relating game theory to more conventional economic theory.

[2] Extensive summaries in English are contained in the two review articles by Arthur W. Marget [5, 1929] and Eveline M. Burns [1, 1929].

[3] It is analyzed as such in our book [20, 1953].

a paper in the *Zeitschrift für Nationalökonomie* "Vollkommene Voraussicht und Wirtschaftliches Gleichgewicht" in which the same illustration of Sherlock Holmes and Moriarty was used once more, but the whole matter of prediction and foresight was put into a wider frame [12, 1935]. That paper, incidentally, intrigued Frank H. Knight so much that he himself translated it and used it in his lectures.[4] I showed that the assumption of perfect foresight leads to paradoxes and is inadmissible for general equilibrium theory, which was thus found critically wanting. After this paper was published, I was invited by Professor Moritz Schlick, the famous philosopher and leader of the so-called "Vienna Circle," to give a discussion of the problems treated in that paper. This I did in a rather lengthy session, and the matter was discussed by many of those present in great detail. To the Vienna Circle belonged, in one way or another, Carnap, Feigl, Frank, Gödel, Hahn, Menger, Popper, Waismann, *etc.* Not all of them were present at this occasion. I went frequently to these meetings as well as to Karl Menger's Colloquium, though I was not a formal member of either group.

I repeated this talk, at Menger's request, in his Colloquium and after the meeting broke up, a mathematician named Eduard Čech came up to me and said that the questions I had raised were identical with those dealt with by John von Neumann in a paper on the Theory of Games published in 1928 [18], the same year that I had published my book on economic forecasting [10, 1928]. Čech, then already a promising mathematician, outlined to me its principal ideas and results and was very eager that I should study this particular work. I intended to do so, but the great burden of work I was carrying at that time as director of the Institute of Business Cycle Research during civil war conditions, with the Nazis threatening, with frequent trips to the League of Nations in Geneva, to Paris and London, *etc.*, made this impossible. Nevertheless, even during those years in the 1930's in Vienna, I managed to read a lot of logic and set theory, *e.g.*, Hilbert-Ackermann, Fraenkel, Hilbert-Bernays, Hahn, Hausdorff, *etc.* I also attempted to approach Kurt Gödel's great work on undecidability, helped and guided by my friend Karl Menger. At the same time Abraham Wald, to whom I had been able to offer a position as statistician in my Institute, gave me special instruction in various fields of mathematics. In those years fall not only the exciting work of Wald on the Walrasian equations but also the publication of Menger's great papers on the theory of returns [9, 1936] and the St. Petersburg Paradox [7, (1934) 1967]. Furthermore, in 1934 Menger's book on the logic of ethics was published [8, (1934) 1974], which I took

[4]This translation is published in [22, 1976].

as the appropriate occasion for a discussion of the role of modern logic for the social sciences [13, 1936].

A little later Karl Menger urged me to attend a meeting of his Colloquium in which John von Neumann, on one of his trips from America to Europe, would speak and present a theory of the expanding economy, which he did in 1937. Unfortunately, precisely during those days when he was in Vienna, I had to be in Geneva to attend the League of Nations committee meetings and so we did not meet.

In January 1938, I left for the United States, invited by the Carnegie Endowment for International Peace as a Visiting Professor at four American universities. I had hopes that I might have a chance to go to Princeton to meet von Neumann, then Professor at the Institute for Advanced Study. I did go to Princeton, but saw only the economists Frank Fetter and Frank Graham and did not encounter von Neumann. The Nazis took over in Vienna in March 1938. I was dismissed as "politically unbearable" from the University as well as from my Institute, which I had left in the hands of my deputy who emerged as a Nazi. The Institute was taken over by him and Wagemann, then the head of the Berlin Institute for Business Cycle Research. While in the United States, I received a number of calls from various American universities to join their faculties. I accepted the invitation from Princeton University for a three-year appointment as Class of 1913 Lecturer in Political Economy. Half my salary for these first three years was paid by the Rockefeller Foundation, which had for a long time supported my Institute in Vienna. The principal reason for my wanting to go to Princeton was the possibility that I might become acquainted with von Neumann and the hope that this would be a great stimulus for my future work.

Von Neumann and I met soon after the University opened. It is curious that years later neither of us could ever remember where we met the first time, but we did remember where we met the second time: I gave an after-luncheon talk on the 1st of February 1939 on business cycles at the Nassau Club, and he was there with Niels Bohr, Oswald Veblen, and others. Both he and Bohr invited me that afternoon for tea at Fine Hall, and we sat several hours talking about games and experiments. This was the first time that we had a talk on games, and the occasion was heightened by Bohr's presence. The disturbance of experiments by the observer was, of course, one of the famous problems raised by Niels Bohr for quantum mechanics. These talks were taken up again with both at Weyl's house—he had also entered my life. This circle still widened when I encountered Einstein for the first time at dinner with Bohr at von Neumann's house, and I recall vividly Einstein's discussion of the priority of theory over experiment, the preeminence of conceptualization, and the deep puzzle of intuition. In many later meetings he would often return to these and related issues.

Von Neumann and I had many other very animated and wide-ranging discussions. There was an instantaneous meeting of minds and a spontaneous empathy between us. I mentioned to him that I was greatly interested in studying both his papers, the one on game theory and the one he had delivered in Vienna on the expanding economy. We quickly exchanged reprints, I giving him in particular my work on perfect foresight. Von Neumann told me that he had done no work on game theory since 1928 or on the expanding economy model. He may have thought one way or the other about it, but never in any systematic way, nor had he put down anything on paper.

I now began to study his paper on game theory seriously. This was no easy matter because parts of the mathematics were new to me, in particular the entire topic of the fixed point theorem. The paper on the expanding economy was also difficult for the same reason. This quickly led to many conversations with Johnny, as I shall from now on call him. I remember vividly my great intellectual excitement, in fact also the emotional involvement with the theory of games, which he had developed in 1928. I saw what was meant and what tremendous possibilities there existed.

So I decided that I would write a paper showing economists the essence and significance of game theory as it then existed, and I set out to produce such a paper. Many further talks with Johnny took place. Our contacts were frequent and, for the rest of our lives, always of the most friendly and intellectually exciting nature. When my paper was well under way, he offered to read the manuscript. He did so and then remarked that it was too short and therefore would not be intelligible to those who had not studied the theory to the extent that I had done. Besides, he and I had already begun to discuss many further possibilities and developments of the theory. So I expanded the paper. When he saw that already greatly enlarged new version, on a memorable day in my bachelor quarters at 12 Nassau Street, which was then the Princeton Bank and Trust Building, he suggested: "Why don't we write this paper together?" I was quite overcome by this suggestion. Our many meetings had already opened up an entire new world toward which I had reached out for years, and here was Johnny wanting to work with me, both of us pushing into a vast new field, never doubting its challenge, difficulty, and promise. Years later I read what Hilbert said when, while in Göttingen, he encountered Minkovsky, who became his closest friend: "Er war mir ein Geschenk des Himmels!" Here was my gift from Heaven.

This was in the fall of 1940. As our paper progressed, we thought—always sitting together and writing jointly—that perhaps the paper might have to be still a little longer. Johnny said it might have to be published in two parts, one piece being too long for a scientific journal. I said that wouldn't matter to me at all; on the contrary, we should be as detailed as the subject matter required. So we began to work and as our work progressed, Johnny said: "You know, this will hardly do as a paper, not even in two parts. Perhaps we should make a small pamphlet out of it, and it could be published in the *Annals of Mathematic Studies*, which Marston Morse is editing." He, Johnny, could probably persuade Marston to accept it, and it might run up to perhaps 100 pages.

As we continued to work, Johnny said: "Why don't we go to the Princeton University Press and ask them whether they would be interested in such a pamphlet?" Datus Smith, then director of The Press, was very accommodating and an agreement was quickly made on the basis of which we committed ourselves to hand over an appropriate manuscript in due course, but I do not recall whether any specific date was set for this. At any rate, a pamphlet of about 100 pages was envisaged. After we had signed this statement, we felt quite relieved and cheered up in a funny sense. Now we really started to write. We completely forgot about any restriction to 100 pages and worked, thought, and discussed endlessly.

Often we went on long walks during which we talked over various games and the whole developing theory. We drove occasionally to the seashore and walked up and down the boardwalk at Sea Girt, in particular, discussing matters. At Christmas in 1940 we were in New Orleans where I had to give a paper on unemployment at a session of the American Economic Association. Later I was vacationing in Biloxi with Johnny and his wife Klari. Again, day after day was spent in discussing the theory. Among the things that came up very early was that we were in need of a number for the pay-off matrices. We had the choice of merely putting in a number, calling it money, and making money equal for both participants and unrestrictedly transferable. I was not very happy about this, knowing the importance of the utility concept, and I insisted that we do more. At first, we were intending merely to postulate a numerical utility, but then I said that, as I knew my fellow economists, they would find this impossible to accept and old-fashioned, in view of the predominance of indifference curve analysis, which neither of us liked.

So later we decided we would settle on thinking about a numerical utility. It did not take us long to construct the axioms on which the present theory is based that gave us a firm utility concept, that of an expected utility, numerical up to a linear transformation. We did not publish the proof of the existence of this number on the basis of our axioms in the first edition of our book, which appeared in 1944 (though we had that proof, of course). In Vienna I had given courses on risk, expectations, and the time element in value theory and also had published on some of these questions. Regarding risk, Karl Menger's important paper of 1934 on the St. Petersburg Paradox [7, (1934) 1967] played a great role. Johnny also read my paper on the time element in value theory [11, 1934] and urged me to expand it into a book (which I never did). He said that the problems discussed there were both very important and would offer great mathematical difficulties. But the construction of axioms for our expected utility came quite naturally. I recall vividly how Johnny rose from our table when we had set down the axioms and called out in astonishment: "Ja hat denn das niemand gesehen?" ("But didn't anyone see that?") Incidentally, we always spoke German even while writing in English, which later caused a knowledgeable reader ironically to remark that the whole book is written ". . . in such nice professorial German."

It was largely my doing that this utility theory was developed, and I am very content with what happened then and also later, although I still feel that the ultimate theory of utility is much more complicated than what we did. We were, of course, aware of the difficulty with the logical foundations of probability theory. We decided we would base our arguments on the classical frequency approach to probability, but we included a footnote saying that one could axiomatize utility and probability together and introduce a subjective notion of probability. This was done later by others.

Johnny's paper on the expanding economy [19, 1937], the one I had been unable to hear him deliver in Vienna, was another area of thought which occupied me. I now also studied this paper very carefully and was immediately convinced of its extraordinary importance. I persuaded Johnny to give a talk on it to the general economics seminar of the Department, which he did. This took place in the old Pyne Library where at that time we had a seminar room. The talk was fairly well attended, but I was dismayed at the total lack of response it found. There were then hardly any mathematical economists at Princeton, let alone persons who could have been receptive to the fundamentally new ideas he propounded.

Even while we were beginning to work on game theory, I also wrote my paper, "Professor Hicks on Value and Capital" published in *The Journal of Political Economy* [14, 1941; 22, 1976]. In that paper I referred to Johnny's model of the expanding economy, and I emphasized the fact that in economics one is confronted essentially with inequalities and not with equations. It is, as far as I know, the first mention of his paper anywhere, certainly in an economic periodical or book. Incidentally, in my paper there is also a strong reference to the fundamental work done by

Abraham Wald (also in the 1930's at Menger's Colloquium) on the Walrasian system, a work again not taken up by Hicks. Johnny read my paper on Hicks carefully and was in full agreement with what I said. He even made one or two annotations to the manuscript. This paper on Hicks has been completely bypassed in the literature, perhaps due to the wartime conditions, perhaps due to its irreverent nature. Von Neumann's view of books on mathematical economics written up to that time (of course, excepting Wald and Menger) and even somewhat later was: "You know, Oskar, if these books are unearthed sometime a few hundred years hence, people will not believe that they were written in our time. Rather they will think that they are about contemporary with Newton, so primitive is their mathematics. Economics is simply still a million miles away from the state in which an advanced science is, such as physics." He repeated this kind of remark even in later years and also with regard to the description of an economy or society.

The topic of the expanding economy had never left me, and Johnny and I talked many times about it. From the beginning I did not like one very restrictive assumption of his beautiful model, which is not its linearity, but the condition that every good produced must enter in the next phase into the production process of every other good produced, no matter in how small a quantity. This would only be true for highly aggregated conditions. In 1956 when he was already very sick with cancer, fortunately I could tell Johnny that J. G. Kemeny, Gerald L. Thompson, and I had succeeded in removing it, generalizing his model substantially, which pleased him very much. This was to become known as the "KMT model," which also established firmly the relationship of the model to game theory. In both, the fundamental minimax theorem is of the essence; KMT showed why this had to be the case and—unexpectedly—that game theory can also be used as a mathematical technique, as a calculus, in addition to being a true model. Many further generalizations were possible beyond the KMT model, an indication of the power of Johnny's original ideas. I also told Johnny that Thompson and I were planning more work along these lines. A series of papers containing further extensions and generalizations led to a book with Thompson, which I found gratifying and which is a major piece in our lives: *Mathematical Theory of Expanding and Contracting Economies* [17, 1976].

We worked intensively through 1941–42. Johnny, upon my urging, had given some lectures at Princeton University on game theory, mostly dealing with two-person theory, although some of our new results on *n*-person game theory also began to appear in his lectures. These were not terribly well attended, I do not know for what reason but, after all, we were already in the midst of war, not an ideal time for work of our kind! At any rate, the lectures helped focus our thoughts and advanced to some extent our manuscript.

There were endless meetings either at my apartment over the bank or at 26 Westcott Road, where Johnny lived with his wife Klari and his daughter Marina (now Mrs. Marina von Neumann Whitman). We wrote virtually everything together and in the manuscript there are sometimes long passages written by one or the other and also passages in which the handwriting changes two or three times on the same page. We spent most afternoons together, consuming quantities of coffee, and Klari was often rather distressed by our perpetual collaboration and incessant conversations. She was at that time collecting elephants made of ivory, glass, and all sorts of other material. At one point she teased us by saying that she would have nothing more to do with the ominous book, which grew larger and larger and consumed more and more of our time if it didn't also have an elephant in it. So we promised we would happily put an elephant in the book: anyone who opens the pages can find a diagram showing an elephant if he knows that he should look for one.

At this point I note a curious incident that shows how chance can influence the direction of scientific work. At the time when we were about to write down a new proof for Johnny's famous minimax theorem, originally developed in 1928, I went out for a walk on a brilliant, snowy cold winter day. I went towards the Institute for Advanced Study and since I was cold, I walked into the library, looking around idly. I picked up E. Borel's *Traité du Calcul des Probabilités,* and there I saw in it suddenly a paper by Jean Ville [23, 1938] dealing with Johnny's 1928 paper. There, in restating Johnny's minimax theory, instead of using Brouwer's fixed point theorem, he gave a more elementary proof (Johnny's two earlier proofs were definitely not elementary). I had not known of Ville's work; so I phoned Johnny to whom this also was news. We met immediately and quickly saw that the best approach was to proceed by considerations of convexity. Thus was developed the "theorem of the alternative for matrices," built upon the theorem of the supporting hyperplane. These were ideas that had occurred nowhere else in any piece on mathematical economics. From this date on (*i.e.*, from 1944 on) stems the introduction of methods of convex bodies into the modern literature of economics, in particular via linear programming (an offspring of game theory).[5]

It is curious to think how different much of our mathematical treatment of game theory would have been, had it not been for this winter

[5] The notion of convexity is used in a fundamental manner in von Neumann's 1932 model of the expanding economy [19, 1937; 17, 1976].

day walk, spotting Borel's book, opening it, and expecting to find nothing there about games of strategy. Of course, eventually convexity considerations would have appeared somewhere in the literature, probably with a delay of years. Many writings in other fields would have been retarded. We were, of course, happy that we could proceed in a more elementary manner, although we had to explain even those underlying mathematics in quite some detail. We were both firmly convinced that one should always try to proceed in as elementary a manner as possible and not display advanced mathematics when not needed. (Fixed point theorems reappeared, however, in a fair number of later writings by mathematical economists, sometimes unnecessarily complicating their arguments.)

In 1942 Johnny moved to Washington, but our manuscript was already very far advanced. The war had broken out, and he went to a research office of the Navy. At the time I was teaching something like twelve or fourteen hours a week. I had not received any "teaching credits" while working on the manuscript. At that time, it was not customary, and I had not asked for it. Neither did we get any financial support of any sort while we were working. While Johnny was in Washington, I frequently went there, sometimes staying at his house, and on weekends we worked furiously in order to finish and to make sure that this big work would really be done before a trip to England. When he left for a brief trip, he was equipped, to Klari's and my great amusement, with a heavy fur coat, a steel helmet, and a volume of the Cambridge Ancient History under his arm. (In spite of our heavy work load, we found time to read and exchange these volumes as well as to study Thucydides, books we discussed extensively when not working on "the next chapter.") Occasionally he came to Princeton and our work continued there deep into the night. He came again at Christmas 1942, and at that time we actually managed to write the last few pages. There were things that we had in mind that we wanted to put into the book, in particular more about economic applications, but we omitted these because the manuscript had already grown enormously. We had said most of the things that were really important to us to say, and time was running out. At Christmas, we put the last touches to the manuscript and also wrote the preface dated January 1943, and it was in the very first days of January that we made the work final.

During the whole period of our collaboration, each day after we met—and we wrote by longhand, of course—I would type two copies of what we had written, put in the formulas, and the next day, or whenever the next occasion arose, I would give one set to Johnny and keep the other; on that basis, we always had a somewhat orderly typed manuscript in front of us. This was a rather big chore, but I did it with pleasure and satisfaction. We had no secretarial or financial help. We did all on our own. Since I was a bachelor at that time, I had breakfast across the street at the Nassau Club, and Johnny, who usually got up early while his wife liked to sleep longer, came almost every day to join me at the Club. He was always wide awake in the morning, and even at the breakfast table we would start talking about what was to be done, if possible the same day in the afternoon. These breakfast meetings continued for many years, even after I was married in 1948, though less frequently.

After the manuscript had been completed, it was clear that we would have to go to Princeton University Press to explain that this was a somewhat bigger matter than the 100-page "pamphlet." The people at The Press were quite overwhelmed seeing a manuscript of about 1200 typed pages full of graphs and uninhibited mathematical notations. They were generous and said they would try very hard to publish the book (during World War II!), but could some subsidy be obtained? First there was the need to produce a clean manuscript; everything had to be retyped and all formulas had to be put into the new copy. We finally managed to get the enormous grant of $500 each from Princeton University and the Institute of Advanced Study for retyping. This was done and then a Japanese "enemy alien," a young mathematician, put in all the formulas from the original manuscript. Johnny remarked in his usual manner that it is the fate of enemy aliens who are mathematicians to be punished for being enemies by having to put other people's formulas into manuscripts.

The retyping and readying of the manuscript, with all the diagrams properly drawn—including the set-theoretical elephant!—done by a draftsman at the National Bureau of Economic Research, Mr. Forman, took quite some time. But the manuscript went to the printer in 1943. Then came the year-long process of typesetting, proofreading, *etc.*, which I shall not further describe here.

We had to settle on a title for our book. For a while we were thinking of calling it *General Theory of Rational Behavior*, but dropped this idea and similar ones quickly. They were not descriptive enough of our work, and we reverted to the title originally considered, *Theory of Games and Economic Behavior*, although, as mentioned, we knew perfectly well of the wider applicability of game theory to politics, sociology, *etc.* That settled, Johnny wanted to have our names listed alphabetically. I absolutely refused to entertain this proposal, and after some struggle he gave in.

The Press accepted the manuscript without any question. It never went to any referee, but they wanted some subsidy not only because of an increase in the cost of production, but also because they thought there was considerable risk involved. I went to a friend, a well-known

American, who made an anonymous donation to the University, which was turned over to The University Press. It was not a very large amount, but it turned the tide, and The Press undertook publication of the book without further hesitation. It came out on the 18th of September 1944. They certainly lost no money.

What were our expectations regarding the fate of our book? Clearly, we were convinced that it represented first of all a fundamental break with conventional economics: we demonstrated that one is not confronted with ordinary maximum or minimum problems (no matter what side conditions!), but with conceptually different situations. Though this becomes intuitively quite easily accessible for ordinary exchange, for monopoly, oligopoly, *etc.*, the phenomenon is all-pervading. The theory—which I shall not even begin to describe here—deals in a new manner even with such things as substitution, complementarity, superadditivity of value, exploitation, discrimination, social "stratification," symmetry in organizations, power and privilege of players, *etc.* Thus the scope of the book extends far beyond economics, reaching into political science and sociology, but economics was of more immediate concern and interest to us. We also knew that there would be much resistance, both because of the basic orientation of the work and because of the unconventional mathematics that extend throughout the book. We took care to explain as much of the latter as feasible, but we knew that the book made great demands on the reader on both counts. Johnny said to me repeatedly that we ought to publish some further joint papers, and we certainly had ideas of what could and should be done. (One on symmetric solutions of general n-person games was distributed as a RAND Research Memorandum [21, 1961].) Otherwise, he said, the book would be "a dead duck." Even so, he did not expect a rapid acceptance, rather we would have to wait for another generation. This view was shared by some of our friends, especially by Wolfgang Pauli and Hermann Weyl.

Matters turned out in some ways quite differently. In 1945 and 1946 there appeared two very fine expositions of our efforts by Leonid Hurwicz [4, 1945] and Jacob Marschak [6, 1946] as well as a long review by Abraham Wald [24, 1947], who in 1945 had already laid a new theory of the foundations of statistical estimation based on the theory of the zero-sum two-person game.

Then, in March of 1946 a long and generally intelligently written article about our book appeared on the front page of a Sunday edition of *The New York Times.* This caused a minor sensation and as a consequence the book was sold out quickly, so that a second edition had to be issued, which appeared in 1947. We added a substantial appendix, giving the proof that our system of axioms for a numerical utility, set forth in the first edition, indeed gave the desired result. That theory has now penetrated into most advanced texts of economic theory and will surely in the long run completely replace conventional indifference curve analysis. We also added some further observations on utility, in particular regarding such topics as partial orderings, non-Archimedean orderings, and the question of a specific utility of gambling. In 1953 we published a third edition, enlarged only by a new preface.

As mentioned above we wrote only one of the planned joint papers. We also had other plans. For example, we were both convinced that the then current methods for time-series analysis of economic data were totally inadequate, that the widespread hostility towards Fourier analysis was unjustified, and that better methods based on Fourier series could be developed. But since we wanted to compute on a large scale, we postponed our work time and again until the electronic computer Johnny was then designing would be available. It never came to that: in 1955 Johnny was stricken by cancer, discovered too late, and after much suffering he died in Washington on 8 February 1957. (I did not give up this plan, but worked towards the development and application of spectral analysis; the fruit of this is my book with C. W. J. Granger, *Predictability of Stock Market Prices* [3, 1970], preceded by various papers.)

Johnny took an active interest also in some other work of mine, such as my investigations of the errors in economic statistics and the general problem of description. My paper "Demand Theory Reconsidered" [15, 1948], pleased him particularly, and he observed that it offered great mathematical difficulties and, as on other occasions, that it would take considerable time for my results to be accepted—as was, indeed, the case.

During the last years, Johnny worked not only on the computer design but was deeply concerned with a theory of automata. We discussed those problems intensively. On many walks through the streets of Princeton, often late at night, he would discuss in great detail especially the possibility of designing a self-reproducing automaton. One problem, in particular, came up time and again: what would the "mouth" of an automaton have to look like if it were confronted with parts of its own kind in order to recognize the parts of which it is made and how to put a duplicate of itself together. These discussions as well as those regarding the future of large scale computations were, of course, restricted by the fact that there existed at that time only bulky and energy-devouring vacuum tubes and as yet no transistors.

By now, our book on game theory has been translated into German, Japanese, and Russian, and translations into Spanish and Italian are in progress. Several international conferences have been held in various countries including the Soviet Union and an *International Journal of Game Theory* has been started, first appearing in 1971. A bibliography on game theory prepared in Vienna lists up to 1970 over 6200 publications, among them dozens of books in many languages.

It is clear that the account given above refers mainly to outward events connected with the writing of this book, and not all of them are here recorded. A great deal would have to be said about what actually went on intellectually between the two of us, but that is another story, which may be told at another time; this would also involve more specific references to the theory itself. This period was, of course, the time of my most intensive intellectual activity ever. How could it have been otherwise: the closest possible lasting association in work and friendship with one of the truly great mathematicians of this century, a man whose genius communicated itself even to those who had only brief encounters with him.

We did an enormous amount of work in a very short time, but it was unceasing pleasure and never a time of drudgery. There was great excitement on both sides. There was the joy of discovery as we went along, with complete immersion in our task. Looking back, it seems strange that we could have found the time and energy, given our other duties and activities. Nor was it merely a time of work, since we also had intensive social contacts, with each other as well as with common friends. Never did a shadow fall on our relationship during this whole period, or in fact, for the rest of our lives. My brief obituary of Johnny in the *Economic Journal* [16, 1958] was written while very much still under the impression of the recency of his death. But there would be so much more to say.

References

1. Burns, Eveline M. "Statistics and Economic Forecasting," *J. Amer. Statist. Assoc.*, June 1929, *24*(166), pp. 152–63.

2. Edgeworth, Francis Y. *Mathematical psychics.* London: Kegan Paul, 1881.

3. Granger, Clive W. J. and Morgenstern, Oskar. *Predictability of stock market prices.* Lexington, Mass.: Heath, Lexington Books, 1970.

4. Hurwicz, Leonid. "The Theory of Economic Behavior," *Amer. Econ. Rev.*, Dec. 1945, *35*(5), pp. 909–25.

5. Marget, Arthur W. "Morgenstern on the Methodology of Economic Forecasting," *J. Polit. Econ.*, June 1929, *37*(3), pp. 312–39.

6. Marschak, Jacob. "Neumann's and Morgenstern's New Approach to Static Economics," *J. Polit. Econ.*, April 1946, *54*(2), pp. 97–115.

7. Menger, Karl. "Das Unsicherheitsmoment in der Wertlehre," *Z. Nationalökon.*, 1934, *5*(4), pp. 459–85. Published in English as: "The Role of Uncertainty in Economics," in *Essays in mathematical economics in honor of Oskar Morgenstern.* Edited by Martin Shubik. Princeton, N.J.: Princeton University Press, 1967, pp. 211–31.

8. ———. *Moral, Wille und Weltgestaltung.* Vienna: Springer, 1934. Published in English as *Morality, decision and social organization.* Dordrecht, Holland: Reidel, 1974.

9. ———. "Bemerkungen zu den Ertragsgesetzen," Z. *Nationalökon.*, 1936, 7(1), pp. 25–56.

10. Morgenstern, Oskar. *Wirtschaftsprognose, eine Untersuchung ihrer Voraussetzungen und Möglichkeiten.* Vienna: Springer Verlag, 1928.

11. ———. "Das Zeitmoment in der Wertlehre," Z. *Nationalökon.*, Sept. 1934, *5*(4), pp. 433–58. Published in English as "The Time Moment in Value Theory," in Schotter [22, 1976].

12. ———. "Vollkommene Voraussicht und wirtschaftliches Gleichgewicht," *Z. Nationalökon.*, August 1935, *6*(3), pp. 337–57. Published in English as "Perfect Foresight and Economic Equilibrium," in Schotter [22, 1976].

13. ———. "Logistik und Sozialwissenschaften," Z. *Nationalökon.*, March 1936, 7(1), pp. 1–24. Published in English as "Logic and Social Science," in Schotter [22, 1976].

14. ———. "Professor Hicks on Value and Capital," *J. Polit. Econ.*, June 1941, *49*(3), pp. 361–93. Reprinted in Schotter [22, 1976].

15. ———. "Demand Theory Reconsidered," *Quart. J. Econ.*, Feb. 1948, *62*, pp. 165–201. Reprinted in Schotter [22, 1976].

16. ———. "John von Neumann, 1903–1957," *Econ. J.*, March 1958, *68*, pp. 170–74. Reprinted in Schotter [22, 1976].

17. ——— and Thompson, Gerald L. *Mathematical theory of expanding and contracting economies.* Lexington, Mass.: Heath, Lexington Books, 1976.

18. von Neumann, John. "Zur Theorie der Gesellschaftsspiele," *Math. Annalen*, 1928, *100*, pp. 295–320.

19. ———. "Über ein ökonomisches Gleichungssystem und eine Verallgemeinerung des Brouwer'schen Fixpunktsatzes," *Ergebnisse eines Math. Kolloquiums*, 1937, *8*, pp. 73–83. Published in English as "A Model of General Economic Equilibrium," in Morgenstern and Thompson [17, 1976].

20. ——— and Morgenstern, Oskar. *Theory of games and economic behavior.* Princeton, N.J.: Princeton University Press, 1944. Third Edition, 1953.

21. ——— and Morgenstern, Oskar. "Symmetric Solutions of Some General *n*-Person Games," RAND Corporation, P-2169, March 2, 1961.

22. Schotter, Andrew, ed. *Selected economic writings of Oskar Morgenstern.* New York: New York University Press, 1976.

23. Ville, Jean. "Sur la Théorie Générale des Jeux où intervient l'Habilité des Joueurs," in *Traité du Calcul des Probabilités et de ses Applications.* Volume IV. Edited by Emile Borel *et al.* Paris: Gautier-Villars, 1938, pp. 105–13.

24. Wald, Abraham. "*Theory of Games and Economic Behavior* by John von Neumann and Oskar Morgenstern," *Rev. Econ. Statist*, 1947, *29*(1), pp. 47–52.

25. Whitehead, Alfred North. *Science and the modern world.* New York: Macmillan, 1925.

26. Wittgenstein, Ludwig. *Tractatus logico-philosophicus.* "Original in final number of Ostwald's *Annalen der Naturphilosophie*," 1921. English edition with Index: London: Routledge & Kegan Paul, 1955.

Index

CREDITS

Review, by Herbert A. Simon. Reprinted from *The American Journal of Sociology* (May 1945) 50 (6): 558–60.

Review, by Arthur H. Copeland. Reprinted by permission of the American Mathematical Society from *Bulletin of the American Mathematical Society* (July 1945) 51 (7): 498–504.

Review (The Theory of Economic Behavior), by Leonid Hurwicz. Reprinted by permission of the American Economic Association from *The American Economic Review* (December 1945) 35 (5): 909–25.

Review, by T. Barna. Reprinted from *Economica* (May 1946) n.s. 13 (50): 136–38.

Review, by Walter A. Rosenblith. Reprinted with permission from *Psychometrika* (March 1951) 16 (1): 141–46.

Heads I Win, and Tails, You Lose, by Paul Samuelson. Reprinted from *Book Week* by permission of Paul Samuelson.

Big D, by Paul Crume. Reprinted with permission from *Dallas Morning News* (December 5, 1957).

Mathematics of Games and Economics, by E. Rowland. Reprinted with permission from *Nature* (February 16, 1946) 157: 172–73. Copyright © 1946 Macmillan Publishers Ltd.

Theory of Games, by Claude Chevalley. Reprinted from *View* (March 1945).

Mathematical Theory of Poker Is Applied to Business Problems, by Will Lissner. Reprinted with permission from *The New York Times* (March 10, 1946). Copyright © 1946 The New York Times Co.

A Theory of Strategy, by John McDonald. Reprinted with permission from *Fortune* (1949): 100–110. Copyright © 1949 Time Inc.

The Collaboration between Oskar Morgenstern and John von Neumann on the Theory of Games, by Oskar Morgenstern. Reprinted by permission of American Economic Association from *Journal of Economic Literature* (September 1976) 14 (3): 805–16.